International Review of
Cytology

A Survey of
Cell Biology

STRUCTURAL AND FUNCTIONAL
ORGANIZATION OF THE NUCLEAR MATRIX

VOLUME 162A

International Review of A Survey of
Cytology Cell Biology

Edited by

Ronald Berezney
Department of Biological Science
State University of New York at Buffalo
Buffalo, New York

Kwang W. Jeon
Department of Zoology
University of Tennessee
Knoxville, Tennessee

STRUCTURAL AND FUNCTIONAL ORGANIZATION OF THE NUCLEAR MATRIX

VOLUME 162A

Academic Press
San Diego   New York   Boston   London   Sydney   Tokyo   Toronto

Academic Press, Inc.
A Division of Harcourt Brace & Company
525 B Street, Suite 1900, San Diego, California 92101-4495

United Kingdom Edition published by
Academic Press Limited
24-28 Oval Road, London NW1 7DX

International Standard Serial Number: 0074-7696

International Standard Book Number: 0-12-364565-4

PRINTED IN THE UNITED STATES OF AMERICA
95 96 97 98 99 00 EB 9 8 7 6 5 4 3 2 1

CONTENTS

The Nuclear Matrix: A Structural Mileu for Nuclear Genomic Function

Ronald Berezney, Michael J. Mortillaro, Hong Ma, Xiangyun Wei, and Jagath Samarabandu

The Architectural Organization of Nuclear Metabolism

Jeffrey A. Nickerson, Benjamin J. Blencowe, and Sheldon Penman

v

The Structural Basis of Nuclear Function

Dean A. Jackson and Peter R. Cook

Nuclear Domains and the Nuclear Matrix

Roel van Driel, Derick G. Wansink, Bas van Steensel, Marjolein A. Grande, Wouter Schul, and Luitzen de Jong

The Nuclear Matrix and the Regulation of Chromatin Organization and Function

James R. Davie

Contributions of Nuclear Architecture to Transcriptional Control

Gary S. Stein, André J. van Wijnen, Janet Stein, Jane B. Lian, and Martin Montecino

Chromatin Domains and Prediction of MAR Sequences

Teni Boulikas

Scaffold/Matrix-Attached Regions: Structural Properties Creating Transcriptionally Active Loci

J. Bode, T. Schlake, M. Ríos-Ramírez, C. Mielke, M. Stengert, V. Kay, and D. Klehr-Wirth

Origins of Replication and the Nuclear Matrix: The DHFR Domain as a Paradigm

P. A. Dijkwel and J. L. Hamlin

The Nuclear Matrix and Virus Function

W. Deppert and R. Schirmbeck

The Nuclear Matrix as a Site of Anticancer Drug Action

D. J. Fernandes and C. V. Catapano

CONTRIBUTORS

Numbers in parentheses indicate the pages on which the authors' contributions begin.

Ronald Berezney (1), *Department of Biological Sciences, State University of New York at Buffalo, Buffalo, New York 14260*

Benjamin J. Blencowe (67), *Department of Biology and Center for Cancer Research, Massachusetts Institute of Technology, Cambridge, Massachusetts 02139*

J. Bode (389), *Gesellschaft für Biotechnologische Forschung m.b.H, Genetik von Eukaryonten, D-38124 Braunschweig, Germany*

Teni Boulikas (279), *Institute of Molecular Medical Sciences, Palo Alto, California 94306*

C. V. Catapano (539), *Department of Experimental Oncology, Hollings Cancer Center, Medical University of South Carolina, Charleston, South Carolina 29425*

Peter R. Cook (125), *CRC Nuclear Structure and Function Research Group, Sir William Dunn School of Pathology, University of Oxford, Oxford OX1 3RE, United Kingdom*

James R. Davie (191), *Department of Biochemistry and Molecular Biology, Faculty of Medicine, University of Manitoba, Winnipeg, Manitoba, Canada R3E OW3*

Luitzen de Jong (151), *E. C. Slater Instituut, University of Amsterdam, 1018 TV Amsterdam, The Netherlands*

W. Deppert (485), *Heinrich-Pette-Institut für Experimentelle Virologie und Immunologie an der Universität Hamburg, D-20251 Hamburg, Germany*

P. A. Dijkwel (455), *Department of Biochemistry, University of Virginia, School of Medicine, Charlottesville, Virginia 22908*

D. J. Fernandes (539), *Department of Experimental Oncology, Hollings Cancer Center, Medical University of South Carolina, Charleston, South Carolina 29425*

Marjolein A. Grande (151), *E.C. Slater Instituut, University of Amsterdam, 1018 TV Amsterdam, The Netherlands*

J. L. Hamlin (455), *Department of Biochemistry, University of Virginia, School of Medicine, Charlottesville, Virginia 22908*

Dean A. Jackson (125), *CRC Nuclear Structure and Function Research Group, Sir William Dunn School of Pathology, University of Oxford, Oxford OX13RE, United Kingdom*

V. Kay (389), *Gesellschaft für Biotechnologische Forschung m.b.H, Genetik von Eukaryonten, D-38124 Braunschweig, Germany*

D. Klehr-Wirth (389), *Gesellschaft für Biotechnologische Forschung m.b.H., Genetik von Eukaryonten, D-38124 Braunschweig, Germany*

Jane B. Lian (251), *Department of Cell Biology and Cancer Center, University of Massachusetts Medical Center, Worcester, Massachusetts 01655*

Hong Ma (1), *Department of Biological Sciences, State University of New York at Buffalo, Buffalo, New York 14260*

C. Mielke (389), *Gesellschaft für Biotechnologische Forschung m.b.H, Genetik von Eukaryonten, D-38124 Braunschweig, Germany*

Martin Montecino (251), *Department of Cell Biology and Cancer Center, University of Massachusetts Medical Center, Worcester, Massachusetts 01655*

Michael J. Mortillaro (1), *Department of Biological Sciences, State University of New York at Buffalo, Buffalo, New York 14260*

Jeffrey A. Nickerson (67), *Department of Biology, Massachusetts Institute of Technology, Cambridge, Massachusetts 02139*

Sheldon Penman (67), *Department of Biology, Massachusetts Institute of Technology, Cambridge, Massachusetts 02139*

M.Ríos-Ramírez (389), *Gesellschaft für Biotechnologische Forschung m.b.H, Genetik von Eukaryonten, D-38124 Braunschweig, Germany*

Jagath Samarabandu (1), *Department of Biological Sciences, State University of New York at Buffalo, Buffalo, New York 14260*

R. Schirmbeck (485), *Institut für Mikrobiologie, Abt. Bakteriologie, Universität Ulm, D-89089 Ulm, Germany*

T. Schlake (389), *Gesellschaft für Biotechnologische Forschung m.b.H, Genetik von Eukaryonten, D-38124 Braunschweig, Germany*

Wouter Schul (151), *E.C. Slater Instituut, University of Amsterdam, 1018 TV Amsterdam, The Netherlands*

Gary S. Stein (251), *Department of Cell Biology and Cancer Center, University of Massachusetts Medical Center, Worcester, Massachusetts 01655*

Janet Stein (251), *Department of Cell Biology and Cancer Center, University of Massachusetts Medical Center, Worcester, Massachusetts 01655*

M. Stengert (389), *Gesellschaft für Biotechnologische Forschung m.b.H, Genetik von Eukaryonten, D-38124 Braunschweig, Germany*

Roel van Driel (151), *E.C. Slater Instituut, University of Amsterdam, 1018 TV Amsterdam, The Netherlands*

Bas van Steensel (151), *E.C. Slater Instituut, University of Amsterdam, 1018 TV Amsterdam, The Netherlands*

André J. van Wijnen (251), *Department of Cell Biology and Cancer Center, University of Massachusetts Medical Center, Worcester, Massachusetts 01655*

Derick G. Wansink (151), *E.C. Slater Instituut, University of Amsterdam, 1018 TV Amsterdam, The Netherlands*

Xiangyun Wei (1), *Department of Biological Sciences, State University of New York at Buffalo, Buffalo, New York 14260*

PREFACE

Research on the nuclear matrix has grown enormously since Berezney and Coffey first reported its isolation and initial characterization in 1974. Since then, more than 1000 papers have been published on the subject by numerous workers around the world, yet there has been no book devoted to reviewing the major developments in this growing field. One of our aims in producing Volume 162, Parts A and B, has been to fill this gap.

We invited many of the world's leading experts in the field to contribute to these volumes, and we were delighted with their positive responses. The chapters cover a variety of topics including isolation of the nuclear matrix, its morphology and correlation with the nuclear structure observed *in situ,* structural and functional domains of the nuclear matrix and its components, and biochemistry and molecular biology of the matrix proteins and associated DNA and RNA. Chapters also discuss functional properties associated with the nuclear matrix such as DNA replication, transcription, RNA splicing, transcription regulation, intranuclear, and nucleocytoplasmic transport and targeting, cell cycle regulation, mitotic regulation steroid hormone receptors, and viral and cancer drug associations. While these chapters deal with a broad range of topics, they are all interrelated in considering the genome organization, function, and regulation in relation to the nuclear architectural parameters. Thus, there are overlaps among some chapters, especially at the conceptual level where authors have been encouraged to "speak their peace" about the nuclear matrix and the progress made in understanding the functional organization of the cell nucleus. This is healthy for a field still in its relatively early stage of development but which shows every sign of entering a stage of explosive growth.

Among the more exciting developments is the recent application of high-resolution three-dimensional microscopy and computer-based image analysis in association with fluorescence techniques to label sites of genomic function inside the cell nucleus. Another development is the recent breakthrough in cloning and identifying genes for nuclear matrix-associated proteins. Combination of these approaches will give us insight into understanding how individual proteins are arranged and interact with each other and in association with the genetic information and expression. This could lead to important advances in correlating molecular details of the genome and its expression with the hierarchy of organization and function characteristic of all living organisms. We hope that these volumes will not only be informative but will also stimulate the interest of many to study the exciting subject of the nuclear matrix.

We sincerely thank our authors, whose willingness to write accounts of recent and cutting-edge developments in the nuclear matrix field in a timely and scholarly fashion has made this project successful. We are also very grateful to the editors and production staff members at Academic Press for their understanding, patience, skill, and devotion to producing the high-quality text and illustrations including many color plates.

Ronald Berezney
Kwang W. Jeon

LIST OF ABBREVIATIONS FOR THE VOLUME

a.a. amino acid
ADP adenosine diphosphate
AFB anticarrot fibrillar bundle antibody
AFM atomic force microscopy
AR rat androgen
araATP arabinofuranosyladenosine 5′-triphosphate
araCTP arabinofuranosylcytosine 5′-triphosphate
ARBP attachment region-binding protein
ARS autonomously replicating sequences
ASTP ATP-stimulated translocation promoter
BAA bromoacetaldehyde
BrdU bromodeoxyuridine
BrUTP 5-bromouridine-5-triphosphate
CAA chloroacetaldehyde
CB coiled body
CCNU chloroethyl-cyclohexl-nitrosourea
CHAT choline acetyltransferase gene
CHO Chinese hamster cell

DAPI 4,6-diamidino-2-phenylindole
DBSF DNA binding stimulatory factor
DFC dense fibrillar component
DHFR dihydrofolate reductase
DHRR dihydrofolate reductase
DHS DNase-hypersensitive site
DNase I-HS DNase 1-hypersensitive site
DRB 5,6-dichloro-(B-D-ribofuranosyl)-benzimidazole
dsDNA double-stranded DNA
dsRNA double-stranded RNA
DTT dithiothreitol
DUE DNA-unwinding element
EBNA1 Epstein-Barr virus nuclear antigen 1
EDTA ethylenediaminetetracetic acid
EM electron microscopy
ER endoplasmic reticulum

ERc estrogen receptor
FA actin filaments
FaraATP arabinofuranosyl-2-fluroadosine 5'-triphosphate
FC fibrillar centers
FISH fluorescence in situ hybridization
FRT FLP recognition target sites
GAPDH glyceraldehyde-3-phosphate dehydrogenase
GC granular component
GlcNac N-acetylglucosamine
GR glucocorticoid receptor
GRE glucocorticoid responsive elements
HD homeodomain
HDase histone deacetylase
HMG high mobility group
HnRNP heterogeneous nuclear ribonucleoprotein
huIFN-β human interferon beta
ICG interchromatin granule
ICS intermediate Cairns structures
IF intermediate filaments
IFA anti-intermediate filament antibody
IFN interferon
Ig immunoglobulin
IG interchromatin granule
IMPDH IMP dehydrogenase
INCENP inner centromere protein
IR inverted repeat

LAP lamin-associated protein
LBR lamin B receptor
LCR locus control region
LCS latest Cairns structures
LIS lithium 3,5-diiodosalicylate
MAb monoclonal antibody
M-AP mitogen-activated protein
MAP microtubule-associated protein
MAR matrix attachment region
MDa megadalton = 10^6 daltons
MKLP mitotic kinesin-like protein
MMTV mouse mammary tumor virus
MNase micrococcal nuclease
MPE methidiumpropyl-EDTA-iron(II)
MPF matrix protein filaments
M$_r$ molecular radius
MT microtubule
MTOC MT organizing centers
MVM minute virus of mice
N/A neutral/alkaline
NAP nucleotide protein
NaTT sodium tetrathionate
NE nuclear envelope
NEL nuclear envelope lattice
NEM N-ethylmaleimide
NLS nuclear localization signal
NM nuclear matrix
nmDNA nuclear matrix-associated DNA

N/N neutral/neutral
NOR nucleolar organizer region
NPC nuclear pore complex
NU in situ nuclei
NuM nucleolar-matrix
NuMA nuclear matrix mitotic apparatus
nup nucleoporin
OBR origin of bidirectional replication
OC osteocalcin
O-glcNAc O-linked *N*-acetyglucosamine moieties
ORC origin recognition complex
ORI origin of replication
PCN perichromonucleolin
PCR polymerase chain reaction
PF perichromatin fibril
PFGE pulse field gel electrophoresis
PG perichromatin granules
PKA protein kinase A
PKC protein kinase C
PMSF phenylmethylsulfonyl-fluoride
PNB prenucleolar body
PPB preprophase band
PPIase proline isomerase
PR progesterone receptor
pRb retinoblastoma protein
PRE progesterone response element
R receptor
RAF receptor accessory factor

RAN BP1 Ran-binding protein-1
RAP repressor-activator binding protein
Rb retinoblastoma
RBF receptor binding factor
rDNA ribosomal DNA
RNM RNA-binding motif
RNP ribonucleoprotein
RP multidrug resistance-associated protein
RPA replication protein A
RRE Rev-responsive element
rRNA ribosomal RNA
R-WGA rhodamine-conjugated wheat germ agglutinin
SAR scaffold attachment region
SAF scaffold attachment factor
SAF-A scaffold attachment factor A
SATB1 special AT-rich sequence binding protein 1
scs special chromatin structures
SDS–PAGE SDS–polyacrylamide gel electrophoresis
SMI1 suppress MAR inhibition
snRNP small nuclear ribonucleoprotein
snoRNA small nucleolar RNA
SMC stability of microchromosome
S/MAR scaffold/matrix-attached regions
SR steroid-receptor
ssDNA single-stranded DNA

SSB single-stranded binding protein
STEM scanning transmission EM
SV40 simian virus 40
T-antigen large tumor antigen
TEM transmission electron microscopy
TF transcription factor
Topo II topoisomerase II
TP terminal protein

Tpr translocated promoter region
TRAP T3 receptor auxiliary protein
TRE thyroid hormone response element
TSA trichostatin A
VDRE vitamin D responsive element
VZV varicella-zoster virus
WAP whey acidic protein
WGA wheat germ agglutinin

The Nuclear Matrix: A Structural Milieu for Genomic Function

Ronald Berezney, Michael J. Mortillaro, Hong Ma, Xiangyun Wei, and Jagath Samarabandu

Department of Biological Sciences, State University of New York at Buffalo, Buffalo, New York 14260

While significant progress has been made in elucidating molecular properties of specific genes and their regulation, our understanding of how the whole genome is coordinated has lagged behind. To understand how the genome functions as a coordinated whole, we must understand how the nucleus is put together and functions as a whole. An important step in that direction occurred with the isolation and characterization of the nuclear matrix. Aside from the plethora of functional properties associated with these isolated nuclear structures, they have enabled the first direct examination and molecular cloning of specific nuclear matrix proteins. The isolated nuclear matrix can be used for providing an *in vitro* model for understanding nuclear matrix organization in whole cells. Recent development of high-resolution and three-dimensional approaches for visualizing domains of genomic organization and function *in situ* has provided corroborative evidence for the nuclear matrix as the site of organization for replication, transcription, and post-transcriptional processing. As more is learned about these *in situ* functional sites, appropriate experiments could be designed to test molecular mechanisms with the *in vitro* nuclear matrix systems. This is illustrated in this chapter by the studies of nuclear matrix-associated DNA replication which have evolved from biochemical studies of *in vitro* nuclear matrix systems toward three-dimensional computer image analysis of replication sites for individual genes.

KEY WORDS: Cell nucleus, Nuclear matrix, Chromatin, RNP (ribonucleoproteins), Nucleolus, Interchromatinic regions, Interchromatin granules (ICGs), Matrix protein filaments (MPFs), Nuclear lamina, Nuclear pore complexes, Nuclear matrins, Matrin 3, Matrin cyp (cyclophilin), Matrin p250, Transcription, RNA splicing, DNA replication, DNA polymerase α, DNA primase, DNA replication sites, Clustersomes, Laser scanning confocal microscopy, Multidimensional computer image analysis, Fluorescence *in situ* hybridization (FISH), Chromosome painting.

I. Introduction

> Thus the task is, not so much to see what no one has seen yet; but to think
> what nobody has thought yet, about that what everybody sees.
>
> *Schopenhauer*

It is generally recognized that the original definition of the nuclear matrix
was a biochemical one in which the major proteins composing this residual
nuclear structure were first identified (Berezney and Coffey, 1974). While
this is true, it is only half of the story. The other half is the morphological
correspondence of the structures visible in the isolated nuclear matrix compared to intact nuclei. As Berezney and Coffey (1974) stated: "The similarities . . . are striking and indicate that the structure of the protein matrix
exists in isolated nuclei, and it is not a result of the treatments with salt,
detergents and enzymes. In addition, similar non-chromatin framework
structures can be detected in the nuclei of a variety of whole cells using
the Bernhard technique." The major conceptual contribution of this study,
therefore, was the idea of a higher order framework structure or "matrix"
which corresponded to identifiable *in situ* nuclear structures and which was
composed of discrete proteins (Berezney and Coffey, 1974). This was soon
followed, however, by the findings of DNA replication (Berezney and
Coffey, 1975) and RNA transcripts (Faiferman and Pogo, 1975; Herman
et al., 1976, 1978; Miller *et al.*, 1978; Herlan *et al.*, 1979) associated with the
nuclear matrix and led to the view of the nuclear matrix as a dynamic
structural system for genomic function (Berezney and Coffey, 1976, 1977;
Berezney, 1979).

In this chapter we present an overall perspective of the nuclear matrix field.
In so doing we will first review the earlier literature of nonchromatin nuclear
structure observed in intact cells and tissue. It is the concept of a nuclear
matrix derived from the initial isolation studies (Berezney and Coffey, 1974,
1977; Comings and Okada, 1976) that actually led to reinterpreting these
earlier ultrastructural studies in terms of an *in situ* nuclear matrix (Berezney,
1984). This is followed by selective perspectives concerning nuclear matrix
isolation, nuclear matrix proteins, and functional properties associated with
the nuclear matrix. Finally, we summarize research progress on DNA replication associated with the nuclear matrix which has been ongoing since 1975
(Berezney and Coffey, 1975) and has led us to recent three-dimensional studies of DNA replication sites using laser-scanning confocal microscopy and
multidimensional computer image analysis.

Throughout the article, emphasis will be placed on the experimental
results from our laboratory with selective supporting references from the
many other studies in this field. We apologize to the authors of the literally

hundreds of other pertinent studies which cannot be directly referenced or discussed in this brief overview of the field.

II. The *in Situ* Nuclear Matrix

A. A Historic Perspective

Research into the organizational, functional, and regulatory properties of the mammalian genome has accelerated in recent years in correlation with the enormous advances in molecular biology and genetics and their associated technologies. Despite this progress our understanding of the architecture of the cell nucleus and how the genome is organized and functions within this architectural context is still in its infancy. Indeed, until relatively recently, our view of the cell nucleus was highly influenced by the classical light microscopy studies of the late 19th and early 20th centuries which revealed three main structural components in the cell nucleus: (1) the chromatin, which actually corresponded to only the dense heterochromatin; (2) the nucleolus; and (3) nucleoplasm, nuclear sap, or karyolymph whose transparent appearance led to the belief that this was a clear fluid or gel.

The simple view of the nucleus as a homogeneous bag containing chromatin and the specialized nucleolar structure immersed in a sea of nucleoplasm has served as a reasonable starting point for investigation. It put the emphasis on the genetic material: the chromatin and the precise chromosomal structures visible during mitosis and meiosis. Indeed, cytological observations in the early 1900s in conjunction with genetic studies provided firm evidence that the cell nucleus is the repository for the genetic information contained within the chromosomes. It was, therefore, natural to consider that the stained chromatin network observed in the cell nucleus by standard transmission light microscopy represented a more amorphous and unraveled form of the chromosomes.

This concept of the interphase nucleus was subjected to major criticism following the application of electron microscopy to the cell nucleus from the late 1950s to the mid-1970s. These studies revealed for the first time heterogeneous structures in the interchromatinic regions between condensed chromatin and led Fawcett (1966) to prophetically state: "The chromatin and nucleolus are dispersed in a matrix traditionally called the *nuclear sap* or *karyolymph*. . . . Although these terms are still in use, they seem inappropriate designations for the congeries of submicroscopic granules of varying size and density that comprise the interchromosomal material of the interphase nucleus. . . . The terms *nuclear matrix* or *nuclear ground substance* are preferable."

As those of us who have managed to survive working in the area of nuclear architecture and function for the past 10 to 20 years readily know, most molecular and cell biologists who were studying the eukaryotic genome did not heed the prophet's voice and stuck with the bag of chromatin model—at least initially! Why was this the case? There are at least two reasons. First, as Max Planck once put it: "A new truth does not triumph by convincing its opponents and making them see the light, but rather because its opponents eventually die and a new generation grows up and is familiar with it."

Second, the "new truth" was too complicated and did not seem to add anything significant to our understanding of genetic processes other than mere speculation. What was clearly lacking in this new electron microscopic view of the cell nucleus was a global perspective of the overall nuclear architecture and how this might relate to genomic organization and function. Despite the significant progress in defining the details of individual granules and fibrils in the nucleus, the relationship of these structures to genomic functions was still unclear and, most importantly, there was no concept of how this might all fit together into an organized framework. As we shall see, the time became "ripe" for such a concept by the mid-1970s.

B. Defining Structural Components of the *in Situ* Nuclear Matrix

1. The Nuclear Matrix in Whole Cells

Figure 1a is a thin-sectioned electron micrograph which illustrates the basic morphology of the *in situ* rat liver nucleus. Dense chromatin is found along the nuclear periphery, the nucleolar periphery, and at other regions in the nuclear interior. Between the dense chromatin areas are the interchromatinic regions which contain at least two main structural components: the diffuse or euchromatin and the nonchromatin *nuclear matrix*. It is here that the distinction becomes vague. Although early studies reported the presence of presumptive nonchromatin granules (i.e., the *interchromatin* and *perichromatin granules*) and proteinaceous fibrils of 30–50 Å (Bernhard and Granboulan, 1963; Smetana *et al.,* 1971), it was generally difficult to distinguish between diffuse chromatin fibers sectioned in various directions and so-called nonchromatin granules and fibers.

An important breakthrough occurred with the development of the EDTA regressive staining procedure (Bernhard, 1969; Monneron and Bernhard, 1969). With this technique it was possible to preferentially bleach the DNA-containing chromatin structures while maintaining significant contrast of the nonchromatin (especially RNA-containing) structures. A remarkably

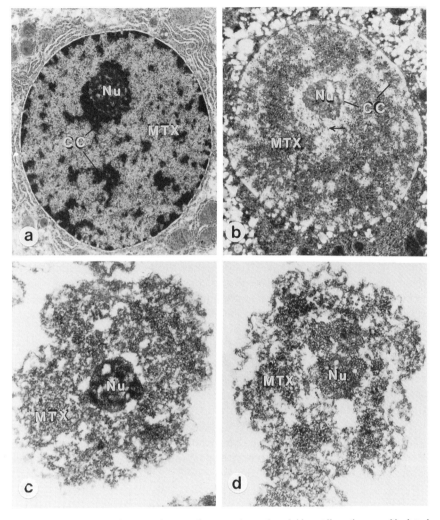

FIG. 1 Thin-sectioned electron micrograph comparison of nuclei in rat liver tissue and isolated nuclear matrix by standard staining with uranyl acetate and lead citrate and EDTA regressive staining. CC, condensed chromatin (heterochromatin); MTX, matrix region; NU, nucleolus. Note the similarity in structure of the isolated matrix and the *in situ* matrix. (a) Rat liver, standard staining; (b) rat liver, EDTA regressive staining; (c) isolated rat liver nuclear matrix, standard staining; (d) isolated rat liver nuclear matrix, EDTA regressive staining.

new view of the cell nucleus was revealed (Fig. 1b). An extensive nonchromatin fibrogranular nuclear matrix is observed within the interchromatinic regions under conditions wherein the dense chromatin regions show little or no staining. The nuclear pore complexes are especially densely stained (see arrows in Fig. 2). Moreover, channels of nuclear matrix material appear to radiate from the nuclear interior to the nuclear pore complexes. This was also observed in the earlier studies of Monneron and Bernhard (1969) and Franke and Falk (1970) and may correspond to passageways for regulated transport of RNA transcripts to the nuclear pores (Blobel, 1985).

2. Chromatin Domains

Aside from the nuclear matrix, the interchomatinic regions contain the diffuse or euchromatin. The studies of Derenzini *et al.* (1977, 1978) have enabled clear visualization of this diffuse chromatin structure inside the interchromatinic regions. Although unstructured, soluble proteins are certainly a component, it is an oversimplification to characterize this region of the nucleus as transcriptionally active chromatin immersed in a homogeneous nuclear sap. Indeed, combined studies of EDTA regressive staining for RNP (ribonucleoprotein) structures and the Feulgen-like osmium–ammine reaction revealed that the diffuse chromatin was only a minor component of the visible structure in the interchromatinic region. The major component were the RNP-rich nonchromatin fibrils and granules that compose the nuclear matrix.

Since the euchromatin contains the actively functioning DNA of the genome, this suggests a potentially close *in situ* association between actively replicating and transcribing chromatin and nuclear matrix components inside the functioning cell nucleus. At least one mode of interaction between the nuclear matrix and chromatin is at the DNA loop attachment sites. Other levels of association between chromatin and nuclear matrix components are intriguing possibilities and will require more detailed investigations of specific nuclear matrix proteins and their potential interactions with chromatin (see Chapter 5, Davie, 1995).

3. Ribonucleoprotein Domains

Electron microscopic studies of structures observed within the interchromatinic regions of the nucleus such as *interchromatin granules, perichromatin granules,* and *fibrils* stress the RNP nature of these components (Bernhard and Granbonlan, 1963; Monneron and Bernhard, 1969), and the EDTA regressive staining pattern of the nucleus in generally interpreted as an enhancement of these presumptive RNP domains (Bernhard, 1969; Monneron and Bernhard, 1969). High-resolution electron microscopic autoradiography in correlation with EDTA regressive staining supports this interpretation (Puvion and Moyne, 1981; Fakan and Bernhard, 1971). Newly transcribed extranucleolar RNA is preferentially localized along the bor-

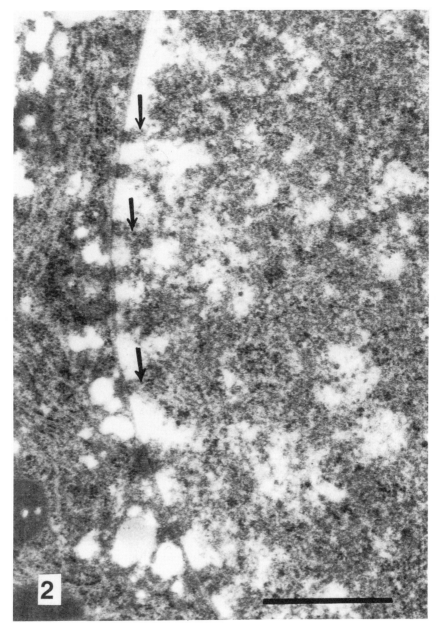

FIG. 2 Thin-sectioned electron micrograph of an *in situ* rat liver nucleus in the area of the nuclear pore complex by the EDTA regressive staining procedure of Bernhard. In this micrograph the interior *in situ* matrix appears contiguous with the nuclear pore complexes of the surrounding nuclear envelope. Arrows indicate the "channels" of matrix material radiating toward the nuclear pore complexes. Bar = 1 μm. (From Berezney, 1984.)

ders (where the perichromatin fibrils are concentrated) and within the interior of the interchromatinic regions (where the interchromatin granules are concentrated). Pulse–chase experiments have led to the idea that once released from the chromatin template, the newly transcribed RNA migrates through the interchromatinic region as RNP particles or fibrils toward the nuclear periphery (Fakan and Bernhard, 1973; Nash *et al.*, 1975; Fakan *et al.*, 1976). This suggests that the *in situ* matrix contains a network of RNP particles and/or fibrils extending from the sites of transcription to final release through the nuclear pore complexes. The strands or channels of nuclear matrix which lead from the interior to the nuclear pore complexes (see Fig. 2) may, therefore, represent dynamic assembly lines for the coordinate transcription, splicing, processing, and transport of RNA.

C. Functional Topography of the Nuclear Matrix

Our current knowledge of *in situ* nuclear structure leads us to view the mammalian cell nucleus as a three-dimensional mosaic complex of condensed chromatin, interchromatinic regions, nucleolar compartments, and a surrounding double-membraned nuclear envelope containing nuclear pore complexes (Fig. 3). The nuclear matrix is depicted as a dynamic fibrogranular structure in the interchromatininic region. At least two main structural domains are postulated within the *in situ* matrix: chromatin domains and RNP domains. This dual aspect of nuclear matrix organization inside the nucleus has important implications. It immediately suggests heterogeneity of the matrix at both the organizational and functional levels. The chromatin domains of the matrix (e.g., DNA loop attachement sites) are presumably quite different from RNP domains with respect to individual components and their organization.

From a functional point of view the matrix can be divided into chromatin domains of replication and/or transcription versus domains of post-transcriptional RNA processing. At the level of replication and transcription the matrix may provide a structural milieu for the expression of functional loops of chromatin and the organization of active replicational and/or transcriptional complexes. The potential continuous nature of the matrix system also suggests that these two distinct structural (chromatin and RNP structures) and functional (DNA replication, transcription, and RNA processing and transport) domains may be interlinked. In this manner the matrix may serve an important integrative and regulatory role for cascade and assembly line events.

This raises a commonly addressed question in this field. Is the nuclear matrix an independent skeletal structure in the nucleus which determines

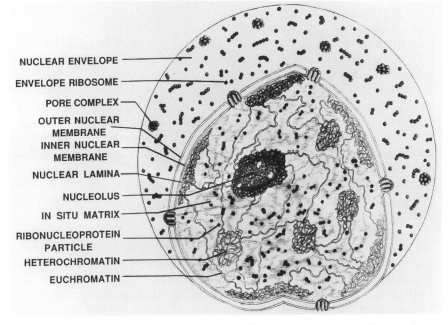

NUCLEAR ENVELOPE

ENVELOPE RIBOSOME

PORE COMPLEX

OUTER NUCLEAR
MEMBRANE

INNER NUCLEAR
MEMBRANE

NUCLEAR LAMINA

NUCLEOLUS

IN SITU MATRIX

RIBONUCLEOPROTEIN
PARTICLE

HETEROCHROMATIN

EUCHROMATIN

FIG. 3 Schematic model of a typical cell nucleus. The nucleus is surrounded by a double-membraned nuclear envelope containing nuclear pore complexes. Ribosome-like structures are found on the surface of the outer nuclear membrane as individual particles and "polysome-like" arrays. The chromatin in the nuclear interior is interpreted as a continuous system of condensed (heterochromatin) and diffuse (euchromatin) regions. The nonchromatin regions of the nuclear interior are simplified to contain the nucleolus, RNP (ribonucleoprotein) particles, and an *in situ* matrix forming a diffuse network which associates with the chromatin and nucleoli in the interior and the nuclear pore complexes at the periphery. The peripherally localized matrix may correspond to the nuclear lamina often observed in close association with the inner nuclear membrane. Drawn by L. A. Buchholtz and from Berezney, R. (1979) *In* "The Cell Nucleus," Vol. 7, (H. Busch, ed.), p. 413. Academic Press, New York.

three-dimensional organization and function of chromatin and RNPs? Or, is the matrix the "phenotypic expression" of nuclear functions with no independent existence devoid of function? It is a fundamental principle of all biology that structure and function, although commonly studied as separate entities, are actually two sides of the same coin: the fundamental biological processes of the living state (von Bertalanffy, 1952). From this perspective, rather than the cell "making" the nuclear matrix and the matrix driving nuclear functions, it is likely that the matrix and nuclear functions both make and drive each other in a unified symbiosis at the macromolecular level.

D. Dynamic Properties

The dynamic properties of nuclear architecture have been previously reviewed (Berezney, 1979). The cell nucleus responds to physiological changes be they cell cycle related, or induced externally or internally via hormones, drugs, or other environmental signals—by a variety of specific morphological alterations. One of the most common alterations involves change in the size of the nucleus. This has been strikingly observed in nuclear transplantation and cell fusion experiments where enormous enlargements have been measured in cell nuclei activated to synthesize DNA (Graham *et al.,* 1966; Harris, 1968, 1970). Moreover, the cell nucleus is constantly changing its size during the cell cycle. The HeLa cell nucleus, for example, doubles in volume between G_1 and the end of S-phase (Maul *et al.,* 1972).

In addition to changes in overall size, treatment of cells with drugs which inhibit various biosynthetic processes often results in striking perturbations in nuclear shape and internal morphology (Berezney, 1979). Many of these drugs are carcinogens. Indeed, it has been known since the pioneering studies by the 19th century cytopathologists that cancer cells are often characterized by the altered shape and internal morphology of their cell nuclei. This is strikingly illustrated in Fig. 4 which shows the irregular shape and morphology of a nucleus in a rat liver hepatoma cell (compare with Fig. 1).

By applying appropriate directional forces to cells, Ingber and colleagues (1994) have been able to observe coordinate changes in overall cell and nuclear shape in relation to mechanical signal transduction. Altering cell shape affected growth, indicating a close relationship of genomic function to the three-dimensional organization of the cell nucleus and the dynamic nature of the nuclear as well as cytoplasmic architecture.

In summary, a variety of studies indicate that the cell nucleus is a dynamic structure capable of rapid and striking changes in organization. We propose that the basis for this is a nuclear matrix structure which is in a state of constant flux and highly dependent on nuclear function. In this view (see Fig. 3 for a static interpretation) an overall nuclear organization is maintained despite the constant alterations in interactions at more local levels. It is these changes at the local level (e.g., replication, transcription, or RNA splicing sites), however, which accumulate and lead to changes in the overall nuclear architecture.

III. The Isolated Nuclear Matrix

If the *in situ* nuclear matrix is indeed a real structure of the cell nucleus, then it should be possible to isolate this structure. In 1974, Berezney and

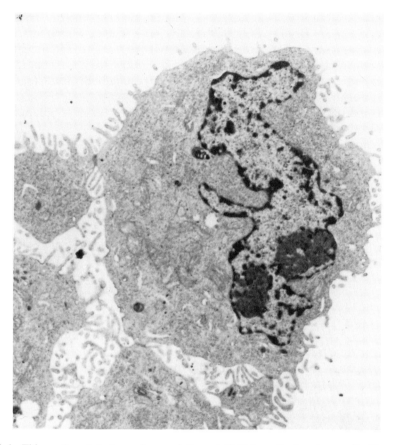

FIG. 4 Thin-sectioned electron micrograph through Zajdela ascites hepatoma cells. Note the bizarre shape of the nucleus.

Coffey, following on previous studies of residual nuclear structures (see Berezney and Coffey, 1976, and Berezney, 1979, for references), reported the first characterization of nuclear matrix isolated from rat liver tissue. Subsequent studies from this group (Berezney and Coffey, 1977) and many others (for references, see Berezney, 1979, 1984; Shaper *et al.,* 1978) have led to the characterization of nuclear matrices from a wide variety of eukaryotic cells from unicellular organisms to humans.

A. Isolation Procedures

Most procedures for nuclear matrix isolation are based on the original protocols reported by Berezney and Coffey (1974, 1977). A procedure currently used in our laboratory for isolating rat liver nuclear matrix is shown in Fig. 5. Briefly, morphologically intact nuclei are isolated and subjected to a series of treatments involving nuclease digestion, salt extractions, and detergent (Triton X-100) extraction. A key point is that morphologically recognizable nuclear structures are maintained throughout the extraction protocol, including the final nuclear fraction, despite the removal of most chromatin and protein and disruption of the nuclear membranes with detergent (Fig. 5).

A major modification for nuclear isolation was introduced by Laemmli's group (Mirkovitch *et al.*, 1984) who used the chaotropic agent and detergent LIS (lithium 3,5-diiodosalicylate) instead of salt solutions for extraction. This preparation has been termed the *nuclear scaffold* to distinguish it from its salt-extracted counterpart and has been widely used for the study of specific DNA sequences associated with the nuclear matrix (see Chapters 8 and 7; Bode *et al.*, 1995; Boulikas, 1995). In a comprehensive study of the HeLa nuclear matrix it was found that nuclear matrices prepared by the LIS method including heat stabilization were remarkably similar to their salt-extracted counterparts in polypeptide profiles and both ultrastructure and biochemical composition (Belgrader *et al.*, 1991a). In contrast, nuclear matrices prepared by salt extraction in the absence of heat or chemical stabilization showed a typical nuclear matrix morphology and chemical composition, while the corresponding LIS-prepared nuclear matrix was highly disrupted and virtually devoid of internal nuclear matrix structure (Belgrader *et al.*, 1991a).

Other preparations of these types of structures have been termed *nucleoskeletons, nuclear ghosts, nuclear cages,* and *residual nuclear structures,* to name just a few. The term *nuclear matrix,* however, is used by the vast majority of the over 1000 papers that have been published in this field. Cook and his associates use the term *nucleoid* for nuclear matrices which have intact, supercoiled DNA associated with them (McCready *et al.*, 1980). The term *DNA-rich nuclear matrices* has also been used to describe these preparations (Berezney and Buchholtz, 1981a). Cook's group has also developed a novel approach to studying nuclear matrix organization. They embed cells in agarose beads, and nuclease digest the chromatin with restriction enzymes and electroelute the chromatin at isotonic salt concentrations (Jackson and Cook, 1985). They call these preparations *nucleoskeletons.*

Nuclear matrices isolated from a wide range of organisms and cell types have a number of characteristic features which are summarized in Table I. Isolated nuclear matrices maintain many of the major architectural features of the intact nucleus despite the removal of 75–90% of the total nuclear protein and virtually all of the chromatin. The isolated matrices also contain large amounts of tightly bound RNA, lesser amounts of DNA (dependent on the degree of nuclease digestion) and only trace amounts

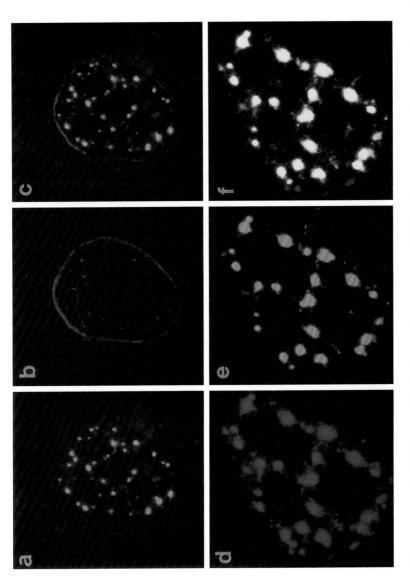

PLATE 1 (Berezney *et al.*, Fig. 15). In mouse 3T3 fibroblasts anti-matrin cyp (a, FITC-labeled) decorates splicing factor rich nuclear speckles and anti-lamin A/C (b, Texas-red labeled) defines the nuclear periphery. (c) Together these antibodies show a predominantly nuclear localization for matrin cyp. Double labeling with B3 (d, Texas-red labeled) and anti-matrin cyp (e, FITC labeled) demonstrates that nuclear speckles are recognized by B3 (f). As seen in (d) B3 also labels structures that are not nuclear speckles, including a fine punctate pattern that is dispersed throughout the nucleus and large domains that frequently are adjacent to nuclear speckles. These may correspond to sites of active transcription.

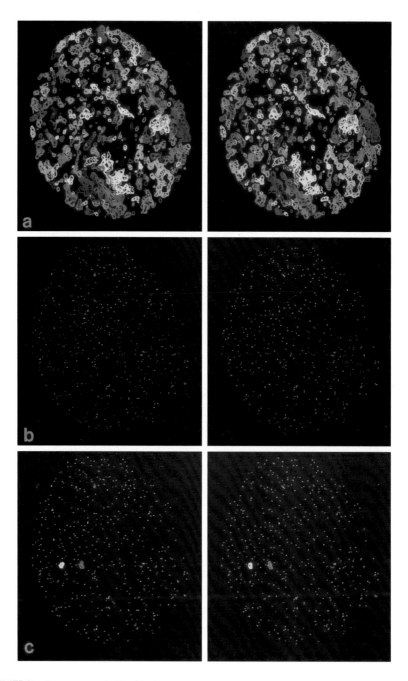

PLATE 2 (Berezney *et al.*, Fig. 25). 3-D reconstruction replication site outlines obtained from the image analysis software. Original image contained 25 sections (0.3 μm intervals). (a) Stereo pair of 3-D contours of the replication sites. Each of the 528 sites is assigned a label as indicated by different colors; (b) stereo pair of the centers of gravity for the contours shown in (a); (c) sites #282 and #368 each with volumes 0.123 and 0.102 μm^3, respectively, shown as selected through the visualization software.

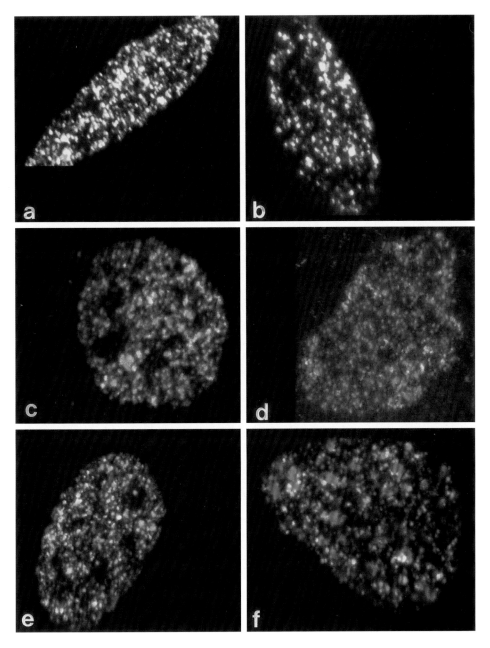

PLATE 3   (Berezney *et al.*, Fig. 26). "Pulse-chase-pulse experiment." Replication sites in 3T3 fibroblasts first labeled for 2 min with CldU (FITC secondary antibody, green sites), chased for 0 min–8 hr and pulsed again for 5 min with IdU (Texas Red secondary antibody, red sites). (a) Simultaneously pulsed; (b) chased for 15 min; (c) chased for 2 hr; (d) chased for 4 hr; (e) chased for 6 hr; (f) chased for 6 hr.

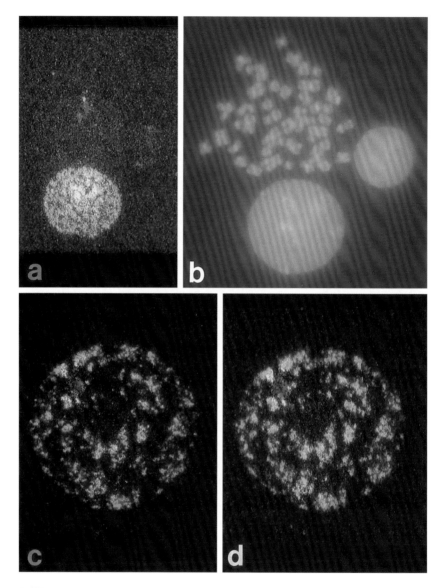

PLATE 4   (Berezney *et al.*, Fig. 27). Replication sites and human chromosome 11 painting. RPMI 7666 human lymphoblasts were labeled for 60 min with 30 μ*M*/ml BrdU before harvest for *in situ* hybridization of a human chromosome 11 painting system (BRL). Probe for identification of human chromosome 11 by FISH was directly labeled with Spectrum Orange. After hybridization, the slides were incubated with anti-BrdU antibody for 30 min followed by incubation with FITC-conjugated secondary antibody against anti-BrdU. (a) Human chromosome 11 was labeled as red color and the S-phase nucleus was labeled with BrdU as green color. Hybridized chromosome 11 was shown in a metaphase (top left); in a S-phase nucleus (in green) and a non-S-phase nucleus without BrdU labeling (the smaller nucleus on right). (b) All chromosomes were stained by DAPI; (c and d) stereo pair of 3-D reconstruction of a S-phase nucleus labeled with BrdU (in green) and human chromosome 11 domains (in red). Specific replication sites of chromosome 11 (yellow sites) can be discerned among the large population of genomic DNA replication (green sites).

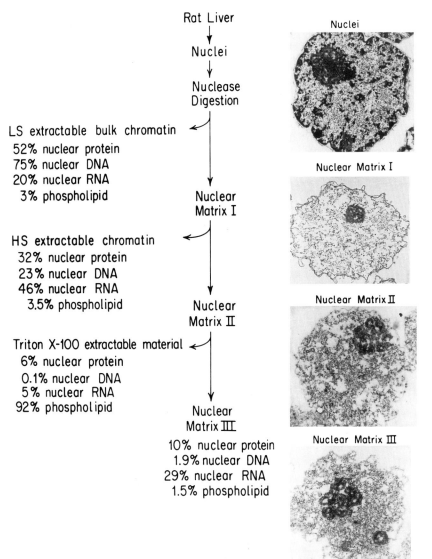

Rat Liver
↓
Nuclei
↓
Nuclease
Digestion

LS extractable bulk chromatin
52% nuclear protein
75% nuclear DNA
20% nuclear RNA
3% phospholipid

Nuclear
Matrix I

HS extractable chromatin
32% nuclear protein
23% nuclear DNA
46% nuclear RNA
3.5% phospholipid

Nuclear
Matrix II

Triton X-100 extractable material
6% nuclear protein
0.1% nuclear DNA
5% nuclear RNA
92% phospholipid

Nuclear
Matrix III
10% nuclear protein
1.9% nuclear DNA
29% nuclear RNA
1.5% phospholipid

Nuclei

Nuclear Matrix I

Nuclear Matrix II

Nuclear Matrix III

FIG. 5   Outline for the preparation of rat liver nuclear matrix. Isolated nuclei from rat liver tissue are sequentially extracted with DNase, low salt concentrations (LS), high salt concentrations (HS), and Triton X-100. Despite the considerable extraction of DNA, protein, RNA, and phospholipid during these steps, the final nuclear matrix (nuclear matrix III, bottom photo) still maintains the major architectural features of the intact nucleus (top photo). The terminology nuclear matrix I, II, and III is explained in more detail elsewhere (Berezney, 1979, 1984). From Berezney, 1984.

TABLE I

Properties of Isolated Nuclear Matrix

Isolation

Wide range from lower eukaryotes to humans.

Structure

Tripartite structure consisting of pore complex lamina, residual nucleoli, fibrogranular
internal matrix.

Composition

Contains 10–25% of the total nuclear protein and tightly bound DNA and RNA.

Polypeptides

Heterogeneous profile of Nonhistone proteins including cell type and differentiation state-
specific proteins as well as common proteins termed the nuclear matrins. Nuclear lamina
has a simpler profile with lamins A, B, and C (60–70 kDa) predominating.

Cell cycle

The cell cycle dynamics of the nuclear matrix in interphase cells remain to be elucidated. It
is likely that components of the interphase nuclear matrix (e.g., the DNA attachment sites)
are preserved in mitotic cells as the chromosome scaffold.

Functions

While the precise role(s) of the nuclear matrix in nuclear function remains to be elucidated,
its proposed role as a site for the organization and regulation of replication, transcription,
and RNA splicing and processing is supported by the vast array of functional properties
associated with isolated nuclear matrix.

---

of lipids if nonionic detergent extraction (e.g., Triton X-100) is performed.
The major macromolecular component is protein which includes a multitude of different proteins with an enrichment of the higher-molecular-
weight nonhistone proteins in the nucleus and a depletion of many of
the lower-molecular-weight ones, especially the histones (Berezney, 1979,
1984). Three of the major proteins are lamins A, B, and C which migrate
between 60 and 70 kDa on SDS–polyacrylamide gel electrophoresis and
are the most abundant components of the surrounding residual nuclear
envelope or *nuclear lamina.*

Laemmli and co-workers (Adolph *et al.,* 1977) first demonstrated that
isolated chromosomes extracted with nuclease and high salt maintain a
residual protein chromosomal structure termed the *chromosome scaffold.*
Further studies showed that the chromosomal DNA loops are attached to
the scaffold structure (Paulson and Laemmli, 1977). Since the DNA loops
are attached to the nuclear matrix in interphase cells, it has been widely
suggested that at least certain components of the interphase nuclear matrix
(e.g., the DNA attachment sites) are maintained in mitotic cells as components of the chromosome scaffold. Despite this widely held belief, our
knowledge of the precise relationships between the proteins composing the
interphase matrix versus the chromosome scaffold is very limited.

B. Morphological Properties of Isolated Nuclear Matrix

Three main structural regions typically compose the isolated nuclear matrix including a surrounding residual nuclear envelope or nuclear lamina containing morphologically recognizable nuclear pore complexes, residual components of nucleoli, and an extensive fibrogranular internal matrix. The latter structure is believed to represent residual components of the *in situ* nuclear matrix structures observed in whole cells (Berezney and Coffey, 1977). Using EDTA regressive staining as a criterion, a similarity is seen between the fibrogranular internal matrix of the isolated rat liver nuclear matrix and the *in situ* nuclear matrix visualized in intact tissue (Figs. 1b and 1d). The virtual identity of all the electron microscopic images seen following EDTA regressive staining versus conventional fixation and staining (Figs. 1c and 1d) further supports the derivation of the internal matrix from the *in situ* nuclear matrix structure in intact liver tissue.

A major consideration in preparing nuclear matrices which optimally resemble the *in situ* nuclear matrix is the procedure used for nuclei isolation. Procedures involving exposure of nuclei to hypotonic solution often lead to major perturbations of internal nuclear structure. These structural alterations will then be maintained in the isolated nuclear matrix. Unfortunately hypotonic swelling of cells is an important step for isolating pure and intact nuclei from most cells grown in culture. Some conditions of nuclei isolated via hypotonic swelling, however, appear to work better than others as illustrated in the study of Belgrader *et al.* (1991a) for isolation of the HeLa nucleus and nuclear matrix. In general, however, study of nuclear matrix morphology of cells grown in culture is best performed by direct extraction of cells grown on coverslips. The laboratories of Deppert (Staufenbiel and Deppert, 1984) and Penman (Fey *et al.*, 1986; Nickerson *et al.*, 1990) have pioneered the development of these so-called *in situ* nuclear matrix preparations which are particularly valuable for electron microscopic studies as well as for immunolocalization studies in the nucleus of intact cells versus *in situ* prepared nuclear matrices. A diagrammatic illustration of a typical *in situ* matrix preparation is shown in Fig. 6. Numerous studies have documented that such *in situ* preparations offer the advantage of better maintenance of nuclear morphology with biochemical and functional properties very similar to those of nuclear matrices prepared from isolated nuclei.

Perhaps the most widely characterized nucleus at the electron microscopic level in both intact tissue and as isolated nuclei is the rat liver nucleus. Procedures have been developed using isotonic and hypertonic sucrose solutions which maintain the ultrastructure of the isolated rat liver nucleus, while allowing for a high degree of purification (Blobel and Potter, 1966; Berezney et al., 1972; Berezney, 1974). This is illustrated in Fig. 7 where the ultrastructure of a typical isolated rat liver nucleus is strikingly similar

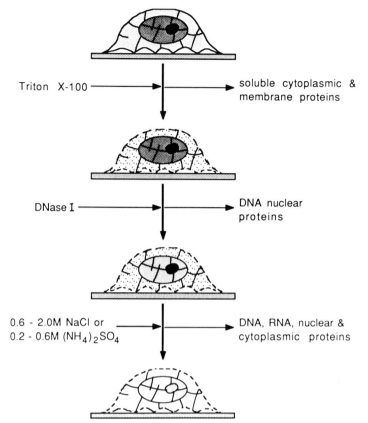

Triton X-100 ⟶ soluble cytoplasmic & membrane proteins

DNase I ⟶ DNA nuclear proteins

0.6 - 2.0M NaCl or
0.2 - 0.6M (NH$_4$)$_2$SO$_4$ ⟶ DNA, RNA, nuclear & cytoplasmic proteins

FIG. 6   Nuclear matrix prepared by *in situ* extraction of cells. Whole cells grown on coverslips can be directly extracted to isolate the nuclear matrix (bottom). Remnants of the cell surface and cytoskeleton help to anchor this *in situ* nuclear matrix to the coverslips. These preparations are especially valuable for cytochemical and immunolocalization studies but are also useful for biochemical studies.

to the *in situ* nuclear morphology observed in rat liver tissue. Aside from the maintenance of the general disposition of condensed chromatin and interchromatinic regions, there is a remarkable maintenance of interchromatin granules or ICGs (including the characteristic clusters of ICGs) and the less discrete appearing fibrous material which is, nonetheless, characteristic of the *in situ* matrix (Berezney and Coffey, 1977).

Preparation of nuclear matrices from such morphologically well-maintained rat liver nuclei enabled us to directly compare the structural details at the electron microscopic level of the isolated nuclear matrix in relation to the nucleus *in situ*. Figures 8 and 9 show the tripartite structure (nuclear lamina, residual nucleus, and internal matrix) at lower and higher

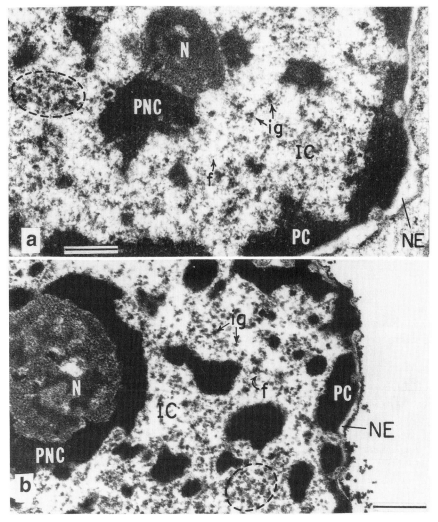

FIG. 7 Electron microscope sections of rat liver nuclei *in situ* in liver tissue (a) and in the isolated nuclear fraction (b). The isolated rat liver nuclei maintain many of the structural features characteristic of nuclei *in situ*. These include: NE, nuclear envelope; N, nucleolus; PNC, perinucleolar condensed chromatin; PC, peripheral condensed chromatin; IC, interchromatinic areas. The interchromatinic areas contain electron-dense particles 150–250 Å in diameter termed interchromatin granules (ig) as well as less electron-dense fibrous material (f). Clusters of interchromatin granules are enclosed by a broken line. ×44,000. (From Berezney and Coffey, 1977.)

magnification. Large empty regions, especially surrounding the nucleoli, suggest that these regions were formerly occupied by condensed chromatin domains. The fibrogranular structure of the isolated nuclear matrix is similar to the fibrogranular structures of the *in situ* matrix. This includes the appearance of discrete ICGs (often appearing as clusters in the isolated nuclear matrix) as well as the less discrete and less densely staining matrix protein filaments (MPFs) (Berezney and Coffey, 1977).

While the fibrous component of the *in situ* and isolated nuclear matrix are difficult to visualize by standard thin-sectioning electron microscopy, the poor quality of imaging does not mean that they do not exist or that they are not important. Indeed, these wispy and poorly staining MPF structures (30–50 Å in diameter) have been consistently identified in this region of the nucleus (Bernhard and Granboulan, 1963) but have been virtually ignored in most studies of nuclear organization. In one of the very few studies, however, Smetana *et al.* (1971) suggested that the fine filaments (MPFs) are protein structures since they were digested with pepsin. In contrast the ICGs were not affected by pepsin but predigestion with pepsin made these granules sensitive to ribonuclease digestion indicating the RNP nature of the ICGs.

The classical studies of Monneron and Bernhard (1969) using EDTA regressive staining also identified 30- to 50-Å fibers as a basic component of various extrachromatinic structures. These include perichromatin fibers and granules, coiled bodies, fibers associated with the nuclear pore complexes, and fibers which interconnect ICGs. Monneron and Bernhard (1969) further stressed the similar staining properties of the fibers and suggested a structural linkage between perichromatin fibers and ICGs.

These ill-defined and hard to visualize filaments which are maintained in appropriately prepared nuclear matrix preparation may, therefore, represent a fundamental structural component of the nuclear matrix. Indeed, Comings and Okada (1976) first termed the MPFs *matrixin* which was later called *matricin* (Berezney, 1980).

FIG. 8 (a) Electron microscope section through the nuclear protein matrix revealing the internal structural components of the matrix. RN, residual nucleolus; IM, internal matrix framework; RE, residual nuclear envelope layer. Note the empty spaces surrounding the residual nucleoli and along the periphery (arrows). These may correspond to regions previously occupied by the perinucleolar and peripheral condensed chromatin in untreated nuclei (compare with Figs. 7a and 7b). ×21,000. (b) Higher magnification electron microscope section of the nuclear protein matrix in the region of the residual nuclear envelope. A close association of the internal matrix framework (IM) with the residual nuclear envelope (RE) is evident (white arrows). A residual nuclear pore complex structure is projecting through the residual nuclear envelope layer (black arrows). Regions of the internal matrix framework (enclosed in broken line) resemble clusters of interchromatin granules seen in both isolated and in situ nuclei (compare with the regions enclosed by broken lines in Figs. 7a and 7b). ×50,000. (From Berezney and Coffey, 1977.)

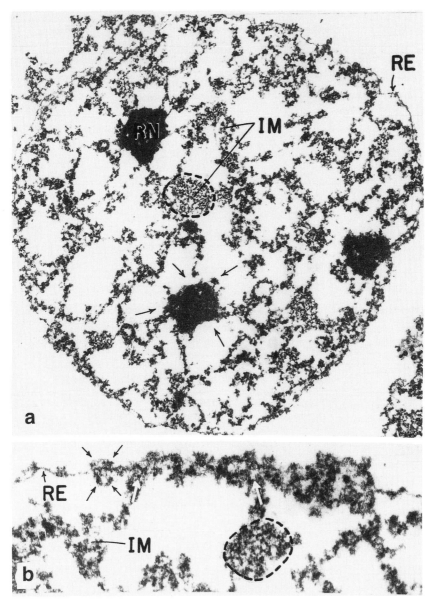

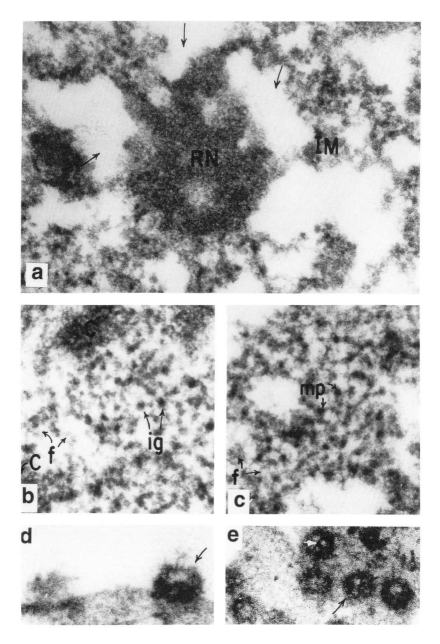

While standard thin-sectioning electron microscopy and EDTA regressive staining enable the visualization of a fibrogranular structure in whole cells and isolated nuclear structures, the structural information obtained is limited. In particular, each section is only a very thin slice through the entire nucleus. Thus the potential three-dimensional organization formed by the ICG's and the MPFs is not addressed. Moreover, the structural information obtained about the MPFs is extremely limited as the tracing of individual filaments would be severely compromised by the thinness of the sections.

As a step toward improving this situation, Penman and co-workers have used whole-mount and resinless thick-section electron microscopy to study the nuclear matrix (Capco et al., 1832; Fey et al., 1986; Nickerson et al., 1990). A complex three-dimensional network of filaments with associated granular structures is observed. Further fractionation revealed a three-dimensional network of core filaments which the Penman group has interpreted as the core framework structure around which other components are assembled in the complete nuclear matrix structure (He et al., 1990). The core filaments averaged 80–100 Å in diameter. Jackson and Cook (1988) have also reported a filament system in the nucleoskeleton obtained without high salt exraction.

We have observed similar approximately 100-Å extended filaments in more spread out regions of whole-mount preparations of isolated rat liver nuclear matrix (Fig. 10). As summarized above, conventional thin-sectioning electron microscopy has demonstrated that the internal matrix

---

FIG. 9 (a) High magnification of the nuclear protein matrix in the area of the residual nucleolus. The residual nucleolus, RN appears continuous with the internal matrix framework (IM). Empty spaces (arrows) surrounding the residual nucleolus may correspond to regions previously occupied by the perinucleolar condensed chromatin (compare with Figs. 7a and 7b). Note that these empty areas are bordered by the nucleolus and the internal matrix framework. ×102,000. (b) High magnification section of a region in the interior of an isolated nucleus. Distinct areas of condensed chromatin, C, are seen in the upper-center and lower-left regions of the micrograph. The regions between these condensed chromatin areas are the interchromatinic regions which contain electron-dense interchromatin granules (ig) and less electron-dense fibrous structures (f). ×116,000. (c) High magnification of a section through the internal matrix framework of the nuclear protein matrix. This residual framework structure consists of electron-dense matrix particles (mp) and matrix fibers (f) which bear a close similarity to the interchromatinic structures of isolated as well as in situ liver nuclei (compare with Figs. 9b and 7a and 7b). ×116,000. (d and e) High magnification sections through the residual nuclear envelope layer of the nuclear protein matrix. Distinct residual nuclear pore complex structures are observed which still retain their characteristic annular structure (arrows). Central granules are often visible in tangential sections through the residual pore complex structures (white arrow in e). ×158,000. (From Berezney and Coffey, 1977.)

is composed predominantly of ICGs that are enmeshed in less electron-dense, fibrous-like material. Filaments with a minimum diameter of 30–50 Å and termed MPFs can be observed extending from this material. The possible identity of the MPFs with the core filaments identified by

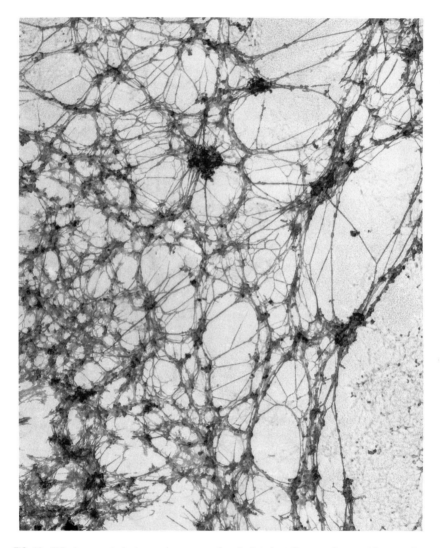

FIG. 10   Whole mount electron microscopy of an isolated rat liver nuclear matrix spread on an aqueous surface reveals an overall fibrous network structure. The specimen was critically point dried and rotary shadowed with platinum–palladium. The delicate matrix lacework is considerably disrupted in the absence of critically point drying.

Penman and co-workers (He *et al.*, 1990) is an important area of future investigation.

In earlier studies of nuclear subfractionation some investigators found that procedures related to those used for nuclear matrix isolation (nuclease, salt, and detergent) could also lead to so-called "empty" nuclear matrices which contained the surrounding nuclear lamina with nuclear pore complexes but were devoid of internal matrix structure. After some initial confusion it became apparent that the internal nuclear matrix is much more sensitive to extraction than the surrounding nuclear lamina (Kaufman *et al.*, 1981). This has led to more optimized preparations for both nuclear matrix with well-preserved internal matrix structure and nuclear lamina free of internal matrix components (Smith *et al.*, 1984; Belgrader *et al.*, 1991a). Figure 11 illustrates this point. If nuclei are digested with RNase A and extracted for nuclear matrix in the presence of sulfhydryl reducing agents such as dithiothreitol, the internal matrix is destabilized and empty matrices consisting exclusively of nuclear lamina are obtained. Preparation of nuclear matrix in the absence of RNase and dithiothreitol leads to typical tripartite matrices with elaborate internal matrix structure.

IV. Nuclear Matrix Proteins

Nuclear matrix proteins are the nonhistone proteins which comprise the nuclear matrix following nuclease, salt, and detergent extraction of isolated cell nuclei (see preceding section on nuclear matrix isolation). While virtually all known nuclear functions are associated with this proteinaceous nucleoskeletal structure (see following section on functional properties of nuclear matrices), our knowledge of the proteins which comprise this intriguing nucleoskeletal structure is very limited. There is no doubt, however, that a detailed molecular analysis of the individual nuclear matrix proteins is of paramount importance for deciphering the structural organization and molecular properties of nuclear matrix structure and the associated functions.

A. General Properties of Nuclear Matrix Proteins

Initial studies of the nuclear matrix proteins suggested that they represent a major component of the acidic nonhistone proteins of the cell nucleus (Berezney and Coffey, 1977). This conclusion was based on a combination of classical fractionation procedures (the nuclear matrix proteins are predominantly soluble in alkaline) and total amino acid analysis which revealed

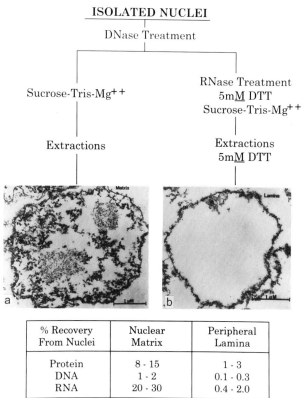

FIG. 11   Scheme for the preparation of nuclear matrix (a) and the peripheral nuclear lamina (b) from isolated rat liver nuclei. Note the typical tripartite structure of the isolated nuclear matrix with the surrounding nuclear lamina and an internal structure consisting of residual nucleoli and a fibrogranular matrix. This characteristic internal structure is lacking in the isolated nuclear lamina.

a remarkable similarity to the previously characterized residual acidic nuclear protein fraction extracted from mammalian cell nuclei (Steele and Busch, 1963). More recent two-dimensional PAGE (polyacrylamide gel electrophoresis) studies, however, have clearly indicated that there are many proteins in the nuclear matrix with overall basic charge as well as a large population of acidic ones (Peters and Comings, 1980; Nakayasu and Berezney, 1991; see Fig. 12).

The first reports of nuclear matrix proteins separated on one-dimensional SDS–PAGE, stressed the predominance of three major proteins between 60 and 70 kDa (Berezney and Coffey, 1974, 1977; Comings and Okada, 1976). It was apparent from those 1-D gels, however, that while the 60- to

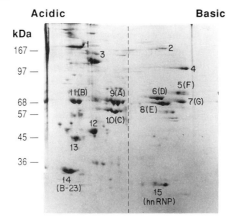

FIG. 12 Two-dimensional polyacrylamide gel electrophoresis (PAGE) of rat liver nuclear matrix proteins. Total rat liver matrix protein was run on a nonequilibrium pH gradient (first dimension), and on SDS–PAGE in the second dimension. The major proteins detected with Coomassie blue staining were numerically labeled (1–15) including one minor spot (protein 2) and another spot (protein 1) which often stained less intensely. The major proteins included lamins A, B, and C, the nucleolar protein B-23, and residual hnRNP proteins. The remaining eight major components appear to represent hitherto uncharacterized major proteins of the cell nucleus. Because these proteins stained the internal fibrogranular matrix by immunofluorescence, they have been termed the *nuclear matrins* to distinguish them from the nuclear lamins along the nuclear periphery (Nakayasu and Berezney, 1991). A group of nonlamin proteins which migrated as more basic components but in the same molecular weight range (60–75 kDa) as the lamins are also labeled matrins D, E, F, and G. Molecular weight markers are indicated in kilodaltons. (From Nakayasu and Berezney, 1991.)

70-kDa triplet was a major component, many other protein bands were present especially at molecular weight from approximately 50,000 to greater than 200,000 (Berezney and Coffey, 1977). At about the same time, Aaronson and Blobel (1975) reported that the nuclear pore complex–lamina fraction was also composed of three major proteins between 60 and 70 kDa which were ultimately termed *nuclear lamins* A, B, and C by Gerace and Blobel (1980).

A controversy then developed as to whether the 60- to 70-kDa proteins were strictly components of the surrounding nuclear lamina or also present in the internal nuclear matrix (Berezney, 1980; Gerace and Blobel, 1980). The final resolution is a good lesson in science. It turned out that both groups were right. Nuclear lamins A, B, and C are major components of the 60- to 70-kDa proteins but there are also other proteins in this molecular weight range which are present in nuclear matrix preparations, but not present in purified nuclear lamina. This was only apparent on two-dimensional SDS–PAGE (Berezney, 1984). Study of these other 60- to

70-kDa proteins has now become a major focus of our research (see Section IV,B). To make the story complete, recent reports now indicate that nuclear lamins are also found in the nuclear matrix interior (Goldman *et al.*, 1992; Bridger *et al.*, 1993; Moir *et al.*, 1994, 1995). This may be yet another lesson in science!

In summary, previous studies of nuclear matrix proteins using one-dimensional SDS–PAGE, while useful for providing an initial indication of the overall polypeptide profile of the nuclear matrix, are extremely limited due to the enormous complexity of the protein composition. This was not fully realized until two-dimensional gels were run. Thus attempts to identify similarities and differences among the polypeptide profiles obtained from nuclear matrices of different species, cellular origins, cell cycle stages, subnuclear fractionation, or physiological states should be regarded as preliminary results which must be extended to two-dimensional analysis (Berezney *et al.*, 1979a,b; Berezney, 1980; Zbarsky, 1981; Detke and Keller, 1982; Lebkowski and Laemmli, 1982; Sevaljevic *et al.*, 1982; Konstantinovic and Sevaljevic, 1983; Song and Adolph, 1983; Pieck *et al.*, 1985; Stuurman *et al.*, 1989).

One problem obtaining pure nuclear matrices is that nuclear matrices prepared from tissue culture cells are invariably contaminated with large amounts of cytoskeletal proteins, particularly the intermediate filament proteins (Capco *et al.*, 1982; Staufenbiel and Deppert, 1983; Verheijen *et al.*, 1986; Belgrader *et al.*, 1991a). Naturally, any *in situ* nuclear matrix preparations would behave in this way. Fey and Penman (1988) have circumvented this problem by an extraction procedure which separates intermediate filament proteins from the true nuclear matrix components. In contrast, the polypeptide profiles of nuclei isolated from tissues such as rat liver are likely to largely reflect the true nuclear proteins.

Two-dimensional analysis of nuclear matrix proteins performed by several different groups all stress the high degree of complexity of these polypeptide profiles (Peters *et al.*, 1982; Fey and Penman, 1988; Stuurman *et al.*, 1990; Nakayasu and Berezney, 1991). Using [^{35}S]methionine labeling for detection, Fey and Penman (1988) have detected over 200 proteins in the nuclear matrix. Stuurman *et al.* (1990) have also found enormous complexity in the two-dimensional profiles with the sensitive silver procedure. Despite this complexity, these studies are already providing valuable information. For example, the total nuclear matrix proteins can be separated into two major classes: those which are found in a variety of cell lines (common matrix proteins) and those which are cell type, hormonal, or differentiation state-dependent (Fey and Penman, 1988; Dworetzky *et al.*, 1990; Getzenberg and Coffey, 1990; Stuurman *et al.*, 1990).

The advantage of using 2-D PAGE to identify specific nuclear matrix proteins is further illustrated by the recent use of this approach to define differences in nuclear matrix proteins between normal and cancer cells grown in culture or in tumor tissues (Getzenberg et al., 1991; Khanuja et al., 1993; Partin et al., 1993; Pienta and Lehr, 1993a; Bidwell et al., 1994; Keesee et al., 1994). The potential application of these findings to the diagnosis, prognosis, and possibly treatment of human cancers is worthy of further studies (see e.g., Pienta and Lehr, 1993b).

Overall our knowledge of the proteins which comprise the nuclear matrix structure is very limited. One general property that needs further investigation is the phosphorylation of nuclear matrix proteins. Previous studies indicate that many of the nuclear matrix proteins are phosphorylated (Allen et al., 1977; Henry and Hodge, 1983). There are only a few reports, however, dealing with the phosphorylation or biosynthesis and turnover of specific nuclear matrix proteins in relation to the cell cycle and its regulation (Milavetz and Edwards, 1986; Halikowski and Liew, 1987; Zhelev et al., 1990).

The nuclear matrix itself contains a variety of protein kinases (see Table III) at least some of which are capable of phosphorylating endogenous nuclear matrix protein (Sahyoun et al., 1984; Capitani et al., 1987; Sikorska et al., 1988; Tawfic and Ahmed, 1994). Payraste et al. (1992) has reported differential localization of phosphoinositide kinases, diacylglycerol kinase, and phospholipase C in the nuclear matrix. Since protein kinase C is also present (Capitani et al., 1987), a complete regulatory signal transduction pathway may exist in the cell nucleus with the nuclear matrix serving an important role in arranging and sequestering the individual components of the regulatory components. In this regard, Coghlan et al. (1994) found a nuclear matrix-associated protein (AKAP 95) that demonstrated high-affinity binding sites for the regulatory subunit of type II cAMP-dependent protein kinase. It is very likely that these finding represent the tip of the iceberg in terms of regulatory factors associated with the nuclear matrix. Loidl and Eberharter (1995) present more details of regulatory proteins associated with the nuclear matrix the companion to this volume.

Use of antibodies raised against nuclear matrix proteins offers great promise. In specific studies, however, it has been difficult to identify the specific nuclear matrix antigen(s) responsible for the immunodecoration or extend the work beyond the initial characterization (Chaly et al., 1984; Lehner et al., 1986; Noaillac-Depeyre et al., 1987; Turner and Franchi, 1987). Progress is being made, however, because the groups of Smith et al. (1985) and Penman (Nickerson et al., 1992; Wan et al., 1994) have gone beyond the initial identification to demonstrate potential roles of their respective nuclear matrix-associated proteins in RNA splicing (Smith et al., 1989; Blencowe et al., 1994).

One very significant development is the recent demonstration by Brinkley and co-workers (Zeng *et al.*, 1994a,b; He *et al.*, 1995) that NuMA (nuclear mitotic apparatus protein) is a nuclear matrix protein. The propensity of this large protein (>200 kDa) to form a long coiled-coil structure (Compton *et al.*, 1992; Yang *et al.*, 1992) suggests an important role of NuMA in the organization of the nuclear matrix—possibly as a major component of the proposed nuclear matrix core filaments (He *et al.*, 1990).

Since the isolated nuclear matrix is composed of three major structural domains, it is important to determine the protein composition of each of these apparent "separate but connected" territories. Initial studies suggested that the polypeptide profile of the residual nucleolus or *nucleolar matrix* was distinct from the total nuclear matrix (Berezney and Coffey, 1977). This has been confirmed and extended by more recent studies of the nucleolar matrix (Olson and Thompson, 1983; Olson *et al.*, 1986; Shiomi *et al.*, 1986; Ochs and Smetana, 1991).

Studies of the nuclear lamin proteins have progressed extremely well. The nuclear lamins are now a bona fide class of intermediate filament proteins (McKeon *et al.*, 1986) and the nuclear envelope-associated LAPs (lamin-associated proteins) provides a basis for beginning to study specific protein–protein associations in this system (Foisner and Gerace, 1993). In this volume Moir *et al.* (1995) present a detailed account of the nuclear lamin proteins.

While basic features of the structure of the nuclear pore complex have been known for decades, the study of the proteins which compose this discrete structure has defied analysis for many years. Progress seemed imminent once Aaronson and Blobel (1975) successfully isolated the pore complex–lamina fraction, but the inability to separate the pore complex proteins from the tightly associated lamina (the predominating component) was a major stumbling block. In the mid-1980s, however, a breakthrough occurred when the combined extraction with moderate salt concentrations and Triton X-100 led to the differential solubilization of nuclear pore proteins from the extraction resistant lamina (Gerace *et al.*, 1984). Subsequent generation of monoclonal antibodies to these soluble pore complex-containing fractions has opened up a whole new area of research and has resulted in the characterization of numerous nuclear pore complex proteins (see the companion volume for details, Bastos *et al.*, 1995). One surprising finding was that a significant number of the nuclear pore complex proteins or *nucleoporins* were O-linked glycoproteins (Park *et al.*, 1987; Bastos *et al.*, 1995). The recent purification of nuclear pore complexes from yeast (Rout and Blobel, 1993) should further accelerate progress in this field.

Despite the need to determine the major protein components of the internal nuclear matrix and their structural organization, progress in identifying and characterizing these proteins has lagged behind. One area of significant development in the past few years, however, is the MAR binding proteins (see this volume Boulikas, 1995; Davie, 1995) which are presumably involved in anchoring the chromatin loop domains to the nuclear matrix (see the companion volume Razin *et al.,* 1995).

B. The Nuclear Matrins

Studies in our laboratory are concentrating on the major proteins of the nuclear matrix which are common at least among mammalian cells. Using a two-dimensional PAGE system, we have detected in rat liver nuclear matrix about one dozen major coomassie blue-stained proteins along with over 50 minor spots (Fig. 12). Polyclonal antibodies were then generated to individual matrix proteins excised from the two-dimensional gels. Antibodies to known nuclear proteins revealed that five of the major Coomassie blue-stained proteins correspond to lamins A, B, and C, the nucleolar protein B-23 and core hnRNP proteins. The remaining eight proteins were termed *nuclear matrins* because they are components of the internal nuclear matrix as revealed by immunofluorescence and are distinct from the nuclear lamins as indicated by 1-D and 2-D peptide maps (Nakayasu and Berezney, 1991). A survey of the literature showed no definite relationship among these proteins (matrins 3,4,D,E,F, G, 12, and 13) to any other known nuclear proteins (Nakayasu and Berezney, 1991). A summary of these major nuclear matrix proteins including their identification as nuclear lamins or matrins is presented in Table II.

Within the eight matrins examined, six formed three "pairs" of related proteins based on antibody cross-reactivity and peptide mapping (matrins D–E, F–G, and 12–13). This suggests that the nuclear matrins may compose a broad family of hitherto undefined proteins in the nucleus with potential subfamilies indicated by the various protein pair homologs. This leads us to propose that the term nuclear matrins be adopted as a general designation of nuclear matrix-associated proteins (determined via immunoblots) which are also components of the internal nuclear matrix (determined via immunofluorescence in whole cells versus cells extracted for nuclear matrix).

We are currently studying the molecular and functional properties of the individual nuclear matrins. As an initial step we have screened nuclear matrix proteins for DNA binding activity on Southwestern blots. Using one/dimensional SDS–PAGE we have shown that the nuclear matrix is enriched in the higher-molecular-weight DNA binding proteins found in total rat liver nuclear matrix proteins (Fig. 13; Hakes and Berezney, 1991). Approximately one dozen major DNA binding proteins with apparent molecular weights exceeding 40,000 were detected on the 1-D blots. Further studies indicated that these proteins preferentially bound DNA when com-

TABLE II

Summary of Major Coomassie Blue-Stained Nuclear Matrix Proteins Derived from Two-Dimensional PAGE

Nuclear matrix protein designation	Approximate kDa	Acidic or basic[b]	Protein identity[c]
1[a]	190	A	Matrin
3	125	A	Matrin
4	105	B	Matrin
5 (F)	77	B	Matrin
6 (D)	72	B	Matrin
7 (G)	68	B	Matrin
8 (F)	66	B	Matrin
9 (A)	68	A	Lamin
10 (B)	66	A	Lamin
11 (C)	62	A	Larnin
12	48	A	Matrin
13	42	A	Matrin
14	34	A	B-23
15	30	B	hnRNP

[a] Designation of protein 1 as a major matrix protein is tentative since it often stains as a more minor component.

[b] Designation as acidic (A) or basic (B) is based on migration across a nonequilibrium pH gradient gel. The dotted lines across the gels indicate the position of pH 7.0. Additional studies with equilibrium pH gradients have verified this general characterization (data not shown).

[c] The identity of the matrix proteins was evaluated by 2-D immunoblots to known nuclear proteins. Those proteins in the nuclear matrix fraction which were not identified as previously known protein are termed the *nuclear matrins* to distinguish them from the well-known nuclear lamins of the nuclear periphery.

peted with excess RNA (Hakes and Berezney, 1991). Two-dimensional Southwestern blots were then performed to identify the specific DNA binding proteins (Fig. 14). Approximately 12 distinct spots were detected including lamins A and C (but not B), matrins D,E,F,G, and 4 (but not 3), and an unidentified protein of about 48 kDa.

As a step toward the further characterization of the nuclear matrins we have been screening λ gt11 cDNA expression libraries with our polyclonal antibodies to these matrins. As outlined in the following sections we have succeeded in cloning two authentic nuclear matrins. In addition we will present recent and surprising results on a high-molecular-weight nuclear matrix-associated protein (matrin p250) which appears to be identical to a hyperphosphorylated form of the large subunit of RNA polymerase II.

It is likely that the next few years will see the elucidation of many of the nuclear matrix proteins using this molecular cloning approach. This will provide fundamental information about a family of proteins which are of obvious significance for nuclear organization and likely function but that have, until recently, defied analysis.

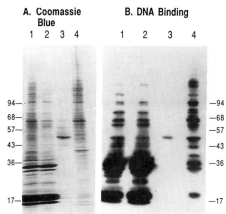

A. Coomassie Blue   **B. DNA Binding**

FIG. 13 Identification of DNA binding proteins in the nuclear matrix by one-dimensional Southwestern blots. Total rat liver nuclear proteins (lanes 1), nuclear matrix proteins (lanes 4), and the proteins from the salt (lanes 2) and Triton X-100 (lanes 3) extraction steps of the nuclear matrix isolation procedure were separated on 5–18% SDS polyacrylamide gradient gels and either Coomassie blue stained (A) or electrophoretically transformed to nitrocellulose paper and probed with labeled genomic DNA (Southwestern blot). Note the enrichment in higher-molecular-weight DNA binding proteins in the nuclear matrix. The positions of molecular weight markers are shown in kilodaltons (Hakes and Berezney, 1991).

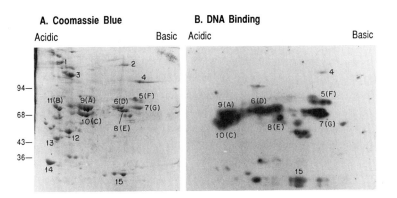

A. Coomassie Blue   **B. DNA Binding**

FIG. 14 Identification of DNA binding proteins in the nuclear matrix by two-dimensional Southwestern blots. Total nuclear matrix proteins were separated on two-dimensional gels, stained with Coomassie blue (A), or transferred to nitrocellulose paper for DNA binding with labeled genomic DNA (B). The major DNA binding proteins identified on the two-dimensional Southwesterns were lamins A and C and matrins D, E, F, and G and an unidentified protein migrating at about 48 kDa (Hakes and Berezney, 1991).

1. Matrin 3

Matrin 3 was identified as an acidic 125-kDa nuclear matrix protein that stained the interior of mammialian cell nuclei in a complex fibrous and granular pattern (Nakayasu and Berezney, 1991). Both human and rat cDNA sequences were obtained which code for matrin 3. Analysis of these sequences indicated that matrin 3 was conserved between species, but shows little homology with other reported proteins, giving few clues to possible nuclear functions (Belgrader *et al.*, 1991b). Antibodies raised against a recombinant matrin 3 protein were able to specifically recognize a 125-kDa protein that was exclusively present in the nucleus and highly enriched in the nuclear matrix. Consistent with the over 40 possible phosphorylation site found in matrin 3, two-dimensional immunoblot analysis revealed that anti-matrin 3 antibodies stained several spots of acidic isoelectric points. Additionally, these antibodies stained mammalian nuclei and *in situ* nuclear matrices in similar fibrogranular patterns throughout the nuclear interior (Belgrader *et al.*, 1991b). Because matrin 3 is an evolutionarily conserved and highly abundant nuclear matrix protein, determination of its function and regulation are of great importance to understanding its structural contributions to the nuclear matrix.

Elucidating the genomic organization of matrin 3 will provide a basis for future work characterizing the regulation and function of this internal nuclear matrix protein (Somanathan *et al.*, 1995). Southern blot analysis of inbred rats demonstrated that there were three separate regions in the rat genome that show high sequence identity throughout the entire rat matrin 3 cDNA (Somanathan *et al.*, 1995; Mortillaro and Berezney, 1993). Attempts to isolate partial gene sequences by low-stringency polymerase chain reaction of rat genomic DNA successfully identified a region from two separate genes that belong to the matrin 3 gene family. The first of the fragments, termed matrin 3A, encodes a sequence identical to a corresponding fragment from the matrin 3 cDNA. The second fragment, matrin 3B, is a novel sequence with high nucleotide and amino acid identity to matrin 3A which may indicate a functionally related role for the matrin 3B gene product (Somanathan *et al.*, 1995). Additionally, analyses of matrin 3A clones isolated from a rat genomic DNA library revealed an inserted sequence just proximal to the putative nuclear localization signal. This inserted sequence contains a short open reading frame, showing significant homology to EF hand divalent cation binding domains which have been shown to be alternatively spliced in other proteins (Perret *et al.*, 1988; Sorimachi *et al.*, 1993). Consistent with this possibility is the isolation of a matrin 3 cDNA sequence that appears to contain an alternatively spliced region encoding a similar EF hand domain (Somanathan *et al.*, 1995). Our results in combination with recent reports of calcium as a regulator of nuclear transport (Gerber and Gerace, 1995) have led us to postulate that matrin 3 may be among those nuclear proteins whose transport is regulated by calcium.

2. Matrin cyp

Recently a 100-kDa nuclear matrix protein, termed matrin cyp (cyclophilin), was identified, cloned and characterized (Mortillaro and Berezney 1995). To briefly summarize our findings, the derived amino acid sequence of matrin cyp contains three interesting domains: (i) a 170-amino acid domain that shows high identity with a family of ubiquitous proteins called cyclophilins. These proteins have been demonstrated to bind the immunosuppressant drug, cyclosporin A, and catalyze conversion of *cis*-proline to *trans*-proline, which is important for protein folding (Fischer *et al.*, 1989; Gething and Sambrook, 1992); (ii) an acidic, serine-rich region similar to the type of nuclear localization signal binding (NLS-binding) domains initially described for Nopp140 (Meier and Blobel 1992); and (iii) a series of serine–arginine (SR) repeats throughout the carboxyl half of the protein which also are also present in several splicing factors (Zahler *et al.*, 1993) and have been demonstrated to target proteins to splicing-factor-rich nuclear speckles (Li and Bingham 1991).

Antibodies raised against matrin cyp were used in immunoblot analysis to determine subcellular location. A single band of 100 kDa was recognized that was located within the nucleus and enriched in the nuclear matrix. Two-dimensional immunoblots revealed several major spots of various isoelectric points at about 100 kDa. This is suggestive of different phosphorylation states and is not surprising considering the high degree of phosphorylation associated with NLS-binding domains and SR repeats (Gui *et al.*, 1994; Meier and Blobel, 1993).

Immunofluorescent experiments corroborated immunoblot analysis, in that, laser scanning confocal microscopy localized the majority of anti-matrin cyp-specific signal within the nucleus and predominantly to large bright foci (Figs. 15a–15c, Color Plate 1). However, unlike the immunoblot analysis, a low level of anti-matrin cyp stained the cytoplasm in a diffuse pattern. To more carefully examine this difference, a technique was developed to quantify immunofluorescence data. The results demonstrated that while most of the high-affinity anti-matrin cyp-specific antigens were present in the nucleus, a small but discrete amount was also in the cytoplasm that would otherwise be undetected via immunoblot analyses. Further evidence that matrin cyp was a nuclear matrix protein was provided by immunofluorescence miscroscopy which showed the staining pattern of antimatrin cyp was maintained in cells extracted for *in situ* nuclear matrices.

Because the cDNA contained numerous SR repeats, and the bright foci observed in the preceding immunofluorescence studies were reminiscent of splicing factor-rich nuclear speckles (Spector, 1984), double labeling experiments were performed with anti-matrin cyp and anti-splicing factor-specific antibodies. Our results demonstrated that matrin cyp and splicing factors, indeed, colocalize at speckles within the nucleus (Mortillaro and Berezney, 1995). This colocalization, plus evidence that some splicing factors interact through SR domains (Wu and Maniatis 1993), raised the possibility that matrin cyp may be functioning in pre-mRNA splicing. However, although matrin cyp was present in HeLa splicing extracts, anti-matrin cyp was incapable of immunoprecipitating splicing complexes *in vitro,* nor could anti-matrin cyp or cyclosporin A inhibit splicing *in vitro* (Mortillaro, Blencowe, Sharp, and Berezney, in preparation). These results indicated no role for matrin cyp in the core pre-mRNA splicing processes associated with the soluble *in vitro* system. Since the nuclear matrix has been demonstrated to increase splicing efficiency of associated pre-mRNA (Zeitlin *et al.,* 1989), it remains to be resolved whether matrin cyp affects splicing when associated with this structure.

Knowledge that proline isomerase (PPIase) activity of cyclophilins is important for the proper folding and functioning of specific cellular proteins (Gething and Sambrook, 1992) led us to determine if the cyclophilin domain found in matrin cyp could function as a proline isomerase and possibly indicate a "foldase" function *in vivo.* To this end a fusion protein containing the cyclophilin domain was recently constructed, purified, and shown to exhibit, *in vitro,* a cyclosporin A-inhibited PPIase activity with a rate constant and cyclosporin A inhibition curve similar to those reported for other cyclophilin proteins (Mortillaro and Berezney, in preparation). Evidence that functioning cyclophilin proteins are present in the nuclear matrix was further indicated when PPIase activity, capable of being inhibited by cyclosporin A, was detected in solubilized nuclear matrix proteins. It is interesting to note that a nucleolar, NLS- and KF506-binding proline isomerase has been implicated as a nucleolar/cytoplasmic shuttle (Shan *et al.,* 1994). The observations that matrin cyp contains a functioning cyclophilin domain, a putative NLS-binding domain, and is present in both the splicing factor-rich nuclear speckles and cytoplasm, leads us to speculate that matrin cyp may associate, via its cyclophilin and NLS-binding domains, with newly synthesized nuclear proteins in the cytoplasm, and by acting as a molecular chaperone, assists in targeting these proteins to the nuclear speckles (Fig. 16).

3. Matrin p250 and RNA Polymerase II-LS

A mouse monoclonal antibody, B3, raised against total rat liver nuclear matrix proteins, recognizes a protein doublet of 250 kDa termed p250 (Nakayasu and Berzney, 1987). Subcellular fractionation followed by immunoblot analysis identified p250 as a nuclear matrix-enriched protein (Fig.

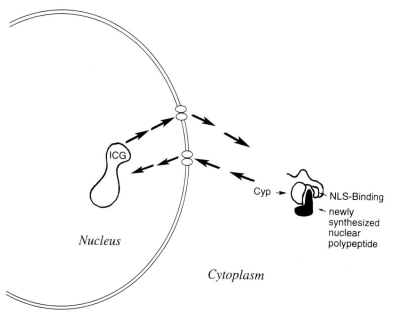

FIG. 16 A schematic model of the putative nuclear speckle/cytoplasmic shuttle function ascribed to matrin cyp (see text for description). Briefly, the cyclophilin (cyp) and nuclear localization signal binding domain (NLS-binding), found on matrin cyp, may bind to a cytoplasmically located nuclear protein (probably recently translated), assist in folding this protein, and then target this protein to the nuclear speckles (icg). Matrin cyp could then release this protein and cycle back into the cytoplasm to bind another protein.

17). In an initial attempt to determine which antigens were bound by B3, antibodies against known nuclear proteins of 240 to 250 kDa were tested for the ability to comigrate with p250. These experiments indicated that p250 is not NuMA (Compton and Cleveland, 1994) nor the TBP-associated factor p250 (Hisatake *et al.,* 1993), but may be the large subunit of RNA polymerase II, as shown by several anti-RNA polymerase II antibodies (Wei *et al.,* 1995; Wei, Du, Warren, and Berezney, unpublished results).

Few studies have characterized the structural organization of transcription factors, including RNA polymerase II, within the nucleus (Bregman *et al.,* 1995). To gain additional insight into the nuclear architecture of RNA polymerase II, immunofluorescent studies were undertaken and revealed that B3 recognized an elaborate fibrogranular structure within the cell nucleus consisting of 30 to 50 brightly staining foci and hundreds of less intensely staining granules (Wei *et al.,* 1995). As in the case of matrin cyp, these large foci were similar in size, shape, and number to nuclear speckles. Double-labeling three-dimensional analysis indicated that most

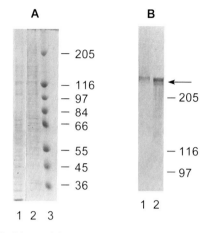

FIG. 17   (A) Coomassie blue staining of purified nuclear proteins, lane 1, nuclear matrix proteins, lane 2, and molecular weight markers, lane 3. (B) Immunoblot analysis demonstrates that a 250-kDa antigen is specifically recognized in purified nuclear proteins, lane 1, and nuclear matrix proteins, lane 2, by the B3 monoclonal antibody. The molecular weight markers are indicated on the right side.

of the bright foci stained by B3 overlapped with splicing factor-rich nuclear speckles (Figs. 15d–15f, Color Plate 1).

To ascertain whether these results had any functional significance to RNA splicing, B3 was tested in *in vitro* splicing reactions. Indeed, B3 was capable of immunoprecipitating antigens associated with splicing complexes *in vitro* (Wei *et al.,* 1995). Although unexpected, this finding may represent evidence that RNA polymerase II is a physical link between splicing and transcription, supporting the hypothesis that the carboxyl-terminal domain of RNA polymerase II is involved in the recruitment of splicing factors to sites of transcription within the nucleus (Greenleaf, 1993).

Further identification and characterization of the major proteins associated with the internal nuclear matrix will contribute to elucidating the complex interactions of functional domains which comprise this structure. Taken together, these studies of matrin cyp, matrin 3, and RNA polymerase II demonstrate that factors associated with the nuclear matrix serve both a structural and functional role and lead us to postulate that the nuclear matrix consists of multiple functional domains whose modular associations constitute its fibrogranular nature.

V. Nuclear Matrix Functions

The nuclear matrix was first identified in whole cells as that region of the nucleus where the actively functioning chromatin is located along with the nonchromatin fibrogranular matrix structures (Berezney and Coffey, 1977; Berezney, 1984). It is, therefore, not surprising that isolated nuclear matrices, which show a structural correspondence to the *in situ* defined structures, have a vast array of functional properties associated with them. Table III summarizes many of these major properties along with sample references.

It is important to stress that while it is no surprise to see this multitude of functional properties ranging from DNA loop attachment sites to DNA replication, to transcriptional associations, to RNA transcripts, to RNA splicing, to viral associations and their associated functions, and to a vast number of regulatory proteins involved in the functioning and regulation of these functions (e.g., steroid hormone receptor receptors, oncogene proteins, tumor suppressers, heat shock proteins, calmodulin binding proteins, protein kinases), the true significance of these associations remain to be determined. Initial results, however, suggest that the isolated nuclear matrix is a potentially powerful *in vitro* approach for studying the molecular biology of higher order nuclear structure and function.

Since only limited studies have been performed on many of these properties, more studies are needed to better define the nature of the associations and the actual role(s) of the nuclear matrix structure in these processes. This is true even for those properties that have been studied in more detail such as DNA loop-attachment sequences (so-called "MAR" or "SAR" sequences, see chapters by Bode *et al.,* 1995; Boulikas, 1995; Moreno Diaz de la Espina, 1995; Razin *et al.,* 1995), active gene sequences, RNA transcripts, viral associations, steroid hormone binding, and DNA replication. What is needed for each functional property is a detailed description of the associated function, what molecular constituents are involved and the relationship of the *in vitro* function to *in situ* associations. Naturally, this last evaluation is most difficult and may require continued studies of the *in vitro* associations until enough is known to plan appropriate experiments at the level of whole cells.

A major development in the past several years which provides optimism for future studies are the multitude of immunofluorescence studies demonstrating a variety of functional domains in the nucleus of intact cells (Spector, 1993) including territories for specific chromosomes (Manuelidis, 1985, 1990; Hadlaczky *et al.,* 1986; Cremer *et al.,* 1993), DNA replication sites (Nakamura *et al.,* 1986; Nakayasu and Berezney, 1989; Fox *et al.,* 1991; Kill *et al.,* 1991; O'Keefe *et al.,* 1992; Neri *et al.,* 1992; Hassan and Cook, 1993; Hozák *et al.,* 1993), and transcriptional and RNA splicing sites (Huang and Spector, 1991; Jackson *et al.,* 1993; Wansink *et al.,* 1993).

In a particularly elegant series of experiments, Lawrence and co-workers (1989) demonstrated that abundantly newly transcribed RNA often forms

TABLE III

Functional Properties Associated with Isolated Nuclear Matrix

DNA loop attachment site sequences	(Gasser and Laemmli, 1987)
DNA binding proteins	(Hakes and Berezney, 1991)
DNA topoisomerase II	(Fernandes and Catapano, 1991)
Replicating DNA	(van der Velden and Wanka, 1987)
Replication origins	(Dijkwel *et al.*, 1986)
DNA polymerase α and primase	(Tubo and Berezney, 1987b)
Other replicative factors	(Tubo and Berezney, 1987a)
Active gene sequences	(Zehnbauer and Vogelstein, 1985)
RNA polymerase II	(Lewis *et al.*, 1984
Transcriptional regulatory proteins	(Feldman and Nevins, 1983)
hn-RNA and RNP	(Verheijen *et al.*, 1988)
Preribosomal RNA and RNP	(Ciejek *et al.*, 1982)
sn-RNA and RNP	(Harris and Smith, 1988)
RNA splicing	(Zeitlin *et al.*, 1987)
Spliceosome-associated proteins	(Blenkowe *et al.*, 1994)
Steroid hormone receptor binding	(Rennie *et al.*, 1983)
Viral DNA and Replication	(Smith *et al.*, 1985)
Viral premessenger RNA	(Mariman *et al.*, 1982)
Viral proteins	(Covey *et al.*, 1984)
Carcinogen binding	(Gupta *et al.*, 1985)
Oncogene proteins	(Eisenman *et al.*, 1985)
Tumor suppressors	(Mancini *et al.*, 1994)
Heat shock proteins	(Reiter and Penman, 1983)
Calmodulin binding proteins	(Bachs and Carafoli, 1987)
HMG-14 and HMG-17 Binding	(Reeves and Chang, 1983)
ADP-ribosylation	(Cardenas-Comna *et al.*, 1987)
Protein phosphorylation	(Allen *et al.*, 1977)
Protein kinase A	(Sikorska *et al.*, 1988)
Protein kinase B	(Sahyoun *et al.*, 1984)
Protein kinase C	(Capitani *et al.*, 1987)
Protein kinase CK2	(Tawfic and Ahmed, 1994)
Histone acetylation and deacetylation	(Hendzel *et al.*, 1991, 1994)
Protein disulfide oxidoreductase	(Altieri *et al.*, 1993)
Reversible size changes	(Wunderlich and Herlan, 1977)

visible "tracks" in the nucleus that presumably extend from sites of transcription toward sites of further processing and transport. This suggested that the newly transcribed RNA was somehow *post-transcriptionally ordered* in the nucleus. The maintenance of these gene transcript track patterns following extraction for nuclear matrix (Xing *et al.*, 1991), implicates the nuclear matrix in this structural organization. Recently, the Lawrence group has extended these studies to show specific orientation of the tracks with respect to sites of transcription and gene splicing for the fibronectin gene using intron-containing and spliced transcripts (Xing *et al.*, 1993).

A major conclusion of these studies is that transcription and RNA splicing may be a coordinated process in which the nuclear matrix plays a fundamental role in the organization. Jiménez-Garcia and Spector (1993) using different approaches also conclude that transcription and splicing may be temporally and spatially linked in the cell nucleus. Similarly, the speckled sites where splicing factors and snRNPs are concentrated (Spector *et al.*, 1983; van Eekelen *et al.*, 1982) and the sites of DNA replication (Nakayasu and Berezney, 1989) are strikingly maintained following extraction of cells on coverslips for nuclear matrix.

The host of regulatory factors which are in the nucleus and nuclear matrix associated (Table III) further supports the model of the nuclear matrix as a dynamic entity involved in structural organization, function, and regulation of the genome (Fig. 3). Additional recent findings include the association of a plethora of transcription factors with nuclear matrix (van Wijnen *et al.*, 1993) and the presumptive role of specific nuclear matrix proteins in RNA splicing *in vitro* (Smith *et al.*, 1989; Blencowe *et al.*, 1994).

The last section of this chapter presents a more detailed description of one of the best-studied functions associated with the nuclear matrix: DNA replication. In particular, recent experiments will be described that are designed to "bridge the gap" between *in vitro* matrix systems and replication *in situ.*

VI. The Nuclear Matrix and DNA Replication

It is known that each enormous molecule of eukaryotic chromosomal DNA is divided into hundreds to thousands of independent subunits of replication termed replicons (Hand, 1978). Replication proceeds bidirectionally within each replicon subunit. Individual replicons are further organized into families or clusters of tandemly repeated subunits which replicate as a unit at particular times in S-phase (Hand, 1975, 1978; Lao and Arrighi, 1981; Fig. 18A). Up to 100 or more replicons may be organized into each replicon cluster with an estimated average size of approximately 25 (Hand, 1978;

Painter and Young, 1976). The numerous reports that specific DNA sequences are duplicated at precise times within the S-phase of eukaryotic
cells (Goldman *et al.*, 1984; Hatton *et al.*, 1988) further support the conclu-

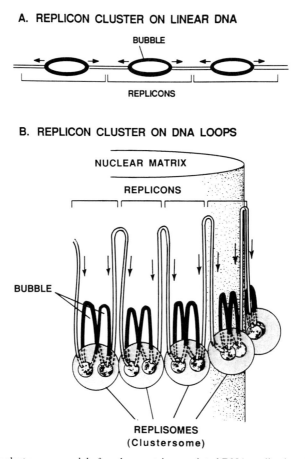

FIG. 18   The clustersome model of nuclear matrix-associated DNA replication. (A) Replicon
cluster on linear DNA. Bidirectional replication along three tandemly arranged replicons in
a hypothetical replicon cluster is illustrated along a linear DNA molecule. The arrows show
the directions of the growing replicational bubbles. Unduplicated DNA is shown in white;
duplicated DNA (replicational bubbles) in black. (B) Replicon cluster on nuclear matrix-
attached DNA loops. The DNA of the same replicon cluster shown in (A) is now arranged
in a series of loops attached to the nuclear matrix at fixed replicational sites. Each of these
sites, known as replisomes, also contains the apparatus for copying the DNA. This occurs
when there is reeling of DNA across the matrix-bound replisomes as shown by the arrows.
Unduplicated DNA loops are shown in white; duplicated DNA loops (replicational bubbles)
in black. Groups of replisomes cluster together to form a higher order assembly for replicon
cluster synthesis termed the clustersome. (From Tubo and Berezney, 1987b.)

sion that replicon cluster synthesis is temporally and spatially regulated along the chromosomal DNA molecule.

While the existence of replicon subunits, their bidirectional replication and organization into functional replicon clusters is well documented, the mechanistic and molecular basis for these fundamental properties remain a long-standing but unsolved mystery. Even less understood is what controls the exquisite spatial and temporal patterns of replicon cluster synthesis during S-phase. It is nothing short of remarkable that the approximately 50,000 to 100,000 individual replicons that comprise the typical mammalian genome are programmed to replicate once and only once in a precisely choreographed process. This all implies a great deal of structural order underlying DNA replication in the cell nucleus. Somehow the molecular details of replication are integrated within the complex three-dimensional organization of the cell nucleus. As summarized below, the key player in this process may be the nuclear matrix.

A. Replicating DNA on the Nuclear Matrix

Numerous studies of *in vivo* replicated DNA associated with isolated nuclear matrix have led to a radically new view of DNA replication inside the cell nucleus (Berezney and Coffey, 1975; Dijkwel *et al.*, 1979; McCready *et al.*, 1980; Pardoll *et al.*, 1980; Berezney and Buchholtz, 1981b; Berezney, 1984). It is envisioned that replicating DNA loops corresponding to individual replicon subunits are bound to the nuclear matrix (see chapters by Bode *et al.*, 1995; Boulikas, 1995; Davie, 1995; Razin *et al.*, 1995 for details of DNA loop attachment sites). Bidirectional replication then occurs by the reeling of DNA at the two ends of the loops through matrix-bound replisomes (Fig. 18B). Topographical organization of the replicating DNA loops and the associated replisomes into functional clusters or clustersomes may then provide the basis for replicon clustering.

Consistent with this clustersome model, DNA polymerase α, primase, and other replicative components have been found associated with isolated nuclear matrix (Table IV and Smith and Berezney, 1980; Jones and Su, 1982; Foster and Collins, 1985; Collins and Chu, 1987; Tubo and Berezney, 1987a,b,c; Paff and Fernandes, 1990) or the nucleoskeleton obtained via electroelution of chromatin fragments following restriction enzyme digestion under isotonic buffer conditions (Jackson and Cook, 1986a,b,c). The *in vitro* synthesis of Okazaki-sized DNA fragments (Smith and Berezney, 1982), density shift experiments which indicate that the matrix-bound synthesis continues replication along *in vivo* initiated DNA strands (Tubo *et al.*, 1985), the striking replicative and prereplicative association of DNA

TABLE IV

Properties of Nuclear Matrix Bound DNA Synthesis

---

1. Replicative-dependent association of DNA polymerase α and primase.
2. Okazaki fragments continue synthesis at *in vivo* forks.
3. ATP-stimulated processive synthesis requires nuclear matrix attachment.
4. Organization into large megacomplexes (100–150S) is replicative dependent.

---

polymerase α, primase, and other replicative components with the nuclear matrix (Smith and Berezney, 1983; Tubo and Berezney, 1987a,b) and the ATP-stimulated processive synthesis by the matrix-bound polymerase (Tubo *et al.*, 1987) all point toward a replicative-related role of these matrix-bound activities.

The clustersome model further predicts that the replicational machinery (replisomes) for a large number of individual replicons is correspondingly clustered at nuclear matrix-bound sites (Fig. 18B). As a step toward testing this aspect of the model, we developed methods to extract the matrix-bound replicational complexes (Fig. 19; Tubo and Berezney, 1987b). Most of the matrix-bound DNA polymerase α and primase activities were released in the form of discrete megacomplexes sedimenting on sucrose gradients at approximately 100S and 150S. In contrast, complexes extracted from nuclei during nuclear matrix preparation, sedimented at about 8–10S which is typical of DNA polymerase–primase complexes purified from cells. The rapid conversion of the megacomplexes into the more typically sized 10S complexes following release from the matrix structure suggested that the megacomplexes were composed of clusters of 10S complexes and might, thus, represent the *in vitro* equivalent of the predicted clustersome (Fig. 19; Tubo and Berezney, 1987b).

B. Visualizing Replication Sites in Mammalian Cells

A model for the arrangement of these putative nuclear matrix bound clustersomes is shown in Fig. 20A. While our biochemical results supported this model (see previous section), the possibilities of rearrangements or aggregations during nuclear and/or nuclear matrix isolation could not be completely ruled out. What was needed was a method to directly visualize the sites of DNA replication in the nuclei of whole cells.

With this in mind, we developed a permeabilized mammalian cell system to study the incorporation of biotin–11-dUTP into newly replicated DNA.

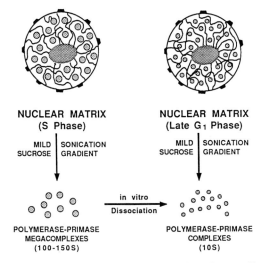

NUCLEAR MATRIX   NUCLEAR MATRIX
(S Phase)   (Late G₁ Phase)

MILD | SONICATION   MILD | SONICATION
SUCROSE | GRADIENT   SUCROSE | GRADIENT

in vitro
Dissociation

POLYMERASE-PRIMASE   POLYMERASE-PRIMASE
MEGACOMPLEXES   COMPLEXES
(100-150S)   (10S)

FIG. 19   Isolation of DNA polymerase–primase megacomplexes from rat liver nuclear matrix. Nuclear matrix was isolated from rat liver nuclei prepared at different times following partial hepatectomy (Tubo and Berezney, 1987b). DNA polymerase α and primase activities were effectively released from the isolated matrices by mild sonication and resolved on sucrose gradients. During active replication in the liver cells (22 hr posthepatectomy), most of the enzyme activity sedimented as large 100–150S complexes (megacomplexes). Just before the onset of replication in the regenerating liver (12 hr posthepatectomy) the DNA polymerase α and primase activities were found predominantly in 10S complexes. The corresponding dissociation of the megacomplexes to 8–10S complexes following release from the matrix structure led to the suggestion that the megacomplexes represent clusters of 10S complexes and thus may represent components of the hypothetical "clustersomes" that we proposed are attached to the nuclear matrix in intact cells (see Tubo and Berezney, 1987b, and Fig. 18).

The sites of biotinylated, newly synthesized DNA were then directly visualized by fluorescence microscopy following reaction with Texes red–streptavidin (Nakayasu and Berezney, 1989). As demonstrated in Fig. 20B, discrete granular sites of replication were observed. The number of replication granules per nucleus (150 to 300) and their size (0.4–0.8 μm in diameter), are consistent with each replication granule being the site of synthesis of a replicon cluster. At any given time in S-phase one would anticipate that thousands of replicons would be active and arranged in up to several hundred clusters.

The discrete nature of the individual replication granules is more apparent at higher magnification (Fig. 21a). Many of the granules appear to have a somewhat elongated or ellipsoid-like shape. In addition, the characteristic size and shape of the individual replication granules remained the same while the fluorescence intensity progressively increased in pulse periods ranging from 2 to 60 min (Nakayasu and Berezney,

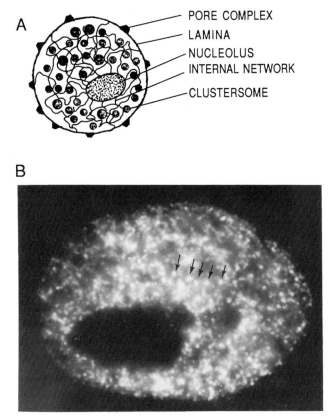

FIG. 20   Schematic diagram of clustersomes attached to the nuclear matrix and direct visualization of replication sites with fluorescence microscopy. A is a schematic diagram of an isolated nuclear matrix with associated, attached clustersomes. The isolated nuclear matrix retains many of the basic architectural landmarks of the intact cell nucleus. In nuclear matrices from cells active in DNA replication, the replication sites are organized into large assemblies termed clustersomes. In B these clustersomes were directly visualized in whole, intact cells by a fluorescence microscopic technique. The individual replication granules or clustersomes (arrows) are distinguished by their intense fluorescence (white granules). Hundreds of clustersomes are detected in each nucleus active in DNA replication.

1989). The size of the individual replication granules, therefore, is not determined by the amount of DNA which is replicated but is rather an inherent organizational property of each replication site. These results strongly support the previously proposed clustersome model. It is proposed, therefore, that each replication granule corresponds to an *in vivo* clustersome (Figure 18).

To study the cell cycle relationships of these replication sites, mouse 3T3 cells were arrested by serum deprivation. Three different motifs

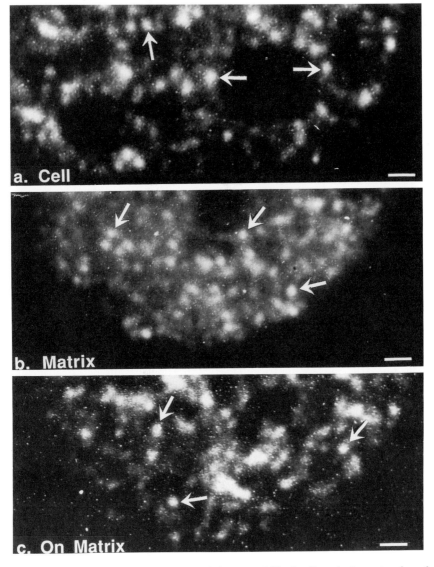

FIG. 21   Replication granules (clustersomes) in permeabilized cells and after extraction of cells for nuclear matrix. (a) PtK-1 cells on coverslips were permeabilized with 0.04% Triton X-100 and incubated with a DNA synthesis medium containing biotin–11-dUTP at 37°C for 5 min (Nakayasu and Berezney, 1989). (b) Following incorporating of biotin–11-dUTP, the cells were extracted for nuclear matrix. (c) PtK-1 cells were first extracted for nuclear matrix. DNA synthesis was then carried out on nuclear matrix-attached DNA fragments in the presence of biotin–11-dUTP. The sites of DNA synthesis were visualized under the fluorescence microscope following incubation with Texas red-conjugated streptavidin. Similar granular sites (clustersomes) were detected in all cases (arrows). Bars correspond to 1 μm.

were observed. The properties of each pattern and its S-phase specific expression are summarized in Table V and illustrated in Fig. 22.

Type I patterns (typical replication granules previously described) were specific for early- to mid-S-phase and found at very low levels during late-S-phase. In contrast, type III patterns over the heterochromatic regions were virtually undetected in early- to mid-S-phase but found as the major pattern (approximately 70%) during late-S-phase. Type II appeared to be a transition between type I and type III patterns with particular replication along the nuclear and nucleolar peripheries and variable amount of type I replication granules throughout the nuclear interior. Type II patterns are a minor component in all stages of S but are particularly prominent during mid- to late-S-phase.

To what extent are the *in situ* nuclear patterns of DNA replication maintained following nuclear matrix isolation? To address this question biotin–dUTP was first incorporated into permeabilized cultured cells (e.g., 3T3 fibroblasts or PtK$_1$ cells) followed by *in situ* extraction for nuclear matrix (Nakayasu and Berezney, 1989). Alternatively, nuclear matrix structures were prepared and followed by *in vitro* incorporation of biotin–dUTP via the nuclear matrix-bound DNA synthesis system. Type I replication granules were observed on the nuclear matrix which were virtually identical in size and number to those in cells (Figs. 22a and 22b).

Identically appearing granules were also detected following DNA synthesis on the short fragments of (circa 1–5 kb) attached to the *in situ* prepared nuclear matrices (Fig. 22c). These results demonstrate that components of the replicational machinery maintain sites on the nuclear matrix which closely correspond to the presumed replicon cluster sites (clustersomes) in intact cells. There is also a remarkable maintenance of type II and III replication patterns on the nuclear matrix (Fig. 22). Thus replicon clusters corresponding to heterochromatin regions in the nucleus are also attached to the nuclear matrix and maintain a similar

TABLE V

Replication Patterns during S-Phase

S-Phase	Replication pattern	Description
Early- to mid-S	Type I	Hundreds of distinct replication granules are distributed over the extranucleolar regions of the nucleus.
Mid- to late-S	Type II	Replication granules are clustered along the nuclear periphery and around individual nucleoli.
Late-S	Type III	Replication sites are concentrated over dense heterochromatin regions of the nucleus.

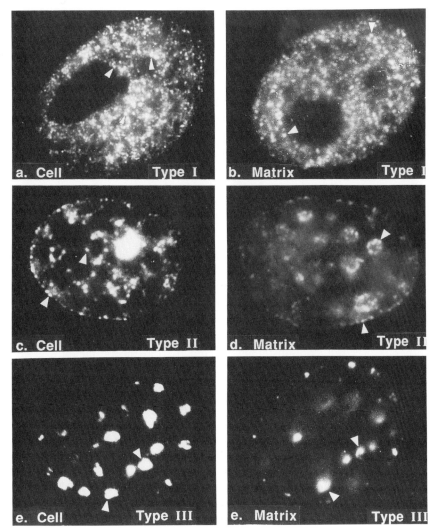

FIG. 22   Visualization of the three types of replicational patterns in 3T3 mouse fibroblasts
and following extraction for nuclear matrix. Permeabilized 3T3 cells were pulsed for 10 min
in the biotin–11-dUTP DNA synthesis medium. Three distinct patterns of replication sites
were detected in the cells and were strikingly maintained following extraction of the cells for
nuclear matrix. (a, b) Type I early-S replication granules (arrowheads) in cell and matrix,
respectively; (c, d) type II mid- to late-S patterns in cell and matrix; characteristic replicational
granules are clustered at heterochromatin regions along the periphery and surrounding the
nucleoli (arrowheads); (e, f) type III late-S patterns in cell and matrix; replication granules
are over the condensed heterochromatin regions (C spots or chromocenters) inside the nucleus.
Individual granules which comprise these stained regions often appear as one large aggregate
(arrowheads) which precisely outline the heterochromatic C spots as defined by intense
Hoechst staining (not shown, see Nakayasu and Berezney, 1989).

size and closely packed configuration despite the removal of over 90% of the total nuclear DNA.

There is a major caveat in the interpretation of the visualization experiments with biotin–dUTP. Permeabilized cells are used for the incorporation, not *in vivo* cells. Although numerous studies have indicated that permeabilized cell systems maintain many basic features of eukaryotic *in vivo* replication there are at least two limitations. One is the apparent inability to ligate DNA of replicon size into larger units or replicated DNA (Berger *et al.,* 1977). Another is that DNA synthesis ends after about 60 min of incorporation (van der Velden and Wanka, 1987). The latter property is due largely to the inability of permeabilized cell systems to initiate DNA synthesis at new replicons. It is conceivable, therefore, that the structural organization for DNA replication sites detected with the biotin–dUTP incorporation differs from that found *in vivo.*

Since we do not know what proportion of the *in vivo* replication sites actually incorporate biotin–dUTP in the permeabilized cells, it is also possible that we have grossly underestimated the actual number of replication sites that are active *in vivo.* For example, if only 10% of replication sites were active in the permeabilized cells, there could be thousands of sites *in vivo.* This would be consistent with each site containing a single replicon rather than a replicon cluster as concluded from the permeabilized cell studies.

To resolve these potential difficulties and obtain direct information on the *in vivo* sites of DNA replication, we incorporated 5-bromodeoxyuridine (BrdU) into cultured cells and performed immunofluorescence localization following reaction with monoclonal antibodies to BrdU and F1TC- or Texas red-conjugated secondary antibodies. Despite the lower sensitivity of this *in vivo* approach, we observed significant similarity in the size and number of replication granules versus those observed in permeabilized cells (Nakayasu and Berezney, 1989). Identically appearing type II and III replication patterns were also observed with the *in vivo* approach. This provides corroborative evidence for the clustersome model and demonstrates that the permeabilized cell approach is a useful approach for studying the organization of replication sites in the cell nucleus. Several other studies of *in vivo* sites of replication in mammalian cells have found similar results (Kill *et al.,* 1991; Hassan and Cook, 1992; Fox *et al.,* 1991; Neri *et al.,* 1992; O'Keefe *et al.,* 1992). In addition, these sites have been visualized with electron microscopy with results consistent with the immunofluorescence patterns (Hozák *et al.,* 1993). Similar sites have been detected following reconstruction of nuclear structures *in vitro* in *Xenopus* egg extracts (Blow and Laskey, 1986).

C. Visualizing Replication Sites in Three Dimensions

A major limitation of these studies is that essentially two-dimensional information is obtained about what is in fact a three-dimensional system.

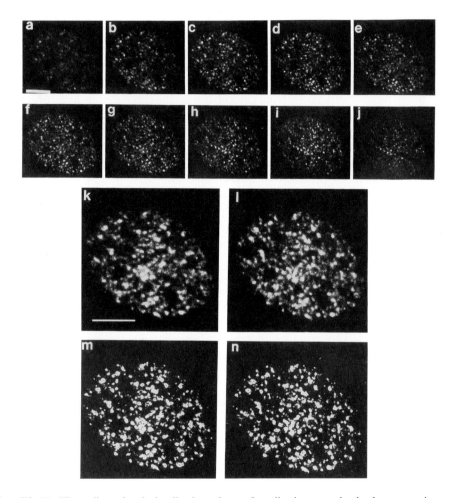

FIG. 23 Three-dimensional visualization of type I replication granules by laser scanning confocal microscopy. (a–j) Ten representative optical sections (0.3-μm intervals) of a total of 17 through the nucleus of 3T3 cells following a 60-min *in vivo* pulse with BrdU and processing for immunofluorescence staining with antibodies to BrdU and FITC-conjugated secondary antibodies; bar denotes 10 μm. (k, l) Stereo pair for the three-dimensionally reconstructed image derived from the 17 individual sections; (m, n). The same three-dimensionally reconstructed nucleus showing the contours of the individual replication granules obtained using a multidimensional image analysis system; bar denotes 10 μm for k–n.

Moreover, the epifluorescence microscopic images are essentially "pseudo two-dimensional" since an indeterminable amount of fluorescence is derived from above and below the actual focal plane on the specimen. To obtain real, three-dimensional information about the spatial organization of the replication sites in the cell nucleus, we used laser scanning confocal fluorescence microscopy following incorporation of BrdU into cultured mammalian cells. Figures 23a to 23j show a series of confocal sections through the nucleus of a 3T3 cell which was replicating its DNA in early to mid-S-phase. A typical type I pattern of replication granules is apparent. Figures 23k and 23l is a stereo pair of the three-dimensionally reconstructed image derived from the complete series of confocal microscopic sections through this nucleus. Aside from demonstrating the three-dimensional organization, the individual sites of replication and their shapes are seen with enhanced clarity and more details with this approach (compare Figs. 21 and 24). Greater heterogeneity in the size and shape of individual clustersomes is observed compared to those visualized with epifluorescence microscopy.

To obtain more precise and quantitative information from these 3-D images we are applying multidimensional computer imaging analysis. In our current approach, the analysis and processing software uses an enhancement algorithm to suppress the background while intensifying the replication sites. A suitable threshold is then used to obtain the contours of individual sections (Duda and Hart, 1973; Ballard and Brown, 1982). Three-dimensional replication sites are built up from these contours by 3-D connected component analysis (Samarabandu *et al.*, 1994). The visualization software renders these volume contours as wire frame diagrams on the screen using different colors to signify individual replication sites (Fig. 25a, Color Plate 2). This greatly enhances the visualization and discrimination of the sites. It is important to note that the x,y, and z coordinates (centers of gravity) are determined for every site and can be displayed in three dimensions (Fig. 25b, Color Plate 2). This enables analysis of specific sites of replication among the hundreds to thousands that are present (Fig. 25c, Color Plate 2).

The segmentation algorithm was further improved by using a spot detection method which can better discriminate closely spaced replication sites.

FIG. 24 Higher magnification of type I replication granules by confocal microscopy. (a) Portion of a typical optical section near the midplane of the nucleus of a 3T3 cell that has been pulsed *in vivo* for 60 min with BrdU and processed for immunofluorescence staining with anti-BrdU antibodies and rhodamine-conjugated secondary antibodies. Arrows point to type I granules. (b and c) Stereo pair of a portion of the three-dimensionally reconstructed image derived from seven sections (0.5-μm intervals) including the section shown in a; bars correspond to 2.2 μm.

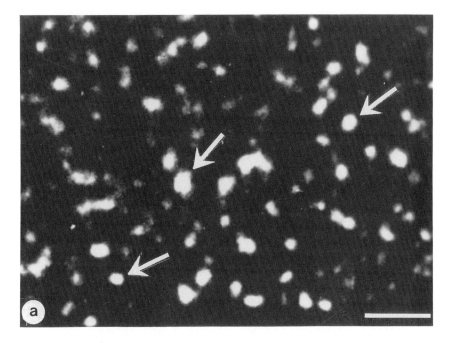

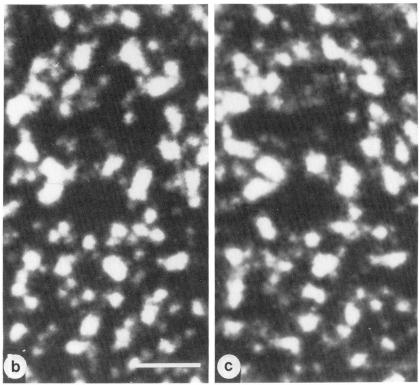

Visualization software is also improved by modeling the replication sites as solid spheres and using ray-tracing algorithms to render these spheres. Higher levels of analysis that are being developed in our laboratory include forming domains of replication sites based on the distance to the nearest neighbor and quantifying various aspects of colocalization of two channel images which shows the DNA replication at different times (Samarabandu et al., 1995).

As a first step in studying the spatiotemporal relationships of DNA replication sites in 3-D, we have performed double-labeling experiments according to the method developed by Aten et al. (1992) at two different times (pulse–chase–pulse) (Berezney et al., 1995). This allows examination of the relationship of early versus later replicated DNA at individual sites. Cells in early S-phase were pulsed for a short time with CldU (green fluorescence), chased for different times, and pulsed again for a short time with IdU (red fluorescence). As a control, simultaneous pulsing with both CldU and IdU resulted in the virtual complete overlap of all the replication sites with the two probes (yellow replication sites, Fig. 26a, Color Plate 3).

Following a 15-min chase, over 50% of the total replication sites were colocalized (yellow sites) and decorated granular replication sites similar to those observed after a short pulse (Fig. 26, Color Plate 3). Later pulsed replication sites (red sites) were consistently observed in juxtaposition to early ones (green or yellow sites). Increasing the chase time between pulses to 1–2 hr resulted in an increasing spatial separation between early and later replication sites (Figs. 26c and 26d, Color Plate 3). Virtually all the green and red sites were separated by a 1-hr chase. This suggests that the average replication site takes less than 1 hr to complete replication. Since the average replicon requires 30–45 min to complete replication in mammalian cells, this implies a reasonable degree of synchrony among the multiple replicons that presumably constitute each replication site (Manders et al., 1992).

Increasing the chase time to 6 hr enabled simultaneous visualization of replication sites labeled in early-S-phase and mid- to late-S-phase. Typical late-S-phase replication around the nuclear periphery (Fig. 26e, Color Plate 3) and at the centromeric hetereochromatin regions (C-spots; Fig. 26f, Color Plate 3) was observed. The maintenance of the early-S-phase type I granular sites postreplicatively in late-S-phase suggests that replication sites in early-S-phase are composed of a higher order domain of chromatin which persists through the S-phase.

An interesting discovery was made when chase times were extended through mitosis into the G$_1$-stage of the next generation of cells. It appears as though mitosis and cell division did little to disrupt the granular pattern of replication site staining, thus providing strong evidence for a tightly regulated maintenance of association between specific regions of the chromatin (Meng and Berezney, 1991). Similar observations were recently published in a study of plant cells (Sparvoli *et al.*, 1994).

Progress has been made in combining dual-color 3-D laser scanning microscopy and computer image analysis with fluorescence *in situ* hybridization (FISH) to identify where specific genes are replicated and ultimately to "map" the DNA sequences at individual replication sites (Berezney, 1995). This would represent a direct test of the "single site-multiple replicon" model and enable us to examine the specificity of the DNA sequences at individual sites of replication. Initial experiments have studied the replication sites of specific chromosomes in RPMI human lymphoblasts. As shown in Figs. 27a and 27c, (Color Plate 4), we were able to detect low levels of chromosome 11 (yellow) replication sites among the multitudes of non-chromosome 11 DNA replication sites (green). Examination of individual sections verified that the yellow sites were due to the specific replication (green) within chromosome 11 regions (red) and were not the result of overlap of individual sections.

In conclusion, the first steps toward defining the macromolecular anatomy of individual replication sites in three dimensions has been taken. Further refinements of the multidimensional analysis programs at both the light and electron microscopic levels and in combination with high-resolution FISH analysis should lead to an increasingly clearer picture of gene replication inside the cell nucleus.

VII. Concluding Remarks

There is growing evidence indicating that genomic organization, function, and regulation are interlinked through nuclear architecture. First are the studies with isolated nuclear matrices demonstrating the association of a vast array of functional and regulatory properties. Second are the fluorescence and electron microscopic studies of the nucleus in whole cells and tissues which show that (i) the chromatin corresponding to individual chromosomes are localized in discrete regions of the nucleus, and (ii) that the genomic functions of replication, transcription, and RNA splicing occupy discrete spatial sites or *domains.*

What is most exciting to the field is that these two lines of evidence are beginning to merge. Recent results demonstrate that the overall organization of sites of replication, transcription, and post-transcriptional RNA transcripts and processing are strikingly maintained in cells extracted for nuclear matrix. These findings corroborate those biochemical results demonstrating functional properties associated with isolated nuclear matrix. They also suggest that *in vitro* nuclear matrix systems represent a valuable new approach for elucidating the relationships of genomic organization, function and regulation in the mammalian cell nucleus.

The nuclear matrix has been a source of study for approximately 20 years. During this time there have been enormous advances in molecular biology and genetics. More extensive application of molecular biology technology to questions concerning the nuclear matrix proteins, their assembly in the nucleus, and their functional and regulatory properties is an important direction for the future.

In one strategy, proteins, or other components known to be involved in genomic processes, can be studied in association with the nuclear matrix. In another approach, the nuclear matrix can be used as a starting point to identify novel proteins important for genomic function and regulation. Finally, both previously known proteins and novel nuclear matrix-associated proteins can be probes for studying the spatial organization of these functional sites in intact cells and nuclear matrix preparations.

Another important direction is computer image analysis. The first steps in applying multidimensional computer imaging to the three-dimensional organization of functional sites in the cell nucleus have already been taken with remarkable results. Realizing the explosive development of computer technology in the past decade, it is awesome to consider what will be possible as we approach the 21st century. At the moment it seems that the use of computer imaging in combination with molecular probes such as specific gene sequences and antibodies to specific nuclear proteins is opening up a whole new field of investigation that we term the *molecular morphology of the genome.*

Acknowledgments

We are extremely grateful to Dr. P. C. Cheng (Department of Electrical and Computer Engineering, SUNY at Buffalo) for the use of the confocal microscopy facilities at the Advanced Microscopy and Imaging Laboratory (AMIL) and his assistance and collaboration on experiments involving this instrumentation. Jim Stamos kindly provided the illustrations and photography. The experiments reported from our laboratory were funded by NIH Grant GM 23922.

References

Aaronson, R. P., and Blobel, G. (1975). Isolation of nuclear pore complexes in association with a lamina. *Proc. Natl. Acad. Sci. U.S.A.* **72,** 1007–1011.

Adolph, K. W., Cheng, S. M., Paulson, J. R., and Laemmli, U. K. (1977). Isolation of protein scaffold from mitotic HeLa cell chromosomes. *Proc. Natl. Acad. Sci. U.S.A.* **74,** 4937–4941.

Allen, S. L., Berezney, R., and Coffey, D. S. (1977). Phosphorylation of nuclear matrix proteins in isolated regenerating rat liver nuclei. *Biochem. Biophys. Res. Commun.* **75,** 111–116.

Altieri, F., Maras, B., Eufemi, M., Ferraro, A., and Turano, C. (1993). Purification of a 57 kDa nuclear matrix protein associated with thiol: protein–disulfide oxidoreductase and phospholipase C activities. *Biochem. Biophys. Res. Commun.* **194,** 992–1000.

Aten, J. A., Bakker, P. J. M., Stap, J., Boschman, G. A., and Veenhof, C. H. N. (1992). DNA double labelling with IdUrd and CldUrd for spatial and temporal analysis of cell proliferation and DNA replication. *Histochem. J.* **24,** 251–259.

Bachs, O., and Carafoli, E. (1987). Calmodulin and calmodulin binding proteins in liver cell nuclei. *J. Biol. Chem.* **262,** 10786–10790.

Ballard, D. H., and Brown, C. M. (1982). "Computer Vision." Prentice–Hall, Englewood Cliffs, NJ.

Bastos, R., Parte, N., and Burke, B. (1995). Nuclear pore complex proteins. *In* "International Review of Cytology" (R. Berezney and K. Jeon, eds.), Vol. 162, in press. Academic Press, San Diego.

Belgrader, P., Siegel, A. J., and Berezney, R. (1991a). A comprehensive study on the isolation and characterization of the HeLa S3 nuclear matrix. *J. Cell Sci.* **98,** 281–291.

Belgrader, P., Dey, R., and Berezney, R. (1991b). Molecular cloning of matrin 3: A 125-kilodalton protein of the nuclear matrix contains an extensive acidic domain. *J. Biol. Chem.* **266,** 9893–9899.

Berezney, R. (1974). Large scale isolation of nuclear membranes from bovine liver. *Methods Cell Biol.* **7,** 205–228.

Berezney, R. (1979). Dynamic properties of the nuclear matrix. *In* "The Cell Nucleus" (Busch, H., ed.), Vol. 7, pp. 413–456. Academic Press, New York.

Berezney, R. (1980). Fractionation of the nuclear matrix. *J. Cell Biol.* **85,** 641–650.

Berezney, R. (1984). Organization and functions of the nuclear matrix. *In* "Chromosomal Nonhistone Proteins" (Hnilica, L. S., ed.), Vol. 4, pp. 119–180. CRC Press, Boca Raton, FL.

Berezney, R. (1995). Replicating the mammalian genome in 3D. *J. Cell. Biochem. Suppl.* **21B,** 118.

Berezney, R., Basler, J., Hughes, B. B., and Kaplan, S. C. (1979a). Isolation and characterization of the nuclear matrix from zajdela ascites hepatoma cells. *Cancer Res.* **39,** 3031–3039.

Berezney, R., Basler, J., Kaplan, S. C., and Hughes, B. B. (1979b). The nuclear matrix of slowly and rapidly proliferating liver cells. *Eur. J. Cell Biol.* **20,** 139–142.

Berezney, R., and Buchholtz, L. B. (1981a). Isolation and characterization of rat liver nuclear matrices containing high molecular weight deoxyribonucleic acid. *Biochemistry* **20,** 4995–5002.

Berezney, R., and Buchholtz, L. B. (1981b). Dynamic association of replicating DNA fragments with the nuclear matrix of regenerating liver. *Exp. Cell Res.* **132,** 1–13.

Berezney, R., and Coffey, D. S. (1974). Identification of a nuclear protein matrix. *Biochem. Biophys. Res. Commun.* **60,** 1410–1417.

Berezney, R., and Coffey, D. S. (1975). The nuclear protein matrix: Association with newly synthesized DNA. *Science* **189,** 291–293.

Berezney, R., and Coffey, D. S. (1976). The nuclear protein matrix: Isolation, structure, and functions. "Advances in Enzyme Regulation" (G. Weber, ed.), Vol. 14, pp. 63–100. Pergamon Press, New York.

Berezney, R., and Coffey, D. S. (1977). Nuclear matrix: Isolation and characterization of a framework structure from rat liver nuclei. *J. Cell. Biol.* **73,** 616–637.

Berezney, R., Ma, H., Wei, X., Meng, C., Samarabandu, J. K., and Cheng, P. C. (1995). Elucidating the higher order assembly of replicating sites in mouse 3T3 cells. *J. Cell. Biochem.* **21B,** 137.

Berezney, R., Macaulay, L. K., and Crane, F. L. (1972). The purification and biochemical characterization of bovine nuclear membranes. *J. Biol. Chem.* **247,** 5549–5561.

Berger, N. A., Petzold, S. J., and Johnson, E. S. (1977). High molecular weight DNA intermediates synthesized by permeabilized L cells. *Biochim. Biophys. Acta* **478**, 44–58.

Bernhard, W. (1969). A new staining procedure for electron microscopical cytology. *J. Ultrastruct. Res.* **27**, 250–265.

Bernhard, W., and Granboulan, N. (1963). The fine structure of the cancer cell nucleus. *Exp. Cell Res.* **9**(Suppl.), 19–53.

Bidwell, J. P., Fey, E. G., van Wijnen, A. J., Penman, S., Stein, J. L., Lian, J. B., and Stein, G. S. (1994). Nuclear matrix proteins distinguish normal diploid osteoblasts from osteosarcoma cells. *Cancer Res.* **54**, 28–32.

Blencowe, B., Nickerson, J. A., Issner, R., Penman, S., and Sharp, P. A. (1994). Association of nuclear matrix antigens with exon-containing splicing complexes. *J. Cell Biol.* **127**, 593–607.

Blobel, G. (1985). Gene gating: A hypothesis. *Proc. Natl. Acad. Sci. U.S.A.* **83**, 8527–8529.

Blobel, G., and Potter, V. R. (1966). Nuclei from rat liver: Isolation method that combines purity with high yield. *Science* **154**, 1662–1664.

Blow, J. J., and Laskey, R. A. (1986). Initiation of DNA replication in nuclei and purified DNA by a cell-free extract of *Xenopus* eggs. *Cell (Cambridge, Mass.)* **47**, 577–587.

Bode, J., Schlake, T., Rios-Ramirez, M., Meike, C., Stengert, M., Kay, V., and Klenk-Wirth, D. (1995). Scaffold/matrix-attached regions (S/MARS): Structural properties of transcriptionally active loci. *In* "International Review of Cytology" (R. Berezney and K. Jeon, eds.), Vol. 162, in press. Academic Press, San Diego.

Boulikas, T. (1995). Chromatin domains and prediction of MAR sequences. *In* "International Review of Cytology" (R. Berezney and K. Jeon eds.), Vol. 162, in press. Academic Press, San Diego.

Bregman, D. B., Du, L., Zee, S., and Warren, S. (1995). Transcription-dependent redistribution of the large subunit of RNA polymerase II to discrete nuclear domains. *J. Cell Biol.* **129**(2), 287–296.

Bridger, J. M., Kill, I. R., O'Farrel, M., and Hutchison, C. J. (1993). Internal lamin structures with G1 nuclei of human fibroblasts. *J. Cell Sci.* **104**, 297–306.

Capco, D. G., Wan, K. M., and Penman, S. (1982). The nuclear matrix: Three-dimensional architecture and protein composition. *Cell (Cambridge, Mass.)* **29**, 847–858.

Capitani, S., Girard, P. R., Mazzei, G. J., Kuo, J. F., Berezney, R., and Manzoli, F. A. (1987). Immunochemical characterization of protein kinase C in rat liver nuclei and subnuclear fractions. *Biochem. Biophys. Res. Commun.* **142**, 367–375.

Cardenas-Corona, M. E., Jacobson, E. L., and Jacobson, M. K. (1987). Endogenous polymers of ADP-ribose are associated with the nuclear matrix. *J. Biol. Chem.* **262**, 14863–14866.

Chaly, N., Bladon, T., Setterfield, G., Little, J. E., Kaplan, J. G., and Brown, D. L. (1984). Changes in distribution of nuclear matrix antigens during the miotic cell cycle. *J. Cell Biol.* **99**, 661–671.

Ciejek, E. M., Nordstro, J. L., Tsai, M. J., and O'Malley, B. W. (1982). Ribonucleic acid precursors are associated with the chick oviduct nuclear matrix. *Biochemistry* **21**, 4945–4953.

Coghlan, V. M., Langeberg, L. K., Fernandez, A., Lamb, N. J. C., and Scott, J. D. (1994). Cloning and characterization of AKAP 95, a nuclear protein that associates with the regulatory subunit of type II cAMP-dependent protein kinase. *J. Biol. Chem.* **269**, 7658–7665.

Collins, J. M., and Chu, A. K. (1987). Binding of the DNA polymerase α–DNA primase complex to the nuclear matrix in HeLa cells. *Biochemistry* **26**, 5600–5607.

Comings, D. E., and Okada, T. A. (1976). Nuclear proteins. III. The fibrillar nature of the nuclear matrix. *Exp. Cell Res.* **103**, 341–360.

Compton, D. A., and Cleveland, D. W. (1994). NuMA, a nuclear protein involved in mitosis and nuclear reformation. *Curr. Opin. Cell Biol.* **6**(3), 343–346.

Compton, D. A., Szilak, I., and Cleveland, D. W. (1992). Primary structure of NuMA, an intranuclear protein that defines a novel pathway for segregation of proteins at mitosis. *J. Cell Biol.* **116**, 1395–1408.

Covey, L., Choi, Y., and Prives, C. (1984). Association of simian virus 40 T antigen with the nuclear matrix of infected and transformed monkey cells. *Mol. Cell. Biol.* **4,** 1384–1392.

Cremer, T., Kurz, A., Zirbel, R., Dietzel, S., Rinke, B., Schröck, E., Speicher, M. R., Mathieu, U., Jauch, A., Emmerich, P., Scherthan, H., Ried, T., Cremer, C., and Lichter, P. (1993). Role of chromosome territories in the functional compartmentalization of cell nucleus. *Cold Spring Harbor Symp. Quant. Biol.* **53,** 777–792.

Davie, J. R. (1995). The nuclear matrix and the regulation of chromatin organization and function. *In* "International Review of Cytology" (R. Berezney and K. Jeon, eds.), Vol. 162, in press. Academic Press, San Diego.

Derenzini, M., Lorenzoni, E., Marinozzi, V., and Barsotti, P. (1977). Ultrastructural cytochemistry of active chromatin in regenerating rat hepatocytes. *J. Ultrastruct. Res.* **59,** 250–262.

Derenzini, M., Novello, F., and Pession-Brizzi, A. (1978). Perichromatin fibrils and chromatin ultrastructural pattern. *Exp. Cell Res.* **112,** 443–454.

Detke, S., and Keller, J. M. (1982). Comparison of the protein present in HeLa cell interphase nucleoskeletons and metaphase chromosome scaffolds. *J. Biol. Chem.* **257,** 3905–3911.

Dijkwel, P. A., Mullenders, L. H. F., and Wanka, F. (1979). Analysis of the attachment of replicating DNA to a nuclear matrix in mammalian interphase nuclei. *Nucleic Acids Res.* **6,** 219–230.

Dijkwel, P. A., Wenink, P. W., and Poddighe, J. (1986). Permanent attachment of replication origins to the nuclear matrix in BHK-cells. *Nucleic Acids Res.* **14,** 3241–3249.

Dworetzky, S. I., Fey, E. G., Penman, S., Lian, J. B., Stein, J. L., and Stein, C. S. (1990). Progressive changes in the protein composition of the nuclear matrix during rat osteoblast differentiation. *Proc. Natl. Acad. Sci. U.S.A.* **87,** 4605–4609.

Duda, R. O., and Hart, P. E. (1973). "Pattern Classification and Scene Analysis." Wiley, New York.

Eisenman, R. N., Tachibana, C. Y., Abrams, H. D., and Hann, S. R. (1985). *v-myc* and *c-myc*-encoded proteins are associated with the nuclear matrix. *Mol. Cell. Biol.* **5,** 114–126.

Faiferman, I., and Pogo, A. O. (1975). Isolation of a nuclear ribonucleoprotein network that contains heterogeneous RNA and is bound to the nuclear envelope. *Biochemistry* **14,** 3808–3816.

Fakan, S., and Bernhard, W. (1971). Localization of rapidly and slowly labeled nuclear RNA as visualized by high resolution autoradiography. *Exp. Cell Res.* **67,** 129–141.

Fakan, S., and Bernhard, W. (1973). Nuclear labeling after prolonged ^{3}H-uridine incorporation as visualized by high resolution autoradiography. *Exp. Cell Res.* **79,** 431–444.

Fakan, S., Puvion, E., and Sphor, G. (1976). Localization and characterization of newly synthesized nuclear RNA in isolated rat hepatocytes. *Exp. Cell Res.* **99,** 155–164.

Fawcett, D. W. (1966). An atlas of fine structure. *In* "The Cell, Its Organelles and Inclusions". Saunders, Philadelphia, PA.

Feldman, L. T., and Nevins, J. R. (1983). Localization of the adenovirus E1 A protein, a positive-acting transcriptional factor, in infected cells. *Mol. Cell. Biol.* **3,** 829–838.

Fernandes, D. J., and Catapano, C. V. (1991). Nuclear matrix targets for anticancer agents. *Cancer Cells* **3,** 134–140.

Fey, E., Krochmalnic, C., and Penman, S. (1986). The nonchromatin substructures of the nucleus: the ribonucleoprotein(RNP) containing and RNP-depleted matrices analyzed by sequential fractionation and resinless section electron microscopy. *J. Cell Biol.* **102,** 1654–1665.

Fey, E., and Penman, S. (1988). Nuclear matrix proteins reflect cell type of origin in cultured human cells. *Proc. Natl. Acad. Sci. U.S.A.* **85,** 121–125.

Fischer, G., Wittmann-Liebold, B., Lang, K., Kiefhaber, T., and Schmid, F. X. (1989). Cyclophilin and peptidyl-prolyl *cis–trans* isomerase are probably identical proteins. *Nature (London)* **337,** 476–478.

Foisner, R., and Gerace, L. (1993). Integral membrane proteins of the nuclear envelope interact with lamins and chromosomes, and binding is modulated by mitotic phosphorylation. *Cell (Cambridge, Mass.)* **73,** 1267–1279.

Foster, K. A., and Collins, J. M. (1988). The interrelation between DNA synthesis rates and DNA polymerases bound to the nuclear matrix in synchronzied HeLa cells. *J. Biol. Chem.* **260,** 4229–4235.

Fox, M. H., Arndt-Jovin, D. J., Jovin, T. M., Baumann, P. H., and Robert-Nicaud, M. (1991). Spatial and temporal distribution of DNA replication sites localized by immunofluorescence and confocal microscopy. *J. Cell Sci.* **99,** 247–255.

Franke, W. W., and Falk, H. (1970). Appearance of nuclear pore complexes after Bernhard's staining procedure. *Histochemie* **24,** 266–278.

Gasser, S. M., and Laemmli, U. K. (1987). A glimpse at chromosomal order. *Trends Genet.* **3,** 16–22.

Gerace, L., and Blobel, G. (1980). The nuclear envelope lamina is reversibly depolymerized during mitosis. *Cell (Cambridge, Mass.)* **19,** 277–287.

Gerace, L., Comeau, C., and Benson, M. (1984). Organization and modulation of nuclear lamina structure. *J. Cell Sci.* **1**(Suppl.), 137–160.

Gerber, U. F., and Gerace, L. (1995). Depletion of calcium from the lumen of endoplasmic reticulum reversibly inhibits passive diffusion and signal-mediated transport into the nucleus. *The J. Cell Biol.* **128,** 5–14.

Gething, M-J., and Sambrook, J. (1992). Protein folding in the cell. *Nature (London)* **355,** 33–45.

Getzenberg, R. H., and Coffey, D. S. (1990). Tissue specificity of the hormonal response in sex accessory tissue is associated with nuclear matrix protein patterns. *Mol. Endocrinol.* **4,** 1336–1342.

Getzenberg, R. H., Pienta, K. J., Huang, E. Y., and Coffey, D. S. (1991). Identification of nuclear matrix proteins in the cancer and normal rat prostate. *Cancer Res.* **51,** 6514–6520.

Goldman, M. A., Holmquist, G. P., Gray, M. C., Caston, L., and Nag, A. (1984). Replication timing of genes and middle repetitive sequences. *Science* **224,** 686–692.

Goldman, A. E., Moir, R. D., Montag-Lowy, M., and Goldman, R. D. (1992). Pathway of incorporation of microinjected lamin A into the nuclear envelope. *J. Cell Biol.* **119,** 725–735.

Graham, C. F., Arms, K., and Gurdon, J. B. (1966). The induction of DNA synthesis by egg cytoplasm. *Dev. Biol.* **14,** 349–359.

Greenleaf, A. L. (1993). Positive patches and negative noodles: linking RNA processing to transcription?. *Trends Biochem. Sci.* **18**(4), 117–122.

Gui, J-F., Lane, W. S., and Fu, X-D. (1994). A serine kinase regulates intracellular localization of splicing factors in the cell cycle. *Nature (London)* **369,** 678–682.

Gupta, R. C., Dighe, N. R., Randerath, K., and Smith, H. C. (1985). Distribution of initial and persistent 2-acetylaminofluorene-induced DNA adducts within DNA loops. *Proc. Natl. Acad. Sci. U.S.A.* **82,** 6605–6608.

Hadlaczky, G., Went, M., and Ringertz, N. R. (1986). Direct evidence for the non-random localization of mammalian chromosomes in the interphase nucleus. *Exp. Cell Res.* **167,** 1–15.

Hakes, D. J., and Berezney, R. (1991). DNA-binding properties of the nuclear matrix and individual nuclear matrix proteins: Evidence for salt resistant DNA binding sites. *J. Biol. Chem.* **266,** 11131–11140.

Halikowski, M. J., and Liew, C. (1987). Identification of a phosphoprotein in the nuclear matrix by monoclonal antibodies. *Biochem. J.* **241,** 693–697.

Hand, R. (1975). Regulation of DNA replication on subchromosomal units of mammalian cells. *J. Cell Biol.* **64,** 89–97.

Hand, R. (1978). Eucaryotic DNA: Organization of the genome for replication. *Cell (Cambridge, Mass.)* **15,** 315–325.

Harris, H. (1968). "Nucleus and Cytoplasm." Clarendon Press, Oxford.

Harris, H. (1970). "Cell Fusion." Clarendon Press, Oxford.

Harris, S. G., and Smith, H. C. (1988). SnRNP core protein enrichment in the nuclear matrix. *Biochem. Biophys. Res. Commun.* **152**, 1383–1387.

Hassan, A. B., and Cook, P. R. (1993). Visualization of replication sites in unfixed human cells. *J. Cell Sci.* **105**, 541–550.

Hatton, K. S., Dhar, V., Gahn, T. A., Brown, E. H., Mager, D., and Schildkraut, C. L. (1988). Temporal order of replication of multigene families reflects chromosomal location and transcriptional activity. *Cancer Cells* **6**, 335–340.

He, D., Nickerson, J. A., and Penman, S. (1990). Core filaments of the nuclear matrix. *J. Cell Biol.* **110**, 569–580.

He, D., Zheng, C., and Brinkley, B. R. (1995). Nuclear matrix proteins as structural and functional components of the mitotic apparatus. *In* "International Review of Cytology" (R. Berezney and K. Jeon, eds.), Vol. 162, in press. Academic Press, San Diego.

Hendzel, M. J., Delcuve, G. P., and Davie, J. R. (1991). Histone deacetylase is a component of the internal matrix. *J. Biol. Chem.* **266**, 21936–21942.

Hendzel, M. J., Sun, J-M., Chen, H. Y., Rattner, J. B., and Davie, J. R. (1994). Histone acetyltransferase is associated with the nuclear matrix. *J. Biol. Chem.* **269**, 22894–22901.

Henry, S. M., and Hodge, L. D. (1983). Nuclear matrix: A cell-cycle-dependent site of increased intranuclear protein phosphorylation. *Eur. J. Biochem.* **133**, 23–29.

Herlan, G., Eckert, W. A., Kaffenberger, W., and Wunderlich, F. (1979). Isolation and characterization of an RNA-containing nuclear matrix from *tetrahymena* macronuclei. *Biochemistry* **18**, 1782–1788.

Herman, R., Weymouth, L., and Penman, S. (1978). Heterogeneous nuclear RNA–protein fibers in chromatin-depleted nuclei. *J. Cell Biol.* **78**, 663–674.

Herman, R., Zieve, G., Williams, J., Lenk, R., and Penman, S. (1976). Cellular skeletons and RNA messages. *Proc. Nucleic Acid Res. Mol. Biol.* **19**, 379–401.

Hisatake, K., Hasegawa, S., Takada, R., Nakatani, Y., Hirikoshi, M., and Roeder, R. G. (1993). The p250 subunit of native TATA box-binding factor TFIID is the cell-cycle regulatory protein CCG1. *Nature* (*London*) **362**, 179–181.

Hozák, P., Hassan, A. B., Jackson, D. A., and Cook, P. R. (1993). Visualization of replication factories attached to a nucleoskeleton. *Cell* (*Cambridge, Mass.*) **73**, 361–373.

Huang, S., and Spector, D. L. (1991). Nascent pre-mRNA transcripts are associated with nuclear regions enriched in splicing factors. *Genes Dev.* **5**, 2288–2302.

Ingber, D. E., Dike, L., Hansen, L., Karp, S., Liley, H., Maniotis, A., McNamee, H., Mooney, D., Plopper, G., and Sims, J. (1994). Cellular tensegrity: exploring how mechanical charges in the cytoskeleton regulate cell growth, migration, and tissue pattern during morphogenesis. *Int. Rev. Cytol.* **150**, 173–224.

Jackson, D. A., and Cook, P. R. (1985). A general method for preparing chromatin containing intact DNA. *EMBO J.* **14**, 913–918.

Jackson, D. A., and Cook, P. R. (1986a). Replication occurs at a nucleoskeleton. *EMBO J.* **5**, 1403–1410.

Jackson, D. A., and Cook, P. R. (1986b). A cell cycle-dependent DNA polymerase activity that replicates intact DNA in chromatin. *J. Mol. Biol.* **192**, 65–67.

Jackson, D. A., and Cook, P. R. (1986c). Different populations of DNA polymerase α in HeLa cells. *J. Mol. Biol.* **192**, 77–86.

Jackson, D. A., and Cook, P. R. (1988). Visualization of a filamentous nucleoskeleton with a 23 nm axial repeat. *EMBO J.* **7**, 3667–3677.

Jackson, D. A., Hassan, A. B., Errington, R. J., and Cook, P. R. (1993). Visualization of focal sites of transcription within human nuclei. *EMBO J.* **12**, 1059–1065.

Jiménez-García, L. F., and Spector, D. L. (1993). In vivo evidence that transcription and splicing are coordinated by a recruiting mechanism. *Cell* (*Cambridge, Mass.*) **73**, 47–59.

Jones, C., and Su, R. T. (1982). DNA polymerase α from the nuclear matrix of cells infected with simian virus 40. *Nucleic Acids Res.* **10,** 5517–5582.

Kaufmann, S. H., Coffey, D. S., and Shaper, J. H. (1981). Considerations in the isolation of rat liver nuclear matrix, nuclear envelope, and pore complex lamina. *Exp. Cell Res.* **132,** 105–123.

Keesee, S. K., Meneghini, M. D., Szaro, R. P., and Wu, Y. J. (1994). Nuclear matrix proteins in human colon cancer. *Proc. Natl. Acad. Sci. U.S.A.* **91,** 1913–1916.

Khanuja, P. S., Lehr, J. E., Soule, H. D., Gehani, S. K., Noto, A. C., Choudhury, S., Chen, R., and Pienta, K. J. (1993). Nuclear matrix proteins in normal and breast cancer cells. *Cancer Res.* **53,** 3394–3398.

Kill, I. R., Bridger, J. M., Campbell, K. H. S., Maldonado-codina, G., and Hutchison, C. J. (1991). The timing of the formation and usage of replicase cluster in S-phase nuclei of human dipoid fibroblasts. *J. Cell Sci.* **100,** 869–876.

Konstantinovic, M., and Sevaljevic, L. (1983). Nuclear matrix from resting and concanavalin A-stimulated human lymphocytes. *Biochim. Biophys. Acta* **762,** 1–8.

Lao, Y. F., and Arrighi, F. F. (1981). Studies of mammalian chromosome replication. II. Evidence for the existence of defined chromosome replicating units. *Chromosoma* **83,** 721–741.

Lawrence, J. B., Singer, R. H., and Marselle, L. M. (1989). Highly localized tracks of specific transcripts within interphase nuclei visualized by in situ hybridization. *Cell (Cambridge, Mass.)* **57,** 493–502.

Lebkowski, J. S., and Laemmli, U. (1982). Non-histone proteins and long-range organization of HeLa interphase DNA. *J. Mol. Biol.* **156,** 325–344.

Lehner, C. F., Eppenberger, H. M., Fakan, S., and Nigg, E. A. (1986). Nuclear substructure antigens monoclonal antibodies against components of nuclear matrix preparations. *Exp. Cell Res.* **162,** 205–219.

Lewis, C. D., Lebkowski, J. S., Daly, A. K., and Laemmli, U. K. (1984). Interphase nuclear matrix and metaphase scaffolding structures. *J. Cell Sci* **1**(Suppl.), 103–122.

Li, H., and Bingham, P. M. (1991). Arginine/serine-rich domains of the su(wa) and tra RNA processing regulators target proteins to a subnuclear compartment implicated in splicing. *Cell* **67,** 335–342.

Loidl, P., and Eberharter, A. (1995). Nuclear matrix and the cell cycle. *In* "International Review of Cytology" (R. Berezney and K. Jeon, eds.), Vol. 162, in press. Academic Press, San Diego.

Mancini, M. A., Shan, B., Nickerson, J. A., Penman, S., and Lee, W. (1994). The retinoblastoma gene product is a cell cycle-dependent, nuclear matrix-associated protein. *Proc. Natl. Acad. Sci. U.S.A.* **91,** 418–422.

Manders, E. M. M., Stap, J., Brakenhoff, G. J., Van Driel, R., and Aten, J. A. (1992). Dynamics of three dimensional replication patterns during the S-phase, analyzed by double labeling and confocal microscopy. *J. Cell Sci.* **103,** 857–862.

Manuelidis, L. (1985). Individual interphase chromosome domains revealed by in situ hybridization. *Hum. Genet.* **71,** 288–293.

Manuelidis, L. (1990). Individual interphase chromosomes. *Science* **250,** 1533–1540.

Mariman, E. C. M., van Eckelen, C. A. G., Reinders, R. J., Berns, A. J. M., and van Venrooij, W. J. (1982). Adenoviral heterogeneous nuclear RNA is associated with the host nuclear matrix during splicing. *J. Mol. Biol.* **154,** 103–119.

Maul, G. G., Maul, H. M., Scogna, J. E., Liebermann, M. W., Stein, G. S., Hsu, B. Y. I., and Borun, T. W. (1972). Time sequence of nuclear pore formation in phytohemagglutinin-stimulated lymphocytes and HeLa cells during the cell cycle. *J. Cell Biol.* **5,** 433–442.

McCready, S. J., Godwin, J., Mason, D. W., Brazell, I. A., and Cook, P. R. (1980). DNA is replicated at the nuclear cage. *J. Cell Sci.* **46,** 365–386.

McKeon, F. D., Kirschner, M. W., and Caput, D. (1986). Homologies in both primary and secondary structure between nuclear envelope and intermediate filament proteins. *Nature (London)* **319,** 463–468.

Meier, U. T., and Blobel, G. (1992). Nopp140 shuttles on tracks between nucleolus and cytoplasm. *Cell* **70,** 127–138.

Meng, C., and Berezney, R. (1991). Replication cluster domains persist throughout the cell cycle of mouse 3T3 cells. *J. Cell Biol.* **115,** 95a.

Milavetz, B. I., and Edwards, D. R. (1986). Synthesis and stability of nuclear matrix proteins in resting and serum-stimulated Swiss 3T3 cells. *J. Cell. Physiol.* **127,** 388–396.

Miller, T. E., Huang, C. Y., and Pogo, A. O. (1978). Rat liver nuclear skeleton and ribonucleoprotein complexes containing hnRNA. *J. Cell Biol.* **76,** 675–691.

Mirkovitch, J., Mirault, M. F., and Laemmli, U. K. (1984). Organization of the higher-order chromatin loop: Specific attachment sites on nuclear scaffold. *Cell (Cambridge, Mass.)* **39,** 223–232.

Moir, R. D., Montag-Lowy, M., and Goldman, R. D. (1994). Dynamic properties of nuclear lamins: lamin B is associated with sites of DNA replication. *J. Cell Biol.* **125,** 1201–1212.

Moir, R. D., Spann, T. P., and Goldman, R. D. (1995). The dynamic properties and possible functions of nuclear lamins, *In* "International Review of Cytology" (R. Berezney and K. Jeon, eds.), Vol 162, in press. Academic Press, San Diego.

Monneron, A., and Bernhard, W. (1969). Fine structural organization of the interphase nucleus in some mammalian cells. *J. Ultrastruct. Res.* **27,** 266–288.

Moreno Diaz de la Espina, S. (1995). Nuclear matrix proteins isolated from plant cells. *In* "International Review of Cytology" (R. Berezney and K. Jeon, eds.), Vol 162, in press. Academic Press, San Diego.

Mortillaro, M. J., and Berezney, R. (1995). Association of a nuclear matrix cyclophilin with splicing factors. *J. Cell. Biochem. Suppl.* **21B,** 142.

Mortillaro, M. J., Vijayaraghavan, S., and Berezney, R. (1993). Evidence that matrin 3 is a member of a complex gene family. *Mol. Biol. Cell* **4,** 81S.

Nakamura, H., Morita, T., and Sato, C. (1986). Structural organization of replication domains during DNA synthetic phase in the mammalian nucleus. *Exp. Cell Res.* **165,** 291–297.

Nakayasu, H., and Berezney, R. (1987). Identification of two high molecular weight internal nuclear matrix proteins with monoclonal antibodies. *J. Cell Biol.* **105,** 70a.

Nakayasu, H., and Berezney, R. (1989). Mapping replicational sites in the eucaryotic cell nucleus. *J. Cell Biol.* **108,** 1–11.

Nakayasu, H., and Berezney, R. (1991). Nuclear matrins: Identification of the major nuclear matrix proteins. *Proc. Natl. Acad. Sci. U.S.A.* **88,** 10312–10316.

Nash, R. E., Puvion, E., and Bernhard, W. (1975). Perichromatin fibrils as components of rapidly labeled extranucleolar RNA. *J. Ultrastruct. Res.* **53,** 395–405.

Neri, L. M., Mazzotti, G., Capitani, S., Maraldi, N. M., Cinti, C., Baldini, N., Rana, R., and Martelli, A. M. (1992). Nuclear matrix-bound replicational sites detected in situ by 5-bromodeoxyuridine. *Histochemistry* **98,** 19–32.

Nickerson, J. A., He, D., Fey, F. G., and Penman, S. (1990). The nuclear matrix. *In* "The Eukaryotic Nucleus. Molecular Biochemistry and Macromolecular Assemblies" (P. R. Strauss and S. H. Wilson, eds.), Vol. 2, pp. 763–782. Telford Press, Caldwell, NJ.

Nickerson, J. A., Krockmalnic, G., Wan, K. M., Turner, C. D., and Penman, S. (1992). A normally masked nuclear matrix antigen that appears at mitosis on cytoskeleton filaments adjoining chromosomes, centrioles and midbodies. *J. Cell Biol.* **116,** 977–987.

Noaillac-Depeyre, J., Azum, M., Geraud, M., Mathieu, C., and Gas, N. (1987). Distribution of nuclear matrix proteins in interphase CHO cells and rearrangements during the cell cycle: An ultrastructural study. *Biol. Cell* **61,** 23–32.

Ochs, R., and Smetana, K. (1991). Detection of fibrillarin in nucleolar remnants and the nucleolar matrix. *Exp. Cell Res.* **197,** 183–190.

O'Keefe, R. T., Henderson, S. C., and Spector, D. L. (1992). Dynamic organization of DNA replication in mammalian cell nuclei: Spatially and temporally defined replication of chromosome-specific α-satellite DNA sequences. *J. Cell Biol.* **116,** 1095–1110.

Olson, M. O., and Thompson, B. A. (1983). Distribution of proteins among chromatin components of nucleoli. *Biochemistry* **22,** 3187–3193.

Olson, M. O., Wallace, M. O., Herrera, A. H., Marshall-Carlson, L., and Hunt, R. C. (1986). Preribosomal ribonucleoprotein particles are a major component of a nucleolar matrix fraction. *Biochemistry* **25,** 484–491.

Paff, M. T., and Fernandes, D. J. (1990). Synthesis and distribution of primer RNA in nuclei of CCRF-CEM leukemia cells. *Biochemistry* **29,** 3442–3450.

Painter, R. B., and Young, B. R. (1976). Formation of nascent DNA molecules during inhibition of replicon initiation in mammalian cells. *Biochim. Biophys. Acta* **418,** 146–153.

Pardoll, D. M., Vogelstein, B., and Coffey, D. S. (1980). A fixed site of DNA replication in eucaryotic cells. *Cell* **19,** 527–536.

Park, M. K., D'Onofrio, M., Willingham, M. C., and Hanover, J. A. (1987). A monoclonal antibody against a family of nuclear pore proteins (nucleoporins): O-linked *N*-acetylglucosamine is part of the immunodeterminant. *Proc. Natl. Acad. Sci. U.S.A.* **84,** 6462–6466.

Partin, A. W., Getzenberg, R. H., Carmichael, M. J., Vindivich, D., Yoo, J., Epstein, J. I., and Coffey, D. S. (1993). Nuclear matrix protein patterns in human benign prostate hyperplasia and prostate cancer. *Cancer Res.* **53,** 744–746.

Paulson, J. R., and Laemmli, U. K. (1977). The structure of histone-depleted metaphase chromosomes. *Cell (Cambridge, Mass.)* **12,** 817–828.

Payrastre, B., Nievers, M., Boonstra, J., Breton, M., Verkleji, A. J., and Van Bergen en Henegouwen, P. M. P. (1992). A differential location of phosphoinositide kinases, diacylglycerol kinase, and phospholipase C in the nuclear matrix. *J. Biol. Chem.* **267,** 5078–5084.

Peick, A. C., van der Velden, H. M., Rijken, A. A., Neis, J. M., and Wanka, F. (1985). Protein composition of the chromosomal scaffold and interphase nuclear matrix. *Chromosoma* **91,** 137–144.

Perret, C., Lomri, N., and Thomasset, M. (1988). Evolution of the EF-hand calcium-binding protein family: Evidence for exon shuffling and intron insertion. *J. Mol. Evol.* **27,** 351–364.

Peters, K. E., and Comings, D. E. (1980). Two-dimensional gel electrophoresis of rat liver nuclear washes, nuclear matrix and hnRNA proteins. *J. Cell Biol.* **86,** 135–155.

Peters, K. F., Okada, T. A., and Comings, D. F. (1982). Chinese hamster nuclear proteins. An electrophoretic analysis of interphase, metaphase and nuclear matrix preparations. *Eur. J. Biochem.* **129,** 221–232.

Pienta, K. J., and Lehr, J. E. (1993a). A common set of nuclear matrix proteins in prostate cancer cells. *Prostate* **23,** 61–67.

Pienta, K. J., and Lehr, J. E. (1993b). Inhibition of prostate cancer growth by estramustine and etoposide: Evidence for interaction at the nuclear matrix. *J. Urol.* **149,** 1622–1625.

Puvion, E., and Moyne, G. (1981). In situ localization of RNA structures. *In* "The Cell Nucleus" (H. Busch, ed.), Vol. 8, pp. 59–115. Academic Press, New York.

Razin, S. V., Gronova, I. T., and Iarovaia, O. V. (1995). Specificity and functional significance of DNA interaction with the nuclear matrix: new approaches to clarify the old questions. *In* "International Review of Cytology" (R. Berezney and K. Jeon, eds.), Vol 162, in press. Academic Press, San Diego.

Reeves, R., and Chang, D. (1983). Investigations of the possible functions for glycosylation in the high mobility group proteins: Evidence for a role in nuclear matrix association. *J. Biol. Chem.* **258,** 679–687.

Reiter, T., and Penman, S. (1983). 'Prompt' heat shock proteins: Translationally regulated synthesis of new proteins associated with the nuclear matrix-intermediate filaments as an early response to heat shock. *Proc. Natl. Acad. Sci. U.S.A.* **80,** 4737–4741.

Rennie, P. S., Bruchovsky, N., and Cheng, H. (1983). Isolation of 3 S androgen receptors from salt resistant fractions and nuclear matrices of prostatic nuclei after mild trypsin digestion. *J. Biol. Chem.* **258,** 7623–7630.

Rout, M. P., and Blobel, G. (1993). Isolation of the yeast nuclear pore complex. *J. Cell Biol.* **123,** 771–783.

Sahyoun, N., LeVine, H., Bronson, D., and Cuatrecasas, P. (1984). Ca(2+)-calmodulin-dependent protein kinase in neuronal nuclei. *J. Biol. Chem.* **269,** 9341–9344.

Samarabandu, J. K., Cheng, P. C., and Berezney, R. (1994). Application of three-dimensional image analysis to the mammalian cell nucleus. *Proc. IEEE Southeastcon '94,* 98–100.

Samarabandu, J. K., Ma, H., Cheng, P. C., and Berezney, R. (1995). Computer aided analysis of DNA replication sites in the mammalian cell. *J. Cell. Biochem. Suppl.* **21B,** 137.

Sevaljevic, L., Petrovic, M., Konstantinovic, M., and Krtolica, K. (1982). Comparative studies of rat liver and sea urchin embryo nuclear matrices: pertial fractional and protein kinase activity distribution. *J. Cell Sci.* **55,** 189–198.

Shan, X., Xue, Z., and Melese, T. (1994). Yeast NPI46 encodes a novel prolyl cis-trans isomerase that is located in the nucleolus. *J. Cell Biol.* **126,** 853–862.

Shaper, J. H., Pardoll, D. M., Kaufmann, S. H., Barrack, F. R., Vogelstein, B., and Coffey, D. S. (1978). The relationship of nuclear matrix to cellular structure and function. *Adv. Enzyme Regul.* **17,** 213–248.

Shiomi, Y., Powers, J., Bolla, R. I., Nguyen, T. V., and Schlessinger, D. (1986). Proteins and RNA in mouse L cell core nucleoli and nucleolar matrix. *Biochemistry* **25,** 5745–5751.

Sikorska, M., Whitfield, J. F., and Walker, P. R. (1988). The regulatory and catalytic subunits of cAMP-dependent protein kinases are associated with transcriptional active chromatin during changes in gene expression. *J. Biol. Chem.* **263,** 3005–3011.

Smetana, K., Lejnar, J., Vlastiborova, A., and Busch, H. (1971). On interchromatinic dense granules of mature human neutrophil granulocytes. *Exp. Cell Res.* **64,** 105–112.

Smith, H. C., and Berezney, R. (1980). DNA polymerase α is tightly bound to the nuclear matrix of actively replicating liver. *Biochem. Biophys. Res. Commun.* **97,** 1541–1547.

Smith, H. C., and Berezney, R. (1982). Nuclear matrix-bound DNA synthesis: An *in vitro* system. *Biochemistry* **21,** 6751–6761.

Smith, H. C., and Berezney, R. (1983). Dynamic domains of DNA polymerase α in regenerating rat liver. *Biochemistry* **22,** 3042–3046.

Smith, H. C., Berezney, R., Brewster, J. M., and Rekosh, D. (1985). Properties of adenoviral DNA bound to the nuclear matrix. *Biochemistry* **24,** 1197–1202.

Smith, H. C., Harris, S. G., Zillmann, M., and Berget, S. M. (1989). Evidence that a nuclear matrix protein participates in pre-mRNA splicing. *Exp. Cell Res.* **182,** 521–533.

Smith, H. C., Puvion, F., Buchholtz, L. A., and Berezney, R. (1984). Spatial distribution of DNA loop attachment and replication sites in the nuclear matrix. *J. Cell Biol.* **99,** 1794–1802.

Somanathan, S., Mortillaro, M. J., and Berezney, R. (1995). Identifying members of the matrin 3 gene family. *J. Cell. Biochem. Suppl.* **21B,** 146.

Song, M. H., and Adolph, K. W. (1983). Phosphorylation of nonhistone proteins during the HeLa cell cycle. *J. Biol. Chem.* **258,** 3309–3319.

Sorimachi, H., Ishiura, S., and Suzuki, K. (1993). A novel tissue-specific calpain species expressed predominantly in the stomach comprises two alternatively spliced products with and without Ca(2+)-binding domain. *J. Biol. Chem.* **268,** 19476–19482.

Sparuoli, E., Levi, M., and Rossi, E. (1994). Replicon clusters may form structurally stable complexes of chromatin and chromosomes. *J. Cell Sci.* **107,** 3097–3103.

Spector, D. (1984). Colocalization of U1 and U2 small nuclear RNPs by immunocytochemistry. *Biol. Cell* **51,** 109–112.

Spector, D. L. (1993). Macromolecular domains within the cell nucleus. *Annu. Rev. Cell Biol.* **9,** 265–315.

Spector, D. L., Schrier, W. H., and Busch, H. (1983). Immunoelectron microscopic localization of SnRNPs. *Bio. Cell* **49,** 1–10.

Staufenbiel, M., and Deppert, W. (1983). Nuclear matrix preparations from liver tissue and from cultured vertebrate cells: Differences in major polypeptides. *Eur. J. Cell Biol.* **31,** 341–348.

Staufenbiel, M., and Deppert, W. (1984). Preparation of nuclear matrices from cultured cells: subfractionation of nuclei in situ. *J. Cell Biol.* **98,** 1886–1894.

Steele, W. J., and Busch, H. (1963). Studies on the acidic nuclear proteins of the Walker tumor and the liver. *Cancer Res.* **23,** 1153–1163.

Stuurman, N., van Driel, L., De Jong, A. M. L., and van Renswoude, J. (1989). The protein composition of the nuclear matrix of murine P19 embryonal carcinoma cells is differentiation-state dependent. *Exp. Cell Res.* **180,** 460–466.

Stuurman, N., Meijne, A. M. L., van der Pol., A. J., de Jong, L., van Driel, R., and van Renswoude, J. (1990). The nuclear matrix from cells of different origin: Evidence for a common set of matrix proteins. *J. Biol. Chem.* **265,** 5460–5465.

Tawfic, S., and Ahmed, K. (1994). Association of casien kinase 2 with nuclear matrix. *J. Biol. Chem.* **269,** 7489–7493.

Tubo, R. A., and Berezney, R. (1987a). Pre-replicative association of multiple replicative enzyme activities with the nuclear matrix during rat liver regeneration. *J. Biol. Chem.* **262,** 1148–1154.

Tubo, R. A., and Berezney, R. (1987b). Identification of 100 and 150S DNA polymerase α–primase megacomplexes solubilized from the nuclear matrix of regenerating rat liver. *J. Biol. Chem.* **263,** 5857–5865.

Tubo, R. A., and Berezney, R. (1987c). Nuclear matrix-bound DNA primase: Elucidation of an RNA primase system in nuclear matrix isolation from regenerating rat liver. *J. Biol. Chem.* **262,** 6637–6642.

Tubo, R. A., Martelli, A. M., and Berezney, R. (1987). Enhanced processivity of nuclear matrix bound DNA polymerase α from regenerating rat liver. *Biochemistry* **26,** 5710–5718.

Tubo, R. A., Smith, H. C., and Berezney, R. (1985). The nuclear matrix continues DNA synthesis at in vivo replicational forks. *Biochim. Biophys. Acta* **825,** 326–334.

Turner, B. M., and Franchi, L. (1987). Identification of protein antigens associated with the nuclear matrix and with clusters of interchromatin granules in both interphase and miotic cells. *J. Cell Sci.* **87,** 269–282.

van der Velden, H. M. W., and Wanka, F. (1987). The nuclear matrix—Its role in the spatial organization and replication of eukaryotic DNA. *Mol. Biol. Rep.* **12,** 69–77.

van Eekelen, C. A. G., Salden, M. H. L., Habets, W. J. A., van de Putte, L. B. A., and van Venrjooij, W. J. (1982). On the existence of an internal nuclear protein structure in HeLa cells. *Exp. Cell Res.* **141,** 181–190.

van Wijnen, A. J., Bidwell, J. P., Fey, E. G., Penman, S., Lian, J. B., Stein, J. L., and Stein, G. S. (1993). Nuclear matrix association of multiple sequence specific DNA binding activities related to SP-1, ATF, CCAAT, C/EBP, Oct-1 and AP-1. *Biochemistry* **32,** 8397–8402.

Verheijen, R., Kuijpers, H., Vooijs, P., van Venrooij, W., and Ramaekers, F. (1986). Protein composition of nuclear matrix preparations from HeLa cells: An immunochemical approach. *J. Cell Sci.* **80,** 103–122.

Verheijen, R., van Venrooij, W. J., and Ramaekers, F. (1988). The nuclear matrix: structure and composition. *J. Cell Sci.* **90,** 11–36.

von Bertalanffy, L. (1952). "Problems of Life." C. A. Watts & Co., London.

Wan, K., Nickerson, J. A., Krockmalnic, G., and Penman, S. (1994). The B1C8 protein is in the dense assemblies of the nuclear matrix and relocates to the spindle and pericentriolar filaments at mitosis. *Proc. Natl. Acad. Sci. U.S.A.* **91,** 594–598.

Wansink, D. G., Schul, W., van der Kraan, I., van Steensel, B., van Driel, R., and de Long, L. (1993). Fluorescent labeling of nascent RNA reveals transcription by RNA polymerase II in domains scattered throughout the nucleus. *J. Cell Biol.* **122,** 283–293.

Wei, X., Mortillaro, M. J., Kim, S., Frego, L., Buchholtz, L., Nakayasu, H., and Berezney, R. (1995). p250 is a novel nuclear matrix protein that colocalizes with splicing factors. *J. Cell. Biochem. Suppl.* **21B,** 142.

Wu, J. Y., and Maniatis, T. (1993). Specific interactions between proteins implicated in splice site selection and regulated alternative splicing. *Cell (Cambridge, Mass.)* **75,** 1061–1070.

Wunderlich, F., and Herlan, G. (1977). A reversibly contractile nuclear matrix. *J. Cell Biol.* **73,** 271–278.

Xing, Y., Johnson, C. V., Dobner, P. R., and Lawrence, J. B. (1993). Higher level organization of individual gene transcription and RNA splicing. *Science* **259,** 1326–1330.

Xing, Y., and Lawrence, J. B. (1991). Preservation of specific RNA distribution within the chromatin-depleted nuclear substructure demonstrated by in situ hybridization coupled with biochemical fractionation. *J. Cell Biol.* **112,** 1055–1063.

Yang, C. H., Lambie, E. J., and Snyder, M. (1992). NuMA: An unusually long coiled-coil related protein in the mammalian nucleus. *J. Cell Biol.* **116,** 1303–1317.

Zahler, A. M., Neugebauer, K. M., Stolk, J. A., and Roth, M. B. (1993). Human SR proteins and isolation of a cDNA encoding SRp75. *Mol. Cell Biol.* **13,** 4023–4028.

Zbarsky, I. B. (1981). Nuclear skeleton structures in some normal and tumor cells. *Mol. Biol. Rep.* **7,** 139–148.

Zehnbauer, B., and Vogelstein, B. (1985). Supercoiled loops in the organization of replication and transcription in eukaryotes. *BioEssays* **2,** 52–54.

Zeitlin, S., Parent, A., Silverstein, S., and Efstratiades, A. (1987). Pre-mRNA splicing and the nuclear matrix. *Mol. Cell. Biol.* **7,** 111–120.

Zeitlin, S., Wilson, R. C., and Efstratiadis, A. (1989). Autonomous splicing and complementation of in vivo-assembled spliceosomes. *J. Cell Biol.* **108,** 765–777.

Zeng, C., He, D., and Brinkley, B. R. (1994a). Localization of NuMA proteins isoforms in the nuclear matrix of mammalian cells. *Cell Motil. Cytoskel.* **29,** 167–176.

Zeng, C., He, D., Berget, S. M., and Brinkley, B. R. (1994b). Nuclear mitotic aparatus proteins: A structural protein interphase between the nucleoskeletons and RNA splicing. *Proc. Natl. Acad. Sci. U.S.A.* **91,** 1505–1509.

Zhelev, N. Z., Todorov, I. T., Philipova, R. N., and Hadjiolov, A. A. (1990). Phosphorylation-related accumulation of the 125K nuclear matrix protein mitotin in human mitotic cells. *J. Cell Sci.* **95,** 59–64.

The Architectural Organization of Nuclear Metabolism

Jeffrey A. Nickerson,* Benjamin J. Blencowe,*,† and Sheldon Penman*
*Department of Biology and †Center for Cancer Research, Massachusetts Institute of Technology, Cambridge, Massachusetts 02139

Nucleic acid metabolism is structurally organized in the nucleus. DNA replication and transcription have been localized to particular nuclear domains. Additional domains have been identified by their morphology or by their composition; for example, by their high concentration of factors involved in RNA splicing. The domain organization of the nucleus is maintained by the nuclear matrix, a nonchromatin nuclear scaffolding that holds most nuclear RNA and organizes chromatin into loops. The nuclear matrix is built on a network of highly branched core filaments that have an average diameter of 10 nm. Many of the intermediates and the regulatory and catalytic factors of nucleic acid metabolism are retained in nuclear matrix preparations, suggesting that nucleic acid synthesis and processing are structure-bound processes in cells. Tissue-specific and malignancy-induced variations in nuclear structure and metabolism may result from altered matrix architecture and composition.

KEY WORDS: Nuclear matrix, RNA metabolism, Nuclear domains, Core filaments, Transcription, RNA splicing, Tissue specificity, Malignancy.

I. Introduction

Nucleic acid metabolism is architecturally organized in the eukaryotic nucleus. Individual catalytic processes and the machinery they require are not uniformly distributed in the nucleus but are structurally constrained to spatial domains. In this review we will argue that the spatial organization of nucleic acids and their metabolism is accomplished by attachments to an intranuclear scaffolding called the nuclear matrix, a structure connected to but distinct from the chromatin.

Many of the reactions, functional components, and intermediates in nuclear metabolism have been identified and characterized by the combined application of biochemical and molecular–genetic approaches. In particular, the development of *in vitro* systems that faithfully perform DNA replication, transcription, the processing of pre-mRNA has allowed the detailed dissection of these processes. Although powerful for the characterization of essential components, a major limitation of *in vitro* systems is their apparent inability to reproduce many of the important kinetic and regulatory features that are observed *in vivo*. These limitations of *in vitro* systems suggest that important features have been lost. One missing element is the architecture of the intact nucleus whose structural components may integrate and coordinate gene expression.

Studies on the spatial distribution of metabolic processes and components have revealed functional domains in the nucleus. The largest of these domains are the nucleoli, sites of ribosomal RNA synthesis and partial ribosome assembly, although many smaller nonnucleolar domains have also been identified. There are domains identified by the processes occurring there such as DNA replication or transcription and there are domains that have been identified by their composition such as the several domains highly enriched in RNA splicing factors.

The structural basis for this segregation of metabolic components has received relatively little attention, perhaps because it cannot be analyzed in a well-defined *in vitro* system. Virtually all of the domain organization of the nucleus remains after the removal of chromatin, demonstrating that a nonchromatin scaffolding within the nucleus is the architectural basis for domain organization. It seems very likely, therefore, that this scaffolding, the nuclear matrix, forms the organizational framework for nuclear metabolism.

The link between nuclear architecture and gene expression may also be an important clue for understanding phenotype-specific gene expression and the alteration of phenotype in malignancy. While regulation of the immediate response of genes to environmental signals is increasingly well understood, far less is known about the mechanisms that set the stable patterns of gene expression in differentiated cells. Structural features of the nucleus, including size, shape, and internal organization, vary with cell type and are radically altered by malignancy. Coincident with these changes are large changes in gene expression. We will argue that phenotype- and malignancy-specific alterations in nuclear matrix structure and composition are the link between these phenomena. The nonchromatin scaffolding of the nucleus may maintain the long-term, tissue-specific states of gene expression that are altered in malignant cells.

II. Functional Domains of the Nucleus

The largest and best studied of nuclear domains are the nucleoli, where ribosomal RNA is synthesized, processed, and assembled into ribosomal subunits. Structural and functional properties of the nucleolus have been extensively reviewed elsewhere (Fischer *et al.*, 1991; Scheer *et al.*, 1993), and will only be mentioned where pertinent here. Several types of nonnucleolar matrix domains have been defined based on criteria such as size, number per nucleus, and presence of specific nuclear constituents. These include structures involved in DNA replication, distinct matrix-associated structures involved in the metabolism of pre-mRNA including "transcript foci," "speckled domains," "coiled bodies," and "tracks." Additional domains, distinct from the above and which are less well characterized, include those that are specifically enriched for the product of the PML (promyelocyte) gene.

All of the nuclear domains listed above appear to be structurally organized by the matrix. Removal of chromatin by nuclease digestion and salt extraction does not appear to alter the number and distribution of these structures, as judged by a comparison of nuclei stained with probes specific for domain-specific constituents before and after nuclear extraction. A striking feature of nuclear organization that is emphasized by a comparison of these nuclear structures is the high level of specificity of sorting for different components, often related to their functions. This implies the existence of a remarkable degree of intranuclear targeting that must be mediated by distinct sets of signals and receptors, analogous to organelle targeting in the cytoplasm. With a few exceptions that we will discuss, the signals mediating domain-specific intranuclear targeting of proteins are not known. It should also be emphasized that most factors specifying the spatial organization of domains within the nuclear matrix structure remain to be identified.

The remainder of this section reviews some of the known properties of each class of nuclear domain. Current information and ideas concerning the functional roles of these matrix-associated structures will also be discussed.

A. DNA Replication Foci

The organization and function of the DNA replication apparatus has been the subject of several recent reviews (Laskey *et al.*, 1989; Cook, 1991). We will therefore only highlight a few of the more recent findings that are pertinent to the nuclear organization of this process. Eukaryote DNA replication takes place in a few hundred discrete replication foci, each of which is thought to contain a portion of the estimated 50,000 replication origins activated during S-phase (Nakamura et al., 1986; Nakayasu and Berezney, 1989). The size and distribution of these foci change in an ordered fashion during S phase of the cell cycle (D'Andrea et al., 1983; Goldman et al., 1984; Hatton et al., 1988; O'Keefe et al., 1992). Not only are factors involved in DNA replication concentrated in these foci, but also factors involved in DNA modification and cell cycle control. To date, factors that have been detected in replication foci include DNA polymerase alpha (Bensch et al., 1982), replication protein A (70-kDa subunit) (Cardoso et al., 1993), DNA ligase (Lasko et al., 1990), proliferating cell nuclear antigen (PCNA) (Bravo and Macdonald-Bravo, 1987), DNA methyltransferase (Leonhardt et al., 1992), and cyclin A and cyclin dependent kinase 2 (cdk2) (Cardoso et al., 1993).

Labeling of cells with 5'-bromodeoxyuridine triphosphate (BrdUTP), followed by detection using anti-BrdU antibodies, demonstrates that DNA synthesis occurs at replication foci (Nakamura et al., 1986; Nakayasu and Berezney, 1989; O'Keefe et al., 1992). In other studies using biotinylated dUTP to pulse label DNA, it was found that at later times of incubation labeling would spread outside these foci into regions of adjacent chromatin, indicating that DNA may be replicated as it passes through the replication foci (Hozák et al., 1993). After removal of chromatin and extraction of soluble nuclear material, remnant DNA and some of the factors in replication foci retain their specific spatial arrangement, indicating that these structures are attached to the nuclear matrix (Pardoll et al., 1980; Berezney et al., 1982; Foster and Collins, 1985; Dijkwel et al., 1986; Jackson and Cook, 1986; Razin, 1987; Tubo and Berezney, 1987a,b; Nakayasu and Berezney, 1989; Vaughn et al., 1990). Recently, resinless section microscopy has been used to visualize these "replication factories" within the matrix-filament network (Hozák et al., 1993).

The concentration of many replication origins within a limited number of foci may help to coordinate the simultaneous "one-off" firing of origins that occurs in S phase. Similarly, the concentration of specific cell cycle proteins, including cyclin A and cdk2, may serve to coordinate cell cycle events with DNA replication. It should be noted that the detection of these cell cycle proteins in replication foci provided the first evidence for the association between cell cycle regulatory proteins and components of the DNA replication apparatus (Cardoso et al., 1993). This study emphasizes the importance of "nuclear cartography" in providing clues for deciphering possible physical interactions that take place between components of coordinated nuclear processes.

B. Transcription Foci

Pulse-labeling of cells with 5'-bromouridine 5'-triphosphate reveals that transcription by RNA polymerase II occurs in over 100 well-defined areas distributed throughout the nucleoplasm, excluding nucleoli (Jackson et al., 1993; Wansink et al., 1993). Consistent with previous observations that the earliest pulse-labeled transcripts associate with the nuclear matrix (see Section VII,A), the number and even nuclear distribution of the transcript domains do not change following removal of chromatin and extraction of the nuclei in 0.25 M salt. As will be discussed in more detail later, these sites are generally not coincident with the sites corresponding to the larger and less abundant "speckled" or coiled body domains, where splicing factors are concentrated (Wansink et al., 1993). The transcript domains are, however, found in proximity to speckle domains and may occasionally overlap.

C. Coiled Bodies

There are approximately one to five coiled bodies per cell nucleus. These structures are highly conserved and have been detected in organisms as diverse as mammals and plants (Lamond and Carmo-Fonseca, 1993). Coiled bodies were initially identified by the Spanish cytologist Ramon y Cajal at the turn of the century (1903, 1909), and subsequently rediscovered several times during the past three decades (Hardin et al., 1969; Monneron and Bernhard, 1969; Seite et al., 1982; Lafarga et al., 1983; Carmo-Fonseca et al., 1992). An important advance in the characterization of the coiled body was the discovery of autoimmune patient sera which selectively label these structures. These autoimmune patient sera recognize a protein of 80-kDa, p80 coilin, which is highly concentrated in coiled bodies, although it is also detected at a relatively low concentration throughout the nucleoplasm, excluding nucleoli (Andrade et al., 1991; Raska et al., 1990, 1991). A p80 coilin cDNA clone has been isolated (Andrade et al., 1991; Chan et al., 1994), which bears some similarity to a protein concentrated in snRNP-rich structures, "snurposomes," in the amphibian oocyte (Tuma et al., 1993). Coiled bodies have been identified as sites of concentration of each of the major snRNAs involved in pre-mRNA splicing and also of several snRNP-associated proteins (Eliceiri and Ryerse, 1984; Fakan et al., 1984; Carmo-Fonseca et al., 1991a,b; Raska et al., 1991; Huang and Spector, 1992; Blencowe et al., 1993). In addition to snRNP components, coiled bodies also contain elevated concentrations of both the 65- and 35-kDa subunits of the non-snRNP splicing factor U2AF (Carmo-Fonseca, et al., 1991a; Zamore and Green, 1991; Zhang et al., 1992).

Recently, it was found that antibodies to p80 coilin give essentially the same immunofluorescence staining pattern on matrix preparations as compared to unextracted cells, indicating that the coiled body structure and p80 coilin are associated with the nuclear matrix (our unpublished findings). In these studies it appeared that p80 coilin was largely retained with the matrix following the removal of chromatin. Consistent with this, it was previously shown that U2 snRNA remains associated with coiled bodies after removal of chromatin and extraction of nuclei in a low salt buffer; unlike p80 coilin, however, the majority of U2 was lost upon extraction at higher salt (Carmo-Fonseca *et al.*, 1991a). These studies could indicate that p80 coilin, as a matrix component, may play a role in the organization and/ or formation of coiled bodies.

The function of coiled bodies is not known. Although these structures contain spliceosomal snRNP components and also a subset of non-snRNP splicing factors, they do not appear to be sites of pre-mRNA splicing. They do not show any preferential labeling with [^{3}H]uridine in pulse-labeling experiments, nor are they detected by *in situ* hybridization with oligo (dT) (Huang *et al.*, 1994). Furthermore, they do not appear to contain non-snRNP essential splicing factors belonging to the mAb104-reactive SR protein family (Carmo-Fonseca *et al.*, 1991b; Raska *et al.*, 1991; Spector *et al.*, 1991). These observations have led to speculation that coiled bodies may be the sites of splicing factor assembly, storage, and/or recycling. However, the presence of coiled bodies does appear to be linked in some way with the transcriptional activity of a cell. For example, conditions that inhibit transcription result in the disassembly of the coiled body, causing a loss of snRNP association and also an altered distribution of p80 coilin (Carmo-Fonseca *et al.*, 1992; Blencowe *et al.*, 1993). As will be discussed in more detail later, it is not clear whether changes such as these reflect a change in functional requirement related to transcript metabolism, or rather, are indirect due to disruption of nuclear organization caused by the loss of structural RNA. Nevertheless, the relative prominence of coiled bodies in rapidly dividing cells, or cells that are undergoing high rates of transcription, suggests that these structures may form in response to increased levels of gene expression (Spector *et al.*, 1992; Carmo-Fonseca *et al.*, 1993).

In addition to snRNP and non-snRNP splicing factors, coiled bodies also contain components of minor nucleoplasmic snRNP particles including U11 and U12 snRNAs (Matera and Ward, 1993). Coiled bodies also contain several factors that are concentrated in nucleoli, including the U3 small nucleolar RNP-associated protein fibrillarin (Raska *et al.*, 1991; Carmo-Fonseca *et al.*, 1993), and two nucleolar proteins Nopp140 and NAP57, which are associated with each other in a complex (Meier and Blobel, 1992, 1994). The accumulation of nucleolar components in coiled bodies suggests a possible structural relationship between these two nuclear compartments. Consistent with this notion, coiled bodies are frequently found adjacent to nucleoli. Visualization by electron microscopy of thin sections stained with gold-conjugated anti-p80 antibodies show that coiled bodies often appear to form continuous interactions with the fibrogranular material of the nucleolus (Hardin *et al.*, 1969; Seite *et al.*, 1982; Lafarga *et al.*, 1983; Carmo-Fonseca *et al.*, 1993). Different-sized structures labeling with p80 antibodies appear at the nucleolar periphery, indicating a possible morphogenesis of coiled bodies at the nucleolus. It is not clear, however, whether these p80-rich structures are partially formed coiled bodies merging and/or emerging from the nucleolus. Real-time video microscopy of cells with fluorescent-tagged p80 may help to track the "life cycle" of coiled bodies in relation to other nuclear structures.

The absence of detectable rRNAs and ribosomal proteins in coiled bodies argues against these structures participating directly in the maturation of rRNA, or in the formation of ribosomal subunits (Raska *et al.*, 1991; Carmo-Fonseca *et al.*, 1993). It is also unlikely that coiled bodies are the sites of snRNA transcription and maturation. Although not examined for all snRNAs concentrated in coiled bodies, it has been shown that the U2 gene cluster does not colocalize with coiled bodies (Matera and Ward, 1993). Furthermore, coiled bodies contain the m3G cap structure (Carmo-Fonseca *et al.*, 1993), which is normally only acquired on snRNAs during their assembly with snRNP proteins in the cytoplasm (Hamm *et al.*, 1990).

The accumulation of certain nucleolar components in coiled bodies suggests possible functions of these structures besides their speculated role in metabolism of splicing factors. For example, it is possible that coiled bodies function in the shuttling of factors imported through the nuclear pore to the nucleoli. As a "transport organelle," the coiled body could perform a function in the maintenance of appropriate concentrations of factors at different sites in the nucleus. This would allow a mechanism for the rapid redistribution of factors according to their specific requirements in processes including, for example, pre-mRNA splicing in the nucleoplasm and rRNA processing and ribosomal subunit assembly in the nucleolus. This model would account for the apparently dynamic nature of the coiled body in response to the metabolic state of the cell. Needless to say, much work must be done to determine the function of these intriguing matrix-associated structures in gene expression.

D. Speckled Domains

Mammalian nuclei typically contain 20–50 domains that are primarily enriched in factors involved in pre-mRNA splicing. Probes specific for factors in these structures give a characteristic "speckled" staining pattern at the level of immunofluorescence. Speckled domains are nonrandomly distributed in nuclei of cultured cells, being concentrated in a plane just below the midline and parallel to the growth surface (Carter et al., 1993). In conventional resin-embedded thin sections visualized by electron microscopy, the speckled pattern appears to correspond to a network of interconnecting structures, often referred to as interchromatin granules and perichromatin fibrils (Perraud et al., 1979; Spector et al., 1983, 1991; Fakan et al., 1984; see later discussion). These structures are preferentially stained by an EDTA-regressive method (Bernhard, 1969), and are also decorated by gold-conjugated antibodies specific for factors concentrated in speckles seen by immunofluorescence staining (Spector, 1993). In resinless sections of nuclear matrix preparations, structures that correspond to interchromatin granules and perichromatin fibrils appear as dense bodies that are enmeshed in the extensive network of matrix core filaments (Nickerson and Penman, 1992). Consistent with the fact that multiple distinct macromolecular structures are associated with the matrix, only a subset of these dense structures is specifically immunogold-labeled with speckle-staining antibodies (Nickerson et al., 1992; Blencowe et al., 1994; Wan et al., 1994).

Nuclear speckles were first detected by the staining patterns of autoimmune patient sera that recognize protein or RNA components of snRNPs (Perraud et al., 1979, Lerner et al., 1981; Spector et al., 1983). More recently, speckled domains (and coiled bodies) have been simultaneously detected with nuclease-stable antisense probes specific for each snRNA involved in splicing, in combination with mAbs specific for snRNP proteins and the snRNA trimethylguanosine (m3G) cap structure (Carmo-Fonseca et al., 1992; Huang and Spector, 1992). These studies indicate, but do not prove, that fully assembled snRNPs are concentrated in these structures. In addition to their concentration in speckles and coiled bodies, components of snRNPs are present at lower concentrations throughout the nucleoplasm, excluding nucleoli.

In addition to snRNP components, speckled domains are also highly enriched in non-snRNP splicing factors belonging to the Ser-Arg (SR) family (Fu and Maniatis, 1990; Roth et al., 1990; Spector et al., 1991). A group of SR proteins share a common phosphoepitope structure detected by a monoclonal antibody, mAb104 (Zahler et al., 1992). These proteins have one or two RNA recognition motifs (RRMs) at the N-terminus, and a domain rich in serine-arginine di-amino acid repeats at the C-terminus. Antibodies that recognize one or more proteins belonging to the mAb104-reactive family show pronounced speckled staining patterns that precisely coincide with the speckled staining patterns obtained with antibodies to snRNPs. Other proteins that contain an SR domain, which are not of the mAb104 class, are also concentrated in speckles. These include the U1 snRNP-associated 70-kDa protein (Verheijen et al., 1986), a 64-kDa autoantigen with sequence similarities to the 65-kDa subunit of U2AF (Imai et *al.,* 1993), and two different *Drosophila* alternative splicing factors, *suppressor of white apricot (su(w^a)* and *transformer (tra)* (Li and Bingham, 1991).

Mutational analyses have shown that the SR domains of *su(w^a)* and *tra* are essential for targeting of these proteins to speckles (Li and Bingham, 1991). Furthermore, fusion of either of these SR domains to β-galactosidase localized the enzyme to the speckled region, demonstrating that these SR domains are both required and sufficient for targeting to speckled domains. Fusion of an SR domain onto B-galactosidase also rendered it more resistant to nuclear extraction, consistent with a role for the SR domain in targeting to a matrix-associated structure. It should be noted, however, that not all SR domains are constitutive for targeting to matrix speckles. As mentioned earlier, both subunits of the heterodimeric non-snRNP splicing factor U2AF contain an SR domain, although in contrast with other SR proteins, U2AF is detected concentrated in coiled bodies and not in speckles (Carmo-Fonseca *et al.,* 1991a; Zamore and Green, 1991; Zhang *et al.,* 1992). Specific structural features that result in the differential targeting of SR proteins are not known.

The speckled domains, which are enriched in splicing factors, appear to be centered within larger domains containing high concentrations of poly(A)$^+$ RNA (Carter *et al.,* 1993). This has been established by the simultaneous use of *in situ* hybridization with an oligo (dT) probe and fluorescent staining with antibodies against either SR proteins or snRNP-associated proteins (Carter *et al.,* 1993). These results suggest that speckled domains are larger than first thought and that these larger domains have an internal organization consisting of an outer region highly enriched in poly(A)$^+$ transcripts and an inner core with high concentrations of splicing factors. When using probes for splicing factors, it is principally the inner core that is detected.

E. Function of Speckled Domains

The concentration of many different splicing factors in speckled domains suggests that these structures could be the sites in the nucleus where splicing takes place. However, some studies have argued that the majority of splicing does not take place in these structures, but rather at sites associated with perichromatin fibrils located at the boundaries of speckles (Fakan and Puvion, 1980; Spector, 1993). This view is primarily based on studies of the localization of nascent transcripts, and also observations on how the distributions of splicing factors change after treatments that affect transcription and splicing. Interpretation of these experiments has sometimes assumed that the majority of pre-mRNA splicing takes place cotranscriptionally. If this is the case, then splicing and transcription would be largely coincident at the resolution of light microscopy. It appears, based on an analysis of Miller spreads, that some splices are completed cotranscription-ally (Beyer and Osheim, 1988) while for other transcripts splice site selection is cotranscriptional but splicing occurs after transcript release from the DNA template (Osheim and Beyer, 1991). For these latter transcripts, splicing could occur at a distance from the gene large enough to be observed by light microscopy.

In early studies on the localization of RNA synthesis, sites of nascent transcription were detected by autoradiography of embedded thin sections from cells labeled with a short pulse of tritiated uridine. After a short pulse, the majority of labeling was detected at the periphery of speckles, coincident with the perichromatin fibril region (Fakan and Bernhard, 1971; Fakan *et al.*, 1976; Fakan and Noblis, 1978). If labeling times were extended, however, an increased frequency of labeling was detected in the interchromatin gran-ule regions that correspond to speckled domains seen by immunofluores-cence. More recently, sites of nascent transcription have been localized by labeling cells with a short pulse of bromouridine triphosphate (BrUTP), followed by detection with a fluorescently tagged antibody specific for BrUTP (Jackson *et al.*, 1993; Wansink *et al.*, 1993). A relatively small percentage of transcript foci detected by this approach was coincident with the speckles detected by antibodies to splicing factors. A larger percentage of foci was present in regions adjacent to splicing factor-containing speckles. Interpretation of these total transcript localization studies is, however, com-plicated by the fact, as discussed further in Section VI, that most hnRNA transcripts do not serveas a precursor for cytoplasmic mRNA and it is not known whether these transcripts require intron removal. The localization of bulk transcription to separated sites from speckled domains does not prove that splicing occurs at these sites. Only localization of specific, intron-containing transcripts can clarify this issue.

Recently, this technique was used to compare the nuclear localization of unspliced and corresponding spliced mRNAs from human β-actin genes (Zhang *et al.*, 1994). It was reported that both spliced and unspliced RNAs colocalized with each other at two transcription foci per cell which, in a population of fibroblasts, appeared to be randomly distributed with respect to speckled domains. According to this report, only 5 cells containing a total of 10 foci were examined. A similar experiment of β-actin transcript localization in fibroblasts, employing a much larger sample size, has been performed (Xing, 1993). Examination of 155 β-actin transcript domains showed that about 90% were adjacent to or overlapping speckled domains enriched in splicing factors. The sample size in this study has subsequently been increased to 350 transcript domains with similar results (J. Lawrence, personal communication). The larger sample size (350 signals vs 10) used in this latter study suggests that the Zhang *et al.* (1994) results may not be entirely representative.

Some of the studies described above concluded that most nascent transcription, and by inference, splicing, occurs at sites located outside of the speckled domains of the nucleus. This view appears to be supported by studies involving manipulations of cells that redistribute splicing factors. Treatments and growth conditions that inhibit transcription and splicing cause the apparent "movement" of splicing factors from regions of nascent transcription (perichromatin fibril region) "into" what resemble rearranged speckled structures. For example, inhibition of transcription with drugs (Spector *et al.*, 1983; Carmo-Fonseca *et al.*, 1992) or by growing cells containing a temperature-sensitive RNA polymerase II subunit at the restrictive temperature (Huang *et al.*, 1994), results in the appearance of fewer and larger speckled structures. A similar rearrangement is seen during the later stages of differentiation of a murine erythroleukemia cell line induced to terminally differentiate (Antoniou *et al.*, 1993). At 6 days post-induction, when transcription levels are greatly reduced, snRNPs are no longer concentrated in 20–50 speckles, or coiled bodies, but are almost entirely concentrated in several large speckle-like structures.

In another study, microinjection into cells of oligonucleotides complimentary to snRNAs, or antibodies to snRNPs that inhibit splicing *in vitro,* also resulted in a similar rearrangement of speckles (O'Keefe *et al.,* 1994). It is important to note, however, that the microinjected oligonucleotides and antibodies used in this study caused a reduction in transcription levels *in vivo.* Consequently, it was not clear whether the redistribution of splicing factors was a direct consequence of the inhibition of snRNP activity, or rather, was an indirect result of reduced transcription. Nevertheless, one interpretation from all of these studies is that splicing factors, when no longer required for the processing of nascent transcripts, move into speckled structures, which may therefore correspond to sites of storage.

While this view is consistent with the findings, another possibility, which is not mutually exclusive, is that the redistribution of splicing factors is a result of altered nuclear structure following the inhibition of transcription. As discussed below, inhibition of transcription is known to collapse chromatin architecture and compromise the structural integrity of the nuclear matrix, possibly through the loss of structural RNA (Nickerson and Penman, 1989, reviewed in Nickerson and Penman, 1992). Consequently, it is likely that treatments that affect transcription lead to a reorganization of speckles and the factors that are contained within them. This could explain the observation that significantly fewer and larger speckles are observed in transcriptionally inhibited cells, rather than larger speckles of the same number. This apparent coalescing of speckles could result from the collapsing together of splicing factors that are normally spatially separated by nuclear structure. For example, members of the SR family, which are highly phosphorylated RNA-binding proteins enriched in speckles, could potentially cluster in such a fashion.

Despite the relative absence of detectable pulse labeled nascent transcripts in speckles, studies on the localization of poly(A)$^+$ transcripts using hybridization *in situ* with oligo (dT) have shown that a significant fraction of nuclear poly(A)$^+$ RNA is in fact concentrated in these structures and is also in larger domains surrounding them (Carter *et al.,* 1991, 1993). A recent study suggests, however, that much of this poly(A)$^+$ RNA may not in fact correspond to precursor RNA destined for export, but rather to a population of RNA that remains stably associated with the nucleus (Huang *et al.,* 1994). In this study, the majority of the poly(A)$^+$ signal appeared to remain in the nucleus and become associated with the rearranged speckled structures following the inhibition of RNA polymerase II transcription. Interestingly, it was found that the majority of the poly(A)$^+$ RNA signal associated with speckles in nondrug-treated cells was retained in matrix preparations following DNase I treatment and salt extraction of nuclei.

These data were used to argue that some of the poly(A)$^+$ RNA retained in speckles after inhibition of transcription may not correspond to functional precursors of mRNA, but is a pool of stable nuclear RNA (Huang *et al.,* 1994). The data are, however, contradicted by a second study employing *in situ* hybridization and electron microscopy to detect poly(A)$^+$ transcripts. In this work, 80% of the detectable poly(A)$^+$ in the interchromatin granule clusters (speckles) was removed in the 1 hr following actinomycin D inhibition of transcription (Visa *et al.,* 1993). It is difficult to evaluate this apparent discrepancy in results. In addition, both of these studies are complicated by the observation that inhibition of transcription itself may stop the processing and transport of preexisting precursor mRNA out of the nucleus. This was initially shown in cells where, first, hnRNA was pulse labeled and then further transcription was inhibited by actinomycin D (Herman and Penman, 1977). Actinomycin blocked the appearance of labeled mRNA in the cytoplasm and decreased the rate of hnRNA decay.

Further, as the authors of the Huang *et al.* (1994) study point out, a complication in quantitating the poly(A)$^+$ signal is that poly (A)$^+$ tail lengthening is associated with the inhibition of transcription. It is possible that some of the poly(A)$^+$ RNA signal in speckles represents a subpopulation of export-competent transcripts. As will be discussed in the following section, recent studies involving the detection of specific transcripts by *in situ* hybridization, support the view that certain pre-mRNAs are associated with speckled structures prior to processing and export.

F. RNA Transcript Tracks and Domains

Recent studies employing fluorescence *in situ* hybridization for the detection of specific genes and corresponding transcripts have resulted in a more detailed picture of transcript localization within the cell nucleus. Using probes that distinguish the localization of a particular gene from its corresponding transcript, and also intron from exon sequences within the transcript, it was shown that certain viral and cellular pre-mRNAs are localized within the confines of curvilinear tracks (Lawrence et al., 1989; Huang and Spector, 1991; Xing et al., 1993). Significantly, both the quantitative and qualitative appearances of individual RNA tracks were preserved during a matrix preparation consisting of removal of 95% of chromatin and also bulk protein and phospholipid material (Xing and Lawrence, 1991). These tracks emerge from a single focus of transcription, coincident with the gene, and then usually extend out toward the nuclear periphery. Consistent with splicing taking place within the track, intron sequences are only detected in a relatively small portion of the track, whereas exon sequences are detected throughout the track (Xing et al., 1993). Significantly, tracks defined by some transcripts are found to intersect a speckled domain, whereas other tracks do not.

In more recent studies, Lawrence and co-workers have localized additional pre-mRNA transcripts in the nucleus, all of which are spliced (Xing, 1993; J. Lawrence, personal communication). Many of these transcripts do not appear as tracks but, rather, as discrete nonlinear domains. For most of these transcripts the domain was consistently overlapping or directly adjacent to an SC-35-containing speckle. A smaller number of transcripts were localized in domains separated from the speckle. Analysis of many cells showed that the localization of a particular transcript was similar in a large majority of cells. These studies indicate that splicing, at least for some pre-mRNAs, may involve a direct interaction with speckled domains and are consistent with an earlier model Lawrence has proposed (Carter et al., 1993; Lawrence et al., 1993; Xing et al., 1993). In this model, transcription of most pre-mRNAs takes place in a larger poly(A)$^+$ containing speckled domain that has a core enriched in splicing factors. During or following transcription in this poly(A)$^+$domain, transcripts move toward the core of the domain that contains splicing factors. This model emphasizes a role for the core of the speckled domain in pre-mRNA splicing.

G. PML Bodies

Mammalian nuclei contain several other types of domains that are spatially and compositionally distinct from those already mentioned. Similar to the other nuclear structures, one of these was initially detected by a class of autoimmune patient sera that recognize antigens concentrated within 10–20 nuclear matrix-associated foci (Ascoli and Maul, 1991). These domains were recently demonstrated to contain the product of the promyelocytic (PML) gene (Dyck et al., 1994; Weis et al., 1994). While the function of

PML is not known, chromosomal translocations that fuse PML to a portion of the retinoic acid α receptor (RARα) result in the onset of acute promyelocytic leukemia (APL) (de Thé *et al.*, 1990; Borrow *et al.*, 1990; Longo *et al.*, 1990). In APL cells, PML bodies are disrupted and the PML-RARα fusion is detected in a few hundred dot-like structures giving the nucleus a microparticulate appearance when stained with anti-PML antibodies (Dyck *et al.*, 1994; Weis *et al.*, 1994). Remarkably, treatment of APL cells with retinoic acid leads to a restoration of PML bodies. This is of particular significance since the administration of retinoic acid to APL patients leads to a remission of symptoms (Huang *et al.*, 1988). Thus it appears that the disruption of a particular nuclear matrix-associated domain is somehow linked to oncogenesis in APL patients.

Biochemical studies on the PML-RARα indicate that the presence of PML in the fusion protein does not alter the ability of RARα portion to bind and activate RA-responsive genes (de Thé *et al.*, 1991; Kastner *et al.*, 1992). By contrast, it has been found that the effects of expression of PML-RARα on the integrity of PML bodies is directly correlated with the ability of the fusion protein to form a heterodimer with wild-type PML (Dyck *et al.*, 1994). This has suggested that modulation of the activity of PML, which presumably normally forms homodimers in PML bodies, could result in oncogenesis. In this scenario, the PML-RARα would act as a dominant negative oncoprotein as a consequence of its interaction with wild-type PML.

These studies pose the critical question as to what function PML bodies play in the regulation of gene expression. Pertinent to this question is the structure of the PML protein and what its cellular function might be. PML contains a cysteine-rich region containing a RING finger motif (Hanson *et al.*, 1991) that has been found in over 40 other proteins, some of which may regulate gene expression (Freemont *et al.*, 1991; Kakizuka *et al.*, 1991). While it is only possible to speculate at this stage, it is has been considered that the PML protein may be involved in the regulation of transcription and/or processing of specific genes involved in the differentiation of promyelocytes. It is interesting in this regard that one of the immediate early products of herpes simplex virus (HSV), ICP0, which is required to activate both HSV and cellular genes during infection, also contains a RING finger motif. ICP0 is transiently concentrated during infection in PML bodies after which it appears to cause disruption of these nuclear structures (Maul *et al.*, 1993; Maul and Everett, 1994) .

The presence of ICP0 and PML in PML bodies indicates that these nuclear structures may concentrate proteins that are required for regulating the activity of specific genes. By determining the target genes of PML and ICP0, and whether these target genes colocalize with PML bodies, may help to address this question. The purification and characterization of the composition of PML bodies should also help to elucidate the function of these structures. The identification of antisera that react with additional proteins enriched in PML bodies seems promising in this regard.

The observation that many of the components that function in nuclear metabolism are constrained to specific loci in the nucleus is inconsistent with any view of the nucleus as a soluble system. The segregation of nuclear metabolic components into domains requires an underlying substructure for maintenance and support. As we will describe in more detail, much of the domain organization of the nucleus resists the removal of chromatin. A nonchromatin scaffolding within the nucleus may therefore be the architectural support for domain organization and chromatin organization.

III. The Nuclear Matrix: History and Definition

Our ideas about cell architecture are shaped by the technology available for examining cells. Early observations of nuclear structure were made by light microscopy and provided little evidence for the existence of an internal, nonchromatin skeleton. The material in which chromatin and nucleoli were suspended appeared to be a translucent gel and this material was called the nuclear sap or karyolymph (Fawcett, 1966). The application of thin section electron microscopy to the study of nuclear organization revealed a much more highly structured nucleus than had been imagined. Some of the structures observed correspond to the functional domains we have already discussed and others that we will describe are architectural elements important for maintaining nuclear organization.

A. The Fibrogranular Ribonucleoprotein (RNP) Network of the Intact Nucleus

When examined by thin section electron microscopy, the nucleus clearly contained two distinct but overlapping nucleic acid-containing structures. The first was the very darkly staining chromatin, a DNA-containing structure organized in dense regions of transcriptionally silent heterochromatin and the less compact transcriptionally active euchromatin.

A fibrogranular network was observed in the interchromatin space between patches of condensed heterochromatin. The use of EDTA-regressive staining on embedded thin sections showed that this material was not chromatin but did contain RNA (Bernhard, 1969; Monneron and Bernhard,

1969; Petrov and Bernhard, 1971). This RNA-, or more precisely RNP-containing network was well distributed throughout the nuclear interior, though excluded from regions of dense heterochromatin. We contend that this RNP-containing structure, visualized in unfractionated nuclei by this technique, corresponds to a large part of the nuclear matrix. While certain nuclear matrix isolation protocols are controversial, the existence of an RNA-containing network distinct from the chromatin is not.

The RNP network selectively stained by the EDTA-regressive method consists of both granules and irregular fibers named according to their spatial relationship to the chromatin. Thus they are called peri- and inter-chromatin granules and fibers. As early as 1978, it was clear that RNP fibrils of this network were retained in the nucleus following the removal of chromatin, although they could be destroyed by a subsequent RNase A digestion (Herman et al., 1978).

Pulse labeling of RNA labels the perichromatin fibrils first, suggesting a directed movement of newly transcribed nonnucleolar RNA through the network (Bachellerie et al., 1975; Fakan et al., 1976; Fakan and Hughes, 1989). The kinetics of RNA labeling and the presence of actively transcribed chromatin suggest the region surrounding interchromatin granule clusters is the major location of hnRNA transcription. Both the interchromatin granule clusters and the perichromatin fibrils defined by the EDTA-regressive technique are candidates for being involved in hnRNA splicing to mRNA. As we have discussed in Sections II,D and II,E, both structures contain polyadenylated RNA, hnRNP proteins, snRNPs, and non-snRNP splicing factors such as SR-proteins (Fakan et al., 1984; Puvion et al., 1984; Spector et al., 1991; Visa et al., 1993). The highly structured nuclear interior imaged by electron microscopy led Fawcett to recognize that terms such as nuclear sap were no longer appropriate for describing the nonchromatin material in the nucleus and to suggest nuclear matrix as a preferable substitute (Fawcett, 1966). Later, Berezney called the nonchromatin structures seen by electron microscopy in the unfractionated nucleus the "in situ nuclear matrix" (Berezney, 1984).

Although many names have subsequently been used to identify this structure, we prefer just nuclear matrix because it was suggested in advance of any matrix-isolation protocol and, therefore, in its original context, was the first term used to describe nonchromatin nuclear substructure. The term nuclear matrix should therefore be used only when referring to a cell structure and not be limited to the product of a particular isolation technique. It is important in this regard that different matrix isolation protocols must be judged according to how well they maintain the morphology and composition of the matrix structure.

B. Development of Matrix Isolation Protocols

Biochemical fractionation protocols demonstrated the existence of insoluble nuclear components and formed the basis for the first matrix isolation procedures. Studies from as early as 1942 (Mayer and Gulick, 1942) identified a subfraction of nuclear proteins that resisted extraction with high-ionic-strength solutions. These proteins were relatively insoluble, and although apparently of interest, this property made them difficult to analyze biochemically. Similar high-ionic-strength extractions would later be used in conjunction with nuclease digestions to isolate the nuclear matrix, largely free of chromatin. The biochemical analysis of the matrix would, however, continue to be plagued by the insolubility of its structural components. Insolubility continues to be a great technical challenge and has slowed the biochemical analysis of nuclear structure.

In their pioneering studies, Berezney and Coffey (1974, 1975, 1977) developed a protocol employing nuclease digestion and elevated ionic strength for the isolation of the nuclear matrix. In their procedure, chromatin was removed from isolated rat liver nuclei by DNase I digestion and removal of cut-chromatin by 2 M NaCl extraction. This was a relatively severe treatment and it insufficiently preserved nuclear matrix ultrastructure.

Since the early work of Berezney and Coffey, many nuclear matrix isolation protocols have been employed. Some are variants on the Berezney–Coffey method—using different salts, ionic strengths, nuclear isolation procedures, or enzymes (for example, Hodge *et al.,* 1977; Adolph, 1980; Kaufman *et al.,* 1981; Capco *et al.,* 1982; Stuurman *et al.,* 1990). Others remove chromatin from the nucleus by more novel means—using detergents such as LIS (Mirkovitch *et al.,* 1984) or using electrophoretic fields (Jackson and Cook, 1985, 1986, 1988). One very important feature of matrix isolation protocols is whether RNA is removed by RNase digestion (for example, Berezney and Coffey, 1974; Kaufman *et al.* 1981) or preserved by the use of RNase inhibitors (Fey *et al.,* 1986; He *et al.,* 1990). As we will describe, the RNA plays an important role in nuclear matrix morphology and nuclear architecture. The nuclear matrix and the organization of chromatin are much altered by the removal of RNA.

C. Comparison of Nuclear Matrix Protocols

The great diversity of nuclear matrix protocols poses an important question. How are we to evaluate the relative merits of different protocols? The two standards that have been used for evaluation are biochemical composition and morphological preservation. Ideally for both approaches, the isolated nuclear matrix should be compared to the matrix in an unfractionated cell, and if it were possible, in the living cell.

Unfortunately, there are at present no methods for determining matrix composition in an intact nucleus so the biochemical fidelity of isolated matrix preparations cannot be determined. Still, the appeal of biochemical criteria is so great that the attempt has sometimes been made. For example, if one assumes that *protein X* is a component of the nuclear matrix, then one might evaluate isolation protocols by their relative retention of *X*. This may be a useful approach to characterizing the apparent binding of factors to the matrix, for example by determining the salt lability of binding, but it cannot validate a protocol. If *protein X* is in fact matrix-associated in the intact cell, then this is a valid approach. If not, this approach leads to artifact, and we are unable to experimentally distinguish between the two results.

The use of "stabilization" protocols to determine composition is even more fraught with interpretational uncertainty. Some proteins have been reported not to be in matrix fractions unless heat or sodium tetrathionate stabilization has been performed. It is very difficult to make a judgment about such experiments. They may be stabilizing a physiologically significant binding or they may be creating an artifact. Stabilization protocols may, of course, be appropriate and useful when used to increase the yield of molecules already established to bind the matrix (Belgrader *et al.,* 1991b) but not in attempts to determine matrix composition or to evaluate matrix protocols.

The second approach for evaluating matrix isolation protocols is to assess morphological preservation. It is possible, if imperfectly, to compare the structural preservation of matrix preparations made according to different protocols. We can determine whether the spatial distribution of specific matrix components is altered by the extraction protocol. There are different ways, all employing microscopy, in which we can look for architectural rearrangement. At the lower resolution of light microscopy we can compare the distribution of specific nuclear antigens before and after extraction using fluorescently tagged antibodies or nucleic acid probes. A more sophisticated, and so far unused, version of this strategy would be to introduce a fluorescently tagged nuclear matrix antigen into the living cell and follow its distribution throughout the fractionation protocol. A good fractionation protocol should preserve the distribution of the antigen at each step in the procedure.

Electron microscopy gives a much finer resolution picture of the matrix. It is possible that a fractionation procedure will leave the gross features of matrix architecture untouched but still degrade the ultrastructure. Only electron microscopy can detect such a change. A good approach to assessing nuclear matrix isolation procedures uses resinless section electron microscopy as a guide (Capco *et al.,* 1982; Fey *et al.,* 1986). This microscopy technique, in contrast to conventional thin-section microscopy, allows the whole structure to be imaged at high contrast and in three dimensions (Capco *et al.*, 1984; Penman, 1995).

Using resinless section microscopy to evaluate fractionation procedures, Penman and co-workers found relatively gentle conditions that effectively removed chromatin and soluble material from nuclei while leaving an apparently well-preserved nuclear structure (Capco *et al.*, 1982; Fey *et al.*, 1986; He *et al.*, 1990). In this nuclear matrix preparation procedure, whole cells were extracted first with Triton X-100 in a buffer of physiological pH and ionic strength. This effectively removed membranes and soluble proteins. A stronger double detergent combination of Tween 40 and deoxycholate in a low-ionic-strength buffer was then used to remove the cytoskeleton, except for the intermediate filaments. The remaining membrane-free nuclei, still connected at their surfaces to the intermediate filaments, were then digested with DNase I before extraction with 0.25 M ammonium sulfate to remove chromatin. This digestion–extraction step removed more than 97% of nuclear DNA and essentially all the histones. The resulting structure was termed an RNA-containing nuclear matrix because it retained about 75% of nuclear RNA. In these experiments, ultrastructural preservation was judged by criteria such as these: Did the structure uniformly fill the nuclear space? Was the nuclear lamina visible and intact? Was the internal matrix attached to the lamina? Were there structures of the right size and distribution resembling nucleolar remnants preserved in the structure? Was fine structure still visible? These criteria were appropriate and the nuclear matrix structure uncovered by the DNase I–0.25 M ammonium sulfate procedure was better preserved than that afforded by previous protocols. The procedure does not allow, however, the direct comparison of matrix ultrastructure before and after the extraction since the matrix in intact nuclei is hidden by a much larger mass of chromatin.

Variants of the Penman protocol have been developed that are equivalent as judged by ultrastructural preservation and protein composition (Fey *et al.*, 1986; Fey and Penman, 1988). We believe that this preparation retains the overall features of nuclear matrix morphology and was an improvement on previous efforts, although some of the ultrastructure may be altered or lost and some components may be stripped from the structure. There remains a need for further matrix isolation procedure development.

The very high degree of RNA retention in the nuclear matrix suggests a method for judging the morphological preservation afforded by any isolation protocol. We might image the distribution of this matrix-attached RNA by electron microscopy using the EDTA regressive staining method of Bernhard (1969) both before and after the fractionation protocol. We think that this would be a more objective ultrastructural approach to evaluating matrix isolation procedures. We believe that the extensive RNP network throughout the nucleus is a large and important part of the matrix. A good matrix isolation protocol should effectively remove chromatin while preserving the RNP network in as native a form as possible. EDTA regressive staining was employed by Berezney (1984) to evaluate an early matrix isolation protocol but since then has rarely been used for this purpose.

D. Definitions of the Nuclear Matrix

The nuclear matrix has often been "operationally defined" in research reports. This approach is inadequate for defining a cellular structure. The nuclear matrix is usually and most easily studied after the removal of chromatin, but as we have discussed above, the matrix is a structure that can be seen in the unextracted, undigested nucleus without the use of any fractionation protocol or "operation." Attempts to operationally define the nuclear matrix in terms of specific buffer conditions, detergents, pHs, and nuclease digestion conditions fuel the criticism that the matrix is an artifact, whose very existence is highly dependent on experimental conditions. In addition, because there are so many matrix isolation protocols, there would be many operationally defined nuclear matrices, all differing somewhat in morphology and composition.

No important cellular structure or organelle can appropriately be defined operationally. A mitochondrion, for example, can be observed in the intact cell. Intensive study of mitochondria may require the partial isolation of this organelle, but the technique used is not definitive. The technique used never yields an organelle that is absolutely complete and native. All of this is equally true for the cytoskeleton as it is for the nuclear matrix. The techniques used in matrix isolation may yield structures with differing degrees of morphological and compositional preservation, but no single technique can perfectly define the matrix that exists within the living nucleus. The great challenge is to develop protocols that preserve matrix structure so that the isolated preparation resembles the *in vivo* matrix as closely as possible.

IV. Ultrastructure of the Nuclear Matrix

The nuclear matrix consists of two parts: the nuclear lamina, which is a protein shell primarily constructed of the lamin proteins A, B, and C (Gerace *et al.,* 1984; Fisher *et al.,* 1986; Franke, 1987; Krohne *et al.,* 1987; Georgatos *et al.,* 1994), and the internal nuclear matrix. The nuclear matrix is part of a larger structure, being connected to an extensive network of intermediate filaments at the surface of the nuclear lamina. This can be seen most clearly in a whole mount electron micrograph (Figs. 1 and 2). The nuclear matrix and intermediate filaments are integrated into a single cell-wide structure that retains the overall geometry and appearance of the intact cell. We have referred to this architectural skeleton as the nuclear matrix–intermediate filament complex.

The intermediate filaments of adjacent cells join at desmosomes (Fig. 2). This suggests that the nuclear matrix–intermediate filament scaffold may be the fundamental support of tissue architecture. The intermediate filaments, by mechanically coupling the nucleus to the exterior of the cell, may stabilize the form and position of the nucleus. This may explain the observation that nuclear shape is coupled to cell shape (Hansen and Ingber,

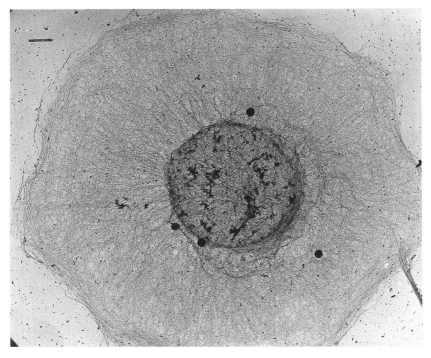

FIG. 1   The nuclear matrix–intermediate filament scaffold. Colon cells grown on nickel grids were extracted with 0.5% Triton X-100 to remove soluble proteins. Chromatin was removed by DNase I digestion and 0.25 M ammonium sulfate extraction. The resulting nuclear matrix–intermediate filament scaffold was fixed, dehydrated, and critical point dried. In this whole mount micrograph cell has not been sectioned and we are viewing the entire structure. The web formed by the intermediate filaments is connected to the internal nuclear matrix through connections to the nuclear lamina. The internal nuclear matrix is not visible but is masked by the dense lamina. Bar: 2 μm.

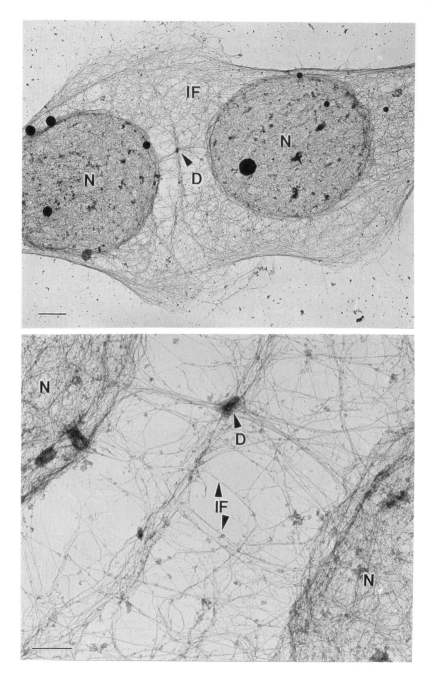

1992; Sims *et al.*, 1992; Ingber, 1993). This architectural arrangement of cells may also couple internal nuclear structures and nuclear metabolism to cell shape and tissue form. Ingber and colleagues have proposed cellular tensegrity models for mechanical signal transduction (Ingber and Folkman, 1989; Ingber *et al.*, 1994) that explain the observed effects of cell shape on growth and gene expression (Folkman and Moscona, 1978; Ben-Ze'ev *et al.*, 1980; Ingber *et al.*, 1994). This model requires a cell structure in which nuclear structure is mechanically linked through the cytoskeleton to the exterior of the cell.

This tissue-wide architectural scaffolding may have evolved to meet the needs of tissue architecture as well as cell structure and we would be surprised to find such a cell organization existing in early unicellular eukaryotes. The oldest eukaryote we have examined so far is the dinoflaggelate. These organisms appear to lack an intermediate filament scaffolding attached to the nucleus (J. A. Nickerson and S. Penman, unpublished observations).

In nuclear matrix preparations the nuclear envelope, a double membrane structure covering the lamina (Nigg, 1989), has been removed by detergent and other extractions. However, the nuclear pores that normally transverse the envelope are not removed by this treatment and remain embedded in the nuclear lamina (Fig. 3). The attachment of pores to a nuclear lamina was first proposed by Aaronson and Blobel (1974) and the structure is often referred to as the pore–lamina complex (Gerace *et al.*, 1978, 1984; Fisher et al., 1982). The intermediate filaments appear to be joined to the lamina at pores by filamentous cross-bridges (Carmo-Fonseca *et al.*, 1987) whose protein composition has not been identified. This RNA-containing nuclear matrix is composed of a network of thick, polymorphic, and irregular fibers bounded by the meshwork of the nuclear lamina (Fey *et al.*, 1986). The fibers suspend larger masses in the nuclear interior. Some of these may correspond to the structural remnants of nucleoli. Some stretches of thinner and more uniform filaments can be seen along the fibers. As we shall discuss, the nuclear matrix is built on a core of branched 10-nm

---

FIG. 2   The intermediate filaments of adjacent cells are connected at desmosomes. Colon cells were grown for whole mount electron microscopy on nickel grids. Cells were extracted in 0.5% Triton X-100 before chromatin was removed with DNase I and 0.25 *M* ammonium sulfate to uncover the nuclear matrix–intermediate filament scaffold. The structure was then fixed, dehydrated, and critical point dried. (Top) Whole mount electron micrograph of two adjacent colon cells. The intermediate filaments (IF) are seen connected to the nucleus (N) and the desmosomes (D) at the cell periphery. Bar: 1 μm. (Bottom) A higher magnification view of the same colon cells showing that the intermediate filaments (IF) of adjacent cells are interconnected at desmosomes (D). Bar: 0.5 μm.

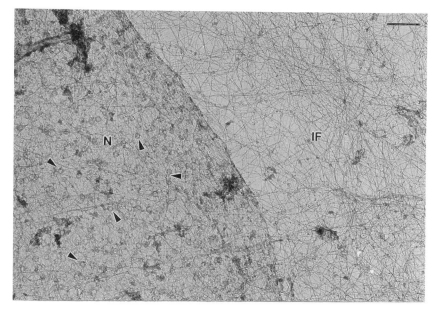

FIG. 3.   The nuclear lamina has embedded nuclear pores. Colon cells grown for whole mount electron microscopy on nickel grids were first extracted with 0.5% Triton X-100. Chromatin was removed by DNase I digestion and 0.25 *M* ammonium sulfate extraction. The resulting nuclear matrix–intermediate filament scaffold was fixed, dehydrated, and critical point dried. The intermediate filaments (IF) are connected to the nuclear lamina (N) through connections to the nuclear lamina, which displays numerous pores (arrowheads). In the unextracted nucleus (not shown) the nuclear envelope covers the lamina leaving only the pores exposed. Bar: 0.5 μm.

filaments. We believe that the structure reported in Fey *et al.* (1986) retains most features of nuclear matrix morphology, but some elements of fine structure may be lost or inadequately preserved. We are working to develop methods of matrix isolation that may afford better preservation of matrix ultrastructure at this later stage of extraction.

When we image the nuclear matrix in extracted and fixed cells, we are examining a static structure. The nuclear matrix in the living cell might, however, be dynamic. Recent studies have shown that the outer shell on the matrix, the nuclear lamina, is much more dynamic than had been imagined. Foci of lamin proteins exist in the interior of the nucleus and can relocate to and reintegrate into the lamina (Goldman *et al.*, 1992; Bridger *et al.*, 1993). Some of the internal foci of lamin B are only present for short times during S-phase and colocalize with sites of DNA replication (Moir *et al.*, 1994). Dynamic rearrangements of the major structural armatures have not been experimentally demonstrated, except of course at mitosis when the whole nucleus disassembles and reforms. Only when the major structural proteins of the matrix have been identified will we be able to tag them in the living cell and follow their redistributions. Studies on matrix dynamism as well as other important questions of matrix structure will, therefore, require a better biochemical identification and characterization of its structural components.

The nuclear matrix is not only a proteinaceous scaffolding but is also a structure of RNA. In a preparation made with RNase inhibitors, the matrix retains more than 70% of total nuclear protein and less than 10% of total nuclear protein (Fey *et al.*, 1986; Fey and Penman, 1988; He *et al.*, 1990). Compared to the unfractionated nucleus, therefore, the matrix is more than sevenfold enriched in nuclear RNA. As we will discuss in greater detail in Section VI, nuclear RNA plays a crucial role in the structure of the nucleus and of the nuclear matrix. Digesting the matrix with RNase A or pretreating cells with transcription inhibitors before matrix isolation causes the matrix to rearrange and the chromatin to collapse (Nickerson *et al.*, 1989).

V. The Core Filaments of the Nuclear Matrix

The RNA-containing nuclear matrix is a complex structure built on a simpler core structure, an underlying network of highly branched 10-nm filaments that we have called the core filaments of the nuclear matrix (He *et al.*, 1990). After removal of chromatin by DNase I digestion and 0.25 *M* ammonium sulfate extraction to expose the nuclear matrix (Capco *et al.*, 1982; Fey *et al.*, 1986), we can further extract with 2 *M* NaCl to remove additional nuclear matrix proteins to reveal an underlying network of core filaments (Fig. 5). Resinless section electron microscopy, which combines the superior three-dimensional imaging of the whole mount with the ability to section, provides a superior view of the nuclear core filament structure as shown in Figs. 4 and 5. So far, we have found these core filaments in HeLa, MCF-7, SiHa, and CaSki cells as well as in rat hepatocytes and we believe them to be universal. They are thinner and more uniform than the thick fibers of the complete matrix, having an average diameter of 10 nm. Many nuclei fractionated in this way have large, dark masses with a number and spatial distribution in the nucleus and with a granular morphology, consistent with their being remnant nucleoli. These presumptive nucleolar remnants are less electron dense than those seen before 2 *M* NaCl extraction and appear to be held in space by the core filaments that are anchored on the nuclear lamina. Core filaments larger than the average diameter of 10 nm emanate from these masses. When seen at higher magnification, the nuclear core filaments in human cells are heterogeneous in appearance.

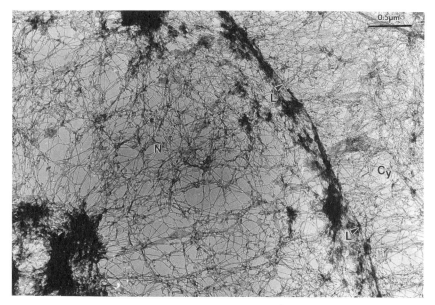

FIG. 4.   Ultrastructure of the HeLa cell nuclear matrix core filament network seen in a resinless section. HeLa cells grown on Mylar coverslips were extracted with 0.5% Triton X-100 before chromatin was removed by DNase I digestion and extracted first with 0.25 M ammonium sulfate and then with 2 M NaCl. Following fixation, the fractionated cells were sectioned and prepared for resinless section electron microscopy (Capco *et al.*, 1984). The nuclear matrix consists of the nuclear lamina (L), which anchors a network of 10-nm filaments (N). The nuclear matrix core filaments suspend masses of different sizes in the nuclear interior, the largest of which may be remnant nucleoli. The lamina is seen to integrate the nuclear matrix core filaments and the intermediate filaments of the cytoskeleton (Cy) into a single structure. Bar: 0.5 μm.

There is some range in sizes with two larger classes of filaments having mean diameters of 9 and 13 nm. The filaments are organized in a network that is highly branched and filaments in this network anastomose smoothly with no obvious junction structures (Fig. 5).

Nuclear filaments have also been reported by Jackson and Cook (1988), who use a very different procedure to partially remove chromatin under isotonic conditions. Following the removal of soluble proteins in detergent, they remove chromatin from agarose-encapsulated cells by *Hae*III digestion and electrophoresis. This procedure removes chromatin incompletely; about 25% of the DNA is retained compared to less than 1% in the Penman procedure described previously (He *et al.*, 1990). The remaining chromatin appears to cover the core filaments, which are seen only in scattered uncovered regions. The filaments revealed by the Penman procedure have a similar appearance but without the regular striations seen following the

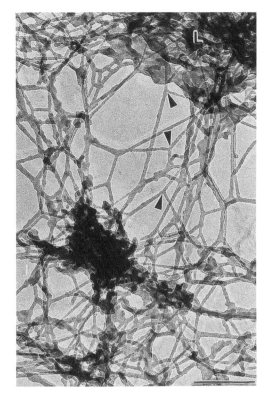

FIG. 5.   Core filaments of the HeLa cell nuclear matrix. HeLa cells grown in suspension were detergent-extracted to remove soluble proteins and then digested with pure DNase I and extracted with 0.25 M ammonium sulfate to remove chromatin and uncover the nuclear matrix. The matrix was then further extracted with 2 M NaCl to reveal the core filaments of the nuclear matrix. The structures were fixed, sectioned, and prepared for electron microscopy by the resinless technique (Capco *et al.,* 1984). The core filament network, shown at high magnification, is connected to the nuclear lamina (L). The web of core filaments is intimately connected to and suspends a dense mass. Some similar masses stain with antibodies against RNA splicing factors and these may correspond to speckled domains. The arrowheads point to the smooth junctions between core filaments. Bar: 0.2 μm.

Jackson and Cook (1988) procedure. The Penman procedure is somewhat simpler, more rapid, and results in a greater release of chromatin. The demonstration of 10-nm core filaments uncovered by two very different procedures increases our confidence that these filaments form the basic support for nuclear matrix structure.

When uncovering nuclear matrix core filaments by the Penman procedure either a stepwise increase in ionic strength, first 0.25 M ammonium sulfate and then 2 M NaCl, or the gradual increase in NaCl concentration over several minutes is necessary for revealing well-preserved core filaments (He *et al.*, 1990). Even with this gentle, stepwise increase in salt concentration, some heterogeneity of morphology is observed. We attribute this in part to the use of DNase I which may damage core filament networks. When DNase I is used in the fractionation protocol, extensive core filaments are only observed by resinless section electron microscopy in some cells (He *et al.*, 1990). In others, the nuclear interior appears to be empty or to contain collapsed masses. This difference could be due either to the partial removal of filaments by DNase I or by the rearrangement of the filament network by DNase I into one region of the nucleus so that some sections are empty. When restriction endonucleases are used for DNA digestion, almost all cells have core filament webs that are uniformly distributed throughout the nuclear interior. A similar sensitivity of 10-nm nuclear filaments to DNase I digestion was observed by Jackson and Cook (1988). The reason for this enzyme-specific difference in core filament preservation is unknown. A small amount of DNA may be required for core filament stability. In our hands the use of restriction enzymes, a combination of *Hae*III and *Pst*I, removes somewhat less DNA (95%) than does DNase I (99%) from the nucleus, leaving more DNA with the core filament network. Alternatively, an activity in DNase I unrelated to its DNA hydrolysis properties may destabilize core filaments. In either case, the actual removal of core filaments probably occurs after fractionation—during fixation and preparation for electron microcopy. We have found no evidence for aggregation or excessive removal of core filament components by immunofluorescent microscopy with antibodies against various proteins retained with the core filaments.

DNase I used during nuclear fractionation persists in core filament preparations and can be observed as a major protein on silver-stained two dimensional electrophoresis gels. This persistence may be related to the very tight actin-binding properties of DNase I (Lazarides and Lindberg, 1974; Hitchcock *et al.*, 1976; Mannherz *et al.*, 1980). Actin is a major protein in nuclear matrix protein preparations (Nakayasu and Ueda, 1983, 1986; Fey and Penman, 1988). Actin-binding might be the mechanism for core filament destabilization by DNase I. Whatever the effect of the enzyme on core filaments, the persistence of DNase I in preparations is a troubling phenomenon. DNase I causes the depolymerization of any filamentous actin (Hitchcock *et al.*, 1976), making it difficult to evaluate any role that F-actin might play in nuclear matrix architecture. This is especially important for understanding the role of the nuclear matrix in RNA metabolism since many studies have suggested a relationship between nuclear actin and RNA (Nakayasu and Ueda, 1985; Schröder *et al.*, 1987a,b; Sahlas *et al.*, 1993). In addition, actin-binding inhibits the enzymatic activity of DNase I (Lazarides and Lindberg, 1974). This inhibition may require the use of larger amounts of enzyme in the removal of chromatin. It would be a technical advance if we could replace DNase I with a highly active deoxyribonuclease that removed DNA more completely than restriction enzymes, did not bind actin, did not persist in matrix preparations, and lacked RNase activity.

The composition of the core filaments remains to be determined. Several nuclear matrix proteins have been characterized biochemically and cloned, but so far, none has been shown to form 10-nm filaments similar to the core filaments. These include several major proteins of the rat liver nuclear matrix that have been called matrins (Belgrader et al., 1991a; Hakes and Berezney, 1991; Nakayasu and Berezney, 1991). Jackson and Cook have noted the similarity of core filaments to the intermediate filaments of the cytoskeleton (Jackson and Cook, 1988) but so far no intermediate filament protein has been detected in these structures. We have used several antibodies against members of the intermediate filament family proteins to stain core filament preparations by immunogold electron microscopy. All antibodies tested, including those against keratins, vimentin, lamins A/C, and lamin B, failed to decorate core filaments (J. A. Nickerson and S. Penman, unpublished results).

One nuclear matrix protein that might form filamentous structures is NuMA (nuclear mitotic apparatus protein). NuMA was originally identified as a high-molecular-weight nuclear matrix protein that redistributed to the poles of the spindle at mitosis (Lydersen and Pettijohn, 1980). At least three isoforms of NuMA are generated in cells by alternative splicing of a common pre-mRNA (Tang et al., 1993). These NuMAs may have different distributions in interphase cells but all associate with spindle poles at mitosis (Tang et al., 1994). The NuMA-l form is uniformly distributed throughout the nucleus, though excluded from nucleoli. It has been reported that NuMA-m and NuMA-s are not nuclear but are detected in the centrosomal region of the cytoplasm (Tang et al., 1994). Secondary structure predictions for the sequence of the 236-kD NuMA-l have revealed a central α-helical region similar to the coiled–coiled domains of proteins such as actin, intermediate filament proteins, lamins, and myosin heavy chains (Tousson et al., 1991; Compton et al., 1992; Yang et al., 1992). This predicted structure suggests that NuMA may be capable of forming filaments. It is this property that makes it a good candidate for a core filament structural protein.

It has been reported that in immunogold labeling experiments one anti-NuMA antibody decorated a small fraction of the 10-nm filaments of the core filament network (Zeng et al., 1994). In our own experiments, using two different anti-NuMA monoclonal antibodies, we were unable to extensively decorate core filaments (J. A. Nickerson and S. Penman, unpublished observations). Most of the NuMA protein was retained with the core filament structure and occasional gold beads were seen on core filaments, but extensive decoration of core filaments was not observed. The reason for this discrepancy in results is unknown. It could be a difference in epitope availability, cross-reacting proteins, staining protocol, or in NuMA isotype.

There is, however, an important similarity between these studies; in both sets of experiments anti-NuMA antibodies did not extensively label the majority of core filaments. While NuMA might be a structural component of core filaments, it is not detectable in a majority of filaments. We believe there are other proteins, as yet unknown, that can form 10-nm nuclear filaments. Identifying and characterizing these proteins is an important goal for matrix research.

Disruption of NuMA function has a profound effect on nuclear formation, consistent with an important structural role for NuMA in the matrix (Compton and Cleveland, 1993). Microinjection of anti-NuMA antibodies or expression of truncated forms of NuMA leads to the formation of many micronuclei following mitosis, instead of two daughter nuclei. Interestingly, a similar phenotype has been observed when cells carrying a temperature-sensitive allele of the RCC1 (regulator of chromatin condensation) protein are grown at the restrictive temperature. Remarkably, transfection of wild-type NuMA into this cell line results in rescue of the micronucleate phenotype, indicating that NuMA and RCC1 may have overlapping functions related to the formation of nuclear structure (Compton and Cleveland, 1993). It will be interesting to determine whether RCC1 is also a component of the nuclear matrix.

We have been assuming in this discussion that the core filaments are constructed only of protein. So far, even this has not been experimentally demonstrated. The observation that core filaments are more stable following restriction enzyme digestion than they are following DNase I digestion could be interpreted to mean that DNA, or some small fragment of DNA, plays a structural role in core filament integrity (He *et al.*, 1990). The use of DNase I during core filament preparation leaves a smaller fraction of DNA with the filament network than does restriction enzyme digestion (1% vs 5%). As discussed previously DNase I digestion results in a greater heterogeneity of ultrastructure from nucleus to nucleus, with some apparent loss of core filaments.

In our earliest observations of core filaments (He *et al.*, 1990) it appeared that RNase A digestion was removing the core filament network from the nucleus. These early electron microscopic results have not, in subsequent experiments, been generally repeatable (J. A. Nickerson and S. Penman, unpublished observations) and are inconsistent with other experimental observations. Several proteins that are supported in space by the core filament network are not removed from the nucleus by RNase A or RNase T1 digestion of core filament preparations (J. A. Nickerson and S. Penman, unpublished observations, and Zeng *et al.*, 1994). Their distribution, as judged by immunofluorescence microscopy, is sometimes altered but not radically enough to think that the entire core filament web is collapsing. It seems, at this time, that the collapse in overall nuclear organization and chromatin architecture caused by RNase A (Nickerson *et al.*, 1989) is a destruction of the linkage between chromatin and the core filaments and not a collapse of the core filament structure itself.

VI. Role for RNA in Nuclear Structure

Detergent extraction of cells removes very little nuclear RNA. Nuclear matrix core filament preparations that have been exposed to long DNase I digestions at 25–35° C retain about 70% of nuclear RNA (He *et al.*, 1990). The 30% of RNA released may be attibuted to an artifactual release during the procedure and not represent genuinely soluble RNA. This RNA retained with the matrix contains about equal amounts of hnRNA and rRNA. Most of the rRNA in core filament preparations is of precursor forms and not the fully processed 18 and 28S forms present in ribosomes.

Most of the hnRNA in matrix and core filament preparations consists of very high molecular weight species. It is important to note that a substantial proportion of hnRNA in the nucleus does not serve as a precursor for cytoplasmic mRNA. Most hnRNA molecules synthesized in the nucleus never appear in polysomal mRNA (Harpold *et al.* 1981; Salditt-Georgieff *et al.*, 1981). This was established by counting the number of capped, polyadenylated hnRNA transcripts synthesized and comparing this with the number of capped, polyadenylated mRNA molecules reaching cytoplasmic fractions. Some of this excess nuclear hnRNA may consist of transcribed repetitive elements, very unlikely to be coding for protein sequences. Transcribed repetitive elements have been identified in sea urchin embryos (Costantini *et al.*, 1978; Scheller *et al.*, 1978), in rat liver (Savage *et al.*, 1978), and in plants (Wu *et al.*, 1994). The *Hin*dIII 1.9-kb repeat is actively transcribed in human cells (Manuelidis, 1982). *Hin*dIII transcripts have great size diversity but are largely retained with the nuclear matrix (Nickerson *et al.*, 1989). Few Hind III repeat transcripts are found in cytoplasmic RNA fractions.

Two lines of evidence suggest a critical role for RNA in nuclear structure and organization. First, growth of cells in the presence of inhibitors such as actinomycin D or 5,6-dichloro1-β-ribofuranosylbenzimidazole at levels required to inhibit all transcription results in a major rearrangement and clumping of chromatin and also disruption of the RNA-containing nuclear matrix. Second, treatment of matrix preparations with RNase A led to an alteration in matrix morphology. If RNase treatment is performed on permeabilized cells, massive rearrangements in chromatin occur, similar to the effects when cells are grown in the presence of actinomycin (Derenzini *et al.*, 1980; Nickerson *et al.*, 1989). Actinomycin D and RNase A effects on chromatin architecture are shown in Fig. 6.

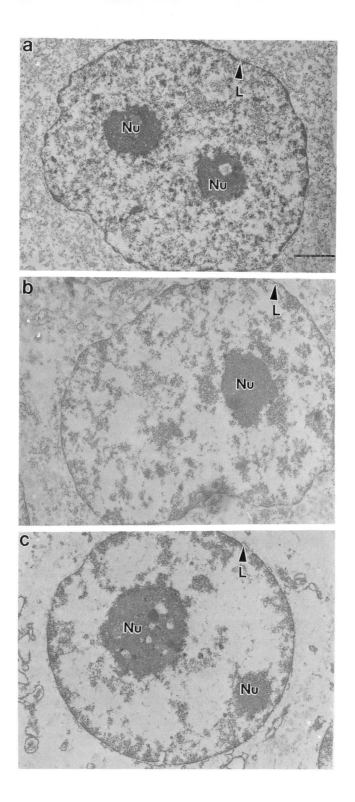

These experiments established that intact RNA and a continuing synthesis of RNA are required for normal nuclear and chromatin architecture. They did not, however, identify the RNA or class of RNAs playing a structural role in nuclear organization. The best candidates for structural RNA may be found among the large fraction of hnRNA which is not a precursor to cytoplasmic message and which may not code for proteins. A recent paper may change our concepts of "noncoding" DNA sequences and of RNA transcripts from those sequences (Mantegna *et al.*, 1994). A statistical linguistic analysis of GenBank sequences that are not protein coding demonstrated that those sequences have properties characteristic of both natural and artificial languages. These results are consistent with the existence of biological information in noncoding sequences. Nucleotide sequences may carry important biological information independent of their ability to direct protein synthesis and some of this information may be expressed structurally through RNA transcripts.

One RNA that may belong to the structural class of transcripts is encoded by the X-inactivation (XIST) gene, expressed exclusively from the inactive X chromosome in females. The 17-kb-long XIST RNA is a spliced transcript with eight exons, yet it does not contain any significant ORFs and is therefore unlikely to code for a protein (Brown *et al.*, 1992). Fluorescence *in situ* hybridization experiments demonstrate that XIST RNA is almost entirely nuclear and is associated with the X-inactivation-associated Barr body. More recent studies have shown that XIST RNA is tightly associated with the nuclear matrix, raising the intriguing possibility that it may function as a matrix-associated structural RNA in the regulation of X-chromosome inactivation (J. Lawrence, personal communication).

VII. Architectural Organization of RNA Metabolism on the Nuclear Matrix

If we seek to better understand the architectural organization of RNA metabolism in the nucleus, we require a more detailed characterization of

---

FIG. 6. Chromatin architecture is rearranged by inhibition of transcription and by removal of RNA. Treatment of Hela cells with actinomycin D (panel b) or treatment of permeabilized nuclei with ribonuclease A (panel c) causes the rearrangement of chromatin. Control Hela cells are shown in panel a. All of the micrographs in this figure are of Epon-embedded sections that were poststained with uranyl acetate and lead citrate, and all are shown at the same magnification. The experimental protocol is described in Nickerson *et al.* (1989). All cells were permeabilized before processing. This procedure allowed the RNase A treatment used for panel c. In control experiments, comparing permeabilized with unextracted cells, no difference in chromatin distribution was observed. The actinomycin D-induced collapse of chromatin shown in panel b can be seen as clearly in unextracted cells. Bar: 1 μm.

the interaction between the nuclear matrix and specific nuclear RNAs. Further, we need to know whether the enzymatic and regulatory elements of that metabolism are soluble components acting upon matrix-attached nucleic acids or whether, instead, they are themselves bound to the matrix.

Little nuclear RNA is soluble; treatment of cells with detergents to remove membranes and allow soluble components to diffuse away does not remove RNA from nuclei (Lenk *et al.*, 1977; Herman *et al.*, 1978). This detergent-resistant RNA includes hnRNA, more than 70% of which remains with the structure throughout a nuclear matrix isolation procedure (Herman *et al.*, 1978; Miller *et al.*, 1978; van Eekelen *et al.*, 1981; Long and Ochs, 1983; Long and Schrier, 1983; He *et al.*, 1990). When specific message sequences are examined, both the intron-containing hnRNA precursor and the fully processed mRNA in the nucleus are retained with the matrix (Mariman *et al.*, 1982; Schröder *et al.*, 1987a,b; Xing and Lawrence, 1991). This very high degree of hnRNA retention by the nuclear matrix suggests that the structure might bind RNA during its synthesis, processing, and export.

Few molecular details are known about RNA retention with the nuclear matrix. Several studies have suggested that nuclear actin is involved, at least indirectly, in RNA binding (Nakayasu and Ueda, 1985; Schröder *et al.*, 1987a,b; Sahlas *et al.*, 1993). Actin is a major protein in nuclear matrix protein preparations (Nakayasu and Ueda, 1983, 1986; Fey and Penman, 1988). It has been reported that the ovalbumin pre-mRNA can be released from isolated nuclear matrices by treatment with the actin-binding drug cytochlasin B or with antiactin antibodies, while the fully processed RNA can be released by treatment with ATP (Schröder *et al.*, 1987a,b). This is an intriguing result but molecular characterization of the possible relationship between RNA and nuclear actin remains to be reported.

A. Transcription

Several different studies have provided evidence for a role for the nuclear matrix in transcription. Active RNA polymerase, nascent RNA transcripts, and actively transcribed genes are all retained in the nucleus following the removal of chromatin (Herman, *et al.*, 1978; Jackson and Cook, 1985). While all nuclear matrix isolation protocols remove DNA, a small fraction always remains with the matrix after isolation. The sequence content of this DNA has been studied and found to be enriched in actively transcribed sequences. Global measurements using end-labeled polyadenylated mRNA to quantify matrix-associated DNA have shown that transcribed sequences resist detachment from the matrix (Jackson and Cook, 1985; Thorburn *et al.*, 1988). The transcribed ovalbumin gene, but not the untranscribed β-

globin gene, is enriched in the matrix DNA of chicken oviduct cells (Robinson et al., 1982; Ciejek et al., 1983). Ribosomal DNA is enriched in matrix fractions only when being actively transcribed (Pardoll and Vogelstein, 1980; Keppel, 1986).

The association of actively transcribed sequences with the nuclear matrix may reflect the involvement of the nuclear matrix in organizing chromatin. Chromatin is organized into loops of heterogeneous size but with an average size of 100 kb. These loops are constrained and attached at their bases to the nuclear matrix (Vogelstein et al., 1980; Gasser and Laemmli, 1987; Zlatanova and Van Holde, 1992). In a nuclear matrix isolation procedure these loops will be digested with nuclease and DNA that is not attached to the matrix will be removed by the subsequent salt extraction. The closer a particular piece of DNA is to the attached base of the loops, the more likely it is to remain attached to the matrix. This suggests that actively transcribed genes are located near the matrix-attached base of the chromatin loops, while inactive sequences are located on the loop fringes.

The DNA sequences that bind chromatin loops to the matrix, matrix attachment regions (MARs), have been localized to specific sequences flanking mouse immunoglobulin genes (Cockerill and Garrard, 1986), the chicken lysosome gene (Phi-Van and Strätling, 1988), the human interferon-β gene (Bode and Maass, 1988), the human β-globin gene (Jarman and Higgs, 1988), and the Chinese hamster dihydrofolate reductase gene (Käs and Chasin, 1987) as well as various Drosophila genes (Mirkovitch et al., 1984; Gasser and Laemmli, 1987). Most MARs (sometimes called scaffold attachment regions or SARs) are A–T-rich sequences. These sequence elements are conserved in evolution and have high concentration sequences similar to topoisomerase II cleavage sites. The common feature of MAR sequences is not a shared sequence element but a shared physical characteristic, the ability to relax base-pairing under torsional stress (Kohwi-Shigematsu and Kohwi, 1990; Bode et al., 1992). Incorporation of MAR sequences into DNA constructs used for transfection of foreign genes increases the rates of transcription from those genes in stable transfectants by 10- to 30-fold (Stief et al., 1989; Phi-Van et al., 1990; Klehr et al., 1991). This effect is independent of the orientation of the inserted MAR sequence. Thus, matrix attachments of genes in the regions flanking their coding sequences may be important for efficient transcription.

The earliest detectable RNA transcripts in uridine pulse-labeled cells are already attached to the matrix (Herman and et al., 1978; Jackson and Cook, 1985). This attachment of RNA to the matrix may occur during transcript elongation. This is suggested by studies on SV40 transcript elongation in isolated nuclei (Abulafia et al., 1984). Prematurely terminated transcripts arising naturally or following 5,6-dichloro1-β-ribofuranosylbenzimidazole pretreatment are not associated with the matrix. Transcripts elongated beyond the 94 bases at the premature termination site are matrix-attached. Nuclear matrices isolated from SV-40-infected cells still contain nascent RNA transcripts. These matrices, following the removal of chromatin, contain only 2 to 6% of viral DNA and yet are capable in an elongation assay of synthesizing more than 35% of the viral RNA synthesized in isolated nuclei (Abulafia *et al.*, 1984).

Stein and colleagues have identified nuclear matrix proteins that bind to histone H4 gene MARs in a sequence-specific way (Dworetzky *et al.*, 1992). This study demonstrated that isolated nuclear matrix proteins provide a pool of proteins that may be enriched in sequence-specific DNA binding proteins. More recently, they have identified tissue-specific nuclear matrix proteins in osteoblasts and osteosarcoma cells that bind to the promoter of the developmentally regulated osteocalcin gene (Dworetzky *et al.*, 1992; Bidwell *et al.*, 1993). One of these factors is a cell-growth-regulated family of proteins, NMP-1, that is related to the transcription factor ATF. NMP-1 consists of at least two proteins with sizes of 43 and 54 kDa. It is partially retained and enriched in nuclear matrix fractions. A second factor, NMP-2, is a 38-kDa cell-type-specific nuclear matrix protein that binds promoter elements resembling the consensus sequence for C/EBP binding and is entirely retained with the nuclear matrix following cell fractionation.

Gel shift assays using HeLa nuclear matrix fractions have identified DNA binding activities with the sequence-specificity of CCAAT, OCT-1, and SP-1 (van Wijnen *et al.*, 1993). Much of the OCT-1 and SP-1 activities is removed with nonmatrix fractions, but a substantial fraction is retained with the matrix. In contrast, most of the CCAAT activity was retained with the matrix. Further biochemical characterization of these activities is required to determine the identity of the nuclear matrix proteins involved. A transcription factor NF1 activity is retained with the nuclear matrices of several chicken tissues, including imature and mature erythrocytes (Sun *et al.*, 1994). Only some members of the NF1 family of proteins are retained with the matrix while others are completely extracted by the matrix preparation technique.

The tissue-specific pituitary transcription factor Pit-1 is retained with the nuclear matrix following the removal of DNA (M. Mancini and Z. D. Sharp, personal communication). Pit-1 is a developmentally regulated member of the POU-homeodomain class of DNA binding proteins. Immunofluorescent localization in unextracted pituitary cells showed a nonhomogeneous distribution suggesting that this factor is partitioned within the nucleus. More than 90% of the Pit-1 is resistant to a 0.5% Triton X-100 extraction demonstrating the existence of an insoluble nuclear pool. Only 10–20% of Pit-1 is solubilized in the standard Dingham 0.45 M NaCl soluble nuclear extract used for *in vitro* transcription and binding assays (Dingham *et al.*, 1983). Further experiments from Z. Dave Sharp's laboratory have shown that about 25% of the Pit-1 is retained with the nuclear matrix core filament network following fractionation by the Penman protocol (He *et al.*, 1990). Immunofluorescent localization of Pit-1 in either unfractionated nuclei or in the core matrix shows a punctate distribution; however, Pit-1 is conspicuously absent in speckled regions of the nucleus enriched in RNA splicing factors. Deletion mutagenisis and epitope-tagged expression of Pit-1 molecules show an interdependant involvement of both the homeodomain and the POU-specific domain in subnuclear targeting. A truncated protein containing mostly the POU domain has a nuclear distribution similar to wild-type Pit-1 while a point mutation in the homoeodomain of Pit-1 causes an abnormal distribution of Pit-1 in the nucleus.

Based on these studies, we expect that additional nuclear matrix proteins will be identified that bind to specific regulatory sequences in genes. The attachment of transcribed DNA, nascent RNA transcripts, and important transcription regulatory proteins to the nuclear matrix suggests a role for this structure in RNA synthesis. The precise molecular mechanisms of matrix involvement require further study, but several models may help to direct our thinking. As early as 1987 Avri Ben-Ze'ev proposed a fixed-site transcription model in which the transcription complex was stationary on the nuclear matrix and nucleic acids were reeled through the site. These fixed-sites would be analogous to the fixed DNA-replication sites in S-phase nuclei (Nakamura *et al.*, 1986; Nakayasu and Berezney, 1989; O'Keefe *et al.*, 1992). This model was expanded by Cook who has proposed that fixed RNA polymerase sites may tie chromatin fibers into loops (Cook, 1994). This would assign RNA polymerase both a catalytic function in transcription and a structural role in the organization of the chromatin.

A second and complimentary model by Stein and colleagues (1994) has proposed that the attachment of transcription factors to adjacent sites on the matrix aids in the regulated formation of transcription complexes. Prepositioning factors and gene regulatory sequences at the same locus may maximize the probability of productive interactions. This might be especially important for low abundance transcription regulators. Each factor must find a short segment, typically 65 Å in length, along 2 m of DNA. Bringing together two low abundance factors on adjacent sites simultaneously would be difficult in a soluble system. However, in a structured system, transcription factors could be prepositioned near each other and the sequence elements to which they bind. This is a much more efficient arrangement than if factors were required to find each other by diffusion. It remains to be seen, however, whether nature is as efficient as our models.

B. RNA Splicing

Many previous studies have provided evidence that the nuclear matrix is the structure on which the major steps in RNA processing take place. The earliest pulse-labeled RNA transcripts that can be detected are associated with the matrix (Herman *et al.*, 1978; Jackson and Cook, 1985) and these transcripts are the most likely to still contain introns. Unspliced and spliced mRNAs of specific genes have been detected in biochemical fractions of nuclear matrices suggesting that pre-mRNA processing could occur on this structure (Mariman *et al.*, 1982; Schröder *et al.*, 1987a,b; Xing and Lawrence, 1991). More direct support for this was provided by Zeitlin *et al.* (1987, 1989) who showed that a β-globin pre-mRNA associated with isolated matrix is rapidly processed following the addition of a soluble nuclear fraction. Taken together with other studies showing that multiple spliceso-mal components, including snRNPs and also non-snRNP splicing factors, remain associated with the nuclear matrix following nuclear extraction, it is likely that the majority if not all splicing takes place on this structure. Based on these observations and also the discovery that specific transcripts and processing tracks are associated with the nuclear matrix, it has been proposed that the matrix provides a solid support on which both RNA splicing and transport takes place.

C. Characterization of Nuclear Matrix Proteins Involved in Pre-mRNA Processing

It is important that we begin to identify nuclear matrix proteins that organize the RNA splicing apparatus on the matrix. One promising experimental approach is to screen antibodies against individual nuclear matrix proteins in an *in vitro* splicing assay to detect matrix antigens that are associated with splicing complexes (Blencowe *et al.*, 1994). We have screened monoclonal antibodies against nuclear matrix proteins that stain nuclei in the speckled pattern in this system. We were able to identify matrix antigens associated with a specific subset of splicing complexes assembled *in vitro*. Three of the mAbs that were characterized efficiently immunoprecipitate splicing complexes, with a distinct preference for complexes containing exon se-quences. Remarkably, these mAbs preferentially immunoprecipitate the exon–product complex but not complexes containing the lariat product after the second step of splicing. Two of the antibodies completely inhibited pre-mRNA splicing *in vitro*. However, none appeared to recognize factors stably associated with splicing snRNPs.

The three antibodies that precipitated splicing complexes predominantly reacted with distinct high-molecular-weight phosphoproteins, which share multiple properties common to members of the SR family. Similar to the known SR proteins, all of three of the high molecular weight antigens selectively precipitated with SR proteins in the presence of high Mg^{2+} concentrations. Two of the high-molecular-weight phosphoproteins were also found to share epitopes with individual defined SR proteins. Furthermore, it was found that the immunoprecipitation and immunostaining properties of the anti-NM mAbs were dependent on the recognition of epitopes on SR proteins. These studies suggest the existence of a class of high-molecular-weight proteins concentrated in speckled structures that may be related to the SR family of non-snRNP-splicing factors. The apparent specificity of these antigens for RNAs containing exon sequences suggests that they could play a fundamental role in exon recognition during the processing of pre-mRNA.

This system provides a powerful approach for identifying matrix proteins involved in RNA splicing and for characterizing their molecular interactions with the specific complexes involved. For this approach to work, however, the matrix protein must be sufficiently soluble in the HeLa nuclear extract required by the assay and an antibody must be available. Solubility would seem to be the greatest obstacle; yet, two of the matrix proteins in our studies that were strongly retained in the nuclear matrix during nuclear fractionation (Nickerson *et al.*, 1992; Wan *et al.*, 1994) were present in an HeLa nuclear extract in sufficient quantity to be detected on Western blots (Blencowe *et al.*, 1994). It would be interesting to know whether the matrix proteins solubilized in nuclear extract are modified in some way or, alternatively, whether they were actually extracted from mitotic cells present in the unsynchronized HeLa cells used as starting material.

VIII. Nuclear Matrix and Phenotype

In higher eukaryotes, every cell contains the complete genome of the organism but expresses only a precisely defined subset of genes. It is the expression of this set of genes that determines and defines the stable cell phenotype. Phenotype-specific genes are expressed at precisely the right place and moment in development and can remain stably turned on for the life of the organism. The mechanism responsible for this remains one of the great puzzles of biology.

The pattern of phenotype-specific gene expression seems mostly determined at the level of transcription, although phenotype-specific alternative splicing mechanisms have been found (Paul *et al.*, 1986; Latchman, 1990; Saga *et al.*, 1990). As we have discussed in Section VII, there is abundant evidence that both transcription and RNA splicing are architecturally organized on an intranuclear scaffolding.

Any cell structure or system responsible for coordinating phenotype-specific gene expression should have a cell-type-specific composition or structure. While chromatin proteins are largely invariant from cell to cell, the protein composition of the nuclear matrix is dependent on the differentiation state and varies from cell type to cell type. This was first observed by Fey and Penman (1988) in a study of the nuclear matrix protein composition of human cell lines. Examining nuclear matrix proteins by two-dimensional gel electrophoresis, they were able to identify nuclear matrix proteins that were present in every cell type and others that were present only in cells derived from one tissue type. These observations were extended by Stuurman (Stuurman et al., 1989, 1990) who found a similar dependence of nuclear matrix composition cell type in murine embryonal cells and tissues. Additional studies by Getzenberg and Coffey (1990, 1991) have shown that different cell types within the rat prostate express distinct, though overlapping, sets of nuclear matrix proteins.

Developmentally regulated changes in nuclear matrix protein composition have also been observed. The fetal rat calvarial osteoblast has become an important in vitro developmental system for studying phenotype-specific gene expression (Stein et al., 1990). In culture, fetal rat osteoblasts differentiate through three well defined developmental stages, each with a unique and characteristic pattern of gene activity. Each developmental stage has a characteristic pattern of nuclear matrix protein composition (Dworetzky et al., 1990). As discussed in Section VII, at least two of these osteoblast nuclear matrix proteins bind regulatory elements in the developmentally regulated osteocalcin gene (Dworetzky et al., 1992; Bidwell et al., 1993). Since the nuclear matrix is the only nuclear component with a marked phenotype-specific composition and since the matrix is involved in gene expression, it is reasonable to think that the nuclear matrix is involved in the regulation and coordination of gene expression that specifies phenotype.

IX. The Nuclear Matrix in Malignancy

A. Alterations in Nuclear Structure and Matrix Composition

Alterations in nuclear structure size, shape, and organization accompany malignancy (Kamel et al., 1990; Underwood, 1990). These changes, while diagnostically significant, are not understood. Since the nuclear matrix determines nuclear form, it is not surprising that malignancy-related alterations in nuclear structure should reflect a change in nuclear matrix composition and structure (Getzenberg and Coffey, 1991; Pienta and Coffey, 1992). The relationship in malignancy between aberrant nuclear structure, nuclear matrix composition, and abnormal cell behavior is inadequately understood and should be a research target of high priority.

As described previously in Section VII, nuclear matrix protein composition reflects cell type and state of differentiation (Fey and Penman, 1988; Stuurman *et al.*, 1989; Dworetzky *et al.*, 1990). Malignant transformation induces changes in all of these properties. Recently, several laboratories have reported that certain tumors have nuclear matrix proteins that are not present in the corresponding normal tissue. In the first such study it was reported that human prostate tumors have a nuclear matrix protein not present in normal prostate tissue or in benign prostatic hyperplasia (Partin *et al.*, 1993). Several malignancy-specific nuclear matrix proteins in human infiltrating ductal carcinoma have been identified that are not present in normal breast tissue (Khanuja *et al.*, 1993). Six nuclear matrix proteins have been reported in human colon adenocarcinoma tumor samples that were absent in normal colon tissue (Keesee *et al.*, 1994). Such malignancy-specific nuclear matrix proteins, if their existence can be confirmed, might have considerable diagnostic value and may help to explain the changes in nuclear structure that accompany malignancy.

These results must, however, be considered preliminary. To date, no study has unambiguously identified a unique polypeptide in the matrix of malignant cells that is not present in normal tissue. All of the early studies have relied on two-dimensional gel analyses of isolated nuclear matrix proteins and have all identified unique protein spots. The question of what these spots represent is still an open one. These unique spots could be unique polypeptides, or alternatively, they could be malignancy-specific post-translational modifications of polypeptides that are also present in normal tissue. Similarly, it is possible that they could be polypeptides that are present in normal tissue but do not fractionate with the matrix. Distinguishing among these possibilities will require a more detailed biochemical analysis than has so far been reported.

The prospect for developing diagnostic reagents based on matrix proteins depends on the results of such an analysis. If malignancy-specific spots represent unique polypeptides synthesized in malignant cells, then it will be easy to develop antibody and nucleotide probes and assays for clinical use. If, however, these spots represent unique post-translational modifications of shared proteins, or if they result from a relocalization of shared proteins to the matrix in malignant cells, then diagnostic assays will be much more difficult to develop. For these reasons, it is very important that preliminary studies that identify apparently unique spots be followed quickly by more in-depth analyses. It is not only important for the field of matrix research, it is important for clinical practice.

The ability to distinguish between malignant and normal tissue is readily and routinely accomplished by microscopic analysis of tissues and cells. In certain cases, matrix-based diagnostics might be an important adjunct to current pathology techniques. A more pressing need, however, is for prognostic markers and assays. Nuclear morphology has been used as a correlate of prognosis in many breast cancer histopathology systems (Elston and Ellis, 1991). Nuclear features of prognostic value are increased size, abnormal shape, and a rearranged internal organization. Nuclear shape used alone, in the absence of other criteria, has great prognostic power. Computer-assisted nuclear morphometry has been used for predicting prognosis (Theissig et al., 1991) with moderate success. For prostate tumors, nuclear morphometry provides a more accurate prediction of prognosis than histologic scoring systems (Partin et al., 1992). Alterations in nuclear morphology contain, therefore, prognostic information. We propose that changes in nuclear matrix structure and protein composition cause the prognostic changes in overall nuclear morphology. Nuclear matrix proteins might therefore be important targets for the identification of prognostic markers. No prognostic changes in matrix protein composition have yet been reported, but we consider the search for such proteins an important and plausible goal.

The finding of malignancy-specific nuclear matrix proteins may at first seem puzzling. As we have seen, nuclear matrix protein composition changes with phenotype. This has been most clearly established in the developing rat calvarial osteoblast. Studies on rat osteosarcoma cells suggest that these tumor cells are expressing matrix proteins normally expressed only at earlier developmental stages in the osteoblast lineage (Bidwell et al., 1994). Thus, the changes in matrix protein composition that are reported in preliminary studies of human tumor tissue may eventually be explained as a regression to earlier stages in the development of the normal tissue.

B. Tumorigenesis

Changes in the nuclear matrix may not merely reflect the malignant phenotype, but may also be involved in the process of malignant progression. The best evidence for this so far comes from tumor virology. Several viral oncoproteins share three common properties: the ability to transform cells, high affinity binding to the nuclear matrix, and a specific association with the hypophosphorylated form of the retinoblastoma tumor suppressor protein, $p110^{RB}$. So far this group of transforming proteins includes the SV40 large T antigen (Schirmbeck and Deppert, 1987, 1989), the adenovirus E1A protein (Chatterjee and Flint, 1986), and the human papilloma virus protein E7 (Greenfield et al., 1991).

The E7 protein of human papilloma virus type 16 has been of the greatest interest to matrix research because, unlike other viral proteins that partition in several nuclear fractions, the 21-kDa E7 protein is entirely bound to the isolated matrix in papilloma virus type 16-transformed cells such as the

CaSki and SiHa cervical carcinoma cell lines (Greenfield *et al.*, 1991). This localization of E7 in the nuclear matrix was established by microscopy immunostaining, by Western blots of isolated nuclear matrix proteins, and by immunoprecipitation of E7 from nuclear matrix preparations. All the detectable E7 in cervical carcinoma cells is present in the nuclear matrix fraction. The E7 protein of cancer-causing strains of human papilloma virus, together with the E6 protein, is necessary for the transformation of cells (Hawley-Nelson *et al.*, 1989; Münger *et al.*, 1989).

The E7 oncoprotein of human papillomavirus, like the other viral-transforming proteins that associate with the matrix, binds to the tumor suppressor protein, $p110^{RB}$. This suggested that $p110^{RB}$ might also have a nuclear matrix association. The work of Mancini *et al.* (1994) demonstrated, in CV-1 cells containing no papilloma virus, that the hypophosphorylated form of $p110^{RB}$ binds to the nuclear matrix in the early G1 phase of the cell cycle (Mancini *et al.*, 1994). In early G1, $p110^{RB}$ is least phosphorylated and it is the hypophosporylated form that is both capable of pausing the cell cycle in G1 (Goodrich *et al.*, 1991) and is stably associated with the matrix. In addition, the hypophosphorylated form of $p110^{RB}$ binds the papillomavirus E7 protein oncoprotein (Imai *et al.*, 1991). As the cell cycle progresses, $p110^{RB}$ becomes hyperphosphorylated (Chen *et al.*, 1989) and is no longer stably associated with the matrix. Based on these results, we proposed that the important interaction between E7 and $p110^{RB}$ takes place on the nuclear matrix during early G1.

Essentially all of the $p110^{RB}$ that is not released from the cell in early G1 by a simple detergent extraction remains associated with the structure throughout the digestions and salt extractions required to uncover the core filament network (Mancini *et al.*, 1994). By treating detergent-extracted WI38 fibroblasts with a variety of protein kinases, Mittnacht *et al.* (1994) were able to show that phosphorylation with cyclin-dependent kinases liberated $p110^{RB}$ from the structure and into the soluble phase.

Although $p110^{RB}$ is capable of binding lamins A and C (Mancini *et al.*, 1994), most of the $p110^{RB}$ in early G1 is bound to internal nuclear matrix sites and not to the lamina. The recent identification of internal foci of lamin in G1 (Goldman *et al.*, 1992; Bridger *et al.*, 1993) suggests that some portion of this internal nuclear $p110^{RB}$ could be lamin-associated, although it is now known that other proteins of the internal nuclear matrix bind $p110^{RB}$. The best characterized of these is an 84-kDa protein, p84, identified by yeast two-hybrid screening by Mancini and Durfee (Durfee *et al.*, 1994). p84 binds to the amino terminal end of $p100^{RB}$, is entirely retained in nuclear matrix preparations, and, intriguingly, is principally located in the splicing speckled domains of the nucleus. While p84 may concentrate some fraction of $p100^{RB}$ to the splicing speckles during G1, the metabolic consequences of this are so far unknown.

These studies link an important tumor suppressor protein, and the viral-transforming proteins that affect its activity, to the nuclear matrix. They show a direct relationship between p110[RB]-dependent cell cycle control and the nuclear matrix and, tantalyzingly, they suggest a role for the matrix in cell transformation and malignant progression. A more complete biochemical characterization of these phenomena and a better understanding of their significance for cell cycle and growth control should be important goals for future research.

X. Concluding Remarks

Biochemical and molecular genetic analysis has successfully allowed the identification and characterization of molecules and reactions important in nucleic acid metabolism. These approachs, however, provide little information about the architectural organization of metabolism in the cell. Indeed, the cell is far more than a vessel holding a solution of interacting molecules. The substructure within the cell imposes an architectural ordering on metabolism and provides an important level of regulation.

A more complete understanding of nucleic acid metabolism in the living cell will require the combined application of biochemistry and structural cell biology. The studies of cell metabolism and of cell architecture have not often been conducted together, in part, because such different technical skills are required. And, perhaps, different philosophies and temperaments are required. Still, neither approach can alone answer all the important questions we can now pose about nuclear metabolism. It is clear that continuing cooperation, collaboration, and communication between disciplines will be essential if we are to characterize and understand the architecture of nuclear metabolism.

Acknowledgments

We are especially grateful to Gabriela Krockmalnic for taking most of the electron micrographs presented in this chapter and for her help in preparing the figures. We thank Phillip A. Sharp for helpful discussions and for his critical reading of the text. We appreciate the efforts of Katherine Mimi Wan, Edward G. Fey, and Robbyn Issner who all contributed experiments that are discussed. We thank Kerstin Bohmann and Angus I. Lamond for the gift of p80 coilin antiserum used for the experiments described in Section II,C. Jeanne B. Lawrence generously allowed us to describe some recent and unpublished experiments from her laboratory. Our discussions with her were an important influence on this work. We thank Gary Stein, Z. Dave Sharp, and Michael Mancini for reading parts of the text and for allowing us to cite unpublished results. B.J.B. acknowledges fellowship support from the Human Frontiers

Science Program Organization and from the USDoD Breast Cancer Research Program. We would like to thank all those whose work we cite here. Even more, we acknowledge the contributions of those whose work we would have cited, had we more space.

References

Aaronson, P. P., and Blobel, G. (1974). On the attachment of the nuclear pore complex. *J. Cell Biol.* **62,** 746–754.

Abulafia, R., Ben-Ze'ev, A., Hay, N., and Aloni, Y. (1984). Control of late simian virus 40 transcription by the attenuation mechanism and transcriptionally active ternary complexes are associated with the nuclear matrix. *J. Mol. Biol.* **172,** 467–487.

Adolph, K. (1980). Organization of chromosomes in HeLa cells: Isolation of histone-depleted nuclei and nuclear scaffolds. *J. Cell Sci.* **42,** 291–304.

Andrade, L. E., Chan, E. K., Raska, I., Peebles, C. L., Roos, G., and Tan, E. M. (1991). Human autoantibody to a novel protein of the nuclear coiled body: Immunological characterization and cDNA cloning of p80-coilin. *J. Exp. Med.* **173,** 1407–1419.

Antoniou, M., Carmo-Fonseca, M., Ferreira, J., and Lamond, A. I. (1993). Nuclear organization of splicing snRNPs during differentiation of murine erythroleukaemia cells in vitro. *J. Cell Biol.* **123,** 1055–1068.

Ascoli, C. A., and Maul, G. G. (1991). Identification of a novel nuclear domain. *J. Cell Biol.* **112,** 785–795.

Bachellerie, J. P., Puvion, E., and Zalta, J. P. (1975). Ultrastructural organization and biochemical characterization of chromatin–RNA–protein complexes isolated from mammalian cell nuclei. *Eur. J. Biochem.* **58,** 327–337.

Belgrader, P., Dey, R., and Berezney, R. (1991a). Molecular cloning of matrin 3. A 125-kilodalton protein of the nuclear matrix contains an extensive acidic domain. *J. Biol. Chem.* **266,** 9893–9899.

Belgrader, P., Siegel, A. J., and Berezney, R. (1991b). A comprehensive study on the isolation and characterization of the HeLa S3 nuclear matrix. *J. Cell Sci.* **98,** 281–291.

Bensch, K., Tanaka, S., Hu, S.-Z., Wang, T., and Korn, D. (1982). Intracellular localization of human DNA polymerase α with monoclonal antibodies. *J. Biol. Chem.* **257,** 8391–8396.

Ben-Ze'ev, A. (1987). The complex cellular networks in the control of SV40 gene expression. *In* "Molecular Aspects of Papoviruses" (Y. Aloni, ed.), pp. 239–268. Martin Nijhoff, Boston.

Ben-Ze'ev, A., Farmer, S., and Penman, S. (1980). Protein synthesis requires cell-surface contact while nuclear events respond to cell shape in anchorage-dependent fibroblasts. *Cell* (*Cambridge, Mass.*) **21,** 365–372.

Berezney, R. (1979). Dynamic properties of the nuclear matrix. *In* "The Cell Nucleus" (H. Busch, ed.), vol. 7, pp. 413–456. Academic Press, Orlando, Florida.

Berezney, R. (1984). Organization and functions of the nuclear matrix. *In* Chromosomal Nonhistone Proteins, (L. S. Hnilica, ed.), vol. IV, pp. 119–180. CRC Press, Boca Raton, Florida.

Berezney, R., and Coffey, D. S. (1974). Identification of a nuclear protein matrix. *Biochem. Biophys. Res. Commun.* **60,** 1410–1417.

Berezney, R., and Coffey, D. S. (1975). Nuclear protein matrix: Association with newly synthesized DNA. *Science* **189,** 291–292.

Berezney, R., and Coffey, D. S. (1977). Nuclear matrix: Isolation and characterization of a framework structure from rat liver nuclei. *J. Cell Biol.* **73,** 616–637.

Berezney, R., Basler, J., Buchholtz, L. A., Smith, H. C., and Siegel, A. J. (1982). Nuclear matrix organization and DNA replication. *In* "The Nuclear Envelope and Nuclear Matrix" (G. G. Maul, ed.), pp. 183–197. Alan R. Liss, New York.

Bernhard, W. (1969). A new procedure for electron microscopical cytology. *J. Ultrastruct. Res.* **27**, 250–265.

Beyer, A. L., and Osheim, Y. N. (1988). Splice site selection, rate of splicing, and alternative splicing on nascent transcripts. *Genes Dev.* **2**, 754–765.

Bidwell, J. P., Van-Wijnen, A. J., Fey, E. G., Dworetzky, S., Penman, S., Stein, J. L., Lian, J. B., and Stein, G. S. (1993). Osteocalcin gene promoter-binding factors are tissue-specific nuclear matrix components. *Proc. Natl. Acad. Sci. U.S.A.* **90**, 3162–3166.

Bidwell, J. P., Fey, E. G., van Wijnen, A. J., Penman, S., Stein, J. L., Lian, J. B., and Stein, G. S. (1994). Nuclear matrix proteins distinguish normal diploid osteoblasts from osteosarcoma cells. *Cancer Res.* **54**, 28–32.

Blencowe, B. J., Carmo-Fonseca, M., Behrens, S.-E., Lührmann, R., and Lamond, A. I. (1993). Interaction of the human autoantigen p150 with splicing snRNPs. *J. Cell Sci.* **105**, 685–697.

Blencowe, B. J., Nickerson, J. A., Issner, R., Penman, S., and Sharp, P. A. (1994). Association of nuclear matrix antigens with exon-containing splicing complexes. *J. Cell Biol.* **127**, 593–607.

Bode, J., and Maass, K. (1988). Chromatin domain surrounding the human interferon-β gene as defined by scaffold-attached regions. *Biochemistry* **27**, 4706–4711.

Bode, J., Kohwi, Y., Dickinson, L., Joh, T., Klehr, D., Mielke, C., and Kohwi-Shigematsu, T. (1992). Biological significance of unwinding capability of nuclear matrix-associating DNAs. *Science* **255**, 195–197.

Borrow, J., Goddard, A. D., Sheer, D., and Solomon, E. (1990). Molecular analysis of acute promyelocytic leukemia breakpoint cluster region on chromosome 17. *Science* **249**, 1577–1580.

Bravo, R., and Macdonald-Bravo, H. (1987). Existence of two populations of cyclin/proliferating cell nuclear antigen during the cell cycle: Association with DNA replication sites. *J. Cell Biol.* **105**, 1549–1554.

Bridger, J. M., Kill, I. R., O'Farrell, M., and Hutchinson, C. J. (1993). Internal lamin structures within G1 nuclei of human dermal fibroblasts. *J. Cell Sci.* **104**, 297–306.

Brown, C. J., Hendrich, B. D., Rupert, J. L., Lafreniere, R. G., Xing, Y., Lawrence, J. B., and Willard, H. F. (1992). The human *xist* gene: Analysis of a 17 kb inactive x-specific RNA that contains conserved repeats and is highly localized within the nucleus. *Cell (Cambridge, Mass.)* **71**, 527–542.

Capco, D. G., Wan, K. M., and Penman, S. (1982). The nuclear matrix: Three-dimensional architecture and protein composition. *Cell (Cambridge, Mass.)* **29**, 847–858.

Capco, D. G., Krocmalnic, G., and Penman, S. (1984). A new method for preparing embedment-free sections for transmission electron microscopy: Applications to the cytoskeletal framework and other three-dimensional networks. *J. Cell Biol.* **98**, 1878–1885.

Cardoso, M. C., Leonhargt, H., and Nadal-Ginard, B. (1993). Reversal of terminal differentiation and control of DNA replication: Cyclin A and Cdk2 specifically localize at subnuclear sites of DNA replication. *Cell (Cambridge, Mass.)* **74**, 979–992.

Carmo-Fonseca, M., Cidadao, A. J., and David-Ferreira, J. F. (1987). Filamentous cross-bridges link intermediate filaments to the nuclear pore complexes. *Eur. J. Cell Biol.* **45**, 282–290.

Carmo-Fonseca, M., Tollervey, D., Pepperkok, R., Barabino, S. M. L., Merdes, A., Brunner, C., Zamore, P. D., Green, M. R., Hurt, E., and Lamond, A. I. (1991a). Mammalian nuclei contain foci which are highly enriched in components of the pre-mRNA splicing machinery. *EMBO J.* **10**, 195–206.

Carmo-Fonseca, M., Pepperkok, R., Sproat, B. S., Ansorge, W., Swanson, M. S., and Lamond, A. I. (1991b). In vivo detection of snRNP-rich organelles in the nuclei of mammalian cells. *EMBO J.* **10**, 1863–1873.

Carmo-Fonseca, M., Pepperkok, R., Carvalho, M. T., and Lamond A. I. (1992). Transcription-dependent colocalization of the U1, U2, U4/U6, and U5 snRNPs in coiled bodies. *J. Cell Biol.* **117**, 1–14.

Carmo-Fonseca, M., Ferreira, J., and Lamond, A. I. (1993). Assembly of snRNP-containing coiled bodies is regulated in interphase and mitosis—evidence that the coiled body is a kinetic nuclear structure. *J. Cell Biol.* **120**, 841–852.

Carter, K. C., Taneja, K. L., and Lawrence, J. B. (1991). Discrete nuclear domains of poly(A) RNA and their relationship to the functional organization of the nucleus. *J. Cell Biol.* **115**, 1191–1202.

Carter, K. C., Bowman, D., Carrington, W., Fogarty, K., McNeil, A., Fay, F. S., and Lawrence, J. B. (1993). A three-dimensional view of precursor messenger RNA metabolism within the mammalian nucleus. *Science* **259**, 1330–1335.

Chatterjee, P. K., and Flint, S. J. (1986). Partition of E1A proteins between soluble and structural fractions of adenovirus-infected and -transformed cells. *J. Virol.* **60**, 1018–1026.

Chan, E. K. L., Takano, S., Andrade, L. E. C., Hamel, J. C., and Matera, A. G. (1994). Structure, expression, and chromosomal location of human p80-coilin gene. *Nucleic Acids Res.* **22**, 4462–4469.

Chen, P.-L., Scully, P., Shew, J.-Y., Wang, J. Y. J., and Lee, W.-H. (1989). Phosphorylation of the retinoblastoma gene product is modulated during the cell cycle and cellular differentiation. *Cell (Cambridge, Mass.)* **58**, 1193–1198.

Ciejek, E. M., Tsai, M.-J., and O'Malley, B. W. (1983). Actively transcribed genes are associated with the nuclear matrix. *Nature (London)* **306**, 607–609.

Cockerill, P. N., and Garrard, W. T. (1986). Chromosomal loop anchorage of the kappa immunoglobulin gene occurs next to the enhancer in a region containing topoisomerase II sites. *Cell (Cambridge, Mass.)* **44**, 273–282.

Compton, D. A., and Cleveland, D. W. (1993). NuMA is required for the proper completion of mitosis. *J. Cell Biol.* **120**, 947–957.

Compton, D. A., Szilak, I., and Cleveland, D. W. (1992). Primary structure of NuMA, an intracellular protein that defines a novel pathway for segregation of proteins at mitosis. *J. Cell Biol.* **116**, 1395–1408.

Cook, P. R. (1991). The nucleoskeleton and the topology of replication. *Cell (Cambridge, Mass.)* **66**, 627–635.

Cook, P. R. (1994). RNA polymerase: Structural determinant of the chromatin loop and the chromosome. *BioEssays* **16**, 425–430.

Costantini, F. D., Schelle, R. H., Britten, R. J., and Davidson, E. H. (1978). Repetitive sequence transcripts in the mature sea urchin oocyte. *Cell (Cambridge, Mass.)* **15**, 173–187.

D'Andrea, A. D., Tantravahi, U., LaLande, M., Perle, M. A., and Latt, S. A. (1983). High resolution analysis of the timing of replication of specific DNA sequences during S phase of mammalian cells. *Nucleic Acids Res.* **11**, 4753–4774.

Derenzini, M., Pession-Brizzi, A., and Novello, F. (1980). Chromatin condensation in isolated rat hepatocyte nuclei induced by ribonuclease treatment. *Experientia* **36**, 181–182.

de Thé, H., Chomienne, C., Lanotte, M., Degos, L., and Dejean, A. (1990). The t(15;17) translocation of acute promyelocytic leukaemia fuses the retinoic acid receptor alpha gene to a novel transcribed locus. *Nature (London)* **347**, 558–561.

de Thé, H., Lavau, C., Marchio, A., Chomienne, C., Degos, L., and Dejean, A. (1991). The PML-RARα fusion mRNA generated by the t(15;17) translocation of acute promyelocytic leukemia encodes a functionally altered RAR. *Cell (Cambridge, Mass.)* **66**, 675–684.

Dijkwel, P. A., Wenink, P. W., and Poddighe, J. (1986). Permanent attachment of replication origins to the nuclear matrix in BHK-cells. *Nucleic Acids Res.* **14**, 3241–3249.

Dingham, J. D., Lebovitz, R. M., and Roeder, R. G. (1983). Accurate transcription initiation by RNA polymerase II in a soluble extract from isolated mammalian nuclei. *Nucleic Acids Res.* **11**, 1475–1489.

Durfee, T., Mancini, M. A., Jones, D., Elledge, S. J., and Lee, W.-H. (1994). The amino-termina region of the retinoblastoma gene product binds a novel nuclear matrix protein that co-localizes to centers for RNA processing. *J. Cell Biol.* **127**, 609–622.

Dworetzky, S. I., Fey, E. G., Penman, S., Lian, J. B., Stein, J. L., and Stein, G. S. (1990). Progressive changes in the protein composition of the nuclear matrix during osteoblast differentiation. *Proc. Natl. Acad. Sci. U.S.A.* **87,** 4605–4609.

Dworetzky, S. I., Wright, K. L., Fey, E. G., Penman, S., Lian, J. B., Stein, J. L., and Stein, G. S. (1992). Sequence-specific DNA binding proteins are components of a nuclear matrix attachment site. *Proc. Natl. Acad. Sci. U.S.A.* **89,** 4178–4182.

Dyck, J., Maul, G. G., Miller, W. H., Chen, J. D., Kakizuka, A., and Evans, R. (1994). A novel macromolecular structure is a target of the promyelocyte–retinoic acid receptor oncoprotein. *Cell (Cambridge, Mass.)* **76,** 333–343.

Eliceiri, G. L., and Ryerse, J. S. (1984). Detection of intranuclear clusters of Sm antigens with monoclonal anti-Sm antibodies by immunoelectron microscopy. *J. Cell. Physiol.* **121,** 449–451.

Elston, C. W., and Ellis, I. O. (1991). Pathological prognostic factors in breast cancer. I. The value of histological grade in breast cancer. Experience from a large study with long-term follow-up. *Histopathology* **19,** 403–410.

Fakan, S., and Bernhard, W. (1971). Localization of repidly and slowly labelled nuclear RNA as visualized by high resolution autoradiography. *Exp. Cell Res.* **67,** 129–141.

Fakan, S., and Nobis, P. (1978). Ultrastructural localization of transcription sites and of RNA distribution during the cell cycle of synchronized CHO cells. *Exp. Cell Res.* **113,** 327–337.

Fakan, S., and Puvion, E. (1980). The ultrastructural visualization of nucleolar and extranucleolar RNA synthesis and distribution. *Int. Rev. of Cytol.* **65,** 255–299.

Fakan, S., and Hughes, M. E. (1989). Fine structural ribonucleoprotein components of the cell nucleus visualized after spreading and high resolution autoradiography. *Chromosoma* **98,** 242–249.

Fakan, S., Puvion, E., and Spohr, G. (1976). Localization and characterization of newly synthesized nuclear RNA in isolated rat hepatocytes. *Exp. Cell Res.* **99,** 155–164.

Fakan, S., Lesser, G., and Martin, T. E. (1984). Ultrastructural distribution of nuclear ribonucleoproteins as visualized by immunocytochemistry on thin sections. *J. Cell Biol.* **98,** 358–363.

Fawcett, D. W. (1966). "An Atlas of Fine Structure: The Cell, Its Organelles and Inclusions," pp. 2–3. Saunders, Philadelphia.

Fey, E. G., and Penman, S. (1988). Nuclear matrix proteins reflect cell type of origin in cultured human cells. *Proc. Natl. Acad. Sci. U.S.A.* **85,** 121–125.

Fey, E. G., Krochmalnic, G., and Penman, S. (1986). The non-chromatin substructures of the nucleus: The ribonucleoprotein (RNP)-containing and RNP-depleted matrices analyzed by sequential fractionation and resinless section electron microscopy. *J. Cell Biol.* **102,** 1654–1665.

Fischer, D., Weisenberger, D., and Scheer, U. (1991). Assigning functions to nucleolar structures. *Chromosoma* **101,** 133–140.

Fisher, D. Z., Chaudhary, N., and Blobel, G. (1986). cDNA sequencing of nuclear lamins A and C reveals primary and secondary structural homology to intermediate filament proteins. *Proc. Natl. Acad. Sci. U.S.A.* **83,** 6450–6454.

Fisher, P. A., Berrios, M., and Blobel, G. (1982). Isolation and characterization of a proteinaceous subnuclear fraction composed of nuclear matrix, peripheral lamina and nuclear pore complexes from embryos of *Drosophila melanogaster. J. Cell Biol.* **92,** 674–686.

Folkman, J., and Moscona, A. (1978). Role of cell shape in growth control. *Nature (London)* **273,** 345–349.

Foster, K. A., and Collins, J. M. (1985). The interrelation between DNA synthesis rates and DNA polymerases bound to the nuclear matrix in synchronized HeLa cells. *J. Biol. Chem.* **260,** 4229–4235.

Franke, W. W. (1987). Nuclear lamins and cytoplasmic intermediate filament proteins: A growing multigene family. *Cell (Cambridge, Mass.)* **48,** 3–4.

Freemont, P. S., Hanson, I. M., and Trowsdale, J. (1991). A novel cysteine-rich motif. *Cell (Cambridge, Mass.)* **64,** 483–484.

Fu, X. D., and Maniatis, T. (1990). Factor required for mammalian spliceosome assembly is localized to discrete regions in the nucleus. *Nature (London)* **343**, 437–441.

Gasser, S. M., and Laemmli, U. K. (1987). A glimpse at chromosomal order. *Trends Gene.* **3**, 16–22.

Georgatos, S. D., Meier, J., and Simos, G. (1994). Lamins and lamin-associated proteins. *Curr. Opin. Cell Biol.* **6**, 347–353.

Gerace, L., Blum, A., and Blobel, G. (1978). Immunocytochemical localization of the major polypeptides of the nuclear pore complex–lamina fraction–interphase and mitotic distribution. *J. Cell Biol.* **79**, 546–566.

Gerace, L., Comeau, C., and Benson, M. (1984). Organization and modulation of the nuclear lamina structure. *J. Cell Sci., Suppl.* **1**, 137–160.

Getzenberg, R. H., and Coffey, D. S. (1990). Tissue specificity of the hormonal response in sex accessory tissues is associated with nuclear matrix protein patterns. *Mol. Endocrinol.* **4**, 1336–1342.

Getzenberg, R. H., and Coffey, D. S. (1991). Identification of nuclear matrix proteins in the cancerous and normal rat prostate. *Cancer Res.* **51**, 6514–6520.

Goldman, A. E., Moir, R. D., Montag-Lowy, M., Stewart, M., and Goldman, R. D. (1992). Pathway of incorporation of microinjected lamin A into the nuclear envelope. *J. Cell Biol.* **119**, 725–735.

Goldman, M. A., Holmquist, G. P., Gray, M. C., Caston, L. A., and Nag, A. (1984). Replication timing of genes and middle repetitive sequences. *Science* **224**, 686–692.

Goodrich, D. W., Wang, P. W., Qian, Y.-W., Lee, E., and Lee, W.-H. (1991). The retinoblastoma gene product regulates progression through the G1 phase of the cell cycle. *Cell (Cambridge, Mass.)* **67**, 293–302.

Greenfield, I., Nickerson, J., Penman, S., and Stanley, M. (1991). Human papillomavirus 16 E7 protein is associated with the nuclear matrix. *Proc. Natl. Acad. Sci. U.S.A.* **88**, 11217–11221.

Hakes, D. J., and Berezney, R. (1991). Molecular cloning of matrin F/G: A DNA binding protein of the nuclear matrix that contains putative zinc finger motifs. *Proc. Natl. Acad. Sci. U.S.A.* **88**, 6186–6190.

Hamm, J., Darzynkiewicz, E., Tahara, S. M., and Mattaj, I. W. (1990). The trimethylguanosine cap structure of U1 snRNA is a component of a bipartite nuclear targeting signal. *Cell (Cambridge, Mass.)* **62**, 569–577.

Hansen, L. R., and Ingber, D. E. (1992). Regulation of nucleocytoplasmic transport by mechanical forces transmitted through the cytoskeleton. *In* "Nuclear Trafficking" (C. Feldherr, Ed.), pp. 71–78. Academic Press, San Diego, CA.

Hanson, I. M., Poustka, A., and Trowsdale, J. (1991). New genes in the class II region of the major histocompatibility complex. *Genomics* **10**, 417–424.

Hardin, J. H., Spicer, S. S., and Greene, W. B. (1969). The paranucleolar structure, accessory body of Cajal, sex chromatin, and related structures in nuclei of rat trigeminal neurons: A cytochemical and ultrastructural study. *Anat. Rec.* **164**, 403–432.

Harpold, M. M., Wilson, M. C., and Darnell, J. E., Jr. (1981). Chinese hamster polyadenylated messenger ribonucleic acid: Relationship to non-polyadenylated sequences and relative conservation during messenger ribonucleic acid processing. *Mol. Cell. Biol.* **1**, 188–198.

Hatton, K. S., Dhar, V., Gahn, T. A., Brown, E. H., Mager, D., and Schildkraut, C. L. (1988). Temporal order of replication of multigene families reflects chromosomal location and transcriptional activity. *Cancer Cells* **6**, 335–340.

Hawley-Nelson, P., Vousden, K. H., Hubbert, N. L., Lowy, D. R., and Schiller, J. T. (1989). HPV16 E6 and E7 proteins cooperate to immortalize human foreskin keratinocytes. *EMBO J.* **8**, 3905–3910.

He, D., Nickerson, J. A., and Penman, S. (1990). The core filaments of the nuclear matrix. *J. Cell Biol.* **110**, 569–580.

Herman, R. C., and Penman, S. (1977). Multiple decay rates of heterogeneous nuclear RNA in HeLa cells. *Biochemistry* **16,** 3460–3465.

Herman, R. C., Weymouth, L., and Penman, S. (1978). Heterogeneous nuclear RNA-protein fibers in chromatin-depleted nuclei. *J. Cell Biol.* **78,** 663–674.

Hitchcock, S. E., Carlsson, L., and Lindberg, U. (1976). Depolymerization of F-actin by deoxyribonuclease I. *Cell (Cambridge, Mass.)* **7,** 531–542.

Hodge, L. D., Mancini, P., Davis, F. M., and Heywood, P. (1977). Nuclear matrix of HeLa S3 cells. Polypeptide composition during adenovirus infection and in phases of the cell cycle. *J. Cell Biol.* **72,** 194–208.

Hozák, P., Hassan, A. B., Jackson, D. A., and Cook, P. R. (1993). Visualization of replication factories attached to a nucleoskeleton. *Cell (Cambridge, Mass.)* **73,** 361–373.

Huang, M.-E., Ye, Y.-C., Chen, S.-R., Chai, J.-R., Zhoa, L., Gu, L.-J., and Wang, Z. Y. (1988). Use of all-trans retinoic acid in the treatment of acute promyelocytic leukemia. *Blood* **72,** 567–572.

Huang, S., and Spector, D. L. (1991). Nascent pre-mRNA transcripts are associated with nuclear regions enriched in splicing factors. *Genes Dev.* **5,** 2288–2302.

Huang, S., and Spector, D. L. (1992). U1 and U2 small nuclear RNAs are present in nuclear speckles. *Proc. Natl. Acad. Sci. U. S. A.* **89,** 305–308.

Huang, S., Deerinck, T. J., Ellisman, M. H., and Spector, D. L. (1994). *In vivo* analysis of the stability and transport of poly(A)⁺ RNA. *J. Cell Biol.* **126,** 877–899.

Imai, H., Chan, E. K. L., Kiyosawa, K., Fu, X.-D., and Tan, E. M. (1993). Novel nuclear autoantigen with splicing factor motifs identified with antibody from hepatocellular carcinoma. *J. Clin. Invest.* **92,** 2419–2462.

Imai, Y., Matsushima, Y., Sugimura, T., and Terada, M. (1991). Purification and characterization of human papillomavirus type 16 E7 protein with preferential binding capacity to the underphosphorylated form of the retinoblastoma gene product. *J. Virol.* **65,** 4966–4972.

Ingber, D. E. (1993). Cellular tensegrity: Defining new rules of biological design that govern the cytoskeleton. *J. Cell Sci.* **104,** 467–475.

Ingber, D. E., and Folkman, J. (1989). Tension and compression as basic determinants of cell form and function: Utilization of a cellular tensegrity mechanism. *In* "Cell Shape Determinants: Regulation and Regulatory Role" (W. Stein and F. Bronner, Eds.), pp. 1–32. Academic Press, San Diego, CA.

Ingber, D. E., Dike, L., Hansen, L., Karp, S., Liley, H., Maniotis, A., McNamee, H., Mooney, D., Plopper, G., Sims, J., and Wang, N. (1994). Cellular tensegrity: Exploring how mechanical changes in the cytoskeleton regulate cell growth, migration, and tissue pattern during morphogenesis. *Int. Rev. Cytol.* **150,** 173–224.

Jackson, D. A., and Cook, P. R. (1985). Transcription occurs at a nucleoskeleton. *EMBO J.* **4,** 919–925.

Jackson, D. A., and Cook, P. R. (1986). Replication occurs at a nucleoskeleton. *EMBO J.* **5,** 1403–1410.

Jackson, D. A., and Cook, P. R. (1988). Visualization of a filamentous nucleoskeleton with a 23nm axial repeat. *EMBO J.* **7,** 3667–3678.

Jackson, D. A., Hassan, A. B., Errington, R. J., and Cook, P. R. (1993). Visualization of focal sites of transcription within human nuclei. *EMBO J.* **12,** 1059–1065.

Jarman, A. P., and Higgs, D. R. (1988). Nuclear scaffold attachment sites in the human globin gene complexes. *EMBO J.* **7,** 3337–3344.

Kakizuka, A., Miller, W. H. Jr., Umesono, K., Warrell, R. P. Jr., Frankel, S. R., Murty, V. V., Dmitrovsky, E., and Evans, R. M. (1991). Chromosomal translocation t(15;17) in human acute promyelocytic leukemia fuses RAR alpha with a novel putative transcription factor, PML. *Cell (Cambridge, Mass.)* **66,** 663–674.

Kamel, H. M., Kirk, J., and Toner, P. G. (1990). Ultrastructural pathology of the nucleus. *Curr. Top. Pathol.* **82,** 17–89.

Käs, E., and Chasin, L. A. (1987). Anchorage of the Chinese hamster dihydrofolate reductase gene to the nuclear scaffold occurs in an intragenic region. *J. Mol. Biol.* **198,** 677–692.

Kastner, P., Perez, A., Lutz, Y., Rochette-Egly, C., Gaub, M., Durand, B., Lanotte, M., Berger, R., Longo, L., and Chambon, P. (1992). Structure, localization and transcriptional properties of two classes of retinoic acid receptor fusion proteins in acute promyelocytic leukemia (APL): Structural similarities with a new family of oncoproteins. *EMBO J.* **11,** 629–642.

Kaufman, S. H., Coffey, D. S., and Shaper, J. H. (1981). Considerations in the isolation of rat liver nuclear matrix, nuclear envelope, and pore complex lamina. *Exp. Cell Res.* **132,** 105–123.

Keesee, S. K., Meneghini, M., Szaro, R. P., and Wu, Y. -J. (1994). Nuclear matrix proteins in colon cancer. *Proc. Natl. Acad. Sci. U.S.A.* **91,** 1913–1916.

Keppel, F. (1986). Transcribed human ribosomal RNA genes are attached to the nuclear matrix. *J. Mol. Biol.* **187,** 15–21.

Khanuja, P. S., Lehr, J. E., Soule, H. D., Gehani, S. K., Noto, A. C., Choudhury, S., Chen, R., and Pienta, K. J. (1993). Nuclear matrix proteins in normal and breast cancer cells. *Cancer Res.* **53,** 3394–3398.

Klehr, D., Maass, K., and Bode, J. (1991). Scaffold-attached regions from the human inteferon-beta domain can be used to enhance the stable expression of genes under the control of various promoters. *Biochemistry* **30,** 1264–1270.

Kohwi-Shigematsu, T., and Kohwi, Y. (1990). Torsional stress stabilizes extended base unpairing in suppressor sites flanking immunoglobulin heavy chain enhancer. *Biochemistry* **29,** 9551–9560.

Krohne, G., Wolin, S. L., McKeon, F. D., Franke, W. W., and Kirschner, M. W. (1987). Nuclear lamin LI of *Xenopus laevis*: cDNA cloning, amino acid sequence and binding specificity of a member of the lamin B subfamily. *EMBO J.* **6,** 3801–3808.

Lafarga, M., Hervas, J. P., Santa-Cruz, M. C., Villegas, J., and Crespo, D. (1983). The "accessory body" of Cajal in the neuronal nucleus. A light and electron microscopic approach. *Anat. Embryol.* **166,** 19–30.

Lamond, A. I., and Carmo-Fonseca, M. (1993). The coiled body. *Trends Cell Biol.* **3,** 198–204.

Laskey, R. A., Fairman, M. P., and Blow, J. J. (1989). S phase of the cell cycle. *Science* **246,** 609–614.

Lasko, D. D., Tomkinson, A. E., and Lindahl, T. (1990). Biosynthesis and intracellular localization of DNA ligase I. *J. Biol. Chem.* **265,** 12618–12622.

Latchman, D. S. (1990). Cell-type specific splicing factors and the regulation of alternative RNA splicing (review). *New Biol.* **2,** 297–303.

Lawrence, J. B., Singer, R. H., and Marselle, L. M. (1989). Highly localized tracks of specific transcripts within interphase nuclei visualized by in situ hybridization. *Cell (Cambridge, Mass.)* 493–502.

Lawrence, J. B., Carter, K. C., and Xing, X. (1993). Probing functional organization within the nucleus: Is genomic structure integrated with RNA metabolism? *Cold Spring Harbor Symp. Quant. Biol.* **58,** 807–818.

Lazarides, E., and Lindberg, U. (1974). Actin is the naturally occurring inhibitor of deoxyribonuclease I. *Proc. Natl. Acad. Sci. U.S.A.* **71,** 4742–4746.

Lenk, R., Ransom, L., Kaufmann, Y., and Penman, S. (1977). A cytoskeletal structure with associated polyribosomes obtained from HeLa cells. *Cell (Cambridge, Mass.)* **10,** 67–78.

Leonhardt, H., Page, A. W., Weier, H. U., and Bestor, T. H. (1992). A targeting sequence directs DNA methyltransferase to sites of DNA replication in mammalian nuclei. *Cell (Cambridge, Mass.)* **71,** 865–873.

Lerner, E. A., Lerner, M. R., Janeway, L. A., and Steitz, J. A. (1981). Monoclonal antibodies to nucleic acid containing cellular constituents: Probes for molecular biology and autoimmune disease. *Proc. Natl. Acad. Sci. U. S. A.* **78,** 2737–2741.

Li, H., and Bingham, P. M. (1991). Arginine/serine-rich domains of the su(wa) and tra RNA processing regulators target proteins to a subnuclear compartment implicated in splicing. *Cell (Cambridge, Mass.)* **67,** 335–342.

Long, B. H., and Ochs, R. L. (1983). Nuclear matrix, hnRNA, and snRNA in Friend erythroleukemia nuclei depleted of chromatin by low ionic strength EDTA. *Biol. Cell* **48,** 89–98.

Long, B. H., and Schrier, W. H. (1983). Isolation from Friend erythroleukemia cells of an RNase-sensitive nuclear matrix fibril fraction containing hnRNA and snRNA. *Biol. Cell* **48,** 99–108.

Longo, L., Pandolfi, P. P., Biondi, A., Rombaldi, A., Mencarelli, A., Lo Coco, F., Direrio, D., Pegoraro, L., Avanzi, G., Tabilio, A., Zangrilli, D., Alcalay, M., Donti, E., Grignani, F., and Pelicci, P. G. (1990). Rearrangements and abberant expression of the retinoic acid receptor α gene in acute myelocytic leukemias. *J. Exp. Med.* **172,** 1571–1575.

Lydersen, B. K., and Pettijohn, D. E. (1980). Human-specific nuclear protein that associates with the polar region of the mitotic apparatus: Distribution in a human/hamster hybrid cell. *Cell (Cambridge, Mass.)* **22,** 489–499.

Mancini, M. A., Shan, B., Nickerson, J. A., Penman, S., and Lee, W.-H. (1994). The retinoblastoma gene product is a cell cycle-dependent nuclear matrix-associated protein. *Proc. Natl. Acad. Sci. U.S.A.* **91,** 418–422.

Mannherz, H. G., Goody, R. S., Konrad, M., and Nowak, E. (1980). The interaction of bovine pancreatic deoxyribonuclease I and skeletal muscle actin. *Eur. J. Biochem.* **104,** 367–379.

Mantegna, R. N., Buldyrev, S. V., Goldberger, A. L., Havlin, S., Peng, C. -K., Simons, M., and Stenley, H. E. (1994). Linguistic features of noncoding DNA sequences. *Phys. Rev. Lett.* **73,** 3169–3172.

Manuelidis, L. (1982). Nucleotide sequence definition of a major human repeated DNA, the Hind III1.9 kb family. *Nucleic Acids Res.* **10,** 3211–3219.

Mariman, E. C. M., van Eekelen, C. A. G., Reinders, R. J., Berns, A. J. M., and van Venrooij, W. J. (1982). Adenoviral heterogeneous nuclear RNA is associated with the host nuclear matrix during splicing. *J. Mol. Biol.* **154,** 103–119.

Matera, A. G., and Ward, D. C. (1993). Nucleoplasmic organization of small nuclear ribonucleoproteins in cultured human cells. *J. Cell Biol.* **121,** 715–727.

Maul, G. G., and Everett, R. D. (1994). The nuclear location of PML, a cellular member of the C3HC4 zinc-finger binding domain protein family, is rearranged during herpes simplex infection by the C3HC4 viral protein ICPO. *J. Gen. Virol.* **75,** 1223–1233.

Maul, G. G., Guldner, H. H., and Spivack, J. G. (1993). Modification of discrete nuclear domains induced by herpes simplex virus type 1 immediate early gene 1 product (ICP0). *J. Gen. Virol.* **74,** 2679–2690.

Mayer, D. T., and Gulick, A. (1942). The nature of the proteins of cellular nuclei. *J. Biol. Chem.* **46,** 433–440.

Meier, U. T., and Blobel, G. (1992). Nopp140 shuttles on tracks between nucleolus and cytoplasm. *Cell (Cambridge, Mass.)* **70,** 127–138.

Meier, U. T., and Blobel, G. (1994). NAP57, a mammalian nucleolar protein with a putative homolog in yeast and bacteria. *J. Cell Biol.* **127,** 1505–1514.

Miller, T. E., Huang, C. Y., and Pogo, A. O. (1978). Rat liver nuclear skeleton and ribonucleoprotein complexes containing hnRNA. *J. Cell Biol.* .**76,** 675–691.

Mirkovitch, J., Mirault, M.-E., and Laemmli, U. K. (1984). Organization of the higher-order chromatin loop: Specific DNA attachment sites on nuclear scaffold. *Cell (Cambridge, Mass.)* **39,** 223–232.

Mittnacht, S., Lees, J. A., Desaj, D., Harlow, E., Morgan, D. O., and Weinberg, R. A. (1994). Distinct sub-populations of the retinoblastoma protein show a distinct pattern of phosphorylation. *EMBO J.* **13,** 118–127.

Moir, R. D., Montag-Lowy, M., and Goldman, R. D. (1994). Dynamic properties of nuclear lamins: Lamin B is associated with sites of DNA replication. *J. Cell Biol.* **125,** 1201–1212.

Monneron, A., and Bernhard, W. (1969). Fine structural organization of the interphase nucleus in some mammalian cells. *J. Ultrastruct. Res.* **27,** 266–288.

Münger, K., Phelps, W. C., Bubb, V., Howley, P. M., and Schlegel, R. (1989). The E6 and E7 genes of the human papillomavirus type 16 together are necessary and sufficient for transformation of primary human keratinocytes. *J. Virol.* **63,** 4417–4421.

Nakamura, H., Morita, T., and Sato, C. (1986). Structural organization of replicon domains during DNA synthetic phase in the mammalian nucleus. *Exp. Cell Res.* **165**, 291–297.

Nakayasu, H., and Berezney, R. (1989). Mapping replication sites in the eukaryotic cell nucleus. *J. Cell Biol.* **108**, 1–11.

Nakayasu, H., and Berezney, R. (1991). Nuclear matrins: Identification of the major nuclear matrix proteins. *Proc. Natl. Acad. Sci. U.S.A.* **88**, 10312–10316.

Nakayasu, H., and Ueda, K. (1983). Association of actin with the nuclear matrix from bovine lymphocytes. *Exp. Cell Res.* **143**, 55–62.

Nakayasu, H., and Ueda, K. (1985). Association of rapidly-labelled RNAs with actin in nuclear matrix from mouse L5178Y cells. *Exp. Cell Res.* **160**, 319–330.

Nakayasu, H., and Ueda, K. (1986). Preferential association of acidic actin with nuclei and nuclear matrix from mouse leukemia L5178Y cells. *Exp. Cell Res.* **163**, 327–336.

Nickerson, J. A., and Penman, S. (1992). The nuclear matrix: Structure and involvement in gene expression. *In* "Molecular and Cellular Approaches to the Control of Proliferation and Differentiation" (G. Stein and J. Lian eds.), pp. 343–380. Academic Press, San Diego, CA.

Nickerson, J. A., Krochmalnic, G., Wan, K. M., and Penman, S. (1989). Chromatin architecture and nuclear RNA. *Proc. Natl. Acad. Sci. U.S.A.* **86**, 177–181.

Nickerson, J. A., Krockmalnic, G., Wan, K. M., Turner, C. D., and Penman, S. (1992). A normally masked nuclear matrix antigen that appears at mitosis on cytoskeleton filaments adjoining chromosomes, centrioles and midbodies. *J. Cell Biol.* **116**, 977–987.

Nigg, E. A. (1989). The nuclear envelope. *Curr. Opinion Cell Biol.* **1**, 435–440.

O'Keefe, R. T., Henderson, S. C., and Spector, D. L. (1992). Dynamic organization of DNA replication in mammalian cell nuclei: Spatially and temporally defined replication of chromosome specific a-satellite DNA sequences. *J. Cell Biol.* **116**, 1095–1100.

O'Keefe, R. T., Mayeda, A., Sadowski, C., Krainer, A. R., and Spector, D. L. (1994). Disruption of pre-mRNA splicing in vivo results in reorganization of splicing factors. *J. Cell Biol.* **124**, 249–260.

Osheim, Y. N., and Beyer, A. L. (1991). EM analysis of Drosophila chorion genes: amplification, transcription termination and RNA splicing. *Electron Microsc. Rev.* **4**, 111–128.

Pardoll, D. M., and Vogelstein, B. (1980). Sequence analysis of nuclear matrix associated DNA from rat liver. *Exp. Cell Res.* **128**, 466–470.

Pardoll, D. M., Vogelstein, B., and Coffey, D. S. (1980). A fixed site of DNA replication in eukaryotic cells. *Cell (Cambridge, Mass.)* **19**, 527–536.

Partin, A. W., Steinberg, G. D., Pitcock, R. V., Wu, L., Piantadosi, S., Coffey, D. S., and Epstein, J. I. (1992). Use of nuclear morphometry, Gleason histologic scoring, clinical stage, and age to predict disease-free survival among patients with prostate cancer. *Cancer (Philadelphia)* **70**, 161–168.

Partin, A. W., Getzenberg, R. H., CarMichael, M. J., Vindivich, D., Yoo, J., Epstein, J. I., and Coffey, D. S. (1993). Nuclear matrix protein patterns in human benign prostatic hyperplasia and prostate cancer. *Cancer Res.* **53**, 744–6.

Paul, J. I., Schwarzbauer, J. E., Tamkun, J. W., and Hynes, R. O. (1986). Cell-type-specific fibronectin subunits generated by alternative splicing. *J. Biol. Chem.* **261**, 12258–12265.

Penman, S. (1995). Rethinking cell structure. *Proc. Natl. Acad. Sci. U.S.A.,* **92**, 5251–5257.

Perraud, M., Gioud, M., and Monier, J. C. (1979). Intranuclear structures of monkey kidney cells recognised by immunofluorescence and immuno-electron microscopy using anti-ribonucleoprotein antibodies. *Ann. Immunol. (Paris)* **130C**, 635–647.

Petrov, P., and Bernhard, W. (1971). Experimentally induced changes of extranucleolar ribonucleoprotein components of the interphase nucleus. *J. Ultrastruct. Res.* **35**, 386–402.

Phi-Van, L., and Strätling, W. H. (1988). The matrix attachment regions of the chicken lysozyme gene co-map with the boundaries of the chromatin domain. *EMBO J.* **7**, 655–664.

Phi-Van, L., von Kries, J. P., Ostertag, W., and Strätling, W. H. (1990). The chicken lysozyme 5' matrix attachment region increases transcription from a heterologous promoter in heterologous cells and dampens positional effects on the expression of transfected genes. *Mol. Cell. Biol.* **10**, 2302–2307.

Pienta, K. J., and Coffey, D. S. (1992). Nuclear-cytoskeletal interactions: Evidence for physical connections between the nucleus and cell periphery and their alteration by transformation. *J. Cell. Biochem.* **49**, 357–365.

Puvion, E., Virion, A., Assens, C., Leduc, E. H., and Jeanteur, P. (1984). Immunocytochemical identification of nuclear structures containing snRNPs in isolated rat liver cells. *J. Ultrastruct. Res.* **87**, 180–189.

Ramón y Cajál, S. (1903). Un sencillo metodo de coloracion selectiva del reticulo protoplasmico y sus efectos en los diversos organos nerviosos. *Trab. Lab. Invest. Biol.* **2**, 129–221.

Ramón y Cajál, S. (1909). "Histologie du système nerveux de l'homme et des vertébrés," Vol. 1 and 2. (Masson, Paris reprinted by Consejo Superior de Investigaciones Cientificas, Madrid, 1952).

Raska, I., Ochs, R. L., Andrade, L. E., Chan, E. K., Burlingame, R., Peebles, C., Gruol, D., and Tan, E. M. (1990). Association between the nucleolus and the coiled body. *J. Struct. Biol.* **104**, 120–127.

Raska, I., Andrade, L. E., Ochs, R. L., Chan, E. K., Chang, C. M., Roos, G., and Tan, E. M. (1991). Immunological and ultrastructural studies of the nuclear coiled body with autoimmune antibodies. *Exp. Cell Res.* **195**, 27–37.

Razin, S. V. (1987). DNA interactions with the nuclear matrix and spatial organization of replication and transcription. *BioEssays* **6**, 19–23.

Robinson, S. I., Nelkin, B. D., and Vogelstein, B. (1982). The ovalbumin gene is associated with the nuclear matrix of chicken oviduct cells. *Cell (Cambridge, Mass.)* **28**, 99–106.

Roth, M. B., Murphy, C., and Gall, J. G. (1990). A monoclonal antibody that recognizes a phosphorylated epitope stains lampbrush chromosome loops and small granules in the amphibian germinal vesicle. *J. Cell Biol.* **111**, 2217–2223.

Saga, Y., Lee, J. S., Saraiya, C., and Boyse, E. A. (1990). Regulation of alternative splicing in the generation of isoforms of the mouse Ly-5 (CD45) glycoprotein. *Proc. Natl. Acad. Sci. U.S.A.* **87**, 3728–3732.

Sahlas, D. J., Milankov, K., Park, P. C., and De Boni, U. (1993). Distribution of snRNPs, splicing factor SC-35 and actin in interphase nuclei: Immunocytochemical evidence for differential distribution during changes in functional states. *J. Cell Sci.* **105**, 347–357.

Salditt-Georgieff, M., Harpold, M. M., Wilson, M. C., and Darnell, J. E., Jr. (1981). Large heterogeneous nuclear ribonucleic acid has three times as many 5' caps as polyadenylic acid segments, and most caps do not enter polyribosomes. *Mol. Cell. Biol.* **1**, 179–187.

Savage, M. J., Sala-Trepat, J. M., and Bonner, J. (1978). Measurement of the complexity and diversity of poly(adenylic acid) containing messenger RNA from rat liver. *Biochemistry* **17**, 462–467.

Scheer, U., Thiry, M., and Goessens, G. (1993). Structure, function and assembly of the nucleolus. *Trends Cell Biol.* **3**, 236–241.

Scheller, R. H., Costantini, F. D., Kozlowski, M. R., Britten, R. J., and Davidson, E. H. (1978). Specific representation of cloned repetitive DNA sequences in sea urchin RNAs. *Cell (Cambridge, Mass.)* **15**, 189–203.

Schirmbeck, R., and Deppert, W. (1987). Specific interaction of simian virus large T antigen with cellular chromatin and nuclear matrix during the course of infection. *J. Virology* **61**, 3561–3569.

Schirmbeck, R., and Deppert, W. (1989). Nuclear subcompartmentalization of simian virus 40 large T antigen: Evidence for in vivo regulation of biochemical activities. *J. Virology* **63**, 2308–2316.

Schröder, H. C., Trölltsch, D., Wenger, R., Bachmann, M., Diehl-Seifert, B., and Müller, W. E. G. (1987a). Cytochalasin B selectively releases ovalbumin mRNA precursors but not the mature ovalbumin mRNA from hen oviduct nuclear matrix. *Eur. J. Biochem.* **167**, 239–245.

Schröder, H. C., Trölltsch, D., Friese, U., Bachmann, M., and Müller, W. E. G. (1987b). Mature mRNA is selectively released from the nuclear matrix by an ATP/dATP-dependent mechanism sensitive to topoisomerase inhibitors. *J. Biol. Chem.* **262**, 8917–8925.

Seite, R., Pebusque, M. J., and Vio-Cigna, M. (1982). Argyrophilic proteins on coiled bodies in sympathetic neurons identified by Ag-NOR procedure. *Biol. Cell* **46**, 97–100.

Sims, J. R., Karp, S., and Ingber, D. E. (1992). Altering the cellular mechanical force balance results in integrated changes in cell, cytoskeletal and nuclear shape. *J. Cell Sci.* **103**, 1215–1222.

Spector, D. L. (1993). Nuclear organization of pre-mRNA processing. *Curr. Opin. Cell Biol.* **5**, 442–448.

Spector, D. L., Schrier, W. H., and Busch, H. (1983). Immunoelectron microscopic localization of snRNPs. *Biol. Cell.* **49**, 1–10.

Spector, D. L., Fu, X.-D., and Maniatis, T. (1991). Associations between distinct pre-mRNA splicing components and the cell nucleus. *EMBO J.* **10**, 3467–3481.

Spector, D. L., Lark, G., and Huang, S. (1992). Differences in snRNP localization between transformed and non-transformed cells. *Mol. Biol. Cell.* **3**, 555–569.

Stein, G. S., Lian, J. B., and Owen, T. A. (1990). Bone cell differentiation: A functionally coupled relationship between expression of cell-growth- and tissue-specific genes. *Curr. Opin. Cell Biol.* **2**, 1018–1027.

Stein, G. S., van Wijnen, A., Stein, J. L., Lian, J. B., Bidwell, J. P., and Montecino, M. (1994). Nuclear architecture supports integration of physiologically regulatory signals for transcription of cell growth and tissue-specific genes during osteoblast differentiation. *J. Cell. Biochem.* **55**, 4–15.

Stief, A., Winter, D. M., Strätling, W. H., and Sippel, A. E. (1989). A nuclear attachment element mediates elevated and position-independent gene activity. *Nature (London)* **341**, 343–345.

Stuurman, N., van Driel, R., de Jong, L., Meijne, A. M., and van Renswoude, J. (1989). The protein composition of the nuclear matrix of murine P19 embryonal carcinoma cells is differentiation-stage dependent. *Exp. Cell Res.* **180**, 460–466.

Stuurman, N., Meijne, A. M., van der Pol, A. J., de Jong, L., van Driel, R., and van Renswoude, J. (1990). The nuclear matrix from cells of different origin. Evidence for a common set of matrix proteins. *J. Biol. Chem.* **265**, 5460–5465.

Sun, J.-M., Chen, H. Y., and Davie, J. R. (1994). Nuclear factor 1 is a component of the nuclear matrix. *J. Cell. Biochem.* **55**, 252–263.

Tang, T. K., Tang, C. C., Chen, Y., and Wu, C. (1993). Nuclear proteins of the bovine esophageal epithelium. II. The NuMA gene gives rise to multiple mRNAs and gene products reactive with the monoclonal antibody W1. *J. Cell Sci.* **104**, 249–260.

Tang, T. K., Tang, C. C., Chao, Y., and Wu, C. (1994). Nuclear mitotic apparatus protein (NuMA): Spindle association, nuclear targeting and differential subcellular localization of varoius NuMA isoforms. *J. Cell Sci.* **107**, 1389–1402.

Theissig, F., Dimmer, V., Haroske, G., Kunze, K. D., and Meyer, W. (1991). Use of nuclear image cytometry, histopathological grading and DNA cytometry to make breast cancer prognosis more objective. *Anal. Cell. Pathol.* **3**, 351–360.

Thorburn, A., Moore, R., and Knowland, J. (1988). Attachment of transcriptionally active sequences to the nucleoskeleton under isotonic conditions. *Nucleic Acids Res.* **16**, 7183.

Tousson, A., Zeng, C., Brinkley, B. R., and Valdivia, M. M. (1991). Centrophilin: A novel mitotic spindle protein involved in mictotubule nucleation. *J. Cell Biol.* **112**, 427–440.

Tubo, R. A., and Berezney, R. (1987a). Identification of 100 and 150S DNA polymerase-primase megacomplexes solubilized from the nuclear matrix of regenerating rat liver. *J. Biol. Chem.* **262,** 5857–5865.

Tubo, R. A., and Berezney, R. (1987b). Nuclear matrix-bound DNA primase. Elucidation of an RNA priming system in nuclear matrix isolated from regenerating rat liver. *J. Biol. Chem.* **262,** 6637–6642.

Tuma, R., Stolk, J. A., and Roth, M. B. (1993). Identification and characterization of a sphere organelle protein. *J. Cell Biol.* **122,** 767–773.

Underwood, J. C. (1990). Nuclear morphology and grading of tumours. *Curr. Top. Pathol.* **82,** 1–15.

van Eekelen, C. A. G., and van Venrooij, W. J. (1981). HnRNA and its attachment to a nuclear protein matrix. *J. Cell Biol.* **88,** 554–563.

van Wijnen, A. J., Bidwell, J. P., Fey, E. G., Penman, S., Lian, J. B., Stein, J. L., and Stein, G. S. (1993). Nuclear matrix association of multiple sequence specific DNA abinding activities related to SP-1, ATF, CCAAT, C/EBP, OCT-1, and AP-1. *Biochemistry* **32,** 8397–8402.

Vaughn, J. P., Dijkwel, P. A., Mullenders, L. H. F., and Hamlin, J. L. (1990). Replication forks are associated with the nuclear matrix. *Nucleic Acids Res.* **18,** 1965–1969.

Verheijen, R., Kuijpers, H., Vooijs, P., van Venrooij, W., and Ramaekers, F. (1986). Distribution of the 70k U1 RNA-associated protein during interphase and mitosis: Correlation with other U RNP particles and proteins of the nuclear matrix. *J. Cell Sci.* **86,** 173–190.

Visa, N., Puvion-Dutilleul, F., Harper, F., Bachellerie, J. P., and Puvion, E. (1993). Intranuclear distribution of poly(A) RNA determined by electron microscope in situ hybridization. *Exp. Cell Res.* **208,** 19–34.

Vogelstein, B., Pardoll, D. M., and Coffey, D. S. (1980). Supercoiled loops and eucaryotic DNA replication. *Cell (Cambridge, Mass.)* **22,** 79–85.

Wan, K., Nickerson, J. A., Krockmalnic, G., and Penman, S. (1994). The B1C8 protein is in the dense assemblies of the nuclear matrix and relocates to the spindle and pericentriolar filaments at mitosis. *Proc. Natl. Acad. Sci. U.S.A.* **91,** 594–598.

Wansink, D. G., Schul, W., van der Kraan, I., van Steensel, B., van Driel, R., and de Jong, L. (1993). Fluorescent labeling of nascent RNA reveals transcription by RNA polymerase II in domains scattered throughout the nucleus. *J. Cell Biol.* **122,** 283–293.

Weis, K., Rambaud, S., Lavau, C., Jansen, J., Carvalho, T., Carmo-Fonseca, M., Lamond, A., and Dejean, A. (1994). Retinoic acid regulates aberrant nuclear localization of PML-RARα in acute promyelocytic leukemia cells. *Cell (Cambridge, Mass.)* **76,** 345–356.

Wu, T., Wang, Y., and Wu, R. (1994). Transcribed repetitive DNA sequences in telomeric regions of rice (*Oryza sativa*). *Plant Mol. Biol.* **26,** 363–375.

Xing, Y. G. (1993). Visualization of individual gene transcription and splicing. Ph.D. Thesis, University of Massachusetts Medical School, Worcester.

Xing, Y. G., and Lawrence, J. B. (1991). Preservation of specific RNA distribution within the chromatin-depleted nuclear substructure demonstrated by *in situ* hybridization coupled with biochemical fractionation. *J. Cell Biol.* **112,** 1055–1063.

Xing, Y. G., Johnson, C. V., Dobner, P. R., and Lawrence, J. B. (1993). Higher level organization of individual gene transcription and splicing. *Science* **259,** 1326–1330.

Yang, C. H., Lambie, E. J., and Snyder, M. (1992). NuMA: An unusually long coiled-coil related protein in the mammalian nucleus. *J. Cell Biol.* **116,** 1303–1317.

Zahler, A. M., Lane, W. S., Stolk, J. A., and Roth, M. B. (1992). SR proteins: a conserved family of pre-mRNA splicing factors. *Genes Dev.* **6,** 837–847.

Zamore, P. D., and Green, M. R. (1991). Biochemical characterization of U2 snRNP auxiliary factor: An essential pre-mRNA splicing factor with a novel intranuclear distribution. *EMBO J.* **10,** 207–214.

Zeitlin, S., Parent, A., Silverstein, S., and Efstratiadis, A. (1987). Pre-mRNA splicing and the nuclear matrix. *Mol. Cell Biol.* **7**, 111–120.

Zeitlin, S., Wilson, R. C., and Efstradiadis, A. (1989). Autonomous splicing and complementation of in vivo assembled splicosomes. *J. Cell Biol.* **108**, 765–777.

Zeng, C., He, D., and Brinkley, B. R. (1994). Localization of NuMA protein isoforms in the nuclear matrix of mammalian cells. *Cell Motil. Cytoskel.* **29**, 167–176.

Zhang, G., Taneja, K. L., Singer, R. H., and Green, M. R. (1994). Localization of pre-mRNA splicing in mammalian nuclei. *Nature (London)* **372**, 809–812.

Zhang, M., Zamore, P. D., Carmo-Fonseca, M., Lamond, A. I., and Green, M. R. (1992). Cloning and intracellular localization of the U2 small nuclear ribonucleoprotein auxiliary factor small subunit. *Proc. Natl. Acad. Sci. U.S.A.* **89**, 8769–8773.

Zlatanova, J. S., and Van Holde, K. E. (1992). Chromatin loops and transcriptional regulation. *CRC Crit. Rev. Eukaryotic Gene Express.* **2**, 211–224.

The Structural Basis of Nuclear Function

Dean A. Jackson and Peter R. Cook
CRC Nuclear Structure and Function Research Group, Sir William Dunn School
of Pathology, University of Oxford, Oxford OX1 3RE, United Kingdom

Most models for transcription and replication involve polymerases that track along the template. We review here experiments that suggest an alternative in which polymerization occurs as the template slides past a polymerase fixed to a large structure in the eukaryotic nucleus—a "factory" attached to a nucleoskeleton. This means that higher-order structure dictates how and when DNA is replicated or transcribed.

KEY WORDS: Nucleoskeleton, DNA replication, Transcription, Repair, Chromatin loop.

I. Introduction

The cytoplasm is a complicated cellular compartment, with different regions dedicated to different functions. Various organelles and subcompartments are organized around different skeletons, and complex mechanisms direct molecular traffic between these subcompartments and surrounding membranes. In contrast with this complexity, the nucleus has traditionally been viewed as relatively unstructured, with few major compartments (e.g., the envelope, heterochromatin, euchromatin, and nucleoli). It is now being realized that the structure of the nucleus is as complex as that of the cytoplasm, with different skeletons and subcompartments, each with its own particular function. For example, the major nuclear processes of replication, the repair of damage in DNA, transcription, RNA processing, and transport, all take place in specific subcompartments. Moreover, complex controls manage traffic between these subcompartments and the cytoplasm. Therefore, nuclear architecture profoundly affects gene function. We review here its roles in replication, transcription, and the repair of damage in DNA. Work using conditions close to the physiological and that maintain as much of the functional integrity as is conveniently possible will be emphasized. The nuclear matrix and scaffold as described in some other chapters in these volumes are commonly isolated under less physiological conditions and hence may display more perturbed functional properties.

The appreciation of the role played by large structures in nuclei has led to a reevaluation of the mechanism of polymerization. The traditional view involved polymerases that tracked along the template (Figs. 1A–C and 2A). (The term polymerase is used here to describe the cluster of many different polypeptides that form an active complex in which the polymerizing subunits are present only as minor components.) This view was sensible if polymerases were small relative to the template and if they acted alone. However, recent experiments show that many DNA polymerases are concentrated together into large structures, "factories", that dwarf the template. Therefore we imagine that several templates slide past a number of DNA polymerases fixed within factories; the template moves rather than the enzyme (Figs. 1D–F). Evidence is also accumulating that transcription occurs similarly (Fig. 2B). Indeed, logic suggests that if the RNA polymerase did track along the helical template, the resulting transcript would be entangled about the template (Fig. 2B,upper); however, no satisfactory mechanism has yet been suggested as to how it might be untangled prior to export to the cytoplasm. A sensible alternative is to immobilize the polymerase;

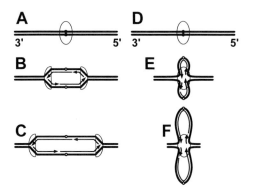

FIG. 1   Models for replication involving (A–C) mobile and (D–F) immobile polymerases. A–C. The conventional model involves (A) binding of the polymerizing complex (oval) to the origin (black dots), before the complex splits into two and (B,C) the two halves track along the template (thick lines) as nascent DNA (thin gray lines) is made. Arrows indicate direction of growth of nascent chains. D–F. The alternative model involves passage of the template through a fixed complex and extrusion of nascent DNA. The origin is shown here detaching from the complex, but it may remain attached. [From Hozák *et al.* (1995a) with permission of Oxford University Press.]

POLYMERASE TRANSLOCATES AND ROTATES

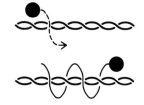

DNA TRANSLOCATES AND ROTATES

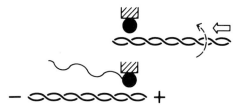

FIG. 2   Models for transcription involving (top) mobile and (bottom) immobile polymerases. Top. The polymerase (solid circle) tracks (arrow) along one of the two helical strands to generate a transcript. This transcript is entangled about the template and must be untangled, but no satisfactory mechanism has yet been suggested with regard to how this might be achieved [discussed by Cook and Gove (1992) and Cook (1994)]. Bottom. DNA rotates (arrow) as it slides (large arrow) past an immobile polymerase (solid circle attached to hatched area); the transcript is extruded as template rotation generates positive and negative supercoils (indicated by + and −) which must be removed by a topoisomerase. The movements involved are similar to driving a screw (the DNA) through a fixed nut (the polymerase). [From Cook (1994) with permission of ICSU Press.]

then an unentangled transcript can be extruded as the template moves both rotationally and laterally through the fixed polymerizing site (Fig. 2B, lower). This means that the position of a sequence in three-dimensional space relative to the fixed polymerases will dictate how easily it can engage a polymerizing complex (i.e., attach) so that replication or transcription can begin. As a result, higher-order structure controls both processes. Although discussion here concentrates on eukaryotic enzymes, it seems that prokaryotic enzymes will work in much the same way, as the catalytic domains of DNA polymerases are structurally related to the eukaryotic enzymes (Braithwaite and Ito, 1993).

A. Artifacts

Isotonic salt concentrations are not generally used during nuclear fractionation—or during polymerase assay—because they cause chromatin to aggregate into an unworkable mess and, quite naturally, more tractable conditions are used instead. However, chromatin is poised in a metastable state and quite small changes in tonicity dramatically affect structure. For example, one-tenth the physiological salt concentration—usually used to isolate nuclei and chromatin—destroys the 30-nm fiber, extracts a quarter of nuclear protein including DNA polymerases and generates a new attachment of the fiber to the substructure for every one that preexists. Often residual aggregation is suppressed by adding "stabilizing" cations like magnesium ions, spermine, or spermidine, but these generate further artifactual attachments. Then it is not surprising that slight differences in isolation conditions generate very different structures, each with its own characteristic set of sequences associated with a different subset of proteins. For example, matrix attached regions (MARs) are bound to different proteins depending on the precise method of isolation, scaffold attached regions (SARs) are often specifically associated with topoisomerase II, and transcribed sequences are bound to "cages." Skeptics point to the fact that even those in the field cannot agree on which sequences are associated with which proteins in a particular substructure and naturally suggest that some, or all, are isolation artifacts with no counterparts *in vivo* (Cook, 1988; Jack and Eggert, 1992).

Against this background, it is not surprising that convincing evidence for a role of larger structures in nuclear functions was only obtained with the use of more physiological conditions that preserved many nuclear functions.

B. Physiological Conditions

The practical problems caused by chromatin aggregation under physiological conditions can be overcome if cells are encapsulated in agarose microbeads (50–150-μm diameter) before lysis. Agarose is permeable to small molecules so encapsulated cells continue to grow in standard tissue-culture media. When they are permeabilized with a mild detergent in a "physiological" buffer most soluble cytoplasmic proteins and RNA diffuse out to leave the cytoskeleton and associated material, as well as the nucleus (Fig. 3A–C; Jackson and Cook, 1985a; Jackson *et al.*, 1988). The agarose protects these cell remnants and, importantly, the encapsulated nuclei can be manipulated freely without aggregation while they remain accessible to probes like antibodies and enzymes. As the template remains intact and as essentially all the replicative and transcriptional activity of the living cell is retained, it seems unlikely that polymerases could have aggregated after lysing the cell. Note that almost all attachments to be discussed below involve active polymerases.

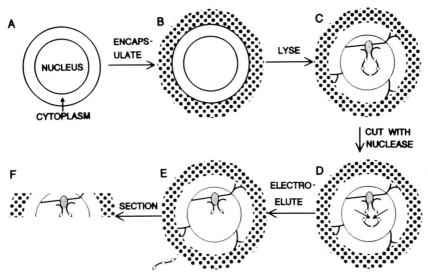

FIG. 3   Procedure for visualizing nucleoskeletons and associated structures. (A) Cells are (B) encapsulated in agarose (dots) and (C) lysed to leave a cytoskeleton, nuclear lamina (dotted circle), and nucleoskeleton (straight line) to which is attached a replication factory (gray oval) and a DNA loop covered with nucleosomes (open circles). Dense chromatin obscures the nucleoskeleton but can be removed by (D) cutting the chromatin fiber with a restriction endonuclease, and then (E) removing any unattached fragments electrophoretically. (F) After cutting thick sections, the nucleoskeleton and associated factory can now be seen in the electron microscope. [From Cook (1994) with permission of ICSU Press.]

II. Nucleoskeletons and Chromatin Loops

Most cell biologists now agree that chromosomal DNA is looped into domains by attachment of the chromatin fiber to a nucleoskeleton. Fortunately, meiotic lampbrush chromosomes of living newt cells provide undisputable proof of chromatin loops attached to a skeleton core (Callan, 1977). Unfortunately, ever since the initial demonstration of individual superhelical domains in "nucleoids" from interphase cells (Cook and Brazell, 1975), there has been little agreement as to what the molecular basis of the skeleton is and what the sequences involved in the attachments might be. As outlined above, controversy has centered on whether the attachments seen *in vitro* are also found *in vivo*.

It is important to recall that the existence of the cytoskeleton was once controversial, largely because it proved so difficult to see; the widths of its individual elements are well below the resolution of the light microscope (i.e., ~200 nm) and the sections used for electron microscopy are usually too thin to contain extended regions of the skeleton. That controversy evaporated once (1) individual elements of the skeleton were seen by fluorescence microscopy after immunolabeling and (2) complete skeletons were imaged by electron microscopy of whole mounts. So perhaps it is not surprising that the nucleoskeleton has proven equally difficult to image, especially since it is obscured by dense chromatin, which also prevents access of antibodies to it.

We have visualized a nucleoskeleton using the approach illustrated in Fig. 3. Encapsulated cells were lysed, the chromatin fiber cut with restriction endonucleases, and then most obscuring chromatin removed by electrophoresis. Electron microscopy of thick (resinless) sections (Capco *et al.,* 1982; He *et al.,* 1990) then revealed residual clumps of chromatin attached to a "diffuse skeleton" that ramified throughout the nucleus (Fig. 4). This network is morphologically (and functionally) very complex, but its "core filaments" had the axial repeat typical of the intermediate-filament family

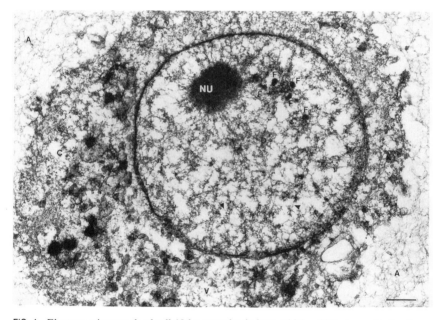

FIG. 4   Electron micrograph of cell 10 hr postmitosis from which ~90% chromatin has been removed. Encapsulated cells were permeabilized, incubated with biotin-dUTP, treated with nucleases, chromatin eluted, incorporated biotin immunolabeled with 5-nm gold particles, and a 500-nm resinless section prepared. Agarose (A) surrounds cytoplasmic (C) and nuclear remnants where residual clumps of chromatin are attached to a diffuse network that ramifies from nucleolus (NU) to lamina. Gold particles, which are not visible at this magnification, were concentrated in replication factories (F). Bar: 1 μm. [From Hozák *et al.* (1994b) with permission of the Company of Biologists Ltd.]

of proteins (Jackson and Cook, 1988). Immunofluorescence—using various antibodies, including one that specifically recognized an epitope in lamin A between amino acids 598 and 611, and cell types (i.e., HeLa, HEp-2, and SW13)—revealed internal "speckles" when the obscuring chromatin was removed (Hozák *et al.,* 1995b). Electron microscopy showed that all three antibodies labeled nodes on the diffuse nucleoskeleton, although the core filaments were weakly labeled (Hozák *et al.,* 1995b). One anti-lamin B2 antibody reacted weakly with foci on this diffuse skeleton in a minority of cells. Moreover, electron microscopy using (conventional) thin sections from which no chromatin was removed revealed as much lamin A in the interior as at the periphery.

This result was surprising. It is widely assumed that the nuclear lamins—as their name indicates—are confined to the nuclear periphery. However, they are occasionally found internally, for example during G1 phase, in certain pathological states, when mutated, or when overexpressed (e.g., Cardenas *et al.,* 1990; Gill *et al.,* 1990; Bader *et al.,* 1991; Beven *et al.,* 1991; Kitten and Nigg, 1991; Eckelt *et al.,* 1992; Goldman *et al.,* 1992; Lutz *et al.,* 1992; Mirzayan *et al.,* 1992; Bridger *et al.,* 1993). This internal location has been attributed to an accidental mislocation of proteins destined for the periphery, but could equally arise from an overconcentration at a normal site. The results described above suggest that the lamins have been mis-named; they have both a peripheral and an internal location.

We have also measured loop size using physiological conditions. Most chromatin was removed as before by treatment with an endonuclease followed by electrophoresis to leave residual clumps of chromatin attached to the nucleoskeleton (Fig. 3). An average loop size of 86 kb (in HeLa cells) was then deduced from the percentage of chromatin remaining and the size of the attached fragments (Jackson *et al.,* 1990). No large changes in loop size could be detected at different stages in the cell cycle. Therefore the fundamental attachments probably persist during the gross structural changes occurring during mitosis. Loop size measured in this way is, of course, an average; for a discussion, see Jackson *et al.,* (1990).

III. Attached Polymerases

Models involving tracking DNA and RNA polymerases can be distinguished from those involving immobile (i.e., attached) enzymes by cutting encapsulated chromatin into fragments of <10 kb with an endonuclease and then removing electrophoretically any unattached fragments (Fig. 3; Jackson and Cook, 1985a,b, 1986a,b; Jackson *et al.,* 1988). If polymerases are attached (either directly to the skeleton or indirectly through a larger structure which is in turn attached to the skeleton), they should remain in the agarose bead after electroelution; if unattached, they should electroelute from the bead with the eluting chromatin. Cutting chromatin in unsynchronized and permeabilized HeLa cells, followed by electroelution of >75% of the chromatin, hardly reduced DNA or RNA polymerizing activity. As very large chromatin fragments (i.e., containing ~150 kb DNA) can escape from beads, this polymerizing activity must be attached. Nascent DNA and RNA—whether labeled *in vivo* or *in vitro* by short incubations with [^{32}P]TTP or [^{32}P]UTP—also resisted elution. These results are simply explained if polymerases are attached, directly or indirectly, to a very large structure like a skeleton.

A. Replication Factories

Seeing is believing; the direct visualization that replication takes place at a limited number of discrete sites in the nucleus has perhaps provided the most persuasive evidence of a role for an underlying structure. In the original experiment, rat fibroblasts in S-phase were incubated with bromodeoxyuridine; then sites where the analogue had been incorporated were labeled using fluorescently tagged antibodies directed against the analogue. These sites were not diffusely spread throughout nuclei but concentrated in ~150 foci (Nakamura *et al.*, 1986). Subsequently it was found that early during S-phase the foci were small and discrete; later they became larger (Nakayasu and Berezney, 1989; Fox *et al.*, 1991; Kill *et al.*, 1991; Humbert and Usson, 1992; Manders *et al.*, 1992) when centromeric and other heterochromatic regions were replicated (O'Keefe *et al.*, 1992). Permeabilized mammalian cells (Fig. 5; Bravo and Macdonald-Bravo, 1987; Hozák *et al.*, 1993) or demembranated frog sperm in egg extracts (e.g., Blow and Laskey, 1986, 1988; Hutchinson *et al.*, 1987, 1988; Mills *et al.*, 1989) incorporate biotin-labeled dUTP into analogous foci, visualized in this case with fluorescently labeled streptavidin or the appropriate antibodies. These foci are not fixation artifacts because similar foci are seen after incorporation of fluorescein-dUTP into permeabilized, but unfixed, cells (Hassan and Cook, 1993). The foci remain even when most chromatin is removed (Nakayasu and Berezney, 1989; Hozák *et al.*, 1993), implying that they are attached to an underlying structure.

Under the conditions used in these experiments, a single replication fork could not possibly incorporate sufficient labeled analogue to be detected, so many forks must be active in each focus; indeed, simple calculations (based on the number of foci, the rate of fork progression, the spacing between forks, the size of the genome, and the length of S-phase) show that ~40 forks must be active in each early S-phase focus in a human cell.

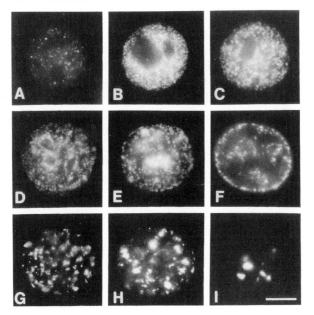

FIG. 5   Fluorescence micrographs of replication patterns found at different stages of S-phase. Synchronized HeLa cells were encapsulated in agarose, permeabilized, incubated with biotin-dUTP, and incorporation sites indirectly immunolabeled. Fluorescence marks replication sites, which change in number and distribution as cells progress from (A) early to (I) late S-phase. Bar: 5 μm. [From Hozák *et al.* (1994b) with permission of the Company of Biologists Ltd.]

This is simply explained if the 40 forks are attached to an underlying structure. Even if polymerases do track, some mechanism must confine the 40 tracking enzymes to one small part of the nucleus and, presumably, this would involve attachment of 40 templates to an underlying structure. So whatever the model, we must invoke some role for a substructure. However, as early S-phase foci have diameters of 100–200 nm, a size that is at the limit of resolution by light microscopy, it would be difficult to use fluorescence microscopy to characterize their relationship to such a substructure.

Many attempts have been made to localize replication sites by electron microscopy. Early autoradiography showed that living cells incorporated [³H]thymidine over periods of many minutes into dispersed chromatin, close to variably sized masses of condensed chromatin (Bouteille *et al.*, 1974; Fakan, 1978). But as the pathlength of the ß-particles emitted by ³H is many tens, even hundreds, of nanometers long, this approach could not provide sufficient resolution for precise localization. Moreover, nascent DNA is made at a rate of about 25 nucleotides per second so that most of the incorporated label would have had plenty of time to move away from its synthetic site. The same is true of experiments involving incubations of 5 min or more with bromodeoxyuridine rather than [³H]thymidine (e.g., Mazzotti *et al.,* 1990; O'Keefe *et al.,* 1992).

Synthetic sites can be immunolabeled with gold particles to a much higher resolution after incubating permeabilized cells with a suboptimal concentration of biotin-dUTP to ensure that the incorporated analogue remains close to the polymerization site. Gold particles then lie within 20 nm of the incorporated biotin (i.e., connected through an antibody bridge). In the first such experiments, most obscuring chromatin was removed as described in Fig. 3 before 400-nm-thick (resinless) sections were viewed in the electron microscope. Electron-dense bodies were scattered along the diffuse nucleoskeleton; they were present in the same numbers as the foci seen by light microscopy, and during early S-phase they were relatively constant in size (100–300-nm diameter). After elongating nascent DNA by ~500 nucleotides, gold particles were associated mainly with these electron-dense bodies (Fig. 6). As the incubation time was progressively increased, longer pieces of DNA were made and gold particles were found progressively further away from the dense bodies. This implies that nascent DNA is extruded from the dense body as templates pass through it.

As cells progress through S-phase, these structures seen by electron microscopy change in numbers, size, shape, and distribution just like foci seen by light microscopy (Fig. 5). Replication foci/factories contain proteins specifically involved in synthesis (e.g., DNA polymerase α, PCNA, and DNA methyltransferase (Leonhardt *et al.,* 1992; Hozák *et al.,* 1993) as well as others that might be involved in regulation [e.g., cyclin A, cdk2, and RPA70, but not cyclin B1 or cdc2 (Adachi and Laemmli, 1992; Cardoso *et al.,* 1993; Sobczak-Thepot *et al.,* 1993)]. As even the smallest contain ~40 active forks and associated leading- and lagging-strand polymerases, it seemed appropriate to call them replication factories. Fragments of such factories were probably isolated by Tubo and Berezney (1987) when they isolated particles containing many polymerases by lightly disrupting somatic

FIG. 6   Replication factories. Encapsulated cells were permeabilized with streptolysin, incubated with biotin-dUTP for 2.5 min, treated with nucleases, ~90% chromatin eluted, and sites of biotin incorporation immunolabeled with 5-nm gold particles. [From Hozák *et al.* (1993) with permission of Cell Press.] A. Seven replication factories (F1–7). NL: nucleolus. L: nuclear lamina. Of the 180 gold particles in the nuclear region 72% (not visible at this magnification) were in factories, indicating that factories were the site of DNA synthesis. Bar: 0.5 μm. B. Higher-power view of F4 and F5. Arrowheads point to the only extrafactory particles. Bar: 0.2 μm. C. Underexposure and further 2× magnification of F4 and F5 to show labeling; arrowheads indicate some of the 30 gold particles. Bar: 0.1 μm.

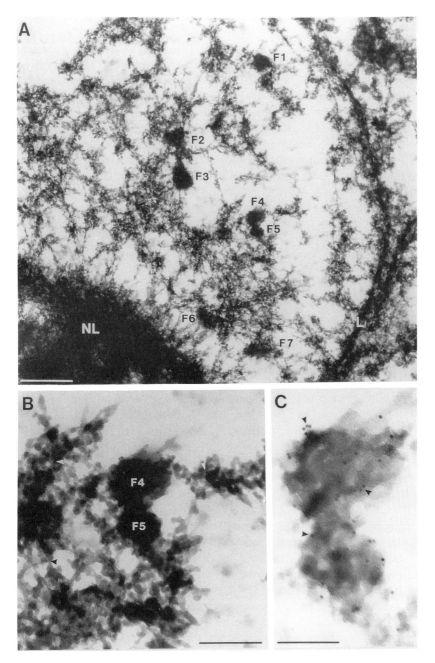

nuclei, while more vigorous procedures give the simpler isolates usually studied by biochemists.

The factories seen in thick resinless sections correspond to a subset of the nuclear "bodies" that have been seen over the years in conventional (thin) sections (Bouteille *et al.*, 1974; Brasch and Ochs, 1992); they are similarly labeled with biotin-dUTP and PCNA and they change in number, shape, and distribution in much the same way (Hozák *et al.*, 1994b).

Although most replication takes place in factories, there is some extrafactory synthesis that increases as cells progress through S-phase (Hozák *et al.*, 1994b). There are special topological problems associated with replicating the last few base-pairs between two replicons (Sundin and Varshavsky, 1980, 1981) so it is attractive to suppose that the extrafactory labeling reflects a tidying-up duplication of hitherto unreplicated DNA.

B. Transcription Foci: RNA Polymerase II

Are many RNA polymerases similarly attached in large factories? Results using the light microscope—though not decisive—suggest that they might be (Jackson *et al.*, 1993; see also Wansink *et al.*, 1993). Encapsulated and permeabilized HeLa cells were incubated with Br-UTP; then 300–500 fluorescent foci were immunolabeled using an antibody against Br-RNA (Fig. 7). The foci contain RNA polymerase II and Sm antigen, a component of the splicing apparatus. This ties the synthetic site into a solid-phase network that might play a role in post-transcriptional events like splicing and transport (e.g., Lawrence *et al.*, 1989; Carmo-Fonseca *et al.*, 1991; Carter *et al.*, 1993). Rough calculations again suggest each transcription focus contains ~50 active RNA polymerases and many templates. α-Amanitin, an inhibitor of RNA polymerase II, prevents incorporation into these foci. These foci remain after removing most chromatin as described in Fig. 3 (Jackson *et al.*, 1993), confirming that synthetic sites are attached to an underlying skeleton.

We might have expected tracking polymerases to be spread throughout open chromatin, but their apparent concentration into foci suggests they do not have the freedom to track everywhere. The simplest explanation of all these results is that many RNA polymerases are attached in factories like the DNA polymerases. However, this conclusion comes with a caution. Estimates of the numbers of RNA polymerases per focus are necessarily less exact than those for DNA polymerases, simply because we do not know the total number of elongating RNA polymerases or foci in a cell. Foci are so variable in size and intensity and close to the resolution afforded by light microscopy that it remains possible that many have intensities below the level of detection by fluorescence microscopy. Consider an extreme case in which ~25,000 polymerase II molecules (Cox, 1976) are in transcription units randomly distributed throughout euchromatin; this inevitably means

5 min.   10 min.

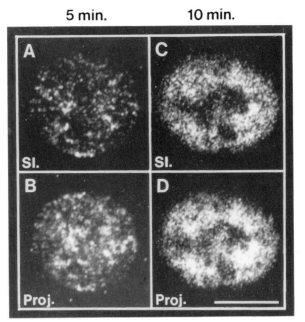

FIG. 7   Transcription sites visualized by confocal microscopy. HeLa cells were permeabilized, incubated with Br-UTP for (A,B) 5 or (C,D) 10 min to extend nascent RNA chains by ~200 and ~400 nucleotides, respectively, and sites containing Br-RNA indirectly immunolabeled. Nine optical slices were taken through a typical nucleus from each sample; A and C show a central slice (Sl.) and B and D the projections (Proj.) of the nine sections onto a single plane. Transcription sites are not spread throughout nuclei, but concentrated in foci or factories. Bar: 5 μm. [From Jackson et al. (1993) with permission of Oxford University Press.]

that there will be local variations in concentration, which—if above a critical threshold—will be detected. By judicious choice of threshold, number of polymerases per transcription unit, and volume occupied by nascent RNA, simulations show that such random distributions can appear as apparently discrete foci (discussed by Jackson et al., 1993). Unfortunately, we have not yet been able to identify a factory by electron microscopy that might correspond to a polymerase II focus seen by light microscopy. However, the nucleolus, which is suspended from the skeleton described above, has a well-characterized ultrastructure and may provide us with a general model for the structure of transcription factories.

C. Nucleolar Factories: RNA Polymerase I

Sites of nucleolar transcription can be seen by light microscopy after incubating permeabilized HeLa cells with α-amanitin (to inhibit RNA polymerase II) and Br-UTP before immunolabeling any incorporated analogue; ~25 discrete nucleolar foci are then visible (Jackson et al., 1993). Again, these nucleolar foci remain after removing most chromatin as described in Fig. 3, confirming that synthetic sites are attached to an underlying structure.

Nucleoli have a well-defined ultrastructure. Several fibrillar centers, which equal the number of polymerase I foci described above, are each surrounded by a dense fibrillar component which is, in turn, embedded in the granular component. In which of these different nucleolar compartments transcription occurs has long been controversial (Jordan, 1991); several factors have contributed to this controversy, including the low resolution offered by autoradiography after incorporation of [^{3}H]uridine and the length of the labeling times required. We have recently resolved this controversy using permeabilized cells and a pulse of Br-UTP short enough to ensure that most transcripts remain at their site of synthesis. After immunogold labeling, gold particles were found over the dense fibrillar component (especially its border with the fibrillar center), indicating that the dense fibrillar component was the synthetic site (Hozák et al., 1994a).

We imagine, then, that the nucleolus is built around fibrillar centers attached to the skeleton; the fibrillar centers store the polymerases, topoisomerases, and other proteins required for transcription. Polymerases directly organize this structure since its formation in mammalian cells is prevented by microinjecting antibodies to the enzyme (Benavente et al., 1987) and yeast mutants with a deleted gene for the second largest subunit of the polymerase assemble several mininucleolar bodies rather than a normal crescent-shaped structure (Oakes et al., 1993). About ~6 active cistrons (each ~5 μm long and packed with ~100 active polymerases) are associated with each fibrillar center in a human nucleolus (see Haaf et al., 1991). Active polymerases—which resist elution in the experiment illustrated in Fig. 3 (Dickinson et al., 1990)—lie on the surface of the fibrillar center and transcription occurs as a transcription unit slides end-on through them over the surface while the nascent rRNA is extruded into the dense fibrillar component. As a promoter emerges from one polymerase, it can soon engage another on the surface. On termination (i.e., when the 3' end of the cistron has slid past a polymerase), the nascent transcript in the dense fibrillar component condenses into the granular component where it completes its maturation. Therefore the dense fibrillar component apparently slides over the surface of the fibrillar center, one end advancing while the other is converted into the granular component and newly inactive enzymes are recycled through the fibrillar center to the growing end of the dense fibrillar component.

This model can be reconciled with the beautiful images of spreads of ribosomal DNA (for a review, see Miller, 1981), if the hypotonic treatment used disperses the granular component and strips transcription units from the surface of the fibrillar center, before spreading gives the well-known "Christmas trees."

It is then attractive to suppose that attached RNA polymerase II molecules are also concentrated in analogous factories, with different templates sliding through polymerases on the surface; nascent transcripts would again be extruded through neighboring processing sites, perhaps then to pass along the skeleton to the cytoplasm (see also Lawrence *et al.*, 1989; Meier and Blobel, 1992).

IV. Replication and Transcription

The location of replication sites relative to transcription sites has been analyzed by incubating permeabilized cells from different stages of the cell cycle with both biotin-dUTP and Br-UTP (Hassan *et al.*, 1994; see also Wansink *et al.*, 1994). During G1 phase, there are ~300 transcription foci in a HeLa nucleus that aggregate on entry into S-phase into ~150 foci; these colocalize with sites of replication. Within about 30 min, many sites solely engaged in transcription reemerge, but the sites involved in replication remain transcriptionally active. Even late during S-phase—when deep heterochromatin is being duplicated—the replication sites remain transcriptionally active (Hassan *et al.*, 1994; but see Wansink *et al.*, 1994). This colocalization of sites of replication and transcription at the G1/S border suggests that transcription sites might seed assembly of replication factories (Hassan and Cook, 1994).

V. Repair Replication

The repair of damage induced in DNA by ultraviolet (UV) light involves excision of damage and then repair synthesis to fill the gap. Therefore sites where this repair takes place can be immunolabeled after incorporation of biotin-dUTP; again sites are not diffusely spread but concentrated in discrete foci at sites that did not reflect the DNA distribution (Jackson *et al.*, 1994a,b). This pattern changed with time; initially intense repair took place at transcriptionally active sites but when transcription became inhibited by the effects of UV irradiation it continued at sites with little transcription. Repair synthesis *in vitro* also occurs in the absence of transcription, showing that the two processes need not be tightly coupled. Repair sites generally contained a high concentration of proliferating cell nuclear antigen but not the tumor suppressor protein, p53.

The repair activity seems not to be as closely associated with the skeleton as the S-phase activity; after treatment with an endonuclease as in Fig. 3, most repaired DNA and the repair foci are removed from beads with the chromatin fragments. However, as electroelution destroys repair activity (but not the S-phase activity), repaired DNA might be attached *in vivo* through a polymerase that was removed by the procedure. So this approach has not allowed us to determine decisively whether repair sites are associated with a skeleton *in vivo*.

VI. Attachment Sites in a Model Loop

It is formally possible that many polymerases could track around a cluster of loops and so resist elution in the experiment illustrated in Fig. 3, if loops were too small to be cut. However, experiments with a model loop confirm that active polymerases resist elution even when the chromatin fiber is shredded (Jackson and Cook, 1993). They also allowed us to define the molecular ties at the bases of the loops.

A plasmid that possessed the SV_{40} origin of replication plus two promoters (i.e., the SV_{40}-early and human $\alpha1$-globin promoters) was transfected into cos7 cells. There it replicated over 2–3 days to give many thousands of minichromosomes. About 1200 minichromosomes per cell resisted elution (i.e., were attached) while any extra copies that were transcriptionally and replicationally inactive could be eluted; this suggests that there are a saturable number of binding sites available to the plasmid. After cutting each attached minichromosome into ~400-bp fragments, ~90% of the contour length could be eluted; the remaining fragment or two were associated with a promoter or an engaged polymerase that was still able to run on along the remaining attached fragment.

These results were surprising and give us a very dynamic view of loop structure. Active minichromosomes might have been attached through a common sequence like an origin of replication or a SAR, but no single sequence was uniquely responsible for attachment; rather different parts of each minichromosome were attached at one or, at most, two points. Moreover, even though both transcription units were highly active, they only contained one polymerase. So we imagine that inert minichromosomes attach to a factory to become active, initially at one of the two promoters. Then a polymerase engages and the minichromosome becomes attached solely through the sequence at the polymerizing site. Finally the template dissociates, but its proximity to binding sites in the factory will mean that it can compete effectively with other minichromosomes for those binding sites. Therefore, at any moment individual templates will be at different stages in this cycle; no one point is always attached and different points have different probabilities of attachment.

VII. Mitotic Chromosomes

The central structural feature of the interphase chromosome is then the chromatin loop, which is tied at its base through an engaged RNA polymerase (or promoter). Does this basic structural feature persist through the gross rearrangements that occur during mitosis? It seems that it does: the contour length of the loops—measured in a physiological buffer—remains unchanged (Jackson *et al.*, 1990) and both RNA polymerases I and II remain bound quantitatively (Matsui *et al.*, 1979), even though nascent transcripts abort (Shermoen and O'Farrell, 1991).

Nucleolar factories may also provide a detailed model for the structural reorganization of mitosis. It is known that as HEp-2 cells enter mitosis, the huge stores of the polymerase I transcription factor, UBF, remain bound at 6–8 of the 10 nucleolar organizing regions (NORs). Subsequently these local concentrations are symmetrically partitioned among daughter cells (Roussel *et al.*, 1993). Here the remnants of the interphase factory—the fibrillar center and its surroundings—persist into metaphase as visible entities, the NORs. However, most nucleolar material—especially the maturing transcripts—are lost from the NORs to form a dense mass of proteins and RNA that coats the surface of the mitotic chromosome (Hernandez-Verdun and Gautier, 1994). It is then attractive to suppose that analogous RNA polymerase II and III factories collapse onto each other to form the axial chromomeres of prophase; subsequent condensation of differently sized loops onto those factories would generate the bands typical of metaphase (Craig and Bickmore, 1993; Cook, 1994). Note that if the vestiges of these factories are segregated symmetrically to daughter cells like UBF, attachments involving RNA polymerases will be inherited by those daughters; as a result, the activity of particular genes will also be inherited (Cook, 1973, 1974).

Like the lamin exoskeleton, the lamin-containing endoskeleton would also depolymerize during mitosis to allow chromosome segregation. On entry into G1 phase, it would repolymerize between factories to provide structural—and so functional—contiguity between factories and perhaps a solid phase for transcript movement. This nucleoskeleton would then play an essential role in integrating nuclear space, in much the same way that cytoplasmic intermediate filaments integrate cytoplasmic space (Lazarides, 1980).

VIII. Relationship to Studies on Other Structures

As the history of subnuclear structures seems to be a history of artifacts, it is foolish to claim that any preparation is free of artifact. However, the attachments of the chromatin fiber to the substructure seen under physiological conditions involve active polymerases that retain the activity that they had *in vivo;* if they had been generated during isolation, we would expect them to lose activity. How, then, do the studies described above relate to those involving other subnuclear structures that have led to different conclusions?

We believe that many of the differences result from the generation of new attachments during isolation. For example, if encapsulated cells are lysed, treated in the ways commonly used to isolate the substructures before returning them to a physiological buffer, and then the loop size measured, we find that different procedures give strikingly different results (Jackson *et al.,* 1990). Exposure to the mild hypotonic conditions generally used to isolate nuclei—usually the first step in the isolation of the substructures— roughly halves loop size, the precise value depending on the detailed ionic constitution of the isolation buffer. This reduction in loop size is not reversed by restoring the tonicity to normal and means that for every attachment existing *in vivo,* about one new attachment is created *in vitro* as nuclei are prepared. High concentrations of magnesium ions also irreversibly fix the chromatin into small loops (i.e., create new attachments). (Note that magnesium ions are chelated with an equal concentration of triphosphate in our physiological buffers.) Subsequent exposure to 2 *M* NaCl, like that used to prepare nuclear matrices, increases loop size from this lower value. Another popular procedure uses the detergent lithium diiodosalicylate (LIS) to generate scaffolds (Mirkovitch *et al.,* 1984). Lysis first increases loop size to ~100 kb, before a stabilization step reduces it to 15 kb. So here, five of six attachments are probably created by the procedure! As so many attachments are being created and destroyed during the isolation of matrices and scaffolds, it is then very difficult to be certain which of the ones studied *in vitro* actually preexisted *in vivo.*

Nucleoids are isolated directly from cells by extraction in 2 *M* NaCl, which increases loop size to 123 kb, so they probably lose only a few attachments. Therefore, their domain structure is very like that found using physiological conditions, with replicating and transcribed sequences (especially promoters and enhancers) being the most closely associated with the substructure (Jackson *et al.,* 1984). It is also reassuring that two completely different sets of conditions (i.e., direct lysis in 2 *M* NaCl to give nucleoids and lysis in physiological buffers) yield structures that have loops of roughly the same size, tied at their bases to the same skeletal structure through engaged RNA polymerases or promoters/enhancers, with active DNA and RNA polymerases attached to the skeleton and that the harsher treatment leads to slightly fewer attachments.

The finding of a lamin endoskeleton is consistent with observations of a distribution only at the periphery, if the dense chromatin generally prevents antibody access to the interior. Note that lamins are frequently associated with the interior of matrices (e.g., Capco *et al.*, 1982; Staufenbiel and Deppert, 1982; Fey *et al.*, 1984; Ludérus *et al.*, 1992) which, like the diffuse nucleoskeleton seen under physiological conditions, contains ribonucleoproteins and, perhaps, RNA (e.g., Long *et al.*, 1979; Fey *et al.*, 1986). It is usually assumed that this internal matrix cannot contain any lamins, and this may be the reason that attempts to isolate it and determine its structure have been so unsuccessful (Jack and Eggert, 1992); however, this failure is easily explicable if the assumption is incorrect. Intermediate filaments (which include the lamins) would then play a role in integrating both cytoplasmic and nuclear space. As lamins also bind DNA and/or chromatin (e.g., Burke, 1990; Glass and Gerace, 1990; Shoeman and Traub, 1990; Höger *et al.*, 1991; Yuan *et al.*, 1991; Glass *et al.*, 1993) and have been implicated in replication (e.g., Jenkins *et al.*, 1993), these results provide a physical basis for additional lamin functions within nuclei (e.g., Traub and Shoeman, 1994).

IX. Conclusions

The results described above suggest that the chromatin loop—the basic structural component of the interphase chromosome—is attached through polymerases and promoters to an underlying lamin-containing endoskeleton that ramifies throughout the nuclear interior. Both RNA and DNA polymerases are attached, either directly or indirectly through a factory, to this skeleton. As a result, loops are dynamic and polymerases static: points of attachment of loops continually change as templates are reeled through the fixed polymerases.

In HeLa cells, replication begins in several hundred small factories with diameters of 100–200 nm. As cells progress through S-phase many of the factories probably become inactive and are quickly dismantled, while others grow and fuse to give the large factories with long axes of 600 nm or more seen late during S-phase. These factories correspond to a subset of the dense structures known to electron microscopists as nuclear bodies.

Transcription of ribosomal genes also occurs in dense nucleolar factories in regions known as the dense fibrillar component surrounding the fibrillar centers. Although analogous factories involving RNA polymerases II and

III have not yet been visualized at the ultrastructural level, discrete focal sites of transcription can be seen by light microscopy suggesting that they, too, will be concentrated within factories.

Immobilization of polymerases forces us to redraw our models for replication and transcription (Cook, 1989, 1991, 1993; Hozák *et al.*, 1995a). It also suggests new ways in which these processes might be controlled, for example at the levels of factory assembly and attachment of the chromatin fiber to the factory.

Acknowledgments

We thank the Cancer Research Campaign and the Wellcome Trust for support.

References

Adachi, Y., and Laemmli, U. K. (1992). Identification of nuclear pre-replication centres poised for DNA synthesis in *Xenopus* egg extracts: Immunolocalization study of replication protein A. *J. Cell Biol.* **119,** 1–15.

Bader, B. L., Magin, T. M., Freudenmann, M., Stumpp, S., and Franke, W. W. (1991). Intermediate filaments formed *de novo* from tail-less cytokeratins in the cytoplasm and in the nucleus. *J. Cell Biol.* **115,** 1293–1307.

Benavente, R., Rose, K. M., Reimer, G., Hugle-Dorr, B., and Scheer, U. (1987). Inhibition of nucleolar reformation after microinjection of antibodies to RNA polymerase I into mitotic cells. *J. Cell Biol.* **105,** 1483–1491.

Beven, A., Guan, Y., Peart, J., Cooper, C., and Shaw, P. (1991). Monoclonal antibodies to plant nuclear matrix reveal intermediate filament-related components within the nucleus. *J. Cell Sci.* **98,** 293–302.

Blow, J. J., and Laskey, R. A. (1986). Initiation of DNA replication in nuclei and purified DNA by a cell-free extract of Xenopus eggs. *Cell (Cambridge, Mass.)* **47,** 577–587.

Blow, J. J., and Laskey, R. A. (1988). A role for the nuclear envelope in controlling DNA replication within the cell cycle. *Nature (London)* **332,** 546–548.

Bouteille, M., Laval, M., and Dupuy-Coin, A. M. (1974). Localization of nuclear functions as revealed by ultrastructural autoradiography and cytochemistry. *In* "The Cell Nucleus." (H. Busch, ed.), Vol. 1, pp. 5–71. Academic Press, New York.

Braithwaite, D. K., and Ito, J. (1993). Compilation, alignment, and phylogenetic relationships of DNA polymerases. *Nucleic Acids Res.* **21,** 787–802.

Brasch, K., and Ochs, R. L. (1992). Nuclear bodies (NBs): A newly 'rediscovered' organelle. *Exp. Cell Res.* **202,** 211–223.

Bravo, R., and Macdonald-Bravo, H. (1987). Existence of two populations of cyclin/proliferating cell nuclear antigen during the cell cycle: Association with DNA replication sites. *J. Cell Biol.* **105,** 1549–1554.

Bridger, J. M., Kill, I. R., O'Farrell, M., and Hutchison, C. J. (1993). Internal lamin structures within G1 nuclei of human dermal fibroblasts. *J. Cell Sci.* **104,** 297–306.

Burke, B. (1990). On the cell-free association of lamins A and C with metaphase chromosomes. *Exp. Cell Res.* **186,** 169–176.

Callan, H. G. (1977). Lampbrush chromosomes. *Proc. R. Soc. London, Ser. B* **214,** 417–448.

Capco, D. G., Wan, K. M., and Penman, S. (1982). The nuclear matrix: Three dimensional architecture and protein composition. *Cell (Cambridge, Mass.)* **29**, 847–858.

Cardenas, M. E., Laroche, T., and Gasser, S. M. (1990). The composition and morphology of yeast nuclear scaffolds. *J. Cell Sci.* **96**, 439–450.

Cardoso, M. C., Leonhardt, H., and Nadal-Ginard, B. (1993). Reversal of terminal differentiation and control of DNA replication: Cyclin A and cdk2 specifically localize at subnuclear sites of DNA replication. *Cell (Cambridge, Mass.)* **74**, 979–992.

Carmo-Fonseca, M., Tollervey, D., Pepperkok, R., Barabino, S. M. L., Merdes, A., Brunner, C., Zamore, P. D., Green, M. R., Hurt, E, and Lamond, A. I. (1991). Mammalian nuclei contain foci which are highly enriched in components of the pre-mRNA splicing machinery. *EMBO J.* **10**, 195–206.

Carter, K. C., Bowman, D., Carrington, W., Fogarty, K., McNeil, J. A., Fay, F. S., and Lawrence, J. B. (1993). A three-dimensional view of precursor messenger RNA metabolism within the nucleus. *Science* **259**, 1330–1335.

Cook, P. R. (1973). Hypothesis on differentiation and the inheritance of gene superstructure. *Nature (London)* **245**, 23–25.

Cook, P. R. (1974). On the inheritance of differentiated traits. *Biol. Rev., Cambridge Philos. Soc.* **49**, 51–84.

Cook, P. R. (1988). The nucleoskeleton: Artifact, passive framework or active site? *J. Cell Sci.* **90**, 1–6.

Cook, P. R. (1989). The nucleoskeleton and the topology of transcription. *Eur. J. Biochem.* **185**, 487–501.

Cook, P. R. (1991). The nucleoskeleton and the topology of replication. *Cell (Cambridge, Mass.)* **66**, 627–635.

Cook, P. R. (1993). A model for reverse transcription by a dimeric enzyme. *J. Gen. Virol.* **74**, 691–697.

Cook, P. R. (1994). RNA polymerase: Structural determinant of the chromatin loop and the chromosome. *BioEssays* **16**, 425–430.

Cook, P. R., and Brazell, I. A. (1975). Supercoils in human DNA. *J. Cell Sci.* **19**, 261–279.

Cook. P. R., and Gove, F. (1992). Transcription by an immobilized RNA polymerase from bacteriophage T7 and the topology of transcription. *Nucleic Acids Res.* **20**, 3591–3598.

Cox, R. F. (1976). Quantitation of elongating form A and B RNA polymerases in chick oviduct nuclei and effects of estradiol. *Cell (Cambrige, Mass.)* **7**, 455–465.

Craig, J. M., and Bickmore, W. A. (1993). Chromosome bands—flavours to savour. *BioEssays* **15**, 349–354.

Dickinson, P., Cook, P. R., and Jackson, D. A. (1990). Active RNA polymerase I is fixed within the nucleus of HeLa cells. *EMBO J.* **9**, 2207–2214.

Eckelt, A., Hermann, H., and Franke, W. W. (1992). Assembly of a tail-less mutant of the intermediate filament protein, vimentin, *in vitro* and *in vivo*. *Eur. J. Cell Biol.* **58**, 319–330.

Fakan, S. (1978). High resolution autoradiographic studies on chromatin functions. *In* "The Cell Nucleus" (H. Busch, ed.), vol 5, pp. 3–53. Academic Press, New York.

Fey, E. G., Wan, K. M., and Penman, S. (1984). Epithelial cytoskeletal framework and nuclear matrix-intermediate filament scaffold: Three dimensional organisation and protein composition. *J. Cell Biol.* **98**, 1973–1984.

Fey, E. G., Krochmalnic, G., and Penman, S. (1986). The nonchromatin substructures of the nucleus: The ribonucleoprotein (RNP)-containing and RNP-depleted matrices analyzed by sequential fractionation and resinless section microscopy. *J. Cell Biol.* **102**, 1654–1665.

Fox, M. H., Arndt-Jovin, D. J., Jovin, T. M., Baumann, P. H., and Robert-Nicoud, M. (1991). Spatial and temporal distribution of DNA replication sites localized by immunofluorescence and confocal microscopy in mouse fibroblasts. *J. Cell Sci.* **99**, 247–253.

Gill, S. R., Wong, P. C., Monteiro, M. J., and Cleveland, D. W. (1990). Assembly properties of dominant and recessive mutations in the small mouse neurofilament (NF-L) subunit. *J. Cell Biol.* **111**, 2005–2019.

Glass, C. A., Glass, J. R., Taniura, H., Hasel, K. W., Blevitt, J. M., and Gerace, L. (1993). The α-helical rod domain of lamins A and C contains a chromatin binding site. *EMBO J.* **12,** 4413–4424.

Glass, J. R., and Gerace, L. (1990). Lamins A and C bind and assemble at the surface of mitotic chromosomes. *J. Cell Biol.* **111,** 1047–1057.

Goldman, A. E., Moir, R. D., Montag-Lowy, M., Stewart, M., and Goldman, R. D. (1992). Pathway of incorporation of microinjected lamin A into the nuclear envelope. *J. Cell Biol.* **119,** 725–735.

Haaf, T., Hayman, D. L., and Schmid, M. (1991). Quantitative determination of rDNA transcription units in vertebrate cells. *Exp. Cell Res.* **193,** 78–86.

Hassan, A. B., and Cook, P. R. (1993). Visualization of replication sites in unfixed human cells. *J. Cell Sci.* **105,** 541–550.

Hassan, A. B., and Cook, P. R. (1994). Does transcription by RNA polymerase play a direct role in the initiation of replication? *J. Cell Sci.* **107,** 1381–1387.

Hassan, A. B., Errington, R. J., White, N. S., Jackson, D. A., and Cook, P. R. (1994). Replication and transcription sites are colocalized in human cells. *J. Cell Sci.* **107,** 425–434.

He, D., Nickerson, J. A., and Penman, S. (1990). Core filaments of the nuclear matrix. *J. Cell Biol.* **110,** 569–580.

Hernandez-Verdun, D., and Gautier, T. (1994). The chromosome perphery during mitosis. *BioEssays* **16,** 179–185.

Höger, T. H., Krohne, G., and Kleinschmidt, J. A. (1991). Interaction of *Xenopus* lamins A and L$_{II}$ with chromatin *in vitro* mediated by a sequence element in the carboxyterminal domain. *Exp. Cell Res.* **197,** 280–289.

Hozák, P., Hassan, A. B., Jackson, D. A., and Cook, P. R. (1993). Visualization of replication factories attached to a nucleoskeleton. *Cell (Cambridge, Mass.)* **73,** 361–373.

Hozák, P., Cook, P. R., Schöfer, C., Mosgöller, W., and Wachtler, F. (1994a). Site of transcription of ribosomal RNA and intra-nucleolar structure in HeLa cells. *J. Cell Sci.* **107,** 639–648.

Hozák, P., Jackson, D. A., and Cook, P. R. (1994b). Replication factories and nuclear bodies: The ultrastructural characterization of replication sites during the cell cycle. *J. Cell Sci.* **107,** 2191–2202.

Hozák, P., Jackson, D. A., and Cook, P. R. (1995a). Role of nuclear structure in DNA replication. *In* "Eukaryotic DNA Replication—Frontiers in Molecular Biology" (J. Blow, ed.). IRL Press, Oxford (in press).

Hozák, P., Sasseville, A. M.-J., Raymond, Y., and Cook, P. R. (1995b). Lamin proteins form an internal nucleoskeleton as well as a peripheral lamina in human cells. *J. Cell Sci.* **108,** 635–644.

Humbert, C., and Usson, Y. (1992). Eukaryotic DNA replication is a topographically ordered process. *Cytometry* **13,** 603–614.

Hutchison, C. J., Cox, R., Drepaul, R. S., Gomperts, M., and Ford, C. C. (1987). Periodic DNA synthesis in cell-free extracts of *Xenopus* eggs. *EMBO J.* **6,** 2003–2010.

Hutchison, C. J., Cox, R., and Ford, C. C. (1988). The control of DNA replication in a cell-free extract that recapitulates a basic cell cycle *in vitro*. *Development (Cambridge, UK)* **103,** 553–566.

Jacks, R. S., and Eggert, H. (1992). The elusive nuclear matrix. *Eur. J. Biochem.* **209,** 503–509.

Jackson, D. A., and Cook, P. R. (1985a). A general method for preparing chromatin containing intact DNA. *EMBO J.* **4,** 913–918.

Jackson, D. A., and Cook, P. R. (1985b). Transcription occurs at a nucleoskeleton. *EMBO J.* **4,** 919–925.

Jackson, D. A., and Cook, P. R. (1986a). Replication occurs at a nucleoskeleton. *EMBO J.* **5,** 1403–1410.

Jackson, D. A., and Cook, P. R. (1986b). A cell-cycle-dependent DNA polymerase activity that replicates intact DNA in chromatin. *J. Mol. Biol.* **192,** 65–76.

Jackson, D. A., and Cook, P. R. (1988). Visualization of a filamentous nucleoskeleton with a 23 nm axial repeat. *EMBO J.* **7**, 3667–3677.

Jackson, D. A., and Cook, P. R. (1993). Transcriptionally-active minichromosomes are attached transiently in nuclei through transcription units. *J. Cell Sci.* **105**, 1143–1150.

Jackson, D. A., McCready, S. J., and Cook, P. R. (1984). Replication and transcription depend on attachment of DNA to the nuclear cage. *J. Cell Sci. Suppl.* **1**, 59–79.

Jackson, D. A., Yuan, J., and Cook, P. R. (1988). A gentle method for preparing cyto- and nucleo-skeletons and associated chromatin. *J. Cell Sci.* **90**, 365–378.

Jackson, D. A., Dickinson, P., and Cook, P. R. (1990). The size of chromatin loops in HeLa cells. *EMBO J.* **9**, 567–571.

Jackson, D. A., Hassan, A. B., Errington, R. J., and Cook, P. R. (1993). Visualization of focal sites of transcription within human nuclei. *EMBO J.* **12**, 1059–1065.

Jackson, D. A., Balajee, A. S., Mullenders, L., and Cook, P. R. (1994a). Sites in human nuclei where DNA damaged by ultra-violet light is repaired: Visualization and localization relative to the nucleoskeleton. *J. Cell Sci.* **107**, 1745–1752.

Jackson, D. A., Hassan, A. B., Errington, R. J., and Cook, P. R. (1994b). Sites in human nuclei were damage induced by ultra-violet light is repaired: Localization relative to transcription sites and concentrations of proliferating cell nuclear antigen and the tumour suppressor protein, p53. *J. Cell Sci.* **107**, 1753–1760.

Jenkins, H., Holman, T., Lyon, C., Lane, B., Stick, R. and Hutchison, C. (1993). Nuclei that lack a lamina accumulate karyophilic proteins and assemble a nuclear matrix. *J. Cell Sci.* **106**, 275–285.

Jordan, E. G. (1991). Interpreting nucleolar structure: Where are the transcribing genes? *J. Cell Sci.* **98**, 437–442.

Kill, I. R., Bridger, J. M., Campbell, K. H. S., Maldonado-Codina, G., and Hutchison, C. J. (1991). The timing of the formation and usage of replicase clusters in S-phase nuclei of human diploid fibroblasts. *J. Cell Sci.* **100**, 869–876.

Kitten, G. T., and Nigg, E. A. (1991). The CaaX motif is required for isoprenylation, carboxyl methylation, and nuclear membrane association of lamin B_2. *J. Cell Biol.* **113**, 13–23.

Lawrence, J. B., Singer, R. H., and Marselle, L. M. (1989). Highly localised tracks of specific transcripts within interphase nuclei visualised by in situ hybridisation. *Cell (Cambridge, Mass.)* **57**, 493–502.

Lazarides, E. (1980). Intermediate filaments as mechanical integrators of cellular space. *Nature (London)* **283**, 249–256.

Leonhardt, H., Page, A. W., Weier, H.-U., and Bestor, T. H. (1992). A DNA targeting sequence directs DNA methyltransferase to sites of DNA replication in mammalian nuclei. *Cell (Cambridge, Mass.)* **71**, 865–873.

Long, B. H., Huang, C.-Y., and Pogo, A. O. (1979). Isolation and characterization of the nuclear matrix in Friend erythroleukaemia cells: Chromatin and hnRNA interactions with the nuclear matrix. *Cell (Cambridges, Mass.)* **18**, 1079–1090.

Ludérus, M. E. E., de Graaf, A., Mattia, E., den Blaauwen, J. L., Grande, M. A., de Jong, L., and van Driel, R. (1992). Binding of matrix attachment regions to lamin B_1. *Cell (Cambridge, Mass.)* **70**, 949–959.

Lutz, R. J., Trujillo, M. A., Denham, K. S., Wenger, L., and Sinensky, M. (1992). Nucleoplasmic localization of prelamin A: Implications for prenylation-dependent lamin A assembly into the nuclear lamina. *Proc. Natl, Acad. Sci. U.S.A.* **89**, 3000–3004.

Manders, E. M. M., Stap J., Brakenhoff, G. J., van Driel, R., and Aten, J. A. (1992). Dynamics of three-dimensional replication patterns during the S-phase, analysed by double labelling of DNA by confocal microscopy. *J. Cell Sci.* **103**, 857–862.

Matsui, S., Weinfeld, H., and Sandberg, A. (1979). Quantitative conservation of chromatin-bound RNA polymerases I and II in mitosis. *J. Cell Biol.* **80**, 451–464.

Mazzotti, G., Rizzoli, R., Galanzi, A., Papa, S., Vitale, M., Falconi, M., Neri, L. M., Zini, N., and Maraldi, N. M. (1990). High-resolution detection of newly-synthesized DNA by anti-bromodeoxyuridine antibodies identifies specific chromatin domains. *J. Histochem. Cytochem.* **38**, 13–22.

Meier, U. T., and Blobel, G. (1992). Nopp140 shuttles on tracks between nucleolus and cytoplasm. *Cell (Cambridge, Mass.)* **70**, 127–138.

Miller, O. L. (1981). The nucleolus, chromosomes, and visualization of genetic activity. *J. Cell Biol.* **91**, 15s–27s.

Mills, A. D., Blow, J. J., White, J. G., Amos, W. B., Wilcock, D., and Laskey, R. A. (1989). Replication occurs at discrete foci spaced throughout nuclei replicating in vitro. *J. Cell Sci.* **94**, 471–477.

Mirkovitch, J., Mirault, M.E., and Laemmli, U. K. (1984). Organisation of the higher-order chromatin loop: Specific DNA attachment sites on nuclear scaffold. *Cell* **39**, 223–232.

Mirzayan, C., Copeland, C. S., and Snyder, M. (1992). The *NUF1* gene encodes an essential coiled-coil related protein that is a potential component of the yeast nucleoskeleton. *J. Cell. Biol.* **116**, 1319–1332.

Nakamura, H., Morita, T., and Sato, C. (1986). Structural organization of replicon domains during DNA synthetic phase in the mammalian nucleus. *Exp. Cell Res.* **165**, 291–297.

Nakayasu, H., and Berezney, R. (1989). Mapping replication sites in the eukaryotic cell nucleus. *J. Cell Biol.* **108**, 1–11.

Oakes, M., Nogi, Y., Clark, M. W., and Nomura, M. (1993). Structural alterations of the nucleolus in mutants of *Saccharomyes cerevisiae* defective in RNA polymerase I. *Mol. Cell. Biol.* **13**, 2441–2455.

O'Keefe, R. T., Henderson, S. C., and Spector, D. L. (1993). Dynamic organization of DNA replication in mammalian cell nuclei: Spatially and temporally defined replication of chromosome-specific α-satellite sequences. *J. Cell Biol.* **116**, 1095–1110.

Roussel, P., Andre, C., Masson, C., Geraud, G., and Hernandez-Verdun, D. (1993). Localization of the RNA polymerase I transcription factor hUBF during the cell cycle. *J. Cell. Sci.* **104**, 327–337.

Shermoen, A. W., and O'Farrell, P. H. (1991). Progression of the cell cycle through mitosis leads to the abortion of nascent transcripts. *Cell* **67**, 303–310.

Shoeman, R. L., and Traub, P. (1990). The in vitro DNA-binding properties of purified nuclear lamin proteins and vimentin. *J. Biol. Chem.* **265**, 9055–9061.

Sobczak-Thepot, J., Harper, F., Florentin, Y., Zindy, F., Brechot, C., and Puvion, E. (1993). Localization of cyclin A at sites of cellular DNA replication. *Exp. Cell Res.* **206**, 43–48.

Staufenbiel, M., and Deppert, W. (1982). Intermediate filament systems are collapsed onto the nuclear surface after isolation of nuclei from tissue culture cells. *Exp. Cell Res.* **138**, 207–124.

Sundin, O., and Varshavsky, A. (1980). Terminal stages of SV40 DNA replication proceed via multiply intertwined catenated dimers. *Cell (Cambridge, Mass.)* **21**, 103–114.

Sundin, O., and Varshavsky, A. (1981). Arrest of segregation leads to accumulation of highly intertwined catenated dimers: Dissection of the final stages of SV40 DNA replication. *Cell (Cambridge, Mass.)* **25**, 659–669.

Traub, P., and Shoeman, R. L. (1994). Intermediate filament and related proteins: Potential activators of nucleosomes during transcription initiation and elongation? *BioEssays* **16**, 349–355.

Tubo, R. A., and Berezney, R. (1987). Identification of 100 and 150S DNA polymerase alpha-primase megacomplexes solubilized from the nuclear matrix of regenerating rat liver. *J. Biol. Chem.* **262**, 5857–5865.

Wansink, D. G., Schul, W., van der Kraan, I., van Steensel, B., van Driel, R., and de Jong, L. (1993). Fluorescent labelling of nascent RNA reveals transcription by RNA polymerase II in domains scattered throughout the nucleus. *J. Cell Biol.* **122**, 283–293.

Wansink, D. G., Manders, E. E. M., van der Kraan, I., Aten, J. A., van Driel, R., and de Jong, L. (1993). RNA polymerase II transcription is concentrated outside replication domains throughout S-phase. *J. Cell Sci.* **107,** 1449–1456.

Yuan, J., Simos, G., Blobel, G., and Georgatos, S. D. (1991). Binding of lamin A to polynucleosomes. *J. Biol. Chem.* **266,** 9211–9215.

Nuclear Domains and the Nuclear Matrix

Roel van Driel, Derick G. Wansink, Bas van Steensel, Marjolein A. Grande,
Wouter Schul, and Luitzen de Jong
E. C. Slater Instituut, University of Amsterdam, 1018
TV Amsterdam, The Netherlands

This overview describes the spatial distribution of several enzymatic machineries and functions in the interphase nucleus. Three general observations can be made. First, many components of the different nuclear machineries are distributed in the nucleus in a characteristic way for each component. They are often found concentrated in specific domains. Second, nuclear machineries for the synthesis and processing of RNA and DNA are associated with an insoluble nuclear structure, called nuclear matrix. Evidently, handling of DNA and RNA is done by immobilized enzyme systems. Finally, the nucleus seems to be divided in two major compartments. One is occupied by compact chromosomes, the other compartment is the space between the chromosomes. In the latter, transcription takes place at the surface of chromosomal domains and it houses the splicing machinery. The relevance of nuclear organization for efficient gene expression is discussed.

KEY WORDS: Interphase nucleus, Transcription, Replication, RNA processing and splicing, Chromatin structure, Chromosomes, Nuclear body.

I. Introduction

The nucleus is the main store of genetic information in eukaryotic cells. Many important processes take place in the nucleus, including replication, transcription, and RNA processing. These processes are precisely controlled by an intricate network of protein factors and are the end points of a variety of signal transduction pathways. For many years, investigators have focused their attention on nuclear processes, such as the molecular mechanisms of control of replication, transcription and RNA processing, and the structure of chromatin. Therefore, it is remarkable that so little attention has been paid to the organelle where it all happens: the nucleus. Paging through different text books on cell biology of the last two decades shows that nuclear structure and functional organization is dealt with only in the most rudimentary form. We know much more about the structure and functional organization of the other major cell compartment, the cytoplasm, than about the nucleus. One reason for this is that the approaches that have been successful for analysis of the cytoplasmic compartment, with its wide variety of membrane-bound organelles, have failed to give us detailed information about the interphase nucleus. Classical electron microscopy techniques have not contributed very much to our understanding of the functional organization of the nucleus, despite the fact that some nuclear compartments have been identified this way, like interchromatin granules and nuclear bodies (Puvion *et al.,* 1994). Another reason why not much attention has been paid to structural aspects of the nucleus is that until recently our understanding of molecular mechanisms that control transcription did not involve other nuclear components than cis-acting sequence elements and trans-acting protein factors. It is only in recent years becoming clear that structural aspects of chromatin and the nucleus are important factors in the control of gene expression (Felsenfeld, 1992).

We first describe what is known about the distribution of various molecular machineries in the nucleus, i.e., those for transcription, RNA processing, and replication. In addition we will briefly portray some nuclear compartments whose function is still unknown. The distribution of chromosomes also will be discussed briefly. In the final section these observations will be integrated into a primitive model. One nuclear compartment, the nucleolus, is not discussed in this overview, but has recently been reviewed elsewhere (Schwarzacher and Wachtler, 1993; Scheer and Weisenberger, 1994).

Three striking general observations concerning nuclear structure can be made from what is discussed in this overview. First, many components of different nuclear machineries show a characteristic, nonhomogeneous distribution. They are often found concentrated in specific domains, while the concentration of these components is low between the domains. This results in typical punctate distributions of nuclear antigens in immunofluorescent labeling experiments. If these punctate distributions are compared for different nuclear components in double labeling experiments, they often do not coincide. The number of domains per nucleus for a particular nuclear component varies widely for different proteins. Some nuclear components are concentrated in only one or at most a few domains (e.g., nuclear bodies; see Section V), whereas for other nuclear factors the number of domains can be over 1000 per nucleus (e.g., clusters of glucocorticoid receptors; see Section II,A). A second general observation is that the nucleus is divided into two major compartments. One is composed of compact chromosomes and the other is the interchromosome space. In the latter compartment transcription takes place at the surface of chromosomes and here the splicing machinery is located. Finally, most components of the nuclear machineries for synthesis of DNA and RNA and processing of RNA do not diffuse freely in the nucleus. Rather, they are associated with the nuclear structure that is called nuclear matrix (also called a nuclear scaffold or karyoskeleton). Evidently, handling of genetic material (i.e., DNA and RNA) is carried out by immobilized enzyme systems.

It is important to realize that we do not yet understand the structure of the fibrogranular, intranuclear matrix in any detail at the molecular level (Stuurman *et al.*, 1992b). So far it is primarily an operationally defined structure that can be visualized by electron microscopic techniques. Also, this intranuclear framework is a structure to which almost all important nuclear machineries are firmly attached. Doubts about the *in vivo* structure and function of the matrix emanate from its instability under certain conditions and from the fact that due to differences in isolation procedures the protein and enzyme composition is found to be different by different investigators (de Jong *et al.*, 1990; Belgrader *et al.*, 1991; van Driel *et al.*, 1991). For each of the nuclear compartments and components discussed below we briefly indicate their interaction with the nuclear matrix. In all cases the precise role of this intranuclear framework remains an enigma that can only be solved if we can unravel its molecular structure.

II. Transcription Machinery and Transcripts

The controlled transcription of genes is one of the major functions of the nucleus. The result of this process determines the properties of the cell, including its differentiation state. Where in the interphase nucleus are actively transcribed genes located? One way to locate the sites of active transcription is to specifically label nascent RNA. This has been done by the incorporation of tritiated nucleotides, followed by high-resolution autoradiography in combination with electron microscopy. Recently, the incorporation of bromouridine (BrU) in RNA has been used to label pre-mRNA. BrU is recognized by specific antibodies that are used for indirect fluorescent labeling or immunogold labeling. Once the distribution of nascent RNA in the nucleus has been determined, and therefore the sites of active transcription, it is important to compare these with the distribution of different components of the transcription and RNA processing machineries, such as RNA polymerases, transcription factors, and proteins involved in splicing. This comparison, as will be discussed, gives insight into the functional organization of the nucleus with respect to the handling of RNA.

A. Sites of Transcription

1. Nascent Pre-mRNA Distribution

A direct way to identify sites in the nucleus where transcription takes place is to visualize nascent RNA. Recently, two groups have independently developed essentially the same procedure to fluorescently label nascent RNA, predominantly pre-mRNA (Jackson *et al.,* 1993; Wansink *et al.,* 1993). During run-on transcription in a permeabilized cell system in the presence of BrUTP (bromouridine triphosphate), or after microinjection of BrUTP into living cells, BrU (bromouridine) is incorporated into nascent RNA. RNA polymerase II incorporates BrUTP with almost the same efficiency in RNA as UTP (Wansink *et al.,* 1993). Subsequently, RNA molecules that contain BrU can be visualized by indirect immunofluorescent labeling using antibodies that recognize this BrU. Labeled RNA is found at a high local concentration in a large number of domains of irregular shape and heterogeneous size. These domains are scattered throughout the nucleus, but are not found in nucleoli [green image in Figs. 1a and b (Color Plate 5) and red image in Figs. 2a–d (Color Plate 6)]. What is labeled is predominantly pre-mRNA (Wansink *et al.,* 1993). Close inspection of high-resolution images obtained by confocal scanning microscopy gives the impression that a network is formed in the nucleoplasm of interconnected domains with high local concentration of nascent RNA (D. G. Wansink, unpublished), similar to what has been reported previously for a component of the splicing machinery (Spector, 1990). Unfortunately, the brominated pre-mRNA is not a substrate for the splicing machinery in the cell (Sierakowska *et al.,* 1989; Wansink *et al.,* 1994b). This makes it impossible to follow the transport of these RNA molecules to the cytoplasm and their concomitant processing.

What is the physiological significance of this distribution of nascent pre-mRNA in the nucleus? Since the number of domains (many hundreds per nucleus) is much smaller than the number of active genes, each fluorescent domain probably contains several transcriptionally active genes. Labeling of nascent pre-mRNA with BrU has shown that active genes are distributed rather evenly throughout the nucleoplasm (except in nucleoli), instead of being concentrated in specific parts of the nucleus, e.g., the nuclear periphery, as has been suggested based on *in situ* nick-translation (Hutchison and Weintraub, 1985; de Graaf *et al.,* 1990).

Two different electron microscopy methods have been used to visualize *in situ* nascent RNA. In one method nascent RNA is radioactively labeled with a short pulse of [^{3}H]uridine, followed by high-resolution *in situ* autoradiography and electron microscopy (Fakan and Puvion, 1980). An alternate, more recent procedure is a modification of the method described above for fluorescent labeling of RNA. Here colloidal gold-labeled antibodies are used against BrU, which is incorporated into nascent RNA molecules. Subsequently, the material is embedded and sectioned (de Graaf *et al.,* 1992; Wansink, 1994). Generally, such preembedment labeling protocols result in excellent labeling efficiencies, but poor preservation of structure. In contrast, immunolabeling on (cryo)sections often results in good preservation of nuclear structure but quite low efficiency of labeling of RNA in the nucleoplasm (Dundr and Raska, 1993). Autoradiography of [^{3}H]uridine-labeled RNA and immunogold labeling of BrU-RNA show that nascent RNA is predominantly found in the interchromatin region, at the surface of electron-dense domains that are rich in chromatin. High-resolution autoradiography associates nascent RNA with perichromatin fibrils (Fakan and Bernhard, 1971; Fakan *et al.,* 1976; Fakan and Nobis, 1978). These fibrils are filamentous structures that contain hnRNA (Bachellerie *et al.,* 1975), hnRNP antigens (Fakan *et al.,* 1984), and RNA polymerase II (Spector *et al.,* 1993; Fakan and Puvion, 1980; Fakan, 1994). The number of perichromatin fibrils in a nucleus correlates with the transcriptional activity of the cell. For instance, the number of perichromatin fibrils markedly decreases after inhibition of transcription with RNA synthesis inhibitors, like α-amanitin (Petrov and Sekeris, 1971). Also, at the onset of transcription during early mammalian embryogenesis the perichromatin fibril concentration rapidly increases (Ordatchenko and Fakan, 1980). These observations support the notion that perichromatin fibrils are a complex of nascent RNA and RNA packaging proteins (hnRNPs). The data indicate that active genes are located at or near the surface of chromatin domains. Nascent pre-mRNA is extruded directly into the interchromatin space, forming fibrillar structures of transcripts packaged as hnRNPs. In these fibrils RNA processing may already have started.

2. RNA Polymerase II

The eukaryotic cell employs three different RNA polymerases, each transcribing different sets of genes. RNA polymerase II (RPII) is responsible for the synthesis of pre-mRNA and snRNA, whereas RNA polymerase I and III are involved in the transcription of ribosomal genes, tRNA genes, and other small RNA genes. Immunolabeling with an antibody against RPII reveals a diffuse distribution of the enzyme in the nucleoplasm (Bona *et al.,* 1981; Jimenez-Garcia and Spector, 1993). The nucleolus excludes RPII. We have found by confocal scanning microscopy that the enzyme is distributed in a fine punctate pattern throughout the nucleus (D. G. Wansink, unpublished). This distribution appears similar to that of nascent pre-mRNA (see previous section).

3. Transcription Factors

Remarkably little is known about the distribution of transcription factors in the interphase nucleus. The few studies that have been published so far show that these regulatory proteins occur in clusters, rather than being diffusely distributed in the nucleoplasm. This is true for the glucocorticoid receptor (GR) in a variety of human and rat tissues and cultured cells (such as HeLa, NRK, T24) and in rat hippocampus tissue (van Steensel *et al.*, 1995). Indirect immunofluorescent labeling of the liganded GR in these nuclei reveals a punctate distribution, which has been interpreted as evidence that this transcription factor is present in the nucleus in many hundreds of clusters of 10 to 100 receptor molecules each (van Steensel *et al.*, 1995). The clusters are evenly distributed throughout the nucleoplasm and are excluded from the nucleoli (Fig. 1e). Upon close inspection of confocal images the clusters seem to be interconnected, apparently forming an intranuclear reticulum. Another transcription factor, the mineralocorticoid receptor (MR), shows a very similar, clustered distribution in hippocampus nuclei (van Steensel *et al.*, 1995). Interestingly, the distributions overlap partially. Certain clusters contain both steroid receptors, whereas others contain either the GR or the MR (van Steensel *et al.*, 1995). This observation parallels our knowledge about the function of the two steroid receptors in hippocampus cells. Certain genes are controlled by the two receptors simultaneously in an antagonistic or synergistic way, whereas other functions involve either of the two receptors alone (Joëls and De Kloet, 1994). It is attractive to speculate that those genes that require the interplay of both receptors are associated with domains that contain both receptors.

A careful analysis by double labeling of nascent pre-mRNA and of the GR in T24 (bladder carcinoma) cells showed that, although both distributions at first sight are similar, the patterns are not related. This indicates that a large fraction of the GR molecules is localized at sites at which little or no transcription is going on (van Steensel *et al.*, 1995). The function of the GR molecules that are clustered at sites where no transcription can be observed remains elusive. Conceivably, these receptor clusters represent storage sites. Alternatively, they are involved in inhibition of transcription rather than transcription activation.

Other transcription factor-like proteins also occur in clusters in the nucleoplasm. This kind of intranuclear distribution has been described for the retinoblastoma tumor suppressor gene product (Mancini *et al.*, 1994) and p53 (Jackson *et al.*, 1994). Also, the PML-RAR fusion protein in acute promyelocytic leukemia cells is distributed in a clustered fashion (Dyck *et al.*, 1994; Koken *et al.*, 1994; Weis et al., 1994) (see Section V,B). Why transcription factors occur as clusters in the nucleus and what the molecular basis is of clustering remains unknown.

B. Distribution of Transcripts in the Nucleus

What route is followed by (pre-)mRNAs after synthesis? One striking observation is that part of the nuclear (pre-)mRNA population is retained inside the nucleus and is probably degraded in the nucleus without ever being transported to the cytoplasm (Harpold et al., 1979, 1981; Salditt-Georgieff et al., 1981; Salditt-Georgieff and Darnell, 1982). An analysis of the localization of specific transcripts in the nucleus by in situ hybridization has shown that different RNAs show different types of distributions. Specific RNA species display an elongated, tracklike, or punctate distribution in the nucleoplasm (Lawrence et al., 1989; Raap et al., 1991; Sui and Spector. 1991; Xing and Lawrence, 1991, 1993; Sibon et al., 1993, 1994; Xing et al., 1993). Tracks have been interpreted as pathways along which the RNA is transported from its site of synthesis to the nuclear envelope (Lawrence et al., 1989; Rosbash and Singer, 1993; Kramer et al., 1994). Some RNAs accumulate near or in nucleoli (Bond and Wold, 1993; Sibon et al., 1994). It is unknown whether these distributions of transcripts reflect a steady state situation of RNA molecules that move to the cytoplasm, or if they should be considered as storage sites of RNA. This uncertainty makes an interpretation of the distributions in terms of RNA transport difficult.

C. The Nuclear Matrix and Transcription

Eukaryotic DNA in interphase (Vogelstein et al., 1980) and in mitotic chromosomes (Paulson and Laemmli, 1977) is organized into topologically independent loops of 5 to 200 kb (Jackson et al., 1990). Matrix-associated regions (or scaffold-associated regions, M/SARs) are a class of genomic sequence elements that form the bases of these loops (Gasser and Laemmli, 1987) and anchor the loops to the nuclear matrix (see also Section VI,C). Another type of interaction of DNA to the nuclear matrix depends on gene activity. Fractionation studies consistently show that transcriptionally active DNA is tightly associated with the nuclear matrix, whereas inactive loci are not (Ciejek et al., 1983; Abulafia et al., 1984; Jost and Seldran, 1984; Jackson and Cook, 1985; Buttyan and Olsson, 1986; Roberge et al., 1988; Cook, 1989; Ogata, 1990; Zambetti et al., 1990; Zlatanova and Van Holde, 1992; Jackson et al., 1993; Gerdes et al., 1994). Also, nascent pre-mRNA is specifically enriched in nuclear matrix preparations (Jackson et al., 1993; Wansink et al., 1993). These studies suggest that the spatial distribution of nascent pre-mRNA in the interphase nucleus is determined at least in part by the nuclear matrix structure. Extraction from nuclei of most of the chromatin and up to about 80% of the nuclear proteins, under conditions that the nuclear matrix remains intact, did not result in a major rearrangement of the sites of transcription (Jackson *et al.,* 1993; Wansink *et al.,* 1993) (Fig. 1c). Little is known about the precise nature of the interaction of transcription complexes with the nuclear matrix. However, since various components of the transcription machinery are found associated with the nuclear matrix, it is likely that the transcription complex itself is responsible for the association of active chromatin with the nuclear matrix.

Nuclear matrix preparations exhibit considerable transcriptional activity (Abulafia *et al.,* 1984; Jackson and Cook, 1985; Razin and Yarovaya, 1985; Dickson *et al.,* 1990). Studies of Cook and co-workers (Jackson and Cook, 1985; Cook, 1989) have shown that nuclear matrix preparations, isolated under close-to-physiological conditions by which about 90% of the DNA is removed, retain most of their transcriptional activity. The picture that emerges is that RNA polymerases are immobilized onto the nuclear matrix structure and DNA reels through the immobilized transcription complexes (Cook, 1994).

In addition to RNA polymerases, transcription factors are also associated with the nuclear matrix. For instance, several members of the steroid receptor superfamily are bound to the matrix (Vollmer *et al.,* 1982; Hora *et al.,* 1986; Kaufmann *et al.,* 1986; Kumara-Siri *et al.,* 1986; Barrack, 1987; Swaneck and Fishman, 1988; van Steensel *et al.,* 1991; Bidwell *et al.,* 1994b) (Fig. 1f). This interaction is dependent on the presence of hormone (Kaufmann *et al.,* 1986; Alexander *et al.,* 1987). Moreover, *in vitro* binding studies show that binding of steroid receptors to the nuclear matrix is of high affinity and saturable (Kirsch and Miller-Diener, 1986; Barrack, 1987; Metzger and Korach, 1990; Schuchard *et al.,* 1991). Furthermore, matrix binding is tissue specific (Barrack, 1983; Metzger and Korach, 1990). Van Steensel *et al.* (1994) have presented evidence that specific protein domains of the glucocorticoid receptor and the androgen receptor are involved in matrix interaction. Like transcription sites, the spatial distribution of clusters of glucocorticoid receptors in the interphase nucleus is determined by the nuclear matrix, since this distribution remains essentially unchanged after removal of most of the chromatin and nuclear proteins, under conditions when the nuclear matrix remains intact (van Steensel *et al.,* 1995). Other transcription factors are also associated with the nuclear matrix. This has been shown by Stein and co-workers as well as others in extensive, systematic studies on matrix interaction of a variety of transcription factors, in relation to the activity of specific genes (Dworetzky *et al.,* 1992; Bidwell *et al.,* 1993, 1994b; Stein and Lian, 1993; Stein *et al.,* 1994; van Wijnen *et al.,* 1993). Also several (proto)oncogene products such as myb (Klempnauer, 1988) and myc (Waitz and Liodl, 1991), and tumor suppressor gene products (Waitz and Loidl, 1991; Mancini *et al.,* 1994) interact with the nuclear matrix.

The observations on the interaction of active genes, RNA polymerase activity, transcription factors, and nascent RNA with the nuclear matrix support the notion that a transcription initiation complex is assembled as an immobilized structure on the nuclear matrix. This results in association of actively transcribed genes to the nuclear matrix in addition to interactions that are not transcription-dependent, e.g., via M/SARs (see Section VI,C). The DNA is moved through the immobilized transcription complex, thereby synthesizing nascent RNA that is firmly associated with the matrix via the transcription complex. Most likely, the newly synthesized RNA remains bound to the nuclear matrix, probably until it has been exported from the nucleus.

III. RNA Processing Machinery

Pre-mRNAs undergo several covalent modifications, i.e., attachment of a 7-methylguanosine nucleotide (a "cap") at their 5' end, addition of a poly(A)-tail at the 3' end, and removal of noncoding introns by the splicing machinery. Furthermore, the RNA is packaged by the binding of a set of hnRNP proteins, polypeptides that interact with the poly(A)-tail, and one or more cap-binding proteins. A comparison of the spatial distribution of some enzyme systems that are responsible for these modifications and the distribution of (pre-)mRNAs has given some insight into the functional organization of the nucleus.

Below we will first examine the distribution of components of the splicing machinery in the interphase nucleus and subsequently compare their localization with that of mRNA and its precursors. Evidence will be summarized showing that these intranuclear distributions are highly dynamic and depend on the physiological state of the cell.

A. Localization of the Splicing Machinery

Splicing of pre-mRNA occurs in a large RNA–protein complex, the spliceosome (Lamond, 1993). It consists of small nuclear ribonucleoprotein particles (snRNPs U1, U2, U4, U6, and U5) and several non-snRNP splicing factors (Lamm and Lamond, 1993). Spliceosomes assemble on pre-mRNA molecules, catalyse the removal of introns, and subsequently dissociate from the mature RNA. Many studies have addressed the question of where in the interphase nucleus the splicing components are located (Spector, 1993a,b). Initial work made use of autoimmune sera that contain antibodies against snRNPs (Spector *et al.*, 1983; Spector, 1984; Nyman *et al.*, 1986; Verheijen *et al.*, 1986). More recently, specific antibodies became available against non-snRNP splicing factors, like SC35 (Fu and Maniatis, 1990), SF2

(Harper and Manley, 1991; Krainer *et al.*, 1991; Lamond, 1991; Zuo and Manley, 1994), and U2AF (Ruskin *et al.*, 1988). Labeling studies with these antibodies have complemented those with autoimmune sera against snRNPs (Fu and Maniatis, 1990; Spector *et al.*, 1991). The antibody against the splicing factor SC35 has become a general marker for the analysis of the distribution of splicing machinery in nuclei (Sahlas *et al.*, 1993, Wansink *et al.*, 1993). *In situ* hybridization experiments, using oligonucleotide probes that are complementary to snRNP RNA molecules, have yielded additional information about the localization of snRNPs (Carmo-Fonseca *et al.*, 1992; Huang and Spector, 1992; Matera and Ward, 1993; Ferreira *et al.*, 1994; Puvion-Dutilleul *et al.*, 1994).

Indirect immunofluorescent labeling studies show that nuclei contain several tens (typically between 20 and 50) of domains in which snRNPs and non-snRNP splicing factors are present in high concentrations [Fig. 1a,b (red image: SC35) and Figs. 2e and 2f (green image: U2 snRNP)]. These domains are often referred to as speckles (Huang and Spector, 1992) and contain local high concentrations of all components of the splicing machinery mentioned above, except the splicing factor U2AF. In addition to the local high concentration in domains, a diffuse and much less intense labeling is often observed between the domains.

Several studies show that snRNPs also are found in a small number of foci, which have been identified by electron microscopy as coiled bodies (Monneron and Bernhard, 1969; Carmo-Fonseca *et al.*, 1991a,b, 1992, 1993; Raska *et al.*, 1991; Huang and Spector, 1992; Spector *et al.*, 1992; Zhang *et al.*, 1992) (Fig. 2e). Interestingly, these foci also contain at least one splicing factor that is not found in SC35-rich nuclear domains: U2AF (Ruskin *et al.*, 1988; Carmo-Fonseca *et al.*, 1991a; Zamore and Green, 1991). In addition, U2AF is found in a diffuse pattern throughout the nucleoplasm (Zamore and Green, 1991). For a discussion about the foci/coiled bodies see Section V,A.

At the ultrastructural level a high concentration of snRNPs is found in clusters of interchromatin granules, perichromatin fibrils, and coiled bodies (Spector *et al.*, 1983, 1991, 1993; Fakan *et al.*, 1984; Puvion *et al.*, 1984; Kopency *et al.*, 1991). In addition, U1 snRNA, but not U2 snRNA, is in HeLa cells localized in a zone that is associated with, but does not coincide with clusters of interchromatin granules (Visa *et al.*, 1993). Interchromatin granules are particles with a diameter between 10 and 30 nm, which are rich in RNA and occur in clusters in regions between chromatin domains (Fakan and Puvion, 1980). These clusters correspond to the speckles observed by indirect immunofluorescent labeling (Carmo-Fonseca *et al.*, 1991b; Spector *et al.*, 1991). Neither nascent RNA (Fakan and Bernhard, 1971; Fakan *et al.*, 1976; Fakan, 1986; Wansink *et al.*, 1993), nor RNA polymerase II (Spector *et al.*, 1993) is found in clusters of interchromatin granules. Also, DNA is absent from these structures (Fakan and Hancock, 1974). Evidence so far supports the notion that the clusters of interchromatin granules (speckles in fluorescent images) represent storage and/or assembly sites of the splicing machinery (see following section). Perichromatin fibrils are thought to contain nascent pre-mRNA molecules that are synthesized at the surface of more or less compact chromatin domains (Fakan and Puvion, 1980, Fakan, 1994). The perichromatin fibrils may well be the structures that are responsible for the diffuse fluorescent labeling by antibodies against splicing components between the speckles (Spector *et al.*, 1991; Sui and Spector, 1991).

B. Sites of Pre-mRNA Splicing

Splicing components are found in three different sites in the interphase nucleus: coiled bodies, clusters of interchromatin granules, and probably perichromatin fibrils, corresponding to the foci, the speckles, and the diffuse immunofluorescent labeling, respectively. There is conflicting evidence about the role of clusters of interchromatin granules/speckles in the splicing process. Some observations suggest that splicing takes place in the speckles. First, microinjection of intron-containing pre-mRNA into nuclei results in accumulation of that RNA in the speckles (Wang *et al.*, 1991). Second, *in situ* hybridization shows that poly(A)-RNA is present in high local concentration in the speckles (Carter *et al.*, 1991; Visa *et al.*, 1993). Third, several nascent transcripts are found in close association with the speckles (Sui and Spector, 1991; Xing *et al.*, 1993), suggesting that transcripts may pass through the speckles before being exported to the cytoplasm (Carter *et al.*, 1993).

In contrast, other experiments support the view that clusters of interchromatin granules/speckles are not the sites where splicing takes place. It was found that most pre-mRNA synthesis occurs at sites distributed throughout the nucleoplasm, mostly outside snRNP-rich domains (speckles) (Wansink *et al.*, 1993) (Fig. 1a,b). There is considerable evidence that splicing already starts cotranscriptionally (Osheim *et al.*, 1985; Fakan *et al.*, 1986; Amero *et al.*, 1992; Matunis *et al.*, 1993; Baurèn and Wieslander, 1994). This leads to the conclusion that splicing occurs at least in part outside the speckles (Wansink *et al.*, 1993). Pulse-chase labeling of RNA with [³H]uridine, followed by autoradiography and electron microscopy, shows that the clusters of interchromatin granules become only slowly labeled, suggesting that the bulk of the (pre-)mRNA does not pass through these subnuclear domains (Fakan and Bernhard, 1971; Fakan *et al.*, 1976; Fakan, 1986). Also, inhibition of RNA polymerase II activity results in depletion of nuclear poly(A)-RNA, except in the speckles, showing that this poly(A)-RNA behaves differently from the bulk of the (pre-)mRNA (Huang *et al.*, 1994). Under these conditions splicing components accumulate in interchromatin granules clusters/speckles (Spector *et al.*, 1993). Finally, the clusters of interchromatin granules are not enriched in hnRNP proteins (Fakan *et al.*, 1984; Leser *et al.*, 1989), suggesting that they contain little hnRNA. Together, these observations suggest that the interchromatin granules/speckles are storage and/or assembly sites for splicing components, rather than sites where splicing takes place. If this conclusion is accepted, the strong poly(A)-RNA signal in the interchromatin granules/speckles remains a puzzling observation. Resolving this problem will give important new information about the function of these structures.

Do coiled bodies play a role as splice sites of pre-mRNAs? Since coiled bodies are not found in all types of cells (Spector *et al.*, 1992) (see Section V,A and Lamond and Carmo-Fonseca, 1993), these structures may not be essential for splicing. Furthermore, coiled bodies do not contain detectable amounts of poly(A)-RNA (Visa *et al.*, 1993; Huang *et al.*, 1994). From these observations it seems unlikely that coiled bodies are the sites of pre-mRNA splicing.

Where is pre-mRNA spliced in the nucleus? Probably, splicing is initiated co-transcriptionally in the perichromatin fibrils, where the pre-mRNA is synthesized. These structures contain snRNPs (Fakan *et al.*, 1984; Puvion *et al* 1984; Kopency *et al.*, 1991) and the splicing factor SC35 (Spector *et al.*, 1991), besides nascent pre-mRNA and hnRNP proteins (Fakan *et al.*, 1984). *In situ* hybridization with intron- and exon-specific probes shows that intron and exon sequences of specific transcripts colocalize near the site of pre-mRNA synthesis, as well as up to relatively large distances away from it (Raap *et al.*, 1991; Lawrence *et al.*, 1993; Xing *et al.*, 1993). Interestingly, beyond a certain distance, the intron has disappeared and only exon probes give a signal. This supports the notion that splicing continues during transport toward the nuclear envelope, although other interpretations cannot be excluded (Rosbash and Singer, 1993).

In conclusion, the domains in the nucleus that contain a high local concentration of splicing components, i.e., the clusters of interchromatin granules/speckles and the coiled bodies, most likely are not the sites where most pre-mRNA is spliced. Rather, these domains are sites of storage and/or assembly. Splicing probably is initiated at the site where the pre-mRNA transcript is synthesized and may be completed during transport to the nuclear pore complex.

C. Dynamics of the Distribution of Splicing Components

The distribution of components of the splicing machinery in the nucleus is highly dynamic. Inhibition of RNA polymerase II results in a marked rounding-up and increase in size of the speckles (Carmo-Fonseca *et al.*, 1991b; Spector *et al.*, 1993) [compare Figs. 2e and 2f (green image: U2 snRNP)]. In addition, virus infection of cells induces a dramatic redistribution of splicing components. Upon infection with herpes virus the splicing machinery accumulates in the interchromatin granules cluster (Martin *et al.*, 1987; Phelan *et al.*, 1993). This is most likely due to a reduction of host mRNA synthesis and the expression of viral genes, most of which do not contain introns. In contrast, infection of cells with adenovirus results in an accumulation of snRNPs and SC35 at sites where massive viral pre-mRNA synthesis and splicing take place (Jimenez-Garcia and Spector, 1993; Puvion-Dutilleul *et al.*, 1994). Remarkably, also the inhibition of the splicing reaction itself, by microinjection of oligonucleotides that interfere with the interaction between snRNA and pre-mRNA, lead to accumulation of SC35 in the speckles and rounding-up of these nuclear domains (Spector *et al.*, 1993; O'Keefe *et al.*, 1994).

The picture that emerges is that the nuclear distribution of splicing components is highly dynamic. The data are consistent with the view that clusters of interchromatin granules/speckles are storage sites from which splicing components can be recruited if required somewhere in the nucleus (Jimenez-Garcia and Spector, 1993; Spector *et al.*, 1993). The speckles are rather uniformly distributed throughout the nucleoplasm. This is an arrangement optimally suited to provide transcripts with splicing factors anywhere in the nucleus.

D. How Do Splicing Components Accumulate in Clusters of Interchromatin Granules/Speckles?

The molecular basis of the formation of nuclear domains, like the clusters of interchromatin granules/speckles, is not known. Two questions should be asked: (1) What is the composition and structure of the individual granules in the cluster of interchromatin granules, and (2) What keeps the granules in a cluster? The answer to both questions is unknown. However, recent experiments begin to shed some light on this problem. Many proteins of the splicing machinery are polypeptides that contain a cluster of alternating serine and arginine residues (SR proteins) (Ge *et al.*, 1991; Fu and Maniatis, 1992; Zhang *et al.*, 1992; Zahler *et al.*, 1993a,b; Cavaloc *et al.*, 1994; Crispino *et al.*, 1994). These serine–arginine-rich domains have been shown to target proteins to the interchromatin granules/speckles (Li and Bingham, 1991). Recently, kinases have been identified that specifically phosphorylate serine residues in the serine–arginine-rich domains of SR proteins (Woppmann *et al.*, 1993; Gui *et al.*, 1994). One such kinase, the human SR-protein kinase SRPK1, is involved in the assembly–disassembly of clusters of interchromatin granules/speckles. Phosphorylation of SR proteins results in disassembly of the speckles and inhibits the *in vitro* splicing reaction (Gui *et al.,* 1994). This strongly suggests that the SR motive has an important function in formation of these nuclear domains. Whether SR-protein domains are instrumental in the formation of the individual granules or play a role in the interactions between the granules remains unknown.

Another observation pertains to the structure of the speckles. Nuclear-mitotic apparatus protein (NuMA) is a coiled-coil protein that is part of the nuclear matrix and also participates in spindle formation during mitosis (Compton *et al.,* 1992; Yang *et al.,* 1992; Tang *et al.,* 1994; Zeng *et al.,* 1994b; Compton and Cleveland, 1994). Evidence has been presented that certain isoforms of NuMA localize to the speckles (Zeng *et al.,* 1994a). This suggests that this structural protein has a skeletal function in the speckle domains in the nucleus. Here we begin to see what may be a glimpse of the interactions that are the molecular basis of the formation of a subnuclear domain. A targeting sequence motive, a regulatory protein kinase, and a skeletal protein are the ingredients of what may be a highly controlled assembly system of a nuclear domain.

E. Role of the Nuclear Matrix in RNA Processing

After its synthesis, pre-mRNA as well as the matured RNA molecules remain associated to the nuclear matrix (Mariman *et al.,* 1982; Ross *et al.,* 1982; Schröder *et al.,* 1987; Xing and Lawrence, 1991; Xing *et al.,* 1993). Similarly, the components of the splicing machinery also are associated with the nuclear matrix (Zeitlin *et al.,* 1987). Nuclear matrices have been prepared that contain preassembled spliceosomes, which can splice pre-mRNA efficiently, provided that some soluble splicing factors are added (Zeitlin *et al.,* 1989). Furthermore, clusters of interchromatin granules (speckles), in which snRNPs and various other components of the splicing machinery accumulate, are retained in nuclear matrix preparations (Wansink, 1994; Zeng *et al.,* 1994a). Interestingly, the nuclear matrix protein NuMA (nuclear mitotic apparatus protein) (He *et al.,* 1990; Compton *et al.,* 1992; Yang *et al.,* 1992; Zeng *et al.,* 1994a) associates specifically with splicing complexes *in vitro* and colocalizes with clusters of interchromatin granules. This suggests that NuMA is involved in binding the splicing machinery to the nuclear matrix. This is true for the splicing components that have accumulated in the speckles. However, it is unclear whether NuMa is also involved in immobilizing the splicing machinery that is located between the speckles, i.e., presumably at the sites of transcription (see Section III,B,C,). Importantly, NuMA may be part of the core filaments of the nuclear matrix (He *et al.*, 1990) by creating potential binding sites for splicing components throughout the nucleoplasm (Zeng *et al.*, 1994b).

These data show that, like the transcription machinery, the splicing machinery is associated with the nuclear matrix. It is likely that splicing is carried out by complexes that are immobilized on the matrix.

IV. DNA Synthesis Machinery

The replication process can be visualized during S-phase by *in vivo* incorporation of radioactive labeled thymidine, followed by autoradiography, or by pulse labeling with bromodeoxyuridine (BrdU), a membrane-permeant nucleoside analogue, which after incorporation in DNA can be visualized by using specific antibodies. Alternatively, replicating DNA can be labeled in permeabilized cells by incorporation of dUTP analogues. Analysis of the distribution of the sites of replication during S-phase has shown that replication occurs in specific domains in the nucleus, as is observed in many different cell types and isolated nuclei (Nakamura *et al.*, 1986; Mills *et al.*, 1989; Nakayasu and Berezney, 1989; Mazzotti *et al.*, 1990; Cox and Laskey, 1991; Fox *et al.*, 1991; Kill *et al.*, 1991; Van Dierendonck *et al.*, 1991; Manders *et al.*, 1992; O'Keefe *et al.*, 1992; Hassan and Cook, 1993; Hozák *et al.*, 1993, 1994; Wansink *et al.*, 1994a). The spatial distribution and size of replication sites change as S-phase progresses.

A. Nuclear Distribution of Replication Domains

Fluorescent labeling of nascent DNA shows that replication occurs in a specific spatial pattern, the texture of which depends on the progression of S-phase (Nakayasu and Berezney, 1989; O'Keefe *et al.*, 1992) (green image in Figs. 2a–2d). In early S-phase replication take place in hundreds of small domains scattered throughout the nucleoplasm. Later in S-phase a smaller number of larger replication domains is observed. Based on morphological criteria three (Nakayasu and Berezney, 1989) to five (O'Keefe *et al.*, 1992) different replication patterns can be discriminated, marking the progression through S-phase. As expected, replication domains have been shown to harbor, in addition to nascent DNA, several proteins involved in the replication process. Examples are cdk2 and cyclin/PCNA (Bravo and Macdonald-Bravo, 1987; Raska *et al.*, 1989; Cardoso *et al.*, 1993; Sobczak-Thepot *et al.*, 1993) and DNA methyltransferase (Leonhardt *et al.*, 1992). The time needed to replicate the DNA in one such domain is about 1 hr and is approximately the same for early and late replicating domains (Manders *et al.*, 1992). After replication is completed in one domain, DNA synthesis starts in another domain. Since there is a close correlation between the timing of replication in S-phase of a locus and its transcriptional activity, the replication pattern somehow reflects the distribution of active and inactive chromatin.

B. Relationship between Replication Domains and Transcription Domains

Early replicating DNA represents mainly active chromatin, whereas late in S-phase constitutive heterochromatin is replicated (Goldman, 1988). If a locus is transcriptionally activated its timing of replication may change from late to early (Forrester *et al.*, 1990). However, also late replicating DNA contains actively transcribed genes (Taljanidisz *et al.*, 1989; Bernard *et al.*, 1992), indicating that the relationship between replication timing and gene activity is not very strict. The relationship between transcription and replication is underscored by the fact that a variety of transcription factors play a role in the replication process (Heintz, 1992; DePamphilis, 1993; Schepers *et al.*, 1993; Li and Botchan, 1994; Wolffe, 1994).

The spatial relationship between replication domains and the site of transcription has been investigated in a synchronized cell population (Hassan *et al.*, 1994) and in nonsynchronized cells (Wansink *et al.*, 1994a) by labeling nascent DNA and nascent RNA simultaneously. Results are partially conflicting in that Hassan *et al.* (1994) found that both types of domains at least in part colocalize, particularly at the G1/S-boundary. Unexpectedly, even during late replication a considerable colocalization of both processes still was observed. This is in contrast to the notion that late replicating DNA is mainly heterochromatin, containing only few active genes. Wansink *et al.* (1994a) found no significant colocalization between transcription and replication domains throughout S-phase [Figs. 2a–2d (green image: replication; red image: transcription)]. Statistical analysis shows that the overlap between the red image (transcription) and the green image (replication) seen in Figs. 2a–2d, particularly during early stages of replication, is not more than is expected for two nonrelated, complex 3D images. It was concluded that transcription is temporarily halted in a domain in which replication takes place (Wansink *et al.*, 1994a).

C. Ultrastructure of Replication Domains

At the ultrastructural level replication domains have been identified as ovoid, electron-dense structures with a diameter of 0.1 to 0.3 μm during early S-phase (Hassan and Cook, 1993; Hozák *et al.*, 1993). These structures (named "replication factories" by Cook and co-workers) contain nascent DNA and PCNA (Hozák *et al.*, 1994) and therefore most likely are the sites of DNA synthesis. They may be the same as the rapidly sedimenting polymerase–primase complexes (100 and 150 S) that can be isolated from regenerating liver tissue (Tubo and Berezney, 1987b). In later states of S-phase the electron-dense structures become larger, probably due to aggregation of individual replication factories. Each replication factory may contain about 40 replication forks (Hozák *et al.*, 1994).

D. Replication and the Nuclear Matrix

The results of many studies show that synthesis of DNA takes place at the nuclear matrix (Dijkwel *et al.*, 1979; Cook, 1991). It is shown that nascent DNA, labeled with a short pulse of radioactive nucleotides followed by autoradiography, or with nucleotide analogues that allow immunochemical detection, is closely associated with the nuclear matrix. After a chase, these labeled regions move away from the matrix in halo preparations and do not cofractionate anymore with the nuclear matrix (Berezney and Coffey, 1975; McCready *et al.*, 1980; Vogelstein *et al.*, 1980; Hozák *et al.*, 1993). Studies on the replication complex in regenerating liver support the idea that the replication complex is matrix associated (Tubo and Berezney, 1987a,b,c; Tubo *et al.*, 1987). The spatial distribution of fluorescently labeled replication sites does not change significantly upon removal of most of the chromatin and nuclear proteins, i.e., the isolation of the nuclear matrix fraction (Nakayasu and Berezney, 1989; Berezney, 1991). In agreement with these observations DNA synthesis activity is found associated with the nuclear matrix (Smith and Berezney, 1982; Collins and Chu, 1987; Tubo *et al.*, 1987; Martelli *et al.*, 1994).

A similar picture emerges for replication as observed for RNA synthesis and splicing. The replication machinery is tightly associated with the nuclear matrix. During replication DNA is reeled through immobilized replication factories (Hozák *et al.*, 1994).

V. Nuclear Domains with Unknown Function

On the basis of structural analyses of interphase nuclei of mammalian cells, a variety of nuclear domains has been identified. Some have been described in detail in the previous sections. In addition, several other discrete nuclear domains have been identified, of which the function is still unknown. These structures often belong to the class of nuclear bodies that show up in the electron microscope as electron-dense structures (Bouteille *et al.*, 1967; Dupuy-Coin *et al.*, 1986; Brasch and Ochs, 1992). Two such structures will be discussed in more detail, the coiled bodies and PML bodies.

A. Coiled Bodies

Coiled bodies are spherical structures with a diameter of 0.1 to 1 μm, that are found in many different cell types and organisms including mammals and plants (Brasch and Ochs, 1992; Lamond and Carmo-Fonseca, 1993; Wu *et al.*, 1993). In the fluorescence microscope these structures light up as bright spots, often called "foci," after indirect immunolabeling with antibodies against one of the antigens that are part of coiled bodies [Fig. 1d (red image) and Figs. 2e and 2f (red/yellow image)]. Generally, cells have between 1 and 10 coiled bodies per nucleus. These structures contain components of the splicing machinery, including all snRNPs (Fakan *et al.*, 1984; Carmo-Fonseca *et al.*, 1991a,b, 1992; Raska *et al.*, 1991; Zamore and Green, 1991; Matera and Ward, 1993). However, quantitatively most of the snRNPs are located outside coiled bodies, i.e., in and between the speckles/SC35-rich nuclear domains, described in Section III. However, coiled bodies are probably not the major sites of pre-mRNA splicing, since they contain little or no poly(A)-RNA (Visa *et al.*, 1993; Huang *et al.*, 1994).

A landmark for coiled bodies is p80-coilin, a 80-kDa protein that is predominantly present in these structures (Raska *et al.*, 1990b, 1991; Andrade *et al.*, 1991). The cDNA that codes for coilin has been cloned and partially analyzed, but the function of the protein is still obscure (Andrade *et al.*, 1991). Interestingly, coiled bodies contain at least one splicing factor (U2AF) that is not found in the speckles/SC35-rich nuclear domains (Ruskin *et al.*, 1988; Carmo-Fonseca *et al.*, 1991a; Zamore and Green, 1991). A remarkable observation is that coiled bodies are often found adjacent to or overlapping with other nuclear domains, particularly PML-containing nuclear bodies (see below) and nuclear domains enriched in the RNA cleavage stimulation factor (CstF) (Schul *et al.*, 1995).

Ultrastructurally, coiled bodies are spherical entities containing coiled, threadlike structures with a cross section of about 50 nm (Bouteille *et al.*, 1967, 1974; Monneron and Bernhard, 1969; Moreno Diaz de la Espina *et al.*, 1980).

The function of coiled bodies is unknown. Several observations indicate that it is a highly dynamic nuclear domain (Andrade *et al.*, 1993; Carmo-Fonesca *et al.*, 1993). For instance, if transcription is inhibited, the snRNPs move away from the coiled bodies and concentrate in the speckles/interchromatin granules (Carmo-Fonseca *et al.*, 1992). Conversely, the number of coiled bodies per cell increases under conditions of high levels of transcriptional activity (Brasch and Ochs, 1992). This suggests some role in transcription or RNA processing. Coiled bodies are often found closely associated with nucleoli (Raska et al., 1990a; Raska and Dundr, 1993; Malatesta et al., 1994; Ochs et al., 1994). Therefore, one of its functions may be the processing of nuclolar transcripts (Jimenez-Garcia et al., 1994). A relationship with nucleoli is supported by the fact that the nucleolar protein fibrillarin is present in coiled bodies (Raska et al., 1990a, 1991). Fibrillarin is an evolutionary highly conserved protein that is functionally related to the processing of pre-rRNA (Tollervey et al., 1991, 1993) and can bind to small nucleolar RNAs (Baserga et al., 1991).

Together, these observations support the view that coiled bodies are involved in the processing of RNA. Alternative functions that cannot be ruled out are assembly and regeneration of snRNPs (Lamond and Carmo-Fonseca, 1993; Wu et al., 1993).

B. PML-Containing Nuclear Bodies

Stuurman et al. (1992a) have described nuclear matrix-associated nuclear bodies that are labeled by a monoclonal antibody, which predominantly recognizes a 126-kDa protein. The antibody has been shown to recognize recombinant PML protein, which is associated with acute promyelocytic leukemia (Reddy et al., 1992; Koken et al., 1994). This disease is characterized by a t(15,17) chromosome translocation, resulting in the fusion of the PML gene with the gene that codes for the α-retinoic acid receptor (RARα). Mutant erythropoietic stem cells are blocked in differentiation beyond the promyelocytic stage, resulting in leukemia. This block can be released by the administration of retinoic acid (Warrell et al., 1994). The 5E10 antigen is present in most rat and human cell lines and tissues.

Cultured cells and cells in tissues contain 10 to 50 of these nuclear bodies per nucleus [Fig. 1d (green image)]. In the electron microscope PML-containing nuclear bodies appear as sharply defined, electron-dense, roughly spherical structures with a diameter of about 0.3 μm (Stuurman et al., 1992a). Interestingly, in leukemic cells the PML-RARα protein is dispersed in over 100 domains throughout the nucleoplasm, rather than concentrated in nuclear bodies. A significant number of these domains colocalize with sites of pre-mRNA snythesis (Grande et al., 1995). Administration of retinoic acid, the ligand for both the RARα receptor and the fusion protein, results in the reformation of the nuclear body-like distribution (Dyck et al., 1994; Koken et al., 1994; Weis et al., 1994). Sometimes, the PML antigen is associated with thread-like structures in the nucleus (Stuurman et al., 1992a), as is observed for nuclear bodies more often

(Lane, 1969; Masurovsky *et al.*, 1970; Bouteille *et al.*, 1974). The significance of this observation is unknown.

Several other nuclear proteins have been found to be localized in the PML-containing nuclear bodies. Among them are antigens recognized by sera from patients suffering from autoimmune disease (Szostecki *et al.*, 1987; Ascoli and Maul, 1991; Fusconi *et al.*, 1991). Some of these proteins, including PML itself, contain sequence motives that suggest a transcription factor-like function (Szostecki *et al.*, 1990; Reddy *et al.*, 1992; Xie *et al.*, 1993). These observations may be interpreted as an indication that PML-containing nuclear bodies are involved in the control of transcription. However, it has been shown that neither pre-mRNA synthesis nor splicing components colocalize with these nuclear bodies (Stuurman *et al.*, 1992a; Weis *et al.*, 1994; Grande *et al.*, 1995). Interestingly, PML-containing nuclear bodies dissociate upon infection with herpes simplex virus (Maul *et al.*, 1993; Maul and Everett, 1994). Evidence has been presented that PML-containing nuclear bodies are also found in the cytoplasm (Stuurman *et al.*, 1992a) and that the protein may shuttle between this compartment and the nucleus (Stuurman, 1991). This suggests a function in nucleocytoplasmic transport. The function of PML bodies in the nucleus remains elusive.

C. Nuclear Bodies and the Nuclear Matrix

Little is known about the interaction of coiled bodies with the nuclear matrix. PML bodies, however, are tightly associated with the nuclear matrix (Stuurman *et al.*, 1992a). The function of this interaction has not been explored yet.

VI. Chromosome Domains

So far we have dealt with a variety of nuclear components and functions, except the most important one, the chromatin. How are the chromosomes arranged in the interphase nucleus and how does this arrangement relate to the position of other nuclear components? When after mitosis the nuclei of the daughter cells are assembled, the metaphase chromosomes are thought to decondense, except areas that are recognized as heterochromatin. The arrangement of chromosomes in interphase nuclei is not random, as is clear, for instance, from the formation of nucleoli that contain ribosomal genes located on different chromosomes. It is likely that the dynamic structure of chromatin is an important factor in the control of gene expression (Felsenfeld, 1992). For that reason the static and dynamic structure and arrangement of chromatin in the interphase nucleus is rapidly attracting more attention.

A. Chromosome Territories

In contrast to earlier models that advocated a dispersed distribution of interphase chromosomes (Comings, 1968), more recent data show that chromosomes remain compact structures in the interphase nucleus. Evidence for this can be derived from two types of experiments. First, local damaging of DNA in interphase nuclei by microbeam UV irradiation results in repair in only a few chromosomes. During the repair reaction [³H]thymidine or nucleotide analogues are incorporated into the damaged chromosomes, which can be visualized during the next metaphase by autoradiography (Cremer et al., 1982a,b; Hens et al., 1983) or immunolabeling (Hens et al., 1983; Cremer et al., 1984), respectively. Results show that each microirradiated area (the diameter of the microbeam was 1 to 2 μm at its focal point) contains parts of only a few chromosomes. Conversely, irradiation of individual chromosomes during metaphase and monitoring subsequent repair in interphase showed that the damaged regions remained compact, rather than unfolding extensively (Cremer et al., 1984). These results support the notion that in interphase individual chromosomes occupy compact territories that are not invaded by other chromosomes (Cremer et al., 1993).

More recently, in situ hybridization of complete chromosomes has strengthened this idea. Initially, experiments were carried out on cell hybrids, containing, for instance, one human chromosome in the nucleus of another species, allowing the use of species-specific genomic probes (Manuelidis, 1985; Schardin et al., 1985; Pinkel et al., 1986). Later on, chromosome painting (Lengauer et al., 1990, 1991) allowed the visualization of individual chromosomes in their normal environment, i.e., in the endogenous interphase nucleus. Experiments on animal cells (Rappold et al., 1984; Cremer et al., 1988; Lichter et al., 1988; Pinkel et al., 1988; Lengauer et al., 1991; Bischoff et al., 1993) and on plant cells (Heslop-Harrison and Bennett, 1990; Leitch et al., 1990) show that each chromosome in interphase occupies its own nuclear territory.

In support of this notion, Manuelidis and co-workers (Manuelidis, 1990; Manuelidis and Chen, 1990) have presented an interesting model on higher order chromatin structure. The model allows for limited, local unfolding of chromatin in order to become transcriptionally active. Attempts have been made to quantify this unfolding by estimating the difference in volume between the active and the inactive X chromosome in human amniotic fluid cells (Bischoff et al., 1993). Results indicate that the active X chromosome has about 25% larger volume than the inactive one (Bischoff *et al.,* 1993; Cremer *et al.,* 1993). This shows that unfolding of chromosomes in interphase is rather limited.

These results suggest that the interphase nucleus is filled with compact chromosomes. The spatial arrangement of the chromosomes is probably not fixed and is possibly dynamic. Occasionally a physical interaction between homologous chromosomes is observed (Hiraoka *et al.,* 1993) or suspected (Pirrotta, 1990; Wu, 1993), whereas in other experiments no evidence for pairing has been found (Cremer *et al.,* 1982a,b; Hens *et al.,* 1983). It remains to be seen whether interphase chromosomes occupy a single, impenetrable territory. It may be that a chromosome consists of subdomains, each consisting of a folded chromatin domain, possibly corresponding to a functional chromatin unit.

B. Relationship between Chromosome Domains and Other Nuclear Components

If the chromosomes are forming compact, recognizable domains in the interphase nucleus, what then is their relationship to domains that are enriched in machineries for transcription, RNA processing, and RNA transport? First, transcription appears to take place almost exclusively at the surface of chromatin domains. Autoradiography of tritiated RNA shows that nascent transcripts are associated with perichromatin fibrils (see Section II,A), structures at the surface of chromatin domains. Immunogold labeling of BrU-containing nascent pre-mRNA confirms that RNA synthesis by RNA polymerase II is predominantly localized at the surface of compact chromatin domains (Wansink, 1994; Wansink *et al.,* 1995). Similarly, the active HPV18 viral genome, integrated in chromosome 8 of HeLa cells, is found almost exclusively at the surface of that chromosomal territory (Zirbel *et al.,* 1993). Together these observations suggest that chromatin is folded so that active loci are at the surface of chromosome (sub)domains (Cremer *et al.,* 1993).

The bulk of the machinery for RNA-splicing is found in clusters of interchromatin granules, localized between chromatin-rich areas (Cremer *et al.,* 1993). A quantitative analysis of the relative position of a splicing component and individual chromosomes shows that both distributions exclude each other, i.e., the antigen was almost never found inside a chromosome territory (Zirbel *et al.,* 1993). These observations make it attractive to speculate that RNA synthesis, processing, and transport occur in the probably continuous reticulum that is formed by the regions that are outside chromosome territories (Cremer *et al.,* 1993). An interchromatin channel network has also been described by Bingham and co-workers (Zacher *et al.*, 1993; Kramer *et al.*, 1994).

C. Chromatin and the Nuclear Matrix

Chromatin is associated to the nuclear matrix via at least two types of interactions. One involves A+T-rich sequences, named matrix-associated regions (or scaffold-associated regions, M/SARs) (Mirkovitch *et al.*, 1984; Gasser *et al.*, 1989; van Driel *et al.*, 1991; Zlatanova and Van Holde, 1992). These are assumed to form the bases of DNA loops bound to the nuclear matrix. The other type of interaction between matrix and chromatin is most likely due to the formation of an active transcription complex on the nuclear matrix (see Section II,C). Several proteins, some of them integral components of the nuclear matrix, have been identified as polypeptides that specifically bind M/SAR sequences. Among them are the rat liver nuclear lamins A, B, and C (Ludérus *et al.*, 1992, 1994), a protein that turned out to be identical to hnRNP-U (Romig *et al.*, 1992; von Kries *et al.*, 1994a), a protein with unknown function (ARBP) (von Kries *et al.*, 1991, 1994b) and the thymus-specific protein SATB1 (Dickinson *et al.*, 1992; Nakagomi *et al.*, 1994). In addition, a variety of other proteins have been described that bind preferentially but not exclusively to M/SAR sequences. Examples are histone H1 and topoisomerase II (Izaurralde *et al.*, 1989; Käs *et al.*, 1993; Zhao *et al.*, 1993). M/SAR sequence elements are also often located close to regulatory sequences (Cockerill and Garrard, 1986; Gasser and Laemmli, 1986a; Cockerill and Yuen, 1987; Jarman and Higgs, 1988), are able to protect transgenes from integration site position effects (Stief *et al.*, 1989; Phi-Van *et al.*, 1990; Bonifer *et al.*, 1991; Breyne et al., 1992; Schöffl *et al.*, 1993), and seem to delimit functional chromatin domains (Mirkovitch *et al.*, 1984; Gasser and Laemmli, 1986a,b; Bode and Maass, 1988; Phi-Van and Strätling, 1988; Levy-Wilson and Fortier, 1989). Evidently, M/SAR sequences unite interesting structural and functional properties.

VII. Concluding Remarks

Three main observations dominate our view on the functional organization of the nucleus. First, many nuclear functions and components occupy well-defined domains in the nucleus, instead of being diffusely distributed throughout the nucleoplasm. This is, among others, true for transcription sites, transcription factors, splicing machinery, and replication sites. Second, many components of the molecular machineries for replication, RNA synthesis, RNA splicing, and replication are immobilized by binding to the nuclear matrix. Evidently, all handling of DNA and RNA occurs on matrix-associated enzyme complexes. Third, each chromosome in the interphase nucleus occupies a rather compact domain. This last observation divides nuclear space into two major compartments, i.e., an interchromosomal space and space occupied by compact chromosomes. In this latter compartment transcription occurs at the surface of these chromatin domains. Nascent RNA molecules are extruded into the interchromosomal space. The RNA processing and RNA transport machineries appear confined to this compartment. This leads to the conclusion that here the processing and the transport of RNA molecules take place. It is attractive to speculate that the interchromatin space forms a continuous reticulum. This would fit the reticular spatial distribution of a number of nuclear components, like splicing components (Spector, 1990).

How does the nuclear matrix fit into this picture? Since we know that several nuclear components in the interchromatin space are associated with the nuclear matrix, it is fair to assume that the matrix is situated in the interchromatin space. Such distribution could accommodate the two types of interactions that exist between the nuclear matrix and DNA, i.e., one that is present only if an active transcription complex is formed, and one in which M/SARs are involved and that is transcription independent. These interactions may all occur predominantly at the surface of the chromosome domains. Does the nuclear matrix penetrate into the chromosome territories or is it confined to the interchromosomal space in the interphase nucleus? Visual inspection of electron micrographs of nuclear matrix preparations does not give the impression that large holes are present that may represent chromosome territories (He *et al.,* 1990; Nickerson and Penman, 1992), suggesting that the question should be answered positively. One should realize, however, that a chromosome territory is probably not a solid mass of packed chromatin. Rather, it is likely to consist of subdomains, each embodying a "lump" of more or less compact chromatin. It is attractive to assume that such a lump represents a functional chromatin domain, such as a cluster of coregulated genes.

The molecular mechanisms that underlie the clustering of nuclear components in recognizable domains, like the nucleolus, clusters of interchromatin granules, and nuclear bodies, are an enigma. It is unknown which interactions are responsible for the formation of these subnuclear structures. In the case of the clusters of interchromatin granules it is likely that the interactions between a heterogeneous set of serine-aginine-rich (SR) proteins and the nuclear matrix protein NuMA play a structural role. A second mystery is the structure of the nuclear matrix. Despite many attempts no major structural components have been identified yet.

Summarizing, the emerging picture is that of a nucleus that is a highly structured and probably is a very dynamic organelle. It is structured so that transcription, RNA processing, and RNA transport can be carried out very efficiently because of the small accessible volume for the different machineries (the interchromosomal space). Transcription, RNA processing, and probably also RNA transport are tightly coupled processes, probably allowing substrate channeling and by that creating an efficient transcript handling system.

Acknowledgments

The authors thank Drs. J. A. Aten, B. Humbel, F. F. M. Cremer, J. G. J. Bauman, O. C. M. Sibon, E. M. M. Manders, and A. P. Otte for critical discussions. Part of the work reported here was supported by the Foundation for Life Sciences (S.L.W.) and the Foundation for Medical Science (G.M.W.), which are subsidized by the Netherlands Organization for the Advancement of Scientific Research (N.W.O.).

References

Abulafia, R., Ben-Ze'ev, A., Hay, N., and Aloni, Y. (1984). Control of late simian virus 40 transcription by the attenuation mechanism and transcriptionally active ternary complexes are associated with the nuclear matrix. *J. Mol. Biol.* **172**, 467–487.

Alexander, R. B., Greene, G. L., and Barrack, E. R. (1987). Estrogen receptors in the nuclear matrix: Direct demonstration using monoclonal antireceptor antibody. *Endocrinology (Baltimore)* **120**, 1851–1857.

Amero, S. A., Raychaudhuri, G., Cass, C. L., van Venrooij, W. J., Habets, W. J., Krainer, A. R., and Beyer, A. L. (1992). Independent deposition of heterogeneous nuclear ribonucleoproteins and small nuclear ribonucleoproteins at sites of transcription. *Proc. Natl. Acad. Sci. U.S.A.* **89**, 8409–8413.

Andrade, L. E. C., Chan, E. K. L., Raska, I., Peebles, C. L., Roos, G., and Tan, E. M. (1991). Human autoantibody to a novel protein of the nuclear coiled body—immunological characterization and cDNA cloning of p80-coilin. *J. Exp. Med.* **173**, 1407–1419.

Andrade, L. E. C., Tan, E. M., and Chan, E. K. L. (1993). Immunocytochemical analysis of the coiled body in the cell cycle and during cell proliferation. *Proc. Natl. Acad. Sci. U.S.A.* **90**, 1947–1951.

Ascoli, C. A., and Maul, G. G. (1991). Identification of a novel nuclear domain. *J. Cell Biol.* **112**, 785–795.

Bachellerie, J.-P., Puvion, E., and Zalta, J.-P. (1975). Ultrastructural organization and biochemical characterization of chromatin-RNA-protein complexes isolated from mammalian cell nuclei. *Eur. J. Biochem.* **58**, 327–337.

Barrack, E. R. (1983). The nuclear matrix of the prostate contains acceptor sites for androgen receptors. *Endocrinology (Baltimore)* **113**, 430–432.

Barrack, E. R. (1987). Steroid hormone receptor localization in the nuclear matrix: Interaction with acceptor sites. *J. Steroid Biochem.* **27**, 115–121.

Baserga, S. J., Yang, X. D. W., and Steitz, J. A. (1991). An intact box-C sequence in the U3 snRNA is required for binding of fibrillarin, the protein common to the major family of nucleolar snRNPs. *EMBO J.* **10,** 2645–2651.

Baurèn, G., and Wieslander, L. (1994). Splicing of Balbiani ring 1 gene pre-mRNA occurs simultaneously with transcription. *Cell (Cambridge, Mass.)* **76,** 183–192.

Belgrader, P., Siegel, A. J., and Berezney, R. (1991). A comprehensive study on the isolation and characterization of the HeLa S3 nuclear matrix. *J. Cell Sci.* **98,** 281–291.

Berezney, R. (1991). Visualizing DNA replication sites in the cell nucleus. *Semin. Cell Biol.* **2,** 103–116.

Berezney, R., and Coffey, D. S. (1975). Nuclear protein matrix: Association with newly synthesized DNA. *Science* **189,** 291–293.

Bernard, M., Pallotta, D., and Pierron, G. (1992). Structure and identity of late-replicating and transcriptionally active gene. *Exp. Cell Res.* **201,** 506–513.

Bidwell, J. P., van Wijnen, A. J., Fey, E. G., Dworetzky, S., Penman, S., Stein, J. L., Lian, J. B., and Stein, G. S. (1993). Osteocalcin gene promoter-binding factors are tissue-specific nuclear matrix components. *Proc. Natl. Acad. Sci. U.S.A.* **90,** 3162–3166.

Bidwell, J. P., Fey, E. G., van Wijnen, A. J., Penman, S., Stein, J. L., Lian, J. B., and Stein, G. S. (1994a). Nuclear matrix proteins distinguish normal diploid osteoblasts from osteosarcoma cells. *Cancer Res.* **54,** 28–32.

Bidwell, J. P., van Wijnen, A. J., Fey, E. G., Merriman, H., Penman, S., Stein, J. L., Stein, G. S., and Lian, J. B. (1994b). Subnuclear distribution of the vitamin-D receptor. *J. Cell. Biochem.* **54,** 494–500.

Bischoff, A., Albers, J., Kharboush, I., Stelzer, E., Cremer, T., and Cremer, C. (1993). Differences of size and shape of active and inactive X-chromosome domains in human amniotic fluid cell nuclei. *Microsc. Res. Tech.* **25,** 68–77.

Bode, J., and Maass, K. (1988). Chromatin domain surrounding the human interferon-beta gene as defined by scaffold attached regions. *Biochemistry* **27,** 4706–4711.

Bona, M., Scheer, U., and Bautz, E. K. F. (1981). Antibodies to RNA polymerase II (B) inhibit transcription in lampbrush chromosomes after microinjection into living amphibian oocytes. *J. Mol. Biol.* **151,** 81–99.

Bond, V. C., and Wold, B. (1993). Nucleolar localization of *myc* transcripts. *Mol. Cell. Biol.* **13,** 3221–3230.

Bonifer, C., Yannoutsis, N., Grosveld, F., and Sippel, A. E. (1991). Determination of DNA-elements necessary for macrophage specific and position independent expression of the chicken lysozyme gene in transgenic mice. *Hum. Gene Transfer* **219,** 267–269.

Bouteille, M., Kalifat, S. R., and Delarue, J. (1967). Ultrastructural variations of nuclear bodies in human diseases. *J. Ultrastruct. Res.* **19,** 474–486.

Bouteille, M., Laval, M., and Dudpuy-Coin, A. M. (1974). Localization of nuclear functions as revealed by ultrastructural autoradiography and cytochemistry. *In* "The Cell Nucleus" (H. Busch, Ed.), pp. 3–71. Academic Press, New York.

Brasch, K., and Ochs, R. L. (1992). Nuclear bodies (NBS)—a newly rediscovered organelle. *Exp. Cell Res.* **202,** 211–223.

Bravo, R., and Macdonald-Bravo, H. (1987). Existence of two populations of cyclin/proliferating cell nuclear antigen during the cell cycle: Association with DNA replication sites. *J. Cell Biol.* **105,** 1549–1554.

Breyne, P., van Montagu, M., De Picker, A., and Gheysen, G. (1992). Characterization of a plant scaffold attachment region in a DNA fragment that normalizes transgene expression in tobacco. *Plant Cell* **4,** 463–471.

Buttyan, R., and Olsson, C. A. (1986). Prediction of transcriptional activity based on gene association with the nuclear matrix. *Biochem. Biophys. Res. Commun.* **138,** 1334–1340.

Cardoso, M. C., Leonhardt, H., and Nadal-Ginard, B. (1993). Reversal of terminal differentiation and control of DNA replication: Cyclin A and cdk2 specifically localize at subnuclear sites of DNA replication. *Cell (Cambridge, Mass.)* **74,** 979–992.

Carmo-Fonseca, M., Tollervey, D., Pepperkok, R., Barabino, S. M. L., Merdes, A., Brunner, C., Zamore, P. D., Green, M. R., Hurt, E., and Lamond, A. I. (1991a). Mammalian nuclei contain foci which are highly enriched in components of the pre-mRNA splicing machinery. *EMBO J.* **10,** 195–206.

Carmo-Fonseca, M., Pepperkok, R., Sproat, B. S., Ansorge, W., Swanson, M. S., and Lamond, A. I. (1991b). In vivo detection of snRNP-rich organelles in the nuclei of mammalian cells. *EMBO J.* **10,** 1863–1873.

Carmo-Fonseca, M., Pepperkok, R., Carvalho, M. T., and Lamond, A. I. (1992). Transcription-dependent colocalization of the U1, U2, U4/U6, and U5 snRNPs in coiled bodies *J. Cell Biol.* **117,** 1–14.

Carmo-Fonseca, M., Ferreira, J., and Lamond, A. I. (1993). Assembly of snRNP-containing coiled bodies is regulated in interphase and mitosis—evidence that the coiled body is a kinetic nuclear structure. *J. Cell Biol.* **120,** 841–852.

Carter, K. C., Taneja, K. L., and Lawrence, J. B. (1991). Discrete nuclear domains of poly(A) RNA and their relationship to the functional organization of the nucleus. *J. Cell Biol.* **115,** 1191–1202.

Carter, K. C., Bowman, D., Carrington, W., Fogarty, K., McNeil, J. A., Fay, F. S., and Lawrence, J. B. (1993). A three-dimensional view of precursor messenger RNA metabolism within the mammalian nucleus. *Science* **259,** 1330–1335.

Cavaloc, Y., Popielarz, M., Fuchs, J. P., Gattoni, R., and Stevenin, J. (1994). Characterization and cloning of the human splicing factor 9G8: A novel 35 KDa factor of the serine/arginine protein family. *EMBO J.* **13,** 2639–2649.

Ciejek, E. M., Tsai, M.-J., and O'Malley, B. W. (1983). Actively transcribed genes are associated with the nuclear matrix. *Nature (London)* **306,** 607–609.

Cockerill, P. N., and Garrard, W. T. (1986). Chromosomal loop anchorage of the kappa immunoglobulin gene occurs next to the enhancer in a region containing topoisomerase II sites. *Cell (Cambridge, Mass.)* **44,** 273–282.

Cockerill, P. N., and Yuen, M. H. (1987). The enhancer of the immunoglobulin heavy chain locus is flanked by presumptive chromosomal loop anchorage elements. *J. Biol. Chem.* **262,** 5394–5397.

Collins, J. M., and Chu, A. K. (1987). Binding of the DNA polymerase alpha-DNA primase complex to the nuclear matrix in HeLa cells. *Biochemistry* **26,** 5600–5607.

Comings, D. E. (1968). The rationale for an ordered arrangement of chromatin in the interphase nucleus. *Am. J. Hum. Genet.* **20,** 440–460.

Compton, D. A., and Cleveland, D. W. (1994). NuMA, a nuclear protein involved in mitosis and nuclear reformation. *Curr. Opin. Cell Biol.* **6,** 343–346.

Compton, D. A., Szilak, I., and Cleveland, D. W. (1992). Primary structure of NuMA, an intranuclear protein that defines a novel pathway for segregation of proteins at mitosis. *J. Cell Biol.* **116,** 1395–1408.

Cook, P. R. (1989). The nucleoskeleton and the topology of transcription. *Eur. J. Biochem.* **185,** 487–501.

Cook, P. R. (1991). The nucleoskeleton and the topology of replication. *Cell (Cambridge, Mass.)* **66,** 627–635.

Cook, P. R. (1994). RNA polymerase: Structural determinant of the chromatin loop and the chromosome. *BioEssays* **16,** 425–430.

Cox, L. S., and Laskey, R. A. (1991). DNA replication occurs at discrete sites in pseudonuclei assembled from purified DNA in vitro. *Cell (Cambridge, Mass.)* **66,** 271–275.

Cremer, T., Cremer, C., Bauman, H., Luedtke, E. K., Sperling, K., Teuber, V., and Zorn, C. (1982a). Rabl's model of the interphase chromosome arrangement tested in Chinese hamster cells by premature chromosome condensation and laser-UV-microbeam experiments. *Hum. Genet.* **60,** 46–56.

Cremer, T., Cremer, C., Schneider, T., Bauman, H., Hens, L., and Kirsch-Volders, M. (1982b). Analysis of chromosome positions in the interphase nucleus of Chinese hamster cells by laser-UV-microirradiation experiments. *Hum. Genet.* **62,** 201–209.

Cremer, T., Baumann, H., Nakanishi, K., and Cremer, C. (1984). Correlation between interphase and metaphase chromosome arrangements as studied by laser-UV-microbeam experiments. *Chromosomes Today* **8,** 203–206.

Cremer, T., Tesin, D., Hopman, A. H. N., and Manuelidis, L. (1988). Rapid interphase and metaphase assessment of specific chromosomal changes in neuroectodermal tumor cells by in situ hybridization with chemical modified DNA probes. *Exp. Cell Res.* **176,** 199–220.

Cremer, T., Kurz, A., Zirbel, R., Dietzel, S., Rinke, B., Schrock, E., Speicher, M. R., Mathieu, U., Jauch, A., Emmerich, P., Scherthan, H., Ried, T., Cremer, C., and Lichter, P. (1993). Role of chromosome territories in the functional compartmentalization of the cell nucleus. *Cold Spring Harbor Symp. Quant. Biol.* **58,** 777–792.

Crispino, J. D., Blencowe, B. J., and Sharp, P. A. (1994). Complementation by SR proteins of pre-mRNA splicing reactions depleted of U1 snRNP. *Science* **265,** 1866–1869.

de Graaf, A., van Hemert, F., Linnemans, W. A. M., Brakenhoff, G. J., de Jong, L., van Renswoude, J., and van Driel, R. (1990). Three-dimensional distribution of DNAse I-sensitive chromatin regions in interphase nuclei of embryonal carcinoma cells. *Eur. J. Cell Biol.* **52,** 135–141.

de Graaf, A., Humbel, B. M., Stuurman, N., Bergen en Henegouwen, P. M. P. V., and Verkleij, A. J. (1992). 3-Dimensional immunogold labeling of nuclear matrix proteins in permeabilized cells. *Cell Biol. Int. Rep.* **16,** 827–836.

de Jong, L., van Driel, R., Stuurman, N., Meijne, A. M. L., and van Renswoude, J. (1990). Principles of nuclear organization. *Cell Biol. Int. Rep.* **14,** 1051–1074.

De Pamphilis, M. L. (1993). How transcription factors regulate origins of DNA replication in eukaryotic cells. *Trends Cell Biol.* **3,** 161–167.

Dickinson, L. A., Joh, T., Kohwi, Y., and Kohwi-Shigematsu, T. (1992). A tissue-specific MAR/SAR binding protein with unusual binding site recognition. *Cell (Cambridge, Mass.)* **70,** 631–644.

Dickinson P., Cook, P. R., and Jackson, D. A. (1990). Active RNA polymerase I is fixed within the nucleus of HeLa cells. *EMBO J.* **9,** 2207–2214.

Dijkwel, P. A., Mullenders, L. H. F., and Wanka, F. (1979). Analysis of the attachment of replicating DNA to a nuclear matrix in mammalian interphase nuclei. *Nucleic Acids Res.* **6,** 219–230.

Dundr, M., and Raska, I. (1993). Nonisotopic ultrastructural mapping of transcription sites within the nucleolus. *Exp. Cell Res.* **208,** 275–281.

Dupuy-Coin, A. M., Moens, P., and Bouteille, M. (1986). Three-dimensional analysis of given cell structures: Nucleolus, nucleoskeleton and nuclear inclusions. *Methods Achiev. Exp. Pathol.* **12,** 1–25.

Dworetzky, S. I., Wright, K. L., Fey, E. G., Penman, S., Lian, J. B., Stein, J. L., and Stein, G. S. (1992). Sequence-specific DNA-binding proteins are components of a nuclear matrix-attachment site. *Proc. Natl. Acad. Sci. U.S.A.* **89,** 4178–4182.

Dyck, J. A., Maul, G. G., Miller, W. H., Chen, J. D., Kakizuka, A., and Evans, R. M. (1994). A novel macromolecular structure is a target of the promyelocyte-retinoic acid receptor oncoprotein. *Cell (Cambridge, Mass.)* **76,** 333–343.

Fakan, S. (1986). Structural support for RNA synthesis in the cell nucleus. *Methods Achiev. Exp. Pathol.* **12,** 105–140.

Fakan, S. (1994). Perichromatin fibrils are *in situ* forms of nascent transcripts. *Trends Cell Biol.* **4,** 86–90.

Fakan, S., and Bernhard, W. (1971). Localisation of rapidly and slowly labelled nuclear RNA as visualized by high resolution autoradiography. *Exp. Cell Res.* **67,** 129–141.

Fakan, S., and Hancock, R. (1974). Localization of newly-synthesized DNA in a mammalian cell as visualized by high resolution autoradiography. *Exp. Cell Res.* **83**, 95–102.

Fakan, S., and Nobis, P. (1978). Ultrastructural localization of transcription sites and of RNA distribution during the cell cycle of synchronized cells. *Exp. Cell Res.* **113**, 327–337.

Fakan, S., and Ordatchenko, N. (1980). Ultrastructural organisation of the cell nucleus in early mouse embryos. *Biol. Cell.* **37**, 211–218.

Fakan, S., and Puvion, E. (1980). The ultrastructural visualization of nuclear and extranucleolar RNA synthesis and distribution. *Int. Rev. Cytol.* **65**, 255–299.

Fakan, S., Puvion, E., and Spohr, G. (1976). Localization and characterization of newly synthesized nuclear RNA in isolated rat hepatocytes. *Exp. Cell Res.* **99**, 155–164.

Fakan, S., Leser, G., and Martin, T. E. (1984). Ultrastructural distribution of nuclear ribonucleoproteins as visualized by immunocytochemistry on thin sections. *J. Cell Biol.* **98**, 358–363.

Fakan, S., Leser, G., and Martin, T. E. (1986). Immunoelectron microscope visualization of nuclear ribonucleoprotein antigens within spread transcription complexes. *J. Cell Biol.* **103**, 1153–1157.

Felsenfeld, G. (1992). Chromatin as an essential part of the transcriptional mechanism. *Nature (London)* **355**, 219–224.

Ferreira, J. A., Carmo-Fonseca, M., and Lamond, A. I. (1994). Differential interaction of splicing snRNPs with coiled bodies and interchromatin granules during mitosis and assembly of daughter cell nuclei. *J. Cell Biol.* **126**, 11–23.

Forrester, W. C., Epner, E., Driscoll, M. C., Enver, T., Brice, M., Papayannopoulou, T., and Groudine, M. (1990). A deletion of the human beta-globin locus activation region causes a major alteration in chromatin structure and replication across the entire beta-globin locus. *Genes Dev.* **4**, 1637–1649.

Fox, M. H., Arndt-Jovin, D. J., Jovin, T. M., Baumann, P. H., and Robert-Nicoud, M. (1991). Spatial and temporal distribution of DNA replication sites localized by immunofluorescence and confocal microscopy in mouse fibroblasts. *J. Cell Sci.* **99**, 247–253.

Fu, X. D., and Maniatis, T. (1990). Factor required for mammalian spliceosome assembly is localized to discrete regions in the nucleus. *Nature (London)* **343**, 437–441.

Fu, X. D., and Maniatis, T. (1992). Isolation of a complementary DNA that encodes the mammalian splicing factor SC35. *Science* **256**, 535–538.

Fusconi, M., Cassani, F., Govoni, M., Caselli, A., Farabegoli, F., Lenzi, M., Ballardini, G., Zauli, D., and Bianchi, F. B. (1991). Anti-nuclear antibodies of primary biliary cirrhosis recognize 78–92-kD and 96–100-kD proteins of nuclear bodies. *Clin. Exp. Immunol.* **83**, 291–297.

Gasser, S. M., and Laemmli, U. K. (1986a). Cohabitation of scaffold binding regions with upstream/enhancer elements of three developmentally regulated genes of *D. melanogaster*. *Cell (Cambridge, Mass.)* **46**, 521–530.

Gasser, S. M., and Laemmli, U. K. (1986b). The organisation of chromatin loops: Characterization of a scaffold attachment site. *EMBO J.* **5**, 511–518.

Gasser, S. M., and Laemmli, U.K. (1987). A glimpse at chromosomal order. *Trends Genet.* **3**, 16–22.

Gasser, S. M., Amati, B. B., Cardenas, M. E., and Hofmann, J. F. X. (1989). Studies on scaffold attachment sites and their relation to genome function. *Int. Rev. Cytol.* **119**, 57–96.

Ge, H., Zuo, P., and Manley, J. L. (1991). Primary structure of the human splicing factor ASF reveals similarities with Drosophila regulators. *Cell (Cambridge, Mass.)* **66**, 373–382.

Gerdes, M. G., Carter, K. C., Moen, P. T., and Lawrence, J. B. (1994). Dynamic changes in the higher-level chromatin organization of specific sequences revealed by in situ hybridization to nuclear halos. *J. Cell Biol.* **126**, 289–304.

Goldman, M. A. (1988). The chromatin domain as a unit of gene regulation. *BioEssays* **9**, 50–55.

Gui, J. F., Lane, W. S., and Fu, X. D. (1994). A serine kinase regulates intracellular localization of splicing factors in the cell cycle. *Nature (London)* **369**, 678–682.

Harper, J. E., and Manley, J. L. (1991). A novel protein factor is required for use of distal alternative 5′ splice sites in vitro. *Mol. Cell. Biol.* **11**, 5945–5953.

Harpold, M. M., Evans, R. M., Salditt-Georgieff, M., and Darnell, J. E. (1979). Production of mRNA in Chinese hamster cells: Relationship of the rate of synthesis to the cytoplasmic concentration of nine specific mRNA sequences. *Cell (Cambridge, Mass.)* **17**, 1025–1035.

Harpold, M. M., Wilson, M. C., and Darnell, J. E., Jr. (1981). Chinese hamster polyadenylated messenger ribonucleic acid: Relationship to non-polyadenylated sequences and relative conservation during messenger ribonucleic acid processing. *Mol. Cell. Biol.* **1**, 188–198.

Hassan, A. B., and Cook, P. R. (1993). Visualization of replication sites in unfixed human cells. *J. Cell Sci.* **105**, 541–550.

Hassan, A. B., Errington, R. J., White, N. S., Jackson, D. A., and Cook, P. R. (1994). Replication and transcription sites are colocalized in human cells. *J. Cell Sci.* **107**, 425–434.

He, D., Nickerson, J. A., and Penman, S. (1990). Core filaments of the nuclear matrix. *J. Cell Biol.* **110**, 569–580.

Heintz, N. H. (1992). Transcription factors and the control of DNA replication. *Curr. Opin. Cell Biol.* **4**, 459–467.

Hens, L., Bauman, M., Cremer, T., Sutter, A., Cornelis, J. J., and Cremer, C. (1983). Immunocytochemical localization of chromatin regions UV-microirradiated in S-phase or naphase. *Exp. Cell Res.* **149**, 257–269.

Heslop-Harrison, J. S., and Bennett, M. D. (1990). Nuclear architecture in plants. *Trends Genet.* **6**, 401–405.

Hiraoka, Y., Dernburg, A. F., Parmelee, S. J., Rykowski, M. C., Agard, D. A., and Sedat, J. W. (1993). The onset of homologous chromosome pairing during *Drosophila melanogaster* embryogenesis. *J. Cell Biol.* **120**, 591–600.

Hora, J., Horton, M. J., Toft, D. O., and Spelsberg, T. C. (1986). Nuclease resistance and the enrichment of native nuclear acceptor sites for the avian oviduct progesterone receptor. *Proc. Natl. Acad. Sci. U.S.A.* **83**, 8839–8843.

Hozák, P., Hassan, A. B., Jackson, D. A., and Cook, P. R. (1993). Visualization of replication factories attached to a nucleoskeleton. *Cell (Cambridge, Mass.)* **73**, 361–373.

Hozák, P., Jackson, D. A. and Cook, P. R. (1994). Replication factories and nuclear bodies: The ultrastructural characterization of replication sites during the cell cycle. *J. Cell Sci.* **107**, 2191–2202.

Huang, S., and Spector, D. L. (1992). U1 and U2 small nuclear RNAs are present in nuclear speckles. *Proc. Natl. Acad, Sci U.S.A.* **89**, 305–308.

Huang, S., Deerinck, T. J., Ellisman, M. H., and Spector, D. L. (1994). In vivo analysis of the stability and transport of nuclear poly(A)$^+$RNA. *J. Cell Biol.* **126**, 877–899.

Hutchison, N., and Weintraub, H. (1985). Localization of DNAase 1-sensitive sequences to specific regions of interphase nuclei. *Cell (Cambridge, Mass.)* **43**, 471–482.

Izaurralde, E., Käs, E., and Laemmli, U. K. (1989). Highly preferential nucleation of histone H1 assembly on scaffold-associated regions. *J. Mol. Biol.* **210**, 573–585.

Jackson, D. A., and Cook, P. R. (1985). Transcription occurs at a nucleoskeleton. *EMBO J.* **4**, 919–925.

Jackson, D. A., Dickinson, P., and Cook, P. R. (1990). The size of chromatin loops in HeLa cells. *EMBO J.* **9**, 567–571.

Jackson, D. A., Hassan, A. B., Errington, R. J., and Cook, P. R. (1993). Visualization of focal sites of transcription within human nuclei. *EMBO J.* **12**, 1059–1065.

Jackson, D. A., Hassan, A. B., Errington, R. J., and Cook, P. R. (1994). Sites in human nuclei where damage induced by ultraviolet light is repaired: Localization relative to transcription sites and concentrations of proliferating cell nuclear antigen and the tumour suppressor protein, p53. *J. Cell Sci.* **107**, 1753–1760.

Jarman, A. P., and Higgs, D. R. (1988). Nuclear scaffold attachment sites in the human globin gene complexes. *EMBO J.* **7**, 3337–3344.

Jimenez-Garcia, L. F., and Spector, D. L. (1993). In vivo evidence that transcription and splicing are coordinated by a recruiting mechanism. *Cell (Cambridge, Mass.)* **73**, 47–59.

Jimenez-Garcia, L. F., Segura-Valdez, M. D., Ochs, R. L., Rothblum, L. I., Hannan, R., and Spector. D. L. (1994). Nucleologenesis: U3 snRNA-containing prenucleolar bodies move to sites of active pre-rRNA transcription after mitosis. *Mol. Biol. Cell* **5**, 955–966.

Joëls, M., and De Kloet, E. R. (1994). Mineralocorticoid and glucocorticoid receptors in the brain. Implications for ion permeability and transmitter systems. *Prog. Neurobiol.* **43**, 1–36.

Jost, J. -P., and Seldran, M. (1984). Association of transcriptionally active vitellogenin II gene with the nuclear matrix of chicken liver. *EMBO J.* **3**. 2005–2008.

Käs, E., Poljak, L., Adachi, Y., and Laemmli, U.K. (1993). A model for chromatin opening: Stimulation of topoisomerase-ii and restriction enzyme cleavage of chromatin by distamycin. *EMBO J.* **12**, 115–126.

Kaufmann, S. H., Okret, S., Wikström A. C., Gustafsson, J. A., and Shaper, J. H. (1986). Binding of the glucocorticoid receptor to the rat liver nuclear matrix. *J. Biol. Chem.* **261**, 11962–11967.

Kill, I. R., Bridger, J. M., Campbell, K. H. S., Maldonadocodina, G., and Hutchinson, C. J. (1991). The timing of the formation and usage of replicase clusters in S-phase nuclei of human diploid fibroblasts. *J. Cell Sci.* **100**, 869–876.

Kirsch, T. M., and Miller-Diener, A. (1986). The nuclear matrix is the site of glucocorticoid receptor complex action in the nucleus. *Biochem. Biophys. Res. Commun.* **137**, 640–648.

Klempnauer, K.-H. (1988). Interaction of myb proteins with nuclear matrix in vitro. *Oncogene* **2**, 545–551.

Koken, M. H. M., Puvion-Dutilleul, F., Guillemin, M. C., Viron, A., Linarescruz, G., Stuurman, N., de Jong, L., Szostecki, C., Calvo, F., Chomienne, C., Degos, L., Puvion, E., and de Thé, H. (1994). The t(15,17) translocation alters a nuclear body in a retinoic acid-reversible fashion. *EMBO J.* **13**, 1073–1083.

Kopency, V., Fakan, S., Pavlok, A., Pivko, J., Grafenau, P., Biggiogera, M., Leser, G., and Martin, T. E. (1991). Immunoelectron microscopic localization of smaller nuclear ribonucleoproteins during bovine early embryogenesis. *Mol. Reprod. Dev.* **29**, 209–219.

Krainer, A. R., Mayeda, A., Kozak, D., and Binns, G. (1991). Functional expression of cloned human splicing factor SF2: Homology to RNA-binding proteins, U1 70K, and Drosophila splicing regulators. *Cell (Cambridge, Mass.)* **66**, 383–394.

Kramer, J., Zachar, Z., and Bingham, P. M. (1994). Nuclear pre-mRNA metabolism: Channels and tracks. *Trends Cell Biol.* **4**, 35–37.

Kumara-Siri, M. H., Shapiro, L. E., and Surks, M. I. (1986). Association of the 3,5,3'-triiodo-L-thyronine nuclear receptor with the nuclear matrix of cultured growth hormone-producing rat pituitary tumor cells (GC cells). *J. Biol. Chem.* **261**, 2844–2852.

Lamm, G. M., and Lamond, A. I. (1993). Non-snRNP protein splicing factors. *Biochim. Biophys. Acta* **1173**, 247–265.

Lamond, A. I. (1991). ASF/SF2—A splice site selector. *Trends Biochem. Sci.* **16**, 452–453.

Lamond, A. I. (1993). The spliceosome. *BioEssays* **15**, 595–603.

Lamond, A. I., and Carmo-Fonseca, M. (1993). The coiled body. *Trends Cell Biol.* **3**, 198–204.

Lane, N. J. (1969). Intranuclear fibrillar bodies in actinomycine D-treated oocytes. *J. Cell Biol.* **40**, 286–291.

Lawrence, J. B., Singer, R. H., and Marselle, L. M. (1989). Highly localized tracks of specific transcripts within interphase nuclei visualized by in situ hybridization. *Cell (Cambridge, Mass.)* **57**, 493–502.

Lawrence, J. B., Carter, K. C., and Xing, X. (1993). Probing functional organization within the nucleus: Is genome structure integrated with RNA metabolism? *Cold Spring Harbor Symp. Quant. Biol.* **58**, 807–818.

Leitch, A. R., Mosgoller, W., Schwarzacher, T., Bennett, M. D., and Heslop-Harrison, J. S. (1990). Genomic in situ hybridization to sectioned nuclei shows chromosome domains in grass hybrids. *J. Cell Sci.* **95,** 335–341.

Lengauer, C., Riethman, H., and Cremer, T. (1990). Painting of human chromosomes with probes generated from hybrid cell lines by PCR with Alu and L1 primers. *Hum. Genet.* **86,** 1–6.

Lengauer, C., Eckelt, A., Weith, A., Endlich, N., Ponelies, N., Lichter, P., Greulich, K. O., and Cremer, T. (1991). Painting of defined chromosomal regions by in situ suppression hybridization of libraries from laser-microdissected chromosomes. *Cytogenet. Cell Genet.* **56,** 27–30.

Leonhardt, H., Page, A. W., Weier, H. U., and Bestor, T. H. (1992). A targeting sequence directs DNA methyltransferase to sites of DNA replication in mammalian nuclei. *Cell (Cambridge, Mass.)* **71,** 865–873.

Leser, G. P., Fakan, S., and Martin, T. E. (1989). Ultrastructural distribution of ribonucleoprotein complexes during mitosis: snRNP antigens are contained in mitotic granule clusters. *Eur. J. Cell Biol.* **50,** 376–389.

Levy-Wilson, B., and Fortier, C. (1989). The limits of the DNAse I-sensitive domain of the human apolipoprotein-b gene coincide with the locations of chromosomal anchorage loops and define the 5'-boundary and 3'-boundary of the gene. *J. Biol. Chem.* **264,** 21196–21204.

Li, H., and Bingham, P. M. (1991). Arginine/serine-rich domains of the $su(w^a)$ and *tra* RNA processing regulators target proteins to a subnuclear compartment implicated in splicing. *Cell (Cambridge, Mass.)* **67,** 335–342.

Li, R., and Botchan, M. R. (1994). Acidic transcription factors alleviate nucleosome-mediated repression of DNA replication of bovine papillomavirus type 1. *Proc. Natl. Acad. Sci. U.S.A.* **91,** 7051–7055.

Lichter, P., Cremer, T., Borden, J., Manuelidis, L., and Ward, D. C. (1988). Delineation of individual human chromosomes in metaphase and interphase cells by in situ suppression hybridization using recombinant DNA libraries. *Hum. Genet.* **80,** 224–234.

Ludérus, M. E. E., de Graaf, A., Mattia, E., den Blaauwen, J. L., Grande, M. A., de Jong, L., and van Driel, R. (1992). Binding of matrix attachment regions to lamin-b1. *Cell (Cambridge, Mass.)* **70,** 949–959.

Ludérus, M. E. E., den Blaauwen, J. L., de Smit, O. J. B., Compton, D. A., and van Driel, R. (1994). Binding of matrix attachment regions to lamin polymers involves single-stranded regions and the minor groove. *Mol. Cell. Biol.* **14,** 6297–6305.

Malatesta, M., Zancanaro, C., Martin, T. E., Chan, E. K. L., Amalric, F., Luhrmann, R., Vogel, P., and Fakan, S. (1994). Is the coiled body involved in nucleolar functions. *Exp. Cell Res.* **211,** 415–419.

Mancini, M. A., Shan, B., Nickerson, J. A., Penman, S., and Lee, W. H. (1994). The retinoblastoma gene product is a cell cycle-dependent, nuclear matrix-associated protein. *Proc. Natl. Acad. Sci. U.S.A.* **91,** 418–422.

Manders, E. M. M., Stap, J., Brakenhoff, G. J., van Driel, R., and Aten, J. A. (1992). Dynamics of three-dimensional replication patterns during the S-phase, analysed by double labelling of DNA and confocal microscopy. *J. Cell Sci.* **103,** 857–862.

Manuelidis, L. (1985). Individual interphase domains revealed by in situ hybridization. *Hum. Genet.* **71,** 288–293.

Manuelidis, L. (1990). A view of interphase chromosomes. *Science* **250,** 1533–1540.

Manuelidis, L., and Chen, T. L. (1990). A unified model of eukaryotic chromosomes. *Cytometry* **11,** 8–25.

Mariman, E. C. M., van Eekelen, C. A. G., Reinders, R. J., Berns, A. J. M., and van Venrooij, W. J. (1982). Adenoviral heterogeneous nuclear RNA is associated with the host nuclear matrix during splicing. *J. Mol. Biol.* **154,** 103–119.

Martelli, A. M., Bareggi, R., and Narducci, P. (1994). Catalytic properties of DNA polymerase alpha activity associated with the heat-stabilized nuclear matrix prepared from HeLa S3 cells. *Cell Biochem. Funct.* **12**, 129–135.

Martin, T. E., Barghusen, S. C., Leser, G. P., and Spear, P. G. (1987). Redistribution of nuclear ribonucleoprotein antigens during herpes simplex virus infection. *J. Cell Biol.* **105**, 2069–2082.

Masurovsky, E. B., Benitez, H. H., Kim, S. U., and Murray, M. R. (1970). Origin, development, and nature of intranuclear rodlets and associated bodies in chicken sympathetic neurons. *J. Cell Biol.* **44**, 172–191.

Matera, A. G., and Ward, D. C. (1993). Nucleoplasmic organization of small nuclear ribonucleoproteins in cultured human cells. *J. Cell Biol.* **121**, 715–727.

Matunis, E. L., Matunis, M. J., and Dreyfuss, G. (1993). Association of individual hnRNP proteins and snRNPs with nascent transcrips, *J. Cell Biol.* **121**, 219–228.

Maul, G. G., and Everett, R. D. (1994). The nuclear location of PML, a cellular member of the C3HC4 zinc-binding domain protein family, is rearranged during herpes simplex virus infection by the C3HC4 viral protein ICPO, *J. Gen. Virol.* **75**, 1223–1233.

Maul, G. G., Guldner, H. H., and Spivack, J. G. (1993). Modification of discrete nuclear domains induced by herpes simplex virus type-1 immediate early gene-1 product (ICPO). *J. Gen. Virol.* **74**, 2679–2690.

Mazzotti, G., Rizzoli, R., Galanzi, A., Papa, S., Vitale, M., Falconi, M., Neri, L. M., Zini, N., and Maralda, N. M. (1990). High-resolution detection of newly synthesized anti-bromodeoxyuridine antibodies indentifies specific chromatin domains. *J. Histol. Cytol.* **38**, 13–22.

McCready, S. J., Godwin, J., Mason, D. W., Brazell, I. A., and Cook, P. R. (1980). DNA is replicated at the nuclear cage. *J. Cell Sci.* **46**, 365–386.

Metzger, D. A., and Korach, K. S. (1990). Cell-free interaction of the estrogen receptor with mouse uterine nuclear matrix—Evidence of saturability, specificity, and resistance to KCl extraction. *Endocrinology (Baltimore)* **126**, 2190–2195.

Mills, A. D., Blow, J. J., White, J. G., Amos, W. B., Wilcock, D., and Laskey, R. A. (1989). Replication occurs at discrete foci spaced throughout nuclei replicating in vitro. *J. Cell Sci.* **94**, 471–477.

Mirkovitch, J., Mirault, M.-E., and Laemmli, U. K. (1984). Organization of the higher-order chromatin loop: Specific DNA attachment sites on nuclear scaffold. *Cell (Cambridge, Mass.)* **39**, 223–232.

Monneron, A., and Bernhard, W. (1969). Fine structural organization of the interphase nucleus in some mammalian cells. *J. Ultrastruct. Res.* **27**, 266–288.

Moreno Diaz de la Espina, S., Sanchez Pina, A., Risueno, M. C., Medina, F. J., and Fernandez-Gomez, M. E. (1980). The role of plant coiled bodies in nuclear RNA metabolism. *J. Electron Microsc.* **2**, 240–241.

Nakagomi, K., Kohwi, Y., Dickinson, L. A., and Kohwi-Shigematsu, T. (1994). A novel DNA-binding motif in the nuclear matrix attachment DNA-binding protein SATB1. *Mol. Cell. Biol.* **14**, 1852–1860.

Nakamura, H., Morita, T., and Sato, C. (1986). Structural organization of replicon domains during DNA synthetic phase in the mammalian nucleus. *Exp. Cell. Res.* **165**, 291–297.

Nakayasu, H., and Berezney, R. (1989). Mapping replication sites in the eucaryotic nucleus. *J. Cell Biol.* **108**, 1–11.

Nickerson, J. A., and Penman, S. (1992). Localization of nuclear matrix core filament proteins at interphase and mitosis. *Cell Biol. Int. Rep.* **16**, 811–826.

Nyman, U., Hallman, H., Hadlaczky, G., Pettersson, I., Sharp, G., and Ringertz, N. R. (1986). Intranuclear localization of snRNP antigens. *J. Cell Biol.* **102**, 137–144.

Ochs, R. L., Stein, T. W., and Tan, E. M. (1994). Coiled bodies in the nucleolus of breast cancer cells. *J. Cell Sci.* **107**, 385–399.

Ogata, N. (1990). Preferential association of a transcriptionally active gene with the nuclear matrix of rat fibroblasts transformed by a simian-virus-40-pBr322 recombinant plasmid. *Biochem. J.* **267**, 385–390.

O'Keefe, R. T., Henderson, S. C., and Spector, D. L. (1992). Dynamic organization of DNA replication in mammalian cell nuclei: Spatially and temporally defined replication of chromosome-specific α-satellite DNA sequences. *J. Cell Biol.* **116**, 1095–1110.

O'Keefe, R. T., Mayeda, A., Sadowski, C. L., Krainer, A. R., and Spector, D. L. (1994). Disruption of pre-messenger RNA splicing in vivo results in reorganization of splicing factors. *J. Cell Biol.* **124**, 249–260.

Osheim, Y. N., Miller, O. L., and Beyer, A. L. (1985). RNP particles at the splice junction sequences on Drosophila chorion transcripts. *Cell (Cambridge, Mass.)* **43**, 143–151.

Paulson, J. R., and Laemmli, U. K. (1977). The structure of histon-depleted metaphase chromosomes. *Cell (Cambridge, Mass.)* **12**, 817–828.

Petrov, P., and Sekeris, C. E. (1971). Early action of alpha-amanitin on extranucleolar ribonucleoproteins, as revealed by electron microscopic observation. *Exp. Cell Res.* **69**, 393–401.

Phelan, A., Carmo-Fonseca, M., McLauchlan, J., Lamond, A. I., and Clements, J. B. (1993). A herpes-simplex virus type-1 immediate–early gene product, IE63, regulates a small nuclear ribonucleoprotein distribution. *Proc. Natl. Acad. Sci. U.S.A.* **90**, 9056–9060.

Phi-Van, L., and Strätling, W. H. (1988). The matrix attachment regions of the chicken lysozyme gene co-map with the boundaries of the chromatin domain. *EMBO J.* **7**, 655–664.

Phi-Van, L., von Kries, J. P., Ostertag, W., and Strätling, W. H. (1990). The chicken lysozyme-5′ matrix attachment region increases transcription from a heterologous promoter in heterologous cells and dampens position effects on the expression of transfected genes. *Mol. Cell. Biol.* **10**, 2302–2307.

Pinkel, D., Straume, T., and Gray, J. W. (1986). Cytogenetic analysis using quantitative, high sensitivity fluorescence hybridization. *Proc. Natl. Acad. Sci. U.S.A.* **83**, 2934–2938.

Pinkel, D., Landegent, J., Collins, C., Fuscoe, J., Segraves, R., Lucas, J., and Gray, J. (1988). Fluorescence in situ hybridization with human chromosome-specific libraries: Detection of trisomy 21 and translocations of chromosome 4. *Proc. Natl. Acad. Sci. U.S.A.* **85**, 9138–9142.

Pirrotta, V. (1990). Transvection and long-distance gene regulation. *BioEssays* **12**, 409–414.

Puvion, E., Viron, A., Assens, C., Leduc, E. H., and Jeanteur, P. (1984). Immunocytochemical identification of nuclear structures containing snRNPs in isolated rat liver cells. *J. Ultrastruct. Res.* **87**, 180–189.

Puvion, E., Hernandez-Verdun, D., and Haguenau, F. (1994). The nucleus and the nucleolus. The contribution of French electron microscopists. *Biol. Cell* **80**, 91–95.

Puvion-Dutilleul, F., Bachellerie, J. P., Visa, N., and Puvion, E. (1994). Rearrangements of intranuclear structures involved in RNA processing in response to adenovirus infection. *J. Cell Sci.* **107**, 1457–1468.

Raap, A. K., van de Rijke, F. M., Dirks, R. W., Sol, C. J., Boom, R., and van der Ploeg, M. (1991). Bicolor fluorescence in situ hybridization to intron and exon mRNA sequences. *Exp. Cell Res.* **197**, 319–322.

Rappold, G. A., Cremer, T., Hager, H. D., Davies, K. E., Muller, C. R., and Yang, T. (1984). Sex chromosome positions in human interphase nuclei studied by in situ hybridization with chromosome specific DNA probes. *Hum. Genet.* **67**, 317–322.

Raska, I., and Dundr, M. (1993). Compartmentalization of the cell nucleus: Case of the nucleolus. *In* "Chromosomes Today" (A. T. Sumner and A. C. Chandley, eds.), pp. 101–119. Chapman & Hall, London.

Raska, I., Koberna, K., Jarnik, M., Petrasovicova, V., Bednar, J., Raska, K., and Bravo, R. (1989). Ultrastructural immunolocalization of cyclin/PCNA in synchronized 3T3 cells. *Exp. Cell Res.* **184**, 81–89.

Raska, I., Ochs, R., Andrade, L., and Chan, E. K. (1990a). Association between nucleolus and the coiled body. *J. Struct. Biol.* **104**, 120–127.

Raska, I., Ochs, R. L., and Salamin-Michel, L. (1990b). Immunocytochemistry of the cell nucleus. *Electron Microsc. Rev.* **3,** 301–353.

Raska, I., Andrade, L. E. C., Ochs, R. L., Chan, E. K. L., Chang, C.-M., Roos, G., and Tan, E. M. (1991). Immunological and ultrastructural studies of the nuclear coiled body with autoimmune antibodies. *Exp. Cell Res.* **195,** 27–37.

Razin, S. V., and Yarovaya, O. V. (1985). Initiated complexes of RNA polymerase II are concentrated in the nuclear skeleton associated DNA. *Exp. Cell Res.* **158,** 273–275.

Reddy, B. A., Etkin, L. D., and Freemont, P. S. (1992). A novel zinc finger coiled-coil domain in a family of nuclear proteins. *Trends Biochem. Sci.* **17,** 344–345.

Roberge, M., Dahmus, M. E., and Bradbury, E. M. (1988). Chromosomal loop/nuclear matrix organization of transcriptionally active and inactive RNA polymerases in HeLa nuclei. *J. Mol. Biol.* **201,** 545–555.

Romig, H., Fackelmayer, F. O., Renz, A., Ramsperger, U., and Richter, A. (1992). Characterization of SAF-A, a novel nuclear DNA binding protein from HeLa cells with high affinity for nuclear matrix/scaffold attachment DNA elements. *EMBO J.* **11,** 3431–3440.

Rosbash, M., and Singer, R. H. (1993). RNA travel—Tracks from DNA to cytoplasm. *Cell (Cambridge, Mass.)* **75,** 399–401.

Ross, D. A., Yen, R. W., and Chae, C. B. (1982). Association of globin ribonucleic acid and its precursors with the chicken erythroblast nuclear matrix. *Biochemistry* **21,** 764–771.

Ruskin, B., Zamore, P. D., and Green, M. R. (1988). A factor, U2AF, is required for U2 snRNP binding and splicing complex assembly. *Cell (Cambridge, Mass.)* **52,** 207–219.

Sahlas, D. J., Milankov, K., Park, P. C., and De Boni, U. (1993). Distribution of snRNPs, splicing factor-SC-35 and actin in interphase nuclei—immunocytochemical evidence for differential distribution during changes in functional states. *J. Cell Sci.* **105,** 347–357.

Salditt-Georgieff, M., and Darnell, J. E. (1982). Further evidence that the majority of primary nuclear RNA transcripts in mammalian cells do not contribute to mRNA. *Mol. Cell. Biol.* **2,** 701–707.

Salditt-Georgieff, M., Harpold, M. M., Wilson, M. C., and Darnell, J. E. (1981). Large heterogeneous nuclear ribonucleic acid has three times as many 5′ caps as polyadenylic acid segments, and most caps do not enter polyribosomes. *Mol. Cell. Biol.* **1,** 179–187.

Schardin, M., Hager, H. D., and Lang, M. (1985). Specific staining of human chromosomes in Chinese hamster x man hybrid cell lines demonstrates interphase chromosome territories. *Hum. Genet.* **71,** 281–287.

Scheer, U., and Weisenberger, D. (1994). The nucleolus. *Curr. Opin. Cell Biol.* **6,** 354–359.

Schepers, A., Pich, D., and Hammerschmidt, W. (1993). A transcription factor with homology to the AP-1 family links RNA transcription and DNA replication in the lytic cycle of Epstein–Barr virus. *EMBO J.* **12,** 3921–3929.

Schöffl, F., Schröder, G., Kliem, M., and Rieping, M. (1993). An SAR-sequence containing 395 bp-DNA fragment mediates enhanced, gene–dosage-correlated expression of a chimaeric heat shock gene in transgenic tobacco plants. *Transgenic Res.* **2,** 93–100.

Schröder, H. C., Trolltsch, D., Wenger, R., Bachmann, M., and Diehl-Seifest, B. (1987). Cytochalasin B selectively releases ovalbumin mRNA precursors but not the mature ovalbumin mRNA from hen oviduct nuclear matrix. *Eur. J. Biochem.* **167,** 239–245.

Schuchard, M., Subramaniam, M., Ruesink, T., and Spelsberg, T. C. (1991). Nuclear matrix localization and specific matrix DNA binding by receptor binding factor-1 of the avian oviduct progesterone receptor. *Biochemistry* **30,** 9516–9522.

Schwarzacher, H. G., and Wachtler, F. (1993). The nucleolus. *Anat. Embryol.* **188,** 515–536.

Sibon, O. C. M., Cremers, F. F. M., Boonstra, J., Humbel, B. M., and Verkleij, A. J. (1993). Localisation of EGF-receptor messenger RNA in the nucleus of A431 cells by light microscopy. *Cell Biol. Int. Rep.* **17,** 1–11.

Sibon, O. C. M., Humbel, B. M., de Graaf, A., Verkleij, A. J., and Cremers, F. F. M. (1994). Ultrastructural localization of epidermal growth factor (EGF)-receptor transcripts in the cell nucleus using pre-embedding in situ hybridization in combination with ultra-small gold probes and silver enhancement. *Histochemistry* **101,** 223–232.

Sierakowska, H., Shukla, R. R., Dominski, Z., and Kole, R. (1989). Inhibition of pre-mRNA splicing by fluoro-, 5-chloro-, and 5-bromouridine. *J. Biol. Chem.* **264,** 19185–19191.

Smith, H. C., and Berezney, R. (1982). DNA polymerase is tightly bound to the nuclear matrix of actively replicating liver. *Biochem. Biophys. Res. Commun.* **97,** 1541–1547.

Sobczak-Thepot, J., Harper, F., Florentin, Y., Zindy, F., Brechot, C., and Puvion, E. (1993). Localization of cyclin A at the sites of cellular DNA replication. *Exp. Cell Res.* **206,** 43–48.

Spector, D. L. (1984). Colocalization of U1 and U2 small nuclear RNPs by immunocytochemistry. *Biol. Cell* **51,** 109–112.

Spector, D. L. (1990). Higher order nuclear organization—3-dimensional distribution of small nuclear ribonucleoprotein particles. *Proc. Natl. Acad. Sci. U.S.A.* **87,** 147–151.

Spector, D. L. (1993a). Macromolecular domains within the cell nucleus. *Annu. Rev. Cell Biol.* **9,** 265–315.

Spector, D. L. (1993b). Nuclear organization of pre-mRNA processing. *Curr. Opin. Cell Biol.* **5,** 442–448.

Spector, D. L., Schrier, W. H., and Busch, H. (1983). Immunoelectron microscopic localization of snRNPs. *Biol. Cell* **49,** 1–10.

Spector, D. L., Fu, X.-D., and Maniatis, T. (1991). Associations between distinct pre-mRNA splicing components and the cell nucleus. *EMBO J.* **10,** 3467–3481.

Spector, D. L., Lark, G., and Huang, S. (1992). Differences in snRNP localization between transformed and nontransformed cells. *Mol. Biol. Cell* **3,** 555–569.

Spector, D. L., O'Keefe, R. T., and Jimenez-Garcia, L. F. (1993). Dynamics of transcription and pre-mRNA splicing within the mammalian cell nucleus. *Cold Spring Harbor Symp. Quant. Biol.* **58,** 799–805.

Stein, G. S., and Lian, J. B. (1993). Molecular mechanisms mediating proliferation/differentiation interrelationships during progressive development of the osteoblast phenotype. *Endocr. Rev.* **14,** 424–442.

Stein, G. S., van Wijnen, A. J., Stein, J. L., Lian, J. B., Bidwell, J. P., and Montecino, M. (1994). Nuclear architecture supports integration of physiological regulatory signals for transcription of cell growth and tissue-specific genes during osteoblast differentiation. *J. Cell. Biochem.* **55,** 4–15.

Stief, A., Winter, D. M., Strätling, W. H., and Sippel, A. E. (1989). A nuclear DNA attachment element mediates elevated and position-independent gene activity. *Nature (London)* **341,** 343–345.

Stuurman, N. (1991). Structure and composition of the nuclear matrix. Ph.D. Thesis, University of Amsterdam, The Netherlands.

Stuurman, N., Meijne, A. M. L., van der Pol, A. J., de Jong, L., van Driel, R., and van Renswoude, J. (1990). The nuclear matrix from cells of different origin: Evidence for a common set of matrix proteins. *J. Biol. Chem.* **265,** 5460–5465.

Stuurman, N., de Graaf, A., Floore, A., Josso, A., Humbel, B., de Jong, L., and van Driel, R. (1992a). A monoclonal antibody recognizing nuclear matrix-associated nuclear bodies. *J. Cell Sci.* **101,** 773–784.

Stuurman, N., de Jong, L., and van Driel, R. (1992b). Nuclear frameworks—Concepts and operational definitions. *Cell. Biol. Int. Rep.* **16,** 837–852.

Sui, H., and Spector, D. L. (1991). Nascent pre-messenger RNA transcripts are associated with nuclear regions enriched in splicing factors. *Genes Dev.* **5,** 2288–2302.

Swaneck, G. E., and Fishman, J. (1988). Covalent binding of the endogenous estrogen 16alpha-hydroxyestrone to estradiol receptor in human breast cancer cells: Characterization and intranuclear localization. *Proc. Natl. Acad. Sci. U.S.A.* **85,** 7831–7833.

Szostecki, C., Krippner, H., Penner, E., and Bautz, F. A. (1987). Autoimmune sera recognize a 100 kD nuclear protein antigen (sp-100). *Clin. Exp. Immunol.* **68,** 108–116.

Szostecki, C., Guldner, H. H., Netter, H. J., and Will, H. (1990). Isolation and characterization of cDNA encoding a human nuclear antigen predominantly recognized by autoantibodies from patients with primary biliary cirrhosis. *J. Immunol.* **145**, 4338–4347.

Taljanidisz, J., Popowski, J., and Sarkar, N. (1989). Temporal order of gene replication in Chinese hamster ovary cells. *Mol. Cell. Biol.* **9**, 2881–2889.

Tang, T. K., Tang, C. J. C., Chao, Y. J., and Wu, C. W. (1994). Nuclear mitotic apparatus protein (NuMA): Spindle association, nuclear targeting and differential subcellular localization of various NuMA isoforms. *J. Cell Sci.* **107**, 1389–1402.

Tollervey, D., Lehtonen, H., Carmo-Fonseca, M., and Hurt, E. C. (1991). The small nucleolar RNP protein NOP1 (fibrillarin) is required for pre-rRNA processing in yeast. *EMBO J.* **10**, 573–583.

Tollervey, D., Lehtonen, H., Jansen, R., Kern, H., and Hurt, E. C. (1993). Temperature-sensitive mutations demonstrate roles for yeast fibrillarin in pre-rRNA processing, pre-rRNA methylation, and ribosome assembly. *Cell (Cambridge, Mass.)* **72**, 443–457.

Tubo, R. A., and Berezney, R. (1987a). Pre-replicative association of multiple replicative enzyme activities with the nuclear matrix during rat liver regeneration. *J. Biol. Chem.* **262**, 1148–1154.

Tubo, R. A., and Berezney, R. (1987b). Identification of 100 and 150 S polymerase alpha-primase megacomplexes solubilized from the nuclear matrix of regenerating rat liver. *J. Biol. Chem.* **262**, 5857–5865.

Tubo, R. A., and Berezney, R. (1987c). Nuclear matrix-bound DNA primase: Elucidation of an RNA primary system in nuclear matrix isolated from regenerating rat liver. *J. Biol. Chem.* **262**, 6637–6642.

Tubo, R. A., Martelli, A. M., and Berezney, R. (1987). Enhanced processivity of nuclear matrix bound DNA polymerase alpha from regenerating liver. *Biochemistry* **26**, 5710–5718.

van Dierendonck, J. H., Wijsman, J. H., Keijzer, R., van de Velde, C. J. H., and Cornelisse, C. J. (1991). Cell-cycle related staining patterns of anti-proliferating cell nuclear antigen monoclonal antibodies—Comparison with BrdUrd labeling and Ki-67 staining. *Am. J. Pathol.* **13**, 1165–1172.

van Driel, R., Humbel, B., and de Jong, L. (1991). The nucleus—A black box being opened. *J. Cell. Biochem.* **47**, 311–316.

van Steensel, B., van Haarst, A. D., de Kloet, E. R., and van Driel, R. (1991). Binding of corticosteroid receptors to rat hippocampus nuclear matrix. *FEBS Lett.* **292**, 229–231.

van Steensel, B., Jenster, G., Damm, K., Brinkman, A. O., and van Driel, R. (1995). Domains of the human androgen receptor and glucocorticoid receptor involved in binding to the nuclear matrix. *J. Cell. Biochem.* **57**, 465–478.

van Wijnen, A. J., Bidwell, J. P., Fey, E. G., Penman, S., Lian, J. B., Stein, J. L., and Stein, G. S. (1993). Nuclear matrix association of multiple sequence-specific DNA binding activities related to SP-1, ATF, CCAAT, C/EBP, OCT-1, and AP-1. *Biochemistry* **32**, 8397–8402.

Verheijen, R., Kuijpers, H., Vooijs, P., van Venrooij, W., and Ramaekers, F. (1986). Distribution of the 70k U1 RNA-associated protein during interphase and mitosis. *J. Cell Sci.* **86**, 173–190.

Visa, N., Puvion-Dutilleul, F., Harper, F., Bachellerie, J. P., and Puvion, E. (1993). Intranuclear distribution of poly(A) RNA determined by electron microscope in situ hybridization. *Exp. Cell Res.* **208**, 19–34.

Vogelstein, B., Pardoll, D. M., and Coffey, D. S. (1980). Supercoiled loops and eucaryotic DNA replication. *Cell (Cambridge, Mass.)* **22**, 79–85.

Vollmer, G., Haase, A., and Eisele, M. (1982). Androgen receptor complex binding in murine skeletal muscle nuclei. *Biochem. Biophys. Res. Commun.* **105**, 1554–1560.

von Kries, J. P., Buhrmester, H., and Strätling, W. H. (1991). A matrix/scaffold attachment region binding protein—Identification, purification, and mode of binding. *Cell (Cambridge, Mass.)* **64**, 123–135.

von Kries, J. P., Buck, F., and Strätling, W. H. (1994a). Chicken MAR binding protein p120 is identical to human heterogeneous nuclear ribonucleoprotein (hnRNP) U. *Nucleic Acids Res.* **22**, 1215–1220.

von Kries, J. P., Rosorius, O., Buhrmester, H., and Strätling, W. H. (1994b). Biochemical properties of attachment region binding protein ARBP. *FEBS Lett.* **342**, 185–188.

Waitz, W., and Loidl, P. (1991). Cell cycle dependent association of c-myc protein with the nuclear matrix. *Oncogene* **6**, 29–35.

Wang, J., Cao, L. G., Wang, Y. L., and Pederson, T. (1991). Localization of pre-messenger-RNA at discrete nuclear sites. *Proc. Natl. Acad. Sci. U.S.A.* **88**, 7391–7395.

Wansink, D. G. (1994). Transcription by RNA polymerase II and nuclear architecture. Ph.D. Thesis, University of Amsterdam, The Netherlands.

Wansink, D. G., Schul, W., van der Kraan, I., van Steensel, B., van Driel, R., and de Jong, L. (1993). Fluorescent labeling of nascent RNA reveals transcription by RNA polymerase-II in domains scattered throughout the nucleus. *J. Cell Biol.* **122**, 283–293.

Wansink, D. G., Manders, E. E. M., van der Kraan, I., Aten, J. A., van Driel, R., and de Jong, L. (1994a). RNA polymerase II transcription is concentrated outside replication domains throughout S-phase. *J. Cell Sci.* **107**, 1449–1456.

Wansink, D. G., Nelissen, R. L. H., and de Jong, L. (1994b). In vitro splicing of pre-mRNA containing bromouridine. *Mol. Biol. Rep.* **19**, 109–113.

Warrell, R. P., Maslak, P., Eardley, A., Heller, G., Miller, W. H., and Frankel, S. R. (1994). Treatment of acute promyelocytic leukemia with all-trans retinoic acid: An update of the New York experience. *Leukemia* **8**, 929–933.

Weis, K., Rambaud, S., Lavau, C., Jansen, J., Carvalho, T., Carmo-Fonseca, M., Lamond, A., and Dejean, A. (1994). Retinoic acid regulates aberrant nuclear localization of PML-RAR alpha in acute promyelocytic leukemia cells. *Cell (Cambridge, Mass.)* **76**, 345–356.

Wolffe, A. P. (1994). The role of transcription factors, chromatin structure and DNA replication in 5 S RNA gene regulation. *J. Cell Sci.* **107**, 2055–2063.

Woppmann, A., Will, C. L., Kornstadt, U., Zuo, P., Manley, J. L., and Luhrmann, R. (1993). Identification of an snRNP-associated kinase activity that phosphorylates arginine/serine rich domains typical of splicing factors. *Nucleic Acids Res.* **21**, 2815–2822.

Wu, C. T. (1993). Transvection, nuclear structure, and chromatin proteins. *J. Cell Biol.* **120**, 587–590.

Wu, Z., Murphy, C., Wu, C. H. H., Tsvetkov, A., and Gall, J. G. (1993). Snurposomes and coiled bodies. *Cold Spring Harbor Symp. Quant. Biol.* **58**, 747–754.

Xie, K. W., Lambie, E. J., and Snyder, M. (1993). Nuclear dot antigens may specify transcriptional domains in the nucleus. *Mol. Cell. Biol.* **13**, 6170–6179.

Xing, Y., and Lawrence, J. B. (1991). Preservation of specific RNA distribution within the chromatin-depleted nuclear substructure demonstrated by in situ hybridization coupled with biochemical fractionation. *J. Cell Biol.* **112**, 1055–1063.

Xing, Y., and Lawrence, J. B. (1993). Nuclear RNA tracks: Structural basis for transcription and splicing? *Trends Cell Biol.* **3**, 346–353.

Xing, Y., Johnson, C. V., Dobner, P. R., and Lawrence, J. B. (1993). Higher level organization of individual gene transcription and RNA splicing. *Science* **259**, 1326–1330.

Yang, C. H., Lambie, E. J., and Snyder, M. (1992). NuMA—An unusually long coiled-coil related protein in the mammalian nucleus. *J. Cell Biol.* **116**, 1303–1317.

Zacher, Z., Kramer, J., Mims, I. P., and Bingham, P. M. (1993). Evidence for channeled diffusion of pre-mRNAs during nuclear RNA transport in metazoans. *J. Cell Biol.* **121**, 729–742.

Zahler, A. M., Neugebauer, K. M., Lane, W. S., and Roth, M. B. (1993a). Distinct functions of SR-proteins in alternative pre-messenger RNA splicing. *Science* **260**, 219–222.

Zahler, A. M., Neugebauer, K. M., Stolk, J. A., and Roth, M. B. (1993b). Human SR-proteins and isolation of a cDNA encoding SRp75. *Mol. Cell. Biol.* **13**, 4023–4028.

Zambetti, G., Fey, E. G., Penman, S., Stein, J., and Stein, G. (1990). Multiple types of messenger RNA-cytoskeleton interactions. *J. Cell. Biochem.* **44,** 177–187.

Zamore, P. D., and Green, M. R. (1991). Biochemical characterization of U2 snRNP auxiliary factor: An essential pre-mRNA splicing factor with a novel intranuclear distribution. *EMBO J.* **10,** 207–214.

Zeitlin, S., Parent, A., Silverstein, S., and Efstratiadis, A. (1987). Pre-mRNA splicing and the nuclear matrix. *Mol. Cell. Biol.* **7,** 111–120.

Zeitlin, S., Wilson, R. C., and Efstratiadis, A. (1989). Autonomous splicing and complementation of in vivo-assembled spliceosomes. *J. Cell Biol.* **108,** 765–777.

Zeng, C. Q., He, D. C., Berget, S. M., and Brinkley, B. R. (1994a). Nuclear-mitotic apparatus protein—A structural protein interface between the nucleoskeleton and RNA splicing. *Proc. Natl. Acad. Sci. U.S.A.* **91,** 1505–1509.

Zeng, C. Q., He, D. C., and Brinkley, B. R. (1994b). Localization of NuMA protein isoforms in the nuclear matrix of mammalian cells. *Cell Motil. Cytoskel.* **29,** 167–176.

Zhang, M., Zamore, P. D., Carmo-Fonseca, M., Lamond, A. I., and Green, M. R. (1992). Cloning and intracellular localization of the U2 small nuclear ribonucleoprotein auxiliary factor small subunit. *Proc. Natl. Acad. Sci. U.S.A.* **89,** 8769–8773.

Zhao, K., Käs, E., Gonzalez, E., and Laemmli, U. K. (1993). SAR-dependent mobilization of histone H1 by HMG-I/Y in vitro—HMG-I/Y is enriched in H1-depleted chromatin. *EMBO J.* **12,** 3237–3247.

Zirbel, R. M., Mathieu, U. R., Kurz, A., Cremer, T., and Lichter, P. (1993). Evidence for a nuclear compartment of transcription and splicing located at chromosome domain boundaries. *Chromosome Res.* **1,** 92–106.

Zlatanova, J. S., and van Holde, K. (1992). Chromatin loops and transcriptional regulation. *CRC Crit. Rev. Eukaryotic Gene Express.* **2,** 211–224.

Zuo, P., and Manley, J. L. (1994). The human splicing factor ASF/SF2 can specifically recognize pre-messenger RNA 5′ splice sites. *Proc. Natl. Acad. Sci. U.S.A.* **91,** 3363–3367.

The Nuclear Matrix and the Regulation of Chromatin Organization and Function

James R. Davie
Department of Biochemistry and Molecular Biology, Faculty of Medicine,
University of Manitoba, Winnipeg, Manitoba, Canada R3E OW3

Nuclear DNA is organized into loop domains, with the base of the loop being bound to the nuclear matrix. Loops with transcriptionally active and/or potentially active genes have a DNase I-sensitive chromatin structure, while repressed chromatin loops have a condensed configuration that is essentially invisible to the transcription machinery. Core histone acetylation and torsional stress appear to be responsible for the generation and/or maintenance of the open potentially active chromatin loops. The transcriptionally active region of the loop makes several dynamic attachments with the nuclear matrix and is associated with core histones that are dynamically acetylated. Histone acetyltransferase and deacetylase, which catalyze this rapid acetylation and deacetylation, are bound to the nuclear matrix. Several transcription factors are components of the nuclear matrix. Histone acetyltransferase, deacetylase, and transcription factors may contribute to the dynamic attachment of the active chromatin domains with the nuclear matrix at sites of ongoing transcription.

KEY WORDS: Nuclear matrix, Chromatin organization, Histone modifications, High mobility group proteins.

I. Introduction

Studies on the structure and organization of chromatin have given us a glimpse of the arrangement of DNA in the nucleus. DNA is not packaged in a random heap in the nucleus. Rather, the DNA is organized into a hierarchy of structures, leading to the level of compaction needed to place 2 m of DNA into a nucleus with a diameter of 10 μm. The orderly packaging of DNA in the nucleus plays an important role in the functional aspects of gene regulation. A small percentage of the nuclear DNA is made available to transcription factors and the transcription machinery, while the remainder of the genome is in a state that is essentially invisible to the RNA polymerases. It is becoming increasingly evident that the nuclear matrix has a role in the organization of nuclear DNA and in nuclear processes such as replication, transcription, and RNA splicing. In this chapter, I will discuss our current understanding of the chromatin structure/ function relationships of the many histone isoforms, the enzymes that generate the modified histone isoforms, high mobility group proteins, and the role of the nuclear matrix in the organization and functioning of chromatin.

II. Organization of DNA within the Nucleus

In multicellular eukaryotic cells, nuclear DNA exists as a hierarchy of chromatin structures, resulting in compaction of the nuclear DNA about 10,000-fold (Fig. 1) (Reeves, 1984; Van Holde, 1988; Pienta *et al.*, 1991).

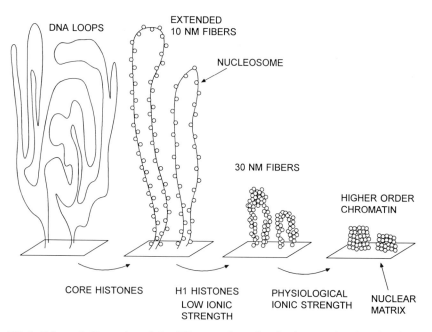

FIG. 1   Schematic illustration of the different orders of packaging postulated to give rise to the organization of DNA in the interphase nucleus. Transcriptionally repressed loops have a condensed chromatin structure, while potentially transcriptionally active loops have a less condensed structure that is sensitive to DNase I digestion.

When chromatin is spread across an electron microscope grid at low ionic strength, a "beads-on-a-string" structure or 10-nm fiber is observed. The beads correspond to the nucleosomes, the repeating structural units in chromatin (DNA packing, 6-fold). The stretch of DNA that joins the nucleosomes is termed the linker DNA. The 10-nm fiber probably occurs rarely in the nucleus and is compacted further into the 30-nm fiber (DNA packing, 40-fold). The 30-nm fiber appears to be folded into loops or domains believed to be anchored by proteins located at the base of the loops to a supporting nuclear structure called the nuclear matrix (DNA packing, 680-fold) (Pienta *et al.*, 1991). The loops range in size from 5 to 200 kb pairs, with an average size of 86 kb (Jackson *et al.*, 1990). Specific DNA sequences (matrix attachment regions or MARs) mediate primary chromatin anchor-

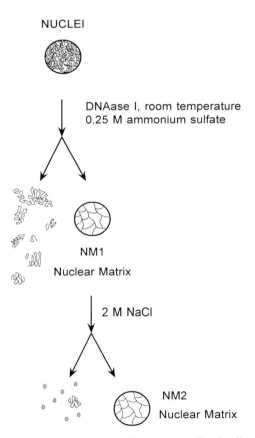

FIG. 2 Experimental procedure to isolate nuclear matrices. For details see Hendzel *et al.* (1994) and Sun *et al.* (1994).

age to the nuclear matrix. Each of these levels of chromatin packaging will be dealt with individually, but first a description of the nuclear matrix.

A. Nuclear Matrix

The nuclear matrix, the nuclear structure that remains following the salt extraction of nuclease treated nuclei, is composed of residual nucleoli, nuclear pore-lamina complex, and internal nuclear matrix (Berezney, 1991). The nuclear matrix defines nuclear shape and provides a structural support from which several nuclear processes such as DNA replication, transcription, and DNA repair are thought to occur (Pienta et al., 1991).

Several different procedures have been used to isolate and analyze the nuclear matrix. The typical nuclear matrix isolation protocol includes a salt extraction of nuclease (usually DNase I)-digested nuclei (Roberge et al., 1988; He et al., 1990; Belgrader et al., 1991a; Sun et al., 1994) (Fig. 2). An alternate procedure is to encapsulate cells or nuclei in agarose prior to nuclease digestion. Instead of extracting the nuclease-digested cells or nuclei with intermediate or high concentrations of salts, the nuclear DNA fragments not associated with the nuclear matrix are removed by electrophoresis (Jackson and Cook, 1985a; Jackson et al., 1988).

The major proteins of the nuclear pore–lamina complex are the nuclear lamins, lamins A (69 kDa), B (67 kDa), and C (62 kDa) (Gerace and Blobel, 1982; Kaufmann et al., 1983). Lamin B is an intrinsic membrane protein (Lebel and Raymond, 1984) and this lamin is attached to intermediate filaments such as peripherin (Djabali et al., 1991) and vimentin (Georgatos and Blobel, 1987). The internal nuclear matrix has a fibrogranular appearance (He et al., 1990; Belgrader et al., 1991a) (Fig. 3). Extraction of the nuclear matrices with high salt (2 M NaCl) may reveal an underlying network of core filaments with diameters of 9 and 13 nm (He et al., 1990). A similar network of 10-nm-diameter core filaments was observed by Jackson and Cook (1988) using electrophoresis to remove HaeIII-digested chromatin. The composition of the core filaments is presently not known.

FIG. 3   Electron micrographic analysis of trout liver nuclear matrices. The 0.25-M ammonium sulfate nuclear matrix preparation (NM1) was fixed with 2.5% glutaraldehyde and analyzed by electron microscopy. NM1 shows a nucleus containing a fibrogranular internal network. N is the residual nucleolus. NPL shows an empty shell of nuclear pore–lamina complex. Each bar: 1μM. [Reprinted with permission, Hendzel et al., 1994 (copyright by the American Society for Biochemistry and Molecular Biology).]

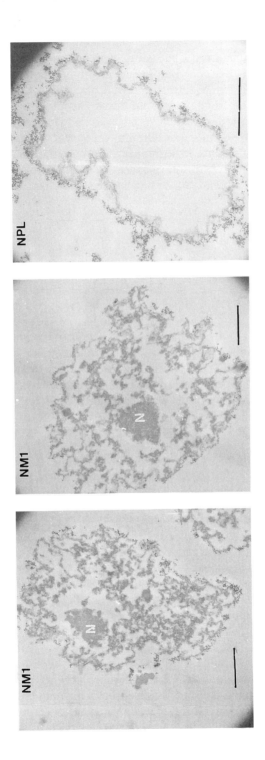

The fibrogranular internal nuclear matrix has a complex protein composition. The proteins that are components of the internal nuclear matrix are called nuclear matrins (Nakayasu and Berezney, 1991). Some of the nuclear matrins have been isolated and the genes encoding these proteins have been cloned (Belgrader *et al.*, 1991b; Hakes and Berezney, 1991). Nuclear matrix proteins are common to different sources of nuclear matrices, while others change in response to hormones, differentiation, and transformation (Fey and Penman, 1988; Stuurman *et al.*, 1989; Dworetzky *et al.*, 1990; Getzenberg and Coffey, 1990; Brancolini and Schneider, 1991; Cupo, 1991).

B. The Nucleosome

The canonical nucleosome core particle is characterized as a wedge-shaped disk, 5.7 nm in height and 11 nm in diameter. It is composed of 1.75 left-handed turns of a DNA superhelix (146 bp) wrapped around a histone octamer core that contains two each of the histones H2A (14.5 kDa), H2B (13.7 kDa), H3 (15.3 kDa), and H4 (11.3 kDa) (Richmond *et al.*, 1984; Arents and Moudrianakis, 1993). The histone octamer is arranged as a tetramer of histones H3 and H4 flanked on either face by a dimer of histones H2A and H2B. The structure of the histone octamer has been determined at 0.31-nm resolution (Arents and Moudrianakis, 1993).

C. Nucleosomal or Core Histones

Multicellular eukaryotes have five classes of histones, histones H2A, H2B, H3, H4, and H1. The nucleosomal histones or core histones (H2A, H2B, H3, and H4) have similar structures (Fig. 4) (Van Holde, 1988; Ramakrishnan, 1994). The distribution of basic amino acids in the nucleosomal histones is asymmetric with the disordered N-terminal portion of the molecule having a high amount of basic amino acid residues. The N-terminal tail of about 30 to 35 amino acids in length is followed by a globular region. In some cases the nucleosomal histones have a short C-terminal tail that is unstructured. Histones H3 and H4 are evolutionarily conserved, as is the globular portion of the histones H2A and H2B. It is the globular portion of these histones that is responsible for the histone-to-histone and histone-to-DNA contacts in the nucleosome. Removal of the histone tails by protease digestion does not affect the structural integrity of the nucleosome. Thus, the N-terminal tails of the nucleosomal histones are not required for nucleosome formation. However, removal of the N-terminal tails prevents

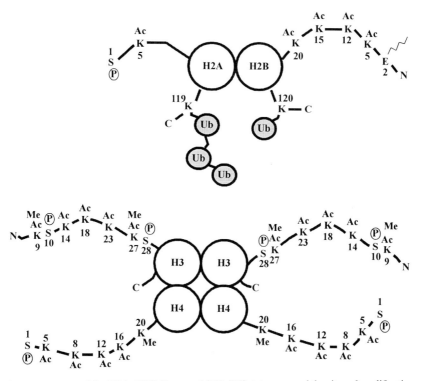

FIG. 4   Structures of the H2A–H2B dimer and (H3–H4)₂ tetramer and the sites of modification. The modifications shown are acetylation (Ac), methylation (Me), phosphorylation (P), ubiquitination (Ub), and poly(ADP-ribosyl)ation (the stepladder structure).

the chromatin fiber from establishing a stable 30-nm fiber in the presence of H1 histones (Allan *et al.*, 1982; Garcia Ramirez *et al.*, 1992).

D. The Extended Form of Chromatin

Chromatin in low ionic strength has been visualized by electron microscopy to have an extended 10-nm fiber, appearing as a beads-on-a-string structure. Recent tapping-mode scanning force microscopy analysis of chicken erythrocyte chromatin fragments at low ionic strength under nondenaturing conditions revealed irregular, three-dimensionally organized nucleosome arrays with a fiber diameter of approximately 30 nm (Leuba *et al.*, 1994). Removal of the H1 histones resulted in a loss of the three-dimensional structure and the generation of the beads-on-a-string form of chromatin.

Thus, histone H1 is required for the condensation of the chromatin at low ionic strength.

E. H1 Histones

The histone H1 has three structural domains, a short, disordered N-terminal domain, a globular central domain, and a long, disordered, highly basic C-terminal domain (Fig. 5). The globular region binds to the exit and entry points of nucleosomal DNA, stabilizing two turns of DNA around the nucleosome and fixing the angle of exit and entrance of the DNA (Van Holde and Zlatanova, 1994). It is currently thought that the N-terminal tail and the positively charged C-terminal tail of H1 bind to the linker DNA or adjacent nucleosomes, contributing to the condensation of the chromatin fiber.

F. Higher Order Chromatin Structure: Role of Histone H1 and Core Histone Tails

Electron microscopic analysis of chromatin folding shows that increasing the concentration of Na^+ to 50 or 60 mM leads to the folding of the chromatin fiber into 30-nm filaments (Fig. 1). Raising the concentration of Na^+ further to 100 mM results in greater condensation of the 30-nm fiber, ending with aggregation and precipitation. The aggregates appear to consist of one or more 30-nm filaments packed tightly together in an end-to-end and side-to-side arrangement (Widom, 1986).

   Stabilization of the 30-nm fiber is dependent on the presence of the H1 histones that bind to the linker DNA and to the core histones (C-terminal portion of histone H2A) (Boulikas et al., 1980). The N-terminal tails of the nucleosomal histones also have a role in the salt-dependent folding of chromatin fragments. Chromatin fragments depleted of H1 histones will fold in 80 mM Na^+. However, chromatin fragments treated with trypsin to remove the nucleosomal histone tails will not (Garcia Ramirez et al., 1992).

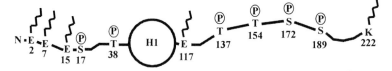

FIG. 5   Structure of histone H1 and sites of modification. The modifications shown are phosphorylation (P) and poly(ADP-ribosyl)ation (the stepladder structure).

Further, trypsin-treated chromatin fragments reconstituted with H1 histones do not fold (Allan *et al.*, 1982). Both the N-terminal tails of the nucleosomal histones and the H1 histones are required to stabilize the folded structure of chromatin at physiological ionic strength.

Several models have been proposed for the 30-nm fiber (Pienta *et al.*, 1991). Based on linker DNA length and the angle between the DNA entering and exiting each nucleosome, Woodcock *et al.* (1993) proposed a model of the chromatin fiber that resembled native structures. This model also was used successfully to predict the structure of chromatin fragments with histone H1 in low ionic strength (Leuba *et al.*, 1994). Graziano *et al.* (1994) used neutron scattering of histone H1 reconstituted chromatin to show that histone H1 is located in the interior of the 30-nm fiber.

G. Loops and Domains

The chromatin fiber is organized into loop domains, with the base of the loop attached to the nuclear matrix (Fig. 1). DNA in interphase nuclei from *Drosophila*, humans, and yeast cells is arranged as negatively supercoiled loops (Pienta *et al.*, 1991). The arrangement of nuclear DNA into a series of loops argued that there exists DNA sequences at the base of the loop that have an affinity for the nuclear matrix. In identifying DNA sequences that bind to the nuclear matrix, two approaches have generally been used. One method (the high salt method) evaluates the ability of labeled naked DNA fragments to bind nuclear matrices in the presence of unlabeled competitor DNA, typically *E. coli* DNA. The nuclear matrices are usually isolated by 2 M NaCl extraction of DNase I-digested nuclei. DNA fragments retained by nuclear matrices under these conditions are operationally defined as matrix attachment regions (MARs) (Cockerill and Garrard, 1986b; Cockerill, 1990). In the second method (the low salt method) nuclei are extracted with a buffer containing 25 mM 3,5-diiodosalicylic acid, lithium salt (LIS) (room temperature) which removes histones. The extracted nuclei are digested with restriction endonucleases (37°C). Southern blot hybridization with labeled probes identifies the sequence of the DNA fragment bound to the nuclear matrix (Mirkovitch *et al.*, 1984). These DNA fragments are operationally defined as scaffold attachment regions (SARs). Both methods yield essentially the same result; that is, a DNA sequence identified as a matrix attachment region (MAR) with the high salt method will be retained to the nuclear matrix prepared by the low salt method (a SAR). I will designate these DNA sequences, MARs or SARs, as MARs.

Matrix attachment regions were first identified in *Drosophila melanogaster* nuclei (Kc cells) (Mirkovitch *et al.*, 1984). Laemmli and colleagues used the low salt method to identify a MAR element that was present in a 167-base-pair DNA fragment located in the nontranscribed region between the histone H1 and histone H3 genes. There exist 100 copies of the histone gene cluster arranged as H3–H4–H2A–H2B–H1 in the haploid genome of *Drosophila*. This finding suggested that the histone gene clusters in *Drosophila* nuclei are organized as a series of 5-kb loops. Using both the low-salt and high-salt methods, Garrard and colleagues identified an intronic MAR element that was located next to an enhancer of the mouse immunoglobulin κ gene (Cockerill and Garrard, 1986b). MARs have now been described for many genes in a wide variety of eukaryotic cells (Gasser *et al.*, 1989; Laemmli *et al.*, 1992). Some MAR sequences have been found near regulatory DNA elements (e.g., enhancers), while others cohabit with regulatory DNA elements (e.g., origins of DNA replication and autonomously regulatory sequences) (Gasser *et al.*, 1989; Farache *et al.*, 1990). The basic mechanism of MAR element attachment to the nuclear matrix appears to be evolutionarily conserved among MAR elements (Cockerill and Garrard, 1986a). However, some MAR sequences do show tissue specificity. For example, the 5′ proximal MAR of the human apolipoprotein B gene domain is associated only with nuclear matrices of cells expressing the gene (Levy-Wilson and Fortier, 1989).

A comparison of the DNA sequences of MAR elements shows that they do not share extensive sequence homology. However, they typically consist of a region of 200 bp or longer of A/T-rich sequence, contain two or more copies of the ATATTT motif, and have topoisomerase II sites. However, not all MAR elements are A/T-rich and have topoisomerase II sites. For example, the 5′ MAR element of the human apolipoprotein B gene domain has neither A/T rich regions nor sites with homology with the topoisomerase II consensus sequence (Levy-Wilson and Fortier, 1989).

Concerns have been raised that MARs identified by the low-salt (LIS extraction) and/or high-salt (2 M NaCl extraction) *in vitro* methods may be artifacts (Eggert and Jack, 1991). Since the isolation of nuclei in low ionic strength buffers and extraction of nuclease-digested nuclei with high salt may lead to rearrangements, it is not clear if all of the operationally defined MARs are nuclear matrix bound *in vivo*. To address these concerns, Cook and colleagues described a method to prepare nuclear matrices and associated chromatin at physiological conditions (Jackson *et al.*, 1988). Cells encapsulated in agarose beads are lysed in a physiological buffer containing 0.5% Triton X-100, digested with a restriction endonuclease at 32°C, and then the detached chromatin is removed by electrophoresis. For HeLa cells the average size of the loops was 86 kb, with the range of loop sizes being 5 to 200 kb (Jackson *et al.*, 1990). The loop sizes were unchanged during the cell cycle. When nuclei were isolated under low ionic strength and subsequently extracted with LIS or 2 M NaCl, the loop size was reduced, suggesting that new attachment sites between the DNA and nuclear matrix were being generated. The report of Eggert and Jack (1991) also shows that an ectopic copy of the *Drosophila* MAR element in the 5' region of the *fushi tarazu* gene was associated with the low-salt (LIS-extracted) nuclear matrix, but the MAR element was removed by electrophoresis from non-treated nuclei. These observations suggest that some MARs may be artifacts.

Another view is that some MARs may not be constitutively bound to the nuclear matrix. The interaction of MARs may be dynamic rather than static. The low-salt and high-salt methods to identify MARs remove the histones so the DNA sequences (e.g., MARs) are no longer in nucleosomes. However, the MAR element in the *Drosophila* nontranscribed histone H1–H3 gene spacer has a highly ordered nucleosomal organization (Worcel *et al.*, 1983). Thus, the position of the MAR element in chromatin (e.g., in nucleosomes or linker DNA) may determine its availability for binding to the nuclear matrix. Further, histone modifications such as histone acetylation that alter nucleosome structure may regulate the binding of MARs to the nuclear matrix (Klehr *et al.*, 1992).

H. Nuclear Proteins That Bind MARs

Cytochemical localization studies as well as other biochemical approaches show that most of the loop attachment sites are to the internal nuclear matrix, with the remainder being bound to the nuclear pore–lamina (Izaurralde *et al.*, 1988; Phi-Van and Strätling, 1988; Zini *et al.*, 1989). Since topoisomerase II is a component of high-salt and low-salt internal nuclear matrices and a major component of metaphase chromosomal scaffold (Earnshaw and Heck, 1985; Gasser *et al.*, 1986; Kaufmann and Shaper, 1991), it seemed reasonable to predict that topoisomerase II is involved in mediating the attachment of MAR elements with the nuclear matrix or chromosome scaffold (Cockerill and Garrard, 1986b; Gasser and Laemmli, 1986; Adachi *et al.*, 1989). However, not all MARs have topoisomerase II sites, and nuclei from nonproliferating cells have very low levels of topoisomerase II (e.g., nuclei from chicken mature erythrocytes and myotubes; Heck and Earnshaw, 1986). Thus, in interphase nuclei other nuclear matrix proteins are likely involved in nuclear matrix–MAR interactions. Von Kries *et al.* (1991) purified a 95-kDa internal nuclear matrix protein called the attachment region-binding protein (ARBP) from chicken oviduct cells. The protein, which is also present in chicken mature erythrocyte nuclei, appears to recognize a structural feature common to MARs. These investigators also isolated another MAR-binding protein, p120, from hen oviduct nuclei (von Kries *et al.*, 1994). The p120 protein was found to be identical to a component of heterogenous nuclear ribonucleoprotein particles, suggesting that p120 has dual roles. SATB1, a MAR binding protein, is expressed predominantly in thymus (Dickinson *et al.,* 1992; Nakagomi *et al.,* 1994). SATB1, which has properties of a transcription factor, binds to a special AT-rich sequence context (ATC sequences) where one strand contains As, Ts, and Cs, but not Gs. RAP-1, a yeast 92-kDa repressor–activator protein associated with the nuclear matrix, binds to the MAR element (Hofmann *et al.,* 1989). SAF-A (scaffold attachment factor A), a 120-kDa abundant nuclear protein that is associated with the nuclear matrix of HeLa cells, binds the MAR element (Romig *et al.,* 1992). Molecules of SAF-A interact with each other and form large aggregates. The affinity of a MAR element for the nuclear matrix is proportional to its binding affinity to SAF-A. Both RAP-1 and SAF-A reconstitute DNA loops *in vitro.* Ludérus *et al.* (1994) have identified lamin A, desmin, and NuMA as proteins that bind specifically to MARs *in vitro.* Lamin B, a major component of the nuclear pore–lamina, also binds to MAR sites (Ludérus *et al.,* 1992). Thus, lamin A and B appear to have roles in the binding of chromatin to the nuclear lamina.

III. Core Histone Variants

There exist multiple forms of the histones H3, H2A, and H2B that differ by one of several amino acids in their primary sequence (Wu *et al.,* 1986). Histone variants generate considerable complexity in the histone octamers of the nucleosomes, and they may be of importance structurally and functionally. For example, in terminally differentiated rat cerebral cortex neurons there are four variants of H2A (H2A.1, H2A.2, H2A.Z, H2A.X), three variants of histone H3 (H3.1, H3.2, H3.3), and two variants of H2B (H2B.1, H2B.2) (Pina and Suau, 1987). The population of nucleosome histone variants change during development, differentiation (e.g., spermatogenesis), and lymphocyte activation (Wu *et al.,* 1986).

Histone H2A.Z, a member of the histone H2A family, appears to be evolutionarily conserved and appears to be a component of transcriptionally active chromatin (Ridsdale and Davie, 1987; White *et al.,* 1988). The histone H2A.Z-like variant hv1 is found in transcriptionally active *Tetrahymena* macronucleus but not in the inactive micronucleus (White *et al.,* 1988). Deletion of the *Drosophila* gene coding for the histone variant H2AvD, which is similar to mammalian H2A.Z, is lethal (van Daal and Elgin, 1992). Contrary to expectations, histone H2A.Z appears to stabilize nucleosome structure (Li *et al.,* 1993).

A novel histone H2A variant, macroH2A, was isolated from rat liver nucleosomes (Pehrson and Fried, 1992). The N-terminal third of macroH2A is 64% identical to H2A, while the remainder of the protein has a segment that resembles a leucine zipper, a dimerization motif found in many transcription factors.

The majority of nucleosomal histone synthesis and incorporation into chromatin is tightly coupled to DNA synthesis during S-phase of the cell cycle (Wu *et al.*, 1986). However, a basal amount of histone synthesis occurs in both G_1- and G_0-phases (Wu *et al.*, 1986). Histones H2A.Z, H2A.X, and H3.3 participate in basal histone synthesis, while histones H2A.1, H2A.2, and H3.2 take part in S-phase histone synthesis. Thus, in G_0- or G_1-phase cells, the proportion of histone H2A variants H2A.X and H2A.Z (G_1 only) and the histone H3 variant (G_0 and G_1) synthesized is increased relative to the S-phase synthesis pattern. Although all four nucleosomal histones are synthesized in the absence of DNA replication, newly synthesized histones H2A and H2B are preferentially incorporated into chromatin by exchanging with nucleosomal histones, preferentially with those associated with transcriptionally active DNA (Hendzel and Davie, 1990). Newly synthesized histones H3 and H4 partition to a greater extent with a soluble histone pool that is not tightly associated with chromatin (Bonner *et al.*, 1988).

IV. Modifications of the Core Histones

The core histones H2A, H2B, H3, and H4 and their variant forms undergo a variety of postsynthetic modifications including acetylation, methylation, ubiquitination, ADP ribosylation, and phosphorylation. Our current understanding of the roles of these modifications in altering chromatin structure/function and of the enzymes catalyzing these modifications will be discussed in the following sections.

A. Histone Acetylation

The histones are modified by two kinds of acetylation. The first kind is amino-terminal acetylation. This type of acetylation occurs shortly after synthesis of the polypeptide and is irreversible. Histones H2A, H4, and H1 usually have acetylated amino terminals, while histones H3 and H2B do not. The second kind of acetylation occurs at the ε-amino group of internal lysines found in the basic N-terminal tails of the histone molecule. This type of histone acetylation is involved in all nuclear processes that require an alteration in the structure of chromatin, including transcription, replication, and repair (Vidali *et al.*, 1988; Turner, 1991). All four of the core histones participate in this kind of acetylation, while H1 histones do not. Histone H2A is modified at one site, while histones H2A.Z, H2B, H3, and H4 are modified at four or five sites (Fig. 4). The sites of acetylation are highly conserved among all eukaryotes. It has been demonstrated directly with an antibody recognizing acetylated core histones that active, but not repressed, genes of chicken erythroid cells are associated with highly acetylated histones (Hebbes *et al.*, 1988).

1. Nonrandom Usage of Acetylation Sites

The order by which the various lysines are acetylated is nonrandom. For mammalian (pig thymus, calf thymus, HeLa cells) and yeast monoacetylated histone H4, lysine 16 is exclusively used (Couppez *et al.*, 1987; Turner *et al.*, 1989; Thorne *et al.*, 1990; Clarke *et al.*, 1993) (Fig. 4). However, in *Tetrahymena* macronucleus, lysine 7 is the preferred site of acetylation in monoacetylated histone H4 (Chicoine *et al.*, 1986). (Note that *Tetrahymena* histone H4 lacks an arginine at position 3 that is found in other histone H4s, off-setting the numbering of sites by one position.) *Drosophila* monoacetylated histone H4 uses lysines at positions 5, 8, or 12 (Munks *et al.*, 1991). From an analysis of the acetylation sites in di- and triacetylated histone H4, acetylation of yeast and mammalian histone H4 follows the order of lysine 16, then lysine 8 or 12, and then lysine 5. Thus, there is a progressive but not sole occupation of sites more N-terminal to lysine 16. For *Tetrahymena* macronuclear histone H4, acetylation follows the order of lysine 7, 4, and then 11 (preferred site, alternatively lysine 15 used). Histone H3 of pig thymus and HeLa cells has a preferred order of site utilization (lysine 14, 23, and then 18) (Thorne *et al.*, 1990). *Tetrahymena* macronuclear histone H3 is acetylated at lysine 9 or 14, then at the alternate site (i.e., lysine 14 or 9) followed by a preference for lysine 18 as the third acetylation site (Chicoine *et al.*, 1986). Pig thymus and HeLa cell histone H2B had a reduced level of specificity in site utilization, with lysines 12 and 15 being used before 5 and 20.

Histone H4 is acetylated in the cytoplasm before being brought into the nucleus, and it is assembled into chromatin as a diacetylated isoform. This form of acetylation is referred to as deposition-related acetylation. In *Tetrahymena*, newly synthesized histone H4 is acetylated at sites lysine 4 and 11. Newly synthesized human histone H4 is acetylated at lysines 5 and 12 (Chicoine *et al.*, 1986). In cuttle fish testis where the histones are being replaced by protamines the monoacetylated isoform of histone H4 is acetylated at lysine 12 and the diacetylated isoform is acetylated at lysines 12 and 5 (Couppez *et al.*, 1987). Thus, replacement- and deposition-related acetylation of histone H4 appear to use the same sites.

Antibodies that distinguish the histone H4 isoforms acetylated at lysines 5, 8, 12, and 16 have been used to determine the distribution of these histone H4 isoforms in the genome (Turner *et al.*, 1992; Jeppesen and Turner, 1993; Sommerville *et al.*, 1993). Immunolabeling studies with these antibodies and human and mouse metaphase chromosomes show that the inactive X chromosome and pericentromeric heterochromatin, transcriptionally inactive regions are underacetylated (Jeppesen and Turner, 1993). Antibodies to the rare (2% of total H4) histone H4 isoform acetylated at lysine 5, which occurs only in tri- and tetraacetylated isoforms, produced the highest differentiation between bands and interbands, with R bands being immunolabeled. The R bands correspond to the parts of the genome that are transcriptionally active during interphase. Thus, the results of these studies suggest that potentially transcriptionally active regions of the genome are associated with highly acetylated histone H4 isoforms, while constitutive and facultative heterochromatin are devoid of acetylated histone H4 (Jeppesen and Turner, 1993).

2. Kinetics of Histone Acetylation and Deacetylation

To study the rate of acetylation–deacetylation, the histone deacetylase inhibitor sodium butyrate has been used. Such studies show that there exist distinct populations of acetylated histones. In hepatoma tissue culture cells, there are two populations of acetylated histone. One population of core histones is characterized by rapid hyperacetylation ($t_{1/2} = 7$ min for monoacetylated histone H4) and rapid deacetylation ($t_{1/2} = 3$ to 7 min) (Covault and Chalkley, 1980). This highly dynamic acetylation–deacetylation is limited to 15% of the core histones. A second population is acetylated ($t_{1/2} = 200$ to 300 min for monoacetylated H4) and deacetylated at a slower rate ($t_{1/2} = 30$ min). The other core histones (H2A, H2B, H3) had two rates of deacetylation, with the acetylated isoforms of histones H2A and H2B having a greater proportion participating in rapid deacetylation than did histones H3 and H4.

About 30% of the histones of chicken erythrocytes are acetylated. Most of the acetylated histones are frozen in this modification state; that is, they are not deacetylated. Among the acetylated histones, however, there is a population of histones that is reversibly acetylated. It has been estimated that 1–2% of the genome of adult chicken mature and immature erythrocytes participates in dynamic acetylation (Zhang and Nelson, 1986). These cells do not replicate and are arrested in the G_0-phase of the cell cycle. In mature erythrocytes, two rates of acetylation are apparent ($t_{1/2} = 12$ and 300 min for monoacetylated histone H4) (Zhang and Nelson, 1988a). Only one rate of acetylation ($t_{1/2} = 12$ min for monoacetylated histone H4) has been observed for the core histones of adult chicken immature erythrocytes (Zhang and Nelson, 1988a; Hendzel and Davie, 1991). However, there are two populations of core histones (H3 and H4) participating in this rapid acetylation. One population of rapidly acetylated histone becomes hyperacetylated (e.g., to the tetraacetylated form of histone H4) in the presence of sodium butyrate. Upon removal of the inhibitor, the hyperacetylated histone species are rapidly deacetylated ($t_{1/2} = 5$ min for both mature and immature cells; Zhang and Nelson, 1988b). The chicken erythrocyte histones that participate in this rapid hyperacetylation and deacetylation are associated with transcriptionally active DNA (Hendzel et al., 1991). The other population of the rapidly acetylated histones (H3 and H4), which is associated with active and competent DNA (Hendzel et al., 1991), achieves only low levels of acetylation (e.g., mono- and diacetylated forms of histone H4) in the presence of sodium butyrate. In the absence of butyrate, these histones are slowly deacetylated ($t_{1/2} = 90$ min for immature cells and 145 min for mature cells). The finding that a small percentage of the genome contains histones that are engaged in rapid acetylation and deacetylation argues that histone acetyltransferase(s) and deacetylase(s) are localized with these regions (see Fig. 14).

The level of histone acetylation changes throughout the cell cycle (Bradbury, 1992). The core histones are all deacetylated in M-phase of the cell cycle, correlating with the condensation of chromatin into metaphase chromosomes (D'Anna et al., 1977; Turner, 1989; Bradbury, 1992). A rapid increase in the level of acetylated histone H4 isoforms occurs in telophase (D'Anna et al., 1977). The cycle of histone H4 acetylation and deacetylation persists within metaphase chromosomes, in regions corresponding to R bands (potentially transcriptionally active regions) (Turner, 1989; Jeppesen and Turner, 1993). This observation suggests that histone acetyltransferases and deacetylases are associated with metaphase chromosomes in potentially transcriptionally active regions, ready to reset in interphase the dynamic acetylation of histones bound to transcriptionally active DNA. Consistent with this idea are the observations that histone acetylation precedes transcription (Clayton et al., 1993). Further, since histone acetyltransferases and deacetylases are associated with the nuclear matrix (Hendzel et al., 1991, 1994), these enzymes may target potentially transcriptionally active regions of the genome to the nuclear matrix as soon as the nuclear matrix reforms and chromosomes become dispersed.

B. Histone Acetylation Alters Higher Order Chromatin and Nucleosome Structure

Transcriptionally active chromatin is associated with H1 histones (Kamakaka and Thomas, 1990; Ridsdale et al., 1990). As the H1 histones play a key role in the stabilization of the higher order folding of the chromatin fiber, what prevents the transcriptionally active chromatin regions from accepting a higher order structure? Reconstitution experiments suggest that core histone acetylation alters the capacity of the H1 histones to condense transcriptionally active/poised chromatin (Ridsdale *et al.*, 1990). Chromatin fragments stripped of H1 histones are soluble in 150 m*M* NaCl, whereas chromatin fragments reconstituted with H1 histones aggregate and precipitate (referred to as exogenously added H1 histone-induced precipitation). Reconstitution experiments revealed that active/poised chromatin fragments are much more resistant than repressed chromatin fragments to exogenously added H1 histone-induced precipitation in 0.15 *M* NaCl. By incubating cells in the presence or absence of sodium butyrate, the acetylation level of dynamically acetylated core histones bound to active/poised DNA can be manipulated. Active/poised chromatin fragments with unacetylated histones were unable to resist exogenously added H1 histone-induced salt precipitation (Ridsdale *et al.*, 1990). Thus, in agreement with the results of others, histone acetylation is responsible for altering the capacity of the H1 histones to condense transcriptionally active/competent chromatin (Annunziato *et al.*, 1988; Perry and Annunziato, 1991; Verreault and Thomas, 1993).

Histone acetylation alters nucleosome structure (Ausio and Van Holde, 1986; Nacheva *et al.*, 1989). Histone acetylation reduces the linking number change per nucleosome, i.e., negative DNA supercoils constrained in unmodified nucleosomes are partially released in nucleosomes with hyperacetylated histones (Norton *et al.*, 1990; Thomsen *et al.*, 1991; but also see Lutter *et al.*, 1992). Acetylation of histones H3 and H4 is responsible for the nucleosome-linking number change (Norton *et al.*, 1990). The ability of histone acetylation to release negative supercoils into a chromatin domain suggests that histone acetyltransferase is functioning as a eukaryotic DNA gyrase (Bradbury, 1992). The liberation of DNA from the nucleosome and/ or change in nucleosome shape as a consequence of acetylation may alter the path of the DNA entering and leaving the nucleosome which in turn may alter the interaction between H1 histones and nucleosomal/linker DNA (Allan *et al.*, 1982; Bauer *et al.*, 1994). These alterations in DNA path and H1 histone–nucleosome interaction may prevent (or alter) the formation of compact higher-order structures.

C. Enzymes Catalyzing Reversible Histone Acetylation

Reversible histone acetylation is catalyzed by histone acetyltransferase and histone deacetylase, with the level of acetylation being governed by the net activities of these two enzymes. Also these enzymes probably have a role in determining the site of acetylation. There are several forms of histone acetyltransferase. Histone acetyltransferase B is a cytoplasmic enzyme that acetylates free histone H4 (Richman *et al.*, 1988; Mingarro *et al.*, 1993). Histone acetyltransferases A and DB (DNA-binding) are bound to chromatin and acetylate all four of the nucleosomal histones when free or within nucleosomes (Böhm *et al.*, 1980; Attisano and Lewis, 1990). Histone deacetylase is a nuclear enzyme (Vidali *et al.*, 1972; Lopez Rodas *et al.*, 1991). As with histone acetyltransferase, there are several forms of histone deacetylases (Brosch *et al.*, 1992; Lopez Rodas *et al.*, 1992). *Physarum polycephalum* has two forms of histone deacetylase (HD) that differ in their substrate specificities (Lopez Rodas *et al.*, 1992).

In unraveling the relationship between histone acetylation and nuclear processes, inhibitors of histone deacetylase have been found to be most helpful. Sodium butyrate has been the most widely used as a reversible, potent inhibitor of histone deacetylase. Mammalian and avian histone deacetylases are inhibited by sodium butyrate at millimolar concentrations (2 to 5 mM) (Candido *et al.*, 1978). However, this inhibitor lacks specificity, has multiple effects on nuclear function, and induces the expression of a histone subtype, H1^0 (Boffa *et al.*, 1978; Candido *et al.*, 1978; D'Anna *et al.*, 1980). Specific and potent histone deacetylase inhibitors, trichostatin A and trapoxin (an antitumor cyclic tetrapeptide), have recently come into use (Yoshida *et al.*, 1990; Kijima *et al.*, 1993). Trichostatin A, but not trapoxin, is a reversible inhibitor of mammalian histone deacetylase. Studies with these three inhibitors show that the reversible nature of histone acetylation has a crucial role in cell cycle control and differential gene expression (Kijima *et al.*, 1993). Trichostatin A and sodium butyrate arrest cells in G$_1$- and G$_2$-phases of the cell cycle (Boffa *et al.*, 1978; Yoshida *et al.*, 1990).

D. Vertebrate Histone Acetyltransferase and Deacetylase Are Associated with the Nuclear Matrix

To determine the nuclear location of the histone acetyltransferase and histone deacetylase activities in chicken erythrocytes, the fractionation protocol shown in Fig. 8 was applied. Chromatin fractions S$_{150}$ and P$_E$ are enriched in transcriptionally active DNA. The P$_E$ fraction consists of chromatin fragments that are bound to the residual nuclear material, the nuclear matrix. Both enzymes were present in S$_{150}$ chromatin fraction, but the P$_E$ fraction, which contains less than 10% of the nuclear DNA and less than 36% of the nuclear protein, retained 79 and 88% of the nuclear histone deacetylase and acetyltransferase activity, respectively (Hendzel *et al.*, 1991, 1994). The demonstration that the enzymes bound to the nuclear skeletons catalyze reversible histone acetylation using as substrate the chromatin fragments attached to the nuclear skeleton suggests that these enzymes are localized near each other (Davie and Hendzel, 1994; Hendzel *et al.*, 1994) (see model in Fig. 14).

We found that 2 *M* NaCl extraction of fraction P_E nuclear skeletons removed the chromatin fragments but not the histone deacetylase activity, suggesting that histone deacetylase is bound to the nuclear matrix. To isolate nuclear matrix, the procedure shown in Fig. 2 was used. Nuclear matrices NM1 and NM2 isolated by this and other procedures and from different sources (chicken erythrocytes, chicken liver, trout liver, trout hepatocellular carcinomas, and hamster liver) retained the majority of the nuclear histone deacetylase and acetyltransferase activities (Hendzel *et al.*, 1991, 1994; Hendzel and Davie, 1992b). For example, the trout liver NM2 nuclear matrices had 76% of the nuclear histone acetyltransferase activity and 77% of the histone deacetylase activity. Consistent with these results, the histone deacetylase of HeLa cells was located in a high-molecular-weight fraction with characteristics of the nuclear matrix (Hay and Candido, 1983a,b).

The nuclear matrix consists of nuclear pore–lamina and internal nuclear matrix. Empty shells of nuclear pore–lamina complexes retained less than 20% of the histone acetyltransferase and deacetylase activities (Hendzel and Davie, 1992b; Hendzel *et al.*, 1994). Nuclear pore–lamina complexes were isolated by the method described by Kaufmann *et al.*, (1983) (Fig. 6).

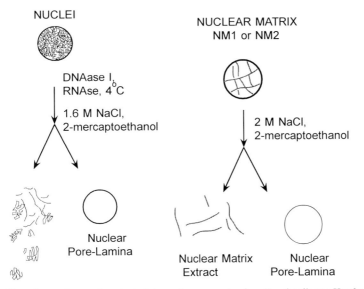

FIG. 6   Experimental procedure to isolate nuclear pore–lamina. For details see Kaufmann *et al.* (1983) and Sun *et al.* (1994).

Salient features of this procedure are nuclease digestions are done at 0°C, a RNase digestion is included, and the nuclease-digested nuclei are extracted with high salt and a reducing agent, 2-mercaptoethanol. These agents destabilize internal nuclear matrix structure (He *et al.*, 1990; Belgrader *et al.*, 1991a; Hendzel and Davie, 1992b). This procedure works well for tissues and erythrocytes (see Fig. 3), but it is not effective in solubilizing the internal nuclear matrix of some tissue culture cells (e.g., human breast cancer cells T-47D-5).

To solubilize the internal nuclear matrix components from chicken mature or immature erythrocyte nuclear matrices, NM1 or NM2 nuclear matrices were extracted with high salt and 2-mercaptoethanol (Sun *et al.*, 1994) (Fig. 6). The structures left after this extraction were empty shells of nuclear pore–lamina. This procedure solubilized the majority (76% or greater) of the histone acetyltransferase and deacetylase activity that was associated with the NM1 or NM2 nuclear matrix. Together these observations show that these enzymes are principally associated with the internal nuclear matrix. Since transcriptionally active DNA is complexed with histones that undergo rapid high acetylation and rapid deacetylation, nuclear matrix-associated histone acetyltransferase and histone deacetylase may mediate an interaction between the nuclear matrix and transcriptionally active chromatin (see Fig. 14).

E. Histone Deacetylase as a Marker Enzyme for the Internal Nuclear Matrix

Lafond and Woodcock (1983) reported that nuclear matrices of chicken mature erythrocytes lack an internal nuclear matrix. The conditions they used to prepare the nuclear matrices for electron microscopy analysis included a DNase I digestion at 4°C and an extraction with 2 *M* NaCl. However, biochemical studies provide evidence for the existence of an internal nuclear matrix in mature erythrocytes (Phi-Van and Strätling, 1988; von Kries *et al.*, 1991). To address this controversy, mature erythrocytes were digested with DNase I at 4°C or at room temperature and extracted sequentially with 0.4 and 2 *M* KCl. The low temperature-digested nuclear matrices had low levels of histone deacetylase activity, while the room temperature-digested nuclear matrices retained greater than 50% of the histone deacetylase activity (Hendzel and Davie, 1992b) and 57% of the histone acetyltransferase activity (Hendzel *et al.*, 1994). Electron microscopic visualization of mature erythrocyte nuclear matrices from room temperature-digested nuclei shows that mature erythrocyte nuclear matrices indeed have an internal nuclear matrix (M. J. Hendzel, J.-M. Sun, H. Y. Chen, J. B. Rattner, and J. R. Davie, unpublished observations, 1994).

These results provide evidence that the stability of the chicken mature erythrocyte internal nuclear matrix is temperature-dependent.

F. Histone Methylation

The core histones H2B, H3, and H4 are modified by methylation. Histones H3 and H4 undergoing methylation in chicken immature erythrocytes are associated with transcriptionally active DNA (Hendzel and Davie, 1989). Histone methylation is a relatively stable modification with a slow turnover rate (Wu *et al.*, 1986). The role of histone methylation in altering nucleosome and/or chromatin structure is not known. In animals and lower eukaryotes but not plants, histone H4 is methylated at lysine 20 (Wu *et al.*, 1986; Van Holde, 1988; Waterborg, 1993b). Lysine residues 9 and 27 of histone H3 are also methylated (Fig. 4). These lysines can be methylated to form N^6-monomethyllysine, N^6-dimethyllysine, and N^6-trimethyllysine (Wu *et al.*, 1986; Waterborg, 1993b). Heat shock of *Drosophila melanogaster* Kc cells induces methylation of histone H2B at N-terminal proline residue (Desrosiers and Tanguay, 1988). Heat shock of these cells also results in methylation of histone H3 at arginine residues.

The level of acetylation for histones H3 and H4 correlates with the extent of methylation (Hendzel and Davie, 1989; Waterborg, 1993a,b). In chicken immature erythrocytes, dynamically acetylated histones are selectively methylated (Hendzel and Davie, 1991). However, the processes of histone methylation and dynamic acetylation are not directly coupled; neither modification predisposes histone H3 or H4 to the other (Hendzel and Davie, 1992a). Histone–lysine methyltransferase is a chromatin-bound enzyme that catalyzes the addition of methyl groups onto the ε-amino groups of chromatin-bound histones H3 and H4 (Hendzel and Davie, 1989).

G. Histone Ubiquitination

In lower and higher eukaryotes, histones H2A and H2B and their variant forms are ubiquitinated (Wu *et al.*, 1986; Van Holde, 1988). One exception is yeast, which lacks ubiquitinated histone H2A (Swerdlow *et al.*, 1990). Histone ubiquitination is a reversible modification. The carboxyl end of ubiquitin, a highly conserved 76 amino acid protein, is attached to the ε-amino group of lysine (Lys-119 in H2A; Lys-120 in H2B; Thorne *et al.*, 1987) (Fig. 4). In higher eukaryotes, histone H2A is typically ubiquitinated to a greater extent than is histone H2B (approximately 10% of H2A versus about 1–2% of H2B). These histones are also polyubiquitinated, with histone H2A having the greater amounts of polyubiquitinated isoforms (Nickel and Davie, 1989; Nickel *et al.*, 1989). The major arrangement of ubiquitin in polyubiquitinated histone H2A is a chain of ubiquitin molecules joined to each other by isopeptide bonds to a ubiquitin molecule that is attached to the ε-amino group of lysine 119 of H2A (Nickel and Davie, 1989).

The ubiquitin ligase system consists of a ubiquitin-activating enzyme, E1, and several ubiquitin-conjugating enzymes, E2s (Hershko, 1988). E1 catalyzes the first step in the conjugation reaction. This enzyme is present in the nucleus (Cook and Chock, 1991; Schwartz *et al.*, 1992). Several E2 enzymes have been shown to catalyze the addition of ubiquitin onto histones (Jentsch *et al.*, 1990). It is not known which E2 enzyme(s) is responsible for this reaction in the nucleus.

Ubiquitinated isoforms of the histones disappear at metaphase and reappear in anaphase (Matsui *et al.*, 1979; Mueller *et al.*, 1985; Raboy *et al.*, 1986). In dividing and nondividing cells, the ubiquitin moiety of the ubiquitinated histones is in rapid equilibrium with a pool of free ubiquitin (Seale, 1981; Wu *et al.*, 1981). The turnover of the ubiquitinated histones is presumably catalyzed by ubiquitin–C-terminal hydrolases (Andersen *et al.*, 1981; Matsui *et al.*, 1982).

Ubiquitinated histone H2B and to a lesser extent ubiquitinated histone H2A are associated with transcriptionally active DNA (Nickel *et al.*, 1989; Davie and Murphy, 1990, 1994). Ubiquitination of histone H2B is the only histone modification that is dependent upon ongoing transcription (Davie and Murphy, 1990, 1994). The COOH terminal sequence of histone H2B, but not H2A, is buried in the nucleosome (Hatch *et al.*, 1983). Thus, the process of transcription may alter nucleosome structure (Morse, 1992; Van Holde *et al.*, 1992) or transiently displace the histone octamer from DNA (Clark and Felsenfeld, 1992; Studitsky *et al.*, 1994), allowing the COOH terminus of histone H2B to become accessible to the enzymes catalyzing the addition of ubiquitin. Another mechanism, which does not require ongoing transcription, is by exchange of newly synthesized ubiquitinated histones H2B and H2A with histones that were in transcriptionally active nucleosomes (Hendzel and Davie, 1990). The introduction of ubiquitinated histone H2B into the nucleosome may result in an alteration in nucleosome and/or higher order chromatin structure.

H. Histone ADP-Ribosylation

Poly(ADP-ribosyl)ation has been implicated in several nuclear processes, including replication, repair, and recombination (Boulikas, 1989, 1990b). The four core histones are modified by adenosine diphospho (ADP) ribosylation, which involves the transfer of the ADP-ribose moiety of NAD^+ to the histone acceptor. One or more ADP-ribose groups may be transferred, resulting in core histones with mono(ADP-ribose) up to highly branched poly(ADP-ribose) residues. The acetylated isoforms of the core histones are preferentially ADP-ribosylated (Boulikas, 1988, 1990a,b; Golderer *et al.*, 1991). About 15 ADP-ribosylated isoforms were observed for histones H3.1, H3.3, H2B.1, and H2B.2 in dimethyl sulfate-treated mouse myeloma cells (Boulikas, 1988). Less than 5% of the core histones are modified by ADP-ribosylation. Further, poly(ADP-ribose) is rapidly turned over, having a half-life of 30 sec to 10 min (Boulikas, 1989, 1990a,b). For histone H2B, the carboxyl group of glutamic acid residue 2 is the site of ADP ribosylation (Ogata *et al.*, 1980) (Fig. 4). The sites of ADP-ribosylation for the other core histones are not known. However, histone H2A has a carboxylate ester ADP-ribose-protein bond, while histones H3 and H4 appear to have arginine-linked ADP-ribose residues (Golderer *et al.*, 1991).

The enzyme catalyzing the addition of ADP-ribose units onto the histones and itself is poly(ADP-ribose) polymerase or synthetase. Poly(ADP-ribose) polymerase is a nuclear, DNA-dependent enzyme that is stimulated by DNA breaks (de Murcia *et al.*, 1988). This property of the enzyme would target its action to sites that have DNA strand breaks (regions of the genome involved in replication, repair, recombination). The enzyme is associated with chromatin areas and perichromatin regions in interphase Chinese hamster ovary cells (Fakan *et al.*, 1988). Both poly(ADP) ribose and poly(ADP-ribose) polymerase are associated with the nuclear matrix (Alvarez Gonzalez and Ringer, 1988; Pedraza Reyes and Alvarez Gonzalez, 1990; but also see Kaufmann *et al.*, 1991). Degradation of the ADP-ribose polymer is catalyzed by the nuclear enzyme poly(ADP-ribose) glycohydrolase and ADP-ribosyl protein lyase.

ADP-ribosylation of the core histones alters nucleosome structure (Huletsky *et al.*, 1989). Realini and Althaus (1992) have put forth the hypothesis that poly(ADP-ribosyl)ation may have a function in histone shuttling. They propose that poly(ADP-ribose) polymerase directed to sites of DNA strand breaks would automodify itself generating multiple ADP-ribose polymers. The polymers would lead to the dissociation of the histones from DNA onto the polymers. The DNA would now be free for processing (e.g., by enzymes involved in excision repair). The action of poly(ADP-ribose) glycohydrolase would degrade the ADP-ribose polymers, leading to the release of the histones that would rebind the DNA. It is possible that histone shuttling takes place on ADP-ribose polymers that are transiently associated with the nuclear matrix.

I. Histone Phosphorylation

All four of the core histones are phosphorylated, with the sites of phosphorylation usually being found in the amino-terminal part of the histones (Van Holde, 1988) (Fig. 4). Phosphorylation of the core histones has been implicated in transcription and chromosome condensation. Histone H3 is phosphorylated to different extents throughout the cell cycle, while histones H2A and H4 are phosphorylated at uniform rates during the cell cycle. Histones H2A.1, H2A.2, and H2A.X are phosphorylated at serine residue 1 (Sung and Dixon, 1970; Pantazis and Bonner, 1981). Histone H2A.Z is not phosphorylated. Phosphorylation of histone H2A occurs in the transcriptionally active macronucleus of *Tetrahymena thermophila*, but not in the transcriptionally inert micronucleus (Allis and Gorovsky, 1981). *Tetrahymena* histone H2A variant hv1 is phosphorylated (Allis *et al.*, 1980). Histone H4, like histone H2A, is phosphorylated at the N-terminal serine residue. Sea urchin sperm-specific histone H2B variants, H2B.1 and H2B.2, are phosphorylated at several sites in the N-terminal domain (H2B.1, two or three sites; H2B.2, four sites). The consensus sequence for the phosphorylation sites was -Ser-Pro-X-Lys/Arg- (X is Thr, Gln, Lys, or Arg). This sequence is recognized by the histone H1 growth-associated kinase or p34[cdc2] kinase (Hill *et al.*, 1990).

Phosphorylation of histone H3 increases dramatically in mitosis, with phosphorylation occurring on serine residue 10 (Shibata *et al.*, 1990). Histone H3 phosphorylation has been linked to the condensation of chromatin, with histone H3 phosphorylation, but not histone H1 phosphorylation, being specific to premature condensation of chromosomes (Ajiro and Nishimoto, 1985). Phosphorylation of histone H3 is also elevated in the G_0–G_1-phase transition. Treatment of G_0-phase-arrested mouse fibroblasts with growth factors, phorbol esters, okadaic acid, or protein synthesis inhibitors leads to the rapid transcriptional activation of the immediate early genes (e.g., c-*fos*, c-*jun*) and the rapid phosphorylation of histone H3 and its variant forms (Mahadevan *et al.*, 1991; Barratt *et al.*, 1994). Serine residues 10 and/or 28 located in the basic N-terminal domain were modified. *In vitro* cAMP-dependent kinase can phosphorylate histone H3 in nucleosomes at serine residues 10 and 28 (Shibata and Ajiro, 1993). Mitogen-activated *rsk*-encoded protein kinases (RSKs or pp90[rsk]) also phosphorylate histone H3 *in vitro* (Chen *et al.*, 1992). Dynamically acetylated histone H3 is preferentially phosphorylated (Barratt *et al.*, 1994). As with histone methylation, acetylation is not a prerequisite for phosphorylation. However, these observations suggest that histone acetyltransferases, histone deacetylases, histone methyltransferases, and mitogen-regulated histone H3 kinase are targeted to specific regions on the nuclear matrix that are engaged in transcription.

V. Histone H1

A. Subtypes

The H1 histones are a heterogenous group of several subtypes that differ in amino acid sequence. Most nuclei (not yeast) typically have more than one histone H1 subtype. The relative amounts of the histone H1 subtypes vary with cell type within a particular species, as well as among various species. For example, mouse tissues contain various levels of histone H1 subtypes, H1a, H1b, H1c, H1d, H1e, and $H1^0$ (Lennox and Cohen, 1983). Recently, Parseghian *et al.* (1994b) put forth a proposal for a coherent nomenclature for the mammalian H1 histones. The expression of the subtypes is differentially regulated throughout development, through the cell cycle, and during differentiation (Cole, 1987; Lennox and Cohen, 1988). Changes in the relative levels of the H1 histone subtypes have been observed in normal and neoplastic cells (Davie and Delcuve, 1991; Giancotti *et al.,* 1993).

Since histone H1 subtypes differ in their abilities to condense DNA and chromatin fragments, it has been proposed that the differential distribution of the H1 histones with chromatin domains may generate chromatin regions with different degrees of compaction. Indirect immunofluorescence studies with histone H1 subtype-specific antibodies have shown that the nuclear location of specific histone H1 subtypes is nonuniform. Rodent histone $H1^0$ co-localized with nucleoli, human H1-3 is found primarily in the nuclear periphery, and human H1-1 is distributed in parallel to the DNA concentration (Breneman *et al.,* 1993; Gorka *et al.,* 1993; Parseghian *et al.,* 1993, 1994a). Antibodies to human histones H1-2 and H1-4 generated a punctate staining pattern, reminiscent of the speckled staining patterns described when the nuclear sites of splicing factors, small nuclear RNAs and RNA synthesis were localized (Huang and Spector, 1992; Jackson *et al.,* 1993; Parseghian *et al.,* 1994a).

B. Modifications of Histone H1

Further heterogeneity in the H1 histone population arises from post-translational modification. The H1 histones are phosphorylated at serine and threonine residues located in the N- and C-terminal domains of the protein (Van Holde, 1988) (Fig. 5). Sea urchin sperm-specific histone H1 and *Tetrahymena* macronuclear histones H1 are phosphorylated in the N-terminal and/or C-terminal domain at several sites that contain the phosphorylation consensus sequence -Ser/Thr-Pro-X-Lys/Arg (Hill *et al.,* 1990; Roth *et al.,* 1991). This sequence corresponds to the $p34^{cdc2}$ kinase consensus motif (Roth *et al.,* 1991). Histone H1 is also phosphorylated by cAMP- or cGMP-dependent kinases, for example at serine residue 37 in rat liver or rabbit thymus histone H1(Van Holde, 1988; Parseghian *et al.,* 1994b).

Immunochemical and biochemical data show that histone H1 phosphorylation increases dramatically as cells progress through the cell cycle (Van Holde, 1988; M. J. Lu *et al.,* 1994). Phosphorylation of H1 begins in G_1, continues at an increasing rate and extent throughout S and G_2, and reaches a maximum in mitosis. The histone H1 subtypes, however, differ in their extent of phosphorylation and the scheduling of some of their phosphorylation during the cell cycle (Hohmann, 1983). Typically most histone H1 subtypes are phosphorylated at several sites in the C-terminal domain during mitosis (Van Holde, 1988). Histone H1 phosphorylation that occurs during cell growth and division is catalyzed by a chromatin-bound enzyme p34^{cdc2} kinase (growth-associated histone H1 kinase) (Chambers and Langan, 1990; Roth et al., 1991). The activity of this kinase is regulated throughout the cell cycle by phosphorylation and interactions with cyclins (Pelech et al., 1990).

Bradbury (1992) proposed that histone phosphorylation drives chromosome condensation. However, several studies suggest that phosphorylation of the H1 histone subtypes decondenses, rather than condenses, chromatin (Roth and Allis, 1992). Reconstitution studies of chromatin with rat thymus histone H1 and phosphorylated histone H1, which was phosphorylated in vitro by p34^{cdc2} kinase to an average of 5.3 phosphates per molecule, showed that phosphorylation of histone H1 did not induce a major structural alteration of chromatin structure. However, at low ionic strength phosphorylation of histone H1 caused a destabilization of chromatin structure (Kaplan et al., 1984).

Recently, Allis and colleagues isolated an antibody that is selective for the highly phosphorylated isoform of Tetrahymena histone H1 (M. J. Lu et al., 1994). This antibody also detected highly phosphorylated HeLa histone H1 subtypes. In indirect immunofluorescence experiments with the antibody, a punctate pattern of nuclear staining was observed for the HeLa that were in G_1-phase of the cell cycle. For cells in S-phase and mitosis, the staining became more diffuse and more intense. The G_1-phase staining pattern was similar to the speckled staining patterns observed when the nuclear sites of splicing factors, small nuclear RNAs, and RNA synthesis were localized (Huang and Spector, 1992; Jackson et al., 1993). These results demonstrate that the phosphorylated isoform of HeLa histone H1 that is recognized by the antibody is nonuniformly distributed in the nuclei of G_1-phase cells, and the possibility exists that the phosphorylated histone subtype is associated with transcriptionally active genes.

Histone H1 is poly(ADP-ribosyl)ated. Histone H1 is ADP-ribosylated at glutamic acid residues 2, 14, and 116 and the C-terminal lysine residue (Parseghian et al., 1994b) (Fig. 5). Note that the position of the glutamic acid may vary with different histone H1 subtypes (Parseghian et al., 1994b). The enzymes catalyzing the addition and removal of the ADP-ribose groups are poly(ADP-ribose) polymerase and poly(ADP-ribose) glycohydrolase/ADP-ribosyl protein lyase, respectively (see previous discussion). Poly(ADP-ribosyl)ated histone H1 is accessible to protein kinases (Wong et al., 1983). Hyper(ADP-ribosyl)ation of histone H1 results in decondensations of the chromatin fiber. Decondensation of chromatin would be one of the steps in presenting the damaged DNA to the enzymes involved in DNA repair, or presenting DNA to be replicated to the DNA polymerases (de Murcia *et al.*, 1988). The highly modified histone H1 remained bound to chromatin, and treatment of the chromatin with poly(ADP-ribose) glyco-hydrolase reversed chromatin decondensation (de Murcia *et al.*, 1988).

VI. High Mobility Group Proteins

The high mobility group (HMG) proteins are associated with transcription-ally active chromatin (Bustin *et al.*, 1990; Reeves, 1992). HMG proteins are a group of nonhistone chromosomal proteins that were so named because of their relatively high mobility on polyacrylamide gels. The HMG proteins can be grouped into three distinct subsets. HMG-1 and HMG-2 proteins (molecular mass about 23 kDa) have two highly structured DNA binding domains (HMG boxes) and an acidic C-terminal tail (Van Holde and Zlatanova, 1994). HMG proteins 1 and 2 stimulate transcription *in vitro* (Tremethick and Molloy, 1986). HMG-14 and HMG-17 (molecular weight, 10,000–12,000) are another class of HMG proteins. HMG proteins 14 and 17 are associated with the nucleosome and are enriched in transcriptionally active chromatin (Postnikov *et al.*, 1991). HMG-14 stimulates RNA poly-merase II elongation on chromatin templates (Ding *et al.*, 1994). The third class of HMG proteins are I, Y, and I-C (molecular weight about 12,000). These proteins, which bind in the minor groove of AT-rich sequences, are preferentially expressed in undifferentiated, neoplastically transformed and rapidly proliferating cells (Manfioletti *et al.*, 1991; Ram *et al.*, 1993). The DNA binding domain of HMG-I, which has been called the "A·T-hook," shares structural similarities with the antitumor agent distamycin A, which competes for HMG-I binding to DNA *in vitro* (Reeves and Nissen, 1990). The N-terminal tails of the core histones must be present for HMG-I to bind to the nucleosome (Reeves and Nissen, 1993).

HMG proteins are susceptible to a number of postsynthetic modifications, including acetylation, phosphorylation, ADP-ribosylation, methylation, and glycosylation. HMG proteins 1, 2, 14, and 17 are reversibly acetylated by what appear to be the same enzymes catalyzing reversible acetylation of the histones (Sterner *et al.*, 1979, 1981). Butyrate inhibition of histone deacetylase increased the levels of the acetylated isoforms of the HMG proteins. The HMG proteins 1, 14, and 17 are acetylated at lysine residues located in their N-terminal regions. HMG-1 and HMG-2 are methylated (N^G,N^G-dimethylarginine) (Boffa *et al.*, 1979). HMG-14 and perhaps HMG-17 are phosphorylated, and their phosphorylation is cell cycle regulated, being greatest in G2/M phase of the cell cycle (Bhorjee, 1981; Walton and Gill, 1983). *In vitro* cAMP-dependent protein kinases, cGMP-dependent protein kinases, and casein kinase II phosphorylated HMG-14 (Walton *et al.*, 1982; Walton and Gill, 1983; Espel *et al.*, 1987). Casein kinase II is a component of the nuclear matrix of rat liver and prostate tissues (Tawfic and Ahmed, 1994). The cAMP- and cGMP-dependent protein kinases phosphorylate HMG-14 at serine residue 6 (major site) and serine residue 24 (minor site) (Walton *et al.*, 1982). The phosphorylation site of casein kinase II is in the acidic COOH-terminal region of HMG-14, possibly serine residue 89 (Walton and Gill, 1983). In neuronal nuclei there is a nuclear matrix-bound Ca^{2+}-calmodulin-dependent protein kinase (56 kDa) that phosphorylates HMG-17 (Sahyoun *et al.*, 1984). HMG-I, Y, and I-C are phosphoproteins. HMG-I and HMG-Y are phosphorylated by p34[cdc2] kinase (metaphase) and by casein kinase II (interphase) (Palvimo and Linnala Kankkunen, 1989; Lund and Laland, 1990; Meijer *et al.*, 1991; Nissen *et al.*, 1991; Reeves *et al.*, 1991). HMG-14 and HMG-17 are poly(ADP-ribosyl)-ated and glycosylated (Reeves *et al.*, 1981). The sugars bound to the HMG proteins through a *N*-glycosidic linkage are *N*-acetylglucosamine, mannose, galactose, glucose, and fucose. Glycosylated HMG proteins 14 and 17 bind to nuclear matrices of mammalian cells, and significantly, removal of the glycosyl side chains greatly diminishes HMG binding to the nuclear matrix (Reeves and Chang, 1983). This observation suggests that glycosylated HMG-14/17 may mediate an interaction between transcriptionally active chromatin and the nuclear matrix.

VII. Chromatin Domains

A. Genes Contained within Nuclease-Sensitive Domains

Transcriptionally active DNA is associated with nucleosomes. Further, it has been demonstrated that RNA polymerases are capable of reading through nucleosomes (Felsenfeld, 1992; Van Holde *et al.*, 1992). However, the transcriptionally active nucleosomes often have an atypical structure (Chen and Allfrey, 1987; Locklear *et al.*, 1990; Walker *et al.*, 1990). The chromatin of transcriptionally active genes differs from the bulk of the genome in susceptibility to digestion by nucleases, including micrococcal nuclease and deoxyribonuclease I (DNase I) (Bloom and Anderson, 1982; Reeves, 1984; Yu and Smith, 1985). Transcriptionally repressed chromatin (e.g., vitellogenin gene) of adult chicken immature erythrocytes generates a typical nucleosome repeat pattern when digested with micrococcal nuclease, while transcriptionally active chromatin (β-globin and histone H5 genes) presents a smear of DNA fragments that on the average are shorter than those from repressed chromatin (Delcuve and Davie, 1989) (Fig. 9). The nucleosome disruption detected by micrococcal nuclease is largely confined to the DNA sequences of the transcribed region (Cohen and Sheffery, 1985; Einck *et al.*, 1986; Strätling *et al.*, 1986). However, the preferential DNase I sensitivity of active genes is not restricted to the coding portion of the gene but extends far upstream and downstream into adjacent nontranscribed DNA sequences before converting to a DNase I-resistant conformation (Reeves, 1984) (Fig. 7). In hen oviduct the ovalbumin DNase I-sensitive domain is 100 kb (Lawson *et al.*, 1982). This domain contains three genes,

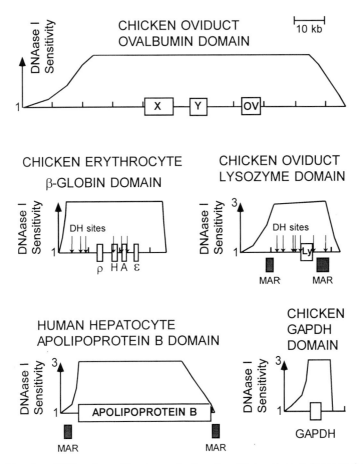

FIG. 7   Mapping of DNase I-sensitive chromatin domains. All domains are drawn to the same scale. DNase I-hypersensitive (DH) sites are indicated by downward-pointing arrows. MAR, matrix attachment regions, are indicated.

the ovalbumin gene and related genes X and Y. The entire domain is resistant to DNase I in tissues where the gene is not expressed. The chicken glyceraldehyde-3-phosphate dehydrogenase (GAPDH) gene, which is constitutively expressed in all cell types, is in a DNase I-sensitive domain approximately 12 kb in length (Alevy *et al.*, 1984). In chicken erythrocytes the β-globin DNase I-sensitive domain is 33 kb in length (Hebbes *et al.*, 1994). The DNase I-sensitive domain of the chicken lysozyme gene in hen oviduct is 25 kb long (Jantzen *et al.*, 1986). The boundaries of the lysozyme gene DNase I-sensitive domain comap with operationally defined MARs (Phi-Van and Strätling, 1988). The human apolipoprotein B gene DNase I-sensitive domain in HepG2 cells has a length of about 47.5 kb (Levy-Wilson and Fortier, 1989). As with the lysozyme DNase I-sensitive domain, the boundaries of the apolipoprotein B domain comap with MARs. These latter studies add credibility to the idea that MAR elements flanking domains anchor the base of the loop to the nuclear matrix. Thus, the DNase I-sensitive chromatin domain probably constitutes a chromosomal loop and, thus, represents a unit of gene regulation (Phi-Van and Strätling, 1988) (see Fig. 1).

The increased nuclease sensitivity of active chromatin is generally considered to indicate some alteration in nucleosome and/or higher order chromatin structure (Felsenfeld, 1992; Van Holde *et al.*, 1992) (see Fig. 1). Several biochemical features distinguish transcriptionally active chromatin from repressed gene chromatin (see Fig. 10), and it is these features that contribute to the altered chromatin structure of transcriptionally active chromatin loop domains. Typically the DNase I sensitivity of transcriptionally active domains is 3-fold greater than that of DNase I-resistant chromatin. However, Villeponteau *et al.* (1984) reported a 20-fold preferential sensitivity of the β-globin gene domain in 12-day embryonic red blood cells. Nicking the DNA of the globin domain with agents such as gamma rays, bleomycin, novobiocin (which induces the release of an endogenous nuclease), or micrococcal nuclease caused a reversal of DNase I sensitivity from 20- to 2–3-fold (Villeponteau *et al.*, 1984, 1986; Villeponteau and Martinson, 1987). These studies provide evidence that the DNase I-sensitive loop domain is under torsional stress and that a single nick in the DNA of the domain is sufficient to dissipate the DNA supercoils and lower the DNase I sensitivity of the domain. The process of transcription generates torsional stress in the domain; positive supercoils are generated ahead of the RNA polymerase and negative supercoils are produced behind it (Liu and Wang, 1987). The positive DNA supercoiling may provide a mechanism for decondensing the chromatin of transcriptionally active loop domains and altering nucleosome structure (Lee and Garrard, 1991a,b).

Several reports have presented evidence that histone acetylation and HMG 14/17 are responsible for the two- to threefold DNase I sensitivity of transcriptionally active chromatin (Reeves, 1984). Chromatin regions containing nucleosomes with hyperacetylated histones are more sensitive to digestion by DNase I than chromatin regions containing unacetylated nucleosomes (Reeves, 1984). Recently, Hebbes *et al.* (1994) demonstrated that acetylated histones were distributed throughout the chicken erythrocyte β-globin DNase I-sensitive domain. The transition of DNA sequences in nucleosomes with acetylated histones to DNA sequences in nuclesomes with unacetylated histones comap with the boundaries of the DNase I-sensitive domain. HMG proteins 14 and 17 are bound to nucleosomes containing transcribed DNA sequences and acetylated histones (Malik *et al.*, 1984; Postnikov *et al.*, 1991). The limited distribution of nucleosomes with HMG 14 and 17 proteins to regions engaged in transcription and not throughout the entire DNase I-sensitive domain raises doubt that these proteins have a principal role in the two- to threefold DNase I sensitivity of active chromatin domains. Thus, torsional stress and histone acetylation appear to be the principal mediators of DNase I sensitivity of active chromatin domains.

B. Fractionation of Chromatin Domains

Employing various chromatin fractionation procedures has shown that transcriptionally active DNA sequences are present in chromatin fragments that are soluble in 25–150 mM NaCl and/or 2–3 mM MgCl$_2$ (or CaCl$_2$) and in chromatin fragments bound to the residual nuclear material (Bloom and Anderson, 1982; Davis *et al.*, 1983; Rocha *et al.*, 1984; Einck *et al.*, 1986; Strätling *et al.*, 1986; Ridsdale and Davie, 1987; Strätling, 1987; Ip *et al.*, 1988; Delcuve and Davie, 1989). The latter fraction will be discussed in a later section. The fractionation procedure used to fractionate chicken erythrocyte chromatin is shown in Fig. 8 (Delcuve and Davie, 1989). The salt-soluble chromatin fragments (fraction S$_{150}$) are principally of mononucleosome size but also have polynucleosome length fragments (Fig. 9). The polynucleosomes soluble in 150 mM NaCl are enriched in transcriptionally active (histone H5, β-globin, histone H2A.Z) and competent DNA sequences (thymidine kinase, glyceraldehyde-3-phosphate dehydrogenase, vimentin, c-*myc*, ε-globin) (Fig. 9). Transcriptionally active sequences are those associated with RNA polymerase II and may be actively transcribed, while transcriptionally competent DNA sequences are those that are not engaged in transcription but are DNase I sensitive. Repressed DNA sequences (e.g., vitellogenin, ovalbumin) are in the salt-soluble mononucleosomes and salt insoluble (P$_{150}$) chromatin fraction. Using a different fractionation procedure, the DNA sequences of the β-globin DNase I-sensitive domain were enriched in the salt (50 mM NaCl/25 mM KCl/2 mM MgCl$_2$)-

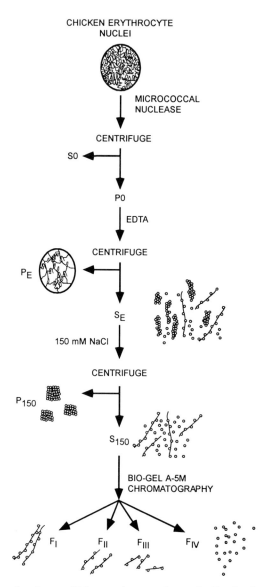

FIG. 8   Procedure to fractionate chicken erythrocyte chromatin. For details see Delcuve and Davie (1989) and Hendzel *et al.* (1991).

soluble chromatin fragments, while a DNA sequence that lies outside of the domain was not (Rocha *et al.*, 1984). Thus, salt solubility of active and competent chromatin fragments correlated with their DNase I sensitivity (Ridsdale *et al.*, 1988). Further, the salt solubility of chromatin fragments

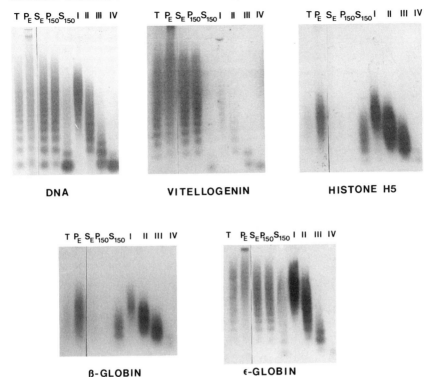

FIG. 9   DNA sequence distribution among erythrocyte chromatin fractions. DNA isolated from each chromatin fraction (see Fig. 8) was run on a 1% agarose gel, stained with ethidium bromide (DNA), or transferred to nitrocellulose and hybridized to radiolabeled probes as indicated. The fractions F_I, F_{II}, F_{III}, and F_{IV} from the Bio-Gel A-5m column (see Fig. 8) are shown as I, II, III, and IV, respectively. [Reprinted with permission, Delcuve and Davie, 1989 (copyright by the Biochemical Society and Portland Press).]

defines the same transcriptionally active β-globin domain as does DNase I (Hebbes *et al.*, 1994).

By size-fractionating the salt-soluble chromatin fragments isolated from adult chicken mature or immature nuclei, it was possible to obtain poly-nucleosomes (fractions F_I, F_{II}, F_{III}, see Fig. 8) that were highly enriched (about 50-fold) in transcriptionally active and competent DNA. The bio-chemically composition of this fraction is markedly different from that of bulk chicken erythrocyte chromatin (Ridsdale and Davie, 1987; Delcuve and Davie, 1989; Hendzel and Davie, 1989; Nickel *et al.*, 1989, 1990; Locklear *et al.*, 1990; Hendzel *et al.*, 1991) (see Fig. 10). The salt-soluble polynucleo-somes are enriched in dynamically acetylated core histones (Fig. 11). The dynamically acetylated histones, which are rapidly hyperacetylated and rapidly deacetylated, are preferentially associated with transcriptionally

ENRICHED IN

HIGHLY ACETYLATED ISOFORMS OF HISTONES H3, H4, H2A, H2A.Z

RAPIDLY ACETYLATED AND DEACETYLATED CORE HISTONES - DYNAMICALLY ACETYLATED HISTONES

METHYLATED HISTONES H3 AND H4

UBIQUITINATED (u) ISOFORMS OF HISTONES H2A AND H2B

HISTONE VARIANTS H2A.Z AND H3.3

NEWLY SYNTHESIZED HISTONES H2A, H2B, H3.3, H4, H2A.Z, uH2A, uH2B

ATYPICAL NUCLEOSOMES - u-SHAPED AND ELONGATED

HISTONE ACETYLTRANSFERASE, DEACETYLASE, METHYLTRANSFERASE

ASSOCIATED WITH

NUCLEAR MATRIX

SENSITIVE TO

DNase I AND MICROCOCCAL NUCLEASE
CHEMICAL CARCINOGENS (AFLATOXIN B1)

ASSOCIATED WITH

HISTONE H1 AND H5

FIG. 10   Biochemical features of transcriptionally active chromatin.

active DNA (Hendzel *et al.*, 1991). The above biochemical features are in general agreement with what others have found for the composition of active DNA-enriched chromatin (Reeves, 1984; Turner, 1991; Ausio, 1992; Morse, 1992; Van Holde *et al.*, 1992). Chicken erythrocyte salt-soluble polynucleosomes are associated with linker histones H1 and H5, but in stoichiometric amounts lower than that found in bulk erythrocyte chromatin (Ridsdale and Davie, 1987). Others have found that H1 histones are associated with transcriptionally active chromatin, but usually at levels lower than that found in bulk, repressed chromatin (Kamakaka and Thomas, 1990; Postnikov *et al.*, 1991; Belikov *et al.*, 1993).

The chicken erythrocyte salt-soluble polynucleosomes are enriched in atypical nucleosomes, referred to as U-shaped nucleosomes and elongated nucleosomes (Locklear *et al.*, 1990). The U-shaped nucleosomes contain

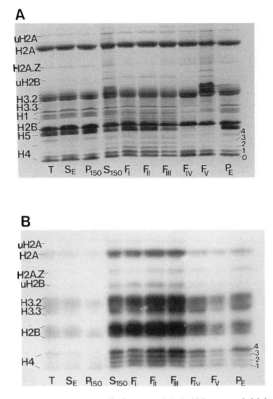

FIG. 11 The chromatin distribution of [³H]acetate-labeled histones of chicken immature erythrocytes. Acid-soluble protein from each chromatin fraction (see Fig. 8) was electrophoretically resolved on 15% polyacrylamide AUT (acetic acid/urea/Triton X-100) gels. Panel A shows Coomassie blue-stained gel pattern, and panel B shows the accompanying fluorogram. The acetylated species of histone H4 are denoted numerically 0, 1, 2, 3, and 4, respectively, representing the un-, mono-, di-, tri-, and tetraacetylated species, respectively. u denotes the ubiquitinated histone isoforms. [Reprinted with permission, Hendzel *et al.*, 1991 (copyright by the American Society for Biochemistry and Molecular Biology).]

the same amount of DNA as canonical nucleosomes, but these particles have 20% less protein mass than that of the typical nucleosome. This reduction of protein mass may be due to a loss of an H2A–H2B dimer. The removal of a histone H2A–H2B dimer may be an intermediate step in transcription through a nucleosome (Baer and Rhodes, 1983; Morse, 1992; Van Holde *et al.*, 1992).

A novel procedure to isolate structurally altered transcriptionally active nucleosomes was designed by Allfrey and colleagues (Walker *et al.*, 1990). Typically the cysteine residues of histone H3, which is usually the only core histone with cysteine residues, are buried in the interior of the nucleosome.

Allfrey and colleagues observed that in structurally altered active nucleosomes the sulfhydryl groups of histone H3 are exposed, allowing isolation of these atypical nucleosomes on organomercurial–agarose columns or magnetic beads (Walker *et al.*, 1990; Chen-Cleland *et al.*, 1993). The affinity purified nucleosomes were enriched in highly acetylated histones that had a high rate of turnover of acetyl groups. Inhibition of RNA polymerase II transcription with α-amanitin resulted in loss of Hg affinity of nucleosomes with active DNA sequences (Boffa *et al.*, 1990). Thus it appears that the process of transcription is required for structural alterations in active nucleosomes to take place. Both torsional stress and histone acetylation may be required to alter the structure of active nucleosomes (Boffa *et al.*, 1990; Ausio, 1992).

C. Transcriptionally Active Chromatin Associated with the Nuclear Matrix

In contrast to the salt solubility properties of transcriptionally active and poised chromatin, the transcribed DNA sequences partition preferentially with the nuclear matrix-attached chromatin fragments. DNA fragments remaining attached to nuclear matrices isolated by extracting nuclei with $2\,M$ NaCl followed by nuclease digestion (DNase I or restriction endonuclease) or shearing are enriched in transcriptionally active sequences (Robinson *et al.*, 1982, 1983; Ciejek *et al.*, 1983; Hentzen *et al.*, 1984; Andreeva *et al.*, 1992). The extraction of the nuclei with $2\,M$ NaCl to remove histones before the DNase I digestion prevented transcriptionally active genes from being preferentially digested. To locate transcriptionally active and inactive sequences in distended DNA loops, Gerdes *et al.* (1994) extracted nuclei with $2\,M$ NaCl and monitored the location of specific DNA sequences by high-resolution fluorescence hybridization. Transcriptionally inactive DNA sequences were on the extended loop, while transcriptionally active DNA sequences remained condensed within the residual nucleus. One criticism about this experimental approach is that RNA polymerase II associated with the transcriptionally active DNA is rendered insoluble by direct high-salt extraction of nuclei (Roberge *et al.*, 1988). Thus, the apparent association of active DNA with the nuclear matrix may be a consequence of RNA polymerase II becoming insoluble. However, other procedures that use low ionic strength or isotonic buffers to liberate nonmatrix chromatin fragments yield the same results as the high-salt procedures; that is, transcriptionally active chromatin is preferentially bound to the nuclear matrix. Residual nuclear material remaining after nuclei are digested with micrococcal nuclease or endogenous nuclease and then extracted with low ionic strength buffers retain transcriptionally active chromatin (Davis *et al.*, 1983; Einck *et al.*, 1986; Strätling *et al.*, 1986; Strätling, 1987; Delcuve and Davie, 1989) (Fig. 9). Such low-salt-extracted nuclear skeletons support transcription (Andreeva *et al.*, 1992). Another experimental procedure that has been used is to encapsulate cells in agarose, digest with restriction endonucleases in physiological buffers, and remove the chromatin fragments not bound to the nucleoskeleton by electrophoresis (Jackson and Cook, 1985b). These nucleoskeletons carry out transcription at *in vivo* rates. Further, transcriptionally active chromatin is retained by the nucleoskeleton (Thorburn *et al.*, 1988; Thorburn and Knowland, 1993). Regardless of the extraction protocol used (high, low, or physiological ionic strength) to isolate the residual nuclear material, studies have shown that inducible genes (e.g., rat liver P450c and P450d genes, *Xenopus* liver estrogen-inducible vitellogenin B2 gene, chick oviduct estrogen-induced lysozyme or ovalbumin genes) are associated with the nuclear matrix only when the gene is transcribed (Ciejek *et al.*, 1983; Robinson *et al.*, 1983; Einck *et al.*, 1986; Strätling *et al.*, 1986; Thorburn and Knowland, 1993). Actively transcribed chromatin regions are thought to be immobilized on the nuclear matrix by multiple dynamic attachment sites (Andreeva *et al.*, 1992).

Transcriptionally active (β-globin and histone H5), but neither competent (ε-globin) nor repressed, DNA sequences are enriched in the residual nuclear material (nucleoskeleton, fraction P_E) of adult chicken immature or mature erythrocytes (Delcuve and Davie, 1989) (Fig. 9). Although the ε-globin gene is part of the DNase I-sensitive β-globin gene domain in chicken erythrocytes, it is not preferentially bound to the nuclear skeleton. Thus, in agreement with the results of others only the transcribed region and nearby flanking DNA sequences (e.g., promoters and enhancers) of the DNase I-sensitive domain are preferentially attached to the nuclear skeleton (Strätling, 1987; Zenk *et al.*, 1990).

The chromatin fragments associated with the nuclear skeletons of chicken immature erythrocytes share several of the biochemical features of salt-soluble polynucleosomes (Fig. 10). The nuclear skeleton (fraction P_E)-bound chromatin fragments are enriched in dynamically acetylated core histones (Fig. 11), methylated histones H3 and H4, and newly synthesized histones. Histone methyltransferase, histone acetyltransferase, and histone deacetylase activities are enriched in the P_E fraction.

D. DNase I Hypersensitive Sites

Within DNase I-sensitive domains are usually found several DNase I-hypersensitive sites (see Fig. 7). These are regions of the chromatin that are 100-fold more sensitive to DNase I attack than is bulk chromatin. DNase I-hypersensitive sites often mark the chromatin fiber for the presence of sequence-specific DNA-binding proteins associated with *cis*-acting regulatory DNA elements (e.g., promoters and enhancers) (Elgin, 1988). For example, the chicken erythrocyte histone H5 gene promoter and downstream enhancer regions have several DNase I-hypersensitive sites, and studies have shown that several transcription factors bind to these regions *in vitro* (Renaud and Ruiz Carrillo, 1986; Sun *et al.*, 1992; Rousseau *et al.*, 1993) (Fig. 12).

E. Transcription Factors and Chromatin

Promoter sequences reconstituted in nucleosomes do not function *in vitro*. To initiate transcription, nucleosomes must be removed from the promoter (Felsenfeld, 1992). Once started, RNA polymerases are able to elongate RNA chains on nucleosomal templates. Thus, for transcription to commence, histone octamers need to be dislodged or prevented from binding to promoter DNA sequences (Felsenfeld, 1992; Becker, 1994; Wallrath *et al.*, 1994). One mechanism to accomplish this clearing of nucleosomes from the promoter has been proposed in the preemptive competition model (Felsenfeld, 1992). In this model, transcription factors preferentially load onto enhancers or locus control regions (LCR), which are rich in factor-binding sites, during replication, blocking histone octamer binding. Transcription factors bound to LCR/enhancers would aid in the loading of factors onto promoter sequences, blocking nucleosomes from forming on promoter sequences. Nucleosomes may aid in the juxtapositioning of LCR/enhancer and promoter elements. Since nucleosomes can govern the path of DNA, strategic positioning of a nucleosome between two distant DNA sequence elements can swing the two DNA sequence elements and their associated factors next to each other (Elgin, 1988; Schild *et al.*, 1993; Van

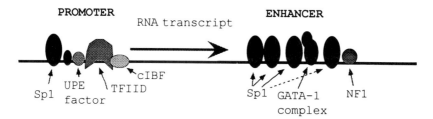

FIG. 12   Transcription factors associated with the chicken erythrocyte histone H5 promoter and 3′ enhancer regions. For details see Sun *et al.* (1992) and Rousseau *et al.* (1993). The regions containing the transcription factors are hypersensitive to DNase I *in situ* (Renaud and Ruiz Carrillo, 1986), suggesting that the promoter and enhancer regions are nucleosome free or associated with atypical nucleosomes.

Holde, 1993; Q. Lu *et al.*, 1994). Another mechanism, the dynamic competition model, which may operate in association with the first mechanism, is the binding of a transcription factor(s) to its DNA binding site in a nucleosome that is present in a LCR/enhancer or promoter region. The binding of the transcription factor to the nucleosome results in the unfolding, sliding, or dissociation of the histone octamer (Felsenfeld, 1992; Workman and Buchman, 1993) (Fig. 13).

Transcription factor binding to nucleosomes associated with enhancers can occur without dissociation of the histone octamer (McPherson *et al.*, 1993) (Fig. 13). For example, the liver albumin enhancer has positioned nucleosomes, and transcription factors (e.g., C/EBP) bind to their recogni-

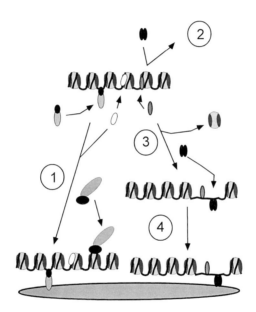

NUCLEAR MATRIX

FIG. 13 Transcription factor–nucleosome and transcription factor–nuclear matrix interactions. Scheme 1 shows two transcription factors binding to chromatin. One factor (the filled oval) binds to its binding site located in the linker DNA. This factor binds to the nuclear matrix and mediates the attachment of chromatin to the nuclear matrix. The other factor (open oval) binds to its binding site in nucleosomal DNA. In neither case does the binding of the factors result in displacement of the histone octamer. In scheme 2 a transcription factor (two solid ovals) cannot bind to its binding site in nucleosomes but once the histone octamer is displaced, it can bind to its recognition site (scheme 3). In scheme 3 a transcription factor (solid oval) binds to its binding site in nucleosomal DNA, which results in the displacement of the histone octamer. Other transcription factors are recruited to this nucleosome-free region (the solid double oval). In this cartoon the solid double oval binds to the nuclear matrix and mediates the attachment between chromatin and the nuclear matrix.

tion sites in nucleosomal DNA. Factors capable of binding to their target sequences in nucleosomes *in vitro* are the glucocorticoid receptors, GAL4, Sp1, TFIIIA, and USF (Archer *et al.*, 1991; Perlmann, 1992; Workman and Kingston, 1992; Lee *et al.*, 1993; Chen *et al.*, 1994; Li *et al.*, 1994). Removal of the N-terminal tails of the core histones or the presence of hyperacetylated histones have been shown to facilitate the access of transcription factors to binding sites in nucleosomes (Lee *et al.*, 1993; Vettese-Dadey *et al.*, 1994). However, not all transcription factors, for example NF1, are capable of binding to recognition sites in nucleosomes (Archer *et al.*, 1991) (Fig. 13). Thus, for these transcription factors binding is possible only when the binding site is in linker DNA or in nucleosome-free regions.

F. Transcription Factors and the Nuclear Matrix

A subset of the nuclear transcription factors is associated with the nuclear matrix (Waitz and Loidl, 1991; Bortell *et al.*, 1992; Dworetzky *et al.*, 1992; Isomura *et al.*, 1992; Thorburn and Knowland, 1993). Sequence-specific DNA-binding proteins that are associated with the nuclear matrix include the estrogen receptor, NMP-1 (possible member of ATF transcription factor family), RAP-1 (repressor–activator binding protein-1), factor F6, c-*Myc*, RFP, Sp1, ATF, CCAAT, C/EBP, Oct1, AP1, and NF1 (Gasser *et al.*, 1989; Waitz and Loidl, 1991; Dworetzky *et al.*, 1992; Isomura *et al.*, 1992; van Wijnen *et al.*, 1993; Vassetzky *et al.*, 1993; Sun *et al.*, 1994). Some factors (Sp1, ATF) are associated with the nuclear matrices of several cell types (e.g., HeLa, a human cervical carcinoma, and ROS 17/2.8, a rat osteosarcoma), while others are found to be cell type specific (e.g., CCAAT, C/EBP). These observations provide evidence that the nuclear matrices of different cells selectively bind specific transcription factors that would contribute to the observed cell-type-specific nuclear matrix protein composition (Cupo, 1991). Such observations further strengthen the hypothesis that the nuclear matrix is involved in transcriptional control (van Wijnen *et al.*, 1993).

To determine whether any of the factors bound to the chicken erythrocyte histone H5 gene promoter and enhancer are components of the chicken mature or immature erythrocyte nuclear matrix, a nuclear matrix was prepared by extracting NM1 or NM2 nuclear matrices with 2 *M* NaCl and 2-mercaptoethanol (Fig. 6). Gel mobility shift and DNase I footprinting assays demonstrated that NF1 but not any of the other transcription factors were components of the erythroid nuclear matrix. Interestingly, purification of the NF1 proteins from nuclei or nuclear matrix extracts (mature erythrocytes) showed that only a subset of the NF1 protein family was bound to the internal nuclear matrix (Sun *et al.*, 1994).

G. Mechanisms for the Attachment of Transcriptionally Active Chromatin to the Nuclear Matrix

The association of transcriptionally active chromatin is a dynamic process involving multiple attachment sites through nuclear matrix components. Components of the nuclear matrix including histone acetyltransferase, histone deacetylase, protein kinases, transcription factors, and glycosylated HMG proteins 14 and 17 may all have roles in mediating the dynamic attachment of transcriptionally active chromatin regions to the nuclear matrix. Further, multiple interactions between transcriptionally active chromatin and the nuclear matrix-bound enzymes and transcription factors would contribute to the insoluble characteristics of transcriptionally active chromatin.

RNA polymerase II transcription localizes to discrete nuclear compartments, called transcription foci, which are thought to contain several transcribed genes per focus (Jackson *et al.*, 1993). With current evidence suggesting that rapidly acetylated and deacetylated core histones are selectively bound to transcriptionally active DNA, it is a reasonable hypothesis that histone acetyltransferases and deacetylases should be targeted to these nuclear compartments to function predominantly on transcribed chromatin. Vertebrate histone acetyltransferase and deacetylase are associated with the nuclear matrix. We propose that the core histones of transcriptionally active nucleosomes are frequently in contact with either the nuclear matrix-bound histone acetyltransferase or deacetylase, resulting in the rapid acetylation–deacetylation that is observed in active chromatin regions (Davie and Hendzel, 1994). Another group of players would be nuclear matrix-associated protein kinases (e.g., casein kinase II) and HMG proteins 14 and I/Y, which are associated with transcriptionally active chromatin. The association of transcriptionally active chromatin with these nuclear matrix-bound enzymes would be among the many dynamic attachments that results in the immobilization of the active chromatin regions onto the nuclear matrix (Fig. 14).

Depending on the gene and the cell type that it is expressed in, one or more transcription factors may contribute to the association of the active chromatin region with the nuclear matrix. Figure 14 shows a model for the transcriptionally active histone H5 gene in adult chicken immature erythrocytes. In this model histone acetyltransferase, deacetylase, and the

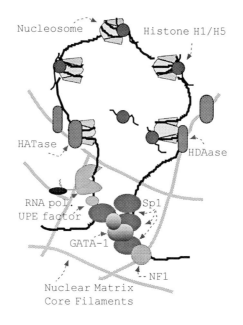

Histone H5 Gene

FIG. 14   A model for the dynamic attachment of histone H5 gene chromatin to the nuclear matrix. In this model the 3' enhancer is positioned next to the 5' promoter through protein–protein interactions, principally those between Sp1s (Sun *et al.*, 1992). Histone acetyltransferase (HATase), histone deacetylase (HDAase), and NF1 are proposed to mediate the dynamic attachments of the histone H5 gene to the nuclear matrix at sites of transcription.

transcription factor NF1 contribute to the attachment of the transcriptionally active histone H5 gene to the nuclear matrix.

H. Formation of Transcriptionally Active Chromatin Domains

The transcriptional activity of a gene that has been integrated into the genome of a transfected cell or transgenic animal or plant is, in general, unpredictable, with the expression of the gene depending on the site of insertion. The identification of MAR elements made it theoretically possible to make DNA constructs which when inserted into a genome would form an independent loop domain. Thus, the gene within the newly formed loop would be insulated from the effects of elements existing near or at the integrated site (position independence). Assuming that the transcription factors driving the expression of the gene are not limiting, one would expect to find an expression level that correlated with the copy number of the integrated transgene. Stief *et al.* (1989) found that expression of an integrated reporter gene (chloramphenicol acetyltransferase) into the genome of chicken promacrophage cells was elevated and independent of chromosomal position when flanked by the 5' MAR element (also called A element) of the chicken lysozyme gene domain. In transient expression assays the presence or absence of the MAR element did not affect the expression of the reporter gene. Thus, the MAR element only exhibits its stimulatory activity when it is in a chromosomal context. This feature of MARs is different from classical enhancers which are stimulatory in transients and stables. Subsequent studies by Phi-Van *et al.* (1990) demonstrated that the chicken lysozyme 5' MAR element enhanced the transcription from a heterologous promoter in heterologous cells, and the expression of the stably transfected gene was copy number dependent. Without the MAR element the expression of the inserted gene was quite variable and independent of copy number. Thus, the MAR element reduced the sensitivity of the expression of the inserted gene to the chromosomal integration site. This chicken MAR element also conferred position-independent hormonal and developmental regulation of the expression of the whey acidic protein gene in transgenic mice (McKnight *et al.*, 1992). However, copy number-dependent expression of the transgene was not observed in the transgenic animals. Similarly, Brooks *et al.* (1994) observed that the human apolipoprotein MAR elements elevated the expression of a transgene in transgenic mice, but the MAR elements did not completely protect the associated transgene from position effects.

Locus control regions play a role in the establishment of transcriptionally active chromatin domains. LCRs are DNA elements that confer position-independent and copy number-dependent expression of linked genes in transgenic mice (Reitman *et al.*, 1990, 1993; Felsenfeld, 1992). The human β globin LCR is located 20 kb 5' of the ε-globin gene. Deletion of the LCR as is the case in $(\gamma\delta\beta)^0$-thalassemias results in the inability of the globin genes to be expressed and in the chromatin structure of the domain becoming resistant to DNase I and late replicating (Forrester *et al.*, 1990). Normally the human β-globin domain is DNase I sensitive and early replicating in erythroid cells. The human β-globin LCR has four prominent hypersensitive sites in cells of the erythroid lineage and many transcription factors bind to these sites (Felsenfeld, 1992). The LCR is configured like a superenhancer. However, LCRs are apparently different from classical enhancers in that the property of position independence can be divorced from that of high-level expression (Reitman *et al.*, 1993). MAR elements are found on both sides of the β-LCR (Jarman and Higgs, 1988). The chicken β-globin 3' enhancer, which cohabits with a MAR (Zenk *et al.*,

1990), functions as an LCR in transgenic mice (Reitman *et al.*, 1990, 1993). The role of the MAR elements in generating the DNase I-sensitive chromatin domain is unknown. However, transcriptional activity of the rearranged immunoglobulin μ heavy chain transgene and formation of a DNase I chromatin structure depended on the intronic MAR elements that flank the μ enhancer (Forrester *et al.*, 1994). It should be noted that the MAR/enhancer/LCR region is not sufficient to form the DNase I-sensitive chromatin domain; a promoter is needed (Jenuwein *et al.*, 1993; Reitman *et al.*, 1993). These studies provide support for the mutual interaction model (Reitman *et al.*, 1993). In this model, the interaction of transcription factors bound to the promoter and enhancer/LCR is a crucial step in opening the chromatin domain.

The studies with transgenes have given us a glimpse of how DNase I-sensitive chromatin domains are formed. A first step appears to be the association of transcription factors with nucleosomes positioned on the enhancer/LCR element (Jenuwein *et al.*, 1993; McPherson *et al.*, 1993) (see Fig. 13). Some of these factors may have affinity for the nuclear matrix, targeting this region to the nuclear matrix. The MAR elements that cohabit or are near the enhancer/LCR may associate with the nuclear matrix, attaching one end of the loop to the matrix. Alternatively, HMG-I may bind to the MAR, preventing nucleosome and/or histone H1 association and establishing a local open or accessible chromatin structure (Reeves, 1992; Zhao *et al.*, 1993; Forrester *et al.*, 1994). Transcription factors bound to the enhancer/LCR and promoter would interact, displacing histone octamers or altering nucleosome structure, forming DNase I-hypersensitive sites and leading to the recruitment of the transcription machinery, the first round of transcription, and further interactions between the domain and the nuclear matrix. At this or an earlier stage histone acetyltransferase and deacetylase are recruited to the potentially active chromatin domain and/or nuclear matrix at the site of domain attachment. Together, histone acetylation and transcription-induced torsional stress would form the DNase I-sensitive chromatin domain.

VIII. Concluding Remarks

Significant progress has been made over the past several years in our understanding of the organization of chromatin in the nucleus. A glimpse of the role of the nuclear matrix in the organization and function of the genome has been gained. However, much needs to be learned about the composition/structure of the nuclear matrix. It appears that several histone modifying enzymes (histone acetyltransferases, deacetylases, methyltransferases, and protein kinases) are colocalized at nuclear sites engaged in transcription. What are the signals by which these enzymes are recruited to these specific nuclear sites? What signals are involved in targeting transcription factors to the nuclear matrix? Progress has been made in identifying the targeting sequences that direct proteins such as DNA methyltransferases and RNA processing regulators to specific nuclear locations (Leonhardt *et al.*, 1992; Li and Bingham, 1991). Are these or other targeting sequences involved in directing nuclear enzymes, transcription factors and splicing factors to the nuclear matrix? Or are modifications of the transcription factors (e.g., glycosylation) involved in the targeting of transcription factors to the nuclear matrix? Also, experiments will have to be designed to test whether the nuclear matrix-bound factors are bound to enhancers/promoters *in situ*. The study of nuclear matrix and chromatin structure/function is presently at an early stage, and the next several years should be a productive period that sheds more light on the mysteries of the organization, function, and formation of chromatin domains in the nucleus.

Acknowledgments

This work was supported from grants from the Medical Research Council of Canada (MT-12147, MA-12283, MT-9186) and a Medical Research of Canada Scientist award.

References

Adachi, Y., Käs, E., and Laemmli, U. K. (1989). Preferential, cooperative binding of DNA topoisomerase II to scaffold-associated regions. *EMBO J.* **8,** 3997–4006.

Ajiro, K., and Nishimoto, T. (1985). Specific site of histone H3 phosphorylation related to the maintenance of premature chromosome condensation. Evidence for catalytically induced interchange of the subunits. *J. Biol. Chem.* **260,** 15379–15381.

Alevy, M. C., Tsai, M. J., and O'Malley, B. W. (1984). DNase I sensitive domain of the gene coding for the glycolytic enzyme glyceraldehyde-3-phosphate dehydrogenase. *Biochemistry* **23,** 2309–2314.

Allan, J., Harborne, N., Rau, D. C., and Gould, H. (1982). Participation of core histone "tails" in the stabilization of the chromatin solenoid. *J. Cell Biol.* **93,** 285–297.

Allis, C. D., and Gorovsky, M. A. (1981). Histone phosphorylation in macro- and micronuclei of *Tetrahymena thermophila. Biochemistry* **20,** 3828–3833.

Allis, C. D., Glover, C. V. C., Bowen, J. K., and Gorovsky, M. A. (1980). Histone variants specific to the transcriptionally active, amitotically dividing macronucleus of the unicellular eucaryote, *Tetrahymena thermophila. Cell (Cambridge, Mass.)* **20,** 609–617.

Alvarez Gonzalez, R., and Ringer, D. P. (1988). Nuclear matrix associated poly(ADP-ribose) metabolism in regenerating rat liver. *FEBS Lett.* **236,** 362–366.

Andersen, M. W., Ballal, N. R., Goldknopf, I. L., and Busch, H. (1981). Protein A24 lyase activity in nucleoli of thioacetamide-treated rat liver releases histone H2A and ubiquitin from conjugated protein A24. *Biochemistry* **20**, 1100–1104.

Andreeva, M., Markova, D., Loidl, P., and Djondjurov, L. (1992). Intranuclear compartmentalization of transcribed and nontranscribed c-myc sequences in Namalva-S cells. *Eur. J. Biochem.* **207**, 887–894.

Annunziato, A. T., Frado, L.-L. Y., Seale, R. L., and Woodcock, C. L. F. (1988). Treatment with sodium butyrate inhibits the complete condensation of interphase chromatin. *Chromosoma* **96**, 132–138.

Archer, T. K., Cordingley, M. G., Wolford, R. G., and Hager, G. L. (1991). Transcription factor access is mediated by accurately positioned nucleosomes on the mouse mammary tumor virus promoter. *Mol. Cell. Biol.* **11**, 688–698.

Arents, G., and Moudrianakis, E. N. (1993). Topography of the histone octamer surface: Repeating structural motifs utilized in the docking of nucleosomal DNA. *Proc. Natl. Acad. Sci. U.S.A.* **90**, 10489–10493.

Attisano, L., and Lewis, P. N. (1990). Purification and characterization of two porcine liver nuclear histone acetyltransferases. *J. Biol. Chem.* **265**, 3949–3955.

Ausio, J. (1992). Structure and dynamics of transcriptionally active chromatin. *J. Cell Sci.* **102**, 1–5.

Ausio, J., and Van Holde, K. E. (1986). Histone hyperacetylation: Its effects on nucleosome conformation and stability. *Biochemistry* **25**, 1421–1428.

Baer, B. W., and Rhodes, D. (1983). Eukaryotic RNA polymerase II binds to nucleosome cores from transcribed genes. *Nature (London)* **301**, 482–488.

Barratt, M. J., Hazzalin, C. A., Cano, E., and Mahadevan, L. C. (1994). Mitogen-stimulated phosphorylation of histone H3 is targeted to a small hyperacetylation-sensitive fraction. *Proc. Natl. Acad. Sci. U.S.A.* **91**, 4781–4785.

Bauer, W. R., Hayes, J. J., White, J. H., and Wolffe, A. P. (1994). Nucleosome structural changes due to acetylation. *J. Mol. Biol.* **236**, 685–690.

Becker, P. B. (1994). The establishment of active promoters in chromatin. *BioEssays* **16**, 541–547.

Belgrader, P., Siegel, A. J., and Berezney, R. (1991a). A comprehensive study on the isolation and characterization of the HeLa S3 nuclear matrix. *J. Cell. Sci.* **98**, 281–291.

Belgrader, P., Dey, R., and Berezney, R. (1991b). Molecular cloning of matrin 3. A 125-kilodalton protein of the nuclear matrix contains an extensive acidic domain. *J. Biol. Chem.* **266**, 9893–9899.

Belikov, S. V., Belgovsky, A. I., Preobrazhenskaya, O. V., Karpov, V. L., and Mirzabekov, A. D. (1993). Two non-histone proteins are associated with the promoter region and histone H1 with the transcribed region of active *hsp-70* genes as revealed by UV-induced DNA–protein crosslinking *in vivo. Nucleic Acids Res.* **21**, 1031–1034.

Berezney, R. (1991). The nuclear matrix: A heuristic model for investigating genomic organization and function in the cell nucleus. *J. Cell. Biochem.* **47**, 109–123.

Bhorjee, J. S. (1981). Differential phosphorylation of nuclear nonhistone high mobility group proteins HMG 14 and HMG 17 during the cell cycle. *Proc. Natl. Acad. Sci. U.S.A.* **78**, 6944–6948.

Bloom, K. S., and Anderson, J. N. (1982). Hormonal regulation of the conformation of the ovalbumin gene in chick oviduct chromatin. *J. Biol. Chem.* **257**, 13018–13027.

Boffa, L. C., Vidali, G., Mann, R. S., and Allfrey, V. G. (1978). Suppression of histone deacetylation *in vivo* and *in vitro* by sodium butyrate. *J. Biol. Chem.* **253**, 3364–3366.

Boffa, L. C., Sterner, R., Vidali, G., and Allfrey, V. G. (1979). Post-synthetic modifications of nuclear proteins: High mobility group proteins are methylated. *Biochem. Biophys. Res. Commun.* **89**, 1322–1327.

Boffa, L.C., Walker, J., Chen, T. A., Sterner, R., Mariani, M. R., and Allfrey, V. G. (1990). Factors affecting nucleosome structure in transcriptionally active chromatin. Histone acetylation, nascent RNA and inhibitors of RNA synthesis. *Eur. J. Biochem.* **194,** 811–823.

Böhm, J., Schlaeger, E.-J., and Knippers, R. (1980). Acetylation of nucleosomal histones *in vitro. Eur. J. Biochem.* **112,** 353–362.

Bonner, W. M., Wu, R. S., Panusz, H. T., and Muneses, C. (1988). Kinetics of accumulation and depletion of soluble newly synthesized histone in the reciprocal regulation of histone and DNA synthesis. *Biochemistry* **27,** 6542–6550.

Bortell, R., Owen, T. A., Bidwell, J. P., Gavazzo, P., Breen, E., van Wijnen, A. J., DeLuca, H. F., Stein, J. L., Lian, J. B., and Stein, G. S. (1992). Vitamin D-responsive protein-DNA interactions at multiple promoter regulatory elements that contribute to the level of rat osteocalcin gene expression. *Proc. Natl. Acad. Sci. U.S.A.* **89,** 6119–6123.

Boulikas, T. (1988). At least 60 ADP-ribosylated variant histones are present in nuclei from dimethylsulfate-treated and untreated cells. *EMBO J.* **7,** 57–67.

Boulikas, T. (1989). DNA strand breaks alter histone ADP-ribosylation. *Proc. Natl. Acad. Sci. U.S.A.* **86,** 3499–3503.

Boulikas, T. (1990a). Studies on protein poly(ADP-ribosylation) using high resolution gel electrophoresis. *J. Biol. Chem.* **265,** 14627–14631.

Boulikas, T. (1990b). Poly(ADP-ribosylated) histones in chromatin replication. *J. Biol. Chem.* **265,** 14638–14647.

Boulikas, T., Wiseman, J. M., and Garrard, W. T. (1980). Points of contact between histone H1 and the histone octamer. *Proc. Natl. Acad. Sci. U.S.A.* **77,** 127–131.

Bradbury, E. M. (1992). Reversible histone modifications and the chromosome cell cycle. *BioEssays* **14,** 9–16.

Brancolini, C., and Schneider, C. (1991). Change in the expression of a nuclear matrix-associated protein is correlated with cellular transformation. *Proc. Natl. Acad. Sci. U.S.A.* **88,** 6936–6940.

Breneman, J. W., Yau, P., Teplitz, R. L., and Bradbury, E. M. (1993). A light microscope study of linker histone distribution in rat metaphase chromosomes and interphase nuclei. *Exp. Cell Res.* **206,** 16–26.

Brooks, A. R., Nagy, B. P., Taylor, S., Simonet, W. S., Taylor, J. M., and Levy-Wilson, B. (1994). Sequences containing the second-intron enhancer are essential for transcription of the human apolipoprotein B gene in the livers of transgenic mice. *Mol. Cell. Biol.* **14,** 2243–2256.

Brosch, G., Georgieva, E. I., Lopez Rodas, G., Lindner, H., and Loidl, P. (1992). Specificity of *Zea mays* histone deacetylase is regulated by phosphorylation. *J. Biol. Chem.* **267,** 20561–20564.

Bustin, M., Lehn, D. A., and Landsman, D. (1990). Structural features of the HMG chromosomal proteins and their genes. *Biochim. Biophys. Acta* **1049,** 231–243.

Candido, E. P. M., Reeves, R., and Davie, J. R. (1978). Sodium butyrate inhibits histone deacetylation in cultured cells. *Cell (Cambridge, Mass.)* **14,** 105–113.

Chambers, T. C., and Langan, T. A. (1990). Purification and characterization of growth-associated H1 histone kinase from Novikoff hepatoma cells. *J. Biol. Chem.* **265,** 16940–16947.

Chen, H., Li, B., and Workman, J. L. (1994). A histone-binding protein, nucleoplasmin, stimulates transcription factor binding to nucleosomes and factor-induced nucleosome disassembly. *EMBO J.* **13,** 380–390.

Chen, R.-H., Sarnecki, C., and Blenis, J. (1992). Nuclear location and regulation of *erk*- and *rsk*-encoded protein kinases. *Mol. Cell. Biol.* **12,** 915–927.

Chen, T. A., and Allfrey, V. G. (1987). Rapid and reversible changes in nucleosome structure accompany the activation, repression, and superinduction of murine fibroblast protooncogenes c-fos and c-myc. *Proc. Natl. Acad. Sci. U.S.A.* **84,** 5252–5256.

Chen-Cleland, T. A., Boffa, L. C., Carpaneto, E. M., Mariani, M. R., Valentin, E., Mendez, E., and Allfrey, V. G. (1993). Recovery of transcriptionally active chromatin restriction fragments by binding to organomercurial-agarose magnetic beads. A rapid and sensitive method for monitoring changes in higher order chromatin structure during gene activation and repression. *J. Biol. Chem.* **268**, 23409–23416.

Chicoine, L. G., Schulman, I.G., Richman, R., Cook, R. G., and Allis, C. D. (1986). Nonrandom utilization of acetylation sites in histones isolated from Tetrahymena. Evidence for functionally distinct H4 acetylation sites. *J. Biol. Chem.* **261**, 1071–1076.

Ciejek, E. M., Tsai, M. J., and O'Malley, B. W. (1983). Actively transcribed genes are associated with the nuclear matrix. *Nature (London)* **306**, 607–609.

Clark, D. J., and Felsenfeld, G. (1992). A nucleosome core is transferred out of the path of a transcribing polymerase. *Cell (Cambridge, Mass.)* **71**, 11–22.

Clarke, D. J., O'Neill, L. P., and Turner, B. M. (1993). Selective use of H4 acetylation sites in the yeast *Saccharomyces cerevisiae. Biochem J.* **294**, 557–561.

Clayton, A. L., Hebbes, T. R., Thorne, A. W., and Crane-Robinson, C. (1993). Histone acetylation and gene induction in human cells. *FEBS Lett.* **336**, 23–26.

Cockerill, P. N. (1990). Nuclear matrix attachment occurs in several regions of the IgH locus. *Nucleic Acids Res.* **18**, 2643–2648.

Cockerill, P. N., and Garrard, W. T. (1986a). Chromosomal loop anchorage sites appear to be evolutionarily conserved. *FEBS Lett.* **204**, 5–7.

Cockerill, P. N., and Garrard, W. T. (1986b). Chromosomal loop anchorage of the kappa immunoglobulin gene occurs next to the enhancer in a region containing topoisomerase II sites. *Cell (Cambridge, Mass.)* **44**, 273–282.

Cohen, R. B., and Sheffery, M. (1985). Nucleosome disruption precedes transcription and is largely limited to the transcribed domain of globin genes in murine erythroleukemia cells. *J. Mol. Biol.* **182**, 109–129.

Cole, R. D. (1987). Microheterogeneity in H1 histones and its consequences. *Int. J. Pept. Protein Res.* **30**, 433–449.

Cook, J. C., and Chock, P. B. (1991). Immunocytochemical localization of ubiquitin-activating enzyme in the cell nucleus. *Biochem. Biophys. Res. Commun.* **174**, 564–571.

Couppez, M., Martin Ponthieu, A., and Sautiere, P. (1987). Histone H4 from cuttlefish testis is sequentially acetylated. Comparison with acetylation of calf thymus histone H4. *J. Biol. Chem.* **262**, 2854–2860.

Covault, J., and Chalkley, R. (1980). The identification of distinct populations of acetylated histone. *J. Biol. Chem.* **255**, 9110–9116.

Cupo, J. F. (1991). Electrophoretic analysis of nuclear matrix proteins and the potential clinical applications. *J. Chromatogr.* **569**, 389–406.

D'Anna, J. A., Tobey, R. A., Barham, S. S., and Gurley, L. R. (1977). A reduction in the degree of H4 acetylation during mitosis in Chinese hamster cells. *Biochem. Biophys. Res. Commun.* **77**, 187–194.

D'Anna, J. A., Tobey, R. A., and Gurley, L. R. (1980). Concentration dependent effects of sodium butyrate in Chinese hamster cells: Cell–cycle progression, inner-histone acetylation, histone H1 dephosphorylation, and induction of an H1-like protein. *Biochemistry* **19**, 2656–2671.

Davie, J. R., and Delcuve, G. P. (1991). Characterization and chromatin distribution of the H1 histones and high-mobility-group non-histone chromosomal proteins of trout liver and hepatocellular carcinoma. *Biochem. J.* **280**, 491–497.

Davie, J. R., and Hendzel, M. J. (1994). Multiple functions of dynamic acetylation. *J. Cell. Biochem.* **55**, 98–105.

Davie, J. R., and Murphy, L. C. (1990). Level of ubiquitinated histone H2B in chromatin is coupled to ongoing transcription. *Biochemistry* **29**, 4752–4757.

Davie, J. R., and Murphy, L. C. (1994). Inhibition of transcription selectively reduces the level of ubiquitinated histone H2B in chromatin. *Biochem. Biophys. Res. Commun.* **203**, 344–350.

Davis, A. H., Reudelhuber, T. L., and Garrard, W. T. (1983). Varigated chromatin structures of mouse ribosomal RNA genes. *J. Mol. Biol.* **167**, 133–155.

Delcuve, G. P., and Davie, J. R. (1989). Chromatin structure of erythroid-specific genes of immature and mature chicken erythrocytes. *Biochem. J.* **263**, 179–186.

de Murcia, G., Huletsky, A., and Poirier, G. G. (1988). Modulation of chromatin structure by poly(ADP-ribosyl)ation. *Biochem. Cell Biol.* **66**, 626–635.

Desrosiers, R., and Tanguay, R. M. (1988). Methylation of *Drosophila* histones at proline, lysine, and arginine residues during heat shock. *J. Biol. Chem.* **263**, 4686–4692.

Dickinson, L. A., Joh, T., Kohwi, Y., and Kohwi Shigematsu, T. (1992). A tissue-specific MAR/SAR DNA-binding protein with unusual binding site recognition. *Cell (Cambridge, Mass.)* **70**, 631–645.

Ding, H.-F., Rimsky, S., Batson, S. C., Bustin, M., and Hansen, U. (1994). Stimulation of RNA polymerase II elongation by chromosomal protein HMG-14. *Science* **265**, 796–799.

Djabali, K., Portier, M. M., Gros, F., Blobel, G., and Georgatos, S. D. (1991). Network antibodies identify nuclear lamin B as a physiological attachment site for peripherin intermediate filaments. *Cell (Cambridge, Mass.)* **64**, 109–121.

Dworetzky, S. I., Fey, E. G., Penman, S., Lian, J. B., Stein, J. L., and Stein, G. S. (1990). Progressive changes in the protein composition of the nuclear matrix during rat osteoblast differentiation. *Proc. Natl. Acad. Sci. U.S.A.* **87**, 4605–4609.

Dworetzky, S. I., Wright, K. L., Fey, E. G., Penman, S., Lian, J. B., Stein, J. L., and Stein, G. S. (1992). Sequence-specific DNA-binding proteins are components of a nuclear matrix-attachment site. *Proc. Natl. Acad. Sci. U.S.A.* **89**, 4178–4182.

Earnshaw, W. C., and Heck, M. M. (1985). Localization of topoisomerase II in mitotic chromosomes. *J. Cell Biol.* **100**, 1716–1725.

Eggert, H., and Jack, R. S. (1991). An ectopic copy of the *Drosophila* ftz associated SAR neither reorganizes local chromatin structure nor hinders elution of a chromatin fragment from isolated nuclei. *EMBO J.* **10**, 1237–1243.

Einck, L., Fagan, J., and Bustin, M. (1986). Chromatin structure of the cytochrome P-450c gene changes following induction. *Biochemistry* **25**, 7062–7068.

Elgin, S. C. (1988). The formation and function of DNase I hypersensitive sites in the process of gene activation. *J. Biol. Chem.* **263**, 19259–19262.

Espel, E., Bernues, J., Guasch, M. D., Querol, E., Plana, M., and Itarte, E. (1987). Phosphorylation of high-mobility-group protein 14 by two specific kinases modifies its interaction with histone oligomers in free solution. *Biochim. Biophys. Acta* **909**, 190–200.

Fakan, S., Leduc, Y., Lamarre, D., Brunet, G., and Poirier, G. G. (1988). Immunoelectron microscopical distribution of poly(ADP-ribose)polymerase in the mammalian cell nucleus. *Exp. Cell Res.* **179**, 517–526.

Farache, G., Razin, S.V., Rzeszowska Wolny, J., Moreau, J., Targa, F. R., and Scherrer, K. (1990). Mapping of structural and transcription-related matrix attachment sites in the alpha-globin gene domain of avian erythroblasts and erythrocytes. *Mol. Cell. Biol.* **10**, 5349–5358.

Felsenfeld, G. (1992). Chromatin as an essential part of the transcriptional mechanism. *Nature (London)* **355**, 219–224.

Fey, E. G., and Penman, S. (1988). Nuclear matrix proteins reflect cell type of origin in cultured human cells. *Proc. Natl. Acad. Sci. U.S.A.* **85**, 121–125.

Forrester, W. C., Epner, E., Driscoll, M. C., Enver, T., Brice, M., Papayannopoulou, T., and Groudine, M. (1990). A deletion of the human β-globin locus activation region causes a major alteration in chromatin structure and replication across the entire β-globin locus. *Genes Dev.* **4**, 1637–1649.

Forrester, W. C., Van Genderen, C., Jenuwein, T., and Grosschedl, R. (1994). Dependence of enhancer-mediated transcription of the immunoglobulin μ gene on nuclear matrix attachment regions. *Science* **265**, 1221–1225.

Garcia Ramirez, M., Dong, F., and Ausio, J. (1992). Role of the histone "tails" in the folding of oligonucleosomes depleted of histone H1. *J. Biol. Chem.* **267,** 19587–19595.

Gasser, S. M., and Laemmli, U. K. (1986). Cohabitation of scaffold binding regions with upstream/enhancer elements of three developmentally regulated genes of *D. melanogaster. Cell (Cambridge, Mass.)* **46,** 521–530.

Gasser, S. M., Laroche, T., Falquet, J., Boy de la Tour, E., and Laemmli, U. K. (1986). Metaphase chromosome structure. Involvement of topoisomerase II. *J. Mol. Biol.* **188,** 613–629.

Gasser, S. M., Amati, B. B., Cardenas, M. E., and Hofmann, J. F. (1989). Studies on scaffold attachment sites and their relation to genome function. *Int. Rev. Cytol.* **119,** 57–96.

Georgatos, S. D., and Blobel, G. (1987). Lamin B constitutes an intermediate filament attachment site at the nuclear envelope. *J. Cell Biol.* **105,** 117–125.

Gerace, L., and Blobel, G. (1982). Nuclear lamina and the structural organization of the nuclear envelope. *Cold Spring Harbor Symp. Quant. Biol.* **46**(2), 967–978.

Gerdes, M. G., Carter, K. C., Moen, P. T., Jr., and Lawrence, J. B. (1994). Dynamic changes in the higher-level chromatin organization of specific sequences revealed by in situ hybridization to nuclear halos. *J. Cell Biol.* **126,** 289–304.

Getzenberg, R. H., and Coffey, D. S. (1990). Tissue specificity of the hormonal response in sex accessory tissues is associated with nuclear matrix protein patterns. *Mol. Endocrinol.* **4,** 1336–1342.

Giancotti, V., Bandiera, A., Ciani, L., Santoro, D., Crane-Robinson, C., Goodwin, G. H., Boiocchi, M., Dolcetti, R., and Casetta, B. (1993). High-mobility-group (HMG) proteins and histone H1 subtypes expression in normal and tumor tissues of mouse. *Eur. J. Biochem.* **213,** 825–832.

Golderer, G., Loidl, P., and Grobner, P. (1991). Cell cycle-dependent ADP-ribosylation of the nuclear matrix. *Eur. J. Cell Biol.* **55,** 183–185.

Gorka, C., Fakan, S., and Lawrence, J. J. (1993). Light and electron microscope immunocytochemical analyses of histone H1^0 distribution in the nucleus of Friend erythroleukemia cells. *Exp. Cell Res.* **205,** 152–158.

Graziano, V., Gerchman, S. E., Schneider, D. K., and Ramakrishnan, V. (1994). Histone H1 is located in the interior of the chromatin 30-nm filament. *Nature (London)* **368,** 351–354.

Hakes, D. J., and Berezney, R. (1991). Molecular cloning of matrin F/G: A DNA binding protein of the nuclear matrix that contains putative zinc finger motifs. *Proc. Natl. Acad. Sci. U.S.A.* **88,** 6186–6190.

Hatch, C. L., Bonner, W. M., and Moudrianakis, E. N. (1983). Minor histone 2A variants and ubiquinated forms in the native H2A:H2B dimer. *Science* **221,** 468–470.

Hay, C. W., and Candido, E. P. (1983a). Histone deacetylase. Association with a nuclease resistant, high molecular weight fraction of HeLa cell chromatin. *J. Biol. Chem.* **258,** 3726–3734.

Hay, C. W., and Candido, E. P. (1983b). Histone deacetylase from HeLa cells: Properties of the high molecular weight complex. *Biochemistry* **22,** 6175–6180.

He, D. C., Nickerson, J. A., and Penman, S. (1990). Core filaments of the nuclear matrix. *J. Cell Biol.* **110,** 569–580.

Hebbes, T. R., Thorne, A. W., and Crane Robinson, C. (1988). A direct link between core histone acetylation and transcriptionally active chromatin. *EMBO J.* **7,** 1395–1402.

Hebbes, T. R., Clayton, A. L., Thorne, A. W., and Crane-Robinson, C. (1994). Core histone hyperacetylation co-maps with generalized DNase I sensitivity in the chicken β-globin chromosomal domain. *EMBO J.* **13,** 1823–1830.

Heck, M. M., and Earnshaw, W. C. (1986). Topoisomerase II: A specific marker for cell proliferation. *J. Cell Biol.* **103,** 2569–2581.

Hendzel, M. J., and Davie, J. R. (1989). Distribution of methylated histones and histone methyltransferases in chicken erythrocyte chromatin. *J. Biol. Chem.* **264,** 19208–19214.

Hendzel, M. J., and Davie, J. R. (1990). Nucleosomal histones of transcriptionally active/competent chromatin preferentially exchange with newly synthesized histones in quiescent chicken erythrocytes. *Biochem. J.* **271**, 67–73.

Hendzel, M. J., and Davie, J. R. (1991). Dynamically acetylated histones of chicken erythrocytes are selectively methylated. *Biochem. J.* **273**, 753–758.

Hendzel, M. J., and Davie, J. R. (1992a). Acetylation and methylation of histones H3 and H4 in chicken immature erythrocytes are not directly coupled. *Biochem. Biophys. Res. Commun.* **185**, 414–419.

Hendzel, M. J., and Davie, J. R. (1992b). Nuclear distribution of histone deacetylase: A marker enzyme for the internal nuclear matrix. *Biochim. Biophys. Acta* **1130**, 307–313.

Hendzel, M. J., Delcuve, G. P., and Davie, J. R. (1991). Histone deacetylase is a component of the internal nuclear matrix. *J. Biol. Chem.* **266**, 21936–21942.

Hendzel, M. J., Sun, J.-M., Chen, H. Y., Rattner, J. B., and Davie, J. R. (1994). Histone acetyltransferase is associated with the nuclear matrix. *J. Biol. Chem.* **269**, 22894–22901.

Hentzen, P. C., Rho, J. H., and Bekhor, I. (1984). Nuclear matrix DNA from chicken erythrocytes contains beta-globin gene sequences. *Proc. Natl. Acad. Sci. U.S.A.* **81**, 304–307.

Hershko, A. (1988). Ubiquitin-mediated protein degradation. *J. Biol. Chem.* **263**, 15237–15240.

Hill, C. S., Packman, L. C., and Thomas, J. O. (1990). Phosphorylation at clustered -Ser-Pro-X-Lys/Arg- motifs in sperm-specific histones H1 and H2B. *EMBO J.* **9**, 805–813.

Hofmann, J. F., Laroche, T., Brand, A. H., and Gasser, S. M. (1989). RAP-1 factor is necessary for DNA loop formation in vitro at the silent mating type locus HML. *Cell (Cambridge, Mass.)* **57**, 725–737.

Hohmann, P. (1983). Phosphorylation of H1 histones. *Mol. Cell. Biochem.* **57**, 81–92.

Huang, S., and Spector, D. L. (1992). U1 and U2 small nuclear RNAs are present in nuclear speckles. *Proc. Natl. Acad. Sci. U.S.A.* **89**, 305–308.

Huletsky, A., de Murcia, G., Muller, S., Hengartner, M., Menard, L., Lamarre, D., and Poirier, G. G. (1989). The effect of poly(ADP-ribosyl)ation on native and H1-depleted chromatin. A role of poly(ADP-ribosyl)ation on core nucleosome structure. *J. Biol. Chem.* **264**, 8878–8886.

Ip, Y. T., Jackson, V., Meier, J., and Chalkley, R. (1988). The separation of transcriptionally engaged genes. *J. Biol. Chem.* **263**, 14044–14052.

Isomura, T., Tamiya Koizumi, K., Suzuki, M., Yoshida, S., Taniguchi, M., Matsuyama, M., Ishigaki, T., Sakuma, S., and Takahashi, M. (1992). RFP is a DNA binding protein associated with the nuclear matrix. *Nucleic Acids Res.* **20**, 5305–5310.

Izaurralde, E., Mirkovitch, J., and Laemmli, U. K. (1988). Interaction of DNA with nuclear scaffolds in vitro. *J. Mol. Biol.* **200**, 111–125.

Jackson, D. A., and Cook, P. R. (1985a). A general method for preparing chromatin containing intact DNA. *EMBO J.* **4**, 913–918.

Jackson, D. A., and Cook, P. R. (1985b). Transcription occurs at a nucleoskeleton. *EMBO J.* **4**, 919–925.

Jackson, D. A., and Cook, P. R. (1988). Visualization of a filamentous nucleoskeleton with a 23 nm axial repeat. *EMBO J.* **7**, 3667–3677.

Jackson, D. A., Yuan, J., and Cook, P. R. (1988). A gentle method for preparing cyto- and nucleo-skeletons and associated chromatin. *J. Cell Sci.* **90**, 365–378.

Jackson, D. A., Dickinson, P., and Cook, P. R. (1990). The size of chromatin loops in HeLa cells. *EMBO J.* **9**, 567–571.

Jackson, D. A., Hassan, A. B., Errington, R. J., and Cook, P. R. (1993). Visualization of focal sites of transcription within human nuclei. *EMBO J.* **12**, 1059–1065.

Jantzen, K., Fritton, H. P., and Igo Kemenes, T. (1986). The DNase I sensitive domain of the chicken lysozyme gene spans 24 kb. *Nucleic Acids Res.* **14**, 6085–6099.

Jarman, A. P., and Higgs, D. R. (1988). Nuclear scaffold attachment sites in the human globin gene complexes. *EMBO J.* **7**, 3337–3344.

Jentsch, S., Seufert, W., Sommer, T., and Reins, H. A. (1990). Ubiquitin-conjugating enzymes: Novel regulators of eukaryotic cells. *Trends Biochem. Sci.* **15,** 195–198.

Jenuwein, T., Forrester, W. C., Qiu, R.-G., and Grosschedl, R. (1993). The immunoglobulin μ enhancer core establishes local factor access in nuclear chromatin independent of transcriptional stimulation. *Genes Dev.* **7,** 2016–2032.

Jeppesen, P., and Turner, B. M. (1993). The inactive X chromosome in female mammals is distinguished by a lack of histone H4 acetylation, a cytogenetic marker for gene expression. *Cell (Cambridge, Mass.)* **74,** 281–289.

Kamakaka, R. T., and Thomas, J. O. (1990). Chromatin structure of transcriptionally competent and repressed genes. *EMBO J.* **9,** 3997–4006.

Kaplan, L. J., Bauer, R., Morrison, E., Langan, T. A., and Fasman, G. D. (1984). The structure of chromatin reconstituted with phosphorylated H1. Circular dichroism and thermal denaturation studies. *J. Biol. Chem.* **259,** 8777–8785.

Kaufmann, S. H., and Shaper, J. H. (1991). Association of topoisomerase II with the hepatoma cell nuclear matrix: The role of intermolecular disulfide bond formation. *Exp. Cell Res.* **192,** 511–523.

Kaufmann, S. H., Gibson, W., and Shaper, J. H. (1983). Characterization of the major polypeptides of the rat liver nuclear envelope. *J. Biol. Chem.* **258,** 2710–2719.

Kaufmann, S. H., Brunet, G., Talbot, B., Lamarr, D., Dumas, C., Shaper, J. H., and Poirier, G. (1991). Association of poly(ADP-ribose) polymerase with the nuclear matrix: The role of intermolecular disulfide bond formation, RNA retention, and cell type. *Exp. Cell Res.* **192,** 524–535.

Kijima, M., Yoshida, M., Sugita, K., Horinouchi, S., and Beppu, T. (1993). Trapoxin, an antitumor cyclic tetrapeptide, is an irreversible inhibitor of mammalian histone deacetylase. *J. Biol. Chem.* **268,** 22429–22435.

Klehr, D., Schlake, T., Maass, K., and Bode, J. (1992). Scaffold-attached regions (SAR elements) mediate transcriptional effects due to butyrate. *Biochemistry* **31,** 3222–3229.

Laemmli, U. K., Käs, E., Poljak, L., and Adachi, Y. (1992). Scaffold-associated regions: cis-acting determinants of chromatin structural loops and functional domains. *Curr. Opin. Genet. Dev.* **2,** 275–285.

Lafond, R. E., and Woodcock, C. L. F. (1983). Status of the nuclear matrix in mature and embryonic chick erythrocyte nuclei. *Exp. Cell Res.* **147,** 31–39.

Lawson, G. M., Knoll, B. J., March, C. J., Woo, S. L., Tsai, M. J., and O'Malley, B. W. (1982). Definition of 5' and 3' structural boundaries of the chromatin domain containing the ovalbumin multigene family. *J. Biol. Chem.* **257,** 1501–1507.

Lebel, S., and Raymond, Y. (1984). Lamin B from rat liver nuclei exists both as a lamina protein and as an intrinsic membrane protein. *J. Biol. Chem.* **259,** 2693–2696.

Lee, D. Y., Hayes, J. J., Pruss, D., and Wolffe, A. P. (1993). A positive role for histone acetylation in transcription factor access to nucleosomal DNA. *Cell (Cambridge, Mass.)* **72,** 73–84.

Lee, M. S., and Garrard, W. T. (1991a). Positive DNA supercoiling generates a chromatin conformation characteristic of highly active genes. *Proc. Natl. Acad. Sci. U.S.A.* **88,** 9675–9679.

Lee, M. S., and Garrard, W. T. (1991b). Transcription-induced nucleosome 'splitting': An underlying structure for DNase I sensitive chromatin. *EMBO J.* **10,** 607–615.

Lennox, R. W., and Cohen, L. H. (1983). The histone H1 complements of dividing and nondividing cells of the mouse. *J. Biol. Chem.* **258,** 262–268.

Lennox, R. W., and Cohen, L. H. (1988). The production of tissue-specific histone complements during development. *Biochem. Cell Biol.* **66,** 636–649.

Leonhardt, H., Page, A. W., Weier, H. U., and Bestor, T. H. (1992). A targeting sequence directs DNA methyltransferase to sites of DNA replication in mammalian nuclei. *Cell (Cambridge, Mass.)* **71,** 865–873.

Leuba, S. H., Yang, G., Robert, C., Van Holde, K. E., Zlatanova, J., and Bustamante, C. (1994). Three-dimensional structure of extended chromatin fibers as revealed by tapping-mode scanning force microscopy. *Proc. Natl. Acad. Sci. U.S.A.* **91,** 11621–11625.

Levy-Wilson, B., and Fortier, C. (1989). The limits of the DNase I-sensitive domain of the human apolipoprotein B gene coincide with the locations of chromosomal anchorage loops and define the 5′ and 3′ boundaries of the gene. *J. Biol. Chem.* **264,** 21196–21204.

Li, B., Adams, C. C., and Workman, J. L. (1994). Nucleosome binding by the constitutive transcription factor Sp1. *J. Biol. Chem.* **269,** 7756–7763.

Li, H., and Bingham, P. M. (1991). Arginine/serine-rich domains of the su(wa) and tra RNA processing regulators target proteins to a subnuclear compartment implicated in splicing. *Cell (Cambridge, Mass.)* **67,** 335–342.

Li, W., Nagaraja, S., Delcuve, G. P., Hendzel, M. J., and Davie, J. R. (1993). Effects of histone acetylation, ubiquitination and variants on nucleosome stability. *Biochem. J.* **296,** 737–744.

Liu, L. F., and Wang, J. C. (1987). Supercoiling of the DNA template during transcription. *Proc. Natl. Acad. Sci. U.S.A.* **84,** 7024–7027.

Locklear, L. J., Ridsdale, J. A., Bazett Jones, D. P., and Davie, J. R. (1990). Ultrastructure of transcriptionally competent chromatin. *Nucleic Acids Res.* **18,** 7015–7024.

Lopez Rodas, G., Tordera, V., Sanchez del Pino, M. M., and Franco, L. (1991). Subcellular localization and nucleosome specificity of yeast histone acetyltransferases. *Biochemistry* **30,** 3728–3732.

Lopez Rodas, G., Brosch, G., Golderer, G., Lindner, H., Grobner, P., and Loidl, P. (1992). Enzymes involved in the dynamic equilibrium of core histone acetylation of Physarum polycephalum. *FEBS Lett.* **296,** 82–86.

Lu, M. J., Dadd, C. A., Mizzen, C. A., Perry, C. A., McLachlan, D. R., Annunziato, A. T., and Allis, C. D. (1994). Generation and characterization of novel antibodies highly selective for phosphorylated linker histone H1 in *Tetrahymena* and HeLa cells. *Chromosoma* **103,** 111–121.

Lu, Q., Wallrath, L. L., and Elgin, S. C. R. (1994). Nucleosome positioning and gene regulation. *J. Cell. Biochem.* **55,** 83–92.

Ludérus, M. E., de Graaf, A., Mattia, E., den Blaauwen, J. L., Grande, M. A., de Jong, L., and van Driel, R. (1992). Binding of matrix attachment regions to lamin B1. *Cell (Cambridge, Mass.)* **70,** 949–959.

Ludérus, M. E. E., den Blaauwen, J. L., de Smit, O. J. B., Compton, D. A., and van Driel, R. (1994). Binding of matrix attachment regions to lamin polymers involves single-stranded regions and the minor groove. *Mol. Cell. Biol.* **14,** 6297–6305.

Lund, T., and Laland, S. G. (1990). The metaphase specific phosphorylation of HMG I. *Biochem. Biophys. Res. Commun.* **171,** 342–347.

Lutter, L. C., Judis, L., and Paretti, R. F. (1992). Effects of histone acetylation on chromatin topology in vivo. *Mol. Cell. Biol.* **12,** 5004–5014.

Mahadevan, L. C., Willis, A. C., and Barratt, M. J. (1991). Rapid histone H3 phosphorylation in response to growth factors, phorbol esters, okadaic acid, and protein synthesis inhibitors. *Cell (Cambridge, Mass.)* **65,** 775–783.

Malik, N., Smulson, M., and Bustin, M. (1984). Enrichment of acetylated histones in poly-nucleosomes containing high mobility group protein 17 revealed by immunoaffinity chromatography. *J. Biol. Chem.* **259,** 699–702.

Manfioletti, G., Giancotti, V., Bandiera, A., Buratti, E., Sautiere, P., Cary, P., Crane Robinson, C., Coles, B., and Goodwin, G. H. (1991). cDNA cloning of the HMGI-C phosphoprotein, a nuclear protein associated with neoplastic and undifferentiated phenotypes. *Nucleic Acids Res.* **19,** 6793–6797.

Matsui, S.-I., Seon, B. K., and Sandberg, A. V. (1979). Disappearance of a structural chromatin protein A24 in mitosis: Implications for molecular basis of chromatin condensation. *Proc. Natl. Acad. Sci. U.S.A.* **76,** 6386–6390.

Matsui, S.-I., Sandberg, A. A., Negoro, S., Seon, B. K., and Goldstein, G. (1982). Isopeptidase: A novel eukaryotic enzyme that cleaves isopeptide bonds. *Proc. Natl. Acad. Sci. U.S.A.* **79,** 1535–1539.

McKnight, R. A., Shamay, A., Sankaran, L., Wall, R. J., and Hennighausen, L. (1992). Matrix-attachment regions can impart position-independent regulation of a tissue-specific gene in transgenic mice. *Proc. Natl. Acad. Sci. U.S.A.* **89,** 6943–6947.

McPherson, C. E., Shim, E.-Y., Friedman, D. S., and Zaret, K. S. (1993). An active tissue-specific enhancer and bound transcription factors existing in a precisely positioned nucleosomal array. *Cell (Cambridge, Mass.)* **75,** 387–398.

Meijer, L., Ostvold, A. C., Walass, S. I., Lund, T., and Laland, S. G. (1991). High-mobility-group proteins P1, I and Y as substrates of the M-phase-specific p34cdc2/cyclincdc13 kinase. *Eur. J. Biochem.* **196,** 557–567.

Mingarro, I., Sendra, R., Salvador, M. L., and Franco, L. (1993). Site specificity of pea histone acetyltransferase B *in vitro. J. Biol. Chem.* **268,** 13248–13252.

Mirkovitch, J., Mirault, M. E., and Laemmli, U. K. (1984). Organization of the higher-order chromatin loop: Specific DNA attachment sites on nuclear scaffold. *Cell (Cambridge, Mass.)* **39,** 223–232.

Morse, R. H. (1992). Transcribed chromatin. *Trends Biochem. Sci.* **17,** 23–26.

Mueller, R. D., Yasuda, H., Hatch, C. L., Bonner, W. M., and Bradbury, E. M. (1985). Identification of ubiquitinated histones 2A and 2B in Physarum polycephalum. Disappearance of these proteins at metaphase and reappearance at anaphase. *J. Biol. Chem.* **260,** 5147–5153.

Munks, R. J., Moore, J., O'Neill, L. P., and Turner, B. M. (1991). Histone H4 acetylation in Drosophila. Frequency of acetylation at different sites defined by immunolabelling with site-specific antibodies. *FEBS Lett.* **284,** 245–248.

Nacheva, G. A., Guschin, D. Y., Preobrazhenskaya, O. V., Karpov, V. L., Ebralidse, K. K., and Mirzabekov, A. D. (1989). Change in the pattern of histone binding to DNA upon transcriptional activation. *Cell (Cambridge, Mass.)* **58,** 27–36.

Nakagomi, K., Kohwi, Y., Dickinson, L. A., and Kohwi-Shigematsu, T. (1994). A novel DNA-binding motif in the nuclear matrix attachment DNA-binding protein SATB1. *Mol. Cell. Biol.* **14,** 1852–1860.

Nakayasu, H., and Berezney, R. (1991). Nuclear matrins: Identification of the major nuclear matrix proteins. *Proc. Natl. Acad. Sci. U.S.A.* **88,** 10312–10316.

Nickel, B. E., and Davie, J. R. (1989). Structure of polyubiquitinated histone H2A. *Biochemistry* **28,** 964–968.

Nickel, B. E., Allis, C. D., and Davie, J. R. (1989). Ubiquitinated histone H2B is preferentially located in transcriptionally active chromatin. *Biochemistry* **28,** 958–963.

Nissen, M. S., Langan, T. A., and Reeves, R. (1991). Phosphorylation by cdc2 kinase modulates DNA binding activity of high mobility group I nonhistone chromatin protein. *J. Biol. Chem.* **266,** 19945–19952.

Norton, V. G., Marvin, K. W., Yau, P., and Bradbury, E. M. (1990). Nucleosome linking number change controlled by acetylation of histones H3 and H4. *J. Biol. Chem.* **265,** 19848–19852.

Ogata, N., Ueda, K., and Hayaishi, O. (1980). ADP-ribosylation of histone H2B: Identification of glutamic acid residue 2 as the modification site. *J. Biol. Chem.* **255,** 7610–7615.

Palvimo, J., and Linnala Kankkunen, A. (1989). Identification of sites on chromosomal protein HMG-I phosphorylated by casein kinase II. *FEBS Lett.* **257,** 101–104.

Pantazis, P., and Bonner, W. T. (1981). Quantitative determination of histone modification: H2A acetylation and phosphorylation. *J. Biol. Chem.* **256,** 4669–4675.

Parseghian, M. H., Clark, R. F., Hauser, L. J., Dvorkin, N., Harris, D. A., and Hamkalo, B. A. (1993). Fractionation of human H1 subtypes and characterization of a subtype-specific antibody exhibiting non-uniform nuclear staining. *Chromosome Res.* **1,** 127–139.

Parseghian, M. H., Harris, D. A., Rishwain, D. R., and Hamkalo, B. A. (1994a). Characterization of a set of antibodies specific for three human histone H1 subtypes. *Chromosoma* **103**, 198–208.

Parseghian, M. H., Henschen, A. H., Krieglstein, K. G., and Hamkalo, B. A. (1994b). A proposal for a coherent mammalian histone H1 nomenclature correlated with amino acid sequences. *Protein Sci.* **3**, 575–587.

Pedraza Reyes, M., and Alvarez Gonzalez, R. (1990). Oligo(3'-deoxy ADP-ribosyl)ation of the nuclear matrix lamins from rat liver utilizing 3'-deoxyNAD as a substrate. *FEBS Lett.* **277**, 88–92.

Pehrson, J. R., and Fried, V. A. (1992). MacroH2A, a core histone containing a large nonhistone region. *Science* **257**, 1398–1400.

Pelech, S. L., Sanghera, J. S., and Daya Makin, M. (1990). Protein kinase cascades in meiotic and mitotic cell cycle control. *Biochem. Cell Biol.* **68**, 1297–1330.

Perlmann, T. (1992). Glucocorticoid receptor DNA-binding specificity is increased by the organization of DNA in nucleosomes. *Proc. Natl. Acad. Sci. U.S.A.* **89**, 3884–3888.

Perry, C. A., and Annunziato, A. T. (1991). Histone acetylation reduces H1-mediated nucleosome interactions during chromatin assembly. *Exp. Cell Res.* **196**, 337–345.

Phi-Van, L., and Strätling, W. H. (1988). The matrix attachment regions of the chicken lysozyme gene co-map with the boundaries of the chromatin domain. *EMBO J.* **7**, 655–664.

Phi-Van, L., von Kries, J. P., Ostertag, W., and Strätling, W. H. (1990). The chicken lysozyme 5' matrix attachment region increases transcription from a heterologous promoter in heterologous cells and dampens position effects on the expression of transfected genes. *Mol. Cell. Biol.* **10**, 2302–2307.

Pienta, K. J., Getzenberg, R. H., and Coffey, D. S. (1991). Cell structure and DNA organization. *CRC Crit. Rev. Eukaryotic Gene Express.* **1**, 355–385.

Pina, B., and Suau, P. (1987). Changes in histones H2A and H3 variant composition in differentiating and mature rat brain cortical neurons. *Dev. Biol.* **123**, 51–58.

Postnikov, Y. V., Shick, V. V., Belyavsky, A. V., Khrapko, K. R., Brodolin, K. L., Nikolskaya, T. A., and Mirzabekov, A. D. (1991). Distribution of high mobility group proteins 1/2, E and 14/17 and linker histones H1 and H5 on transcribed and non-transcribed regions of chicken erythrocyte chromatin. *Nucleic Acids Res.* **19**, 717–725.

Raboy, B., Parag, H. A., and Kulka, R. G. (1986). Conjugation of [125I]ubiquitin to cellular proteins in permeabilized mammalian cells: Comparison of mitotic and interphase cells. *EMBO J.* **5**, 863–869.

Ram, T. G., Reeves, R., and Hosick, H. L. (1993). Elevated high mobility group-I(Y) gene expression is associated with progressive transformation of mouse mammary epithelial cells. *Cancer Res.* **53**, 2655–2660.

Ramakrishnan, V. (1994). Histone structure. *Curr. Opin. Struct. Biol.* **4**, 44–50.

Realini, C. A., and Althaus, F. R. (1992). Histone shuttling by poly(ADP-ribosylation). *J. Biol. Chem.* **267**, 18858–18865.

Reeves, R. (1984). Transcriptionally active chromatin. *Biochim. Biophys. Acta* **782**, 343–393.

Reeves, R. (1992). Chromatin changes during the cell cycle. *Curr. Opin. Cell Biol.* **4**, 413–423.

Reeves, R., and Chang, D. (1983). Investigations of the possible functions for glycosylation in the high mobility group proteins. Evidence for a role in nuclear matrix association. *J. Biol. Chem.* **258**, 679–687.

Reeves, R., and Nissen, M. S. (1990). The A.T-DNA-binding domain of mammalian high mobility group I chromosomal proteins. A novel peptide motif for recognizing DNA structure. *J. Biol. Chem.* **265**, 8573–8582.

Reeves, R., and Nissen, M. S. (1993). Interaction of high mobility group-I(Y) nonhistone proteins with nucleosome core particles. *J. Biol. Chem.* **268**, 21137–21146.

Reeves, R., Chang, D., and Chung, S.-C. (1981). Carbohydrate modifications of the high mobility group proteins. *Proc. Natl. Acad. Sci. U.S.A.* **78**, 6704–6708.

Reeves, R., Langan, T. A., and Nissen, M. S. (1991). Phosphorylation of the DNA-binding domain of nonhistone high-mobility group I protein by cdc2 kinase: Reduction of binding affinity. *Proc. Natl. Acad. Sci. U.S.A.* **88,** 1671–1675.

Reitman, M., Lee, E., Westphal, H., and Felsenfeld, G. (1990). Site-independent expression of the chicken beta A-globin gene in transgenic mice. *Nature (London)* **348,** 749–752.

Reitman, M., Lee, E., Westphal, H., and Felsenfeld, G. (1993). An enhancer/locus control region is not sufficient to open chromatin. *Mol. Cell. Biol.* **13,** 3990–3998.

Renaud, J., and Ruiz Carrillo, A. (1986). Fine analysis of the active H5 gene chromatin of chicken erythroid cells at different stages of differentiation. *J. Mol. Biol.* **189,** 217–226.

Richman, R., Chicoine, L. G., Collini, M. P., Cook, R. G., and Allis, C. D. (1988). Micronuclei and the cytoplasm of growing *Tetrahymena* contain a histone acetylase activity which is highly specific for free histone H4. *J. Cell Biol.* **106,** 1017–1026.

Richmond, T. J., Finch, J. T., Rushton, B., Rhodes, D., and Klug, A. (1984). Structure of the nucleosome core particle at 7 Å resolution. *Nature (London)* **311,** 532–537.

Ridsdale, J. A., and Davie, J. R. (1987). Chicken erythrocyte polynucleosomes which are soluble at physiological ionic strength and contain linker histones are highly enriched in β-globin gene sequences. *Nucleic Acids Res.* **15,** 1081–1096.

Ridsdale, J. A., Rattner, J. B., and Davie, J. R. (1988). Erythroid-specific gene chromatin has an altered association with linker histones. *Nucleic Acids Res.* **16,** 5915–5926.

Ridsdale, J. A., Hendzel, M. J., Delcuve, G. P., and Davie, J. R. (1990). Histone acetylation alters the capacity of the H1 histones to condense transcriptionally active/competent chromatin. *J. Biol. Chem.* **265,** 5150–5156.

Roberge, M., Dahmus, M. E., and Bradbury, E. M. (1988). Chromosomal loop/nuclear matrix organization of transcriptionally active and inactive RNA polymerases in HeLa nuclei. *J. Mol. Biol.* **201,** 545–555.

Robinson, S. I., Nelkin, B. D., and Vogelstein, B. (1982). The ovalbumin gene is associated with the nuclear matrix of chicken oviduct cells. *Cell (Cambridge, Mass.)* **28,** 99–106.

Robinson, S. I., Small, D., Idzerda, R., McKnight, G. S., and Vogelstein, B. (1983). The association of transcriptionally active genes with the nuclear matrix of the chicken oviduct. *Nucleic Acids Res.* **11,** 5113–5130.

Rocha, E., Davie, J. R., Van Holde, K. E., and Weintraub, H. (1984). Differential salt fractionation of active and inactive genomic domains in chicken erythrocyte. *J. Biol. Chem.* **259,** 8558–8563.

Romig, H., Fackelmayer, F. O., Renz, A., Ramsperger, U., and Richter, A. (1992). Characterization of SAF-A, a novel nuclear DNA binding protein from HeLa cells with high affinity for nuclear matrix/scaffold attachment DNA elements. *EMBO J.* **11,** 3431–3440.

Roth, S. Y., and Allis, C. D. (1992). Chromatin condensation: Does histone H1 dephosphorylation play a role? *Trends Biochem. Sci.* **17,** 93–98.

Roth, S. Y., Collini, M. P., Draetta, G., Beach, D., and Allis, C. D. (1991). A cdc2-like kinase phosphorylates histone H1 in the amitotic macronucleus of *Tetrahymena. EMBO J.* **10,** 2069–2075.

Rousseau, S., Asselin, M., Renaud, J., and Ruiz-Carrillo, A. (1993). Transcription of the histone *H5* gene is regulated by three differentiation-specific enhancers. *Mol. Cell. Biol.* **13,** 4904–4917.

Sahyoun, N., LeVine, H., Bronson, D., and Cuatrecasas, P. (1984). Ca2+-calmodulin-dependent protein kinase in neuronal nuclei. *J. Biol. Chem.* **259,** 9341–9344.

Schild, C., Claret, F.-X., Wahli, W., and Wolffe, A. P. (1993). A nucleosome-dependent static loop potentiates estrogen-regulated transcription from the *Xenopus* vitellogenin B1 promoter *in vitro. EMBO J.* **12,** 423–433.

Schwartz, A. L., Trausch, J. S., Ciechanover, A., Slot, J. W., and Geuze, H. (1992). Immunoelectron microscopic localization of the ubiquitin-activating enzyme E1 in HepG2 cells. *Proc. Natl. Acad. Sci. U.S.A.* **89,** 5542–5546.

Seale, R. L. (1981). Rapid turnover of the histone-ubiquitin conjugate, protein A24. *Nucleic Acids Res.* **9,** 3151–3158.

Shibata, K., and Ajiro, K. (1993). Cell cycle-dependent suppressive effect of histone H1 on mitosis-specific H3 phosphorylation. *J. Biol. Chem.* **268,** 18431–18434.

Shibata, K., Inagaki, M., and Ajiro, K. (1990). Mitosis-specific histone H3 phosphorylation in vitro in nucleosome structures. *Eur. J. Biochem.* **192,** 87–93.

Sommerville, J., Baird, J., and Turner, B. M. (1993). Histone H4 acetylation and transcription in amphibian chromatin. *J. Cell Biol.* **120,** 277–290.

Sterner, R., Vidali, G., and Allfrey, V. G. (1979). Studies of acetylation and deacetylation in high mobility group proteins: Identification of the sites of acetylation in HMG 1. *J. Biol. Chem.* **254,** 11577–11583.

Sterner, R., Vidali, G., and Allfrey, V. G. (1981). Studies of acetylation and deacetylation in high mobility group proteins: Identification of the sites of acetylation in high mobility group proteins 14 and 17. *J. Biol. Chem.* **156,** 8892–8895.

Stief, A., Winter, D. M., Strätling, W. H., and Sippel, A. E. (1989). A nuclear DNA attachment element mediates elevated and position-independent gene activity. *Nature (London)* **341,** 343–345.

Strätling, W. H. (1987). Gene-specific differences in the supranucleosomal organization of rat liver chromatin. *Biochemistry* **26,** 7893–7899.

Strätling, W. H., Dolle, A., and Sippel, A. E. (1986). Chromatin structure of the chicken lysozyme gene domain as determined by chromatin fractionation and micrococcal nuclease digestion. *Biochemistry* **25,** 495–502.

Studitsky, V. M., Clark, D. J., and Felsenfeld, G. (1994). A histone octamer can step around a transcribing polymerase without leaving the template. *Cell (Cambridge, Mass.)* **76,** 371–382.

Stuurman, N., van Driel, R., de Jong, L., Meijne, A. M., and van Renswoude, J. (1989). The protein composition of the nuclear matrix of murine P19 embryonal carcinoma cells is differentiation-stage dependent. *Exp. Cell Res.* **180,** 460–466.

Sun, J.-M., Penner, C. G., and Davie, J. R. (1992). Analysis of erythroid nuclear proteins binding to the promoter and enhancer elements of the chicken histone H5 gene. *Nucleic Acids Res* **20,** 6385–6392.

Sun, J.-M., Chen, H. Y., and Davie, J. R. (1994). Nuclear factor 1 is a component of the nuclear matrix. *J. Cell. Biochem.* **55,** 252–263.

Sung, M. T., and Dixon, G. H. (1970). Modification of histones during spermatogenesis in trout: A molecular mechanism for altering histone binding to DNA. *Proc. Natl. Acad. Sci. U.S.A.* **67,** 1616–1623.

Swerdlow, P. S., Schuster, T., and Finley, D. (1990). A conserved sequence in histone H2A which is a ubiquitination site in higher eucaryotes is not required for growth in Saccharomyces cerevisiae. *Mol. Cell. Biol.* **10,** 4905–4911.

Tawfic, S., and Ahmed, K. (1994). Association of casein kinase 2 with nuclear matrix. Possible role in nuclear matrix protein phosphorylation. *J. Biol. Chem.* **269,** 7489–7493.

Thomsen, B., Bendixen, C., and Westergaard, O. (1991). Histone hyperacetylation is accompanied by changes in DNA topology in vivo. *Eur. J. Biochem.* **201,** 107–111.

Thorburn, A., and Knowland, J. (1993). Attachment of vitellogenin genes to the nucleoskeleton accompanies their activation. *Biochem. Biophys. Res. Commun.* **191,** 308–313.

Thorburn, A., Moore, R., and Knowland, J. (1988). Attachment of transcriptionally active DNA sequences to the nucleoskeleton under isotonic conditions. *Nucleic Acids Res.* **16,** 7183.

Thorne, A. W., Sautiere, P., Briand, G., and Crane Robinson, C. (1987). The structure of ubiquitinated histone H2B. *EMBO J.* **6,** 1005–1010.

Thorne, A. W., Kmiciek, D., Mitchelson, K., Sautiere, P., and Crane Robinson, C. (1990). Patterns of histone acetylation. *Eur. J. Biochem.* **193,** 701–713.

Tremethick, D. J., and Molloy, P. L. (1986). High mobility group proteins 1 and 2 stimulate transcription in vitro by RNA polymerases II and III. *J. Biol. Chem.* **261,** 6986–6992.

Turner, B. M. (1989). Acetylation and deacetylation of histone H4 continue through metaphase with depletion of more-acetylated isoforms and altered site usage. *Exp. Cell Res.* **182,** 206–214.

Turner, B. M. (1991). Histone acetylation and control of gene expression. *J. Cell Sci.* **99,** 13–20.

Turner, B. M., O'Neill, L. P., and Allan, I. M. (1989). Histone H4 acetylation in human cells. Frequency of acetylation at different sites defined by immunolabeling with site-specific antibodies. *FEBS Lett.* **253,** 141–145.

Turner, B. M., Birley, A. J., and Lavender, J. (1992). Histone H4 isoforms acetylated at specific lysine residues define individual chromosomes and chromatin domains in Drosophila polytene nuclei. *Cell (Cambridge, Mass.)* **69,** 375–384.

van Daal, A., and Elgin, S. C. (1992). A histone variant, H2AvD, is essential in *Drosophila melanogaster. Mol. Biol. Cell.* **3,** 593–602.

Van Holde, K. E. (1988). "Chromatin," Springer-Verlag, New York.

Van Holde, K. E. (1993). Transcription: The omnipotent nucleosome. *Nature (London)* **362,** 111–112.

Van Holde, K. E., and Zlatanova, J. (1994). Unusual DNA structures, chromatin and transcription. *BioEssays* **16,** 59–68.

Van Holde, K. E., Lohr, D. E., and Robert, C. (1992). What happens to nucleosomes during transcription? *J. Biol. Chem.* **267,** 2837–2840.

van Wijnen, A. J., Bidwell, J. P., Fey, E. G., Penman, S., Lian, J. B., Stein, J. L., and Stein, G. S. (1993). Nuclear matrix association of multiple sequence-specific DNA binding activities related to SP-1, ATF, CCAAT, C/EBP, OCT-1, and AP-1. *Biochemistry* **32,** 8397–8402.

Vassetzky, Y. S., De Moura Gallo, C. V., Bogdanova, A. N., Razin, S. V., and Scherrer, K. (1993). The sequence-specific nuclear matrix binding factor F6 is a chicken GATA-like protein. *Mol. Gen. Genet.* **238,** 309–314.

Verreault, A., and Thomas, J. O. (1993). Chromatin structure of the β-globin chromosomal domain in adult chicken erythrocytes. *Cold Spring Harbor Symp. Quant. Biol.* **58,** 15–24.

Vettese-Dadey, M., Walter, P., Chen, H., Juan, L.-J., and Workman, J. L. (1994). Role of the histone amino termini in facilitated binding of a transcription factor, GAL4-AH, to nucleosome cores. *Mol. Cell. Biol.* **14,** 970–981.

Vidali, G., Boffa, L. C., and Allfrey, V. G. (1972). Properties of an acidic histone-binding protein fraction from cell nuclei: Selective precipitation and deacetylation of histones F2A1 and F3. *J. Biol. Chem.* **247,** 7365–7373.

Vidali, G., Ferrari, N., and Pfeffer, U. (1988). Histone acetylation: A step in gene activation. *Adv. Exp. Med. Biol.* **231,** 583–596.

Villeponteau, B., and Martinson, H. G. (1987). Gamma rays and bleomycin nick DNA and reverse the DNase I sensitivity of beta-globin gene chromatin in vivo. *Mol. Cell. Biol.* **7,** 1917–1924.

Villeponteau, B., Lundell, M., and Martinson, H. (1984). Torsional stress promotes the DNase I sensitivity of active genes. *Cell (Cambridge, Mass.)* **39,** 469–478.

Villeponteau, B., Pribyl, T. M., Grant, M. H., and Martinson, H. G. (1986). Novobiocin induces the in vivo cleavage of active gene sequences in intact cells. *J. Biol. Chem.* **261,** 10359–10365.

von Kries, J. P., Buhrmester, H., and Strätling, W. H. (1991). A matrix/scaffold attachment region binding protein: Identification, purification, and mode of binding. *Cell (Cambridge, Mass.)* **64,** 123–135.

von Kries, J. P., Buck, F., and Strätling, W. H. (1994). Chicken MAR binding protein p120 is identical to human heterogeneous nuclear ribonucleoprotein (hnRNP) U. *Nucleic Acids Res.* **22,** 1215–1220.

Waitz, W., and Loidl, P. (1991). Cell cycle dependent association of c-myc protein with the nuclear matrix. *Oncogene* **6,** 29–35.

Walker, J., Chen, T. A., Sterner, R., Berger, M., Winston, F., and Allfrey, V. G. (1990). Affinity chromatography of mammalian and yeast nucleosomes. Two modes of binding of transcriptionally active mammalian nucleosomes to organomercurial-agarose columns, and contrasting behavior of the active nucleosomes of yeast. *J. Biol. Chem.* **265,** 5736–5746.

Wallrath, L. L., Lu, Q., Granok, H., and Elgin, S. C. R. (1994). Architectural variations of inducible eukaryotic promoters: Preset and remodeling chromatin structures. *BioEssays* **16,** 165–170.

Walton, G. M., and Gill, G. N. (1983). Identity of the in vivo phosphorylation site in high mobility group 14 protein in HeLa cells with the site phosphorylated by casein kinase II in vitro. *J. Biol. Chem.* **258,** 4440–4446.

Walton, G. M., Spiess, J., and Gill, G. N. (1982). Phosphorylation of high mobility group 14 protein by cyclic nucleotide-dependent protein kinases. *J. Biol. Chem.* **257,** 4661–4668.

Waterborg, J. H. (1993a). Histone synthesis and turnover in alfalfa. Fast loss of highly acetylated replacement histone variant H3.2. *J. Biol. Chem.* **268,** 4912–4917.

Waterborg, J. H. (1993b). Dynamic methylation of alfalfa histone H3. *J. Biol. Chem.* **268,** 4918–4921.

White, E. M., Shapiro, D. L., Allis, C. D., and Gorovsky, M. A. (1988). Sequence and properties of the message encoding *Tetrahymena* hv1, a highly evolutionarily conserved histone H2A variant that is associated with active genes. *Nucleic Acids Res.* **16,** 179–198.

Widom, J. (1986). Physicochemical studies of the folding of the 100 Å nucleosome filament into the 300 Å filament. *J. Mol. Biol.* **190,** 411–424.

Wong, M., Miwa, M., Sugimura, T., and Smulson, M. (1983). Relationship between histone H1 poly(adenosine diphosphate ribosylation) and histone H1 phosphorylation using anti-poly(adenosine diphosphate ribose) antibody. *Biochemistry* **22,** 2384–2389.

Woodcock, C. L., Grigoryev, S. A., Horowitz, R. A., and Whitaker, N. (1993). A chromatin folding model that incorporates linker variability generates fibers resembling the native structures. *Proc. Natl. Acad. Sci. USA* **90,** 9021–9025.

Worcel, A., Gargiulo, G., Jessee, B., Udvardy, A., Louis, C., and Schedl, P. (1983). Chromatin fine structure of the histone gene complex of Drosophila melanogaster. *Nucleic Acids Res.* **11,** 421–439.

Workman, J. L., and Buchman, A. R. (1993). Multiple functions of nucleosomes and regulatory factors in transcription. *Trends Biochem. Sci.* **18,** 90–95.

Workman, J. L., and Kingston, R. E. (1992). Nucleosome core displacement in vitro via a metastable transcription factor-nucleosome complex. *Science* **258,** 1780–1784.

Wu, R. S., Kohn, K. W., and Bonner, W. M. (1981). Metabolism of ubiquitinated histones. *J. Biol. Chem.* **256,** 5916–5920.

Wu, R. S., Panusz, H. T., Hatch, C. L., and Bonner, W. M. (1986). Histones and their modifications. *CRC Crit. Rev. Biochem.* **20,** 201–263.

Yoshida, M., Kijima, M., Akita, M., and Beppu, T. (1990). Potent and specific inhibition of mammalian histone deacetylase both in vivo and in vitro by trichostatin A. *J. Biol. Chem.* **265,** 17174–17179.

Yu, J., and Smith, R. D. (1985). Sequential alterations in globin gene chromatin structure during erythroleukemia cell differentiation. *J. Biol. Chem.* **260,** 3035–3040.

Zenk, D. W., Ginder, G. D., and Brotherton, T. W. (1990). A nuclear matrix protein binds very tightly to DNA in the avian beta-globin gene enhancer. *Biochemistry* **29,** 5221–5226.

Zhang, D.-E., and Nelson, D. A. (1986). Histone acetylation in chicken erythrocytes: Estimation of the percentage of sites actively modified. *Biochem. J.* **240,** 857–862.

Zhang, D.-E., and Nelson, D. A. (1988a). Histone acetylation in chicken erythrocytes: Rates of acetylation and evidence that histones in both active and potentially active chromatin are rapidly modified. *Biochem. J.* **250,** 233–240.

Zhang, D.-E., and Nelson, D. A. (1988b). Histone acetylation in chicken erythrocytes: rates of deacetylation in immature and mature red blood cells. *Biochem. J.* **250,** 241–245.

Zhao, K., Käs, E., Gonzalez, E., and Laemmli, U. K. (1993). SAR-dependent mobilization of histone H1 by HMG-I/Y *in vitro*: HMG-I/Y is enriched in H1-depleted chromatin. *EMBO J.* **12,** 3237–3247.

Zini, N., Mazzotti, G., Santi, P., Rizzoli, R., Galanzi, A., Rana, R., and Maraldi, N. M. (1989). Cytochemical localization of DNA loop attachment sites to the nuclear lamina and to the inner nuclear matrix. *Histochemistry* **91,** 199–204.

Contributions of Nuclear Architecture to Transcriptional Control

Gary S. Stein, André J. van Wijnen, Janet Stein, Jane B. Lian, and Martin Montecino
Department of Cell Biology and Cancer Center, University of Massachusetts
Medical Center, Worcester, Massachusetts 01655

Three parameters of nuclear structure contribute to transcriptional control. The linear representation of promoter elements provides competency for physiological responsiveness within the contexts of developmental as well as cell cycle- and phenotype-dependent regulation. Chromatin structure and nucleosome organization reduce distances between independent regulatory elements providing a basis for integrating components of transcriptional control. The nuclear matrix supports gene expression by imposing physical constraints on chromatin related to three-dimensional genomic organization. In addition, the nuclear matrix facilitates gene localization as well as the concentration and targeting of transcription factors. Several lines of evidence are presented that are consistent with involvement of multiple levels of nuclear architecture in cell growth and tissue-specific gene expression during differentiation. Growth factor and steroid hormone responsive modifications in chromatin structure, nucleosome organization, and the nuclear matrix that influence transcription of the cell cycle-regulated histone gene and the bone tissue-specific osteocalcin gene during progressive expression of the osteoblast phenotype are considered.

KEY WORDS: Cell cycle, Osteoblasts, Nucleosomes, Chromatin, Nuclear matrix.

I. Introduction

A fundamental biological paradox is the mechanisms by which, with a limited representation of gene-specific regulatory elements and a low abundance of cognate transactivation factors, sequence-specific interactions occur to support a threshold for initiation of transcription within nuclei of intact cells. Viewed from a quantitative perspective, the regulatory challenge is to account for formation of functional transcription initiation complexes with a nuclear concentration of regulatory sequences that is approximately 20 nucleotides per 2.5 yards of DNA and a similarly restricted level of DNA-binding proteins.

It is becoming increasingly apparent that nuclear architecture provides a basis for support of stringently regulated modulation of cell growth and tissue-specific transcription which is necessary for the onset and progression of differentiation. Here, multiple lines of evidence point to contributions by three levels of nuclear organization to transcriptional control where structural parameters are functionally coupled to regulatory events (Table I). The primary level of gene organization establishes a linear ordering of promoter regulatory elements. This representation of regulatory sequences reflects competency for responsiveness to physiological regulatory signals. However, interspersion of sequences between promoter elements that exhibit coordinate and synergistic activities indicates the requirement of a structural basis for integration of activities at independent regulatory domains. Parameters of chromatin structure and nucleosome organization are a second level of genome architecture that reduce the distance between promoter elements thereby supporting interactions between the modular components of transcriptional control. Each nucleosome (approximately 140 nucleotide base pairs wound around a core complex of 2 each of H3, H4, H2, and H2B histone proteins) contracts linear spacing by sevenfold. Higher order chromatin structure further reduces nucleotide distances between regulatory sequences. Folding of nucleosome arrays into solenoid-type structures provides a potential for interactions that support synergism between promoter elements and responsiveness to multiple signaling path-

TABLE I

Three Levels of Nuclear Organization That Contribute to Transcriptional Control

Structural parameters	Regulatory events
(1) Organization of promoter regulatory elements	(1) Representation of physiologically responsive transcriptional regulatory sequences
(2) Chromatin structure and nucleosome organization	(2) Integration of activities at independent promoter domains Responsiveness to multiple signaling pathways
(3) Composition and organization of nuclear architecture	(3) Gene localization Concentration and targeting of transcription factors Imprinting and modulation of chromatin structure

ways. A third level of nuclear architecture that contributes to transcriptional control is provided by the nuclear matrix. The anastomosing network of fibers and filaments that constitute the nuclear matrix supports the structural properties of the nucleus as a cellular organelle and accommodates structural modifications associated with proliferation, differentiation, and changes necessary to sustain phenotypic requirements of specialized cells. Regulatory functions of the nuclear matrix include but are by no means restricted to gene localization, imposition of physical constraints on chromatin structure that support formation of loop domains, concentration and targeting of transcription factors, RNA processing and transport of gene transcripts, concentration and targeting of transcription factors, as well as imprinting and modifications of chromatin structure. Taken together these components of nuclear architecture facilitate biological requirements for physiologically responsive modifications in gene expression within the context of: (1) homeostatic control involving rapid, short-term, and transient responsiveness; (2) developmental control that is progressive and stage-specific; and (3) differentiation-related control that is associated with long-term phenotypic commitments to gene expression for support of structural and functional properties of cells and tissues (Table II).

We are just beginning to appreciate the significance of nuclear domains in the control of gene expression. However, it is already apparent that local nuclear environments that are generated by the multiple aspects of nuclear structure are intimately tied to developmental expression of cell growth and tissue-specific genes. From a broader perspective it is becoming increasingly evident that, reflecting the diversity of regulatory requirements as well as the phenotype-specific and physiologically responsive representation of nuclear structural proteins, there is a reciprocally functional relationship between nuclear structure and gene expression. Nuclear structure is a primary determinant of transcriptional control and the expressed genes modulate the regulatory components of nuclear architecture. The power of addressing gene expression within the three-dimensional context of nuclear structure would be difficult to overestimate. Membrane-mediated initiation of signaling pathways that ultimately influence transcription have been recognized for some time. Extending the structure–regulation paradigm to nuclear architecture expands the cellular context in which cell structure–gene expression interrelationships are operative.

II. Control of the Cell Cycle-Regulated Histone Genes and the Bone-Specific Osteocalcin Gene during Osteoblast Differentiation

The sequential expression of cell growth and tissue-specific genes during osteoblast differentiation is supported by developmental transcriptional

TABLE II

Contributions of the Nuclear Matrix to Regulation of Gene Expression

---

  I. Involvement of the nuclear matrix in DNA replication (Berezney and Coffey, 1974, 1975,
     1977; Pardoll *et al.*, 1980; Jackson and Cook, 1986; Nakayasu and Berezney, 1989; Vaughn
     et al., 1990; Belgrader *et al.*, 1991)
 II. Involvement of the nuclear matrix in transcriptional control
     A. Biologically relevant modifications in representation of nuclear matrix proteins
        1. Cell type and tissue-specific nuclear matrix proteins (Capco *et al.*, 1982; Fey *et al.*,
           1986, 1991; Fey and Penman, 1988; Nickerson *et al.*, 1990a; He *et al.*, 1990; Pienta
           et al., 1991; Bidwell *et al.*, 1993)
        2. Developmental stage-specific nuclear matrix proteins (Dworetzky *et al.*, 1992)
        3. Tumor-specific nuclear matrix proteins (Getzenberg and Coffey, 1991; Pienta and
           Coffey, 1991; Bidwell *et al.*, 1994a)
        4. Steroid hormone responsive nuclear matrix proteins (Barrack and Coffey, 1983;
           Kumara-Siri *et al.*, 1986; Getzenberg and Coffey, 1990)
        5. Polypeptide hormone responsive nuclear matrix proteins (Bidwell *et al.*, 1994b)
     B. Association of actively transcribed genes with the nuclear matrix (Nelkin *et al.*, 1980;
        Robinson *et al.*, 1982; Ciejek *et al.*, 1983; Mirkovitch *et al.*, 1984; Jackson and Cook,
        1985; Cockerill and Garrard, 1986; Gasser and Laemmli, 1986; Keppel, 1986; Nelson
        et al., 1986; Käs and Chasin, 1987; Bode and Maass, 1988; Jarman and Higgs, 1988;
        Mirkovitch *et al.*, 1988; Phi-Van and Strätling, 1988; Thorburn *et al.*, 1988; Stief *et al.*,
        1989; De Jong *et al.*, 1990; Farache *et al.*, 1990; Phi-Van *et al.*, 1990; Schaack *et al.*,
        1990; Jackson, 1991; von Kries *et al.*, 1991; Dworetzky *et al.*, 1992)
     C. Nuclear matrix localization of transcription factors
        1. Steroid hormone receptors (Barrack and Coffey, 1983; Kumara-Siri *et al.*, 1986;
           Landers and Spelsberg, 1992)
        2. General transcription factors (Dworetzky *et al.*, 1992)
        3. Tissue-specific transcription factors (Bidwell *et al.*, 1993)
        4. Viral transcription factors (Abulafia *et al.*, 1984; (Schaack *et al.*, 1990)
        5. Enhancer binding factors (Zenk *et al.*, 1990)
     D. Selective partitioning of transcription factors between the nuclear matrix and
        nonmatrix nuclear fractions (van Wijnen *et al.*, 1993)
III. Involvement of the nuclear matrix in post-transcriptional control
     A. RNA processing (Herman *et al.*, 1978; Jackson *et al.*, 1981; van Eeklen and van
        Venrooij, 1981; Mariman *et al.*, 1982; Ross *et al.*, 1982; Ben-Ze'ev and Aloni, 1983;
        Schroder *et al.*, 1987a; Zeitlin *et al.*, 1987; Nickerson *et al.*, 1990b; Nickerson and
        Penman, 1992; Xing *et al.*, 1993)
     B. Post-translational modifications of chromosomal proteins (Hendzel *et al.*, 1994)
     C. Phosphorylation (Tawfic and Ahmed, 1994)

---

control (Stein *et al.*, 1990; Stein and Lian, 1993). In both *in vivo* and *in vitro* cultures of normal diploid osteoblasts, proliferating cells express genes that mediate competency for cell growth as well as extracellular matrix biosynthesis. Postproliferatively, genes functionally related to the organization and mineralization of the bone extracellular matrix are expressed (Stein *et al.*, 1990; Stein and Lian, 1993). In this chapter we will focus on regulatory mechanisms controlling transcription of the cell cycle-regulated histone genes in proliferating osteoblasts and those controlling transcription of the bone-specific osteocalcin (OC) gene in mature osteoblasts during extracellular matrix mineralization. Transcription of these cell growth and tissue-specific genes will be considered within the context of regulatory contributions from principal components of nuclear architecture.

A. The Histone Gene Promoter as a Model for the Integration of Regulatory Signals Mediating Cell Cycle Control and Proliferation/Differentiation Interrelationships

The histone gene promoter is a paradigm for cell cycle-mediated transcriptional control (Stein *et al.*, 1992, 1994). Transcription is constitutive throughout the cell cycle, upregulated at the onset of S-phase, and completely suppressed in quiescent cells or following the onset of differentiation (Detke *et al.*, 1979; Plumb *et al.*, 1983a,b; Baumbach *et al.*, 1984). Consequently, activity of the promoter is responsive to regulatory signals that contribute to transcriptional competency for cell cycle progression at the G_1/S-phase transition point and to transcriptional downregulation postproliferatively (Fig. 1). The modularly organized promoter regulatory elements of the histone H4 gene promoter and the cognate transcription factors have been characterized within the context of cell cycle-dependent regulatory parameters (Pauli *et al.*, 1987; van Wijnen *et al.*, 1989, 1992; Ramsey-Ewing *et al.*, 1994). There is a direct indication that the cell cycle regulatory element exhibits phosphorylation-dependent modifications in transcription factor interactions that parallel and are functionally related to cell cycle as well as growth control of histone gene expression. The S-phase transcription factor complexes assembling at the H4 promoter include cdc2, cyclin A, an RB-related protein, and IRF-2 (van Wijnen *et al.*, 1994b; Vaughan *et al.*, 1995), reflecting an integration of phosphorylation-mediated control of histone gene expression, enzymes involved in DNA replication as well as growth stimulation and growth suppression at the G_1/S-phase transition point (Fig. 2).

   This regulatory mechanism is not restricted to cell cycle-dependent transcriptional control of the histone H4 gene. There is an analogous representation and organization of regulatory motifs in the histone H3 and histone H1 gene promoters supporting coordinate control of histone genes which are coexpressed (van den Ent *et al.*, 1994). In a broader biological context there are similarities of histone gene cell cycle regulatory element sequences with those in the proximal promoter of the thymidine kinase gene that exhibits enhanced transcription during S-phase (Dou *et al.*, 1992). The possibility may therefore be considered that genes functionally linked to DNA replication may at least in part be coordinately controlled. Support

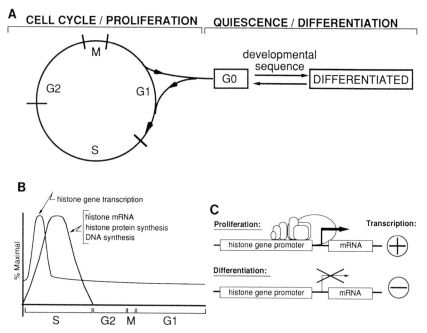

FIG. 1   Regulation of histone H4 gene expression in osteoblasts. (A) Schematic representation
of the cell cycle (G_1, S, G_2, and mitosis), indicating the pathway associated with the postproliferative onset of differentiation initiated following completion of mitosis. (B) Representation
of data defining the principal biochemical parameters of histone gene expression, indicating
a restriction of histone protein synthesis, and the presence of histone mRNA to S-phase cells
(DNA synthesis). Constitutive transcription of histone genes occurs throughout the cell cycle
with an enhanced transcriptional level during the initial 2 hours of S phase. These results
establish the combined contribution of transcription and messenger RNA stability to the
S-phase-specific regulation of histone biosynthesis in proliferating cells, with histone mRNA
levels as the rate limiting step. (C) In contrast, the completion of proliferative activity at
the onset of differentiation is mediated by transcriptional downregulation of histone gene
expression, supported by a parallel decline in rate of transcription and cellular mRNA levels.

for such a mechanism is provided by analogous promoter domains of the
histone and thymidine kinase promoters, which both interact with transcription factor complexes that include cyclins, cyclin-dependent kinases, and
Rb-related proteins.

B. Transcriptional Control of the Bone-Specific Osteocalcin
Gene at the Onset of Extracellular Matrix Mineralization
in Postproliferative Osteoblasts

Influences of promoter regulatory elements that are responsive to basal
and tissue-restricted transactivation factors, steroid hormones, growth fac-

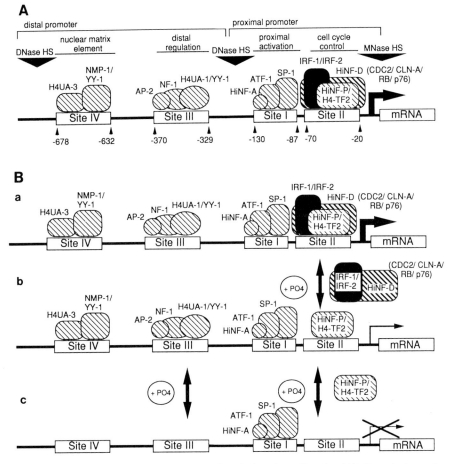

FIG. 2 Regulation of histone gene expression during the cell cycle. (A) Organization of the human histone H4 gene promoter regulatory elements (Sites I–IV) is illustrated. The transcription factors which exhibit sequence-specific interactions with these domains are indicated during the S phase of the cell cycle when the gene is maximally transcribed. Site II contributes to cell cycle regulation of transcription. Site IV binds a nuclear matrix protein complex (NMP-1/YY-1) while the protein–DNA interactions at Sites III and I support general transcriptional enhancement. The Site II complex includes cyclin A, cyclin-dependent kinase cdc2, an RB-related protein, and IRF growth regulatory factors, reflecting integration of phosphorylation-mediated control of histone gene expression. (B) (a) Proliferation of the S phase. Occupancy of the four principal regulatory elements of the histone H4 gene promoter during the S phase of the cell cycle when transcription is maximal is schematically illustrated. (b) Proliferation of the $G_2/M/G_1$ phase. The Site II transcription factor complex is modified by phosphorylation during the G_1/G_2/mitotic periods of the cell cycle resulting in modifications in levels of transcription. Phosphorylation-dependent dissociation of the IRF and HiNF-D (cdc2, cyclin A, and RB-related protein) factors occurs in non-S-phase cells. (c) The complete loss of transcription factor complexes at Sites II, III, and IV occurs following exit from the cell cycle with the onset of differentiation. At this time transcription is completely downregulated.

tors, and other physiologic mediators has provided the basis for understanding regulatory mechanisms contributing to developmental expression of osteocalcin, tissue specificity, and biological activity (Stein *et al.,* 1990; Stein and Lian, 1993; Lian and Stein, 1995). These regulatory elements and cognate transcription factors support postproliferative transcriptional activation and steroid hormone (e.g., vitamin D) enhancement at the onset of extracellular matrix mineralization during osteoblast differentiation (Fig. 3). Thus, the bone-specific osteocalcin gene is to be organized in a manner that supports responsiveness to homeostatic physiologic mediators and developmental expression in relation to bone cell differentiation.

The regulatory sequences illustrated in Fig. 4 have been established in the OC gene promoter and coding region by one or more criteria that includes: (1) demonstration of an influence on transcriptional activity by deletion, substitution, or site-specific mutagenesis *in vitro* and *in vivo;* (2) identification and characterization of sequence-specific regulatory element occupancy by cognate transcription factors *in vitro* and *in vivo;* (3) modifications in protein–DNA interactions as a function of biological activity; and (4) consequential modifications in functional activity following overexpression or suppression of factors that exhibit sequence-specific recognition for regulatory domains. A series of elements contributing to basal expression (Lian *et al.,* 1989) in-

FIG. 3   Reciprocal and functionally coupled relationship between cell growth and differentiation-related gene expression in osteoblasts. (A) These relationships are schematically illustrated as arrows representing changes in expression of cell cycle and cell growth regulated genes as well as genes associated with the regulated and regulatory events associated with the onset and progression of differentiation. The three principal periods of the osteoblast developmental sequence are designated within broken vertical lines (proliferation, matrix maturation, and mineralization). These broken lines indicate two experimentally established transition points in the developmental sequence exhibited by normal diploid osteoblasts during sequential acquisition of the bone cell phenotype—the first at the completion of proliferation when genes associated with extracellular matrix development and maturation are upregulated, and the second at the onset of extracellular matrix mineralization. (B) A series of signalling mechanisms are schematically illustrated whereby the proliferation period supports the biosynthesis of a type I collagen extracellular matrix, which continues to mature and mineralize. The formation of this extracellular matrix downregulates proliferation and extracellular matrix mineralization downregulates the expression of genes associated with the extracellular matrix maturation period. The occupancy of AP-1 sites in the osteocalcin (OC) and alkaline phosphatase (AP) gene promoters by fos-jun and/or related proteins are postulated to suppress both basal and vitamin D-enhanced transcription of phenotypic genes in proliferating osteoblasts. Apoptosis occurs in mature osteoblasts and, together with upregulation of collagenase activity, supports a remodelling of the developing bone extracellular matrix required for development of bone tissuelike organization. (C) Expression of the osteocalcin gene and vitamin D enhancement during the postproliferative period of the osteoblast developmental sequence is shown. H4 histone mRNA is an indication of proliferative activity.

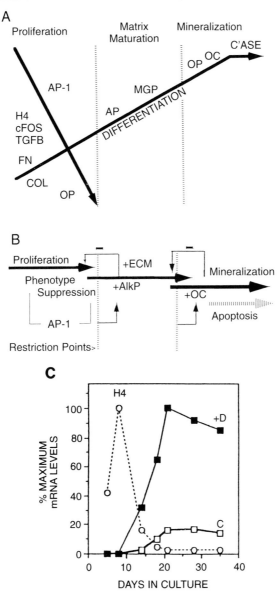

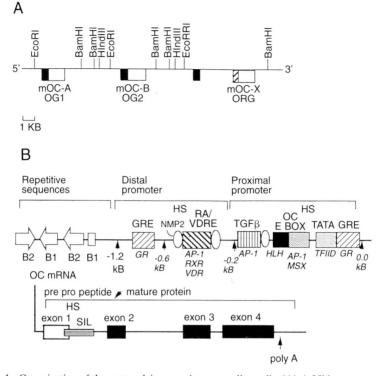

FIG. 4   Organization of the osteocalcin genes in mammalian cells. (A) A 25kb segment of the genome is shown which includes 3 osteocalcin genes designated MOCA, MOCB, and MOCX. MOCA and MOCB are expressed in bone while MOCX is expressed during the late prenatal and early postnatal periods in nonskeletal tissues. (B) The organization of the bone-specific osteocalcin gene is schematically illustrated indicating the regulatory domains within the initial 700 nucleotides 5′ to the transcription start site as well as the exon/intron organization in the mRNA coding region (black boxes) and for distal repetitive elements (open arrows). In the proximal promoter, several classes of transcription factors are represented which bind to key regulatory elements. The OC box is the primary tissue-specific transcriptional element that binds homeodomain proteins (MSX). Fos/jun-related proteins form heterodimers at AP-1 sites or a cryptic tissue-specific complex. HLH proteins bind to the contiguous E box motif. The nuclear matrix protein binding sites for NMP-2 (open ovals; designated A, B and C from left to right) interact with an AML-related transcription factor. Several glucocorticoid response elements (GRE) are indicated as well as the vitamin D response element (VDRE), which is a primary enhancer sequence. Additionally, AP-1 sites are indicated that overlap the TGFβ (TGRE) and the VDRE as well as a silencer element (SIL) in the mRNA coding region and regions of DNAse I hypersensitivity (HS). The combined and integrated activities of overlapping regulatory elements and associated transcription factors provide a mechanism for developmental control of expression during osteoblast growth and differentiation. (C) Osteocalcin gene promoter. Indicated are promoter regulatory domains, cognate transcription factors, as well as consensus elements for AP-1 and NMP-1.

C

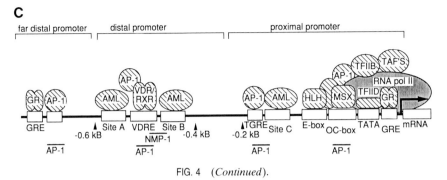

FIG. 4   (*Continued*).

clude a TATA sequence (located at −42 to −39) and the osteocalcin box (OC box), a 24-nucleotide element with a CCAAT motif as a central core, both required for rendering the gene transcribable (Kawaguchi *et al.*, 1992; Towler *et al.*, 1994; Hoffmann *et al.*, 1994). The OC box is a highly conserved regulatory sequence required for basal expression of the rat, mouse, and human OC genes. The osteocalcin box additionally serves a regulatory function in defining the threshold for initiation of transcription and contributes to bone tissue-specific expression of the osteocalcin gene (Heinrichs *et al.*, 1993a, 1995). However, caution must be exercised in attributing tissue-specific transcriptional control to a single element. Contributions of multiple sequences appear to be operative in tissue-specific regulation thereby providing opportunities for expression of the osteocalcin gene in bone under diverse biological circumstances. Multiple glucocorticoid responsive elements (GRE) with sequences that exhibit both strong and weak affinities for glucocorticoid receptor binding have been identified in the proximal promoter (Stromstedt *et al.*, 1991; Heinrichs *et al.*, 1993b; Aslam *et al.*, 1995). Interactions of other transcription factors with the proximal glucocorticoid responsive elements that include NF-IL6 have been reported (Towler and Rodan, 1994) further expanding the potential of the OC gene to be transcriptionally regulated by glucocorticoids. It is reasonable to consider that OC gene GREs may be selectively utilized in a developmentally and/or physiologically responsive manner. The possibility of functional interactions with transcription factors other than glucocorticoid receptors under certain conditions should not be dismissed.

The vitamin D responsive element (VDRE) functions as an enhancer (Kerner *et al.*, 1989; Morrison *et al.*, 1989; Demay *et al.*, 1990; Markose *et al.*, 1990; Terpening *et al.*, 1991). The VDRE transcription factor complex appears to be a target for modifications in vitamin D-mediated transcription by other physiologic factors including TNFα (Kuno *et al.*, 1994) and retinoic acid (Schüle *et al.*, 1990; Kliewer *et al.*, 1992; Bortell *et al.*, 1993; MacDonald *et al.*, 1993; Schrader *et al.*, 1993). Additional regulatory sequences include an NFkB site also reported to be involved in regulation mediated by TNFα (Li and Stashenko, 1993); a series of AP-1 sites (Lian *et al.*, 1991; Ozono *et al.*, 1991; Demay *et al.*, 1992), one of which mediates TGFβ responsiveness (Lian and Stein, 1993; Banerjee *et al.*, 1995); an E box (Tamura and Noda, 1994) that binds HLH containing transcription factor complexes; and a sequence in the proximal promoter that binds a multisubunit complex containing CP1/NF-Y/CBF-like CAAT factor complex. Two osteocalcin gene promoter regulatory domains that exhibit recognition for transcription factors that mediate developmental pattern formation are an MSX binding site within the OC box (Heinrichs *et al.*, 1993a, 1995; Hoffmann *et al.*, 1994; Towler *et al.*, 1994) and an AML-1 site (runt homology) sequence (van Wijnen *et al.*, 1994a; Merriman *et al.*, 1995). These sequences may represent components of regulatory mechanisms that contribute to pattern formation associated with bone tissue organization during initial developmental stages and subsequently during tissue remodeling. Involvement of other homeodomain-related genes that are expressed during skeletal development in control of osteoblast proliferation and differentiation is worthy of consideration. These include but are not restricted to the families of Dlx and Pax (Cohen *et al.*, 1989; Lufkin *et al.*, 1991, 1992; Krumlauf, 1993; Tabin, 1991; Gruss and Walther, 1992; McGinnis and Krumlauf, 1992; Niehrs and DeRobertis, 1992; Simeone *et al.*, 1994) genes. Although a majority of the response elements that have to date been identified reside in the region of the promoter which spans the VDRE domain to the first exxon, upstream sequences that must be further defined may contribute to both basal- and enhancer-mediated control of transcription (Yoon *et al.*, 1988; Terpening *et al.*, 1991; Bortell *et al.*, 1992; Lian and Stein, 1993; Morrison and Eisman, 1993; Owen *et al.*, 1993; Aslam *et al.*, 1994). A GRE residing at -683 to -697 is an example of such an upstream regulatory element (Aslam *et al.*, 1994).

The overlapping and contiguous organization of regulatory elements, as illustrated by the TATA/GRE, E Box/AP-1/CCAAT/homeodomain, and TNFα/VDRE (Kuno *et al.*, 1994) AP-1/VDRE provide a basis for combined activities that support responsiveness to physiologic mediators (Vaishnav *et al.*, 1988; Evans *et al.*, 1990; Nanes *et al.*, 1990, 1991; Owen *et al.*, 1990; Fanti *et al.*, 1992; Li and Stashenko, 1992; Taichman and Hauschka, 1992; Guidon *et al.*, 1993; Jenis *et al.*, 1993; Schedlich *et al.*, 1994). Additionally, hormones modulate binding of transcription factors other than the cognate receptor to nonsteroid regulatory sequences. For example, vitamin D-induced interactions occur at the basal TATA domain (Owen *et al.*, 1993) and 1,25(OH)$_2$D$_3$ upregulates MSX-2 binding to the OC box homeodomain motif as well as supports increased MSX-2 expression (Kawaguchi *et al.*, 1992; Towler *et al.*, 1994; Hoffmann *et al.*, 1994). It is this complexity of OC gene promoter element upregulation that allows for hormone responsiveness in relation to either basal or enhanced levels of expression. The protein–DNA interactions at the principal promoter regulatory elements that mediate levels of osteocalcin gene transcription are schematically summarized in Fig. 4.

III. Nuclear Structure and Cell Cycle Stage-Specific Expression of Histone Genes in Proliferating Osteoblasts

A. Chromatin Structure and Nucleosome Organization

A synergistic contribution of activities by sites I, II, III, and IV H4 histone gene promoter elements to the timing and extent of H4-FO108 gene transcription has been established experimentally (Kroeger *et al.*, 1987; Wright *et al.*, 1992, 1995; Ramsey-Ewing *et al.*, 1994; Birnbaum *et al.*, 1995). The integration of intracellular signals that act independently upon these multiple elements may partly reside in the three-dimensional organization of the promoter within the spatial context of nuclear architecture (Fig. 5).

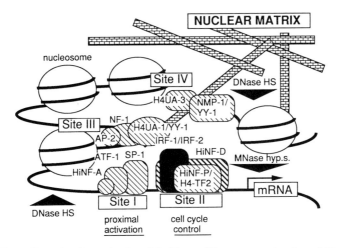

FIG. 5   Three-dimensional organization of the histone H4 gene promoter. A model is schematically presented for the spatial organization of the histone H4 gene promoter based on evidence for nucleosome placement and the interaction of DNA binding sequences with the nuclear matrix. These components of chromatin structure and nuclear architecture restrict mobility of the promoter and impose physical constraints that reduce distances between proximal and distal promoter elements. Such as postulated organization of the histone H4 gene promoter can facilitate cooperative interactions for crosstalk between elements that mediate transcription factor binding and consequently determine the extent to which the gene is transcribed.

The presence of nucleosomes in the H4 promoter when the gene is transcriptionally active (Moreno *et al.,* 1986, 1987; Chrysogelos *et al.,* 1985, 1989) (Figure 6) may serve to increase the proximity of independent regulatory elements, and supports synergistic and/or antagonistic cooperative interactions between histone gene DNA binding activities. In addition, chromatin structure and nucleosomal organization varies as a function of the cell cycle (Chrysogelos *et al.,* 1985, 1989; Moreno *et al.,* 1986, 1987) which may enhance and restrict accessibility of transcription factors, and modulate the extent to which DNA-bound factors are phosphorylated.

Parameters of chromatin structure and nucleosome organization were experimentally established by accessibility of DNA sequences within nuclei of intact cells to a series of nucleases that include micrococcal nuclease for establishing nucleosome placement, DNase I for mapping nuclease-hypersensitive sites, S1 nuclease for determination of single-stranded DNA sequences, and restriction endonucleases to determine protein–DNA interactions within specific sites at single nucleotide resolution (Fig. 6). Findings from these approaches illustrate that nucleosome spacing exhibits cell cycle-dependent variations in response to levels of histone gene expression and/or nuclear organization related to mitotic division. A combined modification is observed in accessibility of histone gene sequences to micrococcal nuclease, DNase I, S1 nuclease, and restriction endonucleases reflecting a remodeling of chromatin architecture that is related to the extent that the histone gene is transcribed to support proliferative activity and cell cycle progression.

B. The Nuclear Matrix

A nuclear matrix attachment site has been identified in the upstream region (-0.8 kb) of the H4-FO108 gene promoter (Dworetzky *et al.,* 1992), which may serve two functions: imposing constraints on chromatin structure and concentrating and localizing transcription factors. Such a role for the nuclear matrix in regulation of histone gene expression is supported by distinct modifications in the composition of nuclear matrix proteins observed when proliferation-specific genes are down-regulated during differentiation (Dworetzky *et al.,* 1990) and, more directly, by the isolation of ATF-related and YY1 transcription factors from the nuclear matrix (Dworetzky *et al.,* 1992; Guo *et al.,* 1995), which interact with site IV of the H4-FO108 gene promoter.

The specific mechanisms by which the 5' histone gene promoter elements and sequence-specific transactivating factors participate in regulating transcription of the histone H4-FO108 gene remain to be determined. However, regulation is unquestionably operative within the context of the complex series of spatial interactions, which are responsive to a broad spectrum of biological signals (Figs. 5 and 6).

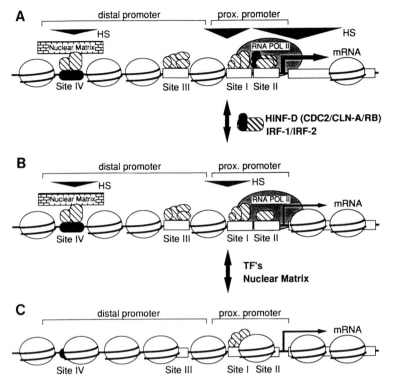

FIG. 6 Schematic illustration of the remodelling of chromatin structure and nucleosome organization which accommodates cell cycle stage-specific and developmental parameters of histone gene promoter architecture to support modifications in level of expression. Placement of nucleosomes and representation as well as magnitude of nuclease hypersensitive sites (solid triangles) are designated. The principal regulatory elements and transcription factors are shown. (A) S-phase, maximal transcription; (B) $G_2/M/G_1$, basal transcription; (C) differentiated, inactive.

IV. Nuclear Structure and Developmental and Steroid Hormone-Responsive Expression of the Osteocalcin Gene during Osteoblast Differentiation

A. Chromatin Structure and Nucleosome Organization

Modifications in parameters of chromatin structure and nucleosome organization parallel both competency for transcription and the extent to which the osteocalcin gene is transcribed. Changes are observed in response to physiological mediators of basal expression and steroid hormone responsiveness. Thus a conceptual and experimental basis is provided for the involvement of nuclear architecture in developmental, homeostatic, and physiologic control of osteocalcin gene expression during establishment and maintenance of bone tissue structure and activity.

In both normal diploid osteoblasts and in osteosarcoma cells basal expression and enhancement of osteocalcin gene transcription are accompanied by two alterations in structural properties of chromatin. Hypersensitivity of sequences flanking the tissue-specific osteocalcin box and the vitamin D-responsive element enhancer domain are observed (Breen *et al.*, 1994; Montecino *et al.*, 1994a,b). Together with modifications in nucleosome placement (Montecino *et al.*, 1994b), a basis for accessibility of transactivation factors to basal and steroid hormone-dependent regulatory sequences can be explained. In early-stage proliferating normal diploid osteoblasts, when the osteocalcin gene is repressed, nucleosomes are placed in the OC box and in VDRE promoter sequences; and nuclease-hypersensitive sites are not present in the vicinity of these regulatory elements. In contrast, when osteocalcin gene expression is transcriptionally upregulated postproliferatively and vitamin D-mediated enhancement of transcription occurs, the osteocalcin box and VDRE become nucleosome free and these regulatory domains are flanked by DNase I-hypersensitive sites. (Figs. 7 and 8)

Functional relationships between structural modifications in chromatin and osteocalcin gene transcription are observed in response to $1,25(OH)_2D_3$ in ROS 17/2.8 osteosarcoma cells that exhibit vitamin D-responsive transcriptional upregulation. There are marked changes in nucleosome placement at the VDRE and OC box as well as DNase I hypersensitivity of sequences flanking these basal and enhancer osteocalcin gene promoter sequences (Breen *et al.*, 1994; Montecino *et al.*, 1994a,b). The complete absence of hypersensitivity and the presence of nucleosomes in the VDRE and osteocalcin box domains of the osteocalcin gene promoter in ROS 24/1 cells which lack the vitamin D receptor additionally corroborate these findings (Breen *et al.*, 1994; Montecino *et al.*, 1994a). (Figs. 7 and 9)

B. The Nuclear Matrix

Involvement of the nuclear matrix in control of osteocalcin gene transcription is provided by several lines of evidence. One of the most compelling is association of a bone-specific nuclear matrix protein designated NMP2 with sequences flanking the VDRE of the osteocalcin gene promoter (Bidwell *et al.*, 1993). Initial characterization of the NMP2 factor has revealed that a component is an AML-1-related transactivation protein which is a runt homology factor associated with developmental pattern formation in *Drosophila* (van Wijnen *et al.*, 1994a; Merriman *et al.*, 1995). These results implicate the nuclear matrix in regulating events that mediate structural properties of the VDRE domain and may also contribute to linkage of

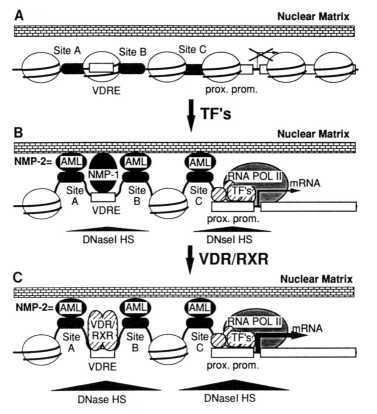

FIG. 7 Schematic representation of the osteocalcin gene promoter organization and occupancy of regulatory elements by cognate transcription factors paralleling and supporting functional relationships to either: (A) suppression of transcription in proliferating osteoblasts, (B) activation of expression in differentiated cells, or (C) enhancement of transcription by vitamin D. The placement of nucleosomes is indicated. Remodelling of chromatin structure in nucleosome organization to support suppression, basal and vitamin D induced transcription of the osteocalcin gene is indicated. The representation and magnitude of DNase I hypersensitive sites are designated by solid triangles and gene-nuclear matrix interactions are shown.

osteocalcin gene expression with modulation of pattern formation requisite for skeletal tissue organization during bone formation and remodeling. Here the possibility of multiple regulatory sequences in the osteocalcin gene promoter that are functionally associated with pattern formation is indicated because of binding by MSX homeodomain proteins at the OC box (Hoffmann *et al.*, 1994; Towler and Rodan, 1994).

It is apparent from available findings that the linear organization of gene regulatory sequences is necessary but insufficient to accommodate the requirements for physiological responsiveness to homeostatic, develop-

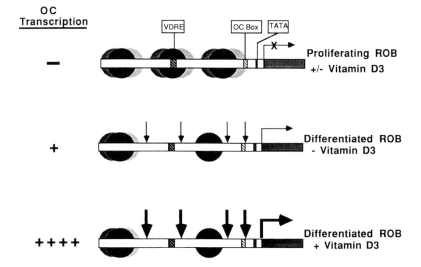

FIG. 8 Developmental remodelling of nucleosome organization in the osteocalcin gene promoter correlates with transcriptional activity during differentiation of normal diploid osteoblasts. The positioning of nucleosomes in the osteocalcin gene promoter was determined by combining DNase I, micrococcal nuclease and restriction endonuclease digestions with indirect endlabelling. The filled circles represent the placement of nucleosomes with the shadows indicating movement within nuclease protected segments. The vertical arrows correspond to the limits of the distal (-600 to -400) and proximal (-170 to -70) DNase I hypersensitive sites, which increase in differentiated normal diploid osteoblasts in response to vitamin D. The VDRE (-465 to -437) and the osteocalcin box (-99 to -77) are designated.

mental, and tissue-related regulatory signals. It would be presumptive to propose a formal model for the three-dimensional organization of the osteocalcin gene promoter. However, the working model presented in Figs. 7 and 10 represents postulated interactions between OC gene promoter elements that reflect the potential for integration of activities by nuclear architecture to support transcriptional control within a three-dimensional context of cell structure and regulatory requirements at the cell and tissue levels.

A functional role of the nuclear matrix in steroid hormone-mediated transcriptional control of the osteocalcin gene is further supported by overlapping binding domains within the VDRE for the VDR and the NMP-1 nuclear matrix protein that we have recently shown to be a YY1 transcription factor. One can speculate that reciprocal interactions of NMP-1 and VDR complexes may contribute to competency of the VDRE to support transcriptional enhancement. Binding of NMP-2 at the VDRE flanking sequence may establish permissiveness for VDR interactions by gene-

OC
Transcription

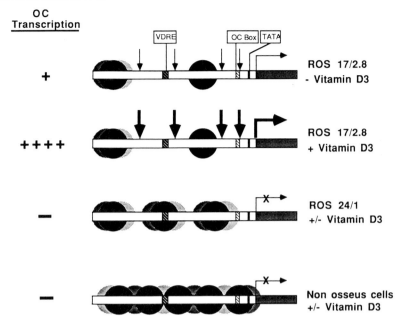

FIG. 9   Schematic representation of nucleosome placement in the rat osteocalcin gene promoter in osteosarcoma cells in relation to osteocalcin gene transcription. The presence and positioning of nucleosomes in the osteocalcin gene promoter of ROS 17/2.8 cells (constitutively expressing osteocalcin) and ROS 24/1 cells (not expressing osteocalcin) were determined by combining nuclease sensitivity (DNase I, micrococcal nuclease and restriction endonucleases) with indirect endlabelling. The filled circles represent positioned nucleosomes and the shadows indicate movement of nucleosomes within protected segments. The vertical arrows correspond to the limits of the distal (-600 to -400) and proximal (-170 to -70) DNase I hypersensitive sites, which are increased following vitamin D treatment. The VDRE (-465 to -437) and the OC box (-99 to -77) are designated within the distal and proximal nuclease hypersensitive domains respectively.

nuclear matrix associations that facilitate conformational modifications in the transcription factor recognition sequences.

V. Conclusions

It is becoming increasingly evident that developmental transcriptional control and modifications in transcription to accommodate homeostatic regulation of cell and tissue function is modulated by the integration of a complex series of physiological regulatory signals. Fidelity of responsiveness necessitates the convergence of activities mediated by multiple regulatory elements

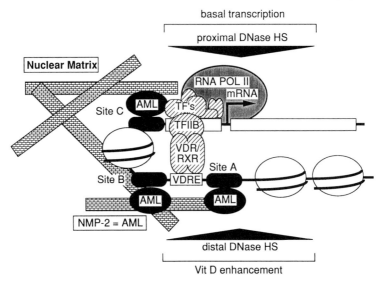

FIG. 10   Three-dimensional organization of the rat osteocalcin gene promoter. A model is schematically presented for the spatial organization of the rat osteocalcin gene promoter based on evidence for nucleosome placement and the interaction of DNA binding sequences with the nuclear matrix. These components of chromatin structure and nuclear architecture restrict mobility of the promoter and impose physical constraints that reduce distances between proximal and distal promoter elements. Such postulated organization of the osteocalcin gene promoter can facilitate cooperative interactions for crosstalk between elements that mediate transcription factor binding and consequently determine the extent to which the gene is transcribed.

of gene promoters. Our current knowledge of promoter organization and the repertoire of transcription factors that mediate activities provides a single dimension map of options for biological control. We are beginning to appreciate the additional structural and functional dimensions provided by chromatin structure, nucleosome organization, and subnuclear localization and targeting of both genes and transcription factors. Particularly exciting is increasing evidence for dynamic modifications in nuclear structure that parallel developmental expression of genes. The extent to which nuclear structure regulates and/or is regulated by modifications in gene expression remains to be experimentally established.

Acknowledgments

Studies reported in this chapter were supported by grants from the National Institutes of Health (GM32010, AR39588, AR42262) and the March of Dimes Birth Defects Foundation.

The authors are appreciative of the editorial assistance of Elizabeth Bronstein in the preparation of this manuscript.

References

Abulafia, R., Ben-Ze'ev, A., Hay, N., and Aloni, Y. (1984). Control of late SV40 transcription by the attenuation mechanism and transcriptionally active ternary complexes are associated with the nuclear matrix. *J. Mol. Biol.* **172**, 467–487.

Aslam, F., Lian, J. B., Stein G. S., Stein, J. L., Litwack, G., van Wijnen, A. J., and Shalhoub, V. (1994). Glucocorticoid responsiveness of the osteocalcin gene by multiple distal and proximal elements. *J. Bone Miner. Res.* **9**, S125.

Aslam, F., Shalhoub, V., van Wijnen, A. J., Banerjee, C., Bortell, R., Shakoori, A. R., Litwack, G., Stein, J. L., Stein, G. S., and Lian, J. B. (1995). Contributions of distal and proximal promoter elements to glucocorticoid regulation of osteocalcin gene transcription. *Mol. Endocrinol.* **9**, 679–690.

Banerjee, C., Stein, J. L., van Wijnen, A. J., Kovary, K., Bravo, R., Lian, J. B., and Stein, G. S. (1995). TGF-β1 response in the rat osteocalcin gene is mediated by an AP-1 binding site. Submitted for publication.

Barrack, E. R., and Coffey, D. S. (1983). Hormone receptors and the nuclear matrix. *In* "Gene Regulation by Steroid Hormones II" (A. K. Roy and J. H. Clark, eds.), pp. 239–266. Springer-Verlag, New York.

Bumbach, L., Marashi, F., Plumb, M., Stein, G. S., and Stein, J. L. (1984). Inhibition of DNA replication coordinately reduces cellular levels of core and H1 histone mRNAs: Requirement for protein synthesis. *Biochemistry* **23**, 1618–1625.

Belgrader, P., Siegel, A. J., and Berezney, R. (1991). A comprehensive study on the isolation and characterization of the HeLa S3 nuclear matrix. *J. Cell Sci.* **98**, 281–291.

Ben-Ze'ev, A., and Aloni, Y. (1983). Processing of SV40 RNA is associated with the nuclear matrix and is not followed by the accumulation of low-molecular weight RNA products. *Virology* **125**, 475–479.

Berezney, R., and Coffey, D. S. (1974). Identification of a nuclear protein matrix. *Biochem. Biophys. Res. Commun.* **60**, 1410–1417.

Berezney, R., and Coffey, D. S. (1975). Nuclear protein matrix: Association with newly synthesized DNA. *Science* **189**, 291–292.

Berezney, R., and Coffey, D. S. (1977). Nuclear matrix: Isolation and characterization of a framework structure from rat liver nuclei. *J. Cell Biol.* **73**, 616–637.

Bidwell, J. P., van Wijnen, A. J., Fey, E. G., Dworetzky, S., Penman, S., Stein, J. L., Lian, J. B., and Stein, G. S. (1993). Osteocalcin gene promoter-binding factors are tissue-specific nuclear matrix components. *Proc. Natl. Acad. Sci. U.S.A.* **90**, 3162–3166.

Bidwell, J. P., Fey, E. G., van Wijnen, A. J., Penman, S., Stein, J. L., Lian, J. B., and Stein, G. S. (1994a). Nuclear matrix proteins distinguish normal diploid osteoblasts from osteosarcoma cells. *Cancer Res.* **54**, 28–32.

Bidwell, J. P., van Wijnen, A. J., Banerjee, C., Fey, E. G., Merriman, H., Penman, S., Stein, J. L., Lian, J. B., and Stein, G. S. (1994b). PTH-responsive modifications in the nuclear matrix of ROS 17/2.8 rat osteosarcoma cells. *Endocrinology* (*Baltimore*) **134**, 1738–1744.

Birnbaum, M. J., van Wijnen, A. J., Odgren, P., Last, T. J., Suske, G., Stein, G. S., and Stein, J. L. (1995). Sp1 trans-activation of cell cycle regulated promoters is selectively repressed by Sp3. *Biochemistry,* in press.

Bode, J., and Maass, K. (1988). Chromatin domain surrounding the human interferon-β gene as defined by scaffold-attached regions. *Biochemistry* **27**, 4706–4711.

Bortell, R., Owen, T. A., Bidwell, J. P., Gavazzo, P., Breen, E., van Wijnen, A. J., DeLuca, H. F., Stein, J. L., Lian, J. B., and Stein, G. S. (1992). Vitamin D-responsive protein-DNA interactions at multiple promoter regulatory elements that contribute to the level of rat osteocalcin gene expression. *Proc. Natl. Acad. Sci. U.S.A.* **89**, 6119–6123.

Bortell, R., Owen, T. A., Shalhoub, V., Heinrichs, A., Aronow, M. A. B., and Stein, G. S. (1993). Constitutive transcription of the osteocalcin gene is osteosarcoma cells is reflected by altered protein-DNA interactions at promoter regulatory elements. *Proc. Natl. Acad. Sci. U.S.A.* **90**, 2300–2304.

Breen, E. C., van Wijnen, A. J., Lian, J. B., Stein, G. S., and Stein, J. L. (1994). In vivo occupancy of the vitamin D responsive element in the osteocalcin gene supports vitamin D dependent transcriptional upregulation in intact cells. *Proc. Natl. Acad. Sci. U.S.A.* **91**, 12902–12906.

Capco, D. G., Wan, K. M., and Penman, S. (1982). The nuclear matrix: Three-dimensional architecture and protein composition. *Cell (Cambridge, Mass.)* **29**, 847–858.

Chrysogelos, S., Riley, D. E., Stein, G. S., and Stein, J. L. (1985). A human histone H4 gene exhibits cell cycle-dependent changes in chromatin structure that correlate with its expression. *Proc. Natl. Acad. Sci. U.S.A.* **82**, 7535–7539.

Chrysogelos, S., Pauli, U., Stein, G. S., and Stein, J. L. (1989). Fine mapping of the chromatin structure of a cell cycle-regulated human H4 histone gene. *J. Biol. Chem.* **264**, 1232–1237.

Ciejek, E. M., Tsai, M.-J., and O'Malley, B. W. (1983). Actively transcribed genes are associated with the nuclear matrix. *Nature (London)* **306**, 607–609.

Cockerill, P. N., and Garrard, W. T. (1986). Chromosomal loop anchorage of the kappa immuno-globulin gene occurs next to the enhancer in a region containing topoisomerase II sites. *Cell (Cambridge, Mass.)* **44**, 273–282.

Cohen, S. M., Broner, G., Kuttner, F., Jurgens, G., and Jackle, H. (1989). *Distal-less* encodes a homeodomain protein required for limb development in *Drosophila. Nature (London)* **338**, 432–434.

De Jong, L., van Driel, R., Stuurman, N., Meijne, A. M. L., and van Renswoude, J. (1990). Principles of nuclear organization. *Cell Biol. Int. Rep.* **14**, 1051–1074.

Demay, M. B., Gerardi, J. M., DeLuca, H. F., and Kronenberg, H. M. (1990). DNA sequences in the rat osteocalcin gene that bind the 1,25-dihydroxyvitamin D_3 receptor and confer responsive to 1,25-dihydroxyvitamin D_3. *Proc. Natl. Acad. Sci. U.S.A.* **87**, 369–373.

Demay, M. B., Kiernan, M. S., DeLuca, H. F., and Kronenberg, H. M. (1992). Characterization of 1,25-dihydroxyvitamin D_3 receptor interactions with target sequences in the rat osteocalcin gene. *Mol. Endocrinol.* **6**, 557–562.

Detke, S., Lichtler, A., Phillips, I., Stein, J. L., and Stein, G. S. (1979). Reassessment of histone gene expression during the cell cycle in human cells by using homologous H4 histone cDNA. *Proc. Natl. Acad. Sci. U.S.A.* **76**, 4995–4999.

Dou, Q., Markell, P. J., and Pardee, A. B. (1992). Retinoblastoma-like protein and cdc2 kinase are in complexes that regulate a G1/S event. *Proc. Natl. Acad. Sci. U.S.A.* **89**, 3256–3260.

Dworetzky, S. I., Fey, E. G., Penman, S., Lian, J. B., Stein, J. L., and Stein, G. S. (1990). Progressive changes in the protein composition of the nuclear matrix during rat osteoblast differentiation. *Proc. Natl. Acad. Sci. U.S.A.* **87**, 4605–4609.

Dworetzky, S. I., Wright, K. L., Fey, E. G., Penman, S., Lian, J. B., Stein, J. L., and Stein, G. S. (1992). Sequence-specific DNA binding proteins are components of a nuclear matrix attachment site. *Proc. Natl. Acad. Sci. U.S.A.* **89**, 4178–4182.

Evans, D. B., Thavarajah, M., and Kanis, J. A. (1990). Involvement of prostaglandin E_2 in the inhibition of osteocalcin synthesis by human osteoblast-like cells in response to cytokines and systemic hormones. *Biochem. Biophys. Res. Commun.* **167**, 194–202.

Fanti, P., Kindy, M. S., Mohapatra, S., Klein, J., Colombo, G., and Malluche, H. H. (1992). Dose-dependent effects of aluminum on osteocalcin synthesis in osteoblast-like ROS 17/2 cells in culture. *Am. J. Physiol.* **263**, E1113–E1118.

Farache, G., Razin, S. V., Rzeszowska-Wolny, J., Moreau, J., Targa, F. R., and Scherrer, K. (1990). Mapping of structural and transcription-related matrix attachment sites in the a-globin gene domain of avian erythroblasts and erythrocytes. *Mol. Cell. Biol.* **10,** 5349–5358.

Fey, E. G., and Penman, S. (1988). Nuclear matrix proteins reflect cell type of origin in cultured human cells. *Proc. Natl. Acad. Sci. U.S.A.* **85,** 121–125.

Fey, E. G., Krochmalnic, G., and Penman, S. (1986). The nonchromatin substructures of the nucleus: The ribonucleoprotein (RNP)-containing and RNP-depleted matrices analyzed by sequential fractionation and resinless section electron microscopy. *J. Cell Biol.* **102,** 1654–1665.

Fey, E., Bangs, P., Sparks, C., and Odgren, P. (1991). The nuclear matrix: Defining structural and functional roles. *CRC Crit. Rev. Eukaryotic Gene Express.* **1,** 127–143.

Gasser, S. M., and Laemmli, U. K. (1986). Cohabitation of scaffold binding regions with upstream/enhancer elements of three developmentally regulated genes of *D. melanogaster. Cell (Cambridge, Mass.)* **46,** 521–530.

Getzenberg, R. H., and Coffey, D. S. (1990). Tissue specificity of the hormonal response in sex accessory tissues is associated with nuclear matrix protein patterns. *Mol. Endocrinol.* **4,** 1336–1342.

Getzenberg, R. H., and Coffey, D. S. (1991). Identification of nuclear matrix proteins in the cancerous and normal rat prostate. *Cancer Res.* **51,** 6514–6520.

Gruss, P., and Walther, C. (1992). Pax in development. *Cell (Cambridge, Mass.)* **69,** 719–722.

Guidon, P. T., Salvatori, R., and Bockman, R. S. (1993). Gallium nitrate regulates rat osteoblast expression of osteocalcin protein and mRNA levels. *J. Bone Miner. Res.* **8,** 103–110.

Guo, B., Odgren, P. R., van Wijnen, A. J., Last, T. J., Fey, E. G., Penman, S., Stein, J. L., Lian, J. B., and Stein, G. S. (1995). The nuclear matrix protein NMP-1 is the transcription factor 44-1. *Proc. Natl. Acad. Sci.,* in press.

He, D., Nickerson, J. A., and Penman, S. (1990). Core filaments of the nuclear matrix. *J. Cell Biol.* **110,** 569.

Heinrichs, A. A. J., Banerjee, C., Bortell, R., Owen, T. A., Stein, J. L., Stein, G. S., and Lian, J. B. (1993a). Identification and characterization of two proximal elements in the rat osteocalcin gene promoter that may confer species-specific regulation. *J. Cell Biochem.* **53,** 240–250.

Heinrichs, A. A. J., Bortell, R., Rahman, S., Stein, J. L., Alnemri, E. S., Litwack, G., Lian, J. B., and Stein, G. S. (1993b). Identification of multiple glucocorticoid receptor binding sites in the rat osteocalcin gene promoter. *Biochemistry* **32,** 11436–11444.

Heinrichs, A. A. J., Bortell, R., Bourke, M., Lian, J. B., Stein, G. S., and Stein, J. L. (1995). Proximal promoter binding protein contributes to developmental, tissue-restricted expression of the rat osteocalcin gene. *J. Cell. Biochem.* **57,** 90–100.

Hendzel, M. J., Sun, J.-M., Chen, H. Y., Rattner, J. B., and Davie, J. R. (1994). Histone acetyltransferase is associated with the nuclear matrix. *J. Biol. Chem.* **269,** 22894–22901.

Herman, R., Weymouth, L., and Penman, S. (1978). Heterogeneous nuclear RNA-protein fibers in chromatin-depleted nuclei. *J. Cell Biol.* **78,** 663–674.

Hoffmann, H. M., Catron, K. M., van Wijnen, A. J., McCabe, L. R., Lian, J. B., Stein, G. S., and Stein, J. L. (1994). Transcriptional control of the tissue-specific developmentally regulated osteocalcin gene requires a binding motif for the MSX-family of homeodomain proteins. *Proc. Natl. Acad. Sci. U.S.A.* **91,** 12887–12891.

Jackson, D. A. (1991). Structure–function relationships in eukaryotic nuclei. *BioEssays* **13,** 1–10.

Jackson, D. A., and Cook, P. R. (1985). Transcription occurs at a nucleoskeleton. *EMBO J.* **4,** 919–925.

Jackson, D. A., and Cook, P. R. (1986). Replication occurs at a nucleoskeleton. *EMBO J.* **5,** 1403–1410.

Jackson, D. A., McCready, S. J., and Cook, P. R. (1981). RNA is synthesized at the nuclear cage. *Nature (London)* **292,** 552–555.

Jarman, A. P., and Higgs, D. R. (1988). Nuclear scaffold attachment sites in the human globin gene complexes. *EMBO J.* **7**, 3337–3344.

Jenis, L. G., Waud, C. E., Stein, G. S., Lian, J. B., and Baran, D. T. (1993). Effect of gallium nitrate in vitro and in normal rats. *J. Cell. Biochem.* **52**, 330–336.

Käs, E., and Chasin, L. A. (1987). Anchorage of the Chinese hamster dihydrofolate reductase gene to the nuclear scaffold occurs in an intragenic region. *J. Mol. Biol.* **198**, 677–692.

Kawaguchi, N., DeLuca, H. F., and Noda, M. (1992). Id gene expression and its suppression by 1,25-dihydroxyvitamin D_3 in rat osteoblastic osteosarcoma cells. *Proc. Natl. Acad. Sci. U.S.A.* **89**, 4569–4572.

Keppel, F. (1986). Transcribed human ribosomal RNA genes are attached to the nuclear matrix. *J. Mol. Biol.* **187**, 15–21.

Kerner, S. A., Scott, R. A., and Pike, J. W. (1989). Sequence elements in the human osteocalcin gene confer basal activation and inducible response to hormonal vitamin D_3. *Proc. Natl. Acad. Sci. U.S.A.* **86**, 4455–4459.

Kliewer, S. A., Umesono, K., Mangelsdorf, D. J., and Evans, R. M. (1992). Retinoic X receptor interacts with nuclear receptors in retinoic acid, thyroid hormone and vitamin D signalling. *Nature (London)* **355**, 441–446.

Kroeger, P., Stewart, C., Schaap, T., van Wijnen, A., Hirshman, J., Helms, S., Stein, G. S., and Stein, J. L. (1987). Proximal and distal regulatory elements that influence in vivo expression of a cell cycle dependent H4 histone gene. *Proc. Natl. Acad. Sci. U.S.A.* **84**, 3982–3986.

Krumlauf, R. (1993). *Hox* genes and pattern formation in the branchial region of the vertebrate head. *Trends Genet.* **9**, 106–112.

Kumara-Siri, M. H., Shapiro, L. E., and Surks, M. I. (1986). Association of the 3,5,3′-triiodo-L-thyronine nuclear receptor with the nuclear matrix of cultured growth hormone-producing rate pituitary tumor cells (GC cells). *J. Biol. Chem.* **261**, 2844–2852.

Kuno, H., Kurian, S. M., Hendy, G. N., White, J., DeLuca, H. F., Evans, C.-O., and Nanes, M. S. (1994). Inhibition of 1,25-dihydroxyvitamin D_3 stimulated osteocalcin gene transcription by tumor necrosis factor-α: Structural determinants within the vitamin D response element. *Endocrinology (Baltimore)* **134**, 2524–2531.

Landers, J. P., and Spelsberg, T. C. (1992). New concepts in steroid hormone action: Transcription factors, proto-oncogenes, and the cascade model for steroid regulation of gene expression. *CRC Crit. Rev. Eukaryotic Gene Express.* **2**, 19–63.

Li, Y., and Stashenko, P. (1992). Proinflammatory cytokines tumor necrosis factor and IL-6, but not IL-1, down-regulate the osteocalcin gene promoter.[1] *J. Immunol.* **148**, 788–794.

Li, Y., and Stashenko, P. (1993). Characterization of a tumor necrosis factor-responsive element which down-regulates the human osteocalcin gene. *Mol. Cell Biol.* **13**, 3714–3721.

Lian, J. B., and Stein, G. S. (1993). Proto-oncogene mediated control of gene expression during osteoblast differentiation. *Ital. J. Miner. Electron. Metab.* **7**, 175–183.

Lian, J. B., and Stein, G. S. (1995). Osteocalcin gene expression: A molecular blueprint for developmental and steroid hormone mediated regulation of osteoblast growth and differentiation. *Endocrin. Rev.* (in press).

Lian, J. B., Stewart, C., Puchacz, E., Mackowiak, S., Shalhoub, V., Collart, D., Zambetti, G., and Stein, G. S. (1989). Structure of the rat osteocalcin gene and regulation of vitamin D-dependent expression. *Proc. Natl. Acad. Sci. U.S.A.* **86**, 1143–1147.

Lian, J. B., Stein, G. S., Bortell, R., and Owen, T. A. (1991). Phenotype suppression: A postulated molecular mechanism for mediating the relationship of proliferation and differentiation by fos/jun interactions at AP-1 sites in steroid responsive promoter elements of tissue-specific genes. *J. Cell. Biochem.* **45**, 9–14.

Lufkin, T., Dierich, A., LeMeur, M., Mark, M., and Chambon, P. (1991). Disruption of the *Hox*-1.6 homeobox gene results in defects in a region corresponding to its rostral domain of expression. *Cell (Cambridge, Mass.)* **66**, 1105–1119.

Lufkin, T., Mark, M., Hart, C. P., LeMeur, M., and Chambon, P. (1992). Homeotic transformation of the occipital bones of the skull by ectopic expression of a homeobox gene. *Nature (London)* **359**, 835–841.

MacDonald, P. N., Dowd, D. R., Nakajima, S., Galligan, M. A., Reeder, M. C., Haussler, C. A., Ozato, K., and Haussler, M. R. (1993). Retinoid X receptors stimulate and 9-*cis* retinoic acid inhibits 1,25-dihydroxyvitamin D$_3$-activated expression of the rat osteocalcin gene. *Mol. Cell. Biol.* **13**, 5907–5917.

Mariman, E. C. M., van Eekelen, C. A. G., Reinders, J., Berns, A. J. M., and van Venrooij, W. J. (1982). Adenoviral heterogeneous nuclear RNA is associated with the host nuclear matrix during splicing. *J. Mol. Biol.* **154**, 103–119.

Markose, E. R., Stein, J. L., Stein, G. S., and Lian, J. B. (1990). Vitamin D-mediated modifications in protein–DNA interactions at two promoter elements of the osteocalcin gene. *Proc. Natl. Acad. Sci. U.S.A.* **87**, 1701–1705.

McGinnis, W., and Krumlauf, R. (1992). Homeobox genes and axial patterning. *Cell (Cambridge, Mass.)* **68**, 283–302.

Merriman, H. L., van Wijnen, A. J., Hiebert, S., Bidwell, J. P., Fey, E., Lian, J. B., Stein, J. L., and Stein, G. S. (1995). The tissue-specific nuclear matrix protein, NMP-2, is a member of the AML/PEB2/runt domain transcription factor family: Interactions with the osteocalcin gene promoter. *Biochemistry*, in press.

Mirkovitch, J., Mirault, M.-E., and Laemmli, U. K. (1984). Organization of the higher-order chromatin loop: Specific DNA attachment sites on nuclear scaffold. *Cell (Cambridge, Mass.)* **39**, 223–232.

Mirkovitch, J., Gasser, S. M., and Laemmli, U. K. (1988). Scaffold attachment of DNA loops in metaphase chromosomes. *J. Mol. Biol.* **200**, 101–109.

Montecino, M., Pockwinse, S., Lian, J. B., Stein, G. S., and Stein, J. L. (1994a). DNase I hypersensitive sites in promoter elements associated with basal and vitamin D dependent transcription of the bone-specific osteocalcin gene. *Biochemistry* **33**, 348–353.

Montecino, M., Lian, J. B., Stein, G. S., and Stein, J. L. (1994b). Specific nucleosomal organization supports developmentally regulated expression of the osteocalcin gene. *J. Bone Miner. Res.* **9**, S352.

Moreno, M. L., Chrysogelos, S. A., Stein, G. S., and Stein, J. L. (1986). Reversible changes in the nucleosomal organization of a human H4 histone gene during the cell cycle. *Biochemistry* **25**, 5364–5370.

Moreno, M. L., Stein, G. S., and Stein, J. S. (1987). Nucleosomal organization of a BPV minichromosome containing a human H4 histone gene. *Mol. Cell. Biochem.* **74**, 173–177.

Morrison, N. A., and Eisman, J. A. (1993). Role of the negative glucocorticoid regulatory element in glucocorticoid repression of the human osteocalcin promoter. *J. Bone Miner. Res.* **8**, 969–975.

Morrison, N. A., Shine, J., Fragonas, J. C., Verkest, V., McMenemy, L., and Eisman, J. A. (1989). 1,25-dihydroxyvitamin D-responsive element and glucocorticoid repression in the osteocalcin gene. *Science* **246**, 1158–1161.

Nakayasu, H., and Berezney, R. (1989). Mapping replicational sites in the eukaryotic nucleus. *J. Cell Biol.* **108**, 1–11.

Nanes, M. S., Rubin, J., Titus, L., Hendy, G. N., and Catherwood, B. D. (1990). Interferon-γ inhibits 1,25-dihydroxyvitamin D$_3$-stimulated synthesis of bone Gla protein in rat osteosarcoma cells by a pretranslational mechanism. *Endocrinology (Baltimore)* **127**, 588–594.

Nanes, M. S., Rubin, J., Titus, L., Hendy, G. N., and Catherwood, B. D. (1991). Tumor necrosis factor alpha inhibits 1,25-dihydroxyvitamin D$_3$-stimulated bone Gla protein synthesis in rat osteosarcoma cells (ROS 17/2.8) by a pretranslational mechanism. *Endocrinology (Baltimore)* **128**, 2577–2582.

Nelkin, B. D., Pardoll, D. M., and Vogelstein, B. (1980). Localization of SV40 genes with supercoiled loop domains. *Nucleic Acids Res.* **8**, 5623–5633.

Nelson, W. G., Pienta, K. J., Barrack, E. R., and Coffey, D. S. (1986). The role of the nuclear matrix in the organization and function of DNA. *Nucleic Acids Res.* **14,** 6433–6451.

Nickerson, J. A., and Penman, S. (1992). The nuclear matrix: Structure and involvement in gene expression. *In* "Molecular and Cellular Approaches to the Control of Proliferation and Differentiation (G. Stein and J. Lian, eds.), pp. 334–380. Academic Press, San Diego, CA.

Nickerson, J. A., Krockmalnic, G., He, D., and Penman, S. (1990a). Immunolocalization in three dimensions: Immunogold staining of cytoskeletal and nuclear matrix proteins in resinless electron microscopy sections. *Proc. Natl. Acad. Sci. U.S.A.* **87,** 2259–2263.

Nickerson, J. A., He, D., Fey, E. G., and Coffey, D. S. (1990b). The nuclear matrix. *In* "The Eukaryotic Nucleus. Molecular Biochemistry and Macromolecular Assemblies" (P. R. Strauss and S. H. Wilson, eds.), p. 763. Telford, Press, Caldwell, NJ.

Niehrs, C., and DeRobertis, E. M. (1992). Vertebrate axis formation. *Curr. Opin. Genet. Dev.* **2,** 550–555.

Owen, T. A., Bortell, R., Yocum, S. A., Smock, S. L., Zhang, M., Abate, C., Shalhoub, V., Aronin, N., Wright, K. L., van Wijnen, A. J., Stein, J. L., Curran, T., Lian, J. B., and Stein, G. S. (1990). Coordinate occupancy of AP-1 sites in the vitamin D responsive and CCAAT box elements by fos-jun in the osteocalcin gene: A model for phenotype suppression of transcription. *Proc. Natl. Acad. Sci. U.S.A.* **87,** 9990–9994.

Owen, T. A., Bortell, R., Shalhoub, V., Heinrichs, A., Stein, J. L., Stein, G. S., and Lian, J. B. (1993). Postproliferative transcription of the rat osteocalcin gene is reflected by vitamin D-responsive developmental modifications in protein–DNA interactions at basal and enhancer promoter elements. *Proc. Natl. Acad. Sci. U.S.A.* **90,** 1503–1507.

Ozono, K., Sone, T., and Pike, J. W. (1991). The genomic mechanism of action of 1,25-dihydroxyvitamin D_3. *J. Bone Miner. Res.* **6,** 1021–1027.

Pardoll, D. M., Vogelstein, B., and Coffey, D. S. (1980). A fixed site of DNA replication in eukaryotic cells. *Cell (Cambridge, Mass.)* **19,** 527–536.

Pauli, U., Chrysogelos, S., Stein, J. L., Stein, G. S., and Nick, H. (1987). Protein-DNA interactions in vivo upstream of a cell cycle regulated human H4 histone gene. *Science* **236,** 1308–1311.

Phi-Van, L., and Strätling, W. H. (1988). The matrix attachment regions of the chicken lysozyme gene co-map with the boundaries of the chromatin domain. *EMBO J.* **7,** 655–664.

Phi-Van, L., von Kries, J. P., Ostertag, W., and Strätling, W. H. (1990). The chicken lysozyme 5′ matrix attachment region increases transcription from a heterologous promoter in heterologous cells and dampens positional effects on the expression of transfected genes. *Mol. Cell. Biol.* **10,** 2302–2307.

Pienta, K. J., and Coffey, D. S. (1991). Correlation of nuclear morphometry with progression of breast cancer. *Cancer (Philadelphia)* **68,** 2012–2016.

Pienta, K. J., Getzenberg, R. H., and Coffey, D. S. (1991). Cell structure and DNA organization. *CRC Crit. Rev. Eukaryotic Gene Express.* **1,** 355–385.

Plumb, M. A., Stein, J. L., and Stein, G. S. (1983a). Coordinate regulation of multiple histone mRNAs during the cell cycle in HeLa cells. *Nucleic Acids Res.* **11,** 2391–2410.

Plumb, M. A., Stein, J. L., and Stein, G. S. (1983b). Influence of DNA synthesis inhibition on the coordinate expression of core human histone genes. *Nucleic Acids Res.* **11,** 7927–7945.

Ramsey-Ewing, A., van Wijnen, A., Stein, G. S., and Stein, J. L. (1994). Delineation of a human histone H4 cell cycle element in vivo: The master switch for H4 gene transcription. *Proc. Natl. Acad. Sci. U.S.A.* **91,** 4475–4479.

Robinson, S. I., Nelkin, B. D., and Vogelstein, B. (1982). The ovalbumin gene is associated with the nuclear matrix of chicken oviduct cells. *Cell (Cambridge, Mass.)* **28,** 99–106.

Ross, D. A., Yen, R. W., and Chae, C. B. (1982). Association of globin ribonucleic acid and its precursors with the chicken erythroblast nuclear matrix. *Biochemistry* **21,** 764–771.

Schaack, J., Ho, W. Y.-W., Friemuth, P., and Shenk, T. (1990). Adenovirus terminal protein mediates both nuclear-matrix association and efficient transcription of adenovirus DNA. *Genes Dev.* **4,** 1197–1208.

Schedlich, L. J., Flanagan, J. L., Crofts, L. A., Gillies, S. A., Goldberg, D., Morrison, N. A., and Eisman, J. A. (1994). Transcriptional activation of the human osteocalcin gene by basic fibroblast growth factor. *J. Bone Miner. Res.* **9**, 143–152.

Schrader, M., Bendik, I., Becker-Andre, M., and Carlberg, C. (1993). Interaction between retinoic acid and vitamin D signaling pathways. *J. Biol. Chem.* **268**, 17830–17836.

Schroder, H. C., Trolltsch, D., Wenger, R., Bachmann, M., Diehl-Seifert, B., and Muller, W. E. G. (1987a). Cytochalasin B selectively releases ovalbumin mRNA precursors but not the mature ovalbumin mRNA from hen oviduct nuclear matrix. *Eur. J. Biochem.* **167**, 239–245.

Schroder, H. C., Trolltsch, D., Friese, U., Bachmann, M., and Muller, W. E. G. (1987b). Mature mRNA is selectively released from the nuclear matrix by an ATP/dATP-dependent mechanism sensitive to topoisomerase inhibitors. *J. Biol. Chem.* **262**, 8917–8925.

Scule, R., Umesono, K., Mangelsdorf, D. J., Bolado, J., Pike, J. W., and Evans, R. M. (1990). Jun-Fos and receptors for vitamins A and D recognize a common response element in the human osteocalcin gene. *Cell (Cambridge, Mass.)* **61**, 497–504.

Simeone, A., Acampora, D., Pannese, M., D'Esposito, M., Stornaiuolo, A., Gulisano, M., Mallamaci, A., Kastury, K., Druck, T., Huebner, K., and Boncinelli, E. (1994). Cloning and characterization of two new members of the vertebrate D1x family. *Proc. Natl. Acad. Sci. U.S.A.* **91**, 2250–2254.

Stein, G. S., and Lian, J. B. (1993). Molecular mechanisms mediating proliferation/differentiation interrelationships during progressive development of the osteoblast phenotype. *Endocr. Rev.* **14**, 424–442.

Stein, G. S., Lian, J. B., and Owen, T. A. (1990). Relationship of cell growth to the regulation of tissue-specific gene expression during osteoblast differentiation. *FASEB J.* **4**, 3111–3123.

Stein, G. S., Stein, J. L., van Wijnen, A. J., and Lian, J. B. (1992). Regulation of histone gene expression. *Curr. Opin. Cell Biol.* **4**, 166–173.

Stein, G. S., Stein, J. L., van Wijnen, A. J., and Lian, J. B. (1994). Histone gene transcription: A model for responsiveness to an integrated series of regulatory signals mediating cell cycle control and proliferation/differentiation interrelationships. *J. Cell. Biochem.* **54**, 393–404.

Stief, A., Winter, D. M., Strätling, W. H., and Sippel, A. E. (1989). A nuclear attachment element mediates elevated and position-independent gene activity. *Nature (London)* **341**, 343–345.

Stromstedt, P. E., Poellinger, L., Gustafsson, J. A., and Carlstedt-Duke, J. (1991). The glucocorticoid receptor binds to a sequence overlapping the TATA box of the human osteocalcin promoter: A potential mechanism for negative regulation. *Mol. Cell. Biol.* **11**, 3379–3383.

Tabin, C. J. (1991). Retinoids, homeoboxes, and growth factors: Towards molecular models for limb development. *Cell (Cambridge, Mass.)* **66**, 199–217.

Taichman, R. S., and Hauschka, P. V. (1992). Effects of interleukin-1 beta and tumor necrosis factor-alpha osteoblastic expression of osteocalcin and mineralized extracellular matrix in vitro. *J. Inflamm.* **16**(6), 587–601.

Tamura, M., and Noda, M. (1994). Identification of a DNA sequence involved in osteoblast-specific gene expression via interaction with helix–loop–helix (HLH)-type transcription factors. *J. Cell Biol.* **126**, 773–782.

Tawfic, S., and Ahmed, K. (1994). Growth stimulus-mediated differential translocation of casein kinase 2 to the nuclear matrix. *J. Biol. Chem.* **269**, 24615–24620.

Terpening, C. M., Haussler, C. A., Jurutka, P. W., Galligan, M. A., Komm, B. S., and Haussler, M. R. (1991). The vitamin D-responsive element in the rat bone Gla protein gene is an imperfect direct repeat that cooperates with other cis-elements in 1,25-dihydroxyvitamin D_3-mediated transcriptional activation. *Mol. Endocrinol.* **5**, 373–385.

Thorburn, A., Moore, R., and Knowland, J. (1988). Attachment of transcriptionally active sequences to the nucleoskeleton under isotonic conditions. *Nucleic Acids Res.* **16**, 7183.

Towler, D. A., and Rodan, G. A. (1994). Cross-talk between glucocorticoid and PTH signaling in the regulation of the rat osteocalcin promoter. *J. Bone Miner. Res.* **9,** S282.

Towler, D. A., Bennett, C. D., and Rodan, G. A. (1994). Activity of the rat osteocalcin basal promoter in osteoblastic cells is dependent upon homeodomain and CP1 binding motifs. *Mol. Endocrinol.* **8,** 614–624.

Vaishnav, R., Beresford, J. N., Gallagher, J. A., and Russell, R. G. G. (1988). Effects of the anabolic steroid stanozolol on cells derived from human bone. *Clin. Sci.* **74,** 455–460.

van den Ent, F. M. I., van Wijnen, A. J., Lian, J. B., Stein, J. L., and Stein, G. S. (1994). Cell cycle controlled histone H1, H3 and H4 genes share unusual arrangements of recognition motifs for HiNF-D supporting a coordinate promoter binding mechanism. *J. Cell. Physiol.* **159,** 515–530.

van Eeklen, C. A. G., and van Venrooij, W. J. (1981). hnRNA and its attachment to a nuclear-protein matrix. *J. Cell Biol.* **88,** 554–563.

van Wijnen, A. J., Wright, K. L., Lian, J. B., Stein, J. L., and Stein, G. S. (1989). Human H4 histone gene transcription requires the proliferation-specific nuclear factor HiNF-D. *J. Biol. Chem.* **264,** 15034–15042.

van Wijnen, A. J., van den Ent, F. M. I., Lian, J. B., Stein, J. L., and Stein, G. S. (1992). Overlapping and CpG methylation-sensitive protein/DNA interactions at the histone H4 transcriptional cell cycle domain: Distinctions between two human H4 gene promoters. *Mol. Cell. Biol.* **12,** 3273–3287.

van Wijnen, A. J., Bidwell, J. P., Fey, Edward G., Penman, S., Lian, J. B., Stein, J. L., and Stein, G. S. (1993). Nuclear matrix association of multiple sequence specific DNA binding activities related to SP-1, ATF, CCAAT, C/EBP, OCT-1 and AP-1. *Biochemistry* **32,** 8397–8402.

van Wijnen, A. J., Merriman, H., Guo, B., Bidwell, J. P., Stein, J. L., Lian, J. B., and Stein, G. S. (1994a). Nuclear matrix interactions with the osteocalcin gene promoter: Multiple binding sites for the runt-homology related protein NMP-2 and for NMP-1, a heteromeric CREB-2 containing transcription factor. *J. Bone Miner. Res.* **9,** S148.

van Wijnen, A. J., Aziz, F., Grana, X., DeLuca, A., Desai, R. K., Jaarsveld, K., Last, T. J., Soprano, K., Giordano, A., Lian, J. B., Stein, J. L., and Stein, G. S. (1994b). Transcription of histone H4, H3 and H1 cell cycle genes: Promoter factor HiNF-D contains CDC2, cyclin A and an RB-related protein. *Proc. Natl. Acad. Sci. U.S.A.* **91,** 12882–12886.

Vaughan, P. S., Aziz, F., van Wijnen, A. J., Wu, S., Soprano, K., Stein, G. S., and Stein, J. L. (1995). Activation of a cell cycle regulated histone gene by the oncogenic transcription factor IRF2. *Nature* (*London*), in press.

Vaughn, J. P., Dijkwel, P. A., Mullenders, L. H. F., and Hamlin, J. L. (1990). Replication forks are associated with the nuclear matrix. *Nucleic Acids Res.* **18,** 1965–1969.

von Kries, J. P., Buhrmester, H., and Strätling, W. H. (1991). A matrix/scaffold attachment region binding mprotein: Identification, purification and mode of binding. *Cell* (*Cambridge, Mass.*) **64,** 123–135.

Wright, R. L., Dell'Orco, R. T., van Wijnen, A. J., Stein, J. L., and Stein, G. S. (1992). Multiple mechanisms regulate the proliferation specific histone gene transcription factor, HiNF-D, in normal human diploid fibroblasts. *Biochemistry* **31,** 2812–2818.

Wright, K. L., Birnbaum, M. J., van Wijnen, A. J., Stein, G. S., and Stein, J. L. (1995). Bipartite structure of the proximal promoter of a human H4 histone gene. *J. Cell. Biochem.,* **58,** 372–379.

Xing, Y., Johnson, C. V., Dobner, P. R., and Lawrence, J. B. (1993). Higher level organization of individual gene transcription and RNA splicing. *Science* **259,** 1326–1330.

Yoon, K., Rutledge, S. J. C., Buenaga, R. F., and Rodan, G. A. (1988). Characterization of the rat osteocalcin gene: Stimulation of promoter activity by 1,25-dihydroxyvitamin D_3. *Biochemistry* **17,** 8521–8526.

Zeitlin, S., Parent, A., Silverstein, S., and Efstratiadis, A. (1987). Pre-mRNA splicing and the nuclear matrix. *Mol. Cell. Biol.* **7,** 111–120.

Zenk, D. W., Ginder, G. D., and Brotherton, T. W. (1990). A nuclear-matrix protein binds very tightly to DNA in the avian β-globin gene enhancer. *Biochemistry* **29,** 5221–5226.

Chromatin Domains and Prediction of MAR Sequences

Teni Boulikas

Institute of Molecular Medical Sciences, Palo Alto, California 94306

Polynuceosomes are constrained into loops or domains and are insulated from the effects of chromatin structure and torsional strain from flanking domains by the cross-complexation of matrix-attached regions (MARs) and matrix proteins. MARs or SARs have an average size of 500 bp, are spaced about every 30 kb, and are control elements maintaining independent realms of gene activity. A fraction of MARs may cohabit with core origins of replication (ORIs) and another fraction might cohabit with transcriptional enhancers. DNA replication, transcription, repair, splicing, and recombination seem to take place on the nuclear matrix. Classical AT-rich MARs have been proposed to anchor the core enhancers and core origins complexed with low abundancy transcription factors to the nuclear matrix via the cooperative binding to MARs of abundant classical matrix proteins (topoisomerase II, histone H1, lamins, SP120, ARBP, SATB1); this creates a unique nuclear microenvironment rich in regulatory proteins able to sustain transcription, replication, repair, and recombination. Theoretical searches and experimental data strongly support a model of activation of MARs and ORIs by transcription factors. A set of 21 characteristics are deduced or proposed for MAR/ORI sequences including their enrichment in inverted repeats, AT tracts, DNA unwinding elements, replication initiator protein sites, homooligonucleotide repeats (i.e., AAA, TTT, CCC), curved DNA, DNase I-hypersensitive sites, nucleosome-free stretches, polypurine stretches, and motifs with a potential for left-handed and triplex structures. We are establishing Banks of ORI and MAR sequences and have undertaken a large project of sequencing a large number of MARs in an effort to determine classes of DNA sequences in these regulatory elements and to understand their role at the origins of replication and transcriptional enhancers.

KEY WORDS: Replication origins, Initiator protein, DNA distortion, Inverted repeat, Enhancer, Transcription factor, Development, Torsional strain, Intrinsic curvature, Nucleosome, Replicon, DNase I-hypersensitive sites, DNA polymorphism.

I. Introduction

The 25 million or so nucleosomes in a single mammalian nucleus are organized into 60,000 chromatin loops (Paulson and Laemmli, 1977) that maintain independent realms of gene activity relaxed in a quantized manner by DNase I nicking (Benyajati and Worcel, 1976). Both interphase chromatin and mitotic chromosomes are organized into loops or domains (Marsden and Laemmli, 1979). This organization is brought about by the anchorage of specific DNA sequence landmarks to a network of protein cross-ties termed nuclear matrix (NM) at interphase (Berezney and Coffey, 1974) or chromosomal scaffolds at mitosis (Mirkovitch *et al.*, 1984). Such landmarks occur within introns of the various genes, as well as within proximal and distal sites flanking the 5' and 3' ends (Hancock and Boulikas, 1982; Mirkovitch *et al.*, 1987; Bodnar, 1988; Verheijen *et al.*, 1988; Cook, 1989; Gasser *et al.*, 1989; de Jong *et al.*, 1990; Garrard, 1990; Berezney, 1991; Bonifer *et al.*, 1991; Fey *et al.*, 1991; Georgiev *et al.*, 1991; Jackson, 1991; Pienta *et al.*, 1991; van Driel *et al.*, 1991; Boulikas, 1992b; Getzenberg, 1994).

Looping of DNA supported the domain chromatin structure (Cook and Brazell, 1975; Paulson and Laemmli, 1977; Hancock; 1982). Native chromatin loops emanating from a central core had been observed in meiotic lumpbrush chromosomes in amphibians (Callan, 1982) and in fixed mitotic or polytene chromosomes (DuPraw, 1965; Björkroth *et al.*, 1988).

SARs or MARs (scaffold- or matrix-attached regions) are supposed to punctuate chromosomal DNA into functional units of topologically constrained loop domains. Thus MARs can be viewed as boundary elements shielding and insulating genes located between them from influences of cis-acting elements and torsional strain from the neighboring chromatin domain (Breyne *et al.*, 1992; Ludérus *et al.*,). In this respect MARS may also represent sequences termed domain boundary elements (Chung *et al.*, 1993), facilitators (Aronow *et al.*, 1995), or special chromatin structures (Kellum and Schedl, 1991) and might be related to locus control regions (LCRs) (Grosveld *et al.*, 1987). Such elements interfere with expression when placed between the promoter and the enhancer region of a gene (Stief *et al.*, 1989).

In addition, MARs and SARs are supposed to aid the cell type-specific expression of genes by their cohabitation and synergistic collaboration with enhancers (Forrester *et al.*, 1994) toward maintaining some domains into condensed inactive structures and other into transcriptionally active, decondensed structures (Boulikas, 1987a,b); topoisomerase II orchestrates the topological looped organization of functional chromatin domains (Berrios *et al.*, 1985; Razin *et al.*, 1991b).

MARs define the borders between chromatin domains each harboring one gene, a segment of a gene, or groups of related genes (Table I). Genes constrained in a single chromatin domain are the human interferon β gene

(Bode and Maass, 1988) and the apolipoprotein B gene (Levy-Wilson and Fortier, 1989). In this case two MARs delineate their 5′ and 3′ ends. Examples of groups of genes included in the same domain are the chicken embryonic π, and the α^D and α^A globin genes (Farache et al., 1990a,b) and the five histone genes in Drosophila (Mirkovitch et al., 1984). Examples of single genes divided over two domains are the mouse immunoglobulin κ light and μ heavy chain genes (Cockerill and Garrard, 1986a; Cockerill et al., 1987), the Chinese hamster dihydrofolate reductase gene (Käs and Chasin, 1987), and the human topoisomerase (TOPO) I gene (Roming et al., 1992). In the latter cases a MAR element was found within an intron.

A fundamental question in understanding MAR function concerns the type of DNA sequences that are used by eukaryotic nuclei to set the borders between chromatin domains while utilizing the same sequences or nearby motifs to initiate replication of the domain and to enhance the transcriptional activity of the genes nested in this particular domain.

Methods for MAR Identification and Matrix Isolation

A. MAR Identification

Several independent methods have been devised to determine the location of MARs within a gene locus.

1. Isolated nuclei are treated with lithium diiodosalicylate (LIS) or 2 M NaCl giving "nuclear halos" where extended chromatin loops puffing out of nuclei are still attached to the matrix; DNA is then cleaved with a restriction enzyme giving soluble loop fragments (~75% of DNA) and insoluble fragments (~25% of DNA). DNA fragments from the two fractions are electrophoresed in equal amounts on agarose gels and blotted with specific gene probes (Mirkovitch et al., 1984).

2. A mixture of DNA restriction fragments from a cloned gene are incubated with isolated nuclear matrix or scaffold proteins and the bound DNA fragments are separated from the unbound DNA by a simple centrifugation step (Cockerill and Garrard, 1986a). Sequences containing enhancers or topoisomerase II sites bind to matrix preparations (Cockerill et al., 1987; Phi-Van and Strätling, 1988); however, the question whether such complexes reflect the status of the genes in vivo remains open (Cook, 1989; Garrard, 1990). The finding that the murine α-globin gene is bound to nuclear matrices in vitro but not in vivo (Avramova and Paneva, 1992) supports the possibility of artifacts during matrix preparation.

3. Micrococcal nuclease (MNase) digestion of chromatin loops in intact

TABLE I

Genes Whose MARs Have Been Determined

Gene	Species	Features	Reference
Histone (intergenic)	*Drosophila*	AT-rich, topo II sites; *Eco*RI–*Hin*fI fragment (657 bp); two protein-binding domains (200 bp) separated by a nuclease-sensitive region (100 bp); recognized by mouse NM	Mirkovitch *et al.* (1984); Cockerill and Garrard (1986a)
HSP70 (intergenic)	*Drosophila*	Topo II sites	Mirkovitch *et al.* (1984)
163-kb locus	*Drosophila* (X chromosome)	Unknown sequence; of the 52 restriction fragments 5 were anchored to the NM giving 4 loops (75, 52, 15, 14 kb), each containing a transcribed sequence	Small and Vogelstein (1985)
α-actin (5′ end)	*Drosophila* Kc cells	Association with NM (25% of DNA in NM) was observed during normal growth (high transcription) and during heat shock (extremely low transcription); NM anchoring was not affected by removal of 99.7% of nascent RNA	Small *et al.* (1985)
Fushi-tarazu	*Drosophila*	5′ and 3′ ends; the 5′ SAR at −4.8 kb is AT-rich; 1.15 kb *Eco*RI fragment from pFKH1	Gasser and Laemmli (1986a); Eggert and Jack (1991)
Sgs-4	*Drosophila*	5′ and 3′ ends; the 5′ SAR (866 bp) encompasses an enhancer	Gasser and Laemmli (1986a); Mirkovitch *et al.* (1986, 1988)
Alcohol dehydrogenase	*Drosophila*	5′ and 3′ ends; near an enhancer	Gasser and Laemmli (1986a); Mirkovitch *et al.* (1986, 1988)
Rosy and Ace loci	*Drosophila*	26- to 112-kb loops over a 320-kb region	Mirkovitch *et al.* (1986)
κ immunoglobulin (Igκ) (intronic)	Mouse	MAR/enhancer; constitutive activity present in different stages of lymphocyte development; in the J–C intron; 70% AT; Topo II sites; (Fig. 1)	Cockerill and Garrard (1986a,b); Sperry *et al.* (1989)
Igκ (intronic)	Rabbit	MAR/enhancer; rabbit κ2 chain gene lacks a MAR which is present in κ1; upstream of the intronic enhancer	Sperry *et al.* (1989)

Igκ	Human	MAR/enhancer; 70% AT; by sequence analysis and homology to mouse IgH MAR	Whitehurst et al. (1992)
Igλ	Chicken	In the J–Cλ region; AT-rich; four topo II sites	Parvari et al. (1990)
Immunoglobulin heavy chain (IgH)	Mouse	MAR/enhancer; one MAR on each side of the J_H–C_μ intronic enhancer region	Cockerill et al. (1987); Cockerill (1990); Kohwi-Shigematsu and Kohwi (1990)
IgH (5′ flank)	Murine	MAR/enhancer; −574 to −425 region of Igμ (149 bp); 72% AT; interacts with IL-5 plus antigen-inducible proteins	Webb et al. (1991)
SV40 ORI	Monkey CV-1	Related to replicating DNA; mediated by large T antigen	Jones and Su (1987); Schirmbeck and Deppert (1989)
SV40 (T antigen-coding region)		Nucleotides 4071–4377; a weak MAR at 3610–4008; topo II sites; recognized by mouse P815 cell nuclear matrix	Pommier et al. (1990)
DHFR	Chinese hamster	Intronic; the sequence is shown in Fig. 1	Käs and Chasin (1987)
DHFR	Chinese hamster	5′ and 3′ flanks; MAR/ORI; near the zone of replication initiation	Dijkwel and Hamlin (1988)
2-Macroglobulin	Rat	Intronic	Ito and Sakaki (1987)
HPRT locus	Human	The sequence is shown in Fig. 1	Sykes et al. (1988)
Interferon-β	Human		Bode and Maass (1988); Mielke et al. (1990); Klehr et al. (1991)
Lysozyme domain	Chicken	At −8.85 and +1.3 kb; known as "A elements"	Phi-Van and Strätling (1988); Stief et al. (1989)
β-globin gene complex	Human	MAR/enhancers; about 8 MARs over a 90-kb region; two MARs are close to enhancers in β-globin gene; absence of MARs over a 140-kb of human α-globin gene complex (−100 to +40 kb) using nuclear matrix from K562 cells	Jarman and Higgs (1988)

TABLE I (*Continued*)

Gene	Species	Features	Reference
γ^A (3′ flank)	Human	750 bp MAR/enhancer; eight regions of TF-DNA interactions (GATA-1, AP-2, CBP-1, Sp1); two of these (I and IV) interact with SATB1 in K562 cells to mediate matrix anchorage	Cunningham *et al.* (1994)
β-major globin (5′ flank)	Murine	−1100 to −300 bp; attachment was not altered by transcriptional activation during induction to terminal differentiation	Greenstein (1988)
α-globin	Murine	First intron and part of 5′ coding region; not interacting with lamins; proximal to DNase I-HS sites; binding occurs only *in vitro* but not *in vivo*	Avramova and Paneva (1992)
β-globin (3′ flanking)	Chicken	MAR/enhancer; 480-bp fragment that spans the 3′ enhancer region between β- and ε-globin genes; facultative MAR activity present in adult reticulocytes but not in thymus (gene is expressed in reticulocytes but not in thymus)	Zenk *et al.* (1990); Brotherton *et al.* (1991)
α-globin (3′ end)	Chicken	Close to the 3′ enhancer; the sequence is shown in Fig. 9	Farache *et al.* (1990a,b)
α-globin (5′ end)	Chicken	MAR/ORI/enhancer; permanent MAR detected in chicken tissues irrespective of α-globin gene transcription; bipartite MAR composed of two small fragments separated by 250 bp of non-MAR sequence; in the 5′ flanking region (about −3.7 to −2 kb) of the chicken π-globin (embryonic) gene; the 1.7-kb fragment contains topo II, NF1, NFIII, Sp1 sites, and a constitutive DNase I-HS site	Farache *et al.* (1990a); Georgiev *et al.* (1991); Razin *et al.* (1991a)
800-kb locus	*Drosophila*	85 restriction fragments showed weak, medium, or strong association with scaffolds from 0–18-hr *Drosophila* embryos	Brun *et al.* (1990); Surdej *et al.* (1990)
Malic enzyme	Chicken	Facultative MAR bound to NM in thymus but not in reticulocytes (gene is expressed in thymus but not in reticulocytes); 5-kb fragment containing the second exon of the gene	Brotherton *et al.* (1991)

284

Gene	Organism	Description	Reference
Apolipoprotein B (5' distal)	Human	Permanent activity; at −5262 to −4048 bp	Levy-Wilson and Fortier (1989)
Apolipoprotein B (5' proximal)	Human	Facultative activity, detected only in HepG2 cells that express the apo B gene; at −2765 to −1801 bp	Levy-Wilson and Fortier (1989)
Apolipoprotein B (3' flank)	Human	Permanent activity; 98% AT; 550 bp, composed of a 30 bp repeat; at +43,186 to +43,850 bp (shown in Fig. 11A)	Levy-Wilson and Fortier (1989)
Ribosomal genes (intergenic)	*Xenopus*	MAR/ORI/promoter; within spacer region between repeats; GC-rich; the homologous region in mouse has ORI activity (Wegner et al., 1989)	Marilley and Gassend-Bonnet (1989)
Ornithine transcarbamylase	Human	>85-kb, X-linked, tissue-specific gene transcribed in liver; lacks CpG clusters; data obtained with halo preparations	Beggs and Migeon (1989)
Blood-clotting factor IX (5' end and intronic)	Human	35-kb, X-linked, tissue-specific gene transcribed in liver; lacks CpG clusters; the strong MAR (4 kb) includes a *KpnI* element in intron 4 and part of exon 5; a weaker MAR includes the 5' end	Beggs and Migeon (1989)
Histone H4	Human	MAR/promoter; 141-bp core element at −730 to −589 of H4 gene including an NMP-1 (ATF) site; facultative MAR associated to NM only during the proliferation period of the osteoblast development	Stein et al. (1991); Dworetzky et al. (1992)
Osteocalcin	Human	MAR/promoter; involves interaction with NMP-1 (ATF) adjacent to the vitamin D-responsive element (−462 to −440); facultative activity detected in mature osteoblasts	Stein et al. (1991)
Proα2(I)collagen (intronic)	Human	MAR (0.9 kb) includes exon 42 (108 bp) and flanking introns; 35-kb gene with 52 exons; completely NM-associated in white cells (gene is inactive) but loosely bound to NM from fibroblasts (gene is active)	Ellis et al. (1988)
Three root-specific genes	Tobacco		Hall et al. (1991)
c-*myc* (3' end and 3' UTR)	Human HL-60 cells	MAR resides within a 1.4-kb region of which a 172-bp subfragment, interacting with p25 nuclear protein, has most MAR activity; p25, interacting with poly(A) signals in 3' UTR, is extracted with low salt and perchloric acid and is thought not to be a matrix protein	Chou et al. (1991)

TABLE I (*Continued*)

Gene	Species	Features	Reference
c-*myc* (5′ flank, exons 1 and 2, and intron 1)	Human	Determined indirectly from the presence of Topo II cleavage sites by use of antitumor drugs (Top II inhibitors) along an 800-kb unit carrying the c-*myc* locus and amplified in a tumor cell line	Gromova *et al.* (1995)
c-*myc* (5′ flank and coding)	Avian	MAR/promoter; both the −2200 to +1194 and −676 to +345 bp genomic fragments were associated to the matrix by slot-blot hybridization of matrix DNA (1–2% of total, DNase I-digested nuclei); no attachment sites in the +1194 to +6000 bp region	Schuchard *et al.* (1991)
Topoisomerase I	Human	Within a 2.9-kb *Bam*HI–*Eco*RI fragment (70% AT, topo II sites) in intron 13 of the gene; the fragment containing the promoter does not display MAR activity; a second MAR is present in intron 1; both include DNase I-HS sites	Romig *et al.* (1992, 1994)
Lectin (−7 kb; several other intergenic in the same locus)	Soybean	The P1 SAR (1.55 kb) at −7 kb contains a 78% AT 550-bp region and a 190-bp 70% AT region separated by 210 bp of 60% AT; six SARs mostly in intergenic regions over a 17-kb locus comprising the lectin and four other genes	Breyne *et al.* (1992)
Plastocyanin (3′ flank)	Pea	Light-inducible gene	Slatter *et al.* (1991)
ST-LS1 (5′ and 3′ flanks)	Potato		Dietz *et al.* (1994)
Agrobacterium T–DNA vector	Petunia	AT-rich, 1.65 kb; T-box consensus; topo II, ATATTT, yeast ARS, direct and inverted repeats; petunia SAR has strong binding affinities for both plant and animal nuclear scaffolds; Ti plasmids integrate into active chromatin; integration occurs close to the SAR site	Dietz *et al.* (1994)
Telomeres	Human	Anchored *via* their TTAGGG repeats; when such repeats are introduced at internal DNA sites do not behave as MARs	de Lange (1992)

286

Gene	Species	Description	Reference
Adh1-S (5' flank)	Maize	688-bp fragment at −1.1 to −0.4 kb of Adh-1; includes three DNase 1-HS sites; the CAAT and TATA boxes are 3' to this MAR and do not exhibit MAR activity *in vitro*	Avramova and Bennetzen (1993)
Adh1 (5' flank)	Maize	−650 to −500-bp region; AT-rich, hyperreactive to OsO_4 both *in vivo* and in supercoiled plasmids	Paul and Ferl (1993)
Histone H5 (3' flank)	Chicken	MAR/3' enhancer (+1146 to +1116, gene ends at +893); only the NF1 binding site (but not Sp1, GATA-1, UPE) is attached to the internal matrix	Sun *et al.* (1994)
Poly(ADP-ribose) polymerase (3' UTR)	Human	Includes most of the 3' untranslated region; TG-rich motifs	Boulikas and Kong (1995)
Choline acetyltransferase (5' UTR)	Human	MAR/ORI/enhancer/promoter; includes the 5' regulatory and 5' UTR region; possesses a GA-rich block of 200 bp	Boulikas *et al.* (1995)
Genes not containing MARs in their vicinity			
α-globin	Murine	Within 1.5 kb upstream and 0.5 kb downstream	Greenstein (1988)
Hypoxanthine phosphoribosyltransferase	Human	CpG islands are excluded from MARs; these genes contain CpG clusters in their 5' end which are methylated in inactive X-chromosomes in mammalian females; using halo preparation (LIS-extracted nuclei) and restriction enzymes	Beggs and Migeon (1989)
Glucose 6-phosphate dehydrogenase	Human		Beggs and Migeon (1989)
P3	Human		Beggs and Migeon (1989)
GdX	Human		Beggs and Migeon (1989)
Phosphoglycerate kinase type I	Human		Beggs and Migeon (1989)
α-galactosidase	Human		Beggs and Migeon (1989)

287

nuclei is used to trim down the loops to the attachment points to the nuclear matrix; MNase is known to be inhibited when it encounters a barrier along the DNA. Soluble nucleosome size classes from the loop are removed and the MAR fragments remaining attached to the nuclear matrix are then cloned and sequenced (Boulikas and Kong, 1993a,b); the genes shielded by these MARs are identified by database searches.

4. Nuclear matrix and loop DNA are hybridized to different genomic probes in order to determine their representation in the matrix DNA. This method has identified the MAR of the chicken α-globin gene (Razin et al., 1986), the MAR of the chicken c-*myc* gene (Schuchard et al., 1991), and has determined that the vitellogenin II gene is anchored preferentially onto the matrix of liver nuclei, a tissue actively expressing this gene encoding for an egg yolk protein, but not onto the matrix of oviduct nuclei where all egg white genes are expressed (Jost and Seldran, 1984).

5. Fishel and co-workers (1993) have used a colony-color assay to screen temperature-sensitive mutants of *Saccharomyces cerevisiae;* this *in vivo* assay for MAR activity is based on the property of MARs to bind transcription factors and on the observation that transcription of the β-galactosidase gene is blocked when a transcription factor binding site is placed in the *GAL1* promoter between the upstream activating sequence (UAS_G) and the TATA box. Yeast cells carrying a MAR-reporter construct are mutagenized and used to identify mutants temperature sensitive for both induction of β-galactosidase and yeast growth. These mutants suppress MAR inhibition and might involve yeast genes encoding defective MAR sequence-binding proteins or defective proteins/enzymes involved in the matrix attachment process. This assay has identified calmodulin and SMI1 proteins as participants of the matrix association process.

B. Matrix Isolation

A great deal of our current knowledge on matrix association of specific proteins or DNA sequences largely depends on the method of isolation of nuclear matrices. Nuclear halos prepared by high salt (2 M NaCl) or LIS treatment of intact nuclei followed by digestion of the expanded loops with restriction endonucleases or DNase I (Mirkovitch et al., 1984) give a nuclear matrix fraction that is most likely different than matrices prepared by nuclease digestion of nuclei followed by salt or LIS extraction of loops (Cockerill and Garrard, 1986a). Nuclei extracted with salt in the presence of EDTA (Cockerill and Garrard, 1986a; Boulikas and Kong, 1993b) or in the presence of Mg^{2+} (Obi et al., 1986; Xu et al., 1994) most likely give different NM preparations.

Treatment of nuclei with sodium tetrathionate that oxidizes the sulfhydryl groups to disulfides stabilizes the nuclear matrix and causes the retention of a higher number of proteins in the NM fraction (Kaufmann et al., 1991;

Stuurman *et al.*, 1992). Stabilization of the nuclear organization with 0.5 m*M* CuSO$_4$ also increases the protein amount in the matrix via matrix metalloproteins but some artifacts are induced by this treatment (Chiu *et al.*, 1993). In contrast to what one might expect, matrix preparations from agarose-encapsulated cells display an astonishing homogeneity with respect to their protein composition irrespective of the method used for matrix isolation (Eberharter *et al.*, 1993).

According to the method of He and co-workers (1990) chromatin is cleaved with nuclease and loop nucleosomes are mildly extracted with 0.25 *M* ammonium sulfate; this moderate ionic strength extraction leaves the NM-intermediate filament network intact and can be further fractionated with 2 *M* NaCl to remove the outer nuclear matrix proteins uncovering a fibrogranular network. Whereas the E7 protein of human papillomavirus 16 is associated with the 0.25 *M* ammonium sulfate-prepared matrix, E7 is lost after 2 *M* NaCl extraction (Greenfield *et al.*, 1991). According to a variation of this method tissue or cells in culture are homogenized in a buffered solution containing 0.5% Triton ×-100 in the presence of RNase inhibitors (vanadyl ribonucleoside) and protease inhibitors (phenylmethyl-sulfonyl fluoride, aprotinin, leupeptin), the nuclear pellet is extracted with 0.25 *M* ammonium sulfate. Chromatin loops are solubilized with DNase I (0.1 mg/ml at 25°C for 45 min); the nuclear matrix pellet is then disassembled in 8 *M* urea, 1% 2-mercaptoethanol, the insoluble carbohydrates and extracellular matrix are removed by ultracentrifugation, and intermediate filaments are reassembled by removal of urea by dialysis preserving nuclear matrix proteins in a soluble form (Fey and Penman, 1988; Khanuja *et al.*, 1993; Keesee *et al.*, 1994). One wonders to what extent results obtained by different methods of matrix isolation can be valid and comparable.

III. Identification of MAR Sequences within Various Genes

Several MAR sequences are known today (Fig. 1; Table I). It was first reported by Mirkovitch and co-workers (1984) that specific matrix attach-

---

FIG. 1 DNA sequences of some MARs and putative factor sites. Each transcription factor site is shown in different font; when sites overlap the new site appearing in the search erases a previous; thus, even one or two nucleotides in different font denote a factor site (see Fig. 9 for details). (A) Mouse κ immunoglobulin gene MAR (365 bp), Blasquez *et al.* (1989a); (B) Chinese hamster dihydrofolate reductase gene SAR-2 (549 bp), Käs and Chasin (1987); (C) Chinese hamster dihydrofolate reductase gene exon 5 MAR (487 bp), Käs and Chasin (1987); (D) HUM hypoxanthine-guanine phosphoribosyltransferase (HPRT) gene MAR (580 bp), Sykes *et al.* (1988).

A

AGCttnngTAtATAATCTTTTAGAGGTAAaatCTACAGCCAGCAAAAAGTCATGGTAAATATTCTTTGACTGAA

CTCTCACTAAACTCCTCTAAattattaTGTCATATTACTCTGGTTAAattaatATaaattGTGACATCACCTTAACT

GGTTAGGTAGGATAttttcTTCATGCAAAAATATGACTAATAAATAATTAGCACAAAAATAATTCCAATACT

TTAATTCTGTGATAgaaaaaTGTTTAACTCAGCTACtaTAAATCCCATAATTTGaaaactAttattAGCTTTTGTG

TTTGACGCTTCCCTGCCAAAGCAAcattaTAAGGACCCCTTTaaaactCTtgaaacTACTTTAG

B

TATACGTAGAATAgtttTCTTCCCTCTGTGTgtttttaAaatagtTACTAATGCTTTCTTGGATCTGCATTTAGG

AGTTATCCTTTCcattaaAaatataAAGctgtttCTTCCAAGGCgacTCCTGGGCGTGGAGCCTACCCTGGGGTA

TggtggtTAAACACAGTGTCACTGTACAATTGTTAGTTATAAattataAATAACTAATTtttAATTatttaaaTTTAAA

CAGTAAATtataatAGCAttttaatgaaaTTAACAttgatTAttatttaAACAACATtaaaaTTTAACATTGTAGAC

ACTGGCTTGGAGTGCTATTAACCAaaacagCGTGAGTGCCAttattaGCAACAACAGGCCAAGatgATTTCCCTGGCTT

ACAGAGAAATGGGGGGCATttttAGGTAATGACAGAGTAATAGGTGTACTGATGGGGGTAGGAGACATGGCTCaa

CAGTTAAGCTCTTGCTAcaCAGTTAttaggACCGATTCAGATCCTAGCACCCACAtaaaaGCAAAAAGGGCA

TCCTGTGAACTTCTCACAGTTCCAACTC

C

GGTACCGTG**ACC**CAATACAC**ACA**TA**A**CCAG**A**C**A**A**A**A**A**AGA**A**taat**ATCTACA**A̲A̲G̲A̲G̲A̲TA*gaataf*G̲A̲A̲G̲A̲A̲A̲**A**c̲a̲

t̲ a̲ a̲ c̲**A̲A̲A̲T**CTtttt**A̲A̲A̲A̲**A̲C̲T̲T̲T̲C̲T̲A̲T̲T̲G̲T̲T̲*TG*A̲G̲C*aacagt*AAGCATGGATT̲ẗ̲ạ̲ạ̲ṯ̲ğ̲CCTAG**TACAAAC**

TTAATtnnnc*AAgaatc*ICTAGT̲C̲T̲C̲T̲T̲TGT̲C̲T̲Ttcaac*ATGGGT*TAAGGAGAGTGATACCTTCTAttttat^TTTA̲A̲A̲G̲

GAAGCCATG**AATCAGC**AGCCA̲TcTCAgactct*TTGTG*ACAAGGATCATG**CCAGGA**Att*TGA̲A̲A̲C̲T̲C̲A̲C̲A̲*

C̲G̲T̲T̲C̲T̲T̲C̲C̲CAGAA*attgat*TGGA*G*R*aatata*AACTTCTCCCAGAGTAAGTACTACA**Attatta**ACTAGTCAG

AAGCACTTTAGAGACTAAT̲TACAGAAGAAg̲g̲g̲gaGCAAGTAAGGACAAAGTACAATG̲n̲n̲cGTGatgaaa

*TTAATACACAGA*ACCATGACTTGactATAATAaaaactA

D

AAGCTT ğğtcaagagty̲-TG̲A̲T̲TTAATICT̲ICTCTAACAGCTTTATCCCTCAGAAGGGAGGAGGCAAGCAAGTTATA

TA̲TGTAG̲t̲t̲a̲t̲ṯ̲T̲C̲T̲A̲A̲G̲A̲C̲T̲G̲ATatgaaa*TTGGAAGAT gaatct*ACtattagC̲T̲T̲T̲A̲A̲T̲Tattttt*ACA̲T̲TT*Aggaat

aTTGC**ATC**AGTAACTcAT*AAT*TTG̲TTTTCTGTTATCCTGAGTT*AACA*CA**AAT**TATCCAAGGAGA̲T̲GGC̲Gc̲a̲t

cac̲*TG*CTTTGAGCTgnn̲TTTGAgaattTTAATGTATCT̲GAATA̲taAAAA̲GGTAAAAATATGCCAA*CTAg*caat

TTCTGCCca**TTCCAGA**AGTT̲GG̲A̲A̲A̲T̲ẗ̲ẗ̲e̲ctattCTAggaattAaataaR̲T̲A̲T̲G̲G̲T̲TTATCTA̲T̲T̲G̲T̲T̲Atacc̲T̲

CTTTTAATTCACATAGCTCATtttatCttttat̲T̲T̲U̲T̲U̲g̲n̲n̲nTTTGAGATGGAGTCTTGCTCGTGTCAC**CAGGCA**

GGAGTGCAGTGA̲Tğc̲a̲a̲t̲CTCCGGCTCA̲C̲T̲C̲T̲A̲C̲CCACCgactcc̲ẗ̲ğğTTCAAGCGATTCTCCT**GCCT**GAGCCT

TCTGAGTAGCTG̲G̲G̲A̲T̲T̲A̲C̲A̲G̲G̲C̲A̲n̲GCACCCACCAC̲GCCCAGC

FIG. 1

291

ment sites occur within the repeated unit of the histone gene complex in *Drosophila*. These sites, occurring in an AT-rich 200-bp intergenic region between the histone H1 and H3 genes, mediate formation of a loop or chromatin domain. Since histone genes are tandemly repeated, these studies suggested that the chromatin of the histone gene locus is organized into a series of 5-kb loops, each loop comprising all five histone genes. These studies were extended by Mirkovitch *et al.* (1986, 1988) and Gasser and Laemmli (1986a) to the 5′ and 3′ flanking regions of *fushi tarazu, Sgs-4,* and the alcohol dehydrogenase single copy genes of *Drosophila*.

A MAR in the intron of the immunoglobulin (Ig) κ light chain gene in the mouse has been identified (Cockerill and Garrard, 1986a,b; Blasquez *et al.*, 1989b). This AT-rich MAR was shown to be in close proximity or to include the intronic tissue-specific enhancer of the gene and to harbor two topoisomerase II consensus cleavage sites. The intronic κ enhancer/MAR element, although playing a minor role in regulating transcription (Xu *et al.*, 1989), greatly promotes somatic hypermutation in the upstream V_κ segment by an error-prone replication (Betz *et al.*, 1994) but also V_κ recombination (Takeda *et al.*, 1993). A second enhancer, about 8.5 kb to the 3′ end of the constant region, was found interacting synergistically with the intronic enhancer to activate the κ gene (Blasquez *et al.*, 1992); whether or not the 3′ enhancer possesses MAR activity has not been investigated.

Two MARs of 350 and 400 bp were found to flank the E_μ intronic enhancer (200-bp) of the immunoglobulin heavy chain μ gene (Cockerill *et al.*, 1987). Each MAR contains a specific inhibitor sequence, IN1 and IN2, which are active in fibroblasts but not in myeloma cells (Wasylyk and Wasylyk, 1986; Imler *et al.*, 1987) and harbor binding sites for transcription factors (Peterson *et al.*, 1986); thus, MARs may contribute to cell type-specific gene expression. The core enhancer between MARs binds a number of transcription factors (Banerji *et al.*, 1983; Gillies *et al.*, 1983). These sequences were the first regulatory regions shown to possess MAR and enhancer activities. Apparently these intronic MARs are transcribed and since they do not impede the passage of RNA polymerase II their occupation by proteins might be a regulated process (Cockerill *et al.*, 1987). The enhancer at the 3′ end of the IgH locus is implicated, like the intronic enhancer, in promoting hypermutation of the V_H segment (Giusti and Manser, 1993; Sohn *et al.*, 1993).

Phi-Van and Strätling (1988) have identified two MARs in the chicken lysozyme gene *in vitro*. These sequences are comprised within two restriction fragments of 2.2 and 3.7 kb, one located 8.85 kb upstream of the transcription start site and the other 1.3 kb downstream of the polyadenylation signal of the gene and thus comap with the boundaries of its chromatin domain. Although these MARs were originally determined not to comprise any of the known enhancers and to contain only one out of seven total

DNase I-hypersensitive sites present within the chicken lysozyme gene locus, they were subsequently shown to have enhancer activity (Stief *et al.*, 1989; Phi-Van *et al.*, 1990). Three topoisomerase II consensus sequences and six T-rich stretches are present within the 5' MAR of the chicken lysozyme gene (Phi-Van and Strätling, 1988).

A 580-bp sequence within the first intron of the human *HPRT* gene has been identified as a MAR (Sykes *et al.*, 1988; Fig. 1). This sequence supports the extrachromosomal replication of plasmids in yeast, displaying indeed three sequence motifs homologous to the *S. cerevisiae* ARS^A_TTTTATATTT^A_T sequence, five motifs homologous to the CTTTTAGC$^{AAA}_{TTT}$ conserved region found to the 3' side of yeast ARS, and three sequences homologous to the *Drosophila* topoisomerase II binding and cleavage consensus sequence. In addition, this 580-bp stretch was shown to harbor intrinsically curved DNA fragments probably due to the presence of a TTTn_6TTTTTTTTTn_5TTTT motif. It was suggested that this MAR sequence might function as an origin of replication *in vivo* (Sykes *et al.*, 1988). Other genes whose attachment regions have been determined are summarized in Table I.

IV. Proteins of the Nuclear Matrix

Among the thousands of nuclear matrix proteins few have been identified and well characterized to date. These can be classified as: (1) Transcription factors with a stringent DNA sequence specificity such as Myb, Myc, RFP, C/EBP, AP-1, Sp1, and NMP-1(ATF) (Table II). (2) DNA-binding proteins of nonstringent sequence specificity exhibiting a cooperative binding and able to distinguish between MAR and non-MAR DNA such as topoisomerase II, histone H1, lamin B1, SP120, SAF-A, and ARBP. These proteins require a minimal length of about 240–350 bp MAR DNA for efficient binding, can cause looping of long DNA included between two distant MARs, prefer AT-rich sequences, are relatively abundant, and can recognize MAR DNA from different genes. (3) Nuclear enzymes such as DNA polymerases, poly(ADP-ribose) polymerase, and histone deacetylase. (4) Matrix adaptors that do not interact with MAR DNA but with other proteins of the matrix. Matrix adaptor proteins may include RB; these molecules might contribute to the formation of large complexes and might function in bridging distant regulatory MAR regions. (5) Other types of proteins (e.g., actin). Matrix proteins can be crosslinked to DNA in intact cells using *cis*-diamminedichloroplatinum (Ferraro *et al.*, 1992).

TABLE II

Proteins of the Nuclear Matrix[a]

Protein	Specificity	Features	Reference
Nonstringent, AT-rich DNA-binding (classical matrix proteins)			
ACBP (yeast) (ARS consensus-binding protein)	$\begin{smallmatrix}T\\A\end{smallmatrix}TTTA\begin{smallmatrix}TA\\CG\end{smallmatrix}TTT\begin{smallmatrix}T\\A\end{smallmatrix}$	Interacts with the ARS element	Hofmann and Gasser (1991)
ARBP (attachment region-binding protein, chicken)	ATTTCA$\begin{smallmatrix}C\\G\end{smallmatrix}$TTGTAAAA	Recognizes a consensus sequence in the MAR of the chicken lysozyme gene locus and MARs from *Drosophila*, mouse, and human genes; binds MAR DNA in a cooperative manner	von Kries et al. (1991)
Histone H1	AT tracts	Involved in gene silencing at a ratio of one molecule H1 per nucleosome over long stretches of chromatin; maintains higher order structures of chromatin	Adachi et al. (1989); Izaurralde et al. (1989)
HMG 1,2	AT tracts	General class transcription factor; recognize single-stranded AT-rich loops of cruciforms	Ivanchenko and Avramova (1992)
Lamin A	Single-stranded AT-rich DNA; narrow minor groove	Proposed to bind single-stranded DNA in active MARs and to bind to a narrow minor groove of inactive MAR structures	Ludérus et al. (1994)
Lamin B		Contacts DNA; localized only in nuclear periphery	Boulikas (1986b); Ludérus et al. (1992)
Matrin 3		Acidic protein of the internal nuclear matrix network of human and rat cells	Belgrader et al. (1991)
Matrin F/G RAP-1 (yeast)	AYCYRTRCAYY$\begin{smallmatrix}A\\T\end{smallmatrix}$	Forms loops with DNA; suggested to mediate the attachment of the HMLa regulatory domains to the scaffold and to repress the expression of genes in this region	Hakes and Berezney (1991) Hofmann et al. (1989)

294

Factor	Binding motif/sequence	Description	Reference
SAF-A (scaffold attachment factor A)	AT-rich motifs	Able to form loops with naked DNA as observed by electron microscopy	Romig et al. (1992)
SATB1 (human)	TTCTAATATAT; ATAATCTTC; TTATTATTTA; TATAAAAA; AAGATTATATA; TTTTAATGAGATAATAA; TATAATCTTC; TATAATCTTC; prefers DNA stretches composed entirely of A, T, and C but not G on the same strand	Tissue-specific; interacts with the minor groove of a special class of AT-rich MARs; makes no contact with DNA bases; binds to the 3' enhancer/MAR of the γ^A-globin gene; has a novel DNA-binding domain	Dickinson et al. (1992); Cunningham et al. (1994); Nakagomi et al. (1994)
SP120	$GTn^{A}_{T}A^{G}_{T}GATTnATnn^{A}_{G}$	120-kDa protein; cooperative binding to 0.4-kb Igκ and ft MAR DNA; reduced binding to smaller MAR fragments in vitro	Tsutsui et al. (1993)
Topoisomerase II	Unknown	Major protein of nuclear matrix; cooperative binding to MAR DNA	Berrios et al. (1985); Razin et al. (1991b)
SMI1 (suppress MAR inhibition)		Evolutionarily conserved (present in yeast, hamster, monkey, and human cells); might encode Igκ intronic MAR sequence-binding protein or other protein involved in the matrix attachment process	Fishel et al. (1993)
Calmodulin; calmodulin-binding proteins	Binds DNA	Calmodulin and at least seven calmodulin-binding proteins are nuclear in animal cells; enriched in nuclear matrix; mutation in the calmodulin gene in yeast suppresses attachment to the Igκ MAR	Bachs et al. (1990); Fishel et al. (1993)
Nuc2+	AT-rich DNA in vitro	Involved in spindle function; mutations arrest cells at metaphase; insoluble in 2 M NaCl; contains a 34-aa repeat proposed to be the matrix-anchoring peptide separated by 40–70 aas from the putative DNA-binding domain	Hirano et al. (1988, 1990)
Nucleolin	Binds RNA and 45-fold higher to T- over A-rich strand of ssDNA	Nucleolin controls rDNA transcription and ribosome assembly; binds MARs from different species; binds to the mutated IgH MAR that has lost its DNA unwinding ability	Dickinson and Kohwi-Shigematsu (1995)

295

TABLE II (*Continued*)

Protein	Specificity	Features	Reference
Transcription factors			
E1A (adenovirus)		Transcription and replication factor sufficient to immortalize primary rodent cells; crosslinked to matrix proteins with oxidation with o-phenanthroline/Cu^{2+}	Chatterjee and Flint (1986)
NF-1 (TGGCA-binding protein), chicken	$YGGMN_{5-6}GCCAA$ (Y = C,T; M = A,C)	By DNase I footprinting in the chicken α-globin gene MAR; a relative of the mammalian nuclear factor 1 (NF-1), within the CR1 repeat	Farache *et al.* (1990a,b)
NF-1	TGGCnnnnnCCCAA	NF-1 was resistant to both 0.25 ammonium sulfate and 2 M NaCl extractions; NM preparations gave NF-1 footprints whereas nuclear extracts gave NF-1, Sp1, and GATA-1 footprints on the enhancer of the H5 gene; present in the internal NM from mature erythrocytes, chicken liver, and trout liver	Sun *et al.* (1994)
ATF family (NMP-1)	TGACGTCCATG	Two proteins of 43 and 54-kDa members of the ATF family were identified by UV crosslinking as integral components of the matrix complex	Stein *et al.* (1991); Dworetzky *et al.* (1992)
Sp1,	GGCGGG		van Wijnen *et al.* (1993)
ATF,	TGACGYAR (Y = T or C; R = G or A)		
CCAAT,	CCAAT		
C/EBP,	CCAAT or TGACG		
AP-1	TGASTMA (S = G or C; M = A or C)		

296

Factor	Binding site	Description	Reference
NMP-1		Regulates histone H4 gene expression (binding at −730 or −589); mediates matrix anchoring of H4 and osteocalcin genes	Stein et al. (1991)
NMP-2 (rat)	Not known	Related to C/EBP; localized exclusively on the nuclear matrix in rat osteosarcoma cells; interacts with the promoter region of the osteocalcin gene	Bidwell et al. (1993)
RFP	GAGGC	Involved in the activation of the *ret* proto-oncogene	Isomura et al. (1992)
T antigen (SV40)		Both replication initiator and helicase; whereas at early times postinfection large T antigen is in the nucleoplasm, it becomes nuclear and is attached to the nuclear matrix with the onset of viral replication	Schirmbeck and Deppert (1987, 1989, 1991)
T antigen (K-128 mutant)		Mutated T antigen at Lys-128, unable to be imported to the nucleus, is found associated with NM at 96 hr pi although cytoplasmic at 24 hr pi; suggested to associate during nuclear envelope disassembly in mitosis	Deppert and von der Weth (1990)
T antigen (polyoma)	GAGGC	The highly phosphorylated and newly synthesized T antigen was matrix bound in mouse 3T6 cells	Buckler-White et al. (1980)
c-Myc	CAcgTG		Waitz and Loidl (1991)
c-Myb	YAACYR		Klempnauer (1988)
Steroid hormone receptors	$AGAACAN_3TGTTCT$ (inverted repeat)	Estradiol binds selectively to matrices of its responsive tissue (chicken liver) and androgen (dihydrotestosterone) binds selectively NM of rat ventral prostate target tissue	Barrack and Coffey (1980); Metzger et al. (1991)
Androgen receptor (AR)	$AGAACAN_3TGTTCT$ (inverted repeat)	AR binds to nuclear matrices from rat Dunning prostate tumor in a manner similar to that of intact nuclei; 1400 sites/nucleus; 50% of AR is extracted with 0.4 M KCl/EDTA	Colvard and Wilson (1984)
Mineralcorticoid receptor (MR)	$AGAACAN_3TGTTCT$ (inverted repeat)	100% of nuclear corticosterone was in NM in rat hippocampus cells *in vivo*; binding was mediated by MR; glucocorticoid receptor was not matrix bound	van Steensel et al. (1991)

TABLE II (*Continued*)

Protein	Specificity	Features	Reference
Estrogen receptor (ER); Progesterone receptor (human)	AGGTCAN$_3$TGACCT AGAACAN$_3$TGTTCT	Complexes of 16α-hydroxyestrone (16α-DHE1, metabolite of estradiol that can covalently attach to ER) with estrogen receptors are localized on the matrix; only 20% of 16α-DHE1 is solubilized with 1 M NaCl; DNase I did not affect extraction but RNase A (10 μg/ml) or sonication solubilized 88% of 16α-DHE1; 16α-hydroxylase responsible for 16α-DHE1 production is elevated in breast cancer patients	Swaneck and Fishman (1988)
Thyroid hormone receptor (ThR)		30–50% of nuclear T3 receptors in cultured GC cells were NM-bound (<1% of DNA in NM)	Kumara-Siri *et al.* (1986)
Tat of HIV	TAGACCAGATCTGAGCCTGGGA (TAR region immediately downstream of CAP site)	Ties the viral genome to the nuclear matrix with a K_a of 7.5×10^7 M^{-1}; 4000 sites/nucleus; requires Zn^{2+}	Müller *et al.* (1990)
E7 of HPV-16	Unknown	E7 is the major transforming protein encoded by human papilloma virus 16 (HPV-16); interacts with retinoblastoma; bound to NM prepared by 0.2 M ammonium sulfate but extracted with 2 M NaCl	Greenfield *et al.* (1991)
Adaptors			
Retinoblastoma (RB)	Interacts with E2F, c-Myc, Cyclins to modulate gene expression and the G1/S transition	Hypophosphorylated RB is NM-bound during early G1; mutant RB in tumor cells does not associate with RB; RB is preferentially localized in peripheral matrix, associated with lamins A and C	Mancini *et al.* (1994)
Progesterone receptor-binding proteins 1 and 2 (RBF-1, RBF-2)		RBF-1 (10 kDa) is abundant in liver matrix but only in trace amounts in the matrix of spleen; progesterone acts on the rapidly regulated nuclear matrix oncogenes c-*myc* and c-*jun*	Schuchard *et al.* (1991)

298

Enzymes

Enzyme	Description		Reference
Histone deacetylase			Hendzel et al. (1991)
Histone acetyltransferase	60–76% of the activity is in internal NM in immature chicken erythrocyte and trout liver nuclei; may mediate a dynamic attachment of active chromatin to the NM		Hendzel et al. (1994)
Casein kinase II (pea)	Remains bound to lamina-NM from pea nuclei after extraction with Triton, EGTA, 0.3 M NaCl, and a pH 10.5 buffer; might phosphorylate lamin A for entry of cells into mitosis		Li and Roux (1992)
Casein kinase II (rat)	NM-association of CKII declines dramatically in androgen-deprived rat prostate		Tawfic and Ahmed (1994)
DNA polymerase α	Four subunits (180, 68, 30, and 20 kDa) involved in lagging strand synthesis; p180 is catalytic; p68 interacts with T antigen to tether the polymerase to the replication fork; p30 and p20 are primases		Smith and Berezney (1980, 1982, 1983); Nishizawa et al. (1984); Yamamoto et al. (1984)
DNA polymerase β	Involved in DNA repair; localized in peripheral matrix		Smith et al. (1984)
Poly(ADP-ribose) polymerase	Stimulated by DNA strand breaks; only 5–26% of enzyme is NM-associated from different cell types; when sodium tetrathionate (NaTT, sulfhydryl crosslinker) is used during matrix isolation 50% of enzyme is in NM; in the absence of NaTT no enzyme was recovered in NM		Fakan et al. (1988); Kaufmann et al. (1991)
Phospholipase C/thiol:protein disulfide oxidoreductase	57 kDa; located in internal matrix		Altieri et al. (1993)
Topoisomerase I	AAAAAGACTT↓AGAAAAATT	By DNA topo I assays in isolated matrix preparations	Nishizawa et al. (1984)

TABLE II (Continued)

Protein	Specificity	Features	Reference
Other proteins			
B23 (numatrin)		Nucleolar 40-kDa phosphoprotein involved in ribosome assembly and processing of rRNA; elevated in tumor cells and in early G1 after stimulation of B lymphocytes with mitogens	Feuerstein et al. (1988)
β-actin; γ-actin; acidic actins		Identified by immunoblotting in mouse leukemia cells; actin filaments are associated with snRNPs	Capco et al. (1982); Nakayasu and Ueda (1986)
Vimentin		The cytoskeleton network extends to nuclei retaining their spatial disposition; observed after complete removal of DNA and RNA in epithelial cells but not in fibroblasts; vimentin is perinuclear	Fey et al. (1984)
hnRNA proteins; intermediate filaments		Internal matrix is a 3D network of thick filaments covered with 20- to 30-nm electron-dense hnRNA particles	Fey et al. (1986)
Ki-67		Monoclonal antibody of prognostic importance recognizing a nuclear antigen expressed in proliferating cells: normal blood lymphocytes are negative whereas PHA stimulation gives a positive nuclear reaction pattern	Verheijen et al. (1989)
Adenovirus terminal protein		55 kDa; covalently attached to the 5' end of Ad DNA; initiates DNA replication; mediates adenovirus anchorage to NM in a mode resistant to 1 M guanidine extraction	Bodnar et al. (1989); Schaack et al. (1990); Fredman and Engler (1993)
Minute virus of mice (MVM) terminal protein (TP)		Mediates attachment of MVM to the NM of MVM-infected mouse A9 cells; TP becomes covalently attached to both 5' termini to initiate replication	Bodnar et al. (1989)

Intermediate filaments	Proposed to stabilize NM irrespective of method of extraction	Beven et al. (1991); Eberharter et al. (1993)
NUF1 (nuclear filament-related)	110-kDa yeast protein, highly charged, resistant to extraction from nuclei with nonionic detergent, salt, DNase, or RNase; predicted from sequence analysis to form a large coiled coil structure (627 aa) flanked by nonhelical N- and C-terminal domains	Mirzayan et al. (1992)
H1B2; B1C8 monoclonal antibody-detected protein	180 kDa; in the dense 11-nm filaments of NM; relocates to the spindle and pericentriolar filaments at mitosis	Wan et al. (1994)
p84	Retinoblastoma-associated; localizes to intranuclear speckles, subnuclear regions of RNA processing; binds to the N-terminal region of hypophosphorylated RB during G1	Durfee et al. (1994)
Ser–Arg family of protein-splicing factors	Specify splice site selection in pre-mRNA; present in nonsmall nuclear ribonucleoprotein particles corresponding to speckles and containing U2, U4/6, U5 snRNPs	Blencowe et al. (1994)
Transcription factors not found in matrix		
Sp1, GATA-1, UPE-binding protein	In 3′ enhancer region of chicken H5 histone gene	Sun et al. (1994)

301

[a] DNA sequence specificity of TFs is from Boulikas (1994a).

A. Transcription Factors (TFs) on the Matrix

A number of transcription factors (TFs) have been found to be enriched in the nuclear matrix (Table II). It is proposed here that the attachment of TFs to the matrix is a process regulated by protein phosphorylation creating a dynamic matrix environment that changes during the cell cycle or from one cell type to another. Several protein transcription factors interact with one another on a transcriptional enhancer sequence germane to a particular gene. The interdigitation and crosstalk between transcription protein factor sites on the regulatory regions of genes allow for the complex process of cell type speciation to be manifested during development. The strong specificity, giving a dissociation constant of a protein factor from the DNA in the order of 10^{-7} to 10^{-15}, depends on the precise interaction of protein factors with one another after locking onto positioned sites on the DNA at optimal distance. TF binding may be modulated by protein phosphorylation, DNA methylation, and its integration into a chromatin environment (Boulikas, 1995d). A critical mass of activator protein seems to be necessary to effectively stimulate the basic transcriptional machinery. Creating polymers of single factor binding sites, which by themselves may have weak or no stimulatory activity, can lead to strong transcriptional stimulation, especially when one single high-affinity binding site is present in the complex (Boulikas, 1994a).

We have proposed that transcription factors are components of the nuclear matrix (Boulikas, 1992b,c, 1993a,b) and that TFs bind to a core ORI or enhancer, part of a composite MAR element; the cooperative binding of classical matrix proteins to the AT-rich MAR juxtaposes control elements (Boulikas, 1995b). Growing evidence independently supports this notion (Dworetzky *et al.*, 1992; Isomura *et al.*, 1992; Bidwell *et al.*, 1993; van Wijnen *et al.*, 1993; Table II).

A number of transcription factors have been shown to be enriched or exclusively located on the nuclear matrix. TFs attached to the matrix include: Sp1, ATF, CCAAT, C/EBP, and AP-1 activities (van Wijnen *et al.*, 1993); Sp1 and ATF were bound to the matrices in a variety of cell types, whereas CCAAT and C/EBP were cell type-specific; the NMP-2 localized exclusively in the nuclear matrix (Bidwell *et al.*, 1993); c-myc (Waitz and Loidl, 1991) and Myb proteins (Klempnauer, 1988); corticosteroid receptors (van Steensel *et al.*, 1991); the transcription factor RFP involved in the activation of the *ret* proto-oncogene (Isomura *et al.*, 1992); and large T antigen; specifically the attachment of SV40 large T antigen in TC7 monkey cells is a regulated process (Schirmbeck and Deppert, 1991) probably involving T antigen phosphorylation (Boulikas, 1995d). The fraction of T antigen anchored to the matrix exhibited the highest ATPase and by consequence helicase activity (Schirmbeck and Deppert, 1987, 1989, 1991). The MAR to the 5' flank (-500 bp) of IgH gene includes a 46-bp enhancer that interacts with IL-5 plus antigen-inducible TFs (Webb *et al.*, 1991).

NF-1, involved in both transcription and replication, was specifically found in the internal nuclear matrix of mature chicken erythrocytes but also in the internal NM of chicken liver and trout liver tissue (Sun *et al.*, 1994); DNase I footprinting analysis clearly demonstrated that proteins able to recognize the NF1 site (position +1116 to +1146) but not the Sp1 and GATA-1 sites (position +1050 to +1110) of the chicken histone H5 gene were present in matrix preparations; on the contrary, nuclear extracts, normally isolated by 0.35 M salt extraction of nuclei, gave footprints on all three TF binding sites. These data conclusively showed that NF1 is a protein of the internal NM, whereas Sp1 and GATA-1 are not (Sun *et al.*, 1994). Since NF1 has binding sites in the regulatory regions of a number of genes including the c-*myc* ORI and adenovirus ORI (Boulikas, 1994a) the studies of Sun and co-workers (1994) suggest that NF1 might mediate the attachment of these regulatory regions to the matrix in a constitutive or phosphorylation-regulated manner. NF1 might need to ineract with other TFs or matrix proteins to accomplish this mission.

The view of the nuclear matrix as a site of nucleation of protein transcription factors agrees with findings showing that transcription takes place on the nuclear matrix (e.g., Robinson *et al.*, 1982; Ciejek *et al.*, 1983), that MARs cohabit with a transcriptional enhancer activity (e.g., Xu *et al.*, 1989; Forrester *et al.*, 1994), that ORIs, believed to be activated by transcription factors (Boulikas, 1995c) are attached to the matrix (see below), and with emerging findings showing sequestering of transcriptional regulators in the matrix (Table II).

B. DNA-Binding Matrix Proteins of Nonstringent Sequence Specificity (Classical Matrix Proteins)

1. SATB1, ARBP, Matrins, Lamins, ACBP, SP120, SMI1, and SAF-A

The tissue-specific human SATB1 protein expressed predominantly in thymus binds to the minor groove with little contact to DNA bases of a special class of AT-rich MARs with high unwinding potential and with A, T, C on one strand, excluding G (Dickinson *et al.*, 1992). Human SATB1 binds strongly and selectively to MARs from different species.

The chicken ARBP protein (attachment region binding protein) recognizes the consensus sequence $\mathrm{ATTTCA}^{\mathrm{C}}_{\mathrm{G}}\mathrm{TTGTAAAA}$ in the MAR of the chicken lysozyme locus (von Kries *et al.*, 1991). This protein is a component of the internal nuclear network, selectively binds to the bases of the chromatin loop harboring the chicken lysozyme locus, but also recognizes MARs from *Drosophila,* mouse, and human genes. ARBP is a relatively abundant protein (100,000 copies per oviduct nucleus) and binds MAR DNA in a cooperative manner; dimerization between ARBP molecules bound to distant sites on the DNA causes looping out *in vitro,* suggesting a role in generating functional chromatin loops (von Kries *et al.,* 1991).

Eight proteins of the internal nuclear matrix have been identified and termed matrins; matrin 3 (845 amino acids) is an acidic DNA-binding protein of the rat internal nuclear matrix network with a 170-amino acid domain in the C-terminus; it displays a 96% identity with human matrin 3 (Belgrader *et al.,* 1991). The presence of an acidic domain suggests either a role of this molecule as a transcriptional transactivator or an interaction with nucleosomal histones in a mode similar to that of nucleolin, HMG1, and nucleoplasmin possessing long acidic domains (Belgrader *et al.,* 1991). Matrin F/G also has been characterized (Hakes and Berezney, 1991).

Lamins A, B, and C form the peripheral nuclear matrix and are mainly responsible for the functional organization of chromatin loops anchored to the periphery of nuclei (Fisher *et al.,* 1986; McKeon *et al.,* 1986; Osman *et al.,* 1990). Lamin B1 exhibits a DNA-binding activity (Boulikas, 1986b; Ludérus *et al.,* 1992); lamin A also binds DNA and displays a preference for single-stranded regions or narrow minor grooves (Ludérus *et al.,* 1994). It is an attractive idea that the preference of lamins for some sequence motifs and a possible interaction of lamins with some type of TFs might organize specific classes of genes to the nuclear periphery. Both lamin A, primarily synthesized in differentiated cells, and lamin B, found almost exclusively in nondifferentiated cells, interact with polynucleosomes on one hand and with nuclear membrane receptors on the other (Yuan *et al.,* 1991; Glass *et al.,* 1993).

The yeast protein ACBP (ARS consensus-binding protein) and at least one other protein bind to the T-rich strand of the ARS (autonomously replicating sequence) consensus (Hofmann and Gasser, 1991); two of these proteins, however, were later shown to be RNA-binding proteins unlikely to play a direct role in DNA replication (Cockell *et al.,* 1994).

Tsutsui and co-workers (1993) have identified a 120-kDa matrix protein, designated SP120, that exhibits a cooperative mode of interaction with the Igκ and the *fushi-tarazu* gene MARs; SP120, like SATB1, is unable to discriminate MARs of different sequences. The affinity of the protein was reduced when MAR subfragments of the 397-bp Igκ MAR, prepared by PCR, were used: filter blotting analysis using renatured SP120 showed that the 176- and 221-bp subfragments had about half, whereas subfragments of 114 and 125 bp had about 10% of the affinity of the full-length 397-bp Igκ MAR. The reduction in the affinity was independent of the position of the subfragment from the MAR sequence (Tsutsui *et al.,* 1993). These *in vitro* studies may not necessarily reflect the *in vivo* situation where SP120 might interact with other proteins of the matrix, TFs, and nucleosomes.

Binding assays of matrix proteins with pBR322 of various topological forms has identified two classes of proteins with respect to their DNA-binding sites: one class was highly specific for supercoiled DNA, whereas the other class lacked this specificity (Tsutsui *et al.,* 1988). When the matrix–DNA was prepared by *Eco*RI digestion of LIS-treated nuclei the preparation had an affinity only for supercoiled DNA; further trimming of isolated matrix DNA loops with MNase generated additional sites on the nuclear matrix which also bound relaxed and linear DNA (Tsutsui *et al.,* 1988).

Fishel and co-workers (1993) have used a colony-color assay to screen yeast cells carrying a MAR-reporter plasmid to identify mutants temperature-sensitive for both induction of β-galactosidase and yeast growth. These mutants suppress MAR inhibition, which is caused by genetically engineering a MAR within the *GAL1* promoter and might involve yeast genes encoding defective MAR sequence-binding proteins or defective proteins/enzymes involved in the matrix attachment process. This assay has identified calmodulin and SMI1 (suppress MAR inhibition) proteins as participants of the matrix association process. The monkey protein, counterpart of the yeast SMI1, from CV-1 cells is required for yeast cell growth and replication at 37°C, shows an intranuclear localization/nucleolar exclusion using immunostaining, and is also present in hamster and HeLa cells; Yeast SMI1 (505 aa, 57 kDa) has a putative KRKVK nuclear localization signal in its N-terminus, a very high concentration of charged amino acids in its C-terminus, and a weak leucine zipper/heptad repeat.

SAF-A (scaffold attachment factor A) was identified from its binding to a 2.9-kb SAR fragment (70% AT, Topo II sites) from intron 13 of the human topoisomerase I gene (Romig *et al.,* 1992). SAF-A (120-kDa, abundant nuclear protein) was able to recognize a number of homologous and heterologous SARs from vertebrate cells and to bind at multiple sites containing AT-rich stretches in a cooperative mode. SAF-A (but not Topo I, II, or SV40 large T) was shown by electron microscopy to be able to bring together distantly spaced SAF-A sites inducing looping of DNA. Proteins homologous to SAF-A were detected in human (HeLa and 293), mouse (lung, liver, heart, kidney), bovine thymus, and *Xenopus* oocytes (Romig *et al.,* 1992). SAF-A binds to RNA albeit with lower affinity than DNA and is identical to hnRNP-U thought to be involved in packaging of hnRNA into RNP particles; SAF-A binds both dsDNA and ssDNA (Fackelmayer *et al.,* 1994).

Half of the MAR-binding sites on the nuclear matrix may originate from proteins like ARBP, SAF-A, SP120, and SATBP, which do not recognize single-stranded DNA and bind cooperatively (Kay and Bode, 1994). The other half, accounting for more than 60,000 sites per nuclear equivalent, seem to recognize single-stranded DNA and are saturated in a hyperbolic manner (Kay and Bode, 1994).

2. Topoisomerases I and II

Topoisomerase II functions both in the relaxation of supercoiled DNA and in the decatenation of intertwined DNA loops (Giaever and Wang, 1988; Sperry *et al.*, 1989). A topoisomerase activity associated with the nuclear matrix (thought to be topoisomerase I) was first reported by Nishizawa and collaborators (1984). Subsequent studies have shown that DNA topoisomerase II is a major nuclear matrix protein component (Berrios *et al.*, 1985; Earnshaw *et al.*, 1985; Mirkovitch *et al.*, 1987; Adachi *et al.*, 1989) and that topoisomerase II-DNA associations at MAR sequences largely contribute to the topological organization of chromatin into loops (Razin *et al.*, 1991b). One dimer of topoisomerase II associates *in vitro* with about 200 bp of SAR DNA (Adachi *et al.*, 1989) in a cooperative mode of binding resembling the binding of other matrix proteins like H1, SP120, and ARBP.

Topoisomerase II, although contributing to a significant extent in interphase and mitotic nuclear matrix structure, is not the only protein involved nor absolutely necessary for matrix formation; topoisomerase II is absent from hen erythrocytes yet hen erythrocyte nuclei retain a nuclear matrix structure (Phi-Van and Strätling, 1988). The levels of topoisomerase II within a particular cell reflect the proliferative state of that cell. Relevant to the presence of large quantities of topoisomerase II in the nuclear matrix might be the role of nuclear matrix as the principal or exclusive site of DNA replication and transcription.

Searches for topoisomerase II binding and cleavage consensus in MARs have predicted or shown the presence of topoisomerase II cleavage sites within the MAR/enhancer region of the mouse κ immunoglobulin gene (Cockerill and Garrard, 1986a; Sperry *et al.*, 1989), in the MAR of SV40 (Pommier *et al.*, 1990); in the 3' MAR (Farache *et al.*, 1990a,b) and in the 5' MAR/ORI (Razin *et al.*, 1991a) of the chicken α-globin gene, and in the 1-kb intergenic region included between H1 and H3 genes of the *Drosophila* histone gene cluster (Käs and Laemmli, 1992).

Topoisomerase I prefers intrinsically curved DNA segments such as the motif AAA<u>AAGACTT</u> ↓ AGAAAAATT (Krogh *et al.*, 1991); the arrow points to the topoisomerase I cleavage site (Bonven *et al.*, 1985). Although topoisomerase II is a major constituent of the nuclear matrix, the evidence for localization of topoisomerase I on the nuclear matrix is weak (Nishizawa *et al.*, 1984). UV-crosslinking in intact *Drosophila* cells followed by identification of the DNA sequences that crosslink to topoisomerase I by immunoprecipitation showed recruitment of topoisomerase I to the transcribed *hsp70* gene (active genes are bound to the matrix) but not to its flanking sequences (Gilmour *et al.*, 1986). Immunolocalization by in situ hybridization on *Drosophila* polytene chromosomes has shown a different localization of topoisomerases I and II (Gilmour and Elgin, 1987). Provided that topoisomerase II is a major constituent of the matrix (Berrios *et al.*, 1985) topoisomerase I might become associated with the matrix in a facultative manner (e.g., during transcription) and with matrix regions remote from the topoisomerase II sites.

3. Histone H1 in Chromatin Loop Structure

Histone H1 locks the two helical turns of the DNA around the nucleosome (Boulikas *et al.*, 1980) and maintains higher-order chromatin structures (Thoma *et al.*, 1979; Fig. 2; see Boulikas, 1993c). Histone H1 is the most abundant repressor of gene activity (Croston *et al.*, 1991; Laybourn and Kadonaga, 1991). Even physically unconnected mononucleosomes are assembled into higher-order chromatin structures by exogenous H1 resembling domain subunits (Boulikas, 1986a). This is in spite of the presence of histone H1 within active genes (Grossbach *et al.*, 1990) but in stoichiometric amounts about twofold lower than in inactive genes believed to break the cooperative interaction between H1 molecules leading to unfolded chromatin (Kamakaka and Thomas, 1990; Postnikov *et al.*, 1991).

One copy of histone H1 per nucleosome promotes higher-order chromatin structures; the 1–72 amino acid domain of histone H1 interacts with three core histones of neighboring nucleosomes, whereas the 73–106 amino acid segment of H1 contacts histone H2A of its "own" nucleosome and locks the two ends of DNA around the particle into a closed conformation (Boulikas *et al.*, 1980; Fig. 2). Thus removal of histone H1 might constitute the first step in chromatin unfolding and in the unlocking of nucleosomes; H1 removal might be accomplished by its poly(ADP-ribosyl)ation (Boulikas, 1988), by its competition by HMG-I/Y (Zhao *et al.*, 1993), or by its displacement by a transcription factor (Laybourn and Kadonaga, 1991). H1 can suppress binding of USF to nucleosomes but only slightly suppress binding of GAL4-AH (Juan *et al.*, 1994).

4. Histone H1 and HMG1 Possess Nuclear Matrix Activities

Histone H1 was found to have an additional astonishing property: It behaves like nuclear matrix proteins in the sense that it binds to the same set of restriction fragments derived from a genomic clone of a gene, thus causing their precipitation (Izzauralde *et al.*, 1989; Adachi *et al.*, 1991; Romig *et al.*, 1992). H1 binds to A tracts in short homopolymeric runs of A or T; these stretches have a narrow minor groove that binds selectively to distamycin. An exciting suggestion is that nuclear matrix could act as a nucleation site

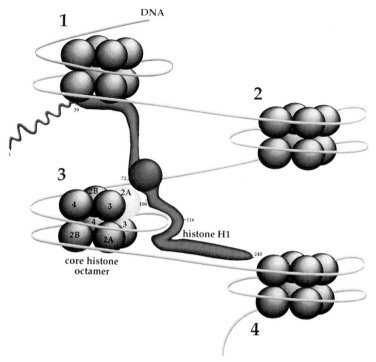

FIG. 2   A model for the involvement of different segments of the histone H1 molecule in
higher order chromatin structure. Second order chromatin structures are shown as part of a
zig-zag dinucleosome helix; the 73–106 amino acid globular segment of H1 contacts H2A in
its "own" nucleosome 3; the 1–72 N-terminal segment of H1 contacts two histones in nucleosome 1, an interaction thought to be responsible for chromatin condensation. From Boulikas (1993c).

for molecules of histones H1 into nonnucleosome organizations and that
shuttling of H1 molecules between matrix and polynucleosomes could control the activity of genes (Adachi *et al.*, 1991). A titration of H1 sites on
the A tracts of MARs by HMG-I/Y might cause redistribution of H1 to
non-MAR sequences (Zhao *et al.*, 1993). This finding places histone H1
and HMG, 1,2, I/Y at the crossroads of nuclear matrix structure and the
dynamic process of the switch on or off of genes. Interactions between
histone H1 and protein transcription factors or nuclear matrix proteins
could participate in the domain chromatin structure and in the remodeling
of the attachment points of chromatin loops to the nuclear matrix believed
to occur during development and the cell cycle.

HMG (high mobility group) 1 and 2 molecules (Ivanchenko and Avramova, 1992) as well as HMG-1/Y (Zhao *et al.*, 1993) were found to possess an

ability similar to H1 in recognizing MAR from loop DNA. HMG nonhistone proteins containing positively and negatively charged amino acid domains are thought to modulate nucleosome and higher-order chromatin structures and to promote gene expression. A number of transcription factors possess an "HMG box" as their DNA-binding domain (Boulikas, 1994a). HMG I, an activator of gene expression, and histone H1 share an 11 amino acid DNA binding domain peptide (TPKRPRGRPKK) called "AT hook," which is directed toward the minor groove of AT-rich sequences (Reeves and Nissen, 1990; Reeves et al., 1991). HMGI(Y) (which does not possess an HMG domain) binds to the minor groove at the NK-κB site in the IFN-β promoter; this interaction occurring with the NF-κB occupying the major groove and HMGI(Y) the minor groove of the same site (GGGAAAT-TCC) facilitates NF-κB loading (Thanos and Maniatis, 1992).

A protein from rat liver nuclei that specifically recognizes cruciform structures suspected to constitute a hallmark of MAR DNA (see below) was identified as HMG 1 (Bianchi et al., 1989, 1992). HMG 1 sequestered within the matrix of nuclei might participate in the stabilization of AT-rich cruciforms in MAR/ORI sequences during DNA replication.

V. MARs as Transcriptional Enhancers

MARs constitute boundaries of independently controlled chromatin units and, under proper conditions, display a moderate transcriptional enhancer activity (Blasquez et al., 1989a; Stief et al., 1989; Xu et al., 1989; Phi-Van et al., 1990; Klehr et al., 1991). However, a number of studies indicate that when a MAR is placed between an enhancer and a promoter element, the transcriptional activity of reporter genes is reduced (Stief et al., 1989). Furthermore, for some unknown reason, two MARs flanking a minidomain do not enhance efficiently the expression of reporter genes in transient transfection experiments but only in stably transfected cell lines (Blasquez et al., 1989a; Stief et al., 1989; Allen et al., 1993; Poljak et al., 1994), in transgenic animals (Xu et al., 1989; Kellum and Schedul, 1991; Forrester et al., 1994), or transgenic plants (Breyne et al., 1992).

The intronic Igμ enhancer (220-bp) is flanked at either side by an AT-rich MAR of 200–300 bp (Cockerill et al., 1987). Deletion of the MARs decreased the abundance of mRNA by a factor of 35 to >1000 in B lymphocytes of transgenic animals (Forrester et al., 1994). Deletion of the MAR from the J–C intron in the immunoglobulin κ gene and introduction of the gene that was defective in matrix attachment, into plasmacytoma cells in culture and transgenic mice led to about three- to fourfold reduction in the mean level of tissue-specific κ gene expression

(Blasquez *et al.*, 1989a; Xu *et al.*, 1989); this MAR vastly enhances the somatic hypermutation of the κ gene contributing to antibody diversity (Betz *et al.*, 1994).

A 2.2-kb fragment including the MAR to the 5′ side of the human interferon-β gene when linked to marker genes was found to increase their expression 10-fold only after integration of the construct into the genome (Stief *et al.*, 1989; Mielke *et al.*, 1990; Phi-Van *et al.*, 1990; Klehr *et al.*, 1991). A 20- to 40-fold stimulation of the CAT reporter gene by the *Drosophila* H1–H3 intergenic or *HSP70* gene MAR occurred only in stably but not transiently transfected HeLa cells; the stimulatory effect of MARs could be blocked by insertion of a 70% GC fragment containing many CpG islands between the MAR and the CAT gene (Poljak *et al.*, 1994).

The soybean P1-SAR but not the mammalian β-globin SAR was shown to reduce position effect variegation in the expression of the reporter β-glucuronidase gene/nopaline synthase promoter in 50 tobacco transgenic plants (Breyne *et al.*, 1992).

Two SARs (from yeast *ARS1*) flanking the β-glucuronidase (GUS) reporter gene under the influence of the 35S promoter of the cauliflower mosaic virus led to a 24-fold increase in expression in stably transformed tobacco cell lines; seven cell lines exhibited an expression level from 100- to 850-fold higher than control cell lines transfected with the nonMAR construct; only a 2-fold increase in expression was observed with the MAR construct in transient expression assays (Allen *et al.*, 1993).

MARS were proposed:

(1) To enhance the expression of reporter genes by formation of a domain and its insulation from the position effect variagation from neighboring domains at the integration site (Breyne *et al.*, 1992; Allen *et al.*, 1993). In this model most of the chromatin in higher eukaryotes (but not in yeast) is in an inactive conformation exerting a negative influence on foreign genes unless insulated with MARs (Allen *et al.*, 1993)

(2) To act by facilitating the displacement of histone H1 from chromatin through interactions with proteins with similar DNA binding motifs, such as HMG-I/Y (Käs *et al.*, 1993; Poljak *et al.*, 1994)

(3) To act as long-range modulators of chromatin structure although lacking any enhancer activity on their own, MARs were unable to stimulate the expression of reporter genes in the absence of the 72-bp repeats of SV40 enhancer (Poljak *et al.*, 1994)

(4) To facilitate communication between enhancers and promoters or to propagate an enhancer-induced local alteration in chromatin structure to a proximal promoter (Forrester *et al.*, 1994)

(5) To become stably unwound under torsional strain leading to transcription-poised chromatin structures (Bode *et al.*, 1992)

(6) To target integration of reporter genes to transcriptionally active chromosomal locations or recruit topoisomerase II to relieve torsional stress and promote transcription (Blasquez *et al.*, 1989a)

(7) To bring together distantly located control elements (ORIs, enhancers) (Boulikas, 1995b; Fig. 3); this model requires an evolutionary

COMPOSITE MAR ELEMENT

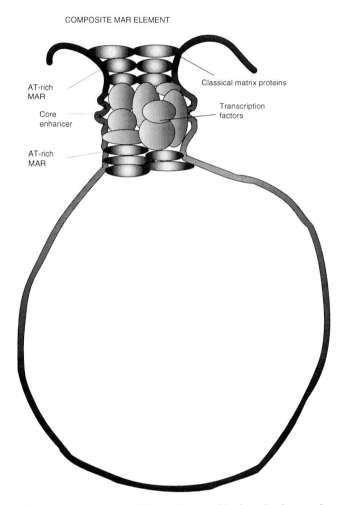

FIG. 3 A model for looping out of DNA and juxtapositioning of enhancers by synergistic interaction of classical MAR proteins (composite MAR model). The core enhancer is flanked by AT-rich sequences of about 300–500 bp able to sequester classical matrix proteins interacting cooperatively; this process, juxtaposing distant MAR elements and causing looping of DNA, brings together on the nuclear matrix two core enhancers or a core enhancer and a core origin of replication (100–200 bp) that cohabit with the AT-rich MAR (along with the transcription factors bound to them) to facilitate transcription and replication.

conservation of a cohabitation mechanism between AT-rich MARs and control elements, something that offers an advantage to the cell by facilitating the regulation of transcription and replication. The strong cooperative interaction between abundant classical MAR proteins (topoisomerase II, histone H1, lamins, SP120, and others, Table II) might bring AT-rich MAR elements together on the nuclear matrix in a mechanism also pulling together the closely juxtaposed enhancer, ORI, and promoter elements loaded with their transcrition factors. Mutual juxtapositioning of distant regulatory elements might be difficult in the absence of AT-rich regions in the core enhancer/ORI flanking regions.

The anchoring of enhancers and ORIs to the matrix might also facilitate their occupancy by the scarce transcription factors mediated by the interaction of TFs with classical matrix proteins and might juxtapose regulatory regions on the matrix, thus creating a unique nuclear microenvironment rich in TFs and enzymes to facilitate nuclear functions (transcription, replication, repair, recombination).

The observation that nuclear matrix shows a distinct preference for supercoiled plasmid forms (Tsutsui et al., 1988; Kay and Bode, 1994) could partly explain the preferential attachment of active but not of inactive genes to the matrix. Active but not inactive genes are under torsional strain (Giaever and Wang, 1988). A number of studies have shown the dependence of matrix anchorage to gene activity (Leibovitch et al., 1983). Precursors to ovalbumin and ovomucoid mRNAs, including various splicing intermediates and rRNA precursors, were found to be exclusively associated with the nuclear matrix from chicken oviduct nuclei (Ciejek et al., 1982, 1983). The transcriptionally active ovalbumin gene was found to be associated with such nuclear matrices in chicken oviduct nuclei but it was not associated with the nuclear matrix in chicken liver cells that do not express ovalbumin (Robinson et al., 1982). Polyoma virus transcripts in nine different transformed cell lines are associated with the nuclear cage (Cook et al., 1982).

Furthermore, a distinct class of sites recognizing single-stranded regions seems to be present on the matrix (Kay and Bode, 1994); single-stranded regions might be preferentially induced in the regulatory regions of active genes only as part of a mechanism for transcriptional activation.

Splicing of HnRNA to mRNA occurs on the matrix (Zeitlin et al., 1987). HnRNA is a component of the internal matrix fiber network linked via the poly(A) tails and secondary double-stranded regions (Herman et al., 1978; Miller et al., 1978; Verheijen et al., 1988). The status of association of HnRNA with the internal matrix is, however, controversial since pure RNPs aggregate in hypotonic conditions (Lothstein et al., 1985).

VI. Replication and Matrix

A. Origins of Replication (ORIs) as MAR Elements

The spatial organization of the eukaryotic DNA replication is faciliated by the anchoring of the replication machinery to the nuclear matrix; elegant experimental approaches have provided strong evidence that origins of replication are associated with the proteinacious insoluble nuclear matrix (Berezney and Coffey, 1975; Pardoll *et al.*, 1980; Berezney and Bucholtz, 1981; Smith and Berezney, 1982; Carri *et al.*, 1986; Razin *et al.*, 1986; Dijkwel and Hamlin, 1988). The activation of a fraction of the 30,000 or so potential ORIs present in a single mammalian nucleus (Martin and Oppenheim, 1977) is tightly linked to the transcriptional activity of neighboring genes.

DNA replication complexes are anchored to the nuclear matrix and the DNA is reeled through these complexes as it is replicated (Pardoll *et al.*, 1980). Replicating forks from HeLa cells (Valenzuela *et al.*, 1983) or from a slime mold (Aelen *et al.*, 1983) were found enriched in the matrix fraction of nuclei. Isolation of nucleoids or matrix halos and visualization of the sites of [³H]thymidine incorporation by autoradiography and mode of elongation during cold chase also favor reeling of DNA loops through the matrix (Mc-Cready *et al.*, 1980; Buongiorno-Nardelli *et al.*, 1982; Dijkwel *et al.*, 1986).

Sequence analysis of *ors,* putative origins of replication of the monkey genome, has shown them to have motifs similar to the consensus of the scaffold attachment region T and the consensus of the minimal autonomously replicating sequence of *S. cerevisiae* (Rao *et al.*, 1990). In yeast, matrix attachment elements coincide with the putative origins of replication and a chromosomal centromere (Amati and Gasser, 1988). *Drosophila* SARs can promote autonomous replication of plasmids in yeast (Amati and Gasser, 1990; Amati *et al.*, 1990). Furthermore, Vaughn and co-workers (1990) showed that replication forks derived from the amplified *DHFR* gene were associated with the nuclear matrix. These studies suggested that the origins of replication in mammalian cells are in close proximity to MARs or that the DNA structure of origins might allow them to act as matrix attachment sites.

Razin and co-workers (1991a) have found nuclear matrix attachment regions in the vicinity of the chicken replication origin of α-globin gene. MAR/ORIs can be found closely associated with promoter or enhancer elements. The promoter region of the *Xenopus laevis* ribosomal RNA gene repeat is a GC-rich 0.5-kb region that anchors the gene repeat to the matrix (Marilley and Gassend-Bonnet, 1989). The GC-rich region from the spacer region of the 44-kb mouse rDNA gene, which is amplified in some cell lines, is used as the origin of bidirectional replication of the locus (Wegner *et al.*, 1989).

The mammalian genome contains more potential ORIs than actually needed for its replication at a particular stage of development. Those remaining active are presumably the ones that sequester TFs among those available in a particular cell type and stage of development. The number of active origins decreases with pattern and cell type determination during development (Fig. 4; Callan, 1974; Sprädling and Orr-Weaver, 1987; Goldman, 1988). A similar change in chromatin loop size was found during *Xenopus* development: the maximum fluorescence halo radius was 9 μm in blastula, 9–15 μm in neurula and gastrula, and about 15 μm in adult blood cells; this suggests structural rearrangement of chromatin loops during gastrulation when the embryonic genome enters a phase of active gene expression (Buongiorno-Nardelli *et al.*, 1982).

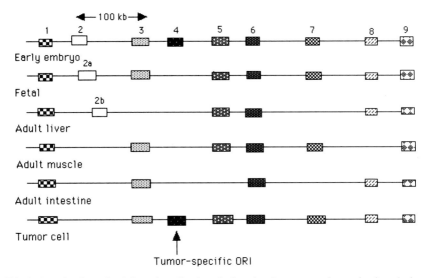

FIG. 4 Inactivation of origins of replication during development and reactivation during transformation. About 30% of ORIs functioning in early developmental stages might undergo inactivation in a program that might be associated with gene inactivation. This schema shows nine hypothetical chromatin replicons initiated from origins 1 to 9; an origin not used in a developmental stage or cell type is indicated by the absence of a box. In cases of developmentally controlled genes like the genes between origins 2 and 3 (for example, this domain may represent the human β-globin gene complex) inactivation of the early embryo-specific ORI 2 may be accompanied by activation of ORI 2a in fetal stage and ORI 2b in adult liver. During tumorigenesis or transformation a number of ORIs presumably representing early developmental stage-specific ORIs become activated (Martin and Oppenheim, 1977). Such ORIs are defined here as tumor cell-specific; these should differ among different tumor cells depending on their cell type and stage of dedifferentiation. From Boulikas (1995c).

As development proceeds, different replication origins, which are supposed to be anchored to the nuclear matrix, might be deactivated with a concomitant decrease in the total number of active origins (Fig. 4). The multiplicity of origins may drive the differential gene expression and cell type formation during embryogenesis (Boulikas, 1995c).

B. A Model of Replication Enhancers

We have proposed a model to explain a number of observations showing stimulation of replication by transcription factors and the presence of *cis*-acting elements distant from the origin of bidirectional replication able to affect origin firing (Boulikas, 1995c). According to this model (Fig. 5) all transcription enhancers are also elements exploited by cells to enhance replication efficiency in a cell type- or developmental stage- and substage of S-phase-specific manner. Transcription is specified by the interaction of the

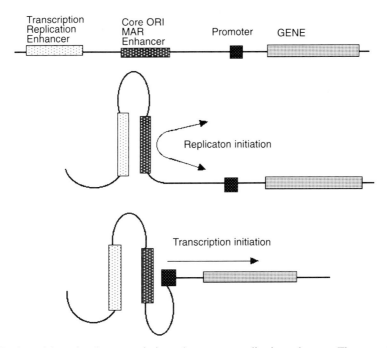

FIG. 5   A model arguing that transcription enhancers are replication enhancers. The transcription/replication enhancer acts upon the core ORI (which is also an enhancer and a matrix-attached region) to stimulate replication. On the other hand, interaction of the promoter element with both enhancers results in transcription initiation. From Boulikas (1995c).

enhancer with the promoter whereas replication is specified by the interaction of the enhancer with the core origin (Fig. 5). This model predicts one large replicon (determined by one core ORI) to be divided into several chromatin loops by the anchoring of every enhancer and core ORI element to the matrix. This idea is compatible with studies in the 240-kb dihydrofolate reductase (*DHFR*) amplicon showing one ORI or zone of replication initiation and about three MARs (Dijkwel and Hamlin, 1988); with the studies of Cockerill (1990) in the IgH locus also showing one ORI and 4–5 MARs; with studies in a 800-kb locus from the *Drosophila* X chromosome that bears about 25 fragments of 1–9 kb with ARS activity and 74 fragments with MAR activity (Brun *et al.*, 1990); and with measurements of replicon sizes from [^{3}H]thymidine autoradiography in microscopic slides and of sizes of chromatin loops from the diameter of chromatin halos. It was concluded that the average replicon length in all species examined is about four times the maximum fluorescence halo radius (that is twice the size of a DNA loop); this gives twice as many MARs as ORIs (Buongiorno-Nardelli *et al.*, 1982).

These studies provide evidence that of the several composite MARs along the genomic DNA only a fraction (about 30%) may cohabit with potential ORIs; this finding is in accord with the replication enhancer model (Fig. 5)—of the several potential MARs a certain fraction may be transcription/replication enhancers and the remaining may be core origins/MARs.

VII. Repair Occurs on the Nuclear Matrix

Repair of cyclobutane dimers and/or (6–4)photoproducts incurred at low doses of UV radiation were found to take place in association with the nuclear matrix (McCready and Cook, 1984; Harless and Hewitt, 1987; Mullenders *et al.*, 1988). Repair of carcinogenic chromium-induced lesions is higher in the matrix (Xu *et al.*, 1994). It is not clear at present whether or not these findings reflect the preferential repair of active over inactive genes (Bohr *et al.*, 1985; Russev and Boulikas, 1992); active genes are attached to the nuclear matrix. At high doses of UV radiation (30 J/m^2) no distinct preferential repair of matrix DNA was found (Mullenders *et al.*, 1983, 1988; Harless and Hewitt, 1987). An issue relevant to this model is also the preferential damage of matrix over loop DNA by mutagens (Obi *et al.*, 1986; Boulikas, 1992a) and chromate oxyanion (CrO$_4^{2-}$)-induced DNA adducts and DNA–protein crosslinks (Xu *et al.*, 1994).

Among the multitude of repair enzymes only poly(ADP-ribose) polymerase (Fakan *et al.*, 1988) and DNA polymerase β (Smith *et al.*, 1984) have been shown to be preferentially localized on the matrix using immunochemical methods. Protein phosphorylation might turn out to play an important role in cell-regulated repair events or in inducing the anchoring of a repair enzyme to the matrix. A number of DNA repair enzymes identified in yeasts and in mammalian cells encode for endonucleases, DNA helicases, or single-stranded DNA (ssDNA)-binding proteins (Sung *et al.*, 1987; Drapkin *et al.*, 1994). We predict that a number of these enzymes are anchored to the matrix either in a constitutive manner as matrix components or in a facultative manner by their association with matrix proteins or lesions of DNA during repair.

Of the three models evoked to explain the preferential repair of active versus inactive genes in rodent cells one is based on the anchoring of active but not of inactive genes to the nuclear matrix. According to this model transcriptionally active parts of the genome are already anchored permanently to the nuclear matrix where all the repair enzymes are to be found and this results in the preferential repair of active over inactive genes (Russev and Boulikas, 1992). In addition, the nuclear matrix displays a preference for binding single-stranded DNA (Kay and Bode, 1994; Ludérus *et al.*, 1994); this property could arise from the sequestering of helicases and ssDNA-binding proteins (Sung *et al.*, 1987; Drapkin *et al.*, 1994) to the nuclear matrix and could explain the anchoring of repairing regions to the matrix.

A mechanism for the transient anchoring of damaged DNA sites to the matrix was proposed as a prerequisite for efficient repair (Fig. 6). Induction of a lesion on a loop site was proposed to be recognized by a matrix protein mediating the anchoring of the site to the matrix thus dividing the loop into two subloops. Nuclear matrix is thus viewed as a nuclear microenvironment enriched not only in DNA and RNA polymerases but also in repair enzymes and protein transcription, replication, and repair factors. Nuclear matrix was proposed to act as a "sink" for all proteins that regulate replication, transcription, repair, and recombination, because of the nucleotide composition of matrix DNA, DNA secondary structures, and the three-dimensional configuration of the underlying superstructure.

Relevant to the repair mechanism is the view of MARs as hotspots of illegitimate recombination (Sperry *et al.*, 1989).

VIII. Rules Diagnostic for the Prediction of MAR/ORI Sequences

MARs and ORIs are shown or proposed to share fundamental sequence characteristics including inverted repeats, intrinsically curved DNA, potential triplex or left-handed DNA stretches, homooligonucleotides, DNase-I-hypersensitivity at the time of their activation by their binding for example

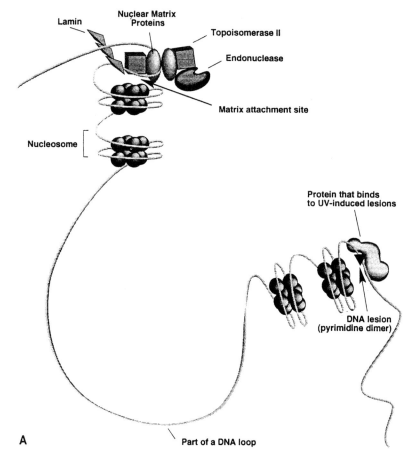

FIG. 6   A schematic diagram showing how a protein might recognize a DNA lesion (A) and by virtue of its binding to the nuclear matrix or its interaction with protein components of the nuclear matrix might mediate the anchoring of a damaged DNA site to the matrix (B). Nuclear matrix is the principal site where DNA repair takes place (Mullenders *et al.*, 1988). Attachment of damaged DNA to the nuclear matrix is suggested here to be the first, essential step for efficient repair. From Boulikas (1992a).

of a transcription/replication factor, and AT tracts (Boulikas, 1993a, 1995a). AT tracts may harbor DNA unwinding elements, binding sites for histone H1, HMG-I/Y, topoisomerase II, homeodomain proteins, and single-stranded DNA-binding proteins. AT tracts with these properties may lie within the core enhancer/origin of replication region or might extend within the classical MAR sequence.

We propose that a significant fraction of ORIs are composite MARs or cohabit with MARs; we call such elements MAR/ORIs. Furthermore, a sig-

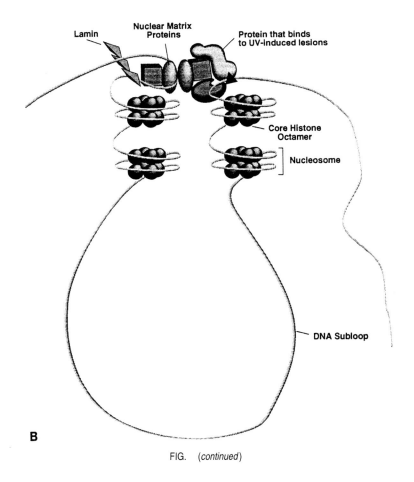

FIG.   (*continued*)

nificant fraction of MAR/ORIs may also be enhancers; these may represent about one-third to one-fourth of composite MARs. Another major fraction of MARs are enhancers only, without ORI function. Here it is attempted to summarize some features that appear to be common to many ORIs (Boulikas, 1995a) and MARs (Boulikas, 1993a) in an effort to pave the way toward identification and definition of DNA sequences able to function both as MARs and ORIs. Some studies show cohabitation rather than coincidence between MARS and ORIs (e.g., Dijkwel and Hamlin, 1988; Razin *et al.*, 1991a); nevertheless, an ORI (or an enhancer) can be considered as a complex element also including AT-rich tracts with a shown affinity for classical matrix proteins. Alternatively, a MAR may not only be composed of the AT-rich segment but also of the core origin (or enhancer) sequences that may not be AT-rich and may bind TFs (Fig. 3). We thus arrive at the notion of composite

MARs such as the MAR/ORI/enhancer in the chicken α-globin 5' flanking region (Razin *et al.*, 1991a) and the intronic Igμ MAR/enhancer (Cockerill *et al.*, 1987).

Of the proposed 21 rules for MAR/ORI elements, rules A–L can be verified by both experimental and computer analysis whereas rules M–U can only be verified by experimental trials for a given MAR sequence. A given MAR may obey only a set, not all, rules. Some rules (A–D) are proposed to be universal and to apply to all MAR/ORI elements.

A. MARs/ORIs Contain AT Tracts

Eukaryotic genomes are punctuated in a systematic way by AT-rich sequences at a periodicity of 10–100 kb (Moreau *et al.*, 1982). At-rich regions of varying size are omnipresent components of origins of replication (reviewed by Boulikas, 1995a). AT-rich stretches might flank the core ORI, as for example in EBV ORI or be included in the core ORI as in SV40 ORI. A replication initiator protein binds to core ORI causing local distortion in the double helix and initiating DNA unwinding in the AT tract. The AT-rich tract may lie between two binding sites of the initiator protein as is the case of the 18-bp AT stretch in the ORIs of HSV1 (Elias *et al.*, 1990).

A major class of matrix attachment sites are known to be AT-rich sequences of about 500 bp in length (Mirkovitch *et al.*, 1984, 1986; Gasser and Laemmli, 1986a,b) resembling in this respect some yeast ORIs (Boulikas, 1995a); these sequences bind to the classical matrix proteins (Table II). SARs were determined to contain multiple copies of the 10-bp AATAAA-TAAA motif (Gasser and Laemmli, 1986a).

Several different functions can be assigned to AT-rich stretches. First, the most conspicuous role of the AT-rich regions is that they facilitate DNA unwinding catalyzed by helicase molecules, torsional strain, or proteins that bind to ssDNA. In cases studied, AT tracts serve as the points for initiation of DNA unwinding caused by the binding of the initiator proteins to the AT-flanking region; single-stranded DNA-binding proteins (as for example replication protein A, RPA) then interact with melted AT tract and attract the DNA polymerase α-primase complex.

Second, they might be involved in the attachment of the ORI to the nuclear matrix in eukaryotic cells where the majority of matrix-attached regions appear to be AT-rich or to possess AT-rich tracts; a DNA unwinding matrix activity is found associated with rather short AT-rich motifs. Indeed, the AATATATTT motif present within the MARs of both IgH and β-interferon genes becomes the nucleation site of a DNA unwinding effect under torsional strain that is propagated to neighboring sequences to include a 200-bp stretch of stably unpaired DNA (Dickinson *et al.*, 1992).

Third, AT-rich stretches in ORIs might interact with HMG 1 and 2 like in the 50-bp AT-rich stretches of the amplification origins located within the nontranscribed spacer of the murine rDNA which function in ORI activation but not in transcription (Wegner *et al.*, 1989). The ORI of the mouse rDNA gene repeat contains the stretch:

TTTTTTAAAATTCTTTAAAATttttATTTTATATTTTTTTAGTTTAGT
TTAGTTTAATTTA(145bp)TATTTTTAAAATTTTTAAAATTAT
ATTTATTTAATTTATTTTTTTTTGTTTTTTT which is composed of two large footprints of HMG 1 separated by a nucleosome-length of 145 bp (Wegner *et al.*, 1989). Mirror symmetries and inverted repeats (shown outlined/double-underlined with the loop in lowercase) are present whereas slippage structures might be formed between or within these AT-rich sequences that can be brought together as two linkers flanking the same nucleosome. HMG1,2 were shown to bind to At-rich loops of cruciforms (Bianchi *et al.*, 1989).

Fourth, AT-rich stretches represent binding sites of a special class of regulatory proteins functioning in replication initiation; for example, yeast ARS elements include the ^T_ATTTA$^{TA}_{CG}$TTT^T_A 11-bp sequence representing the binding site of the OBP protein compex (Bell and Stillman, 1992; Diffley and Cocker, 1992).

Fifth, AT tracts harbor topoisomerase II binding and cleavage sites as well as binding sites for histone H1 both major scaffold proteins (Berrios *et al.*, 1985; Izzauralde *et al.*, 1989; Käs *et al.*, 1989). The binding of SATB1 although directed toward AT-rich tracts nevertheless depends on the presence of a DNA unwinding capacity in these tracts (Dickinson *et al.*, 1992). A number of other matrix proteins display a distinct preference for AT-rich motifs including ARBP, SP120, SAF-A, and lamin B1 (Table II).

Sixth, AT tracts may represent TF sites; a recent compilation of DNA binding sites for transcription factors has revealed the presence of a class of proteins with an AT-rich sequence specificity (Boulikas, 1994a).

Seventh, although not the actual sites of binding of the initiator protein, AT tracts may be the principal site of local distortion of the double helix caused by the binding of the initiator protein to immediate flanking regions in MAR/ORI elements as in SV40 ORI (Mastrangelo *et al.*, 1989; Dean and Hurwitz, 1991) and in yeast (Umek and Kowalski, 1988).

Eighth, the AT-rich region might be part of a palindrome; this appears to be the case for the $(TA)_{16}$ element within the core origin of varicella-zoster virus (VZV), which forms the center of a potential cruciform structure of 46 bp as well as for the HSV $AA(TA)_5TTATTA$ element (see Boulikas, 1995a); as part of a palindrome, the AT tract may still represent the binding site of a transcription or replication factor.

Ninth, AT tracts in their single-stranded form bind a special class of proteins (Bergemann *et al.*, 1992) that might be constitutive matrix components; the easy unwinding of AT tracts might facilitate this interaction

(Bode *et al.*, 1992). This proposal is further supported by the finding that the poly(A) tail in hnRNA is used for its anchoring to the matrix (Herman et al., 1978); although RNA is structurally different from DNA this observation suggests binding of the A-rich strands in AT tracts to DNA in their single-stranded form, a proposal supported by recent experimental data (Ludérus *et al.*, 1994).

B. DNA Unwinding Elements (DUEs) in MAR/ORIs

A DNA unwinding element was described in the origin of *E. coli* as a GATCT*n*TT*n*TTTT element which is thermodynamically unstable as evidenced by its sensitivity to the single-strand-specific mung bean nuclease (Kowalski and Eddy, 1989).

The AATATATTT motif present within the MARs of both IgH (including an enhancer and a negative regulatory element) and β-interferon genes becomes the nucleation site of a DNA unwinding effect under torsional strain that is propagated to neighboring sequences to include a 200-bp stretch of stably unpaired DNA (Kohwi-Shigematsu and Kohwi, 1990; Bode *et al.*, 1992). Mutation of this motif to ACTGCTTT abolishes both the MAR activity of the fragment as well as its DNA unpairing ability. The 25-bp sequence TCTTTAATTTCT**AATATATTT**AGAA from the MAR of the IgH gene when octamerized and inserted into pUC18 plasmid becomes unpaired in supercoiled plasmids under a variety of ionic conditions; the mutated form displays neither MAR activity nor unwinding capability (Bode *et al.*, 1992).

It has been speculated that negative superhelical strain in chromatin loops could be relieved by unwinding of specific DNA sequence motifs in MARs; the thermodynamic energy stored in the underwound stretches of the MAR could be released at a remote site within the chromatin loop and at other times to generate negative supercoiling to relax the positive supercoils generated ahead of the point of transcript elongation (Bode *et al.*, 1992; Klehr *et al.*, 1992).

AT tracts harboring DNA-unwinding elements (DUEs) may be the sites of initiation of DNA replication and might be diagnostic of MAR/ORI elements. DUEs may represent sites of the primary structural distortion upon binding of the replication initiator protein manifested as unwinding, untwisting, bending, or melting; this transition allows the binding of helicase molecules that promote further unwinding of ORI DNA thus allowing the assembly of the replication initiation complex.

An inverted repeat is noted that lies in the DNA unwinding region of the MAR in the human IFN-β gene (Bode *et al.*, 1992) shown outlined and with the loop in lowercase: TTTTATTTTTTTACATATAAATAtattcccTGTT-TTTCTAAAAAAGAAAA; this agrees with observations showing potential cruciform structures in MARs (Boulikas and Kong, 1993a,b). Whether this cruciform plays a crucial role in DNA unwinding or its ORI function, the type of nuclear matrix proteins that might interact preferentially with the unpaired region of the loop, the stem, or the base of the cruciform if any, and their role in propagation of a DNA unwinding to both sides of this cruciform should await further experimentation.

The negative control elements flanking both 5' and 3' ends of the enhancer of the immunoglobulin heavy chain gene convert into single-stranded regions with increased superhelical density (Bode et al., 1992). Are matrix proteins involved in DNA unwinding or is this process entirely dependent on the DNA sequence? The mechanism of origin firing has been best studied in SV40 ORI where an initiator protein (T antigen) causes structural distortion and eventual unwinding of the AT tract flanking the T antigen sites. This suggests that matrix proteins may help stabilize an unwound stretch of an ORI/MAR. DNA unwinding in MARs might be facilitated by the enrichment in DNA unwinding proteins presumed to be single-strand specific proteins.

The observations that yeast ARS, residing within the histone H4 gene, unwind under superhelical strain (Umek and Kowalski, 1988) and that matrix attachment regions in yeast coincide with the ARS elements, pulative origins of replication of the yeast genome (Amati and Gasser, 1988, 1990; Amati et al., 1990; Rao et al., 1990) further supports a DNA unwinding ability residing in MAR/ORI elements.

C. MAR/ORIs May Be Enriched in Inverted Repeats (IRs)

Inverted repeats (palindromic or snapback sequences) are able to convert into cruciform (hairpin) structures upon introduction of torsional strain on the DNA (Panayotatos and Wells, 1981). Only about seven nucleotides in the stem are thought to be able to maintain a cruciform and special computer programs are available to determine inverted repeats and their free energy of formation (Boulikas, 1993b).

Inverted repeats are of two kinds; (1) short (5–20 bp), usually representing the binding sites of replication initiator proteins and transcription factors and (2) long, best demonstrated by a 144-bp perfect inverted repeat in HSV-1 ori$_L$; long palindromes are believed to convert into cruciform structures and to act like efficient sinks of torsional strain thus facilitating unwinding of the double helix at the core origin.

Inverted repeats may represent protein binding sites in the origin of replication. For example, cooperative binding of the UL9 protein occurs on palindromic sequences in HSV ori$_S$ (Elias et al., 1992); the four binding sites of T antigen lie within an imperfect inverted repeat in SV40 ORI

(Borowiec and Hurwitz, 1988; Parsons *et al.*, 1990). The role of a special class of inverted repeat-binding proteins and their function in stabilizing DNA in its cruciform structure is just emerging (Pearson *et al.*, 1994, 1995; Xie and Boulikas, 1995).

Several independent lines of evidence suggest that cruciform structures are enriched in MAR/ORIs. First, early studies have shown that hnRNA, contributing to the fiber network of the matrix, is anchored via double-stranded hnRNA regions corresponding to the inverted repeats in DNA (Herman *et al.*, 1978). Second, the recognition sequence of a class of transcription factors known to be highly enriched in the nuclear matrix, such as steroid hormone receptors (Table II) are perfect inverted repeats (e.g., $AGAACAN_3T$-$GTTCT$). Third, HMG1 that specifically binds to the single-stranded loop of cruciforms (see Bianchi *et al.*, 1989, 1992) is a component of the nuclear matrix (Ivanchenko and Avramova, 1992). Fourth, origins of replication are characterized by an enrichment in cruciforms. These include viral DNA such as SV40 (Hay and DePamphilis, 1982) and polyoma virus (Hendrickson *et al.*, 1987), and mammalian ORI sequences (Frappier and Zannis-Hadjopoulos, 1987; Iguchi-Ariga *et al.*, 1988). An inverted repeat of 144 bp with only two mismatches is present in HSV-1*ori*$_L$, whereas SV40 ORI contains a 7-bp cruciform (Boulikas, 1995a). Fifth, monoclonal antibodies directed against cruciforms occurring in putative origins of replication of the monkey genome were shown to enhance DNA replication in permeabilized monkey cells (Zannis-Hadjopoulos *et al.*, 1988). This was interpreted to be a consequence of stabilization of the inverted repeats into their cruciform configuration by the antibody. Using quantitative fluorescence flow cytometry, $3-5 \times 10^5$ cruciforms/nucleus were estimated for monkey CV-1 and human colon adenocarcinoma SW48 cells throughout S-phase; however, no cruciform-like structures could be detected during G0, G2/M, or in metaphase chromosomes (Ward *et al.*, 1990, 1991). Sixth, S1 nuclease-sensitive sites appear as rodent cells move through G1 phase (Collins *et al.*, 1982). The increase in the number of single-stranded DNA-specific S1 nuclease cutting sites found during the S-phase by Collins (1979) and Collins and co-workers (1982) supports a model of a transient conversion of inverted repeats into cruciform structures during initiation of DNA replication. Seventh, formation of the two inverted repeats within the origin in the R1162 plasmid was found to be essential for activation of replication (Lin and Meyer, 1987). Eighth, MAR sequences found using random cloning of the nuclear matrix DNA from human cells have identified a novel class of non-AT-rich MARs highly enriched in potential cruciform structures (Boulikas and Kong, 1993a,b).

The 22 stable cruciforms along 77.3 kb of human β-globin locus are nonrandomly distributed along DNA (Fig. 7). Their location on the locus map is shown in Fig. 8. They are located close to DNase I-HS sites, at high densities of homeodomain protein-binding sites, and within the MARs of the locus (Jarman and Higgs, 1988). The location of the inverted repeats was proposed to reveal two zones of replication initiation, one in the LCR and the other in the 5′ flank and intron of β- and δ-globin genes (Boulikas, 1993b); subsequent studies provided evidence for one ORI at the predicted site in the 5′ flank of the β-globin gene (Kitsberg et al., 1993).

However, there is one experiment arguing against an enrichment of cruciforms in MARs. In an attempt to detect cruciform structures in the nuclear matrix Ward et al. (1991) have used DNase I digestion of nuclei treated with anticruciform monoclonal antibodies. In this experiment, the monoclonal antibody was expected to stabilize inverted repeats into their cruciform structures and render them sensitive to DNase I. The fraction of the DNA associated with the nuclear matrix was found to be resistant to DNase I, and it was concluded that cruciforms either are not present in the nuclear matrix or are buried in the scaffold, remaining inaccessible to DNase I (Ward et al., 1991).

D. MAR/ORIs May Possess Binding Sites for Protein Transcription Factors (TFs)

A number of studies (Table II; van Wijnen et al., 1993) support the idea that nuclear matrix is a microenvironment for the dwelling of transcription factors. The two MARs (each about 350 bp) that flank both sides of the 220-bp Igμ enhancer (Cockerill et al., 1987) have binding sites for NFμNR (Scheuermann and Chen, 1989) and SATB-1 (Dickinson et al., 1992).

Several studies on origins of replication in viruses, metazoans, and yeast also support a model for their activation by transcription factors (Boulikas, 1995a,c). Large T antigen and AP-1 have strong binding sites in the origin region of SV40 and stimulate both replication and transcription (Hay and DePamphilis, 1982; Murakami et al., 1991). Sp1 stimulates SV40 DNA replication (Guo and DePamphilis, 1992). The transcription factor C/EBP stimulates replication of adenovirus DNA in vitro (Pruijn et al., 1986). Oct-1 can stimulate initiation of adenovirus DNA replication in vitro (O'Neill et al., 1988). The DNA sequence-specific activity RIP-60 might

---

FIG. 7   Potential cruciform structures within the β-globin gene domain, formed under torsional strain on the DNA. These structures are numbered with 1 to 22 according to their position in the locus, which is indicated at the left horizontal line. The figure 6/6 in the palindrome 1 represents 6 matches over 6 nucleotides in the stem. The legend to each palindrome describes its position with respect to the DNase I-HS site and the genes. From Boulikas (1993b).

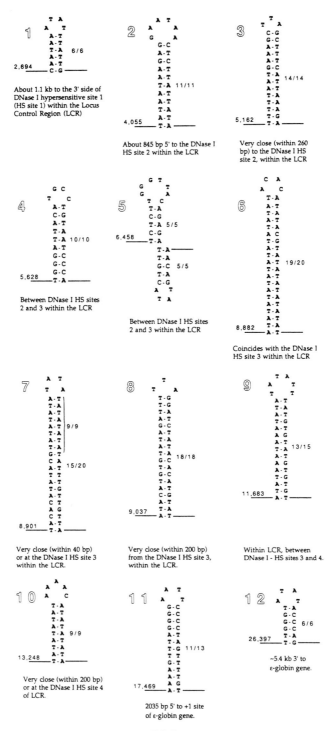

FIG. 7

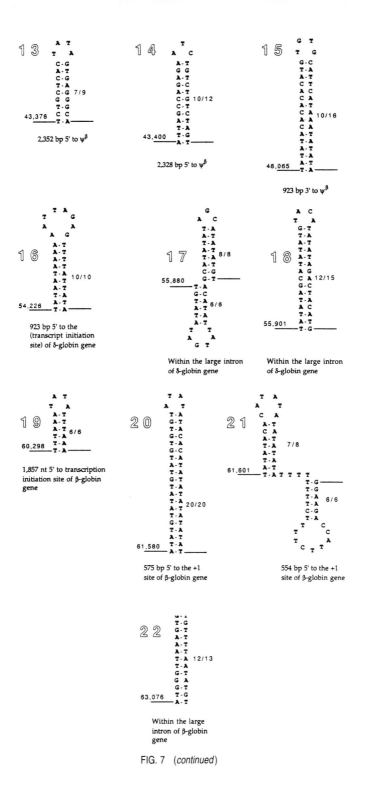

FIG. 7 *(continued)*

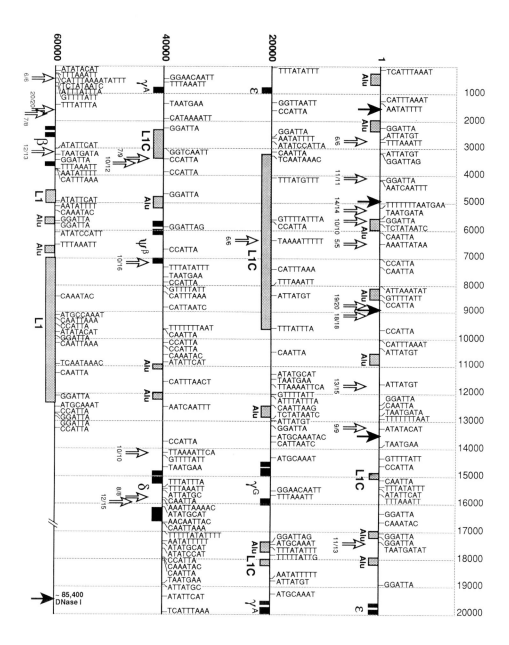

FIG. 8 Homeotic protein binding sites, MARs, ORIs, and palindromes in the human β-globin gene locus. This figure is in scale. Each of the four horizontal lines represents 20 kb of DNA. The ε, γ^G, γ^A, ψ^β, δ, and β gene exons, three in each gene, are shown as solid boxes. The *Alu* and L1 repeats are shown in gray. L1C (L1 complementary) represents an L1 repeat in the opposite orientation. Solid arrows represent the DNase I-hypersensitive sites; open arrows indicate the position of palindromes (see also Fig. 7). The first number before the slash at the base of open arrows represents the number of matches, and the second, the total nucleotides in the stem of each palindrome (see also Fig. 7). The short sequence motifs derived from homeotic protein binding sites, MARs and ORIs of other known genes are shown, with the thin lines representing their exact position on the map. From Boulikas (1993b).

constitute an additional candidate replication initiator protein; RIP-60 binds to the *DHFR* replication origin (Caddle *et al.*, 1990a; Dailey *et al.*, 1990).

The constellation of transcription factors implicated in the control of DNA replication may also include p53, a sequence-specific DNA-binding factor with a GC-rich sequence preference that interacts with the Sp1 site of SV40 ORI (Boulikas, 1994a); however, an additional important mechanism for the involvement of p53 in replication is via its ability to cause G1 arrest or apoptosis in response to DNA damage or inappropriate expression of c-Myc (Wagner *et al.*, 1994), activation of the Cdk inhibitor Cip1 (Di Leonardo *et al.*, 1994), and via its specific interaction with replication initiators such as large T antigen (Thukral *et al.*, 1994).

A number of yeast transcription factors are involved in replication. GAL4 was capable of stimulating polyoma virus DNA replication (Baru *et al.*, 1991). ABF1 may contribute either to repression or activation of transcription, and its binding to several autonomously replicating sequence (ARS) elements may stimulate initiation of replication (Buchman *et al.*, 1988). MCM1 is a transcription activator that also affects DNA replication (Christ and Tye, 1991). RAP1/GRF-I is both a transcription and replication factor (Kimmerly *et al.*, 1988). The origin recognition complex in yeast functions both in ORI firing and in transcription repression (Bell *et al.*, 1993).

c-Myc can substitute for SV40 large T antigen in an *in vitro* SV40 replication system (Classon *et al.*, 1987; Iguchi-Ariga *et al.*, 1987). The ORI/enhancer of the human c-*myc* locus located approximately 2 kb upstream of the transcription start site contains c-Myc binding sites (Iguchi-Ariga *et al.*, 1988). A different activity was shown with DNase I footprinting to protect a region that includes the AATTTTTTTT element of c-*myc* ORI (Saigo *et al.*, 1994).

The murine adenosine deaminase ORI (3993 bp) contains a 377-bp polypurine stretch and putative binding sites for several transcription factors including PUR and RIP-60, EBNA1, as well as Oct-1, C/EBP, and p53 implicated in stimulating DNA replication (Virta-Pearlman *et al.*, 1993). The 5' MAR/ORI in the chicken α-globin gene complex harbors putative binding sites for NF1, NFIII, and Sp1 (Georgiev *et al.*, 1991).

ORIs in multicellular organisms were proposed to be activated by a higher number of protein transcription/replication factors because of their higher complexity over viral and yeast ORIs. This level of complexity might allow their programming during embryogenesis, their differential replication during S-phase intervals, and might be tightly coupled to gene expression. Transcriptional enhancers were proposed to act as replication enhancers to increase initiation of replication from the core origin. Thus the functional origin of replication in our model is composed of the core origin and one or more replication enhancers at variable distances from the core origin (Boulikas, 1995c; Fig. 5).

1. Theoretical Searches for Transcription Factor Sites on MARs

In a study (Boulikas, 1994b) regulatory regions of genes including promoters, enhancers, origins of replication, and MARs were searched for transcription factor sites; all regulatory sequences gave a similar density of transcription factor sites which are illustrated using the 967-bp MAR of the chicken α-globin gene as an example (Fig. 9). A total number of 760 hexanucleotides from factor sites or origins of replication were used for this search. It was found that: (1) The occurrence of protein transcription factor binding sites overall on MAR fragments as well as on the enhancer and promoter regions of genes was only about 1.2 to 1.5 times higher than in random DNA. However, a higher concentration (up to 2.7 times over random sequences) of hexanucleotide factor sites was observed on small stretches of regulatory regions. The nucleotide composition did not alter the appearance of the landscape. (2) Some regulatory protein binding sites were underrepresented whereas others were overrepresented, giving to each MAR or other regulatory region a particular transcription factor flavor. Transcription factor density might be of diagnostic power for the regulatory or nonregulatory function of a stretch of DNA especially when combined with other diagnostic rules. MARs were proposed to constitute important regulatory elements of genes in addition to enhancers, promoters, silencers, locus control regions, and origins of replication.

2. Factor Binding Sites in the MAR of the Chicken α-Globin Gene

Figure 9 suggests that the chicken α-globin gene MAR sequence is a mosaic of 6-nt binding sites of a great number of transcription factors, including the mammalian Oct-3, μ-enhancer-binding protein, CD28, NFκB, Ets-1, SRF, CBF, and the Pu.1 factor related to Ets-1 oncoprotein. In addition, several 6-nt motifs identical to the c-*myc* and bovine papilloma virus origins of replication and from the core sequence of the adenovirus 2 and 4 origin of replication, which are known to interact with protein replication/transcription factors, were found.

An important feature of the factor hexanucleotides on the chicken α-globin gene 3' MAR, shown to cohabit with a transcriptional enhancer (Farache *et al.,* 1990a) is that some factor sites are underrepresented while other factor sites are overrepresented in this MAR sequence. The most striking examples of underrepresented factor/ORI sites are those of SV40-ORI, ETS-1, VDR, TFEB, RAR, and HOX4D. The most overrepresented factor sites are those of CD28, Pu.1, PEA3, and AGIE (Boulikas, 1994b).

CCG **GCTAGG**CAGAGCTGGGCAC *GGTG*T*GGGAA*cAggaaccCTG   **43**
  RFX   Pax-1   AGIE   Pu.1
                       TCF2   NTF

CCTAGGCCCCTCACCCTCTTCCTAC*AATCATAGAAATC*AT   **83**
  CTCF   CTCF   YB-1   CD28   AGIE
           PEA3   HNF-1   DBP   HNF-1

AgaaacaC*A*GAATGG**TTTGGG**TTTGG**AAGG**gaccttACAGC**CC**   **126**
  NF-AT   CBF   LIT-1   PCF   LIT-1
  H1TF1   ELP
  TGT3   TR

CCAGCT*CCACCC*CTGCTGT*GGGCTG*GCT**GCC***CCCACC*A   **164**
AP2   PuF   AP2   ETF   EGR2   PuF
           ARP-1

GCTCAGGCTGC**CCAGAGCCCCTCCATG**GCCTTGGGC**ACATCCAA**   **208**
       UBP-1   ETF   AP2   SRF   AP2
               H4TF1   Ad2,4-ORI

GGATGGGG**CACTCA**CAGCTCC GGGCAGCAGT*GCCAT*CAC*TTC*   **250**
Ets-1   TTF1   H2RII   TGCCA-protein NF-D   EF-1A

*ACT*GCCCTCTATGTGAA*GGA*tttcctcCTC**ACATCT**TACCTAAAT**C**   **296**
PRD1   CTCF   AGIE   Pu.1   SRF
H2RII   NF-AT

tccccTCTTTTGGT**TTAAAA**GC*attccc*T CTCGTCC*TCTCACT*   **339**
H4TF1   Oct-5   MBP-1   Pu.1   AhR
NTF   NFκB   CTCF
  CTCF   PCF

ATCTACCCG*TGTAAAAAGAA*GA*caagaga*AGGCAAGCAT   **378**
GATA   VBP   LSF   AhR   ISGF2   AhR
  Pax-1   CD28   c-myc-ORI

cCTAGAGGAGTT*TAGCAA*gaattt*CCC*att*caa*AATGTACttc   **421**
Pax-1   Pu.1   Oct-3   CD28   PCF   c-myc-ORI   EF-1A
             RFX   NFκB   AGIE   DBP   Pu.1
                   Oct-5   NTF

FIG. 9 Chicken α-globin gene 3' MAR (967 bp). Protein transcription factor binding sites in the chicken α-globin gene matrix attachment region. Different fonts and formats of all possible hexanucleotides from the consensus binding sites of vertebrate protein transcription factors homeodomain protein sites, as well as from vertebrate core sites of origins of replication known to interact with transcription/replication protein factors were used to screen this MAR sequence. Protein factor acronyms are aligned below the hexanucleotide motifs. Four topoisomerase II recognition sites predicted by Farache and co-workers (1990b) are overlined; the region from nucleotides 50 to 334 is the CRI repeat containing a site for the TGGCA-binding protein (overlined), the chicken analog of NF-1 determined by DNase I footprinting (Farache et al., 1990b). From Boulikas (1994b).

ctcATT**TAGCAAAGAAGAAA**ACACAgaatttATCTCCTGAGA   **463**

Oct-3	LSF	AhR	H1TF1	Oct-5	GATA
RFX	CD28	CD28		TGT3	CD28	RVF
		YB-1		NFκB	Pax-1

GCCAATTGT**AA**CATACCA**CTTGAG**ATCCTCTTGAT**AGA**   501

Ad2,4-ORI	Hbox1	TTF1	Pu.1	GATA
YB-1	Hbox1			UBP-1
BPV-ORI				

GAG**ACGAGGG**AGAGCATCACCTGC**TTACAAA**aggttcTTCTAC   **544**

CTCF		Hbox1	NTF	AhR
		DBP	LSF	

GTTTGCATC**TG**ACAC**TGGCA**TTCAGT**A**GGCCAATA GCA   **582**

TGT3	RFX	NFe	ISGF3	YB-1	RFX
NF-W				Ad2,4-ORI	

AAACCCA**TTATTGG**TGCAACCTATAGGTCTTCTgaaaaa   621

RVF	HNF-1	NF-W	AhR	NF-AT
	HOX4D			YB-1
	Ad2,4-ORI			DBP

ATCCTGTCAAAGCAGCT**TGTTTGC**CTTAGCTTGGCcagttTTC   **665**

AGIE	H1TF1		YB-1	IREBF	CD28
	TGT3			ISGF3	YB-1
				NF-AT	

ttcctAAA**TGTCAG**CCTGAACTG**CTCTCTCTG**ATATATCCCAA   **709**

EBP-1	SRF	NFe	UBP-1	AGIE (3 sites)
PCF			UBP-1	
PEA3				
NF-AT				

CCattttt**AA**CCCAACG**CTTTGG**GTTGCTCACAGCTCTTCCAGCTC   **755**

IREBF	Oct-5	CBF	LIT-1	RFX
	YB-1	AGIE	YB-1	H2RII
	HOX4D		CBF	

AGATCCATCTATCT**gaggtt**TTA**AGTAAT**GTA*TTCTC*T**ATTACA**   **799**

GATA	NTF	Oct-5	AAF	ISGF2	AAF
	RVF				Hbox1
	HOX4D	Topo II			

AcAcataagtattgaTGAAGgaaca gTGAATAAT   **834**

H1TF1	VBP	Ad2,4-ORI	Pu.1	PRD1	HNF-1
c-myc-ORI			NF-AT	BPV-ORI (3 sites)
			H2RII	Topo II

GAtgtt gaaccAAGAACtggaatcTGTCACAgaaggaaaAC   **875**

c-myc-ORI	LSF	MBP-1	RFX	PEA3	RFX
		NFκB	H2RII	NF-AT

*AGA*CT***CATCTGAAA***tattcaTTCTTGGCAAAAGCA GTGCTtattc   **920**

NF-E2   RFX   CBF   c-myc-ORI
                       c-myc-ORI   (2 sites)
                                                  Topo II

atctcttaCTTCA ATTAAATGTCCactgttAGTTGTGAATT   **961**
ISGF2   Ad2,4-ORI IINF-1   SRF   NF-AT   Oct-5
         (2 sites)   **Topo II**
         Hbox1

GAAGTT   **967**

Ad2,4-ORI

FIG. (*continued*)

3. Intrinsically Curved DNA Stretches and Positioned Nucleosomes in the α-Globin Gene MAR

MARs are characterized by the presence of intrinsically curved DNA (see below). Figure 10A shows the curvature map of the chicken α-globin gene MAR sequence. Peaks above about 0.3 curvature units on the map (Fig. 10A) are significant curvature points (Shpigelman *et al.,* 1993). The highest curved synthetic or naturally occurring DNA fragments display curvature values of up to 0.6 units in our program. The curvature map shows intrinsically curved DNA points in decreasing order at nucleotide 622 (0.47 units), 429, 447, 576, 583, 490, 459, 409, 662, 875, and 353 (0.35 units).

We have scanned the 967-bp MAR sequence for the preferred sites of positioned nucleosomes using a program that takes into consideration all AA and TT dinucleotides and compares their positions against a given matrix of strongly positioned nucleosomes (Ioshikhes *et al.,* 1992). A strongly positioned nucleosome (score 0.304) around nucleotide 648 occupies the 564–731-bp region of the MAR fragment. The most probable location of five nucleosomes on the 967-bp MAR fragment are shown by arrows in Fig. 10A. Nucleosome positioning is determined by (1) preference of core histone octamers for features on DNA, (2) interactions between neighboring nucleosomes, (3) by transcription/replication regulatory proteins acting as boundaries for positioned nucleosomes, and (4) by the higher order folding of chromatin (Thoma, 1992). The map (Fig. 8) does not take into account the three of four total factors that might determine nucleosome positioning *in vivo;* for example, strongly bound transcription regulatory proteins at precise positions on the MAR fragment could act as boundaries

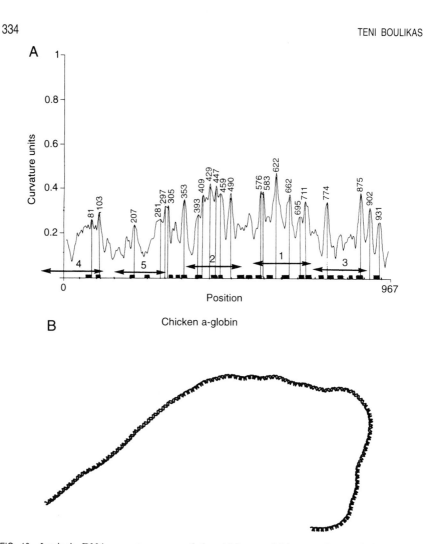

FIG. 10 Intrinsic DNA curvature map of the chicken α-globin gene MAR. (A) A 15-nt window was moved along the sequence. Peaks above 0.2 curvature units denote significant positions of curvature. Numbers denote the nucleotide positions of major peaks on the map. Filled rectangles on the abscissa show the overlapping factor sites. Arrows above the abscissa show possible nucleosome locations. (B) Three-dimensional projection of curvature along the 967-bp fragment. From Boulikas (1994b).

to position two nucleosomes at their flanking regions (Roth *et al.,* 1990; Thoma, 1992). This information can only arise from footprint analysis.

Figure 10A also shows the locations of overlapping protein factor sites on the curvature map (shown on the abscissa as solid rectangles). Only five stretches of clustered factor sites out of 23 total clusters fall within predicted linkers, those at the flanking region of nucleosome 2. The occupancy of a portion of the MAR region by a nucleosome with strong regulatory protein sites in the nucleosome-flanking regions might bring into juxtaposition protein factors bound to the flanking regions at the entry and exit points of the DNA to and from the nucleosome.

E. Replication Initiator Protein Sites in MAR/ORIs

Replication in animal viruses is more often stimulated by a single replication factor also talented to regulate transcription of viral genes. Such replication initiator proteins distort the DNA causing initiation by local unwinding that is propagated to nearby AT-rich sequences. Large T antigen of SV40 binds to the four GAGGC motifs in SV40 ORI (Parsons et al., 1990). EBNA1 (Epstein–Barr virus nuclear antigen 1), interacts with four repeats at the genetically defined replication origin of EBV–oriP; these four sites contain the consensus EBNA1 recognition sequence GATAGCATATGC-TACC (Hsieh et al., 1993). UL9 initiator protein binds to three sites in the origin of HSV-1 ori_s as shown by DNase I and MNase footprinting; interaction of UL9 with sites I and II of the core ORI causes looping of the AT-rich stretch in between (Koff et al., 1991).

So far SV40 has been shown to be attached to NM only in its actively replicating form; both SV40 and polyoma large T antigen is attached to the nuclear matrix of infected cells concomitant with the onset of virus replication (Buckler-White et al., 1980; Schirmbeck and Deppert, 1987). Molecules of T antigen anchored to the matrix are only those that have ATPase/helicase activity (Schirmbeck and Deppert, 1987, 1989, 1991) probably representing molecules phosphorylated at specific positions (reviewed by Boulikas, 1995d).

Bell and Stillman (1992) described the purification from yeast cells of a six-protein complex—the origin recognition complex (ORC)—with molecular weights of 120, 72, 62, 56, 53, and 50 kDa: ORC binds with high sequence specificity to the *ARS1* yeast origin of replication, causing local distortions of the double helix.

The mechanism of ORI activation in metazoans might be more complex than in viruses; candidate replication initiators include c-Myc (Iguchi-Ariga et al., 1988), RIP60 (Caddle et al., 1990a), PUR (Bergemann and Johnson, 1992), ssA-TIBF (Umthun et al., 1994), and a cruciform-binding protein (Pearson et al., 1994); these might be specific for a class of ORIs only. On the other hand RPA, an auxiliary to DNA polymerases α and δ (Tsurimoto and Stillman, 1989; Erdile et al., 1991), might play a broader role in ORI unwinding, acting at all ORIs.

Motifs in MARs that bind replication initiator proteins are not known; the cohabitation of MARs with the ORI of the chicken α-globin gene (Razin *et al.*, 1991b) and the broad ORI zone of the *DHFR* locus in Chinese hamster cells (Vaughn *et al.*, 1990) suggest that some type of metazoan replication initiator may bind near some MARs. In our model a number of ORIs might be included within MARs or MARs might be included within ORIs; this model is supported by studies on the human choline acetyltransferase gene, which is a composite control element comprising a MAR, an ORI, an enhancer, and a promoter—all four coinciding and forming a bipartite MAR/ORI/enhancer/promoter element (Boulikas *et al.*, 1995). By consequence, replication initiators, which have not been unequivocally identified in metazoans as yet, are proposed to bind to the special class of composite MARs possessing ORI activity.

F. MAR/ORIs Enriched in ATTA and ATTTA Motifs May Comprise Recognition and Binding Sites of Homeodomain (HD) Proteins

ATTA motifs may be more abundant in the AT tracts of ORI/MARs and might contribute to the execution of the developmental program of silencing (and less often of activation) of specific ORIs during cell type formation; furthermore, ATTA and ATTTA motifs appear to be enriched in elements controlled by homeodomain proteins and in major classes of ORIs and MARs (Boulikas, 1992c). This observation is supported by studies in other laboratories (Cunningham *et al.*, 1994; Todd *et al.*, 1995). Homeodomain proteins execute the body plan in all organisms studied and mutations in their genes give severe disruptions in body parts (Scott *et al.*, 1989). A statistical treatment of this idea should await the identification of ORI stretches among the sequences of the GenBank.

Figure 11a shows a striking example of this kind; the constitutive MAR to the 3′ side of the human apolipoprotein B gene, which is present in cell types expressing the gene as well as in cell types harboring the gene into an inactive chromatin structure (Levy-Wilson and Fortier, 1989), appears to be a mosaic of homeodomain protein-binding sites entirely composed of a contiguous stretch of 555 bp made of TAAT, TAAAT, ATTA, ATTTTA, TAAAAT, and ATTTA motifs (Levy-Wilson and Fortier, 1989; Boulikas, 1993a). This stretch of DNA needs to interact with unidentified nuclear matrix proteins in order to anchor the 3′ flanking region of the gene to the nuclear matrix. This suggests that nuclear matrix proteins might exist that are able to recognize motifs composed entirely of AT and in particular those containing in their recognition site the ATTA motif; those might include homeodomain proteins, i.e., transcription factors involved in body

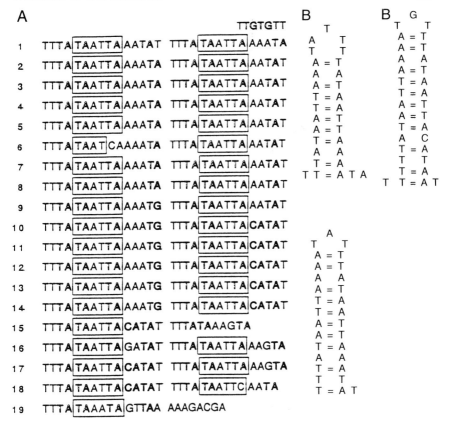

FIG. 11   (A) The constitutive 3′ MAR of the human apolipoprotein B gene is a mosaic of
TAAT, ATTA, TAAAT, ATTTTA, and TAAAAT motifs. The sequence is from Levy-
Wilson and Fortier (1989) but aligned in Boulikas, 1993a). The TAATTA motifs, core elements
of homeotic protein recognition and binding sites, are boxed. The TAATTA boxes are
separated by the TAAAT and ATTTTA motifs. The TA, CA, and TG nucleotides that are
characteristic of kinked DNA at a spacing of 0.5 or 1.0 helical turns between centers and
which are overrepresented in protein binding sites on DNA (see text) are outlined. This MAR
is a 30 nucleotide sequence shown on a single line repeated 18.5 times. Each 30mer is a dimer
of a 15 nucleotide sequence. TA + CA + TG account for 26.6% of all dinucleotides in this
particular MAR compared with 18.75% expected for random DNA. This particular MAR is
a striking example of AT-rich MARs that contain putative homeodomain protein binding
sites with a spacing of 1.5 helical turns and kinked DNA. (B) The best cruciform alignment
of the 30-nt repeating unit in the 3′ MAR of the human apolipoprotein B gene. About 18
energetically favored palindromic structures are predicted to be formed in the 564-bp MAR
sequence. This MAR fragment can also form a great number of slippage structures.

plan formation containing the conserved 61-amino acid homeodomain, although one could imagine binding of other types of proteins (e.g., HMG-I/Y, topoisomerase II, histone H1, SATB1, or other yet unidentified classical matrix proteins; see Tabel II). The proper alignment of the nucleotide sequences (Fig. 11A) shows that this MAR sequence is composed of a 30-nucleotide element, repeated 18.5 times. An ORI function of this MAR sequence has not as yet been investigated.

The 550-bp AT-rich MAR to the 3' border of the human apolipoprotein B gene contains about 18 potential cruciforms (Fig. 11B). These studies support the model that inverted repeats are highly enriched in the attachment regions of chromatin loops to the nuclear matrix and that these sequences in their cruciform configuration play an important role in the initiation of DNA replication (see Section VIII,C).

Figure 8 reveals a nonrandom clustering of putative homeodomain protein sites along the human β-globin gene complex; the map, which is in scale, shows several ATTA-containing motifs in the LCR (-22 to -6kb of ε-globin gene) as well as in the 5' flank of γ^G, 3' flank of ψ^β and in 3' and 5' flanks and large introns of both δ- and β-globin genes. Most of these clusters include DNase I-HS sites, inverted repeats, and the MARs of the complex identified by Jarman and Higgs (1988) and Cunningham and co-workers (1994).

One ATTA motif is not sufficient to specify a homeodomain protein site; instead, several ATTA motifs need to be clustered (Scott et al., 1989). In addition, homeodomain protein sites need to be properly spaced with other transcription factor sites or homeodomain protein sites to allow for the interaction of the proteins with one another and with DNA (Boulikas, 1994a).

It has been proposed that a fraction of homeodomain proteins from insects to mammals are nuclear matrix proteins and that matrix proteins are involved in the switching of genes during development by changing the attachment points of the chromatin loops to the matrix (Boulikas, 1992b,c). According to this hypothesis, genes that are coordinately expressed during development contain MAR-like elements in their upstream regulatory regions to facilitate chromatin loop remodeling; each MAR activity is expected to be developmentally regulated and to bind to specific homeotic proteins expressed at specific developmental stages. This mechanism is proposed to be intimately connected to programs of ORI inactivation presumably involving MAR/ORI remodeling (Fig. 4).

In support of the hypothesis that an enrichment of ATTA motifs may be found in developmentally controlled ORIs, RIP-60, a candidate replication initiation protein with binding sites within the ORIs of the *DHFR* locus, rhodopsin locus, and murine adenosine deaminase locus, recognizes the AT-TATTATTATT or a repeat of the (ATT)$_n$ motif (Dailey et al., 1990). A putative origin of replication from the monkey genome, *ors8*, contains a 186-bp internal fragment that is required for autonomous replication containing three ATTA and five ATTTAT motifs (Todd *et al.*, 1995). A role of the nuclear matrix in gene switch and differential activation of origins of replication in development is supported from changes in protein composition of the matrix during osteoblast differentiation (Dworetzky *et al.*, 1990).

G. Intrinsically Curved DNA in MAR/ORIs

A number of studies on origins of replication in viral and in other genomes sustain the idea that either the origin harbors intrinsically curved DNA motifs or that a severe bent is produced at the origin fragment as a result of its interaction with replication initiator proteins (Mukherjee *et al.*, 1985; Deb *et al.*, 1986; Koepsel and Khan, 1986; Schnos *et al.*, 1988, 1989). Curved DNA is a characteristic element of replication origins of yeast (Snyder *et al.*, 1986), SV40 DNA origin region I (Ryder *et al.*, 1986), and bacteriophage λ (Zahn and Blattner, 1985). SV40 T antigen, a component of the matrix (Table II), prefers intrinsically curved motifs; intrinsically curved DNA has been suggested to play an important role in nuclear processes involving specific protein–DNA interactions such as recombination (Ross *et al.*, 1982), transcription (Bossi and Smith, 1984), and replication (Zahn and Blattner, 1987). Several transcription regulators are known to induce a severe bent upon binding to B-form DNA including the CAP protein on the lac promoter of *E. coli* (Wu and Crothers, 1984), TBP on TATA box (Kim *et al.*, 1993), and others (e.g., Hatfull *et al.*, 1987). Stretches occupied by proteins of this type have been proposed as preferential sites of targeting by antineoplastic drugs (Boulikas, 1992b).

MARs, too, have been proven to possess intrinsically curved DNA from their retarded mobility on agarose gels (Anderson, 1986; Homberger, 1989). In addition, sequence analysis of specific MAR sequences has suggested the presence of stretches of intrinsically curved DNA (Sykes *et al.*, 1988; Boulikas, 1994b; Fig. 10). However, two studies suggested that intrinsically curved DNA may not be essential for matrix binding (Amati *et al.*, 1990; von Kries *et al.*, 1991).

Intrinsically curved DNA is formed by the occurrence of certain dinucleotides with a wedge angle every one-helical turn (10–11 nt) of DNA; optimal curvature will be expected with four to five consecutive A residues, such as in the sequence $AAAAn_6AAAAn_7AAAA$; alternation between AAAA and TTTT stretches does not favor curving of DNA (Trifonov, 1980, 1986, 1991; Trifonov and Sussman, 1980; Wu and Crothers, 1984; Koo *et al.*, 1986; Travers, 1987; Crothers *et al.*, 1990). In addition, motifs such as T_mA_n (but not A_nT_m) or GTTTAAAC (but not GAAATTTC) appear to have a pronounced bend into the major groove and a wide minor groove at the TA region; 5' A_nT_m 3' are essentially straight (Chuprina *et al.*, 1991). In addition to the AA and TT dinucleotides, the pairs AG, GA, GC, CG, CT, and TC when phased correctly cause an appreciable curvature on the DNA due to their large wedge values (Bolshoy *et al.*, 1991). GGGGCCC repeats with a 10–11 nucleotide distance from their centers are also curved (Brukner *et al.*, 1991).

Any DNA sequence can be subjected to an analysis for intrinsically curved DNA motifs using computer programs (Fig. 10).

H. Left-Handed and Triple Helical Structures in MAR/ORI Flanks

We propose that a class of MAR/ORIs may be flanked at one or both sides by DNA sequences that can adopt left-handed or triple helical structures when torsional strain is introduced in the DNA. Such motifs might (1) act as sinks of negative torsional strain that can be released during transcription and replication to relieve positive strain ahead of the advancing RNA and DNA polymerases; (2) serve as binding site for a special class of Z- and triplex-DNA binding proteins of the matrix; and (3) serve for TF binding in their B-DNA form.

Alternating purines/pyrimidines such as TG, TA, CA, or CG repeats adopt a left-handed (Z-DNA) structure under torsional strain (Johnston and Rich, 1985). Crosslinking Z-DNA antibodies with laser radiation at 266 nm for 10 ns has shown precise locations of Z-DNA formation in the c-*myc* gene during transcriptional activation; Z-DNA formation was near the promoter (Wittig *et al.*, 1992). Triple-helical (H-DNA) structures are formed by DNA stretches containing polypurines on one strand and poly-pyrimidines on the other and may include poly(dA), poly(dG), poly(dC), poly(dT), poly(dGA), or poly(dCT); triplex DNA is also formed by stretches with a mirror symmetry and can adopt the triple-helical structure under special conditions of torsional strain, pH, or salt (Johnston, 1988).

Polypurine or polypyrimidine stretches possessing a mirror symmetry have been identified in ORIs. A polypurine repeat of 248 bp is present within the ORI of mouse adenosine deaminase locus; it contains stretches of potential triplex and slippage structures such as in the segment A_5(G-GAAA)$_2$(**GGAA**)$_9$(AG)$_4$AAA(GGAA)$_2$GG(AG)$_5$AAAGAAA(GGAA)$_2$ GTAGA(GAAA)$_2$(GA)$_3$GGGAGA(GGGA)$_2$G(GAAA)$_2$ (Virta-Pearl-man *et al.*, 1993); an extensive mirror symmetry motif is present in the 36-bp segment (GGAA)$_9$ shown in bold.

An examination of the 6.2-kb zone of replication of the amplified *DHFR* locus in CHO cells has identified potential Z-DNA and triple helices by sequence analysis and from their sensitivity to mung bean nuclease (Caddle *et al.*, 1990b). The mouse rDNA ORI contains the potential Z-DNA stretch: $(TG)_8(TA)_2$; the ribosomal protein S14 (*RPS14*) ORI includes a $(CA)_7C$ motif (Boulikas, 1995a). Potential MAR sites in the β-globin gene complex are flanked by potential Z- or H-DNA stretches (Boulikas, 1993b). Z-DNA and H-DNA fall in the immediate flanking region of the maize *Adh1* gene MAR (Avramova and Bennetzen, 1993).

Several functions have been suggested for the interspersed $(TG)_n$ microsatellite, including those of a hotspot of recombination (see Boulikas, 1992b, for references). We suggest that an additional function of $(TG)_n$ as well as of $(GA)_n$ and $(GC)_n$ might be to delineate the borders of matrix anchorage sites.

Z-DNA and triple-helical structures in origins are expected to be prone to mutagens and have been proposed to contribute to the low degree of conservation of some stretches in ORIs flanking elements highly conserved during evolution (see below).

I. A Class of MAR/ORIs May Harbor Kinked DNA

Kinks may be generated on DNA by the presence of TG, CA, or TA dinucleotides separated from each other by 2 to 4 or 9 to 12 nucleotides. Kinked DNA may contain TAn_3TGn_3CA motifs with TA, TG, and CA occurring in any order (McNamara *et al.*, 1990; E. N. Trifonov, personal communication). A study on the sequence specificity of 420 transcription factors has shown that the CA, TA, and TG dinucleotides are overrepresented over other dinucleotides (Trifonov and Brendel, 1987).

We have observed that some MARs and ORIs from published studies and MARs identified in our laboratory may diplay an unusual richness in TA, TG, and CA and proper phasing of these dinucleotides into a kinkable configuration such as the MAR to the 3' side of human apolipoprotein B gene (Fig. 11A).

J. Attachment of Satellite DNA and *Alu* Repeats to the Matrix

Studies on whether repetitive or single-copy DNA is used by the cells to seclude chromatin domains have been controversial. The studies of Razin and collaborators (1978) and Jeppensen and Bankier (1979) on mitotic chromosome scaffolds in L cells, Matsumoto (1981) on bovine cells, and Neuer-Nitsche and co-workers (1988) on mouse cells suggested that AT-rich satellite DNA might be involved in anchoring the genome to the nuclear matrix. Satellite DNA was found to be overrepresented in the

DNA associated with the nuclear matrix in mouse cells (Boulikas, 1986b). Heterochromatin has also been reported associated with the nuclear matrix (Ellison and Howard, 1981). The CR1 repeat makes up about one-third of the MAR to the 3' flank of the chicken α-globin gene complex (Farache *et al.*, 1990a,b). AT-rich satellite DNA was thought to give 10-kb nuclease-resistant fragments that were preferentially interacting with the nuclear lamina in mouse cells (Georgiev *et al.*, 1991). It is not clear at present whether satellite DNA complexed with histones in the form of phased nucleosomes and other nonhistones forms a nuclear substructure onto which additional MAR landmarks of genes might be attached.

Contradictory experiments show that satellite sequences are not over-represented in matrix DNA: melted (single-stranded) matrix DNA fragments reassociated with kinetics virtually identical to those of total nuclear DNA (Pardoll and Vogelstein, 1980; Basler *et al.*, 1981). This means that all repetition frequency classes of DNA (single-copied, middle repetitive, and highly repetitive or satellite DNA) display the same distribution in matrix DNA as in total nuclear DNA. It has also been shown using cloned DNA that the matrix DNA in the mouse is not enriched in satellite sequences (Basler *et al.*, 1981). Blot hybridization experiments suggested that the scaffold attachment sites of the *Drosophila* chromosome are within unique sequences (Mirkovitch *et al.*, 1986). It is unlikely that these discrepancies arise from real differences between repetition classes of DNA sequences anchored to the matrix among organisms. Rather, they arise from differences in the experimental protocols.

The studies of Chimera and Musich (1985) indicate that a subset, not all sequences, of the *Kpn*I long interspersed repetitive DNA (LINE) is associated with the nuclear matrix. One *Kpn*I element is found within the 4-kb intronic MAR of the human blood-clotting factor IX, a gene linked to the X chromosome (Beggs and Migeon, 1989). The recombination junction of a human ring chromosome 21 contains Topo II consensus sequences that lie within two MARs (1.6 and 1.4 kb) separated by a 750-bp region; this fragment contains the 3' end of a LINE and is the recombination junction not exhibiting MAR activity (Sperry *et al.*, 1989).

Some *Alu* sequences may also be MAR elements (Small *et al.*, 1982). Such sequences may only comprise those members of *Alu* that contain a sufficiently high number of ATTA, ATTTA HD and TF protein sites properly distanced between their centers (Boulikas, 1992c). This idea suggests that the discrepancy on whether *Alu* sequences function or not as origins of replication in primate cells (Johnson and Jelinek, 1986; Rao *et al.*, 1990) and may or may not have MAR activity (Small *et al.*, 1982) might depend on the presence of elements proposed here as characteristic of MAR/ORIs in some members of the repeats.

Alu sequences in the human genome are not exactly identical. Preliminary data indicate that only a fraction of the 900,000 total *Alu* sequence repeats, which are dispersed in the human genome, are expected to function as matrix attachment sites (T. Boulikas and J. Jurka, unpublished). For example, only 3 out of the 1825 *Alu* sequences in the GenBank (as of September 1991) had five ATTA motifs 6 had 4 ATTA, and 19 had 3 ATTA motifs known to be HD core sites. Searching for transcription factor sites, replication initiation protein sites, intrinsic curvature, AT tracts, DNA unwinding elements, inverted repeats, and for potential triplex and Z-DNA in *Alu* sequences might provide a stronger suggestion for the potential role of some *Alu* as MAR/ORI elements.

Satellite repeats may show a periodicity characteristic of curved DNA such as the ATT periodicity occurring exactly every 10 nucleotides over long stretches of DNA in nematodes (Teschke *et al.*, 1991) and in the 370-bp repeat of rat satellite (Pech *et al.*, 1979). Individual repeats in these classes of satellite DNA may represent ORI/MAR stretches provided that they divert slightly from the sequence of the repeat and display additional characteristics of ORI/MARs such as TF binding sites. A less likely alternative possibility is that every sequence in a satellite repeat may act as a weak MAR/ORI structure; replication and MAR activity may initiate in this case at any site along the thousands or millions of repeats.

K. MAR/ORI Elements Harbor Topoisomerase II Binding and Cleavage Sites

A number of MARs have been shown to possess topoisomerase II sites including those of the κ Ig gene (Cockerill and Garrard, 1986a), the histone gene repeat (Gasser and Laemmli, 1986b; Käs and Laemmli, 1992), the *DHFR* locus (Käs and Chasin, 1987), and the MAR/ORI in the 5′ flanking region of the chicken α-globin gene (Razin *et al.*, 1986, 1991a). Topoisomerase II with an AT-rich sequence specificity (Table II) is a major protein of the nuclear matrix (Berrios *et al.*, 1985) functioning in replication, transcription, and chromatin assembly. Inhibitors of topoisomerase II find wide application as antineoplastic agents (Boulikas, 1992b). A provocative idea is that such agents target the function of this enzyme at matrix anchorage points. MARs were thought to enhance the transcription level of reporter genes in stably transfected cell lines or transgenic animals by recruiting topoisomerase II (Blasquez *et al.*, 1989a).

However, not all MARs bind strongly to Topo II; Topo II sites are neither a necessary nor a sufficient condition to specify a MAR (Sperry *et al.*, 1989).

L. Short Homo-Oligonucleotide and Other Direct Repeats in MAR/ORIs

A number of simple homo-oligonucleotides like CCC, GGG, TTT, and AAA are overrepresented in ORIs (Boulikas, 1995a). They might be remnants of evolution, products of extensive errors during replication of origins, or elements of intrinsic curvature. In support of this rule the yeast replication factor MCM1 recognizes among other the sequence GATATTTC-CAATTTGGGAAATTTCCCAAATCAGTAAT (Kuo and Grayhack, 1994) containing a number of homotrinucleotides; the putative replication initiator protein RIP-60 in mammals recognizes tandem repeats of ATT (Dailey *et al.*, 1990). Sequences containing similar homooligonucleotide repeats in ORIs may constitute recognition sites of similar or unrelated to MCM1 and RIP-60 replication factors.

MARs also appear to contain a higher number of homooligonucleotide repeats (i.e., AAA, TTT, CCC) than average loop DNA or random RNA sequences (Figs. 1 and 11A). Although difficult to fully evaluate this proposal in a statistical manner since very few of the MAR/ORI elements that might be present in the GenBank sequences are known, one could estimate the number of AAA, TTT, CCC, GGG, and all direct repeats in known MAR and ORI sequences and compare them to those in random DNA of different AT percentage.

M. Potential Origins of Replication Cohabit with MARs

It is proposed that about one-third of all composite MARs harbor core ORIs and the remaining two-thirds of composite MARs represent the habitat of core enhancers. A subset of structural MARs also may exist; in a mammalian nucleus we estimate the presence of about 30,000 ORIs (40,000 ORIs in early embryos and somewhere in-between in dedifferentiated tumor cells) and about 100,000 MARs. Most ORIs are expected to display MAR and enhancer activity in our model (Fig. 5).

Newly replicated DNA is specifically anchored to the nuclear matrix, and origins of replication are permanent MAR elements (Berezney and Coffey, 1975; Pardoll *et al.*, 1980; Smith *et al.*, 1987). Newly synthesized DNA isolated from replication forks by extrusion has been shown to hybridize with nuclear matrix DNA (Razin *et al.*, 1986). New DNA identified by [³H]thymidine pulse labeling has been shown by a number of methods, including autoradiography on microscopic slides, to be localized on the nuclear cage or matrix (McCready *et al.*, 1980; Buongiorno-Nardeli *et al.*, 1982; Aelen *et al.*, 1983; Carri *et al.*, 1986; Dijkwel *et al.*, 1986; Martelli *et al.*, 1992). New DNA has been found to be enriched in the matrix fraction by electron microscopy (Valenzuela *et al.*, 1983). MARs from yeast coincide with putative origins of replication and *Drosophila* MARs can drive the autonomous replication of plasmids in yeast (Amati and Gasser, 1988, 1990; Amati *et al.*, 1990).

DNA polymerase α is a nuclear matrix enzyme (Smith and Berezney, 1980, 1982, 1983; Berezney and Bucholtz, 1981; Nishizawa *et al.*, 1984; Smith *et al.*, 1984; Yamamoto *et al.*, 1984). Two-dimensional ORI mapping techniques on matrix and loop DNA have shown sequestering of the zone of replication of the *DHFR* locus to the matrix (Dijkwel and Hamlin, 1988). Two MARs cohabit with the origin in the 5' flanking region of the chicken α-globin gene (Razin *et al.*, 1991a). Seventy-four subfragments representing a contiguous 800-kb locus from the *Drosophila* X chromosome were examined for ARS activity in yeast and for MAR activity using LIS scaffolds from *Drosophila* embryos; the same set of fragments (about 25) possessing ORI activity also displayed MAR activity (Brun *et al.*, 1990). MARs have been localized close to or at the replication origin of bovine papilloma virus 1 (Adom and Richard-Foy, 1991). Active SV40 minichromosomes are associated with the matrix during transcription and replication; this attachment is mediated at least via T antigen (Jones and Su, 1987; Schirmbeck and Deppert, 1991).

Isolation and cloning of putative origins of replication from monkey cells in culture (Frappier and Zannis-Hadjopoulos, 1987) has shown them to possess sequence homology with MARs (Rao *et al.*, 1990). Origins of replication, nuclear matrix anchorage sites, and homeotic protein recognition and binding sites share the ATTA, ATTTA, and ATTTTA motifs (Boulikas, 1992b). Finally, isolation of a number of matrix anchorage regions from the human genome by random cloning has shown that a significant fraction of these act as origins for autonomous replication of plasmids in transient transfection experiments (Boulikas *et al.*, 1995).

N. Transcriptional Enhancers Cohabit with MARs

A significant number of ·studies have shown that MAR sequences are located near or at enhancer sites (e.g., Cockerill and Garrard, 1986a; Cockerill *et al.*, 1987). In addition, MAR sequences were shown to act like transcriptional enhancers in experiments involving stable transfection of animal cells in culture (Blasquez *et al.*, 1989a; Stief *et al.*, 1989; Phi-Van *et al.*, 1990; Klehr *et al.*, 1991), of plant cell cultures (Allen *et al.*, 1993), transgenic animals (Grosveld *et al.*, 1987; Xu *et al.*, 1989; Kellum and Schedl, 1991; McKnight *et al.*, 1992; Brooks *et al.*, 1994; Forrester *et al.*, 1994; Thompson *et al.*, 1994; Vazquez and Schedl, 1994), or transgenic plants (Breyne *et al.*, 1992). This property of MARs was attributed to an insulation effect of MARs upon reporter genes (Grosveld *et al.*, 1987; Kellum and Schedl, 1991; McKnight *et al.*, 1992; Allen *et al.*, 1993), to recruitment of topoisomerase II (Blasquez *et al.*, 1989a), to a propagation of the enhancer-induced open chromatin conformation to the promoter (Forrester *et al.*, 1994), to absorption of torsional strain by DNA-unwinding elements in MARs (Bode *et al.*, 1992), and to facilitation of displacement of H1 by HMG-I(Y) in the chromatin loop (Poljak *et al.*, 1994).

However, it has been suggested that MARs are void of enhancer activity, unable to act in the absence of a strong enhancer such as the SV40 72-bp repeats; in this model MARs simply amplify (20- to 40-fold) the effect of the strong enhancer by removing histone H1 and opening up chromatin structures (Poljak *et al.*, 1994).

The intronic enhancer of the immunoglobulin μ heavy chain locus (E_μ) is flanked at both sides by MARs; in this case it was thought that MAR and enhancer do not coincide but cohabit (Forrester *et al.*, 1994). In our opinion the entire region is a composite MAR element including transcription factor sites (defined as a classical enhancer) and sites for matrix proteins of nonstringent, AT-rich sequence specificity displaying a cooperative binding to DNA (defined as a classical MAR).

In a different model addressing the activation of origins, the core ORI acts both as an enhancer and as a MAR element for the gene with which it is linked (Fig. 6).

O. DNase I-Hypersensitive (HS) Sites in MAR/ORIs

A number of studies show that MARs can induce DNase I-hypersensitive site (DNase I-HS) sites in chromatin. The DNase I-HS sites in the *Drosophila* histone gene repeat coincide with topoisomerase II cleavage sites and with the MAR sector in the H1–H3 intergenic region (Käs and Laemmli, 1992). The MAR/ORI to the 5' flanking region of the chicken α-globin gene cluster (upstream of π-gene, Farache *et al.*, 1990a) harbors a constitutive DNase I-HS site that is detected in chromatin from many chicken tissues (Weintraub *et al.*, 1981). The formation of an extensive DNase I-HS chromatin conformation in the Igμ intronic enhancer was absolutely dependent upon the presence of the two MARs in its flanks in B lymphocytes from transgenic animals (Forrester *et al.*, 1994). Similar studies in transgenic animals showed that whereas the 95-bp Igμ core enhancer was necessary and sufficient for the accessibility of TF to T7 promoter it required the MAR region to induce DNase I hypersensitivity (Jenuwein *et al.*, 1993). The ORI of SV40 minichromosomes is attached to the matrix and is hypersensitive to DNase I (Schirmbeck and Deppert, 1991). The 1-kb "facilitator" elements probably related to MARs that flank a 200-bp enhancer in the human *ADA* gene are responsible for the DNase I hypersensitivity of the enhancer (Aronow *et al.*, 1995).

The origin of replication in the 5′ nontranscribed spacer of the amplified macronuclear rRNA genes in the protozoan *Tetrahymena thermophila* includes three DNase I-HS sites, direct repeats, DUEs, intrinsically curved DNA, a single-stranded DNA-binding activity, and a double-stranded DNA-binding activity of 105 kDa distinct from the *Tetrahymena* HMG B and C proteins footprinting when probed with hydroxyl radicals to a $(T)_7GGC(A)_{11}$-$C(A)_5TAGTAA$ motif of the type I repeat (Umthun *et al.*, 1994).

The proposal that MAR/ORI elements include or lie close to DNase I-HS sites can be further substantiated by correlating the great number of the mapped DNase I-HS sites in various genes with new data on MAR activity in these regions.

ORI/MARs are proposed here to constitute important control elements, colocalizing with transcriptional enhancers, tightly linked with the transcriptional activity and the temporal manner of replication of genes during the S-phase intervals. Initiator proteins distort the DNA. It is proposed that MAR/ORIs become DNase I-hypersensitive at the time of their activation as a result of binding of transcription and replication factors, as a result of unwinding, likely to be generated from the transcriptional activity in their domain (Bode *et al.*, 1992), a process that generates a single-stranded region, and as a result of removal of nucleosomes, a process that might be a prelude to ORI activation (Hsieh *et al.*, 1993). Experimental evidence shows that the Epstein–Barr virus ORI has a local region of its double helix distorted at the time of activation and is rendered hypersensitive to DNase I, and to some chemicals that preferentially attack single-stranded over double-stranded DNA such as $KMnO_4$ (Frappier and O'Donnell, 1992). However, TFs can bind to precisely positioned nucleosomes and confer DNAse I-hypersensitivity without the ejection of the underlying histone octamer (McPherson *et al.*, 1993).

One additional reason to believe that core ORIs are into DNase I-hypersensitive conformations at the time of their activation emerges from studies in synchronized cells; it has been conjectured that cruciform structures appear at the origin only during their activation in S phase and that these structures disappear at all other stages of the cell cycle (Ward *et al.*, 1990, 1991); the loop of the cruciform is hypersensitive to DNase I.

P. Active MAR/ORIs Might Be Free of Nucleosomes and Might Harbor Poor Nucleosome Positioners

The currently emerging idea that MARs harbor sites for interaction of transcription and replication protein factors with DNA, justifying their role as potential origins of replication and transcriptional enhancers, is appealing. In this case one might expect that nucleosomes will be absent in active MAR/ ORI stretches interacting with transcription/replication factors in a model similar to that of nucleosome-free promoters in active genes (Archer *et al.*, 1991), but nucleosomes might be present in inactive MAR/ORI structures. Studies in β-thalassemias and transgenic mice suggest that LCRs known to act as MAR/boundary elements play a critical role in displacing the nucleosomes over the promoter of the genes in the β-globin complex; MAR/insulators when inserted between the LCR and the promoter might prevent the LCR from disrupting the nucleosomes (Chung *et al.*, 1993).

A number of studies have rather conclusively demonstrated that promoter regions in active genes need to be free of nucleosomes for optimal binding to transcription factors (Archer *et al.*, 1991) and that the presence of a nucleosome can preclude binding of TFs (Venter *et al.*, 1994). Active origins of replication in the few cases examined like that of SV40 were also found to be nucleosome free in transcription-poised minichromosomes (Varshavsky *et al.*, 1979; Saragosti *et al.*, 1980). One might expect that MARs in transcriptionally active genes but not silenced tissue-specific MAR structures ought to have nucleosome-free stretches especially in regions that interact with replication/transcription factors. However, a class of TFs can bind DNA when wrapped around histone octamers (Chao *et al.*, 1980; McPherson *et al.*, 1993).

A DNase I-hypersensitive site at -1730 of the human interferon-β gene transcription start site is flanked in the 5' side by a nucleosome-free domain (Bode *et al.*, 1986) and the entire region forms the 5' border of the 5' MAR of the locus (Bode and Maass, 1988; Mielke *et al.*, 1990).

Bendable DNA arising by low-level periodic signals along the DNA may in part determine the positioning of nucleosomes, which have a tendency to place such intrinsically curved motifs at certain locations within the core particle (Trifonov and Sussman, 1980; Trifonov, 1991; Thoma, 1992). There is a delicate balance between stability of nucleosomes and their functionality; the natural nucleosomes are rather metastable (i.e., they unfold easily by a small energy input facilitating replication and transcription) as opposed to nucleosomes formed on strong nucleosome positioning sequences (Shrader and Crothers, 1989) that can act like immovable objects (E. N. Trifonov, personal communication).

It is proposed that weak or poor nucleosome positioning sequences might have an advantage as MAR/ORIs: in this case transcription factors need not compete strongly with histone octamers for occupying a cognate site in a regulatory region and might be complexed readily with transcription factors. Nucleosome positioning might constitute one of the codes of DNA, overlapping with a number of other codes such as coding sequence, intrinsic DNA curvature, intron/exon junction, transcription factor sites, higher order chromatin folding, and transcription/replication initiation and termination (Trifonov, 1989). Programs for predicting positioned nucleosomes are available (Boulikas, 1994b). Theoretically predicted sites of positioned nucleosomes and factor binding sites on MARs (Fig. 10A) as well as *in vivo* footprinting data might determine possible sites where competition between transcription factors and core histone octamers for a defined sequence of MAR DNA might take place.

Are there nucleosomes integrated within the nuclear matrix structure? Lamins A and C interact nonelectrostatically with polynucleosomes (Yuan *et al.*, 1991; Glass *et al.*, 1993) mediating anchoring to the nuclear envelope (Foisner and Gerace, 1993); this suggests a close proximity of nucleosomes to the peripheral nuclear matrix. How do proximal nucleosomes interact with nuclear matrix components? Are there interactions between classical matrix proteins and matrix transcription factors with both histones and nucleosomal DNA? Are there stretches of MARs occupied by nucleosomes only in some cell types? Are inactive MAR/ORI sequences, i.e., those linked with inactive genes, structured into inactive nucleosome conformations and fall into the constitutive heterochromatin fraction in a cell type-specific mode? Answers to these questions should await further experimentation.

Q. Nuclear Matrix Recognizes Supercoiled DNA

Total nuclear matrix when mixed with plasmid DNA shows a distinct preference for supercoiled forms of plasmids (usually plasmids, as isolated from *E. coli*, have a negative torsional strain) (Tsutsui *et al.*, 1988). The presence of MAR inserts within the supercoiled plasmid does not appear to be required for binding to matrix proteins; however, recognition of the SAR/MAR inserts seems to be impeded in the majority of topoisomers displaying a lower degree of superhelical density over the fully supercoiled molecules as well as in the nicked form (Kay and Bode, 1994). The strong recognition of the supercoils by the matrix has been attributed to topoisomerase II; however, histone H1 and HMG 1,2 display a pronounced preference for supercoiled DNA (Boulikas, 1995e). Competition assays have shown that the sites of the matrix responsible for binding linear MARs or supercoiled DNA are distinct nonoverlapping entities (Kay and Bode, 1994).

During transcription a substantial amount of superhelical density is generated that is positive ahead of and negative behind the advancing transcription complex (Giaever and Wang, 1988). This preference of the matrix proteins for supercoiled DNA could partly explain the preferential attachment of active but not of inactive genes to the matrix or might be a consequence of the anchoring of active genes to the matrix.

R. MAR/ORIs May Interact with Single-Stranded DNA-Binding Proteins

Competition filter assays have shown a preference of mouse nuclear matrix proteins for single-stranded homopolymers with a higher preference for poly(dT) in addition to AT-rich DNA (Comings and Wallack, 1978); single-stranded DNA has been suggested to contribute to the binding of MARs (Gasser et al., 1989). Lamin A binds with high affinity to evolutionarily conserved features on the matrix; of the two types of interaction one involves single-stranded regions and the other a narrow minor groove of double-stranded DNA (Ludérus et al., 1994).

Scaffolding proteins have been shown to have a specificity for single-stranded DNA. Among the sites of the matrix responsible for the recognition of linear MARs there is a distinct class recognizing single-stranded regions; more than 60,000 sites for single-stranded DNA per nuclear equivalent have been estimated to be present in mouse cells using a titration/competition curve analysis. These sites, representing one-half of the total MAR-binding sites on total matrix proteins, were easily competed for by single-stranded DNA; the remaining half of matrix protein sites were thought to originate from proteins like ARBP, SAF-A, SP120, and SATBP which do not recognize single-stranded DNA and bind cooperatively (Kay and Bode, 1994).

An increasing number of studies show a role of single-stranded proteins in origin activation (Boulikas, 1995c). Replication protein A (RPA) is an essential auxiliary protein for both DNA polymerases α and δ (Tsurimoto and Stillman, 1989; Erdile et al., 1991). Yeast proteins that interact with the ARS consensus sequence bind the T-rich strand only (Hofmann and Gasser, 1991); two of these proteins, however, were later shown to be RNA-binding proteins unlikely to play a direct role in DNA replication (Cockell et al., 1994). Other candidate replication initiator proteins that bind only to single-stranded DNA are PUR (28 kDa) recognizing a ss purine-rich element in the c-myc replication zone (Bergemann and Johnson, 1992; Bergemann et al., 1992); MSSP-1 and 2 proteins (sequence specificity TCTCTTA) identical to Scr2 that complements defective cdc2 kinase in S. pombe (Takai et al., 1994); RIP60 (60 kDa) that binds to intrinsically curved ATT repeats in the DHFR ORI (Caddle et al., 1990a; Dailey et al., 1990); ssA-TIBF that recognizes the A-rich strand of the type I repeat sequence in the Tetrahymena rDNA ORI (Umthun et al., 1994); and several others (Yee et al., 1991; Kolluri et al., 1992; Wang et al., 1993; Jacquemin-Sablon et al., 1994). This class of proteins may interact with single-stranded DNA thus stabilizing stretches of the double helix into melted structures. A special class of such proteins might be those that interact with one strand of a potential cruciform in mammalian origins of replication (Xie and Boulikas, 1995).

A number of proteins identified by molecular cloning from the ability of their genes to complement defects in cells deficient in repair like some of the xeroderma pigmentosum complementation groups turned out to be single-stranded DNA binding proteins (SSBPs) and helicases. For example, the *S. cerevisiae RAD3* and the mammalian *ERCC2* and *ERCC3* genes code for helicases (Sung *et al.*, 1987; Drapkin *et al.*, 1994); *ERCC1* encodes for a ssDNA-binding protein (Sung *et al.*, 1987). These proteins might be attached to the matrix and might explain the anchoring of repairing patches of chromatin to the nuclear matrix (Mullenders *et al.*, 1983, 1988; McCready and Cook, 1984; Harless and Hewitt, 1987).

S. Narrow Minor Grooves in MAR/ORIs

A narrow minor groove favors interaction of some matrix proteins like histone H1, topoisomerase II (Adachi *et al.*, 1989; Käs *et al.*, 1989), lamin A (Ludérus *et al.*, 1994), and SATB1 (Dickinson *et al.*, 1992). Distamycin also has a distinct and high sequence preference by establishing H bonds and van der Waals interactions with the floor and the sides of narrower minor grooves in A tracts and effectively competes out H1 from MARs (Zhao *et al.*, 1993). Other matrix proteins that bind DNA without a stringent sequence specificity in a cooperative manner with a preference for AT-rich stretches may indeed fall into this class of proteins with a preference for a narrow minor groove. This rule applies only to the AT-rich class of MARs.

T. Polymorphic and Conserved Sequences in MAR/ORIs

The recognition of MAR DNA signals by matrix proteins from different species has prompted the hypothesis of an evolutionary conservation of MAR-protein interactions (Cockerill and Garrard, 1986b). For example, yeast scaffolds can recognize the intronic MAR of mouse Igκ gene (Cockerill and Garrard, 1986b) and *Drosophila* SARs (Amati *et al.*, 1990); the *Drosophila* histone gene MAR is recognized by mouse nuclear matrices (Cockerill and Garrard, 1986a) and rat liver nuclear matrices (Izaurralde *et al.*, 1988); the chicken lysozyme gene MAR is recognized by mouse cells in transgenic studies (McKnight *et al.*, 1992); tobacco cells can effectively use the yeast *ARS1* MAR element (Allen *et al.*, 1993) and the soybean lectin gene P1-SAR (Breyne *et al.*, 1992); and a pea nuclear scaffold exhibits a high binding affinity for both a petunia 1.65-kb SAR and the 2.2-kb human IFN-β SAR (Dietz *et al.*, 1994). Similar studies have shown recognition of soybean MAR sequences by *Drosophila* scaffolds (Breyne *et al.*, 1992). However, the human 3′ and 5′ MARs from the interferon-β locus were unable to insulate the luciferase reporter gene in differentiated tissues in transgenic mice, although efficiently shielding this gene from position effects only in preimplantation embryos (Thompson *et al.*, 1994).

ORI/MARs are proposed to be made up of highly conserved elements representing protein binding sites, surrounded by hypervariable sequence elements displaying variability not only across species but even among individuals of the same species. The evolutionarily conserved DNA sequences may include the inverted repeats provided they represent protein-binding sites.

The breakpoint of a chromosomal translocation occurs in the intronic κ gene MAR near four strong topoisomerase II cleavage sites; MARs may reside next to long interspersed repetitive elements within recombination junctions (Sperry *et al.*, 1989).

Transgenic mice carrying the κ immunoglobulin gene with a deletion spanning the intronic enhancer/MAR region displayed a drastic drop in the rate of somatic hypermutation in the flanking V region; the mutation frequency was 0.7×10^{-3} mutations per base pair which was the same for PCR-induced background mutations (Betz *et al.*, 1994). Thus MARs might increase the mutation rate of a flanking sequence in the special case of the immunoglobulin genes; somatic mutation rates of the intronic enhancer/ MAR in the κ or μ genes have not been studied in B lymphocytes; the boundaries of somatic mutations are well defined, extending from the promoter to 1 kb downstream from the V(D)J segment that defines the border of the MAR region (Both *et al.*, 1990; Lebécque and Gearhart, 1990). Hotspots of somatic hypermutation are also found in this region (Betz *et al.*, 1993a,b).

Z-DNA, cruciforms, triplex, and slippage structures proposed to be responsible for the variability in some MAR/ORI stretches are more likely to form in the active part of the genome due to torsional strain associated with gene activity (Giaever and Wang, 1988). One might expect a higher hypervariability in active over inactive MAR/ORI structures; this leads to the hypothesis that differences in hypervariability of a special class of MAR/ORI sequences may be found in a cell type-specific manner among individuals of the same species.

1. Proposed Mechanisms for the Vulnerability of MAR/ORIs to Mutagens

Several salient features of the MAR/ORI DNA lead to the idea that these regulatory regions may be more vulnerable to distinct classes of chemical carcinogens. These features are: (1) The AT richness of a class of MAR/ ORIs or the presence of AT tracts and DNA unwinding elements. (2) The enrichment of MAR/ORIs in potential cruciform structures. (3) The enrichment of MAR/ORIs in potential Z-DNA and triple-helical structures. (4) The presence of putative single-stranded DNA-binding proteins in the matrix. ORIs bind DNA helicases and other single strand-specific proteins, and thus MAR/ORI DNA is expected to have more single-stranded stretches of DNA than bulk sequences. (5) The presence of homooligonucleotide and other direct repeats which might form slippage structures exposing single-stranded regions. (6) The presence of curved DNA at the origin/MAR rendering the flanking regions prone to small reactive molecules. (7) The presence of TF sites including replication initiator protein and homeodomain protein sites in MAR/ORIs; the binding of DNA sequence-specific transcription factors to their cognate DNA causes a local distortion to the DNA that is usually detected as a DNase I-hypersensitive site. These features of the nuclear matrix DNA have one property in common—their tendency to expose the bases either by distorting the regular B-form of the DNA double helix or by opening up the two strands of DNA. (8) Finally, a higher mutation rate of MAR over loop DNA could arise from an enrichment of recombinase-binding consensus sequences in MAR/ORIs or from an association of recombinase molecules with the nuclear matrix.

2. Evidence That MAR/ORI DNA Is Prone to Mutagens and Refractory to Repair

The DNA of the nuclear matrix, as site of replication and transcription, has been shown to be damaged preferentially by ionizing radiation and certain alkylating agents (reviewed by Fernandes and Catapano, 1991; Boulikas 1992b). The studies of Obi and co-workers (1986) and Obi and Billett (1991) show an increased vulnerability of nuclear matrix DNA to benzo[a]-pyrene. Nuclear matrix isolated by low and high salt extraction of MNase-digested nuclei in the presence of Mg^{2+} (12% of total DNA) was found to be 3.4-fold enriched in chromium–DNA adducts and DNA–protein crosslinks induced by treatment of CHO cells with carcinogenic chromium (Xu et al., 1994). The frequency of protein–DNA crosslinks after γ-radiation of Chinese hamster cells was 10- to 16-fold higher in the matrix (5% of DNA) than loop DNA consistent with both a preferential formation of crosslinks on the matrix and attachment of γ-radiation-induced lesions to the matrix (Chiu et al., 1986).

Origins of replication also display an enhanced vulnerability to MNNG–induced amplification (Yalkinoglu et al., 1991) and replicating chromating is 3.3 times more prone than mature chromatin DNA to γ-radiation–induced damage (Warters and Childers, 1982). Hypervariability has been shown for the replication origin of mitochondria, herpes simplex virus type 1, and the E. coli (see Boulikas, 1992a, for references).

Examination of the unusual MAR to the 3' end of the human apolipopro-tein B locus which is composed of a 30-bp repeating unit composed almost exclusively of AT (Fig. 11A) has shown several distinct polymorphic forms differing in the number of repeats among humans (Boerwinkle *et al.*, 1989); although all MARs are not ORIs it might be that an ORI includes or flanks this hypervariable region.

The studies showing drastic differences in two-dimensional protein pat-terns in nuclear matrix proteins between normal and tumor cell lines or tissues (Bidwell *et al.*, 1994; Keesee *et al.*, 1994) give further support to an involvement of nuclear matrix in carcinogenesis. A significant difference in the accessibility of DNA loops in nucleoids to DNase I solubilization between a radiosensitive and a radioresistant human cell line was thought to arise from a looser association of DNA with the nuclear matrix in radiosensitive cells (Milner *et al.*, 1993).

The unpaired regions induced in the loop of a cruciform structure, in the loop of the triple helical DNA, or in the single-stranded regions completely expose the aromatic bases and render them vulnerable to chemicals and to S1 nuclease (Naylor *et al.*, 1986; van Belkum *et al.*, 1990). Also the altered stacking of bases in Z-DNA and triple helices renders them more vulnerable to chemicals (Johnston and Rich, 1985; Johnston, 1988). Furthermore, the bases of the DNA at the B–Z junctions (Johnston and Rich, 1985; Nejedly *et al.*, 1985; Kohwi, 1989) and in Z–Z junctions (Johnston *et al.*, 1991) are much more vulnerable to chemicals. Although the AT-rich DNA is attacked by rather few chemicals, such as distamycin, the G-rich and purine-rich stretches, which are proposed to flank the AT-rich motifs in composite nuclear matrix anchorage sites (Boulikas, 1993a,b), are prone to ramifica-tions by a diverse class of agents including DNA methylating agents (such as MNNG, *N*-methyl-*N*-nitrosourea, and dimethyl sulfate) as well as aflatoxin, benzopyrene-7,8-diol-9,10-epoxide, *cis*-diaminedichloroplatinum(II), and others (Boulikas, 1992a).

Z-DNA is refractory to repair enzymes that have been evolved to recog-nize B-DNA as a substrate (Boiteux *et al.*, 1985). Although B- and Z-DNA have the same mean sensitivity toward radiolytic attack (generating hydroxy radicals, OH·, by radiolysis of water) guanine sites are more sensitive and cytosine sites less sensitive in Z- than in B-DNA giving to Z-DNA a radiolytic signature (Tartier *et al.*, 1994). Based on these observations, it might be predicted that potential Z-DNA and potential triple-helical stretches (also called H-DNA) and by consequence a fraction of the nuclear matrix DNA might display a higher mutation rate than the genome overall (Boulikas, 1992b). Thus a class of MARs might be expected to harbor more "initiated" sites in the two-step, initiation and promotion, chemical carcinogenesis model. The nature of the initiating event depends upon the nature of the carcinogen and affects the probability of progression to malignancy. A similar model has proposed that tumor cell heterogeneity originates from an unstable nuclear matrix (Pienta and Ward, 1994).

3. Evolutionary Consequences

Damage and repair are two opposing forces in molecular evolution largely responsible for the sculpturing of today's genomes (Boulikas, 1992a). Since damage has been inferred to be more frequently occurring on the nuclear matrix and damaged DNA in non-B form (triplex, Z-, cruciform) might be repaired at low rates, one can indeed predict a higher mutation rate in MAR over loop DNA. Since, however, the repair process is associated with the nuclear matrix one expects to find some evolutionarily conserved MAR motifs.

U. Level of DNA Methylation in MAR/ORIs

Specific methylation in promoter, enhancer, and other regulatory regions as well as in coding regions of inactive but not of active genes in higher eukaryotes is a main control mechanism for transcriptional inactivation. By consequence, one would expect DNA methylation at CpG to be higher in developmentally inactivated MAR/ORIs. Indeed, a study on the intronic MAR of the human procolagen $\alpha 2(I)$ gene shows one fully methylated site in white blood cells (gene is inactive) but partially methylated in fibroblasts where the gene is active (Ellis *et al.*, 1988). CpG islands seem to be excluded from MARs since a number of X-linked genes containing CpG islands (*HPRT, G6PD*) were found not to attach to the nuclear matrix in halo preparations; this was in contrast to X-linked genes, which lack CpG islands (ornithine transcarbamylase, factor IX), that were associated with the nuclear matrix in human lymphoblast cultures (Beggs and Migeon, 1989). CpG clusters, usually occurring in the 5' region of X-linked genes, are methylated in inactive but not active X chromosomes.

Alkylation of matrix DNA *in vitro* (1 alkylation every 100 bp) resulted in 60% reduction in its interaction with matrix proteins (Tew *et al.*, 1983). A study on DNA methylation in CHO cells, however, using PCR and a genomic sequencing protocol able to pinpoint all 5-methylcytosines, has shown that a defined 127-bp region within the ORI of the *RPS14* locus is densely methylated at almost all C residues, including CpG, CpA, CpT, and CpC dinucleotides, only during replication; similar data were found for a 516-bp region within the ORI 17 kb to the 3' side of the *DHFR* gene locus (Tasheva and Roufa, 1994). Such densely methylated islands could: (1) mediate the association of ORIs to the nuclear matrix, (2) function in licensing particular ORIs to inactive states during execution of specific developmental programs, or (3) be used by cells as signposts to mark previously replicated ORIs (Tasheva and Roufa, 1994). The association of the *RPS14* gene ORI to the nuclear matrix has not been examined.

MARs can affect the level of DNA methylation in neigboring sequences independently of the initiation of transcription; the intronic MAR/ enhancer in the mouse κ immunoglobulin gene is essential for the demethylation of the V_κ loop domain and for increasing its accessibility to recombinases; the κ gene is highly methylated in the postimplantation embryo and in most somatic cells but not in mature B lymphocytes where it undergoes differentiation-specific demethylation; transfection of B cell cultures with the κ gene methylated *in vitro* by *Hpa*II and *Hha*I methylases resulted in specific demethylation in a tissue-specific manner (no demethylation took place in fibroblasts); deletion of the MAR/ enhancer region from the κ construct failed to induce demethylation in B cells (Lichtenstein *et al.*, 1994).

V. Perspectives for MAR/ORI Prognosis

We hope that future studies will be directed at screening gene sequences from databases for inverted repeats, transcription factor binding sites, homeodomain protein sites, replication initiator protein sites, curved motifs, DUEs, Z- and triplex-DNA, Topo II sites, and nucleosome positioning in an effort to predict putative positions of MARs, and by consequence, positions of transcriptional enhancers and origins of replication. Such theoretical data could then be correlated with future and presently available experimental work defining origins of replication, DNase I-HS sites at the chromatin level of the various genes, polymorphic motifs, level of DNA methylation among tissues, protein footprinting studies, and enhancer elements.

The results of combined computer searches for these elements might be of important diagnostic power for MAR/ORI/enhancer/promoter elements. Such efforts might have an important impact on our understanding of the function of stretches of MARs as transcriptional enhancers and origins of replication or of an evolutionary conservation of a cohabitation mechanism between MARs and important control elements. In the latter case the AT-rich MAR with its accessory proteins might aid the cooperative binding between proteins from distantly located enhancers (Fig. 3).

It will be interesting to correlate the cellular function of a gene with the type of transcription factor sites theoretically predicted to be present on its MAR sequence. Such data can further be refined taking into account the possible involvement of a MAR at the origin of replication and/or transcriptional enhancer for the gene with which it is associated.

IX. Other Features of MAR/ORI DNA

A. Sequence Composition of MAR/ORIs

MARs may be classified into different types with respect to their DNA sequence characteristics:

1. AT-rich MAR/ORIs. An AT-rich class of ORIs is known (Boulikas 1995a). Historically, the first identification of MAR sequences (two regions, each 200 bp long, identified between the H1 and H3 histone genes, responsible for the anchorage of the histone gene cluster to the matrix) has shown them to be AT-rich (Mirkovitch *et al.*, 1984; Gasser and Laemmli, 1986b). Other MARs and SARs have been characterized as specific AT-rich DNA sequences (Blasquez *et al.*, 1989b; Levy-Wilson and Fortier, 1989; see Fig. 1) usually harboring topoisomerase II cleavage sites in their vicinity (Cockerill and Garrard, 1986a; Gasser and Laemmli, 1986b; Cockerill *et al.*, 1987; Käs and Chasin, 1987; Farache *et al.*, 1990b; Razin *et al.*, 1991a). A special class of AT-rich MAR harboring binding sites for SATB1 may contain some cytosines (Dickinson *et al.*, 1992; Nakagomi *et al.*, 1994; Cunningham *et al.*, 1994). In addition, AT tracts rather than entire AT-rich stretches may be present in the other sequence classes of MAR/ORIs. We define here composite MAR elements (Fig. 3) as those containing one or two AT-rich blocks (300–500 bp each) at the vicinity of a core ORI or enhancer.

2. Polypurine stretches on MAR/ORIs. This type of MAR has been found within the MAR/ORI/enhancer of the human choline acetyltransferase (*CHAT*) gene (Boulikas *et al.*, 1995), in some stretches of the *RPS14* ORI, in the human c-*myc* ORI, within the ORI of the mouse *ADA* locus (Virta-Pearlman *et al.*, 1993), and within the ORI/promoter of the rat aldolase B gene (Zhao *et al.*, 1994). Our MAR libraries from different mammalian species (T. Boulikas, C. F. Kong, and L. Xie, unpublished) also show a distinct class of MARs containing polypurine blocks of variable sizes. TFs that specifically interact with GA-rich (or TC-rich) DNA have been identified (Boulikas, 1994a); distinct proteins may recognize single strands of GA-rich blocks (Aharoni *et al.*, 1993), but neither the interaction of these motifs with nuclear matrices nor the subnuclear distribution of the proteins involved have been examined.

3. Alternations of polypurine and polypyrimidine tracts. A class of MAR sequences may contain short (5–50 nt) repeats of CT-rich stretches alternating with 5–50 nucleotide-long repeats of GA-rich stretches (Boulikas and Kong, 1993a,b). This type of motif increases the likelihood of occurrence of factor sites whose recognition sequence is composed of a purine-rich and a pyrimidine-rich moiety; these factors form a distinct characteristic group in the constellation of vertebrate transcription factors (Boulikas, 1994a).

4. One example of a GC-rich type of MAR has been identified in the promoter region of ribosomal genes (Marilley and Gassend-Bonnet, 1989). A similar region in the 18S/28S ribosomal repeat of the mouse is an origin of replication (Wegner *et al.,* 1989). GC-rich tracts occur in the origins of SV40, HSV-1 ori$_S$, HSV-1 ori$_L$, and ori$_L$ of pseudo-rabies virus (reviewed by Boulikas, 1995a).

5. A class of MAR/ORIs may be enriched in TG/CA motifs. Such motifs, like TGTTTTG, TGTTTTTTG, and TTTTGGGG, have been identified in abundance in the MAR that includes the 3' untranslated region of the poly(ADP-ribose) polymerase gene; they might arise from 3' untranslated regions of genes and recombinogenic regions of the genome (Boulikas and Kong, 1995).

6. The mixed motif type of MARs are the most abundant in our MAR sequence database; these frequently include AT tracts of 10–30 bp, GA and CT blocks, as well as inverted repeats. However, they cannot be classified to any of the above sequence groups from their overall sequence composition.

B. Genomic Position of MAR/ORIs

Most origins in multicellular organisms fall within upstream or downstream flanking regions of the gene they are linked with. For example the ORI of the human c-*myc* gene is at about −2 kb from the transcription initiation site, the ORI of the Chinese hamster *DHFR* gene is 17 kb to the 3' site, and that of the rDNA gene is in the spacer region between two genes in the gene repeat (Boulikas, 1995c).

However, as more studies are being completed, a more diverse picture with respect to the genomic position of the origins is emerging: the ORI of the S14 ribosomal protein gene in Chinese hamster cells comprises part of the coding and intronic regions (Tasheva and Roufa, 1994), and that of the human *HSP70* gene (−500 to −150) includes part of the promoter region (Taira *et al.,* 1994). The ORI of the rat aldolase B gene of 1 kb in size includes the promoter region (Zhao *et al.,* 1994). The ORI of the human choline acetyltransferase gene includes the entire 5' untranslated region as well as upstream flanking regions (Boulikas *et al.,* 1995). Initiation of DNA replication within a transcribed region has been found by Wu and co-workers (1993).

MARs also differ with respect to their location to coding regions. For example, MARs are found in a 200-bp region between the histone H1 and H3 genes of the histone gene complex in *Drosophila* (Mirkovitch *et al.,* 1984), within the intron of the immunoglobulin κ light chain gene in the mouse (Cockerill and Garrard, 1986a,b), within the first intron of the human *HPRT* gene (Sykes *et al.,* 1988), in intron 13 of the human topoisomerase I gene (Romig *et al.,* 1992), in the far upstream (at a distance of 8.85 kb) and downstream (1.3 kb) positions in the chicken lysozyme gene (Phi-Van and Strätling, 1988), and in the coding (second exon) region of the avian malic enzyme gene (Brotherton *et al.,* 1991) (Table I).

Several codes overlap on the DNA including the nucleotide triplex of coding regions, the factor sites, nucleosome positioning sequences, and exon/intron splice junctions (Trifonov, 1989). The cell has to accommodate these overlapping codes on the same stretch of DNA and to satisfy several structural/sequence requirements for each one including a mechanism for evolutionary conservation. MAR/ORIs also found in coding and in transcribing regions might constitute a novel overlapping code in eukaryotic genomes.

C. Estimates of MAR and ORI Sequences Per Nucleus

From the size of replicons, 20,000–35,000 origins of replication in mammalian cells can be estimated (Martin and Oppenheim, 1977). What fraction of these origins of replication coincide with MARs? About 60,000 total constitutive and developmentally facultative MARs per mammalian haploid genome can be estimated from the average size of 60 kb for a chromatin loop (Paulson and Laemmli, 1977; Hancock and Hughes, 1982; Table III). What fraction of the 60,000 or so MARs are enhancers (or repressors), ORIs, promoters, or structural MARs? The answer to these questions should await the characterization of a more representative sample of MARs from the various eukaryotic genomes.

Our estimate of MAR sequences is about 60,000 permanent MARs per nucleus, which together with the facultative class of MARs may amount to a total of 100,000 or more MARs per mammalian nucleus. This gives an average spacing of one MAR every 30,000 bp of DNA. The fraction of composite MAR elements among MARs is not known. Among the 30,000 ORIs a significant fraction may be composite MARs (MAR/ORIs). At least about 60,000 enhancers might be present in the mammalian genomes; the fraction of these that cohabit with MARs is not known. Two-thirds of composite MARs might be MAR/enhancers and we propose the presence of a smaller number of structural MARs. The promoters of active genes might constitute a facultative MAR/promoter fraction.

Kay and Bode (1994) have estimated about 40,000 binding sites for supercoiled DNA and about 60,000 sites for single-stranded DNA in mouse matrices per nuclear equivalent; competition/saturation binding studies using a particular MAR sequence of 800 bp derived from the upstream position of the human interferon-β locus and containing a DNA unwinding element has shown the presence of about 20,000 accessible binding sites per mouse nucleus (Kay and Bode, 1994).

TABLE III

Sizes of Replicons and Chromatin Loops[a]

Genes or loci/total genomic	Genomic DNA size	Replicon size	Chromatin loop size	Reference
Specific Genes				
Human β-globin	3.4 pg	200 kb	About 8 MARs over a 90-kb stretch	Kitsberg et al. (1993); Jarman and Higgs (1988)
Human α-globin		>140 kb		Jarman and Higgs (1988)
Human IFN-β			14kb	Bode and Maass (1988)
Human Apo B			47.5 kb	Levy-Wilson and Fortier (1989)
Human Apo b (5′ flank)			1.3 kb (enhancer–promoter loop?)	Levy-Wilson and Fortier (1989)
Murine IgH	7 pg	200 kb	20, 30, 20, and >70 kb	Brown et al. (1987); Cockerill (1990)
Chicken lysozyme		Not known	19 kb	Phi-Van and Strätling (1988)
Chicken α-globin (π-, $α^D$, $α^A$ genes)			11.5 kb	Farache et al. (1990a)
Drosophila 163-kb locus (7F) in X chromosome	0.18 pg		Four loops of 75, 52, 15, 14 kb	Small and Vogelstein (1985)
Drosophila 800-kb locus in X chromosome		25 replicons (in vitro), 32 kb average	15–115 kb loops; weak SARs give 11 kb loops	Brun et al. (1990); Surdej et al. (1990)
Drosophila histone genes		Not known	5 kb	Mirkovitch et al. (1984)
Drosophila Sgs-4		Not known	4.7 kb	Mirkovitch et al. (1987)
Drosophila Adh		Not known	5, 3.2, and 4 kb	Mirkovitch et al. (1987)
Drosophila ftz		Not known	11.1 kb	Mirkovitch et al. (1987)
Drosophila 320-kb region (rosy and Ace loci)		Not known	> 78, 43, 112, 26, >62 kb	Mirkovitch et al. (1986, 1987)

360

Average (total genomic)

HeLa human	3.4 pg	5–200 kb (average 86 kb)	Jackson et al. (1990)
Mouse	7 pg	10–180 kb (average 53 kb)	Hancock and Hughes (1982)
Chinese hamster	4.2 pg	30 μm (90 kb)	6.1 μm (9.3 kb)	Buongiorno-Nardelli et al. (1982)
Mouse cultured cells	7 pg	45.4 μm (136 kb)	15 μm (45 kb)	Buongiorno-Nardelli et al. (1982)
Gallus domesticus	1.4 pg	63 μm (189 kb)	14.1 μm (42 kb)	Buongiorno-Nardelli et al. (1982)
Xenopus laevis	3.0 pg	57.5 μm (172 kb) (A6 cultured cells)	14 μm (42 kb) (kidney cultured cells)	Buongiorno-Nardelli et al. (1982)
Xenopus blastula		9 μm (27 kb)	Buongiorno-Nardelli et al. (1982)
Xenopus neurula		9–15 μm (27–45 kb)	Buongiorno-Nardelli et al. (1982)
Xenopus gastrula		9–15 μm (27–45 kb)	Buongiorno-Nardelli et al. (1982)
Xenopus adult blood		15 μm (45 kb)	Buongiorno-Nardelli et al. (1982)
Triturus cristatus liver primary culture	23.3 pg	188 μm (564 kb)	53.7 μm (161 kb)	Buongiorno-Nardelli et al. (1982)
Drosophila		85 kb (DNase I nicking)	Benyajati and Worcel (1976)
Drosophila melanogaster	0.18 pg	13.6 μm (41 kb)	6 μm (18 kb)	Buongiorno-Nardelli et al. (1982)
Physarum polycephalum		9 μm (27 kb)	2.5–3 μm (7.5–9 kb)	Buongiorno-Nardelli et al. (1982)
Pisum sativum	5.8 pg	54.7 μm (164 kb)	14.8 μm (44 kb)	Buongiorno-Nardelli et al. (1982)

[a] Chromatin loop size wherever given in μm (1 μm is about 3 kb) is the maximum fluorescence halo radius estimated by staining HP40-lysed cells on microscope slides with ethidium bromide in high salt buffer (Buongiorno-Nardelli et al., 1982).

Loop sizes from specific loci or average loop sizes measured in nuclei halos are given in Table III.

X. The Project of Random Cloning of MARs

We have undertaken a large-scale sequencing of MARs from the human genome representing ~1% of total DNA. We consider MARs as important control elements and the isolation of MAR DNA as a convenient method to "fish out" important regulatory sequences that function in the regulation of gene expression, in initiation of DNA replication, as well as in setting the borders between chromatin domains (Boulikas and Kong, 1993a,b). The strength of our approach relies in the ability of our method to exclude false functional MAR sequences from MAR libraries (T. Boulikas, unpublished). This avoids the cumbersome situation of cloning any genomic sequence that might adventitiously anchor to the matrix during transcription/replication elongation or during repair. This "fishing expedition" on genomic DNA fragments resistant to solubilization after digestion of chromatin loops with nucleases and associated with the nuclear matrix has revealed the origins of replication for poly(ADP-ribose) polymerase and choline acetyltransferase genes (Boulikas *et al.*, 1995).

Large-scale sequencing of MARs from individual chromosomes is an easily realizable project proposed to aid the Human Genome Project. Indeed, only about 5000 MARs are estimated on the average per human chromosome, each MAR having an average length of 500 bp or about 1 kb for composite MAR elements. On the basis of a large enough number of MAR sequences we expect to gain insights into the structuring of genomes into domains. Such results are expected to have notable implications with respect to our knowledge of the control of initiation of DNA replication and the nature of transcriptional enhancers.

MARs were proposed to sculpture the crossroads of the differential activation of origins (during development and S-phase) and of the control of gene expression and pattern formation in embryogenesis (Boulikas, 1992b). Thus, large-scale random cloning of MARs might advance our knowledge of the control of initiation of DNA replication and function of transcriptional enhancers as well as domain structuring of genomes during development. Furthermore, MAR clones displaying a strong origin of replication or enhancer activity might be useful as gene transfer and expression vehicles for transfection of mammalian cells, in transgenic experiments, or in providing tools for the episomal replication of foreign genes in mammalian cells.

Active genes become attached to the nuclear matrix for their transcription. These findings suggest that the matrix structure of the nucleus plays a profound role in development and cell memory; understanding the dynamics of the nuclear matrix and the molecular mechanisms that govern its morphological and compositional changes during transformation, carcinogenesis, differentiation, and dedifferentiation is imperative for a better look at human cancers and other diseases.

XI. Conclusions

Several conclusions can be reached on the functional significance of MARs:

1. The nuclear matrix appears to constitute a structural nuclear entity governing all nuclear functions. About 30% of composite MARs may harbor ORIs or simple MARs may cohabit with origins of replication of the mammalian genome.

2. About 60% of MARs might colocalize with transcriptional enhancers, possess transcriptional enhancer activity, or insulate genes from neighboring chromatin effects. The cooperative binding of classical matrix proteins may juxtapose control elements (enhancers, promoters, origins of replication) that may cohabit with composite MAR elements.

3. The nuclear matrix is involved in repair and recombination. Thus, nuclear matrix may act as a "sink" for all proteins that regulate replication, transcription, repair, and recombination because of its nucleotide composition and its three-dimensional configuration in the nucleus.

4. MAR sequences as representing the attachment points of chromatin loops to the nuclear matrix define the borders between chromatin domains, each harboring one gene or segment of a gene or groups of related genes. Even upon completion of the human genome project the information on loop domain borders can only arise by mapping MARs along the DNA; this information will be valuable for our understanding of the functional organization of the genome.

5. Intimate contacts between lamins that are attached to the nuclear envelope and peripheral MARs (Boulikas, 1986b; Ludérus et al., 1992, 1994) might participate in the three-dimensional chromatin architecture fixed to the nuclear periphery. Similar anchorage points in the nuclear interior might govern a fixed spatial topological organization of chromatin loops throughout the nuclear architecture (Agard and Sedat, 1983; Hochstrasser et al., 1986).

6. Significant changes in the nuclear matrix protein pattern during osteoblast differentiation suggested that the nuclear matrix might be involved in the developmental decision of cell type formation (Dworetzky et al., 1990), cell memory, and body plan formation in development (Boulikas, 1992c). It

can be speculated that nuclear matrix might be involved in the differential activation of origins of replication during development and substages of S-phase. MARs nest origins of replication, have transcriptional enhancer activity, and via their interaction with protein transcription factors may govern gene switch during development and tissue-specific gene expression.

7. Active genes become associated with the nuclear matrix in a cell type-specific manner that might reflect activation of the promoter by binding of transcription factors and its anchoring to the matrix next to enhancers. Such data are in agreement with earlier studies showing the association of pulse-labeled RNA with the nuclear matrix (e.g., Fakan and Puvion, 1980). Promoters of active genes were proposed to be anchored to the matrix in a facultative manner (Fig. 12).

8. The intronic MAR/enhancer of the immunoglobulin κ gene potentiates somatic hypermutation in a narrow window of B lymphocyte develop-

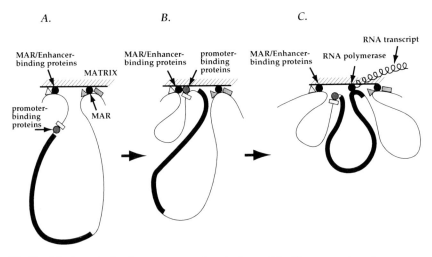

FIG. 12 Attachment of active gene promoters to the matrix. The expression of sequence-specific protein factors interacting with the immediate 5' regulatory region of genes, and the interaction of these proteins with nuclear matrix transcription factors permanently attached to the enhancer presumed to be the MAR distal element of the gene, results in the looping out of the DNA. The enhancer element in this model is either a permanent or facultative matrix attachment site of the chromatin domain, interacting with protein transcription factors. Its juxtaposition with the promoter, which is a facultative or a functional MAR activity detected only in active genes, results in the attachment of the immediate 5' regulatory region of the gene to the matrix enriched in RNA polymerases, DNA unwinding elements, and protein transcription factors. This results in high level expression of the gene. This model agrees with the looping out of DNA after physical contact between the enhancer–protein and the promoter–protein complexes proposed by others (Ptashne, 1988; Müller et al., 1989; Ptashne and Gann, 1990; Dunaway and Ostrander, 1993) and with reeling of the transcription unit through its matrix attachment site (Jackson et al., 1981). From Boulikas (1992d).

ment (Betz *et al.*, 1994), and promotes V_κ–J_κ joining (Takeda *et al.*, 1993) and demethylation in the V_κ loop (Lichtenstein *et al.*, 1994).

9. The nuclear matrix appears to bind preferentially supercoiled over relaxed DNA (Tsutsui *et al.*, 1988) and to actively recruit topoisomerase II (Earnshaw *et al.*, 1985). These observations together with the DNA unwinding properties of MARs (Kohwi-Shigematsu and Kohwi, 1990; Bode *et al.*, 1992) suggest that MARs may relieve superhelical strain generated during transcription by their unwinding; MARs may later release the super-coiling at other stretches of the loop (Bode *et al.*, 1992).

10. It might be possible to predict MAR elements from their DNA sequence. The power of the predictive rules proposed here will increase as more factor recognition sites become available. This knowledge is impor-tant for a better understanding of the function of enhancers and their varying location with respect to the coding region of genes. Of the 21 predictive rules proposed here to be common for origins of replication and MARs, rules 1–5 might be of higher predictive power and are proposed as universal properties of MAR/ORIs. The other rules might be confined to distinct classes of ORI/MAR sequences.

The results of combined computer searches for factor sites, positioned nucleosomes, maxima of intrinsic curvature, inverted repeats, DNA un-winding elements, potential triple-helical and left-handed stretches, topo-isomerase II consensus sites, and homooligonucleotide repeats might be of important diagnostic power for MAR/ORI/enhancer/promoter elements.

11. MARs/SARs can be used to enhance the level of expression of foreign genes introduced into animal or plant cells in both stably transfected cell lines or in transgenic animals and plants; two MARs, one on each side, are needed to flank the reporter gene. Transcription enhancement by MARs is not pronounced in transient transfection experiments.

12. MARs are thought to act by reducing position effect variegation from local chromatin structures at the integration site of foreign genes (Allen *et al.*, 1993; Huber *et al.*, 1994), by recruitment of topoisomerase II by the MAR sequence (Blasquez *et al.*, 1989a), by facilitating the displace-ment of histone H1 repressor from the chromatin loop and its replacement by HMG-I(Y) activator (Poljak *et al.*, 1994), by absorbing torsional strain by virtue of their DNA unwinding ability (Bode *et al.*, 1992), or by propagat-ing the enhancer-induced local alteration in chromatin structure to the promoter (Forrester *et al.*, 1994).

13. AT-rich MARs that may cohabit with core ORI/enhancers are pro-posed to facilitate the interaction of distantly spaced control elements by cooperative interactions of MAR-bound nuclear matrix proteins.

14. Mapping, isolation, and cloning of MAR sequences in a systematic way is proposed to lead to the isolation of a significant portion of the regulatory sequences from the eukaryotic genomes.

Recent studies have shown that large parts of 5′ and 3′ flanking regions of genes are essential for their correct developmental expression in transgenic experiments (Grosveld *et al.*, 1987; Hanna *et al.*, 1993; Bonifer *et al.*, 1994; Gourdon *et al.*, 1994; Huber *et al.*, 1994); the LCR of the β-globin gene complex is 50 kb away from the β-globin gene (Fig. 8); distant sequences required for correct developmental expression of genes may include MARs and other key sequences the exact nature of which remains largely obscure. Indeed, MAR sequences may fall in proximal and distal sites flanking the 5′ and 3′ ends of genes (chicken lysozyme), but also in introns (mouse *Igκ*, human *Topo I*, Chinese hamster *DHFR*), in the intergenic region between related genes (*Drosophila* histone gene complex, soybean lectin, and related genes), in a GC-rich promoter region (*Xenopus* rDNA), and in the 3′ and 5′ untranslated regions of genes (*pADP–Rpol* and *CHAT*) (Table 1). Prediction of MAR sequences along a stretch of a cloned gene might help our understanding of control mechanisms governing transcription and replication during development.

MARs seem to constitute novel regulatory elements of genes in addition to promoters, enhancers, silencers, locus control regions, and origins of replication. The timing of activation of ORI/MAR sequences during S-phase; the differential inactivation or activation of MARs, ORIs, enhancers, and promoters during development intimately connected to gene switch and body plan formation; and the mechanisms of repair, splicing, and recombination may cross on the nuclear matrix.

Acknowledgments

Special thanks to Emile Zuckerkandl, Maria Zannis-Hadjopoulos, Bernhard Hirt, Jerzy Jurka, Lyle Arnold, and Ed Trifonov for discussions or communications; and to present and past members of my group, notably Linda Xie, Geeta Kadambi, Dawn Brooks, C. F. Kong, John Costouros, and Ken Miura for their dedication and contributions to the MAR project.

References

Adachi, Y., Käs, E., and Laemmli, U. K. (1989). Preferential, cooperative binding of DNA toposiomerase II to scaffold-associated regions. *EMBO J.* **8**, 3997–4006.

Adachi, Y., Luke, M., and Laemmli, U. K. (1991). Chromosome assembly in vitro: Topoisomerase II is required for condensation. *Cell (Cambridge, Mass.)* **64**, 137–148.

Adom, J. N., and Richard-Foy, H. (1991). A region immediately adjacent to the origin of replication of bovine papilloma virus type 1 interacts in vitro with the nuclear matrix. *Biochem. Biophys. Res. Commun.* **176**, 479–485.

Aelen, J. M. A., Opstelten, R. J. G., and Walka, F. (1983). Organization of DNA replication in *Physarum polycephalum*. Attachement of origins of replicons and replication forks to the nuclear matrix. *Nucleic Acids Res.* **11**, 1181–1195.

Agard, D. A, and Sedat, J. W. (1983). Three-dimensional architecture of a polytene nucleus *Nature (London)* **302,** 676–681.

Aharoni, A., Baran, N., and Manor, H. (1993). Characterization of a multisubunit human protein which selectively binds single stranded d(GA)$_n$ and d(GT)$_n$ sequence repeats in DNA. *Nucleic Acids Res.* **21,** 5221–5228.

Allen, G. C., Hall, G. E., Jr., Childs, L. C., Weissinger, A. K., Spiker, S., and Thompson, W. F. (1993). Scaffold attachment regions increase reporter gene expression in stably transformed plant cells. *Plant Cell* **5,** 603–613.

Altieri, F., Maras, B., Eufemi, M., Ferraro, A., and Turano, C. (1993). Purification of a 57 kDa nuclear matrix protein associated with thiol:protein–disulfide oxidoreductase and phospholipase C activities. *Biochem. Biophys. Res. Commun.* **194,** 992–1000.

Amati, B. B., and Gasser, S. M. (1988). Chromosomal ARS and CEN elements bind specifically to the yeast nuclear scaffold. *Cell (Cambridge, Mass.)* **54,** 967–978.

Amati, B. B., and Gasser, S. M. (1990). *Drosophila* scaffold-attached regions bind nuclear scaffolds and can function as ARS elements in both budding and fission yeasts. *Mol. Cell. Biol.* **10,** 5442–5454.

Amati, B. B. Pick, L., Laroche., T., and Gasser, S. M. (1990). Nuclear scaffold attachment stimulates, but is not essential for ARS activity in *Saccharomyces cerevisiae:* Analysis of the *Drosophila ftz* SAR. *EMBRO J.* **9,** 4007–4016.

Anderson, J. N. (1986). Detection, sequence patterns and function of unusual DNA structures. *Nucleic Acids Res.* **14,** 8513–8533.

Archer, T. K., Cordingley, M. G., Wolford, R. G., and Hager, G. L. (1991). Transcription factor access is mediated by accurately positioned nucleosomes on the mouse mammary tumor virus promoter. *Mol. Cell. Biol.* **11,** 688–698.

Aronow, B. J., Ebert, C. A., Valerious, M. T., Potter, S. S., Wiginton, D. A., Witte, D. P., and Hutton, J. J. (1995). Dissecting a locus control region: Facilitation of enhancer function by extended enhancer-flanking sequences. *Mol. Cell. Biol.* **15,** 1123–1135.

Avramova, Z., and Bennetzen, J. L. (1993). Isolation of matrices from maize leaf nuclei: Identification of a matrix-binding site adjacent to the *Adh 1* gene. *Plant Mol. Biol.* **22,** 1135–1143.

Avramova, Z., and Paneva, E. (1992). Matrix attachment sites in the murine α-globin gene. *Biochem. Biophys. Res. Commun.* **182,** 78–85.

Bachs O., Lanini, L., Serratosa, J., Coll, M. J., Bastos, R., Aligue, R., Rius, E., and Carafoli, E. (1990). Calmodulin-binding proteins in the nuclei of quiescent and proliferatively activated rat liver cells. *J. Biol. Chem.* **265,** 18595–18600.

Banerji, J., Olson, L., and Schaffner, W. (1983). A lymphocyte-specific cellular enhancer is located downstream of the joining region in immunoglobulin heavy chain genes. *Cell (Cambridge, Mass.)* **33,** 729–740.

Barrack, E. R., and Coffey, D. S. (1980). The specific binding of estrogens and androgens to the nuclear matrix of sex hormone responsive tissues. *J. Biol. Chem.* **255,** 7265–7275.

Baru, M., Shlissel, M., and Manor, H. (1991). The yeast GAL4 protein transactivates the polyomavirus origin of DNA replication in mouse cells. *J. Virol.* **65,** 3496–3503.

Basler, J., Hastie, N. D., Pietras, D., Matsui, S. -I., Sandberg, A. A., and Berezney, R. (1981). Hybridization of nuclear matrix attached deoxyribonucleic acid fragments. *Biochemistry* **20,** 6921–6929.

Beggs, A. H., and Migeon, B. R. (1989). Chromatin loop structure of the human X chromosome: Relevance to X inactivation and CpG clusters. *Mol. Cell. Biol.* **9,** 2322–2331.

Belgrader, P., Dey, R., and Berezney, R. (1991). Molecular cloning of matrin 3. A 125-kiloDalton protein of the nuclear matrix contains an extensive acidic domain. *J. Biol. Chem.* **266,** 9893–9899.

Bell, S. P., and Stillman, B. (1992). ATP-dependent recognition of eukaryotic origins of DNA replication by a multiprotein complex. *Nature (London)* **357,** 128–134.

Bell, S. P., Kobayashi, R., and Stillman, B. (1993). Yeast origin recognition complex functions in transcription silencing and DNA replication. *Science* **262**, 1844–1849.

Benyajati, C., and Worcel, A. (1976). Isolation, characterization, and structure of the folded interphase genome of Drosophila melanogaster. *Cell (Cambridge, Mass.)* **9**, 393–407.

Berezney, R. (1991). The nuclear matrix: A heuristic model for investigating genomic organization and function in the cell nucleus. *J. Cell. Biochem.* **47**, 109–123.

Berezney, R., and Bucholtz, L. A. (1981). Dynamic association of replicating DNA fragments with the nuclear matrix of regenerating liver. *Exp. Cell Res.* **132**, 1–13.

Berezney, R., and Coffey, D. S. (1974). Identification of a nuclear protein matrix. *Biochem. Biophys. Res. Commun.* **60**, 1410–1417.

Berezney, R., and Coffey, D. S. (1975). Nuclear protein matrix: Association with newly synthesized DNA. *Science* **189**, 291–293.

Bergemann, A. D., and Johnson, E. M. (1992). The HeLa Pur factor binds single-stranded DNA at a specific element conserved in gene flanking regions and origins of DNA replication. *Mol. Cell. Biol.* **12**, 1257–1265.

Bergemann, A. D., Ma, Z.-W., and Johnson, E. M. (1992). Sequence of cDNA comprising the human *pur* gene and sequence-specific single-stranded-DNA-binding properties of the encoded protein. *Mol. Cell. Biol.* **12**, 5673–5682.

Berrios, M., Osheroff, N., and Fisher, P. A. (1985). *In situ* localization of DNA topoisomerase II, a major polypeptide component of the *Drosophila* nuclear matrix fraction. *Proc. Natl. Acad. Sci. U.S.A.* **82**, 4142–4146.

Betz, A. G., Neuberger, M. S., and Milstein, C. (1993a). Discriminating intrinsic and antigen-selected mutational hotspots in immunoglobulin V genes. *Immunol. Today* **14**, 405–411.

Betz, A. G., Rada, C., Pannell, R., Milstein, C., and Neuberger, M. S. (1993b). Passenger transgenes reveal intrinsic specificity of the antibody hypermutation mechanism: Clustering, polarity, and specific hot spots. *Proc. Natl. Acad. Sci. U.S.A.* **90**, 2385–2388.

Betz, A. G., Milstein, C., González-Fernández, A., Pannell, R., Larson, T., and Neuberger, M. S. (1994). Elements regulating somatic hypermutation of an immunoglobulin κ gene: Critical role for the intron enhancer/matrix attachment region. *Cell (Cambridge, Mass.)* **77**, 239–248.

Beven, A., Guan, Y., Peart, J., Cooper, C., and Shaw, P. (1991). Monoclonal antibodies to plant nuclear matrix reveal intermediate filament-related components within the nucleus. *J. Cell Sci.* **98**, 293–302.

Bianchi, M. E., Beltrame, M., and Paonessa, G. (1989). Specific recognition of cruciform DNA by nuclear protein HMG1. *Science* **243**, 1056–1059.

Bianchi, M. E., Falciola, L., Ferrari, S., and Lilley, D. M. J. (1992). The DNA binding site of HMG1 protein is composed of two similar segments (HMG boxes), both of which have counterparts in other eukaryotic regulatory proteins. *EMBO J.* **11**, 1055–1063.

Bidwell, J. P., van Wijnen, A. J., Fey, E. G., Dworetzky, S., Penman, S., Stein, J. L., Lian J. B., and Stein, G. S. (1993). Osteocalcin gene promoter-binding factors are tissue-specific nuclear matrix components. *Proc. Natl. Acad. Sci. U.S.A.* **90**, 3162–3166.

Bidwell J. P., Fey, E. G., van Wijnen, A. J., Pennman, S., Stein, J. L., Lian., J. B., and Stein, G. S. (1994). Nuclear matrix proteins distinguish normal diploid osteoblasts from osteosarcoma cells. *Cancer Res.* **54**, 28–32.

Björkroth, B., Ericsson, C., Lamb, M. M., and Daneholt, B. (1988). Structure of the chromatin axis during transcription. *Chromosoma* **96**, 333–340.

Blasquez, V. C., Xu, M., Moses, S. C., and Garrard, W. T. (1989a). Immunoglobulin κ gene expression after stable integration. I. Role of the intronic *MAR* and enhancer in plasmacytoma cells. *J. Biol. Chem.* **264**, 21183–21189.

Blasquez, V. C., Sperry, A. O., Cockerill, P. N., and Garrard, W. T. (1989b). Protein:DNA interactions at chromosomal loop attachement sites. *Genome* **31**, 503–509.

Blasquez, V. C., Hale, M. A., Trevorrow, K. W., and Garrard, W. T. (1992). Immunoglobulin κ gene enhancers synergistically activate gene expression but independently determine chromatin structure. *J. Biol. Chem.* **267**, 23888–23893.

Blencowe B. J., Nickerson, J. A., Issner, R., Penman, S., and Sharp, P. A. (1994). Association of nuclear matrix antigens with exon-containing splicing complexes. *J. Cell Biol.* **127**, 593–607.

Bode, J., and Maass, K. (1988). Chromatin domain surrounding the human interferon-β gene as defined by scaffold-attached regions. *Biochemistry* **27**, 4706–4711.

Bode, J., Pucher, H.-J., and Maass, K. (1986). Chromatin structure and induction-dependent conformational changes in human interferon-β genes in a mouse host cell. *Eur. J. Biochem.* **158**, 393–401.

Bode, J., Kohwi, Y., Dickinson, L., Joh, T., Klehr, D., Mielke, C., and Kohwi-Shigematsu, T. (1992). Biological significance of unwinding capability of nuclear matrix-associating DNAs. *Science* **255**, 195–197.

Bodnar, J. W. (1988). A domain model for eukaryotic DNA organization: A molecular basis for cell differentiation and chromosome evolution. *J. Theor. Biol.* **132**, 479–507.

Bodnar, J. W., Hanson, P. I., Polvino-Bodnar, M., Zempsky, W., and Ward, D. C. (1989). The terminal regions of adenovirus and minute virus of mice DNAs are preferentially associated with the nuclear matrix in infected cells. *J. Virol.* **63**, 4344–4353.

Boerwinkle, E., Xiong, W., Fourest, E., and Chan, L. (1989). Rapid typing of tandemly repeated hypervariable loci by the polymerase chain reaction: Application to the apolipoprotein B 3′ hypervariable region. *Proc. Natl. Acad. Sci U.S.A.* **86**, 212–216.

Bohr, V. A., Smith, C. A., Okumoto, D. S., and Hanawalt, P. C. (1985). DNA repair in an active gene: Removal of pyrimidine dimers from the DHFR gene of CHO cells is much more efficient than in the genome overall. *Cell (Cambridge, Mass.)* **40**, 359–369.

Boiteux, S., deOliveira, R. C., and Laval, J. (1985). The *Escherichia coli* O^6-methylguanine-DNA methyltransferase does not repair promutagenic O^6-methylguanine residues when present in Z-DNA. *J. Biol. Chem.* **260**, 8711–8715.

Bolshoy, A., McNamara, P., Harrington, R. E., and Trifonov, E. N. (1991). Curved DNA without A-A: Experimental estimation of all 16 DNA wedge angles. *Proc. Natl. Acad. Sci. U.S.A.* **88**, 2312–2316.

Bonifer, C., Hecht, A., Saueressig, H., Winter, D. M., and Sippel, A. E. (1991). Dynamic chromatin: The regulatory domain organization of eukaryotic gene loci. *J. Cell. Biochem.* **47**, 99–108.

Bonifer, C., Yannoustsos, N., Kruger, G., Grosveld, F., and Sippel, A. E. (1994). Dissection of the locus control function located on the chicken lysozyme gene domain in transgenic mice. *Nucleic Acids Res.* **22**, 4202–4210.

Bonven, B. J., Gocke, E., and Westergaard, O. (1985). A high affinity topoisomerase I binding sequence is clustered at DNAase I hypersensitive sites in Tetrahymena R-Chromatin. *Cell (Cambridge, Mass.)* **41**, 541–551.

Borowiec, J. A., and Hurwitz, J. (1988). Localized melting and structural changes in the SV40 origin of replication induced by T-antigen. *EMBO J.* **7**, 3149–3158.

Bossi, L., and Smith, D. M. (1984). Conformational change in the DNA associated with an unusual promoter mutation in a tRNA operon of Salmonella. *Cell (Cambridge, Mass.)* **39**, 643–652.

Both, G. W., Taylor, L., Pollard, J. W., and Steele, E. J. (1990). Distribution of mutations around rearranged heavy-chain antibody variable-region genes. *Mol. Cell. Biol.* **10**, 5187–5196.

Boulikas, T. (1986a). Nucleosomes are assembled into discrete size structures by histone H1 *in vitro*. *Biochem. Cell. Biol.* **64**, 463–473.

Boulikas, T. (1986b). Protein-protein and protein-DNA interactions in calf thymus nuclear matrix using cross-linking by ultraviolet irradiation. *Biochem. Cell. Biol.* **64**, 474–484.

Boulikas, T. (1987a). Nuclear envelope and chromatin structure. *Int. Rev. Cytol., Suppl.* **17**, 493–571.

Boulikas, T. (1987b). Functions of chromatin and the expression of genes. *Int. Rev. Cytol. Suppl.* **17,** 599–684.

Boulikas, T. (1988). At least sixty ADP-ribosylated histones are present in nuclei from dimethylsulfate-treated and untreated cells. *EMBO J.* **7,** 57–67.

Boulikas, T. (1992a). Evolutionary consequences of preferential damage and repair of chromatin domains. *J. Mol. Evol.* **35,** 156–180.

Boulikas, T. (1992b). Chromatin and nuclear matrix in development and in carcinogenesis: A theory. *Int. J. Oncol.* **1,** 357–372.

Boulikas, T. (1992c). Homeotic protein binding sites, origins of replication and nuclear matrix anchorage sites share the ATTA and ATTTA motifs. *J. Cell. Biochem.* **50,** 1–13.

Boulikas, T. (1992d). Poly(ADP-ribosyl)ation, repair, chromatin and cancer. *Curr. Perspect. Mol. Cell. Oncol.* **1,** 1–109.

Boulikas, T. (1993a). Nature of DNA sequences at the attachment regions of chromatin loops to the nuclear matrix. *J. Cell. Biochem.* **52,** 14–22.

Boulikas, T. (1993b). Homeodomain protein binding sites, inverted repeats, and nuclear matrix attachment regions along the human β-globin gene complex. *J. Cell. Biochem.* **52,** 23–36.

Boulikas, T. (1993c). Poly(ADP-ribosyl)ation, DNA strand breaks, chromatin and cancer. *Toxicol. Lett.* **67,** 129–150.

Boulikas, T. (1994a). A compilation and classification of DNA binding sites for protein transcription factors from vertebrates. *CRC Crit. Rev. Eukaryotic Gene Express.* **4,** 117–321.

Boulikas, T. (1994b). Transcription factor binding sites in the matrix attachment region of the chicken α-globin gene. *J. Cell. Biochem.* **55,** 513–529.

Boulikas, T. (1995a). Common structural features in origins of replication among all life forms *J. Cell. Biochem.* (in press).

Boulikas, T. (1995b). How enhancers work: Juxtapositioning of DNA control elements by synergistic interaction of MARs. *Int. J. Oncol.* **6,** 1313–1318.

Boulikas, T. (1995c). A model of transcription /replication enhancer. *Oncol. Rep.* **2,** 171–181.

Boulikas, T. (1995d). Phosphorylation of transcription factors and control of the cell cycle. *CRC Crit. Rev. Eukaryotic Gene Express.* **5,** 1–77.

Boulikas, T. (1995e). Interaction of histone H1 and HMG1,2 with supercoiled plasmids. In preparation.

Boulikas, T., and Kong, C. F. (1993a). Multitude of inverted repeats characterize a class of anchorage sites of chromatin loops to the nuclear matrix. *J. Cell. Biochem.* **53,** 1–12.

Boulikas, T., and Kong, C. F. (1993b). A novel class of matrix attached regions (MARs) identified by random cloning and their implications in differentiation and carcinogenesis. *Int. J. Oncol.* **2,** 325–330.

Boulikas, T., and Kong, C. F. (1995). The 3′ untranslated region of the human poly(ADP-ribose) polymerase gene is a matrix attached region. In preparation.

Boulikas, T., Wiseman, J. M., and Garrard, W. T. (1980). Points of contact between histone H1 and the histone octamer. *Proc. Natl. Acad. Sci. U.S.A.* **71,** 127–131.

Boulikas T., Xie, L., Kong, C. F., Todd, A., and Zannis-Hadjopoulos, M. (1995). The 5′ flanking region of the human choline acetyltransferase gene displays enhancer, nuclear matrix attachment, and bipartite origin of replication activities separated by an origin silencer. In preparation.

Breyne, P., Van Montagu, M., Depicker, A., and Gheysen, G. (1992). Characterization of a plant scaffold attachment region in a DNA fragment that normalizes transgene expression in tobacco. *Plant Cell* **4,** 463–471.

Brooks, A. R., Nagy, B. P., Taylor, S., Simonet, W. S., Taylor, J. M., and Levy-Wilson, B. (1994). Sequences containing the second-intron enhancer are essential for transcription of the human apolipoprotein B gene in the livers of transgenic mice. *Mol. Cell. Biol.* **14,** 2243—2256.

Brotherton, T., Zenk, D., Kahanic, S., and Reneker, J. (1991). Avian nuclear matrix proteins bind very tightly to cellular DNA of the β-globin gene enhancer in a tissue-specific fashion. *Biochemistry* **30,** 5845–5850.

Brown, E. H., Iqbal, M. A., Stuart, S., Hatton, K. S., Valinsky, J., and Schildkraut, C. L. (1987). Rate of replication of the murine immunoglobulin heady-chain locus: Evidence that the region is part of a single replicon. *Mol. Cell. Biol.* **7**, 450–457.

Brukner, I., Jurukovski, V., Konstantinović, M., and Savić, A. (1991). Curved DNA without AA/TT dinucleotide step. *Nucleic Acids Res.* **19**, 3549–3551.

Brun, C., Dang, Q., and Miassod, R. (1990). Studies of an 800-kilobase DNA stretch of the *Drosophila* X chromosome: Comapping of a subclass of scaffold-attached regions with sequences able to replicate autonomously in *Saccharomyces cerevisiae. Mol. Cell. Biol.* **10**, 5455–5463.

Buchman, A. R., Kimmerly, W. J., Rine, J., and Kornberg, R. D. (1988). Two DNA-binding factors recognize specific sequences at silencers, upstream activating sequences, autonomously replicating sequences, and telomeres in *Saccaromyces cerevisiae. Mol. Cell. Biol.* **8**, 210–225.

Buckler-White, A. J., Humphrey, G. W., and Pigiet, V. (1980). Association of polyoma T antigen and DNA with the nuclear matrix from lytically infected 3T6 cells. *Cell (Cambridge, Mass.)* **22**, 37–46.

Buongiorno-Nardelli, M., Micheli, G., Carrî, M. T., and Marilley, M. (1982). A relationship between replicon size and supercoiled loop domains in the eukaryotic genome. *Nature (London)* **298**, 100–102.

Caddle M. S., Dailey. L., and Heintz, N. H. (1990a). RIP60, a mammalian origin-binding protein, enhances DNA bending near the dihydrofolate reductase origin of replication. *Mol. Cell. Biol.* **10**, 6236–6243.

Caddle, M. S., Lussier, R. H., and Heintz, N. H. (1990b). Intramolecular DNA triplexes, bent DNA and DNA unwinding elements in the initiation region of an amplified dihydrofolate reductase replicon. *J. Mol. Biol.* **211**, 19–33.

Callan, H. G., (1974). DNA replication in the chromosomes of eukaryotes. *Cold Spring Harbor Symp. Quant. Biol.* **38**, 195–203.

Callan, H. G. (1982). Lampbrush chromosomes *Proc. R. Soc. London, Ser. B* **214**, 417–448.

Capco, D. G., Wan, K. M., and Penman. S. (1982). The nuclear matrix: Three-dimensional architecture and protein composition. *Cell (Cambridge, Mass.)* **29**, 847–858.

Carri, M. T., Micheli, G., Graziano, E., Pace, T., and Buongiorno-Nardelli, M. (1986). The relationship between chromosomal origins of replication and the nuclear matrix during the cell cycle. *Exp. Cell Res.* **164**, 426–436.

Chao, M. V., Gralla, J. D., and Martinson, H. G. (1980). lac operator nucleosomes, 1. Repressor binds specifically to operator within the nucleosome core. *Biochemistry* **19**, 3254–3260.

Chatterjee, P. K., and Flint, S. J. (1986). Partition of E1A proteins between soluble and structural fractions of adenovirus-infected and -transformed cells. *J. Virol.* **60**, 1018–1026.

Chimera, J. A., and Musich, P. R., (1985). The association of the interspersed repetitive *Kpn*I sequences with the nuclear matrix. *J. Biol. Chem.* **260**, 9373–9379.

Chiu, S.-M., Xue, L. -Y., Friedman, L. R., and Oleinick, N. L. (1993). Cooper ion-mediated sensitization of nuclear matrix attachment sites to ionizing radiation. *Biochemistry* **32**, 6214–6219.

Chiu, S.-M., Friedman., Sokany, N. M., Xue, L.-Y., and Oleinick, N. L. (1986). Nuclear matrix proteins are crosslinked to transcriptionally active gene sequences by ionizing radiation. *Radiat. Res.* **107**, 24–38.

Chou, R. H., Churchill, J. R., Mapstone, D. E., and Flubacher, M. M. (1991). Sequence-specific binding of a *c-myc* nuclear-matrix-associated region shows increased nuclear matrix retention after leukemic cell (HL-60) differentiation. *Am. J. Anat.* **191**, 312–320.

Christ, C., and Tye, B.-K. (1991). Functional domains of the yeast transcription/replication factor MCM1. *Genes Dev.* **5**, 751–763.

Chung, J. H., Whiteley, M., and Felsenfeld, G. (1993). A 5′ element of the chicken β-globin domain serves as an insulator in human erythroid cells and protects against position effect in Drosophila. *Cell (Cambridge, Mass.)* **74**, 505–514.

Chuprina, V. P., Fedoroff, O. Y., and Reid, B. R. (1991). New insights into the structure of A$_n$ tracts and B′-B′ bends in DNA. *Biochemistry* **30,** 561–568.

Ciejek, E. M., Nordström, J. L., Tsai, M.-J., and O'Malley, B. W. (1982). Ribonucleic acid precursors are associated with the chick oviduct nuclear matrix. *Biochemistry* **21,** 4945–4953.

Ciejek, E. M., Tsai, M.-J, and O'Malley, B. W. (1983). Actively transcribed genes are associated with the nuclear matrix. *Nature (London)* **306,** 607–609.

Classon, M., Henriksson, M., Sümegi, J., Klein, G., and Hammaskjöld, M. -L. (1987). Elevated *c-myc* expression facilitates the replication of SV40 DNA in human lymphoma cells. *Nature (London)* **330,** 272–274.

Cockell, M., Frutiger, S., Hughes, G. J., and Gasser, S. M. (1994). The yeast protein encoded by PUB1 binds T-rich single stranded DNA. *Nucleic Acids Res.* **22,** 32–40.

Cockerill, P. N. (1990). Nuclear matrix attachment occurs in several regions of the IgH locus. *Nucleic Acids Res.* **18,** 2643–2648.

Cockerill, P. N., and Garrard, W. T. (1986a). Chromosomal loop anchorage of the kappa immunoglobulin gene occurs next to the enhancer in a region containing topoisomerase II sites. *Cell (Cambridge, Mass.)* **44,** 273–282.

Cockerill, P. N., Garrard, W. T. (1986b). Chromosomal loop anchorage sites appear to be evolutionarily conserved. *FEBS Lett.* **204,** 5–7.

Cockerill, P. N., Yuen, M.-H and Garrard, W. T. (1987). The enhancer of the immunogloblin heavy chain locus is flanked by presumptive chromosomal loop anchorage elements. *J. Biol. Chem.* **262,** 5394–5397.

Collins, J. M. (1979). Transient structure of replicative DNA in normal and transformed human fibroblasts. *J. Biol. Chem.* **254,** 10167–10172.

Collins, J. M., Glock, M. S., and Chu, A. K. (1982). Nuclease S$_1$ sensitive sites in parental deoxyribonucleic acid of cold- and temperature-sensitive mammalian cells. *Biochemistry* **21,** 3414–3419.

Colvard, D. S., and Wilson, E. M. (1984). Androgen receptor binding to nuclear matrix in vitro and its inhibition by 8S androgen receptor promoting factor. *Biochemistry* **23,** 3479–3486.

Comings, D. E., and Wallack, A. S. (1978). DNA-binding properties of nuclear matrix proteins. *J. Cell Sci.* **34,** 233–246.

Cook, P. R. (1989). The nucleoskeleton and the topology of transcription. *Eur. J. Biochem.* **185,** 487–501.

Cook, P. R., and Brazell, I. A. (1975). Supercoils in human DNA. *J. Cell Sci.* **19,** 261–279.

Cook, P. R., Lang, J., Hayday, A., Lania, L., Fried, M., Chiswell, D. J., and Wyke, J. A. (1982). Active viral genes in transformed cells lie close to the nuclear cage. *EMBO J.* **1,** 447–452.

Croston, G. E., Kerrigan, L. A., Lira, L. M., Marshak, D. R., and Kadonaga, J. T. (1991). Sequence-specific antirepression of histone H1-mediated inhibition of basal RNA polymerase II transcription. *Science* **251,** 643–649.

Crothers, D. M., Haran, T. E., and Nadeau, J. G. (1990). Intrinsically bent DNA. *J. Biol. Chem.* **265,** 7093–7096.

Cunningham, J. M., Purucker, M. E., Jane, S. M., Safer, B., Vanin, E. F., Ney, P. A., Lowrey, C. H., and Nienhuis, A. W. (1994). The regulatory element 3′ to the $_A\gamma$-globin gene binds to the nuclear matrix and interacts with special A-T-rich binding protein 1 (SATB1), an SAR/MAR-associating region DNA binding protein. *Blood* **84,** 1298–1308.

Dailey, L., Caddle, M. S., Heintz, N., and Heinz, N. H. (1990). Purification of RIP60 and RIP100, mammalian proteins with origin-specific DNA-binding and ATP-dependent DNA helicase activities. *Mol. Cell. Biol.* **10,** 6225–6235.

Dean, F. B., and Hurwitz, J. (1991). Simian virus 40 large T antigen untwists DNA at the origin of DNA replication. *J. Biol. Chem.* **266,** 5062–5071.

Deb, S., DeLucia, A. L., Koff, A., Tsue, S., and Tegtmeyer, P. (1986). The adenine-thymine domain of the simian virus 40 core origin directs DNA bending and coordinately regulates DNA replication. *Mol. Cell. Biol.* **6,** 4578–4584.

de Jong, L., van Driel, R., Stuurman, N., Meijne, A. M. L., and van Renswoude, J. (1990). Principles of nuclear organization. *Cell Biol. Int. Rep.* **14**, 1051–1074.

de Lange, T. (1992). Human telomeres are attached to the nuclear matrix. *EMBO J.* **11**, 717–724.

Deppert, W., and von der Weth, A. (1990). Functional interaction of nuclear transport-defective simian virus 40 large T antigen with chromatin and nuclear matrix. *J. Virol.* **64**, 838–846.

Dickinson, L. A., Joh, T., Kohwi, Y., and Kohwi-Shigematsu T. (1992). A tissue-specific MAR/SAR DNA-binding protein with unusual binding site recognition. *Cell (Cambridge, Mass.)* **70**, 631–645.

Dietz, A., Kay, V., Schlake, T., Landsmann, J., and Bode, J. (1994). A plant scaffold attached region detected close to a T-DNA integration site is active in mammalian cells. *Nucleic Acids Res.* **22**, 2744–2751.

Diffley, J. F. X., and Cocker, J. H. (1992). Protein-DNA interactions at a yeast replication origin. *Nature (London)* **357**, 169–172.

Dijkwel, P. A., and Hamlin, J. L. (1988). Matrix attachment regions are positioned near replication initiation sites, genes, and an interamplicon junction in the amplified dihydrofolate reductase domain of Chinese hamster ovary cells. *Mol. Cell. Biol.* **8**, 5398–5409.

Dijkwel, P. A., Wenink, P. W., and Poddighe, J. (1986). Permanet attachment of replication origins to the nuclar matrix in BHK-cells. *Nucleic Acids Res.* **14**, 3241–3249.

Di Leonardo, A. D., Linke, S. P., Clarkin, K., and Wahl, G. M. (1994). DNA damage triggers a prolonged p53-dependent G_1 arrest and long-term induction of Cip 1 in normal human fibroblasts. *Genes Dev.* **8**, 2540–2551.

Drapkin, R., Reardon, J. T., Ansari, A., Huang, J.-C., Zawel, L., Ahn, K., Sancar, A., and Reinberg, D. (1994). Dual role of TFIIH in DNA excision repair and in transcription by RNA polymerase II. *Nature (London)* **368**, 769–772.

Dunaway, M., and Ostrander, E. A. (1993). Local domains of supercoiling activate a eukaryotic promoter in vivo. *Nature (London)* **361**, 746–748.

DuPraw, E. J. (1965). Macromolecular organization of nuclei and chromosomes: a folded fibre model based on whole-mount electron microscopy. *Nature (London)* **206**, 338–343.

Durfee, T., Mancini, M. A., Jones, D., Elledge, S. J., and Lee, W-H. (1994). The amino terminal region of the retinoblastoma gene product binds a novel nuclear matrix protein that co-localized to centers for RNA processing. *J. Cell Biol.* **127**, 609–622.

Dworetzky, S. I., Fey, E. G., Penman, S., Lian, J. B., Stein, J. L., and Stein, G. S. (1990). Progressive changes in the protein composition of the nuclear matrix during rat osteoblast differentiation. *Proc. Natl. Acad. Sci. U.S.A.* **87**, 4605–4609.

Dworetzky, S. I., Wright, K. L., Fey, E. G., Pennman, S., Lian, J. B., Stein, J. L., and Stein, G. S. (1992). Sequence-specfic DNA-binding proteins are components of a nuclear matrix-attachment site. *Proc. Natl. Acad. Sci. U.S.A.* **89**, 4178–4182.

Earnshaw, W. C., Halligan, B., Cooker, C. A., Heck, M. M. S, and Liu, L. F. (1985). Topoisomerase II is a structural component of mitotic chromosome scaffolds. *J. Cell Biol.* **100**, 1706–1715.

Eberharter, A. Grabher, A., Gstraunthaler, G., and Loidl, P. (1993). Nuclear matrix of the lower eukaryote *Physarum polycerphalum* and the mammalian epithelial *LLC-PK₁* cell line. A comprehensive investigation of different preparation procedures. *Eur. J. Biochem.* **212**, 573–580.

Eggert, H., and Jack, R. S. (1991). An ectopic copy of the *Drosophila ftz* associated SAR neither reorganizes local chromatin structure nor hinders elution of a chromatin fragment from isolated nuclei. *EMBO J.* **10**, 1237–1243.

Elias, P., Gustafsson, C. M., and Hammarsten, O. (1990). The origin binding protein of herpes simplex virus 1 binds cooperatively to the viral origin of replication ori$_s$. *J. Biol. Chem.* **265**, 17167–17173.

Elias, P., Gustafsson, C. M., Hammarsten, O., and Stow, N. D. (1992). Structural elements required for the cooperative binding of the herpes simplex virus origin binding protein to oriS reside in the N-terminal part of the protein. *J. Biol. Chem.* **267**, 17424–17429.

Ellis, G. C., Grobler-Rabie, A. F., Hough, F. S., and Bester, A. J. (1988). Location and methylation pattern of a nuclear matrix associated region in the human PROa2(I) collagen gene. *Biochem. Biophys. Res. Commun.* **157**, 500–506.

Ellison, J. R., and Howard, G. C. (1981). Non-random position of the satellite A-T rich DNA sequences within nuclei in early embryos of *Drosophila virilis. Chromosoma* **83**, 555–561.

Erdile, L. F., Heyers, W.-D., Kolodners, R., and Kelly, T. J. (1991). Characterization of cDNA encoding the 70-kDa single-stranded DNA-binding subunit of human replication protein A and the role of the protein in DNA replication. *J. Biol. Chem.* **266**, 12090–12098.

Fackelmayer, F. O., Dahm, K., Renz, A., Ramsperger, U., and Richter, A. (1994). Nucleic-acid-binding properties of hnRNP-U/SAF-A, a nuclear-matrix protein which binds DNA and RNA *in vivo* and *in vitro. Eur. J. Biochem.* **221**, 749–757.

Fakan, S., and Puvion, E. (1980). The ultrastructural visualization of nucleolar and extranucleolar RNA synthesis and distribution. *Int. Rev. Cytol.* **65**, 255–299.

Fakan, S., Leduc, Y., Lamarre, D., Brunet, G., and Poirier, G. G. (1988). Immunoelectron microscopical distribution of poly(ADP-ribose) polymerase in the mammalian cell nucleus. *Exp. Cell Res.* **179**, 517–526.

Farache, G., Razin, S. V., Targa, F. R., and Scherrer, K. (1990b). Organization of the 3'-boundary of the chicken α globin gene domain and characterization of CR 1-specific protein binding site. *Nucleic Acids Res.* **18**, 401–409.

Farache, G., Razin, S. V., Rzeszowska-Wolny, J., Moreau, J., Targa, F. R., and Scherrer, K. (1990a). Mapping of structural and transcription-related matrix attachment sites in the α-globin gene domain of avian erythroblasts and erythrocytes. *Mol. Cell. Biol.* **10**, 5349–5358.

Fernandes, D. J., and Catapano, C. V. (1991). Nuclear matrix targets for anticancer agents. *Cancer Cells* **3**, 134–140.

Ferraro, A., Grandi, P., Eufemi, M., Altieri, F., and Turano, C. (1992). Crosslinking of nuclear proteins to DNA by cis-diammminedichloroplatinum in intact cells. *FEBS Lett.* **307**, 383–385.

Feuerstein, N., Chan, P. K., and Mond, J. J. (1988). Identification of numatrin, the nuclear matrix protein associated with induction of mitogenesis, as the nucleolar protein B23. Implication for the role of the nucleolus in early transduction of mitogenic signals. *J. Biol. Chem.* **263**, 10608–10612.

Fey, E. G., and Penmann, S. (1988). Nuclear matrix proteins reflect cell type of origin in cultured human cells. *Proc. Natl. Acad. Sci. U.S.A.* **85**, 121–125.

Fey, E. G., Wan, K. M., and Penman, S. (1984). Epithelial cytoskeletal framework and nuclear matrix-intermediate filament scaffold: Three-dimensional organization and protein composition. *J. Cell Biol.* **98**, 1973–1984.

Fey, E. G., Krochmalnic, G., and Penman, S. (1986). The nonchromatin substructures of the nucleus: The ribonucleoprotein containing and RNP-deleted matrices analyzed by sequential fractionation and resinless section electron microscopy. *J. Cell Biol.* **102**, 1654–1665.

Fey, E. G., Bangs, P., Sparks, C. and Odgren, P. (1991). The nuclear matrix: Defining structural and functional roles. *CRC Crit. Rev. Eukaryotic Express.* **1**, 127–143.

Fishel, B. R., Sperry, A. O., and Garrad, W. T. (1993). Yeast calmodulin and a conserved nuclear protein participate in the *in vivo* binding of a matrix association region. *Proc. Natl. Acad. Sci. U.S.A.* **90**, 5623–5627.

Fisher, D. Z., Chaudhary, N., and Blobel, B. (1986). cDNA sequencing of nuclear lamins A and C reveals primary and secondary structural homology to intermediate filament proteins. *Proc. Natl. Acad. Sci. U.S.A.* **83**, 6450–6464.

Foisner, R., and Gerace, L. (1993). Integral membrane proteins of the nuclear envelope interact with lamins and chromosomes, and binding is modulated by mitotic phosphorylation. *Cell (Cambridge, Mass.)* **73**, 1267–1279.

Forrester, W. C., van Genderen, C., Jenuwein, T., and Grosschedl, R. (1994). Dependence of enhancer-mediated transcription of the immunoglobulin μ gene of nuclear matrix attachment regions. *Science* **265**, 1221–1225.

Frappier, L., and O'Donnell, M. (1992). EBNA1 distorts *oriP,* the Epstein-Barr virus latent replication origin. *J. Virol.* **66,** 1786–1790.

Frappier, L., and Zannis-Hadjopoulos, M. (1987). Autonomous replication of plasmids bearing monkey DNA origin-enriched sequences. *Proc. Natl. Acad. Sci. U.S.A.* **84,** 6668–6672.

Fredman, J. N., and Engler, J. A. (1993). Adenovirus precursor to terminal protein interacts with the nuclear matrix in vivo and in vitro. *J. Virol.* **67,** 3384–3395.

Garrard, W. T. (1990). Chromosomal loop organization in eukaryotic genomes. *Nucleic Acids Mol. Biol.* **4,** 163–175.

Gasser, S. M., and Laemmli, U. K. (1986a). Cohabitation of scaffold binding regions with upstream/enhancer elements of three developmentally regulated genes of *D. melanogaster. Cell (Cambridge, Mass.)* **46,** 521–530.

Gasser, S. M., and Laemmli, U. K. (1986b). The organization of chromatin loops: Characterization of a scaffold attachment site. *EMBO J.* **5,** 511–518.

Gasser, S. M., Amati, B. B., Cardenas, M. E., and Hofmann, J. F.-X. (1989). Studies on scaffold attachment sites and their relation to genome function. *Int. Rev. Cytol.* **119,** 57–96.

Georgiev, G. P., Vassetzky, Y. S., Jr., Luchinik, A. N., Chernokhvostov, V. V., and Razin, S. V. (1991). Nuclear skeleton, DNA domains and control of replication and transcription. *Eur. J. Biochem.* **200,** 613–624.

Getzenberg, R. H. (1994). Nuclear matrix and the regulation of gene expression: Tissue specificity. *J. Cell. Biochem.* **55,** 22–31.

Giaever, G. N., and Wang, J. C. (1988). Supercoiling of intracellular DNA can occur in eukaryotic cells. *Cell (Cambridge, Mass.)* **55,** 849–856.

Gillies, S. D., Morrison, S. L., Oi, V. T., and Tonegawa, S. (1983). A tissue-specific transcription enhancer element is located in the major intron of a rearranged immunoglobulin heavy chain gene. *Cell (Cambridge, Mass.)* **33,** 717–728.

Gilmour, D. S., and Elgin, S. C. R. (1987). Association of topoisomerase I with transcriptionally active loci in Drosophila. *NCI Monogr.* **4,** 17–21.

Gilmour, D. S., Pflugfelder, G., Wang, J. C., and Lis, J. T. (1986). Topoisomerase I interacts with transcribed regions in Drosophila cells. *Cell (Cambridge, Mass.)* **44,** 401–407.

Giusti, A. M., and Manser, T. (1993). Hypermutation is observed only in antibody H chain V region transgenes that have recombined with endogenous immunoglobulin H DNA: Implications for the location of cis-acting elements require for somatic mutation. *J. Exp. Med.* **177,** 797–809.

Glass, C. A., Glass, J. R., Taniura, H., Hasel, K. W., Blevitt, J. M., and Gerace, L. (1993). The α-helical rod domain of human lamins A and C contains a chromatin binding site. EMBO J. **12,** 4413–4424.

Goldman, M. A. (1988). The chromatin domain as a unit of gene regulation. *BioEssays* **9,** 50–55.

Gourdon, G., Sharpe, J. A., Wells, D., Wood, W. G., and Higgs, D. R. (1994). Analysis of a 70 kb segment of DNA containing the human ζ and α-globin genes linked to their regulatory element (HS-40) in transgenic mice. *Nucleic Acids Res.* **22,** 4139–4147.

Greenfield, I., Nickerson, J., Penman, S., and Stanley, M. (1991). Human papillomavirus 16 E7 protein is associated with the nuclear matrix. *Proc. Natl. Acad. Sci. U.S.A.* **88,** 11217–11221.

Greenstein, R. J. (1988). Constitutive attachment of murine erythroleukemia cell histone-depleted DNA loops to nuclear scaffolding is found in the β-major but not the α1-globin gene. *DNA* **7,** 601–607.

Gromova, II., Thomsen, B., and Razin, R. V. (1995). Different topoisomerase II antitumor drugs direct similar specific long-range fragmentation of an amplified c-*MYC* gene locus in living cells and in high salt-extracted nuclei. *Proc. Natl. Acad. Sci. U.S.A.* **92,** 102–106.

Grossbach, E. R., Björkroth, B., and Daneholt, B. (1990). Presence of histone H1 on an active Balbiani ring gene. *Cell (Cambridge, Mass.)* **60,** 78–83.

Grosveld, F., van Assendelft, G. B., Greaves, D. R., and Kollias, G. (1987). Position-independent, high-level expression of the human β-globin gene in transgenic mice. *Cell (Cambridge, Mass.)* **51,** 975–985.

Guo, Z.-S., and De Pamphilis, M. L. (1992). Specific transcription factors stimulate simian virus 40 and polyomavirus origins of DNA replication. *Mol. Cell. Biol.* **12,** 2514–2524.

Hakes, D. J., and Berezney, R. (1991). Molecular cloning of matrin F/G: A DNA binding protein of the nuclear matrix that contains putative zinc finger motifs. *Proc. Natl. Acad. Sci. U.S.A.* **88,** 6186–6190.

Hall, G., Jr., Allen, G. C., Loer, D. S., Thompson, W. F., and Spiker, S. (1991). Nuclear scaffolds and scaffold-attachment regions in higher plants. *Proc. Natl. Acad. Sci. U.S.A.* **88,** 9320–9324.

Hancock, R. (1982). Topological organisation of interphase DNA: The nuclear matrix and other skeletal strucures. *Biol. Cell* **46,** 105–122.

Hancock, R., and Boulikas, T. (1982). Functional organization in the nucleus. *Int. Rev. Cytol.* **79,** 165–214.

Hancock, R., and Hughes, M. E. (1982). Organisation of DNA in the interphase nucleus *Biol. Cell* **44,** 201–212.

Hanna, Z., Simard, C., Laperriere, A., and Jolicoeur, P. (1993). Specific expression of the human CD4 gene in mature CD4⁺ CD8⁻ and immature CD4⁺ CD4⁺ T cells and in macrophages of transgenic mice. *Mol. Cell. Biol.* **14,** 1084–1094.

Harless, J., and Hewitt, R. R. (1987). Intranuclear localization of UV-induced DNA repair in human VA13 cells. *Mutat. Res.* **183,** 177–184.

Hatfull, G. F., Noble, S. M., and Grindley, N. D. F. (1987). The γδ resolvase induces an unusual DNA structure at the recombinational crossover point. *Cell (Cambridge, Mass.)* **49,** 103–110.

Hay, R. T., and DePamphilis, M. L. (1982). Initiation of SV40 DNA replication in vivo: Location and structure of 5′ ends of DNA synthesized in the *ori* region. *Cell (Cambridge, Mass.)* **28,** 767–779.

He, D., Nickerson, J. A., and Penman, S. (1990). Core filaments of the nuclear matrix. *J. Cell Biol.* **110,** 569–580.

Hendrickson, E. A., Fritze, C. E., Folk, W. R., and DePamphilis, M. L. (1987). The origin of bidirectional DNA replication in polyoma virus. *EMBO J.* **6,** 2011–2018.

Hendzel, M. J., Delcuve, G. P., and Davie, J. R. (1991). Histone deacetylase is a component of the internal nuclear matrix. *J. Biol. Chem.* **266,** 21936–21942.

Hendzel, M. J., Sun, J. M., Chen, H. Y., Rattner, J. B., and Davie, J. R. (1994). Histone acetyltransferase is associated with the nuclear matrix. *J. Biol. Chem.* **269,** 22894–22901.

Herman, R., Weymouth, L., and Penman, S. (1978). Heterogeneous nuclear RNA-protein fibers in chromatin-depleted nuclei. *J. Cell Biol.* **78,** 663–674.

Hirano, T., Hiraoka, Y., and Yanakida, M. (1988). A temperature-sensitive mutation of the *Schizosaccharomyces pombe* gene *nuc2⁺* that encodes a nuclear-scaffold protein blocks spindle elongation in mitotic anaphase. *J. Cell Biol.* **106,** 1171–1183.

Hirano, T., Kinoshita, N., Morikawa, K., and Yanakida, M. (1990). Snap helix with knob and hole: Essential repeats in S. pombe nuclear protein nuc2⁺. *Cell (Cambridge, Mass.)* **60,** 319–328.

Hochstrasser, M., Mathog, D., Gruenbaum, Y., Saumweber, H., and Sedat, J. W. (1986). Spatial organization of chromosomes in the salivary gland nuclei of *Drosophila melanogaster. J. Cell Biol.* **102,** 112–123.

Hofmann, J. F.-X., and Gasser, S. M. (1991). Identification and purification of a protein that binds the yeast ARS consensus sequence. *Cell (Cambridge, Mass.)* **64,** 951–960.

Hofmann, J. F.-X., Laroche, T., Brand, A. H., and Gasser, S. M. (1989). RAP-1 factor is necessary for DNA loop formation in vitro at the silent mating type locus HML. *Cell (Cambridge, Mass.)* **57,** 725–737.

Homberger, H. P. (1989). Bent DNA is a structural feature of scaffold-attached regions in *Drosophila melanogaster* interphase nuclei. *Chromosoma* **98,** 99–104.

Hsieh, D.-J., Camiolo, S. M., and Yates, J. L. (1993). Constitutive binding of EBNA1 protein to the Epstein-Barr virus replication origin, oriP, with distortion of DNA structure during latent infection. *EMBO J.* **12**, 4933–4944.

Huber, M. C., Bosch, F. X., Sippel, A. E., and Bonifer, C. (1994). Chromosomal position effects in chicken lysozyme gene transgenic mice are correlated with suppression of DNase I hypersensitive site formation. *Nucleic Acids Res.* **22**, 4195–4201.

Iguchi-Ariga, S. M. M., Itani, T., Yamaguchi, M., and Ariga, H. (1987). c-myc protein can be substituted for SV40 T antigen in SV40 DNA replication. *Nucleic Acids Res.* **15**, 4889–4899.

Iguchi-Ariga, S. M. M., Okazaki, T., Itani, T., Ogata, M., Sato, Y., and Ariga, H. (1988). An initiation site of DNA replication with transcriptional enhancer activity present upstream of the c-myc gene. *EMBO J.* **7**, 3135–3142.

Imler, J. L., Lemaire, C., Wasylyk, C., and Wasylyk, B. (1987). Negative regulation contributes to tissue specificity of the immunoglobulin heavy-chain enhancer. *Mol. Cell. Biol.* **7**, 2558–2567.

Ioshikhes, I., Bolshoy, A., and Trifonov, E. N. (1992). Preferred positions of AA and TT dinucleotides in aligned nucleosomal DNA sequences. *J. Biomol. Struct. Dyn.* **9**, 1111–1117.

Isomura, T., Tamiya-Koizumi, K., Suzuki, M., Yoshida, S., Taniguchi, M., Matsuyama, M., Ishigaki, T., Sakuma, S., and Takahashi, M. (1992). RFP is a binding protein associated with the nuclear matrix. *Nucleic Acids Res.* **20**, 5305–5310.

Ito, T., and Sakaki, Y. (1987). Nuclear matrix association regions of rat α_2-macroglobulin gene. *Biochem. Biophys. Res. Commun.* **149**, 449–454.

Ivanchenko, M., and Avramova, Z. (1992). Interaction of MAR-sequences with nuclear matrix proteins. *J. Cell. Biochem.* **50**, 190–200.

Izaurralde, E., Mirkovitch, J., and Laemmli, U. K. (1988). Interaction of DNA with nuclear scaffolds *in vitro. J. Mol. Biol.* **200**, 111–125.

Izaurralde, E., Käs, E., and Laemmli, U. K. (1989). Highly preferential nucleation of histone H1 assembly on scaffold-associated regions. *J. Mol. Biol.* **210**, 573–585.

Jackson, D. A. (1991). Structure-function relationships in eukaryotic nuclei. *BioEssays* **13**, 1–10.

Jackson, D. A., McCready, S. J., and Cook, P. R. (1981). RNA is synthesized at the nuclear cage. *Nature (London)* **292**, 552–555.

Jackson, D. A., Dickinson, P., and Cook, P. R. (1990). The size of chromatin loops in HeLa cells. *EMBO J.* **9**, 567–571.

Jacquemin-Sablon, H., Triqueneaux, G., Deschamps, S., le Maire, M., Doniger, J., and Dautry, F. (1994). Nucleic acid binding and intracellular localization of unr, a protein with five cold shock domains. *Nucleic Acids Res.* **22**, 2643–2650.

Jarman, A. P., and Higgs, D. R. (1988). Nuclear scaffold attachment sites in the human globin gene complexes. *EMBO J.* **7**, 3337–3344.

Jenuwein, T., Forrester, W. O., Qiu, R.-G., and Grosschedl, R. (1993). The immunoglobulin μ enhancer core establishes local factor access in nuclear chromatin independent of transcriptional stimulation. *Genes Dev.* **7**, 2016–2032.

Jeppesen, P. G. N., and Bankier, A. T. (1979). A partial characterization of DNA fragments protected from nuclease degradation in histone-depleted metaphase chromosomes of the Chinese hamster. *Nucleic Acids Res.* **7**, 49–67.

Johnson, E. M., and Jelinek, W. R. (1986). Replication of a plasmid bearing a human Alu-family repeat in monkey COS-7 cells. *Proc. Natl. Acad. Sci. U.S.A.* **83**, 4660–4664.

Johnston, B. H. (1988). The S1-sensitive form of $d(C-T)_n \cdot d(A-G)_n$: Chemical evidence for a three-standard structure in plasmids. *Science* **241**, 1800–1804.

Johnston, B. H., and Rich, A. (1985). Chemical probes of DNA conformation: Detection of Z-DNA at nucleotide resolution. *Cell (Cambridge, Mass.)* **42**, 713–724.

Johnston, B. H., Quigley, G. J., Ellison, M. J., and Rich, A. (1991). The Z-Z Junction: The boundary between two out-of-phase Z-DNA regions. *Biochemistry* **30**, 5257–5263.

Jones, C., and Su, R. T. (1987). Association of viral and plasmid DNA with the nuclear matrix during productive infection. *Biochim. Biophys. Acta* **910**, 52–62.

Jost, J.-P., and Seldran, M. (1984). Association of transcriptionally active vitellogenin II gene with the nuclear matrix of chicken liver. *EMBO J.* **3**, 2005–2008.

Juan, L., Utley, R. T., Adams, C. C., Vettese-Dadey, M., and Workman, J. L. (1994). Differential repression of transcription factor binding by histone H1 is regulated by the core histone amino termini. *EMBO J.* **13**, 6031–6040.

Kamakaka, R. T., and Thomas, J. O. (1990). Chromatin structure of transcriptionally competent and repressed genes. *EMBO J.* **9**, 3997–4006.

Käs, E., and Chasin, L. A. (1987). Anchorage of the Chinese hamster dihydrofolate reductase gene to the nuclear scaffold occurs in an intragenic region. *J. Mol. Biol.* **198**, 677–692.

Käs, E., and Laemmli, U. K. (1992). *In vivo* topoisomerase II cleavage of the Drosophila histone and satellite III repeats: DNA sequence and structural characteristics. *EMBO J.* **11**, 705–716.

Käs, E., Izaurralde, E., and Laemmli, U. K. (1989). Specific inhibition of DNA binding to nuclear scaffolds and histone H1 by distamycin: The role of oligo(dA) · oligo(dT) tracts. *J. Mol. Biol.* **210**, 587–599.

Käs, E., Poljak, L. Adachi, Y., and Laemmli, U. K. (1993). A model for chromatin opening: Stimulation of topoisomerase II and restriction enzyme cleavage of chromatin by distamycin. *EMBO J.* **12**, 115–126.

Kaufmann, S. H., Brunet, G., Talbot, B., Lamarr, D., Dumas, C., Shaper, J. H., and Poirier, G. (1991). Association of poly(ADP-ribose) polymerase with the nuclear matrix: The role of intermolecular disulfide bond formation, RNA retention, and cell type. *Exp. Cell. Res.* **192**, 524–535.

Kay, V., and Bode, J. (1994). Binding-specificity of a nuclear scaffold: Supercoiled, single-stranded, and scaffold-attached-region DNA. *Biochemistry* **33**, 367–374.

Keesee, S. K., Meneghini, M. D., Szaro, R. P., and Wu, Y.-J. (1994). Nuclear matrix proteins in human colon cancer. *Proc. Natl. Acad. Sci. U.S.A.* **91**, 1913–1916.

Kellum, R., and Schedl, P. (1991). A position-effect assay for boundaries of higher order chromosomal domains. *Cell* **64**, 941–950.

Khanuja, P. S., Lehr, J. E., Soule, H. D., Gehani, S. K., Noto, A. C., Choudhury, S., Chen, R., and Pienta, K. J. (1993). Nuclear matrix proteins in normal and breast cancer cells. *Cancer Res.* **53**, 3394–3398.

Kim, J. L., Nikolov, D. B., and Burley, S. K. (1993). Co-crystal structure of TBP recognizing the minor groove of a TATA element. *Nature (London)* **365**, 520–527.

Kimmerly, W., Buchman, A., Kornberg, R., and Rine, J. (1988). Roles of two DNA-binding factors in replication, segregation and transcriptional repression mediated by a yeast silencer. *EMBO J.* **7**, 2241–2253.

Kitsberg, D., Selig, S., Keshet, I., and Cedar, H. (1993). Replication structure of the human β-globin gene domain. *Nature (London)* **366**, 588–590.

Klehr, D., Maass, K., and Bode, J. (1991). Scaffold-attached regions from the human interferon β domain can be used to enhance the stable expression of genes under the control of various promoters. *Biochemistry* **30**, 1264–1270.

Klehr, D., Schlake, T., Maass, K., and Bode, J. (1992). Scaffold-attached regions (SAR elements) mediate transcriptional effects due to butyrate. *Biochemistry* **31**, 3222–3229.

Klempnauer, K.-H. (1988). Interaction of myb proteins with the nuclear matrix in vitro. *Oncogene* **2**, 545–551.

Koepsel, R. R., and Khan, S. A. (1986). Statid and initiator protein-enhanced bending of DNA at a replication origin. *Science* **233**, 1316–1318.

Koff, A., Schwedes, J. F., and Tegtmeyer, P. (1991). Herpes simplex virus origin-binding protein (UL9) loops and distorts the viral replication origin. *J. Virol.* **65**, 3284–3292.

Kohwi, Y. (1989). Non-B-DNA structure: Preferential target for the chemical carcinogen glycidaldehyde. *Carcinogenesis (London)* **10**, 2035–2042.

Kohwi-Shigematsu, T., and Kohwi, Y. (1990). Torsional stress stabilizes extended base unpairing in suppressor sites flanking immunoglobulin heavy chain enhancer. *Biochemistry* **29**, 9551–9560.

Kolluri, R., Torrey, T. A., and Kinniburgh, A. J. (1992). A CT promoter element binding protein: Definition of a double-strand and a novel single-strand DNA-binding motif. *Nucleic Acids Res.* **20**, 111–116.

Koo, H.-S., Wu, H.-M., and Crothers, D. M. (1986). DNA bending at adenine · thymidine tracts. *Nature (London)* **320**, 501–506.

Kowalski, D., and Eddy, M. J. (1989). The DNA unwinding element: A novel, *cis*-acting component that facilitates opening of the *Escherichia coli* replication origin. *EMBO J.* **8**, 4335–4344.

Krogh, S., Mortensen, U. H., Westergaard, O., and Bonven, B. J. (1991). Eukaryotic topoisomerase I-DNA interaction is stabilized by helix curvature. *Nucleic Acids Res.* **19**, 1235–1241.

Kumara-Siri, M. H., Shapiro, L. E., and Surks, M. I. (1986). Association of the 3,5,3′-triiodo-L-thyronine nuclear receptor with the nuclear matrix of cultured growth hormone-producing rat pituitary tumor cells (GC cells). *J. Biol. Chem.* **261**, 2844–2852.

Kuo, M.-H., and Grayhack, E. (1994). A library of yeast genomic MCM1 binding sites contains genes involved in cell cycle control, cell wall and membrane structure, and metabolism. *Mol. Cell. Biol.* **14**, 348–359.

Laybourn, P. J., and Kadonaga, J. T. (1991). Role of nucleosomal cores and histone H1 in regulation of transcription by RNA polymerase II. *Science* **254**, 238–245.

Lebécque, S. G., and Gearhart, P. J. (1990). Boundaries of somatic mutation in rearranged immunoglobulin genes: 5′ boundary is near the promoter, and 3′ boundary is ~1kb from V(D)J gene. *J. Exp. Med.* **172**, 1717–1727.

Leibovitch, S. A., Leibovitch, M. P., Hillion, H., Krug, J., and Harel, J. (1983). A destabilized DNA conformation associated with tightly bound nuclear proteins in active genes of rat myoblast. *Nucleic Acids Res.* **11**, 4035–4047.

Levy-Wilson, B., and Fortier, C. (1989). The limits of the DNase I-sensitive domain of the human apolipoprotein B gene coincide with the locations of chromosomal anchorage loops and define the 5′ and 3′ boundaries of the gene. *J. Biol. Chem.* **264**, 21196–21204.

Lichtenstein, M., Keini, G., Cedar, H., and Bergman, Y. (1994). B cell-specific demethylation: A novel role for the intronic κ chain enhancer sequence. *Cell (Cambridge, Mass.)* **76**, 913–923.

Li, H., and Roux, S. J. (1992). Casein kinase II protein kinase is bound to lamina-matrix and phosphorylates lamin-like protein in isolated pea nuclei. *Proc. Natl. Acad. Sci. U.S.A.* **849**, 8434–8438.

Lin, L.-S., and Meyer, R. J. (1987). DNA synthesis is initiated at two positions within the origin of replication of plasmid R1162. *Nucleic Acids Res.* **15**, 8319–8331.

Lothstein, L., Arenstorf, H. P., Chung, S.-Y., Walker, B. W., Wooley, J. C., and LeStourgeon, W. M. (1985). General organization of protein in HeLa 40S nuclear ribonucleoprotein particles. *J. Cell Biol.* **100**, 1570–1581.

Ludérus, M. E. E., de Graaf, A., Mattia, E., den Blaauwen, J. L., Grande, M. A., de Jong, L., and van Driel, R. (1992). Binding of matrix attachment regions to lamin B₁. *Cell (Cambridge, Mass.)* **70**, 949–959.

Ludérus, M. E. E., den Blaauwen, J. L., de Smith, O. J. B., Compton, D. A., and van Driel, R. (1994). Binding of matrix attachment regions to lamin polymers involves single-stranded regions and the minor groove. *Mol. Cell. Biol.* **14**, 6297–6305.

Mancini, M. A., Shan, B., Nickerson, J. A., Penman, S., and Lee, W.-H. (1994). The retinoblastoma gene product is a cell cycle-dependent, nuclear matrix associated protein. *Proc. Natl. Acad. Sci. U.S.A.* **91**, 418–422.

Marilley, M., and Gassend-Bonnet, G. (1989). Supercoiled loop organization of genomic DNA: A close relationship between loop domains, expression units and replicon organization in rDNA in *Xenopus laevis*. *Exp. Cell Res.* **180**, 475–489.

Marsden, M. P. F., and Laemmli, U. K. (1979). Metaphase chromosome structure: Evidence for a radial loop model. *Cell* (*Cambridge, Mass.*) **17**, 849–858.

Martelli, A. M., Gilmour, R. S., Bareggi, R., and Cocco, L. (1992). The effect of in vitro heat exposure on the recovery of nuclear matrix-bound DNA polymerase α activity during the different phases of the cell cycle in synchronized HeLa S3 cells. *Exp. Cell Res.* **201**, 470–476.

Martin, R. G., and Oppenheim, A. (1977). Initiation points for DNA replication in nontransformed and simian virus 40-transformed Chinese hamster lung cells. *Cell* (*Cambridge, Mass.*) **11**, 859–869.

Mastrangelo, I. A., Hough, P. V. C., Wall, J. S., Dodson, M., Dean, F. B., and Hurwitz, J. (1989). ATP-dependent assembly of double hexamers of SV40 T antigen at the viral origin of DNA replication. *Nature* (*London*) **38**, 658–662.

Matsumoto, L. H. (1981). Enrichment of satellite DNA on the nuclear matrix of bovine cells. *Nature* (*London*) **294**, 481–482.

McCready, S. J., and Cook, P. R. (1984). Lesions induced in DNA by ultraviolet light are repaired at the nuclear cage. *J. Cell Sci.* **70**, 189–196.

McCready, S. J., Godwin, J., Mason, D. W., Brazell, I. A., and Cook, P. R. (1980). DNA is replicated at the nuclear cage. *J. Cell Sci.* **46**, 365–386.

McKeon, F. D., Kirschner, M. W., and Caput, D. (1986). Homologies in both primary and secondary structure between nuclear envelope and intermediate filament proteins. *Nature* (*London*) **319**, 463–468.

McKnight, R. A., Shamay, A., Sankaran, L., Wall, R. J., and Henninghausen, L. (1992). Matrix-attachment regions can impart position-independent regulation of a tissue-specific gene in transgenic mice. *Proc. Natl. Acad. Sci. U.S.A.* **89**, 6943–6947.

McNamara, P. T., Bolshoy, A., Trifonov, E. N., and Harrington, R. E. (1990). Sequence-dependent kinks induced in curved DNA. *J. Biomol. Struct. Dyn.* **8**, 529–538.

McPherson, C. E., Shim, E. Y., Friedman, D. S., and Zaret, K. S. (1993). An active tissue-specific enhancer and bound transcription factor existing in a precisely positioned nucleosomal array. *Cell* (*Cambridge, Mass.*) **75**, 387–398.

Metzger, D. A., Curtis, S., and Korach, K. S. (1991). Diethylstilbestrol metabolites and analogs: Differential ligand effects on estrogen receptor interactions with nuclear matrix sites. *Endocrinology* (*Baltimore*) **128**, 1785–1791.

Mielke, C., Kohwi, Y., Kohwi-Shigematsu, T., and Bode, J. (1990). Hierarchical binding of DNA fragments derived from scaffold-attached regions: Correlation of properties *in vitro* and function *in vivo. Biochemistry* **29**, 7475–7485.

Miller, T. E., Huang, C.-Y., and Pogo, A. O. (1978). Rat liver nuclear skeleton and ribonucleoprotein complexes containing HnRNA. *J. Cell Biol.* **76**, 675–691.

Milner, A. E., Gordon, D. J., Turner, B. M., and Vaughan, A. T. M. (1993). A correlation between DNA-nuclear matrix binding and relative radiosensitivity in two human squamous cell carcinoma cell lines. *Int. J. Radiat. Res.* **63**, 13–20.

Mirkovitch, J., Mirault, M.-E., and Laemmli, U. K. (1984). Organization of the higher-order chromatin loop: Specific DNA attachment sites on nuclear scaffold. *Cell* (*Cambridge, Mass.*) **39**, 223–232.

Mirkovitch, J., Spierer, P., and Laemmli, U. K. (1986). Genes and loops in 320,000 base-pairs of the *Drosophila melanogaster* chromosome. *J. Mol. Biol.* **190**, 255–258.

Mirkovitch, J., Gasser, S. M., and Laemmli, U. K. (1987). Relation of chromosome structure and gene expression. *Philos. Trans. R. Soc. London, Ser. B* **317**, 563–574.

Mirkovitch, J., Gasser, S. M., and Laemmli, U. K. (1988). Scaffold attachment of DNA loops in metaphase chromosomes. *J. Mol. Biol.* **200**, 101–109.

Mirzayan, C., Copeland, C. S., and Snyder, M. (1992). The *NUF1* gene encodes an essential coiled-coil related protein that is a potential component of the yeast nucleoskeleton. *J. Cell Biol.* **116**, 1319–1332.

Moreau, J., Marcaud, L., Maschat, F., Kejzlarova-Lepesant, J., Lepesant, J.-A., and Scherrer, K. (1982). A + T-rich linkers define functional domains in eukaryotic DNA. *Nature (London)* **295,** 260–262.

Mukherjee, S., Patel, I., and Bastia, D. (1985). Conformational changes in a replication origin induced by an initiator protein. *Cell (Cambridge, Mass.)* **43,** 189–197.

Mullenders, L. H. F., van Zeeland, A. A., and Natarajan, A. T. (1983). Analysis of the distribution of DNA repair patches in the DNA-nuclear matrix complex from human cells. *Biochim. Biophys. Acta* **740,** 428–435.

Mullenders, L. H. F., van Kasteran-van Leeuwen, A. C., van Zeeland, A. A., and Natarajan, A. T. (1988). Nuclear matrix associated DNA is preferentially repaired in normal human fibroblasts, exposed to a low dose of ultraviolet light but not in Cockayne's syndrome fibroblasts. *Nucleic Acids Res.* **16,** 10607–10623.

Müller, H.-P., Sogo, J. M., and Schaffner, W. (1989). An enhancer stimulates transcription in trans when attached to the promoter via a protein bridge. *Cell (Cambridge, Mass.)* **58,** 767–777.

Müller, W. E. G., Okamoto, T., Reuter, P., Ugarkovic, D., and Schröder, H. C. (1990). Functional characterization of Tat protein from human immunodeficiency virus. Evidence that Tat links viral RNAs to nuclear matrix. *J. Biol. Chem.* **265,** 3803–3808.

Murakami, Y., Satake, M., Yamaguchi-Iwai, Y., Sakai, M., Maramatsu, M., and Ito, Y. (1991). The nuclear protooncogenes *c-jun* and *c-fos* as regulators of DNA replication. *Proc. Natl. Acad. Sci. U.S.A.* **88,** 3947–3951.

Nagagomi, K., Kohwi, Y., Dickinson, L. A., and Kohwi-Shigematsu, T. (1994). A novel DNA-binding motif in the nuclear matrix attachment DNA-binding protein SATB1. *Mol. Cell. Biol.* **14,** 1852–1860.

Nakayasu, H., and Ueda, K. (1986). Preferential association of acidic actin with nuclei and nuclear matrix from mouse leukemia L5178Y cells. *Exp. Cell Res.* **163,** 327–336.

Naylor, L. H., Lilley, D. M. J., and van de Sande, J. H. (1986). Stress-induced cruciform formation in a cloned d(CATG)$_{10}$ sequence. *EMBO J.* **5,** 2407–2413.

Nejedly, K., Kwinkowski, M., Galazka, G., Klysik, J., and Palecek, E. (1985). Recognition of the structural distortions at the junctions between B and Z segments in negatively supercoiled DNA by osmium tetroxide. *J. Biomol. Struct. Dyn.* **3,** 467–478.

Neuer-Nitsche, B., Lu, X., and Werner, D. (1988). Functional role of a highly repetitive DNA sequence in anchorage of the mouse genome. *Nucleic Acids Res.* **16,** 8351–8360.

Nishizawa, M., Tanabe, K., and Takahashi, T. (1984). DNA polymerases and DNA topoisomerases solubilized from nuclear matrices of regenerating rat livers. *Biochem. Biophys. Res. Commun.* **124,** 917–924.

Obi, F. O., and Billett, M. A. (1991). Preferential binding of the carcinogen benzo[a]pyrene to proteins of the nuclear matrix. *Carcinogenesis (London)* **12,** 481–486.

Obi, F. O., Ryan, A. J., and Billett, M. A. (1986). Preferential binding of the carcinogen benzo[a]pyrene to DNA in active chromatin and the nuclear matrix. *Carcinogenesis (London)* 907–913.

O'Neill, E. A., Fletcher, C., Burrow, C. R., Heintz, N., Roeder, R. G., and Kelly, T. J. (1988). Transcription factor OTF-1 is functionally identical to the DNA replication factor NF-III. *Science* **241,** 1210–1213.

Osman, M. Paz, M., Landesman, Y., Fainsod, A., and Gruenbaum, Y. (1990). Molecular analysis of the *Drosophila* nuclear lamin gene. *Genomics* **8,** 217–224.

Panayotatos, N., and Wells, R. D. (1981). Cruciform structures in supercoiled DNA. *Nature (London)* **289,** 466–470.

Pardoll, D. M., and Vogelstein, B. (1980). Sequence analysis of nuclear matrix associated DNA from rat liver. *Exp. Cell Res.* **128,** 466–470.

Pardoll, D. M., Vogelstein, B., and Coffey, D. S. (1980). A fixed site of DNA replication in eucaryotic cells. *Cell (Cambridge, Mass.)* **19,** 527–536.

Parsons, R., Anderson, M. E., and Tegtmeyer, P. (1990). Three domains in the simian virus 40 core origin orchestrate the binding, melting and DNA helicase activities of T antigen. *J. Virol.* **64,** 509–518.

Parvari, R., Ziv, E., Lantner, F., Heller, D., and Schechter, I. (1990). Somatic diversification of chicken immunoglobulin light chains by point mutations. *Proc. Natl. Acad. Sci. U.S.A.* **87,** 3072–3076.

Paul, A.-N., and Ferl, R. J. (1993). Osmium tetroxide footprinting of a scaffold attachment region in the maize *Adh1* promoter. *Plant Mol. Biol.* **22,** 1145–1151.

Paulson, J. R., and Laemmli, U. K. (1977). The structure of histone-depleted metaphase chromosomes. *Cell (Cambridge, Mass.)* **12,** 817–828.

Pearson, C. E., Ruiz, M. T., Price, G. B., and Zannis-Hadjopoulos, M. (1994). Cruciform DNA-binding protein in HeLa cell extracts. *Biochemistry* **33,** 14185–14196.

Pearson, C. E., Zannis-Hadjopoulos, M., Price, G. B., and Zorbas, H. (1995). A novel type of interaction between cruciform DNA and a cruciform binding protein from HeLa cells. *EMBO J.* **14** 1571–1580.

Pech, M. Igo-Kemenes, T., and Zachau, H. G. (1979). Nucleotide sequence of a highly repetitive component of rat DNA. *Nucleic Acids. Res.* **7,** 417–432.

Peterson, C. L., Orth, K., and Calame, K. L. (1986). Binding in vitro of multiple cellular proteins to immunoglobulin heavy-chain enhancer DNA. *Mol. Cell Biol.* **6,** 4168–4178.

Phi-Van, L., and Strätling, W. H. (1988). The matrix attachment regions of the chicken lysozyme gene co-map with the boundaries of the chromatin domain. *EMBO J.* **7,** 655–664.

Phi-Van, L., von Kries, J. P., Ostertag, W., and Strätling, W. H. (1990). The chicken lysozyme 5' matrix attachment region increases transcription from a heterologous promoter in heterologous cells and dampens position effects on the expression of transfected genes. *Mol. Cell. Biol.* **10,** 2302–2307.

Pienta, K. J., and Ward, W. S. (1994). An unstable nuclear matrix may contribute to genetic instability. *Med. Hypotheses* **42,** 45–52.

Pienta, K. J., Getzenberg, R. H., and Coffey, D. S. (1991). Cell structure and DNA organization. *CRC Crit. Rev. Eukaryotic Gene Express.* **1,** 355–385.

Poljak, L., Seum, C., Mattioni, T., and Laemmli, U. K. (1994). SARs stimulate but do not confer position independent gene expression. *Nucleic Acids Res.* **22,** 4386–4394.

Pommier, Y., Cockerill, P. N., Kohn, K. W., and Garrard, W. T. (1990). Identification within the simian virus 40 genome of a chromosomal loop attachment site that contains topoisomerase II cleavage sites. *J. Virol.* **64,** 419–423.

Postnikov, Y. V., Shick, V. V., Belyavsky, A. V., Khrapko, K. R., Brodolin, K. L., Nikolskaya, T. A., and Mirzabekov, A. D. (1991). Distribution of high mobility group proteins 1/2,E and 14/17 and linker histones H1 and H5 on transcribed and non-transcribed regions of chicken erythrocyte chromatin. *Nucleic Acids Res.* **19,** 717–725.

Pruijn, G. J. M., van Driel, W., and van der Vliet, P. C. (1986). Nuclear factor III, a novel sequence-specific DNA-binding protein from HeLa cells stimulating adenovirus DNA replication. *Nature (London)* **322,** 656–659.

Ptashne, M. (1988). How eukaryotic transcriptional activators work. *Nature (London)* **335,** 683–689.

Ptashne, M., and Gann, A. A. F. (1990). Activators and targets. *Nature (London)* **346,** 329–331.

Rao, B. S., Zannis-Hadjopoulos, M., Price, G. B., Reitman, M., and Martin, R. G. (1990). Sequence similarities among monkey *ori*-enriched (*ors*) fragments. *Gene* **87,** 233–242.

Razin, S. V., Mantieva, V. L., and Georgiev, G. P. (1978). DNA adjacent to attachment points of deoxyribonucleoprotein fibril to chromosomal axial structure is enriched in reiterated base sequences. *Nucleic Acids Res.* **5,** 4737–4751.

Razin, S. V., Kekelidze, M. G., Lukanidin, E. M., Scherrer, K., and Georgiev, G. P. (1986). Replication origins are attached to the nuclear skeleton. *Nucleic Acids Res.* **14,** 8189–8207.

Razin, S. V., Vassetzky, Y. S., and Hancock, R. (1991a). Nuclear matrix attachment regions and topoisomerase II binding and reaction sites in the vicinity of a chicken DNA replication origin. *Biochem. Biophys. Res. Commun.* **177**, 265–270.

Razin, S. V., Petrov, P., and Hancock, R. (1991b). Precise localization of the α-globin gene cluster within one of the 20- to 300-kilobase DNA fragments released by cleavage of chicken chromosomal DNA at topoisomerase II sites *in vivo:* Evidence that the fragments are DNA loops or domains. *Proc. Natl. Acad. Sci. U.S.A.* **88**, 8515–8519.

Reeves, R., and Nissen, M. S. (1990). The A·T-DNA-binding domain of mammalian high mobility group I chromosomal proteins: A novel peptide motif for recognizing DNA structure. *J. Biol. Chem.* **265**, 8573–8582.

Reeves, R., Langan, T. A., and Nissen, M. S. (1991). Phosphorylation of the DNA-binding domain of nonhistone high-mobility group I protein by cdc2 kinase: Reduction of binding affinity. *Proc. Natl. Acad. Sci. U.S.A.* **88**, 1671–1675.

Robinson, S. I., Nelkin, B. D., and Vogelstein, B. (1982). The ovalbumin gene is associated with the nuclear matrix of chicken oviduct cells. *Cell (Cambridge, Mass.)* **28**, 99–106.

Romig, H., Fackelmayer, F. O., Renz, A., Ramsperger, U., and Richter, A. (1992). Characterization of SAF-A, a novel nuclear DNA binding protein from HeLa cells with high affinity for nuclear matrix/scaffold attachment DNA elements. *EMBO J.* **11**, 3431–3440.

Romig, H., Ruff, J., Fackelmayer, F. O., Patil, M. S., and Richter, A. (1994). Characterization of two intronic nuclear-matrix-attachment regions in the human DNA topoisomerase I gene. *Eur. J. Biochem.* **221**, 411–419.

Ross, W., Shulman, M., and Landy, A. (1982). Biochemical analysis of *att*-defective mutants of the phage lambda site-specific recombination system. *J. Mol. Biol.* **156**, 505–529.

Roth, S. Y., Dean, A., and Simpson, R. T. (1990). Yeast a2 repressor positions nucleosomes in TRP1/ARS1 chromatin. *Mol. Cell. Biol.* **10**, 2247–2260.

Russev, G., and Boulikas, T. (1992). Repair of transcriptionally active and inactive genes during S and G2 phases of the cell cycle. *Eur. J. Biochem.* **204**, 267–272.

Ryder, K., Silver, S., DeLucia, A. L., Fanning, E., and Tegtmeyer, P. (1986). An altered DNA conformation in origin region I is a determinant for the binding of SV40 large T antigen. *Cell (Cambridge, Mass.)* **44**, 719–725.

Saigo, K., Yaginuma, K., Hayashi, H., Isono, K., Takada, S., and Koike, K. (1994). Enhanced expression of c-*myc* gene and binding protein to the 2.3 kb upstream region of the P1 promoter in human liver and colon cancers. *Int. J. Oncol.* **5**, 1099–1104.

Saragosti, S., Moyne, G., and Yaniv, M. (1980). Absence of nucleosomes in a fraction of SV40 chromatin between the origin of replication and the region coding for the late leader RNA. *Cell (Cambridge, Mass.)* **20**, 65–73.

Schaack, J., Ho, W. Y.-W., Freimuth, P., and Shenk, T. (1990). Adenovirus terminal protein mediates both nuclear matrix association and efficient transcription of adenovirus DNA. *Genes Dev.* **4**, 1197–1208.

Scheuermann, R. H., and Chen, U. (1989). A developmental-specific factor binds to suppressor sites flanking the immunoglobulin heavy-chain enhancer. *Genes Dev.* **3**, 1255–1266.

Schirmbeck, R., and Deppert, W. (1987). Specific interaction of simian virus 40 large T antigen with cellular chromatin and nuclear matrix during the course of infection. *J. Virol.* **61**, 3561–3569.

Schirmbeck, R., and Deppert, W. (1989). Nuclear subcompartmentalization of simian virus 40 large T antigen: Evidence for in vitro regulation of biochemical activities. *J. Virol.* **63**, 2308–2316.

Schirmbeck, R., and Deppert, W. (1991). Structural topography of simian virus 40 DNA replication. *J. Virology* **65**, 2588.

Schnos, M., Zahn, K., Inman, R. B., and Blattner, F. R. (1988). Initiation protein induced helix destabilization at the I origin: A prepriming step in DNA replication. *Cell (Cambridge, Mass.)* **52**, 385–395.

Schnos, M., Zahn, K., Blattner, F. R., and Inman, R. B. (1989). DNA looping induced by bacteriophage I O protein: Implications for formation of higher order structures at the origin of replication. *Virology* **168,** 370–377.

Schuchard, M., Subramaniam, M., Ruesink, T., and Spelsberg, T. C. (1991). Nuclear matrix localization and specific matrix DNA binding by receptor binding factor 1 of the avian oviduct progesterone receptor. *Biochemistry* **30,** 9516–9522.

Scott, M. P., Tamkun, J. W., and Hartzell, G. W. (1989). The structure and function of the homeodomain. *Biochim. Biophys. Acta* **989,** 25–48.

Shpigelman, E. S., Trifonov, E. N., and Bolshoy, A. (1993). Curvature: Software for the analysis of curved DNA. *Comput. Appl. Biosci.* **9,** 435–440.

Shrader, T. E., and Crothers, D. M. (1989). Artificial nucleosome positioning sequences. *Proc. Natl. Acad. Sci. U.S.A.* **86,** 7418–7422.

Slatter, R. E., Dupree, P., and Gray, G. (1991). A scaffold-associated region is located downstream of the pea plastocyanin gene. *Plant Cell* **3,** 1239–1250.

Small, D., and Vogelstein, B. (1985). The anatomy of supercoiled loops in the *Drosophila* 7F locus. *Nucleic Acids Res.* **13,** 7704–7713.

Small, D., Nelkin, B., and Vogelstein, B. (1982). Nonrandom distribution of repeated DNA sequences with respect to supercoiled loops and the nuclear matrix. *Proc. Natl. Acad. Sci. U.S.A.* **70,** 5911–5915.

Small, D., Nelkin, B., and Vogelstein, B. (1985). The association of transcribed genes with the nuclear matrix of *Drosophila* cells during heat shock. *Nucleic Acids Res.* **13,** 2413–2431.

Smith, H. C., and Berezney, R. (1980). DNA polymerase α is tightly bound to the nuclear matrix of actively replicating liver. *Biochem. Biophys. Res. Commun.* **97,** 1541–1547.

Smith, H. C., and Berezney, R. (1982). Nuclear matrix-bound deoxyribonucleic acid synthesis: An in vitro system. *Biochemistry* **21,** 6751–6761.

Smith, H. C., and Berezney, R. (1983). Dynamic domains of DNA polymerase α in regenerating rat liver. *Biochemistry* **22,** 3042–3046.

Smith, H. C., Puvion, E., Buchholtz, L. A., and Berezney, R. (1984). Spatial distribution of DNA loop attachment and replicational sites in the nuclear matrix. *J. Cell Biol.* **99,** 1794–1802.

Smith, H. C., Ochs, R. L., Lin, D., and Chinault, A. C. (1987). Ultrastructural and biochemical comparisons of nuclear matrices prepared by high salt or LIS extraction. *Mol. Cell. Biol.* **77,** 49–61.

Snyder, M., Buchman, A. R., and Davis, R. W. (1986). Bent DNA at a yeast autonomously replicating sequence. *Nature (London)* **324,** 87–89.

Sohn, J., Gerstein, R. M., Hsieh, C., Lemer, M., and Selsing, E. (1993). Somatic hypermutation of an immunoglobulin μ heavy chain transgene. *J. Exp. Med.* **77,** 493–504.

Sperry, A. O., Blasquez, V. C., and Garrard, W. T. (1989). Dysfunction of chromosomal loop attachment sites: Illegitimate recombination linked to matrix association regions and topoisomerase II. *Proc. Natl. Acad. Sci. U.S.A.* **86,** 5497–5501.

Sprädling, A., and Orr-Weaver, T. (1987). Regulation of DNA replication during *Drosophila* development. *Annu. Rev. Genet.* **21,** 373–403.

Stein, G. S., Lian, J. B., Dworetzky, S. I., Owen, T. A., Bortell, R., Bidwell, J. P., and van Wijnen, A. J. (1991). Regulation of transcription-factor activity during growth and differentiation: Involvement of the nuclear matrix in concentration and localization of promoter binding proteins. *J. Cell. Biochem.* **47,** 300–305.

Stief, A., Winter, D. M., Strätling, W. H., and Sippel, A. E. (1989). A nuclear DNA attachment element mediates elevated and position-independent gene activity. *Nature (London)* **341,** 343–345.

Stuurman, N., Floore, A., Colen, A., de Jong, L., and van Driel, R. (1992). Stabilization of the nuclear matrix by disulfide bridges: Identification of matrix polypeptides that form disulfides. *Exp. Cell Res.* **200,** 285–294.

Sun, J.-M., Chen, H. Y., and Davie, J. R. (1994). Nuclear factor 1 is a component of the nuclear matrix. *J. Cell. Biochem.* **55**, 252–263.

Sung, P., Prakash, L., Matson, S. W., and Prakash, S. (1987). RAD3 protein of *Saccharomyces cerevisiae* is a DNA helicase. *Proc. Natl. Acad. Sci. U.S.A.* **84**, 8951–8955.

Surdej, P., Got, C., Rosset, R., and Miassod, R. (1990). Supragenic loop organization: Mapping in *Drosophila* embryos, of scaffold-associated regions on a 800 kilobase DNA continuum cloned from the 14B-15B first chromosome region. *Nucleic Acids Res.* **18**, 3713–3722.

Swaneck, G. E., and Fishman, J. (1988). Covalent binding of the endogenous estrogen 16α-hydroxyestrone to estradiol receptor in human breast cancer cells: Characterization and intranuclear localization. *Proc. Natl. Acad. Sci. U.S.A.* **85**, 7831–7835.

Sykes, R. C., Lin, D., Hwang, S. J., Framson, P. E., and Chinault, A. C. (1988). Yeast ARS function and nuclear matrix association coincide in a short sequence from the human *HPRT* locus. *Mol. Gen. Genet.* **212**, 301–309.

Taira, T., Iguchi-Ariga, S. M. M., and Ariga, H. (1994). A novel DNA replication origin identified in the human heat shock protein 70 gene promoter. *Mol. Cell. Biol.* **14**, 6386–6397.

Takai, T., Nishita, Y., Iguchi-Ariga, S. M. M., and Ariga, H. (1994). Molecular cloning of MSSP-2, a c-myc gene single-strand binding protein: Characterization of binding specificity and DNA replication activity. *Nucleic Acids Res.* **22**, 5576–5581.

Takeda, S., Zou, Y.-R., Bluethmann, H., Kitamura, D., Muller, U., and Rajewsky, K. (1993). Deletion of the immunoglobulin κ chain intron enhancer abolishes κ chain gene rearrangement in *cis* but not λ chain gene rearrangement in *trans*. *EMBO J.* **12**, 2329–2336.

Tartier, L., Michalik, V., Spotheim-Maurizot, M., Rahmouni, A. R., Sabatter, R., and Charlier, M. (1994). Radiolytic signature of Z-DNA. *Nucleic Acids Res.* **22**, 5565–5570.

Tasheva, E. S., and Roufa, D. J. (1994). Densely methylated DNA islands in mammalian chromosomal replication origins. *Mol. Cell. Biol.* **14**, 5636–5644.

Tawfic, S., and Ahmed, K. (1994). Growth stimulus-mediated differential translocation of casein kinase 2 to the nuclear matrix. Evidence based on androgen action in the prostate. *J. Biol. Chem.* **269**, 24615–24620.

Teschke, C., Solleder, G., and Moritz, K. B. (1991). The highly variable pentameric repeats of the AT-rich germline limited DNA in *Parascaris univalens* are the telomeric repeats of somatic chromosomes. *Nucleic Acids Res.* **19**, 2677–2684.

Tew, K. D., Wang, A. L., and Schein, P. S. (1983). Alkylating agent interactions with the nuclear matrix. *Biochem. Pharmacol.* **32**, 3509–3516.

Thanos, D., and Maniatis, T. (1992). The high mobility group protein HMGI(Y) is required for NF-κB-dependent virus induction of the human IFN-b gene. *Cell* (*Cambridge, Mass.*) **71**, 777–789.

Thoma, F. (1992). Nucleosome positioning. *Biochim. Biophys. Acta* **1130**, 1–19.

Thoma, F., Koller, T., and Klug, A. (1979). Involvement of histone H1 in the organization of the nucleosome and of the salt-dependent superstructures of chromatin. *J. Cell Biol.* **83**, 403–427.

Thompson, E. M., Christians, E., Stinnakre, M. G., and Renard, J. P. (1994). Scaffold attachment regions stimulate HSP70.1 expression in mouse preimplantation embryos but not in differentiated tissues. *Mol. Cell. Biol.* **14**, 4694–4703.

Thukral, S. K., Blain, G. C., Chang, K. K. H., and Fields, S. (1994). Distinct residues of human p53 implicated in binding to DNA simian virus 40 large T antigen, 53BP1, and 53BP2. *Mol. Cell. Biol.* **14**, 8315–8321.

Todd, A., Landry, S., Pearson, C. E., Khoury, V., and Zannis-Hadjopoulos, M. (1995). Deletion analysis of minimal sequence requirements for autonomous replication of *ors8*, a monkey early-replicating DNA sequence. *J. Cell. Biochem.* **57**, 280–289.

Travers, A. A. (1987). DNA bending and nucleosome positioning. *Trends Biochem. Sci.* **12**, 108–112.

Trifonov, E. N. (1980). Sequence-dependent deformational anistropy of chromatin DNA. *Nucleic Acids Res.* **8**, 4041–4053.

Trifonov, E. N. (1986). Curved DNA. *CRC Crit. Rev. Biochem.* **19**, 89–106.

Trifonov, E. N. (1989). The multiple codes of nucleotide sequences. *Bull. Math. Biol.* **51**, 417–432.

Trifonov, E. N. (1991). DNA in profile. *Trends Biochem. Sci.* **16**, 467–470.

Trifonov, E. N., and Brendel, V. (1987). "Gnomic: A Dictionary of Genetic Codes." Springer-Verlag, Berlin.

Trifonov, E. N., and Sussman, J. J. (1980). The pitch of chromatin DNA is reflected in its nucleotide sequence. *Proc. Natl. Acad. Sci. U.S.A.* **77**, 3816–3820.

Tsurimoto, T., and Stillman, B. (1989). Multiple replication factors augment DNA synthesis by the two eukaryotic DNA polymerases, α and δ. *EMBO J.* **8**, 3883–3889.

Tsutsui, K., Tsutsui, K., and Muller, M. T. (1988). The nuclear scaffold exhibits DNA-binding sites selective for supercoiled DNA. *J. Biol. Chem.* **263**, 7235–7241.

Tsutsui, K., Tsutsui, K., Okada, S., Watarai, S., Seki, S., Yasuda, T., and Shohmori, T. (1993). Identification and characterization of a nuclear scaffold protein that binds the matrix attachment region DNA. *J. Biol. Chem.* **268**, 12886–12894.

Umek, R. M., and Kowalsky, D. (1988). The ease of DNA unwinding as a determinant of initiation at yeast replication origins. *Cell (Cambridge, Mass.)* **52**, 559–567.

Umthun, A. R., Hou, Z., Sibenaller, Z. A., Shaiu, W.-L., and Dobbs, D. L. (1994). Identification of DNA-binding proteins that recognize a conserved type I repeat sequence in the replication origin region of *Tetrahymena* rDNA. *Nucleic Acids Res.* **22**, 4432–4440.

Valenzuela, M. S., Mueller, G. C., and Dasgupta, S. (1983). Nuclear matrix-DNA complex resulting from EcoR1 digestion of HeLa nucleoids is enriched for DNA replicating forks. *Nucleic Acids Res.* **11**, 2155–2164.

van Belkum, A., Blommers, M. J. J., van den Elst, H., van Boom, J. H., and Hilbers, C. W. (1990). Biochemical and biophysical studies on the folding of the core region of the origin of replication of bacteriophage M13. *Nucleic Acids Res.* **18**, 4703–4710.

van Driel, R., Humbel, B., and de Jong, L. (1991). The nucleus: A black box being opened. *J. Cell. Biochem.* **47**, 311–316.

van Steensel, B., van Haarst, A. D., de Kloet, E. R., and van Driel, R. (1991). Binding of corticosteroid receptors to rat hippocampus nuclear matrix. *FEBS Lett.* **292**, 229–231.

van Wijnen, A. J., Bidwell, J. P. Fey, E. G., Penman, S., Lian, J. B., Stein, J. L., and Stein, G. S. (1993). Nuclear matrix association of multiple sequence-specific DNA binding activities related to SP-1, ATF, CCAAT, C/EBP, OCT-1, and AP-1. *Biochemistry* **32**, 8397–8402.

Varshavsky, A. J., Sundin, O., and Bohn, M. (1979). A stretch of "late" SV40 viral DNA about 400 bp long which includes the origin of replication is specifically exposed in SV40 minichromosomes. *Cell (Cambridge, Mass.)* **16**, 453–466.

Vaughn, J. P., Dijkwel, P. A., Mullenders, L. H. F., and Hamlin, J. L. (1990). Replication forks are associated with the nuclear matrix. *Nucleic Acids Res.* **18**, 1965–1969.

Vazquez, J., and Schedl, P. (1994). Sequences required for enhancer blocking activity of scs are located within two nuclease-hypersensitive regions. *EMBO J.* **13**, 5984–5993.

Venter, U., Svaren, J., Schmitz, J., Schmid, A., and Horz, W. (1994). A nucleosome precludes binding of the transcription factor Pho4 in vivo to a critical target site in the PHO5 promoter. *EMBO J.* **13**, 4848–4855.

Verheijen, R., van Venrooij, W., and Ramaekers, F. (1988). The nuclear matrix: Structure and composition. *J. Cell Sci.* **90**, 11–36.

Verheijen, R., Kuijpers, H. J. H., Schlingemann, R. O., Boehmer, A. L. M., van Driel, R., Brakenhoff, G. J., and Ramaekers, F. C. S. (1989). Ki-67 detects a nuclear matrix-associated proliferation-related antigen. *J. Cell Sci.* **92**, 123–130.

Virta-Pearlman, V. J., Gunaratne, P. H., and Chinault, A. C. (1993). Analysis of a replication initiation sequence from the adenosine deaminase region of the mouse genome. *Mol. Cell. Biol.* **13,** 5931–5942.

von Kries, J. P., Buhrmester, H., and Strätling, W. H. (1991). A matrix/scaffold attachment region binding protein: Identification, purification and mode of binding. *Cell (Cambridge, Mass.)* **64,** 123–135.

Wagner, A. J., Kokontis, J. M., and Hay, N. (1994). Myc-mediated apoptosis requires wild-type p53 in a manner independent of cell cycle arrest and the ability of p53 to induce p21$^{waf1/cip1}$. *Genes Dev.* **8,** 2817–2830.

Waitz, W., and Loidl, P. (1991). Cell cycle dependent association of c-myc protein with the nuclear matrix. *Oncogene* **6,** 29–35.

Wan, K. M., Nickerson, J. A., Krockmalnic, G., and Penman, S. (1994). The B1C8 protein is in the dense assemblies of the nuclear matrix and relocates to the spindle and perinucleolar filaments at mitosis. *Proc. Natl. Acad. Sci. U.S.A.* **91,** 594–598.

Wang, Z.-Y., Lin, X.-H., Nobuyoshi, M., and Deuel, T. F. (1993). Identification of a single-stranded DNA-binding protein that interacts with an S1 nuclease-sensitive region in the platelet-derived growth factor A-chain gene promoter. *J. Biol. Chem.* **268,** 10681–10685.

Ward, G. K., McKenzie, R., Zannis-Hadjopoulos, M., and Price, G. B. (1990). The dynamic distribution and quantification of DNA cruciforms in eukaryotic nuclei. *Exp. Cell Res.* **188,** 235–246.

Ward, G. K., Shihab-el-Deen, A., Zannis-Hadjopoulos, M., and Price, G. B. (1991). DNA cruciforms and the nuclear supporting structure. *Exp. Cell Res.* **195,** 92–98.

Warters, R. L., and Childers, T. J. (1982). Radiation-induced thymine base damage in replicating chromatin. *Radiat. Res.* **90,** 564–574.

Wasylyk, C., and Wasylyk, B. (1986). The immunoglobulin heavy-chain B-lymphocyte enhancer efficiently stimulates transcription in non-lymphoid cells. *EMBO J.* **5,** 553–560.

Webb, C. F., Das, C., Eneff, K. L., and Tucker, P. W. (1991). Identification of a matrix-associated region 5' of an immunoglobulin heavy chain variable region gene. *Mol. Cell. Biol.* **11,** 5206–5211.

Wegner, M., Zastrow, G., Klavinius, A., Schwender, S., Muller, F., Luksza, H., Hoppe, J., Wienberg, J., and Grummt, F. (1989). Cis-acting sequences from mouse rDNA promote plasmid DNA amplification and persistence in mouse cells: Implication of HMG-I in their function. *Nucleic Acids Res.* **17,** 9909–9932.

Weintraub, H., Larsen, A., and Groudine, M. (1981). α-globin-gene switching during the development of chicken embryos: Expression and chromosome structure. *Cell (Cambridge, Mass.)* **24,** 333–344.

Whitehurst, C., Henney, H. R., Max, E. E., Schroeder, H. W., Jr., Stüber, F., Siminovitch, K. A., and Garrard, W. T. (1992). Nucleotide sequence of the intron of the germline human κ immunoglobulin gene connecting the J and C regions reveals a matrix association region (MAR) next to the enhancer. *Nucleic Acids Res.* **20,** 4929–4930.

Wittig, B., Wölfl, S., Dorbic, T., Vahrson, W., and Rich, A. (1992). Transcription of human c-myc in permeabilized nuclei is associated with formation of Z-DNA in three discrete regions of the gene. *EMBO J.* **11,** 4653–4663.

Wu, C., Zannis-Hadjopoulos, M., and Price, G. B. (1993). *In vivo* activity for initiation of DNA replication resides in a transcribed region of the human genome. *Biochim. Biophys. Acta* **1174,** 258–266.

Wu, H.-M., and Crothers, D. M. (1984). The locus of sequence-directed and protein-induced DNA bending. *Nature (London)* **308,** 509–513.

Xie, L., and Boulikas, T. (1995). A potential replication initiator protein interacts with one strand of a cruciform within the putative origin of replication/MAR/enhancer of the human choline acetyltransferase gene. In preparation.

Xu, J., Manning, F. C. R., and Patierno, S. R. (1994). Preferential formation and repair of chromium-induced DNA adducts and DNA-protein crosslinks in nuclear matrix DNA. *Carcinogenesis* **15**, 1443–1450.

Xu, M., Hammer, R. E., Blasquez, V. C., Jones, S. L., and Garrard, W. T. (1989). Immunoglobulin κ gene expression after stable integration. II. Role of the intronic *MAR* and enhancer in transgenic mice. *J. Biol. Chem.* **264**, 21190–21195.

Yalkinoglu, A. O., Zentgraf, H., and Hübscher, U. (1991). Origin of adeno-associated virus DNA replication is a target of carcinogen-inducible DNA amplification. *J. Virol.* **65**, 3175–3184.

Yamamoto, T., Takahashi, T., and Matsukage, A. (1984). Tight association of DNA polymerase α with granular structures in the nuclear matrix of chick embryo cell: Immunocytochemical detection with monoclonal antibody against DNA polymerase α. *Cell Struct. Funct.* **9**, 83–90.

Yee, H. A., Wong, A. K. C., van de Sande, J. H., and Rattner, J. B. (1991). Identification of novel single-stranded $d(TC)_n$ binding proteins in several mammalian species. *Nucleic Acids Res.* **19**, 949–953.

Yuan, J., Simos, G., Blobel, G., and Georgatos, S. D. (1991). Binding of lamin A to polynucleosomes. *J. Biol. Chem.* **266**, 9211–9215.

Zahn, K., and Blattner, F. R. (1985). Sequence-induced DNA curvature at the bacteriophage λ origin of replication. *Nature* (*London*) **317**, 451–453.

Zahn, K., and Blattner, F. R. (1987). Direct evidence for DNA bending at the lambda replication origin. *Science* **236**, 416–422.

Zannis-Hadjopoulos, M., Frappier, L., Khoury, M., and Price, G. B. (1988). Effect of anti-cruciform DNA monoclonal antibodies on DNA replication. *EMBO J.* **7**, 1837–1844.

Zeitlin, S., Parent, A., Silverstein, S., and Efstratiadis, A. (1987). Pre-mRNA splicing and the nuclear matrix. *Mol. Cell. Biol.* **7**, 111–120.

Zenk, D. W., Ginder, G. D., and Brotherton, T. W. (1990). A nuclear matrix protein binds very tightly to DNA in the avian β-globin gene enhancer. *Biochemistry* **29**, 5221–5226.

Zhao, K., Käs, E., Gonzalez, E., and Laemmli, U. K. (1993). SAR-dependent mobilization of histone H1 by HMG-I/Y in vitro: HMG-I/Y is enriched in H1-depleted chromatin. *EMBO J.* **12**, 3237–3247.

Zhao, Y., Tsutsumi, R., Yamaki, M., Nagatsuka, Y., Ejiri, S., and Tsutsumi, K. (1994). Initiation zone of DNA replication at the aldolase B locus encompasses transcription promoter region. *Nucleic Acids Res.* **22**, 5385–5390.

Scaffold/Matrix-Attached Regions: Structural Properties Creating Transcriptionally Active Loci

J. Bode, T. Schlake, M. Ríos-Ramírez, C. Mielke, M. Stengert, V. Kay, and D. Klehr-Wirth

Gesellschaft für Biotechnologische Forschung m.b.H., Genetik von Eukaryonten, D-38124 Braunschweig, Germany

The expression characteristics of the human interferon-β gene, as part of a long stretch of genomic DNA, led to the discovery of the putative domain bordering elements. The chromatin structure of these elements and their surroundings was determined during the process of gene activation and correlated with their postulated functions. It is shown that these "scaffold-attached regions" (S/MAR elements) have some characteristics in common with and others distinct from enhancers with which they cooperate in various ways. Our model of S/MAR function will focus on their properties of mediating topological changes within the respective domain.

KEY WORDS: Attachment sequence, Chromatin structure, DNA core-unwinding element, FLP recognition target sites (FRT), Insulator elements, Matrix-attached regions (MARs), Metallothionein, Nucleosomes, Nucleosomal positioning, Retroviral integration sites, Scaffold-attached regions.

I. Introduction

Within the nucleus, DNA is packed into nucleosomes as a 10-nm string that is folded further into a compact fiber of 30-nm diameter. Transcriptional competence is critically dependent on local disruptions of this structure in order to permit the recognition of DNA sequence by transcription factors. This function is under the control of a master switch like a locus control region (LCR), the action of which has to be limited to a particular domain (Fig. 17). The nature of the bordering elements that define such a domain is the main subject of this article.

An organization of the eukaryotic genome into topologically independent domains was originally concluded from electron microscopy on histone-depleted metaphase chromosomes that revealed a supporting structure (scaffold) to which DNA was attached in the form of superhelical loops (Paulson and Laemmli, 1977). Later studies confirmed this packaging principle by thin sectioning (Marsden and Laemmli, 1979) and immunofluorescence. A recent refinement showed that the scaffolding is mirror-symmetrically helically coiled to yield compact sister chromatids (Boy de la Tour and Laemmli, 1988). The bands that are well known for intact chromosomes were found to be generated by an organized line-up of AT-rich sequences (the "AT-queue"; Saitoh and Laemmli, 1993, 1994) which are believed to represent the "scaffold-attached regions" (SARs) or matrix-associated regions (MARs) found at the borders of numerous eukaryotic genes. The relative use of terms "SAR" or "MAR" entirely depends on the protocol applied for their characterization which will be dealt with in Section II,C,1, and more fully in Kay and Bode (1995). Since there is reason to believe that these protocols uncover the same functional elements, it has become common to use a consensus term, "S/MAR."

In this chapter, we will review the domain structure of the inducible human interferon-β gene in light of these models and discuss their implication for gene function in interphase nuclei. By excising the bordering elements and using them for constructing prototype minidomains we will try to elucidate the nature of the domain borders and relate them to properties of other chromatin domains.

II. Human Interferon-β: Does It Form a Prototype Domain?

A. Initial Observations: Gene Expression Depending on Flanking Sequences

The interferons (IFNs) are a class of cytokines inducing an antiviral state in many cell types against a wide range of viruses (Bode and Hauser, 1992). All human class I IFN genes (encoding at least 15 type α-IFNs and the single β-IFN) are localized on chromosome 9 and lack an intron—an unusual situation in higher eukaryotes. Virtually all cell types produce one or more class I IFNs upon induction with virus or a synthetic inducer like double-stranded RNA (dsRNA). Because IFNs have numerous functions and, in particular, antiproliferative properties, their synthesis is strictly controlled.

For the study of the induction process, primary human fibroblasts, such as FS-4 cells, which express and secrete human IFN-β (huIFN-β) in response to dsRNA, have played a primary role. Gross *et al.* (1981) isolated and characterized a 36-kb genomic region surrounding and including the single human gene which was then transfected and expressed in murine cells by Hauser *et al.* (1982). For this purpose, a cosmid containing these 36-kb (corresponding to the segment spanning fragments (G–A in Fig. 1) and a plasmid containing the HSV-tk gene as a selection marker, were co-transfected into mouse Ltk$^-$ cells. From the resulting clones pCosIFN-β could be reisolated by cosmid rescue suggesting that some if not all of the cosmid copies are integrated head-to-tail, forming a tandem array.

A number of cell clones have been isolated from these transfections and characterized for the properties regarding gene expression (Table I) and chromatin constitution (see Section II,B). Considering the fact that the expression of the human IFN gene occurs in a heterologous background, completely depending on the mouse induction system and factors, the IFN (calculated on a per-copy basis) was surprisingly high (Table I). This was also true for a parameter called "cotransfer efficiency" marking the percentage of cells that survive the selection procedure (here: HAT selection) and at the same time produce more than a threshold value of huIFN-β (Collins, 1983, 1984; Klehr and Bode, 1988). The exceptional properties of cell lines obtained by the transfer of long segments of genomic DNA became even more unique when they were compared with the related results obtained from shorter segments, other modes of transfer, or episomal vectors.

The 1.8-kb *Eco*RI fragment "F" (spanning DNA between positions -284 and $+1545$ relative to the transcriptional start site, Fig. 1) and its 1.6-kb *Eco*RI–*Hind*III- and 0.9-kb *Eco*RI–*Bgl*II subfragments cover the huIFN-β gene and all its known control sequences (negative and positive regulatory domains, i.e., NRD and PRDI to PRDIV, cf. Du *et al.,* 1993). These fragments have been used extensively to gain information about the induction mechanism of this gene. Some representative examples are reproduced in Table I to give an idea about the interdependence of gene transfer procedures, copy numbers, and nuclear localization on one hand and the induced maximum levels of gene expression on the other. Even this restricted selection of early data which are based on the same induction protocol involving dsRNA allows a number of conclusions that are supported by the evidence that accumulated over a decade.

1. Standard transfection protocols yield high numbers of integrated copies although some reduction can be achieved if template DNA is transfected in a linear rather than supercoiled state and if carrier DNA is omitted (Klehr *et al.,* 1991).

2. Electroporation protocols can be optimized to yield a high proportion of single-copy integration events enabling relevant controls for the work

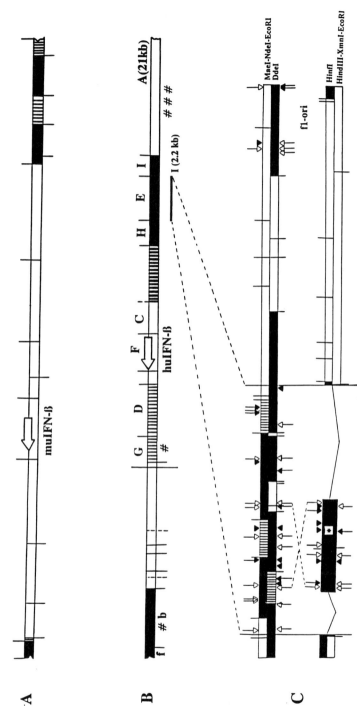

A

muIFN-ß

B

G D   F C   H E I   A(21kb)

fl   #   huIFN-ß   # # #

b   I (2.2 kb)

C

MaeI-NdeI-EcoRI
DdeI

fl-ori

HinfI
HindIII-XmnI-EcoRI

TABLE I

Dependence of IFN-Expression on the Length of Genomic DNA[a]

Cell	Gene transfer	Cotransfer efficiency	Copies per cell	Rel IFN-β yield
FS-4 (diploid fibroblast)	(Endogenous)	—	2	*100%*
Ltk⁻ × IFN[36]	(Co)transfection	>90%		
Clone 12₃			30	10%
Clone 2₄			100	10%
L × IFN[0.9]	(Co)transfection	40%		
Clone 1			80	0.5%
Clone 2			2	3%
L × IFN[0.9]	(Co)electroporation	1%		
Clone 1			1	0.5%
Clone 2			1	0.1%
C127 × BPVIFN[1.6]	Transfection			
				0.5%
				0.5%

[a] Superscripts next to IFN in the first column mark DNA length in kilobases.

to be reported. A low cotransfer efficiency is a logical consequence of the fact that the simultaneous transfer from solution of the physically unlinked plasmids containing the IFN or selection sequences is a rare event.

3. Appropriately constructed, BPV-based episomal vectors maintain a rather constant and high number of (nonintegrated) copies (DiMaio *et al.*, 1982; Goodbourn *et al.*, 1986). However, critical controls revealed long-

FIG. 1 Chromatin domains for β-type interferons: Overview and summary of data. A: The murine IFN-β gene (muIFN-β) is part of a 22-kb domain (J. Bode, unpublished). Solid lines pointing up: EcoRI-sites. Solid lines pointing down: HindIII sites. B: The human IFN-β gene (huIFN-β) is embedded into a 14-kb domain (Bode and Maass, 1988). Solid lines pointing up: EcoRI-sites separating fragments A–H (part of cosmid pCosIFNβ, sorted according to size, see Gross et al., 1981). Solid lines pointing down: BglII fragments. Dashed lines pointing down: HindIII fragments. Regions of repetitive DNA have been marked by "#." C: Common sequence motifs within the scaffold attached region "E" and the adjacent vector sequences (control). Solid arrowheads: ATATT; light arrows: Topo II consensus (*Drosophila:* GTNA/TAT/CATTNATNNG/A); solid arrows: Topo II consensus sequences for which the core (AT/CATTN) is ATATTT. Segment 'E' has been subdivided by the intrinsic *Mae*I sites (lines pointing up) and *Dde*I sites (lines pointing down) and the SAR-properties of the individual subfragments have been compared. The asterisk stands for the core unwinding element within S/MAR fragment E. Black/strong crosshatched/light crosshatched/white regions mark a very strong/significant/minor/lacking affinity for nuclear scaffolds (see Section II,C). Adapted with permission from Mielke et al. (1990). Copyright 1990, American Chemical Society.

term deletion and integration events demonstrating the instability of this expression system (Klehr and Bode, 1988).

4. Short integrated sequences give raise to a large interclone variation of expression levels while long intregrated genomic sequences show a much more uniform high-level expression. On a per-copy basis, the latter are much superior to the BPV episomes for which only a subfraction responds to the induction protocol (Klehr and Bode, 1988).

Most of the studies to follow will be based on the high-copy cell clone 2_4 (Table I).

B. Studies on Chromatin Structure

1. Nucleosome Mapping and DNase I-Hypersensitive Sites

Being one of the most strictly controlled, inducible genes, huIFN-β offers the unique possibility to study the chromatin procedure during the activation process. This has been exploited in some of our early studies (Bode *et al.,* 1986) using indirect-end labeling techniques as indicated in Fig. 2. As an example, genomic DNA that had been reisolated from nuclei treated with a mapping agent (MNase, DNase I, MPE, or BAA), could be cut with *Hinc*II (the site at position +69), sorted according to size on an agarose gel, and visualized (after blotting) by a short, cloned, and radioactively labeled probe covering the sequence between -69 and -14. For this setup, the autoradiograph reveals the structural features upstream of the *Hinc*II reference site. The same genomic DNA, cleaved with *Nco*I (the site at position -14) and processed as before will then disclose the analogous features downstream from the *Nco*I reference site. Using MNase, an enzyme cutting in the DNA linkers between nucleosomes but in a manner that is also dictated by DNA sequence (Hörz and Altenburger, 1981), a dramatic induction-dependent change in accessibility was noted: while there was barely any cutting in the quiescent state, the cuts after induction revealed many features of histone-free DNA. Clearcut nucleosomal patterns could only be derived by another nuclease (DNase I) and a chemical mapping agent, methidiumpropyl–EDTA–iron(II) (MPE).

Endogenous nucleases were already sufficient to reveal the predominant hypersensitive site upstream to the huIFN-β gene at a position, which at that time was unexpectedly remote from the start of transcription (-1730, see lane 1 in Fig. 3). Addition of trace amounts of DNase I intensified this band and higher concentrations of the same enzyme produced a fragmentation pattern reminiscent of a nucleosomal repeat (Fig. 3 and Bode *et al.,* 1986). A notable feature in these analyses was the fact that a region further

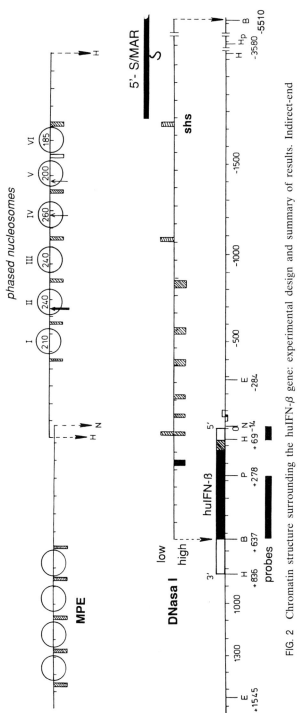

FIG. 2 Chromatin structure surrounding the huIFN-β gene: experimental design and summary of results. Indirect-end labeling analyses were obtained using the indicated probes for DNA isolated from 2$_4$cells (corresponding to fragments G through A in Fig. 1B). A register of six-phased nucleosomes (I–VI) was detected upstream and a similar register of at least four nucleosomes downstream from the huIFN-β gene. The dominant DNase-I sensitive site has been marked (shs, superhypersensitive site) and a subsequently identified scaffold attached region has been included (5' S/MAR).

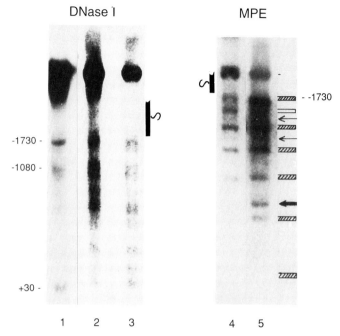

FIG. 3 Chromatin structure upstream from the human interferon-β gene embedded in 36 kb of genomic DNA (fragments A through G in Fig. 1), analyzed as part of the genome of Ltk cells (line 2₄, cf. Table I). Lanes 1–3: Increasing activities of DNase I were added to reveal the superhypersensitive site at −1730 bp (lane 1) and aspects of the nucleosomal organization (lanes 2, 3). Lanes 4, 5: Nucleosomal borders upstream of huIFN-β were visualized using the chemical mapping reagent MPE. MPE was applied to cell nuclei prepared from cells before (lane 4) and after induction of the huIFN-β gene (lane 5). Note a secondary nucleosome register (arrows) and an induction-dependent new nucleosome border (wide arrow).

upstream from position −1730 remained inaccessible to the enzyme under all physiological and experimental conditions.

MPE is a double-stranded DNA (dsDNA)-cleaving reagent which, at this level of resolution, lacks a sequence specificity. The methidium moiety is used to deliver a chelated Fe(II) ion to regions that are able to accommodate the intercalator and strand scission is then initiated by a hydroxy radical generated during a Fenton-type reaction from H_2O_2 (Cartwright and Elgin, 1989). This chemistry forms the basis of the now common hydroxy radical footprinting technique but in the present case the reaction gains added specificity from the fact that the internucleosomal linkers are the primary sites of intercalation and thereby cutting. In contrast to MNase, an obvious and very well-defined ladder of nucleosomes could be visualized upstream and downstream from the gene while the immediate control region and the coding sequences appeared nucleosome free. These analyses (exemplified by lane 4 in Fig. 3) showed that the 100 copies of the 36-kb fragment of DNA surrounding huIFN-β are organized into an almost uniform nucleosomal pattern terminating at the dominant DNase I-hypersensitive site characterized before. Controls on cells transfected by shorter fragments of DNA lacked some or all of these features.

For the sake of simplicity we will at this point introduce the fact that the -1730 site marks the end of the putative chromatin domain, i.e., the beginning of an extended scaffold-attached region (S/MAR) covering 7 kb (Fig. 1). While static nucleosomes around S/MARs have been described only for particular cases and it seems premature to imply them in domain bordering functions (see Section II,B,3), we asked whether the occurrence of a constitutive DNase-I hypersensitive site could serve as a general and convenient marker to define the boundaries of regulatory units in the eukaryotic genome.

Constitutive hypersensitive sites have been documented close to the presumptive end(s) of quite a number of eukaryotic gene domains.

- Sites of preferential DNase I cleavage flank the human β-globin gene cluster, most notably a quadruplet of sites (I–IV, called HS1–HS4 by others) that serve the function of the locus-control region (Fraser *et al.*, 1993). Most of these sites are typical of erythroid cells and hence of an active state of the domain. In addition, two constitutive sites (called "V" and "VI" by Dhar *et al.*, 1990, and located -21.4 kb from the 5' end of the ε gene or 3' from the β-gene, respectively) are detected close to the regions that have been identified as S/MARs by Jarman and Higgs (1988). The S/MAR coinciding with site V acts as a prototype insulator in full accord with its supposed bordering function (Li and Stamatoyannopoulos, 1994a). Other S/MARs within the domain comap with important regulatory regions (see Section V,E).
- A somewhat related anatomy has been described for the α-globin locus from chicken (de Moura Gallo *et al.*, 1991; Razin *et al.*, 1994). Moreover, both S1 nuclease (Targa *et al.*, 1994) and topoisomerase II (Razin *et al.*, 1993) could be used to excise full-domain-length DNA using sites within the MARs reflecting the distinct properties of the attachment sequences.
- Similarly, the active chromatin domain of the chick lysozyme gene has S/MARs at its boundaries and these again occur in close association with hypersensitive sites (Bonifer *et al.*, 1991).
- "Special chromatin structures," (scs and scs') were found in the 87A7 heat shock locus of *Drosophila melanogaster* (Kellum and Schedl, 1992; Vazquez *et al.*, 1993). Both elements are located downstream from the two divergently transcribed 87A7 hsp 70 genes, at the

edge of the chromatin domain that decondenses to form a puff after heat shock. Although scs and scs' are no SARs by strict definition, they have a number of properties in common such as a large nuclease-resistant core spanning a DNA segment that is very AT rich and flanked by DNase I-hypersensitive sites. Moreover, after heat shock, both elements are primary targets for the action of Topo II.

- Using a constitutive hypersensitive site (HS4) 5' to the end of the chicken β-globin cluster as a marker, Chung *et al.* (1993) discovered a putative domain-bordering element that blocks the action of enhancers in a way resembling scs and scs'. Thus far, this situation is analogous to the site V element in the human β-globin locus. However, this element, meanwhile reported to be GC-rich, has no S/MAR activity *in vitro* and if it were matrix-attached it would be so by a different but possibly related mechanism.

This brief list of examples suffices to indicate that constitutive DNase I-hypersensitive sites are quite common (although not unique) markers for a domain border even despite the fact that there may be more than one type of bordering element.

2. Nucleosome Reconstitution

Both transacting factors and particular sequence features of DNA can be responsible for positioning nucleosomes. Certain factors that bind to DNA during the nucleosome assembly process can serve as "bookends" to define an orderly array of nucleosomes. Proteins attaching the domain bordering elements (S/MARs) to the nuclear matrix or scaffold would be excellent candidates to have such a role and this idea was actually the starting point for the studies in this chapter. Alternatively, DNA sequences with an intrinsic curvature or bendability prefer to be accommodated within a nucleosome causing phased arrays of nucleosomes even in the absence of additional proteins (Wolffe, 1994).

In an extended series of experiments, we reconstituted a number of DNA segments from the human interferon domain (Ríos-Ramírez, 1994). To this end, we transferred preformed functional histone octamers from reconstituted core particles to the cloned and end-labeled fragments *in vitro*. Reconstituted complexes were then cleaved by either MNase or MPE and cleavage sites were visualized by autoradiography of the gels used for fragment separation. Some representative results, which include the Fig. 3 data, are summarized in Fig. 4. This overview indicates that in a simple reconstitution involving only DNA and histones, the position of one nucleosome (nucleosome IV in Fig. 2) is invariant whereas other nucleosomes appear to extend the nucleosomal ladder, more or less by a passive spreading mechanism. Analyses on the sequence underlying the "strong" nucleosome show that

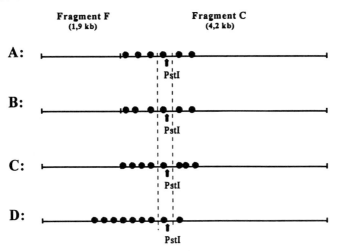

FIG. 4 Correlation of nucleosomal positions *in vivo* and after reconstitution of DNA with core histones. A: Nucleosomes mapped in cell nuclei (see Figs. 2 and 3); B: Nucleosomes reconstituted on long DNA (comprising fragments C, H, and E), using limited amounts of octamers; C: Nucleosomes reconstituted onto fragment C; D: like B but using stoichiometric amounts of octamers.

it is suited to determine a defined rotational setting over a restricted range (see Pina *et al.*, 1990, for a similar analysis on a regulatory nucleosome within the MMTV-LTR). This sequence allows runs of A,T to expose their major grooves whereas runs of G,C are permitted to orient their minor grooves facing out. In conclusion, the positioned nucleosomes close to the upstream domain border are primarily directed by sequence, possibly in a way keeping the regulatory elements nucleosome free. Therefore, the borders themselves have at most an indirect function in preventing the passive spreading of nucleosomes into the regions implicated in scaffold attachment.

Along these lines, we also investigated the nucleosome-organizing capacity of the S/MARs themselves by the same techniques. These S/MARs are the regions that appeared to be tightly protected *in vivo* (Fig. 3). In the present analyses on reconstituted S/MAR–histone complexes (Fig. 5) they exhibit an irregular pattern of cleavage sites indicating the absence of at least positioned nucleosomes. At this point we had to address the question about the nucleosome organizing potential of bordering DNA sequences *in vitro* and *in vivo*, i.e., in a situation where histones and scaffold components compete for the occupancy of S/MARs.

3. Bordering Elements: Are They Nucleosomally Organized?

Early work in the field of nucleosome reconstitution indicated that homopolymeric DNAs such as $d(A)_n(dT)_n$ cannot be reconstituted into nucleo-

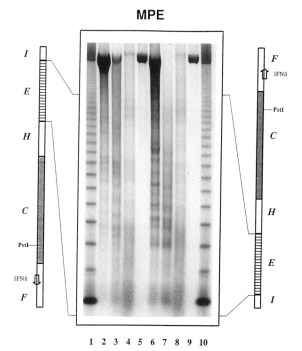

FIG. 5   Lack of positioned nucleosomes on an S/MAR fragment (E). Lanes 1 and 10: 123_n ladders of marker fragments. Lanes 2, 3 and 6, 7: Degradations at 0.25 and 0.5 mM of MPE, respectively. Lanes 4 and 8: Controls degradations on histone-free DNA. Lanes 5 and 9: no MPE addition.

some core particles and this initial observation has been explained by a reduced number of base pairs per turn (Travers, 1987). Only a more recent assay, performed under competitive conditions, demonstrated that this homopolymer reconstitutes as well as heterogenous sequence DNA (Puhl *et al.*, 1991). Since S/MARs contain many A tracts, which by some authors are believed to represent their dominant sequence characteristics (Laemmli *et al.*, 1992), it was not at all clear at the time of our study whether or not S/MARs have a nucleosomal organization. If S/MARs were devoid of nucleosomes, they would be permanently free for an interaction with scaffold proteins like histone H1 or topoisomerase II without any interference arising from nucleosome assembly processes. This would fit well into the conventional concepts implying the existence of constitutive domain borders. The problem was tackled on different levels.

Our initial assays used a chromatin reconstitution procedure as before but applied unlabeled S/MAR- and non-S/MAR fragments. After a nucleosomal ladder was generated by MNase, electrophoresis, and blotting, different probes were applied to visualize the contribution of mono- and oligonucleosomes to various sequences. We show in Fig. 6A that sequences positioning nucleosomes (fragment C) and sequences devoid of such property (fragment E) are indistinguishable if compared by this procedure and the same applies for any of the other fragments (not shown). The various bands that appear during the end-labeling technique in Fig. 5 are hence ascribed to nucleosomes forming various alternative registers.

During the next stage, a related assay was performed on chromatin that had been fragmented by MNase inside nuclei of the 2_4 cell line. The corresponding Southern blots in Fig. 6B demonstrate that the above conclusion can be extended to the properties *in vivo*, at least for a cell line that harbors multiple transfected copies of the genomic region including huIFN-β.

The final level of analyses was based on the endogenous IFN-β gene of human osteosarcoma MG-63 cells, which in the past have been used for a highly efficient production of huIFN-β (Billiau *et al.*, 1977). Since the signal strength obtained from this line did not permit detection by the above analysis scheme, we introduced a new protocol during which mononucleosomal DNA from the huIFN-β genomic range was used as a probe to signal the presence of mononucleosomes on any of the fragments G through I specified in Fig. 1. In Fig. 6C we reproduce a result obtained by hybridizing with such a probe, derived from an intense MNase digest that left only the most stable mononucleosomes. The Southern blot reveals the strongest signals for fragments C (containing the phased nucleosome register and the beginning of the upstream S/MAR) as well as E and I (from the distal end of the 7 kb S/MAR). The interpretation is straightforward: typical S/MAR elements are also organized into nucleosomes *in vivo;* the nucleosomes, although not positioned, are significantly more stable than those from the coding fragment (F) the other fragments from the domain's interior (G and D) and those from the vector sequences (pTZ). If S/MARs were permanently attached, this property would then have to be mediated by the association of scaffold protein with the internucleosomal linkers (see the model by Laemmli *et al.*, 1992).

Is there a precedent for the nucleosomal organization of bordering elements or the adjacent stretches of DNA? Strätling *et al.* (1986) have performed chromatin mapping studies analogous to Fig. 6B for the chicken lysozyme domain and described two classes of nucleosomes for noncoding sequences. While the coding sequences themselves lack any nucleosome repeat, the adjacent nucleosomes from the domain possess a disturbed nucleosomal organization (class I). It should be mentioned that the lysozyme domain contains eight scattered hypersensitive sites marking particular chromatin structures in regions with a now established enhancer func-

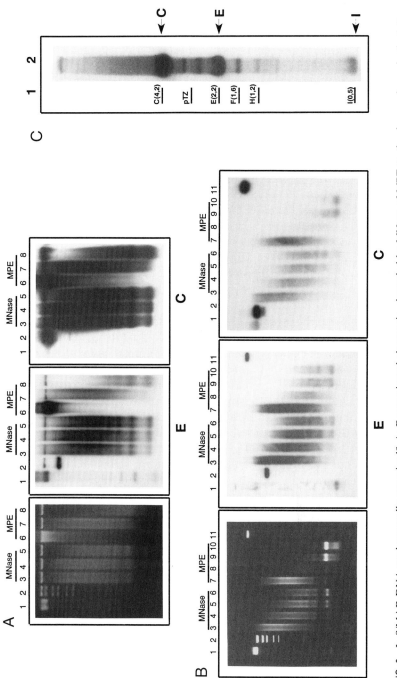

FIG. 6 Is S/MAR-DNA nucleosomally organized? A: Reconstituted chromatin, degraded by MNase of MPE to the degree shown in the left-hand ethidium-stained gels and hybridized to fragments E or C, respectively. B: The corresponding analyses on multicopy integrates in murine 2₄ cells. C: Stable nucleosomes on the huIFN-β domain in human MG63 osteosarcoma cells. A nucleosomal ladder was produced and analyzed on ethidium-stained gels as above. Mononucleosomal DNA was isolated and hybridized to a cloned DNA covering fragments C, H, E, and I; nonhybridizing portions were removed on an agarose gel. Subsequently, the hybridized part was recovered by denaturing gel electrophoresis and labeled by random priming. The resulting probe was used to trace nucleosomes within the S/MAR fragments C, E, and I.

tion. The bordering regions (S/MARs, first termed attachment or "A" elements) generated an extended regular nucleosomal ladder (class II) indicating the existence of stable nucleosomes on the flanking sequences).

Regarding the arrangement of nucleosomes at bordering elements the information is scarce and restricted to special cases. Worcel and co-workers were the first to describe the chromatin organization of the 5-kb histone-gene repeat in *Drosophila* (Worcel *et al.*, 1983) which appeared as a static 200-bp ladder in a region including the attachment complex localized by Gasser and Laemmli (1986b). Two of the 200-bp protected regions met the criteria of a protein–scaffold complex—they resisted the extraction by lithiumdiiodosalicylate (LIS) and formed a barrier to *Exo*III digestion. They were therefore believed to be different from nucleosomes. Altogether, being positioned within a nucleosomal ladder, this 657-bp S/MAR-fragment is too short to serve as a model for nucleosome occupancy and positioning.

S/MARs located 5' to the *Drosophila* hsp70 genes at both loci, 87C1 and 87A7, also fall into regions of static, phased nucleosomes the pattern of which is maintained at phases of transcriptional activity (Mirkovitch *et al.*, 1984; Gasser and Laemmli, 1987). Again, using the nucleosomal ruler, the length of the attachment region (960 bp) is rather short prohibiting any generalization. This particular model is complicated by the fact that, in addition to the S/MAR-sequence on the nontranscribed spacer between the divergent hsp70 genes, putative bordering elements (scs and scs') have been localized downstream from these genes (Kellum and Schedl, 1992, see Section V,C). As mentioned above, although these elements share many properties with S/MARs, they do not meet the *in vitro* criteria that are common for such a classification.

For some S/MARs the MNase cleavage pattern within the linkers of nucleosome-sized particles is matched by the topoisomerase II cleavage pattern produced *in vivo* (Käs and Laemmli, 1992). In addition to the S/MAR separating the histone domains, internucleosomal cleavage is also observed in one of the linker regions of the two nucleosomes spanning satellite III, a centromeric AT-rich and S/MAR-like DNA sequence with a repeat length of 359 bp. Once more, regarding the particular nature of this repetitive DNA, no general conclusions about the nucleosomal architecture of S/MAR-DNA can be derived.

4. Transcription-Dependent Changes

The ordered control of transcription processes requires a subdivision of the genome into domains. Early work on prototype genes like the chicken α- and β-globin domains (Stadler *et al.*, 1980; Weintraub *et al.*, 1981), and the domain of ovalbumin-related genes (Lawson *et al.*, 1982) clearly demonstrated that a region of general DNase I sensitivity spreads beyond the borders of transcribing DNA within cells of the erythroid lineage but not in control tissues. In line with these observations, a housekeeping gene, GAPDH, formed a 12-kb-sensitive domain present in all tissues (Alevy *et al.*, 1984).

Work on the β-globin chromosomal domain in mature chicken erythrocytes showed that, despite a general repression at this stage, the domain maintains its nuclease-sensitive configuration (Verreault and Thomas, 1993). Similarly, when hormone is withdrawn eliminating transcription in hen oviduct, the entire ovalbumin domain remains in a sensitive configuration that is clearly different from the same domain in spleen, liver, or erythrocytes (Lawson *et al.*, 1982). These findings raise questions about both the initiation and maintenance of the open state. A possible mechanism has recently been proposed by Lee and Garrard (1991a,b) who demonstrated the role of positive superhelical strain generated during the first round of transcription. The topological stress arising in front of the polymerase alters the nucleosome structure and these changes may be maintained by the association of additional factors or the modification of histones. In our concluding section (VIII) we will delineate an alternative or additional mechanism in which the domain borders are programmed by changing their association mode during the first round of transcription.

So far, a convincing co-localization of S/MARs with regions of a decreasing, general DNase sensitivity could only be demonstrated in two cases, chicken lysozyme (Bonifer *et al.*, 1991) and human apolipoprotein (Levy-Wilson and Fortier, 1989). For lysozyme, the extension of the open chromatin region has been studied in oviduct tubular gland cells in which the gene is maximally active to produce lysozyme as one of the egg white proteins and in the macrophage lineage where it is controlled by a different set of DNase I-hypersensitive sites and hence *cis*-regulatory elements. In both cases the open domain had the same extension terminating within the attachment regions (S/MARs or A elements, see Sippel *et al.*, 1993). The chromatin domain encompassing the human apolipoprotein gene is defined in a similar way by S/MARs that colocalize with the decay of general DNase I sensitivity in cells where it is expressed. Interestingly, in this case a second 5' anchorage site was found for expressing cells suggesting that, depending on the transcriptional status, a miniloop may be formed from a facultative S/MAR within a constitutive domain.

The organization of the human β-globin locus, which frequently serves as a paradigm, is clearly more complicated than the previous examples. While it is tempting to test for the domain borders in regions where constitutive hypersensitive sites coincide with S/MARs (sites V and VI, see Jarman and Higgs, 1988), only the S/MAR upstream from the ε gene marks a region of decreasing general sensitivity (Forrester *et al.*, 1990; Dillon and Grosveld, 1993). At the other end, the domain stays open past the hypersensitive site VI (21 kb downstream from the β-globin gene) and even past another erythroid-specific developmentally stable hypersensitive site 100 kb 3' to the β gene. The sensitive status is entirely dependent on the locus control region (LCR, hypersensitive sites I–IV) between the ε gene and site V since the Hispanic thalassemia deletion of major parts converts all the mentioned regions to a DNase I-resistant conformation (Forrester *et al.*, 1990).

In all the above examples, transcriptionally competent, tissue-specific, or constitutive genes have been chosen as a model. Type I interferons are clearly different, since they can be expressed in nearly all cell types by the concomitants of a viral infection and since their transcriptional status does not depend on a developmental program. Except from certain transformed cells, the gene is totally silent before induction. After induction, a burst of transcription can be monitored followed by a shutoff regulated at both the transcriptional and post-transcriptional stages (Bode and Hauser, 1992). On the chromatin level, the accessibility to MNase in isolated nuclei is dependent on the active status of the domain (Bode *et al.*, 1986) to an extent that required a small chemical, MPE, to reveal aspects of the nucleosomal organization (Fig. 3). Induction causes a dramatic increase of MPE-mediated cutting close to the upstream end of the domain, in a region that normally accommodates three phased nucleosomes. Whether these nucleosomes are lost or undergo a modification permitting the reagent's intercalation is not known, but the topological consequences would be similar in both cases and will be a major point in the final discussion.

Using the criteria of DNase I or MNase accessibility, our attempts to delineate the extension of the huIFN-β domain for the quiescent gene have been unsuccessful altogether. Based on the "last-cut" approach (Fritton *et al.*, 1988) we followed the time of digestion within nuclei that is required to degrade fragments around the huIFN-β gene to a nonhybridizable size. As a model, we used the endogenous IFN-β copies of the human osteosarcoma cell line MG-63. Prior to induction, the sensitivity of fragments across the domain is indistinguishable. However, after induction, fragments from the domain's interior and from its putative borders behave strikingly different (Fig. 7).

- The coding region shows the expected sensitization which is ascribed to ongoing transcription.
- The immediate upstream reagion from which positioned nucleosomes are lost during phases of gene activity (C_{nuc}) is also sensitized to a significant extent.
- Starting with the putative upstream domain border (close to the *Pst*I-site intrinsic to C) DNA becomes *de*sensitized, i.e., more resistant to nucleolysis. Quite remarkably, this phenomenon is most pronounced in the core of the S/MAR region (as defined in Section II,C).

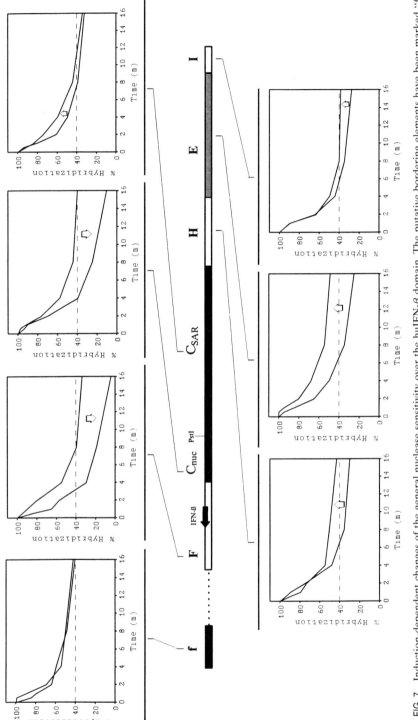

FIG. 7 Induction-dependent changes of the general nuclease sensitivity over the huIFN-β domain. The putative bordering elements have been marked "f" (downstream border) and "C$_{SAR}$-H-E-I" (upstream border; cf Fig. 1B). The coding and promoter regions are part of fragment F; C$_{nuc}$ marks the upstream region containing a register of six-phased nucleosomes. Analyses follow the last-cut approach essentially as described by Fritton *et al.* (1988) but varying the time of exposure to nuclease rather than the enzyme's activity. Traces are reproduced for the quiescent and active gene; the direction of activity-dependent changes is marked by arrows.

- The domain's far downstream end (fragment "f") remains unchanged, which may be a consequence of its larger distance from the site of transcription (Fig. 1), its shielding due to intervening S/MAR sequences (fragments "b" and "G") or to the fact that it is influenced by the positively supercoiled part of a transcribed twin domain (Wu *et al.*, 1988). Note that the sensitivity of fragments b and G cannot be addressed in human cells owing to their content of repetitive sequences (see Fig. 1).

While most mapping techniques involve some degree of perturbance by the preparation of cell nuclei and their incubation in the respective media, bromoacetaldehyde (BAA) is a single-stranded DNA-specific reagent that can be applied to living cells. The reagent marks important regulatory sites that are otherwise sensitive to nuclease S1 (Kohwi-Shigematsu *et al.*, 1983) and has been applied by us in order to follow the induction of the huIFN-β gene(s) (Bode *et al.*, 1986). Surprisingly prior to induction there was no labeling close to the coding region. Instead, labeling started around the dominant DNase I-hypersensitive site and continued, with increasing intensity, in the upstream direction that lateron qualified as a scaffold-attached region (Bode and Maass, 1988). After gene induction, there was an immediate decrease of labeling in this region that followed the time course of transcriptional activation (Fig. 8). On balance, the transcribed region gained reactivity while more extended upstream districts accumulated the label lateron.

Like the MPE-mapping analyses in Fig. 3, these data confirm that all 100 copies of transfected DNA undergo transcription-dependent alterations in synchrony. The initial single-stranded character of protein-associated S/MAR sequences appears to be a more general phenomenon since it was also found in other cases by using OsO_4 as the labeling reagent (Paul and Ferl, 1993). Regarding the unwinding properties of these sequences that were established by subsequent assays *in vitro*, these findings will be a central aspect when we will develop our view of S/MAR functions in the final chapter.

C. The Putative Domain Borders of the Human IFN-β Domain Are Prototype S/MARs

S/MARs were determined by assays *in situ* ("halo-mapping") and *in vitro* ("scaffold-reassociation"). The nuclear matrix is the protein framework involved in the organization of interphase chromatin. Initially, the nuclear matrix was isolated by extracting nuclei with high-salt and DNA disgestion (Berezney and Coffey, 1974). Later modifications of this salt extraction procedure start with an intermediate-salt step for a better preservation of

time post induction (h)

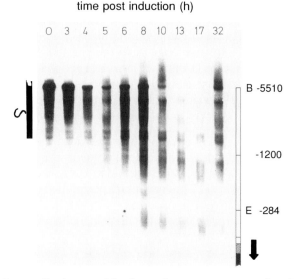

FIG. 8   Induction-specific changes of the chromatin structure upstream from huIFN-β, monitored by treating whole cells with the single-strand specific reagent bromoacetaldehyde. Prior to induction, reactivity is restricted to a region that has subsequently been identified as an S/MAR element (symbol to the left).

morphological features (Belgrader *et al.*, 1991) or to prevent any artifactual precipitation of transcription complexes, which has frequently hampered the discussion of matrix functions (Mirkovitch *et al.*, 1984; Roberge *et al.*, 1988).

An alternative approach for the resolution of artifactual precipitation and rearrangement or sliding phenomena was the replacement of salt by a supposedly mild detergent, LIS, to extract the majority of histones and many nonhistone proteins. Analogous to the residual structures obtained from metaphase chromosomes, the remaining entities were termed nuclear scaffolds to distinguish them from the nuclear matrix structures obtained by salt extraction. Today, LIS extraction is thought to dissociate most transcription-dependent associations explaining the fact that—unlike for the high-salt procedure—an attachment of DNA is detected, for the most part, independent of the transcriptional status.

For the original *in situ* assay, halos (the complex of scaffold proteins and nucleic acids) were degraded with a set of restriction enzymes to trim down to a minimum the DNA regions that are in contact with or released from the protein complex. A visualization of both populations on Southern blots then served to identify SARs (Mirkovitch *et al.*, 1988). Probably motivated by the subsequent finding that cloned DNA fragments containing "attachment regions" bind to complete nuclear scaffolds by competition with the endogenous sites, the term "SAR" was reinterpreted to mean "scaffold-associated regions" (Izaurralde *et al.*, 1988). These observations led us to refine a hybrid procedure in which nuclear halos are preincubated with an appropriate set of restriction enzymes and labeled restriction fragments were added to the resulting scaffolds in the presence of endogenous DNAs (here called the "*in vitro*" procedure). To achieve a specific reassociation, we added *E. coli* genomic DNA at an excess up to 10,000-fold. These stringent conditions permitted a direct comparison of associated and nonassociating DNA by autoradiography and led to the conviction that SAR-type DNA is exclusively found in eukaryotes (Mielke *et al.*, 1990; see Kay and Bode, 1995, for an updated protocol).

Although they do not strictly depend on a stabilization step, specific SAR-scaffold interactions become particularly evident if nuclei are incubated prior to LIS extraction at 37 to 42° for 20 min, resulting in "matrix I" type scaffolds. As any standard matrix preparation, matrix I contains three components, the peripheral nuclear lamina, an internal protein network, and a residual nucleolar structure. SARs associated with a type I matrix are also found to recognize metaphase scaffolds. Since the latter have a much simpler protein composition than interphase, it appeared that the specific scaffold–DNA interaction is limited to the respective set of proteins that does not involve the lamins. Later reinvestigations on empty nuclear shells (so-called matrix II structures), however, showed the lamins to account for at least 50% of binding (Ludérus *et al.*, 1991, 1994) indicating that several components of an interphase matrix or scaffold are likely to contribute.

We have identified the SARs surrounding the huIFN-β domain by the *in situ* and *in vitro* procedures, respectively, both based on LIS-extracted scaffolds (Fig. 9) and have confirmed that essentially the same results are obtained with the high-salt procedure of Cockerill and Garrard (1986) that has been developed for the identification of matrix-associated regions (MARs). All our analyses establish a strong affinity for fragments C (localized proximal to the 5' end of the gene, see Mielke *et al.*, 1990), H, E, and I which form a continuous 7-kb S/MAR element. At least 21 kb of sequences further upstream are totally devoid of any binding potential. Using the same criteria, the downstream domain border was localized to fragments b and f, which cover a 5-kb segment (Fig. 1).

Without wishing to overemphasize the differences between the methods to define SARs (i.e., the detection of endogenous elements by the halo mapping, "*in situ*" method and the "*in vitro*" reassociation approach) and the minor but reproducible differences between the transcriptional states, we should mention that the affinity of fragments D and G is not fully exploited *in situ*. D and G represent a region of intermediate binding

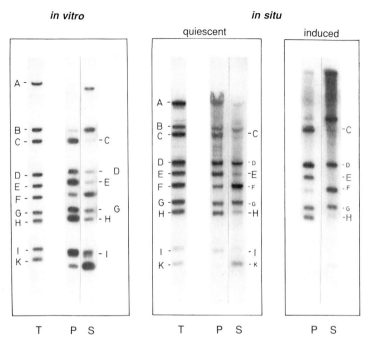

FIG. 9 S/MARs around the huIFN-β gene, analyzed by a scaffold-reassociation assay (*in vitro*) and by a halo-mapping procedure (*in situ*). The *in vitro* assay uses labeled fragments that are allowed to reassociate with a stabilized scaffold obtained by LIS extraction. Attached fragments (P-fraction) and nonattached regions (S-fraction) are separated by centrifugation and visualized by autoradiography. Trace T reflects the total input mixture of fragments. For the *in situ* method, nuclear halos were prepared by LIS extraction of nuclei and stabilization. After extensive washing, halos were degraded by an overnight incubation with *Eco*RI and probed by Southern blotting either before (T) or after the separation of attached (P) and released fragments. Adapted with permission from Bode and Maass (1988). Copyright 1988, American Chemical Society. See also Kay and Bode (1995).

potential and we followed the idea that, as in the case of ovalbumin, the gene's induction could lead to forming a miniloop further supporting the transcriptional potential. While a comparison of data for the quiescent and induced states lends little support to such a mechanism, the coding-region fragment "F" appears to be less attached in the active state. Since the homologous murine gene (muIFN-β) lacks a region of intermediate SAR activity (Fig. 1), the role of a moderate attachment potential must remain unanswered.

Although to a first approximation, the LIS procedure tends to abolish activity-dependent differences of the loop organization (a fact that may account for its reproducibility), some data are available to suggest that the extraction of specific attachment factors may be incomplete. Käs and Chasin (1987) show that the SAR-scaffold interaction of a dihydrofolate reductase gene is lost at metaphase and may hence be mediated by an interaction with the lamins. Levy-Wilson and Fortier (1989) demonstrate for the apolipoprotein B gene that two distal SAR elements are detected in any cell while an additional 5'-proximal SAR, positioned between the promoter and the enhancer, is specific for cells expressing the gene. Similarly, Brotherton *et al.* (1991) demonstrate the tissue-specific attachment of an avian β-globin enhancer element by the *in situ* method using either salt or LIS; as before, this apparently specific interaction is lost during a reassociation (*in vitro*) approach.

III. S/MAR Structural Features Leading to Scaffold Association: Superhelicity, Strand Separation, and Single Strands

Although S/MAR families have been detected by cross-hybridization to the *Drosophila* heat shock S/MAR (Mirkovitch *et al.*, 1984), S/MARs do not generally contain common sequence motifs that would facilitate their recognition. Thus, the S/MAR scaffold interaction cannot be expected to be mediated by DNA-binding proteins that make precise sequence-specific contacts with the DNA bases.

Early analyses (Cockerill and Gerrard, 1986) revealed the presence in "S/MAR elements" (the consensus term characterizing the apparent identity of scaffold- or matrix-associating regions) of several, evenly spaced repeats of a motiv ATATTT which conforms to the core of the loosely defined consensus sequence for *Drosophila* topoisomerase II (GTNA/TAC/TATTNATNNA/G, see Sander and Hsieh, 1985). Although topoisomerase associates preferentially with S/MAR-type sequences (Käs *et al.*, 1993) and cleavage within S/MARs has been exploited to excise complete eukaryotic domains (Razin *et al.*, 1993), the predominant cuts are usually not found at such a consensus (Käs and Laemmli, 1992). This is a consequence of the fact that the enzyme's specificity is primarily guided by topological features, i.e., supercoils (Osheroff *et al.*, 1991) and supercoiling-induced structures like crossovers (Zechiedrich and Osheroff, 1990) and hairpins (Froelich-Ammon *et al.*, 1994).

Other properties that have been implicated in the scaffold-S/MAR recognition are the occurrence of oligo(dA) tracts and of DNA bending. Due to a propeller twist, base pairs in regions of oligo(dA)–oligo(dT) develop a system of additional H-bonds diagonally across the major groove which causes carbonyl oxigens to receive two H-bonds. Thereby these stretches become conformationally more rigid concomitant with a narrowing of the minor groove. Bending arises at the transition to B-type DNA owing to larger than usual roll angles (Nelson *et al.*, 1987). Prominent proteins recognizing these features are again topoisomerase II, histone H1, and HMG I/Y which colocalize with the AT-rich queues (S/MARs) that have recently been identified in metaphase chromosomes (Saitoh and Laemmli, 1993, 1994).

A nonrandom distribution of oligomeric (dA) tracts leads to macroscopically bent DNA if they occur with a 10–11-bp periodicity. Such a feature may attract proteins or protein complexes that are involved in topoisomerization, recombination, transcription or replication, or any process that requires DNA bending and will therefore occur in regions that are either bendable or bent (Travers, 1990).

We have marked the criteria mentioned so far on two pieces of DNA (X and XI, Fig. 10) which are comparable both in length and in base composition. XI covers a segment of mouse mitochondria DNA that contains an origin of replication and promotes a head-to-tail multimeric insertion after transfection into recipient cells (Lutfalla *et al.*, 1985). Although these properties are not unusual for S/MARs (Schlake, 1994), XI displays only minor S/MAR activity *in vitro* (Fig. 10). X is an authentic S/MAR discovered at a tobacco gene. It is one of the elements used to demonstrate the fact that S/MAR activities are conserved across species borders (cf., fragment W in Fig. 15; Mielke *et al.*, 1990; Dietz *et al.*, 1994). An inspection of Fig. 10 shows that for this couple of sequences, the repetitive occurrence of a hexanucleotide, ATATTT, is the most discriminating feature whereas,

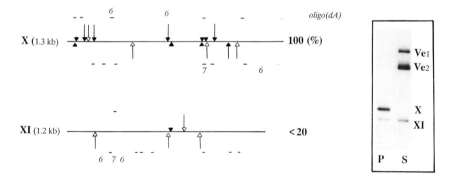

FIG. 10   Sequence motifs implicated in S/MAR function. An authentic S/MAR element from a light-inducible potato gene (X), which shows 100% binding to a scaffold preparation (insert) is compared with an equally AT-rich and similarly sized non-S/MAR DNA from mouse mitochondria (XI). Adapted with permission from Mielke *et al.* (1990). Copyright 1990, American Chemical Society.

obviously, the presence of oligo(dA) tracts, Topo II boxes, and potentially bent regions is insufficient to provide XI with a significant S/MAR activity *in vitro* (Fig. 10) and *in vivo* (Mielke *et al.,* 1990). These conclusions gain added support from a study by von Kries *et al.* (1990) that showed bending, although common in S/MARs (Anderson, 1986; Homberger, 1989) is dispensable for strong matrix binding.

In an attempt to study the contribution of superhelicity to the recognition of S/MAR sequences, we not only confirmed observations by Tsutsui *et al.* (1988) that supercoils are bound, but also made the unexpected observation that an S/MAR-free plasmid (pTZ) associated in a way indistinguishable from an S/MAR plasmid (pCL). Even more importantly, enzymatic activities surviving the LIS procedure converted pTZ and pCL with the same kinetic parameters (Kay and Bode, 1994). These residual activities, i.e., nicking, linearization, and topoisomerization (Fig. 11A) are all reminiscent of the presence of topoisomerase II as is the fact that binding of supercoils requires a threshold value of superhelical density (cf., the transition from the nonbinding to the binding state between α values of -13 and -15, Fig. 11A and Kay and Bode, 1994). The responsible protein was clearly different from the main activity mediating the binding of linear S/MAR fragments as the latter was extracted at LIS concentrations that left the association of supercoils unaffected (Kay and Bode, 1994).

The number of accessible S/MAR sites for its part is a parameter dependent on the degree of extraction as S/MAR sites are successively uncovered between 5 and 8 mM of LIS while they are lost at 12.5 mM detergent concentration (Fig. 11B). Using our standard conditions (5 mM LIS), a nuclear equivalent binds about 20,000 S/MARs of 800 bp (SAR$_{800}$ in Fig. 11B) whereas only 3000–4000 specimens of the parent 2.2-kb fragment can be accommodated. In both cases, the association process occurs in a significant, cooperative manner (Mielke *et al.,* 1990). These parameters hint at a multiple-site attachment mechanism that is clearly different from the association of supercoils (pTZ$_{sc}$) that follows a hyperbolic saturation function and involves a larger population of sites. Subsequently we also confirmed the results by Hakes and Berezney (1991) who found a major population (in their case 150,000 sites per nuclear equivalent) of single-stranded DNA binding sites. An earlier observation by Probst and Herzog (1985) that matrix-associated DNA regions expose single strands had already pointed at a physiological role for this type of specificity.

The above experiments indicate a completely independent association of S/MARs and supercoils and this could be confirmed by the incapability of supercoils (pTZ$_{sc}$) to compete for S/MAR sites (Fig. 11B). Some minor competition monitored at higher concentrations of pCL$_{sc}$ is not in contradiction as it is a consequence of the already mentioned nicking and linearization activities that enable the recognition of the S/MAR-part intrinsic to pCL.

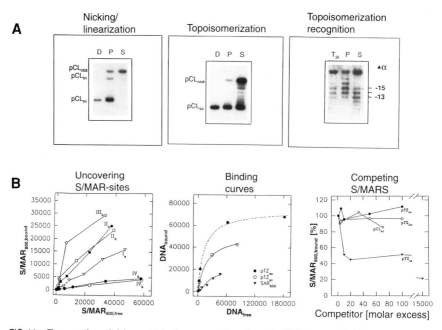

FIG. 11   Enzymatic activities and binding properties of a scaffold. Top row: Nicking, linearizing, and nicking-closing activities become evident if prototype scaffold is supplied with a super-coiled vector (pCL). Topoisomers are recognized only above a threshold superhelical density. Bottom row, left: Scaffolds, prepared by LIS concentrations increasing from 5 (I) to 12.5 m*M* (IV) are tested for their S/MAR-binding capacity in the absence (open symbols) and presence (solid symbols) of an equivalent amount of supercoils. Center: Saturation binding curves have been determined for single-stranded (pTZ$_{ss}$), supercoiled pTZ$_{sc}$), and linear S/MAR-DNA using a scaffold prepared by 5 m*M* LIS-extraction. Right: A 5-m*M* LIS-scaffold saturated with S/MAR$_{800}$ has been treated with the following potential competitors: S/MAR-free plasmid in its supercoiled (pTZ$_{sc}$), linearized (pTZ$_{lin}$) and single stranded form (pTZ$_{ss}$), and supercoiled S/MAR-plasmid (pCL). Adapted with permission from Kay and Bode (1994). Copyright 1994, American Chemical Society.

Finally, the most revealing observation concerned the competition of about half of the S/MAR sites by single-stranded DNA (pTZ$_{ss}$ in Fig. 11B) which suggested to us some overlap of S/MAR and ssDNA sites. It should be mentioned in this context that some of the most prominent S/MAR-binding proteins retained in LIS-scaffolds, the lamins, reflect the same mode of interaction (Ludérus *et al.*, 1994). Another abundant protein, scaffold-attachment factor A (SAF-A, alias hnRNP-U) recognizes both S/MARs and ssDNA although, apparently, at separate sites (Fackelmayer *et al.*, 1994; von Kries *et al.*, 1994).

Intrigued by the facts that single-stranded DNA can interfere with S/MAR binding while S/MARs themselves display single-stranded characteristics by their reactivity towards BAA (bromoacetaldehyde) *in vivo*, we investigated the conditions providing S/MARs with these properties. Since the sites that are reactive toward the unpaired DNA-specific probes BAA or CAA (chloroacetaldehyde) *in vivo* are often reactive toward the same reagents if they are part of a supercoiled plasmid, we tried to map the core unpairing elements by applying this technology to cloned S/MAR vectors. After modification, DNA was digested with a restriction enzyme, labeled on one end, and separated from unmodified portions by electrophoresis. The chemical cleavage at the site of CAA modification was then performed according to a Maxam and Gilbert sequencing reaction by successive treatments with hydrazine and piperidine. This chemistry is routinely used for a cleavage at C-residues and a C-ladder can hence be used to trace the additional cuts occurring on CAA-reacted A-residues (Fig. 12). For the

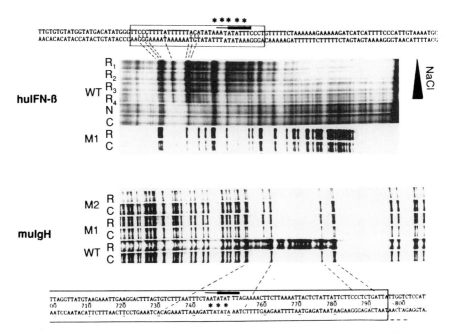

FIG. 12   Core unwinding elements in the huIFN-β upstream S/MAR and the murine immunoglobulin heavy chain enhancer (muIgH). Supercoiled plasmids with an S/MAR insert were reacted with chloroacetaldehyde (CAA). Modified DNAs were subjected to a C-type reaction according to Maxam-Gilbert in order to localize the modified A residues. Traces R: DNAs, reacted with CAA at increasing ionic strength (R_1–R_4). N: DNA subjected to the same chemistry but isolated from the unreacted portion of the gel used for the enrichment of R species. C: Controls from DNAs not pretreated with CAA ("C-ladders"). WT: wild-type sequences as shown. M1 and M2, the corresponding DNAs for which the core-unwinding elements have been destroyed by mutating the positions marked by an asterisk.

two examples probed first, the huIFN-β upstream S/MAR and the S/MAR sequences surrounding the murine IgH enhancer, cleavage occurred within the same AATATATTT motifs, and by increasing the ionic strength the nucleation site could be localized to the A residues of the AT-alternating section. As a control, mutations at these positions were sufficient to largely eliminate the reactivity. Since this dramatic change of unpairing occurs with little decrease of the overall AT content and is also dependent on this sequence being embedded in a particular environment, the design of S/MARs depends on a combination of factors that may involve the repetition of related motifs at certain intervals. At this point it is important to stress the fact that other S/MARs that have subsequently been mapped displayed core unwinding at different motifs as $(GA)_{10}$ and $(GAA)_{43}$ tracts embedded in a generally AT-rich environment (T. Kohwi-Shigematsu and J. Bode, unpublished). These motifs belong to the group of homopurine–homopyrimidine sequences that are common in the regulatory region of eukaryotic genes where they may form hypersensitive sites (Lu *et al.*, 1992) and attract binding factors implicated in rearrangements of the chromatin structure (Lu *et al.*, 1993). Both tracts form a dG·dG·dC-type triple helix exposing one single strand (Kohwi and Kohwi-Shigematsu, 1993, and private communication). Recent analyses by Boulikas and Kong (1993) suggest that S/MARs do not even need a threshold AT content as long as they develop, under torsional stress, single-stranded regions by B–Z-type transitions, the formation of triple helices or cruciforms.

Many of the S/MAR-specific structural characteristics are sensed by two proteins, SATB1 (special-At-rich DNA-binding protein) and nucleolin, which have been isolated via their recognition of an AATATATTT unwinding element as opposed to the respective nonunwinding mutant (Dickinson *et al.*, 1992, and unpublished). SATB1 is the prototype for a new class of DNA-binding proteins in that it recognizes a specific DNA sequence context ("ATC sequences") rather than a sequence consensus that is typical for proteins making base contacts through the major groove (Nakagomi *et al.*, 1994).

While we have demonstrated a critical role for the unwinding potential, for instance of $(AT)_n$-type sequences (Bode *et al.*, 1992), Laemmli and co-workers have stressed the fact that, on linear DNA, binding of prototype matrix proteins like histone H1 and topoisomerase II tends to initiate in the minor groove of A_n tracts and is competed for by distamycin (Adachi *et al.*, 1989; Käs *et al.*, 1989). It is noted that prototype proteins of both groups recognize most naturally occurring S/MARs and removing the unwinding potential of short synthetic oligomers by mutagenesis can reduce but not totally abolish a scaffold–DNA interaction (Mielke *et al.*, 1990; Bode *et al.*, 1992). Ludérus *et al.* (1994) have proposed that both types of interaction may be of relevance for the various physiological states since histone H1 is an important component of heterochromatin whereas proteins stabilizing single-stranded or unwound DNA are in the position to fix an activated state and to prevent rebinding of H1. These ideas will be incorporated into the concluding section (VIII).

IV. S/MARs and Transcriptional Regulation

In addition to the established promoter elements, the function of many specialized eukaryotic genes is supported by other *cis*-acting elements like the enhancers. In contrast to the immediate promoter sequences, enhancers work independent of their position, orientation, and distance from the gene. In this respect they resemble the more evolved LCRs which were first characterized for the human β-globin domain (Dillon and Grosveld, 1993). LCRs were detected as elements coinciding with DNase I-hypersensitive sites and found to confer position-independent expression upon their coordinate gene (van Assendelft *et al.*, 1989). In the following we will compare the properties of these elements with the S/MARs which will be shown to represent a third and separate group of DNA sequences with a stimulating effect on transcription.

A. S/MARs Enhancers and Locus-Control
Regions: General Properties

Figure 13 shows a stepwise reconstruction of the human IFN-β domain. Stable expression assays were performed to show increased gene activities as S/MARs were reattached to the gene. The first step already yielded a factor 15 increase and this correlated with the attachment potential rather than the presence of the dominant hypersensitive site between the white and crosshatched upstream areas. These observations indicated that a genuine, remote enhancer function was not involved.

Next we isolated the segments with an S/MAR activity and created a variety of single-S/MAR constructs containing these segments in their original as well as new positions, if possible in both orientations (Fig. 13, bottom). For all constructs the interferon titer was determined and the clones arising from an individual transfection were counted. Additional controls confirmed that clone numbers reflected the expression of the selector gene (neo[r]). These experiments allowed the following preliminary conclusions:

- Expression levels mirror the S/MAR activity *in vitro* (note that the entire fragment "C" behaves as a strong S/MAR element although

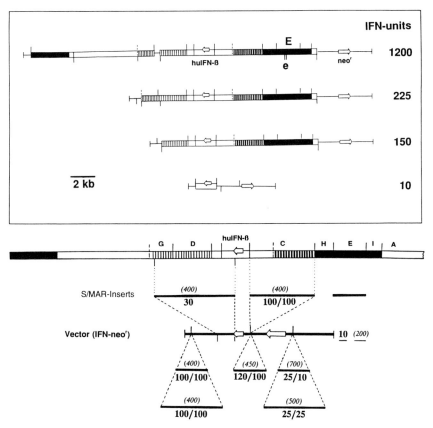

FIG. 13   IFN-β expression in partially and fully reconstituted chromatin domains and their variants. Top: Various portions of the domain have been attached to the gene and interferon units have been measured by a biological assay. Bottom: S/MAR-effects depend on SAR strength but are independent of the origin and orientation of the SAR element. The vector (IFN-neo[r], yielding 200 clones and 10 units of interferon per 10^6 cells subsequent to transfection) is the acceptor plasmid used to accommodate various DNA inserts; all clone numbers and IFN titers refer to a single insert at the respective location. Bold numbers separated by a slash refer to IFN units obtained for either of two possible S/MAR-orientations. Numbers in brackets mark the average number of clones obtained from the respective transfection.

the binding functions are concentrated within the section distal to huIFN-β).

- S/MAR function is independent of S/MAR orientation,
- S/MAR effects are monitored for the proximal gene and not for a more distally located one; see the constructs with either "C" or "E" upstream from neo[r].

After these orienting studies we tackled the problem of S/MAR activities *in vivo* in an extended series of experiments based on luciferase as the marker gene and neo^r as the selection marker. Central properties are summarized in Fig. 14 and some aspects studied in more detail in Figs. 15 and 16. Although the results of Fig. 14A indicate that S/MAR effects can be monitored if marker and selector are physically coupled, observations by Blasquez *et al.* (1989) have suggested that this situation dampens S/MAR functions due to a possible preselection for transcriptionally poised chromatin locations. As an alternative, transfections with both genes on separate vectors were suggested. Since the transfer of separate vectors is possible by transfection (causing multicopy integration events) but hardly by electroporation (enabling the integration of one or a few copies), we compared both approaches to enable specific controls later on (Fig. 14A). It turned out that both methods are similarly suited for the determination of S/MAR effects although there is a difference in the overall expression level, mainly because of differences in copy number. In conclusion, the transcriptional effects of the S/MARs can clearly be traced using the coupled gene transfer approach (Figs. 14B,C) and this will be the only method applied in the following.

A detailed series of tests was designed to clarify the importance of length, composition, origin (plants versus mammals), position, and combination of the S/MARs (single S/MARs versus flanking elements). Figure 15 demonstrates the equivalence of upstream and downstream positions, exemplified by constructs with the upstream S/MAR (E) of the human IFN-β domain. The same is true for both orientations of a S/MAR element (data not shown). In this respect S/MARs behave like enhancers, but in contrast to them they show a cooperative effect when flanking a gene while a cumulative arrangement has at most an additive effect. It should be noted that this conclusion has been derived from constructs E-p-W and EW-p, in which a luciferase marker gene controlled by the SV40 promoter enhancer is attached, in alternative ways, to a human (E) and a potato–S/MAR element (W).

Using a standard construct (S/MAR–promoter–luciferase), Fig. 14C shows that S/MAR action is observed independent of the particular cell line. Although there are some differences regarding the absolute expression levels that may depend on the properties of the promoter within a given cellular background, the expression ratio between non-S/MAR controls and SAR constructs is rather consistent.

Elevated expression levels due to the presence of S/MARs are restricted to the stable expression phase as the analogous experiments with transiently expressed genes failed to reveal such an activity (Fig. 14B). This agrees with the properties of the lysozyme attachment sequences (Stief *et al.,* 1989) or of hypersensitive sites 1, 3, and 4 of the human β-globin LCR (Antoniou and Grosveld, 1990). To exclude the possibility that the apparent lack of an enhancing effect is caused by a supersaturation of the attachment sites

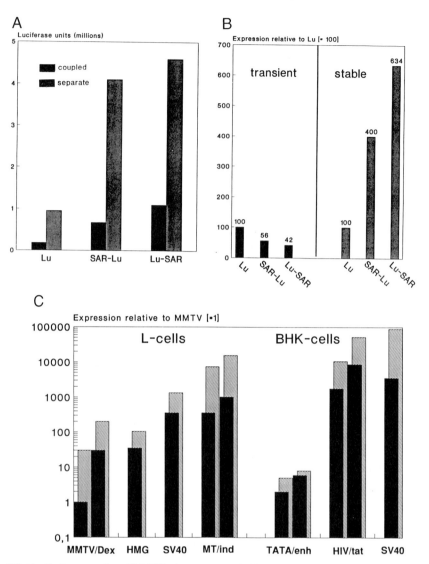

FIG. 14 Basic properties of S/MAR-elements revealed by prototype experiments. A: Uncoupled or linked transfer. The marker gene (luciferase, Lu) and the selector gene (neo^r) have been transfected either on separate vectors (ratio 10:1; crosshatched bars) or as parts of the same vector (solid bars). Both series include a basic construct (Lu) or a S/MAR construct (S/MAR-Lu, Lu-S/MAR, constructs with different S/MARs either in the upstream or downstream positions, respectively). B: Transient or stable expression: using the coupled transfer approach, luciferase expression was either measured during the transient stage (solid bars) or after integration into the genome. C: All promoters profit from the presence of an S/MAR element. Promoters were tested in two different cell lines (mouse-L and BHK). MMTV, mouse mammary tumor virus LTR without/with induction by dexamethasone; HMG, mouse hydroxymethylglutaryl-CoA-reductase; SV40, SV40 promoter–enhancer; MT, mouse metallothionein without/with induction by heavy metals; TATA, T7 promoter functioning as a TATA box, isolated/coupled to CMV enhancer; HIV, HIV-LTR alone/transactivated by Tat-protein.

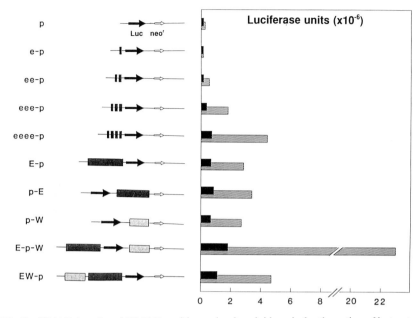

FIG. 15   S/MAR-length and S/MAR-position; role of a minidomain for the action of butyrate. Constructs based on the SV40 promoter/enhancer are shown driving the luciferase (Luc) gene. Black bars mark expression levels in the absence and crosshatched bars expression levels in the presence of 1 mM butyrate. Adapted with permission from Schlake *et al.* (1994). Copyright 1994, American Chemical Society.

existing in the nuclear matrix, we performed a titration in order to compare the relationship between the amount of transfected DNA of non-S/MAR and S/MAR constructs and the respective expression levels: assuming that binding to the nuclear matrix is a prerequisite for transcriptional enhancement, an excess of nonbound S/MAR constructs would lead to a transcriptional level not distinguishable from that of a non-S/MAR construct. Figure 16 clearly indicates and other tests confirm that there is a linear correlation between the two parameters, even at DNA concentrations just sufficient to cause a detectable expression. These results validate our conclusion that the missing S/MAR effect depends on the extrachromosomal state of the construct that prevents the formation of a chromatin structure characteristic of genomic DNA.

B. Cooperation with Promoter Elements

For a deeper insight into the function of S/MARs and the other *cis*-acting elements, it is necessary to investigage their mutual interaction. As an

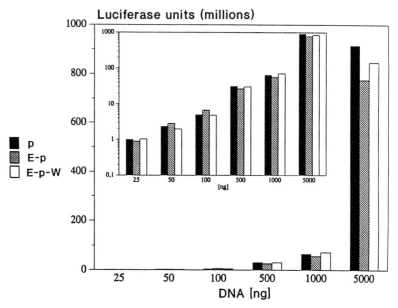

FIG. 16 Lack of S/MAR action on transiently expressed constructs is independent of the transgene density. Transient expression levels for increasing amounts of non-S/MAR and S/MAR constructs were compared. Designation of constructs is as for Fig. 15.

example, the LCR of the β-globin locus is not able to establish a hypersensitivity on its own but requires the interaction with a promoter (Reitman *et al.*, 1993). Additional evidence for an interrelation of elements comes from the chicken lysozyme domain, for which the property of a position-independent expression became evidence only in the presence of the native enhancer (Stief *et al.*, 1989).

For these reasons we studied the S/MAR effects on a luciferase reporter under the control of widely different promoters. The results for some of the most divergent constructs are summarized in Fig. 14C and later on in Table II. Our selection comprises viral and cellular as well as inducible and housekeeping promoters. Furthermore, a construct with the reporter driven by the bacteriophage T7 promoter was used, the TATA-like sequence of which is recognized by the eukaryotic RNA–polymerase II (Sandig *et al.*, 1993). With all propoters a clearcut S/MAR effect was measured (Klehr *et al.*, 1991, 1992; Schlake *et al.*, 1994). Compared to a non-S/MAR construct, apparent transcriptional increases between three- and eightfold were monitored, in this case largely independent on the presence of other *cis*-active elements or transacting factors.

The correlation of increased transcriptional rates and the nature of the promoter elements identified so far shows no obvious link between a

TABLE II

Transcriptional Stimulation of Various Promoters by S/MARs and/or Butyrate (But)[a]

Promoter	CCAAT	GC	TATA	enh	S/MAR	LCR	Fold stimulation		
							S/MAR	But	Both
HIV	+	+	+	+			8×	4×	50×
huIFN-β	−	−	+	+	+		8×	1×	30×
muHMG	−	+	−				3×	5×	20×
muMT	−	+	+	+		+	6×	80×	1900×
T7	−	−	+	−			3×	2×	6×

[a] Promoter boxes and Distal elements (S/MAR, LCR) have been marked by + and − as far as information is available from the literature.

S/MAR effect and any of the regulatory sequences. Neither an enhancer nor a TATA box are a prerequisite for obtaining an elevated expression in the presence of a S/MAR and even a naked TATA box does respond. For the time being, molecular mechanisms for this behavior must remain unresolved although a general explanation on topological grounds will be proposed.

The question arises whether an action that is essentially independent of any further sequences also holds for the other effects that have been ascribed to the S/MARs or whether it is limited to their immediate effects on transcriptional rates. As an example, studies by Yu *et al.* (1994) demonstrate that in case of a S/MAR from the human β-globin domain the condition of a position-independent expression is only observed in combination with an enhancer (see below).

Looking at the interrelation of S/MARs and butyrate, which is the subject of the following paragraph, it will become obvious that differences between the investigated promoters do actually exist that are amplified by this reagent. While in combination with the T7 promoter S/MAR and butyrate act independently of each other, in all other cases there is a more or less pronounced synergism of these two paramaters leading to a 1900-fold stimulation at maximum (Table II). Therefore, even though a "minimal" promoter responds to S/MARs, a group of upstream elements is responsible for the synergistic action of butyrate and S/MARs.

C. Butyrate and Trichostatin A: Epigenetically Active Agents Supporting S/MAR Functions

The fatty acid butyrate influences the degree of histone acetylation when delivered to cultured eukaryotic cells. While there is no effect on the process of acetylation (Vidali *et al.*, 1978), butyrate inhibits the action of histone deacetylases causing a net effect that results in a well-characterized hyperacetylation of histones H3 and H4 (Candido *et al.*, 1978; Sealy and Chalkley, 1978; Boffa *et al.*, 1981). This process triggers the release of the negative superhelicity that is constrained by the presence of nucleosomes (Norton *et al.*, 1989, 1990) and this in turn is able to support the transcription *in vitro* (Hirose and Suzuki, 1988; Mizutani *et al.*, 1991) and *in vivo* (Luchnik *et al.*, 1988). The degree of negative superhelicity at which expression reaches a maximum depends on the nature of the promoter (Hirose and Ohta, 1990; Mizutani *et al.*, 1991).

One particular requirement for these actions of butyrate is that the ends of DNA are fixed to limit the superhelicity to the gene's domain, for instance by attaching S/MARs to the nuclear matrix (Klehr *et al.*, 1992; Schlake *et al.*, 1994). Addressing this question, the present experiments clearly reflect the synergistic action of butyrate and S/MARs (Fig. 15, Table II). While even single S/MARs increase the transcriptional enhancement due to butyrate (Klehr *et al.*, 1992), this becomes more pronounced if two of these elements are present at either side of the gene (Schlake *et al.*, 1994).

If one tries to interpret these results in a straightforward manner, a problem arises because of the multiple effects of butyrate (Kruh, 1982). In addition to the influence on histone acetylation, butyrate inhibits the phosphorylation of histones H1 and H2A and changes the phosphorylation pattern of some nonhistone proteins (Boffa *et al.*, 1981). Moreover, the reagent affects the degree of DNA methylation in a cell-specific manner (Cosgrove and Cox, 1990) as well as the methylation and synthesis of proteins (Boffa *et al.*, 1981). These combined effects can lead to secondary effects like the induction of proteins such as hemoglobin F, methallothioneine I, and albumin (Thomas *et al.*, 1991; McDonagh *et al.*, 1992; Saito *et al.*, 1992) or they inhibit the accumulation of CDC2 and MyoD (Charollais *et al.*, 1990; Johnston *et al.*, 1992). c-myc and c-jun, in contrast, are not affected (Charollais *et al.*, 1990). Finally, butyrate is also known as a potent inhibitor of DNA replication and cell proliferation (Kruh, 1982) and it induces the differentiation of MEL cells (Leder and Leder, 1975) whereas that of muscle cells is blocked (Johnston *et al.*, 1992).

While we favored the contribution of a hyperacetylation of histones (Klehr *et al.*, 1992) as an explanation and excluded the involvement of "*de novo*" protein synthesis (Schlake *et al.*, 1994), the modification of a regulatory protein was suggested by others to account for the actions of butyrate on the responsive elements of the HIV1-LTR and the 5' region of CCP1, respectively (Bohan *et al.*, 1987, 1989; Fregeau *et al.*, 1992). Still another report related the effect of butyrate to the *de novo* synthesis of a protein (Yeivin *et al.*, 1992).

With the advent of a novel agent, Trichostatin A (TSA), and knowledge about its mode of action, we were in the position to confirm the level of butyrate action responsible for its synergism with S/MARs. TSA was originally isolated in 1976 by Tsuji *et al.* and shown later to induce the differentiation of MEL cells and to inhibit the cell cycle of rat fibroblasts (Yoshida *et al.*, 1987; Yoshida and Beppu, 1988). Interestingly, these actions were stereospecific as they were restricted to the natural (R)-enantiomer (Yoshida *et al.*, 1990b). All these secondary effects could unambiguously be linked to the influence of TSA on the acetylation of histones, which again occurs by inhibiting the histone deacetylases in a reversible manner and with a K_i of 3.4 nM (Yoshida *et al.*, 1990a). The identification of a TSA cell line that was resistant to TSA due to a deacetylase mutant supported these ideas. Finally, there is yet another agent, Trapoxin, that mirrors the biological effects of TSA by inhibiting histone deacetylases in a different, irreversible fashion (Kijima *et al.*, 1993).

Using TSA in place of butyrate, we repeated the most representative experiments of Figs. 14 and 15. As anticipated by us, both reagents yielded similar results in the cells that were amenable to TSA (Schlake *et al.*, 1994): an apparent failure of TSA to improve the transcription of S/MAR-constructs in BHK cells was correlated with the absence of hyperacetylated forms of H4 and may be caused by an inability of uptake of the agent or its transport into the nucleus.

As TSA is able to mimic the effect of butyrate in our system and since titration experiments demonstrated that this is true both for the qualitative and the quantitative aspects (Schlake *et al.*, 1994), there is no need to postulate any other function than hyperacetylation to explain the synergism of butyrate and S/MARs.

V. S/MARs as Domain Borders: Model Studies

A. Position Independence

It is widely accepted that the eukaryotic genome consists of 50,000–100,000 looped domains comprising between 5 and 200 kb of DNA. Changes in the superhelicity of a loop, triggered by the action of ethidium bromide, cannot pass the intervening anchor sites, which is interpreted to mean that domains form topologically separate entities (Zehnbauer and Vogelstein, 1986). On a functional level, looped domains qualify to be regulatory units delimiting the action of promoter/enhancer or promoter/LCR complexes.

S/MARs are prime candidates for a function as domain borders as they are able to mediate an attachment to the nuclear matrix *in vitro* and are frequently found in the vicinity of the ends of a domain that are otherwise defined by constitutive DNase I-hypersensitive sites or a decrease of the general DNase I sensitivity (Section II,B,4). This leads to the model in Fig. 17 which defines a transcriptionally (and replicationally) independent domain according to constitutive S/MARs. Inside the domain an LCR is responsible for the chromatin's general accessibility whereas enhancers and promoter elements are only involved in the regulation of their coordinate gene(s)—a picture that may require refinement considering the overlapping functions of enhancers and LCRs (see following). In addition, regulated S/MARs (elements that are only attached transiently) may be part of the domain. Their possible interaction with elements involved in gene regulation will be considered in the concluding chapter.

From a more practical standpoint, a domain is seen as that piece of DNA which after transfer into a recipient cell or tissue and integration into the genome establishes an authentic, high-level expression pattern that is "position-independent," i.e., independent of the site of integration. Different elements are currently considered as candidates to confer this property. For some genes, enhancer-like, dominant sequences serving the function of an LCR are thought to mediate transcriptional patterns and levels independent of the position of integration, even in the absence of true domain borders. These elements do not insulate the domain from influences by

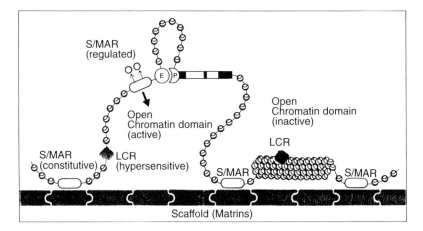

FIG. 17 Current ideas about structure–function relationships in eukaryotic gene domains. A central switch (LCR) determines whether a domain is active or inactivated, for instance as a 30-nm fiber. S/MARs are the bordering elements, which by their association with a scaffold serve as a topological barrier between adjacent domains. Within a domain, topological changes occur as a consequence of the loss of nucleosomes or of transcription. Facultative or regulated S/MAR elements may support the action of an enhancer. These basic features are extended in Fig. 22.

adjacent chromatin structures but rather form extraordinary stable complexes with "their" promoters in a way that overcomes external influences (Dillon and Grosveld, 1993); a possible exception exists for truly heterochromatic localizations. This view is supported by the work of Reitman *et al.* (1993) who demonstrated the importance of promoter sequences and/ or transcription factors for this LCR function that is lost if certain motifs are destroyed. In line with these models, position independence can be restored for such a truncated entity if the boundaries are re-added (Sippel *et al.*, 1993).

Besides the S/MARs, at least one other class of bordering elements seems to exist. The "special chromatin structures" (scs and scs') delimiting the 87A7 locus of *Drosophila* (Uvardy *et al.*, 1985) are both able to shield a gene from position effects (Kellum and Schedl, 1991) and to limit enhancer activities to promoters within the domain (Kellum and Schedl, 1992). scs elements have no enhancing activity by themselves, which makes them different from the prototype attachment elements bordering the chicken lysozyme domain that are S/MARs by definition (Phi-Van and Strätling, 1988; Stief *et al.*, 1989).

For a rigorous determination of an element's potential to mediate position independence, a single copy of the respective test construct would have to be integrated at various genomic loci for which the expression of a reporter gene has to be determined. The potential has to be judged relative to a control missing the element of interest, assuming that neither transgene selects for a particular class of sites. A novel solution for such a test, which would be impossible or exceedingly laborious if based on conventional techniques, will be given in Section VII. For obvious reasons, up to now virtually all studies made use of indirect test systems. These consist of the determination of expression levels as a function of copy numbers or of the shielding of transcriptional units from the action of enhancers.

B. Copy Number Dependence

The most common way of demonstrating a position independence is by determining copy numbers and expression levels: if a transfected and integrated construct establishes a reporter gene expression independent of the surroundings, then to a first approximation, the overall transcription should be proportional to the number of gene copies. Based on this assumption, a linear correlation between the number of stably integrated constructs and level of expression could be demonstrated for the lysozyme-attachment elements (A elements) when flanking a reporter gene. This is not only true for the native domain containing the lysozyme gene and its promoter

(Bonifer *et al.*, 1990) but also for constructs with homologous and heterologous promoters and genes (Stief *et al.*, 1989; Phi-Van *et al.*, 1990). Studies on other S/MAR elements give additional evidence for their potential to confer copy number-dependent expression. As an example, an S/MAR localized in a gene cluster of soybean dampens the influence of the surrounding chromatin at the site of integration (Breyne *et al.*, 1992). Similarly, the 5′ S/MAR of the human β-globin locus, which behaves as a true insulator (Li and Stamatoyannopoulos, 1994a), causes copy-dependent transcription but only in conjunction with an enhancer (Yu *et al.*, 1994).

A situation resembling that of the A elements has been found in the case of the LCRs, but there, a single element is sufficient (Grosveld, 1987; van Assendelft *et al.*, 1989). Dissections of the human β-globin LCR made it clear that at least three out of the four DNase I-hypersensitive sites constituting the LCR confer an expression proportional to the number of integrated copies (Fraser *et al.*, 1990). More recent data suggest a similar action for the fourth site (Fraser *et al.*, 1993).

Although "copy number dependence" is by far the most common assay principle, there are major difficulties inherent in this method (demonstrated convincingly by Poljak *et al.*, 1994). These are at least in part due to the fact that multiple copies do not integrate at random (covering multiple genomic sites of a single cell) but as a single tandem array at a location that is unique for each cell. This situation provokes many unexpected interactions rearrangements, and cellular "defense mechanisms" (Kricker *et al.*, 1992; Dorer and Henikoff, 1994; Kalos and Fournier, 1995). Therefore, it is not all that surprising that even in the case of low copy numbers a true linear correlation with gene transcription may be missing (see the exponential dependence on copy number in the study by Stief *et al.*, 1989). As a laborious alternative, the selection of single-copy clones has been undertaken in order to demonstrate the shielding potential of S/MARs— necessarily on a small database (Kalos and Fournier, 1995). A completely novel approach with the potential to circumvent these problems will be introduced in Section VII).

Even for the human β-globin LCR, which mediates a significant proportionality between copy number and gene expression, this is only true for cells with fewer than 15 copies. This is even more pronounced in plants where interactions between duplicated genes cause a silencing effect termed "cosuppression" (Matzke *et al.*, 1989; Matzke and Matzke, 1990; van der Krol *et al.*, 1990). A striking example is the study of Allen *et al.* (1993) where there is no copy number dependence for stably transformed plant cells and a S/MAR effect that decreases at high copy numbers. On balance, the obvious enhancing activity of an S/MAR for integration events involving less than 50 copies hints at a certain disruption of the co-suppression phenomenon.

C. Enhancer Blocking Assays

A second method that has been used in a few cases to determine the insulating potential of certain elements is the "enhancer blocking" assay that evolved from early findings in yeast that protein-binding sequences placed between the UAS sequences and the TATA box of the GAL1 promoter impede transcription (see Fishel *et al.*, 1993). This test relates to the definition of a domain boundary. A boundary element should restrict the action of an enhancer or LCR to "their" domain and at the same time shield the domain's gene(s) against influences from the exterior. One would therefore expect that an element capable of mediating position independence would also be in the position to eliminate the effect of an enhancer when positioned between it and a promoter.

The first element tested by this approach was the scs sequence from the 87A7 domain of *Drosophila*. Kellum and Schedl (1992) demonstrate convincingly that this "special chromatin structure" insulates a gene from an enhancer while an S/MAR from the interior of the same domain had no comparable effect. Subsequent tests of this type revealed two more elements acting as insulators. These were the hypersensitive site (HS4) at the 5' end of the chicken β-globin cluster (Chung *et al.*, 1993) that has no S/MAR activity *in vitro* and HS5 of the human β-globin locus (Li and Stamatoyannopoulos, 1994a,b), an element coinciding with an S/MAR.

Experiments from the authors' laboratory support the idea that S/MARs can indeed act as domain boundaries and therefore block the effect of an enhancer. As a test construct we used the luciferase gene as a reporter driven by a minimal promoter in order to eliminate any sophisticated interactions. The S/MARs used as putative domain borders were from petunia and the human interferon-β domain. They were chosen because of their activities in binding and transcriptional assays by which they qualified as the two most potent elements available to us. The use of two S/MARs was dictated by the fact that standard transfection techniques were applied that promote the integration of tandem repeats.

Since S/MARs do not stimulate transcription in the transient expression but do so after integration into the genome, we were interested to find whether a possible blocking function was likewise restricted to the stable state. Therefore, the test was performed in parallel for the transient expression and stable phases, respectively. Figure 18 shows that, regardless of the arrangement of the elements, a stimulatory effect due to the enhancer can be observed throughout the transient assays. In contrast, after integration the enhancer can exert its function only if there are no S/MARs or if it is part of the domain defined by the S/MARs.

We have to conclude that so far the results concerning S/MAR elements are contradictory (Kellum and Schedl, 1992, on one side; Chung *et al.*, 1993;

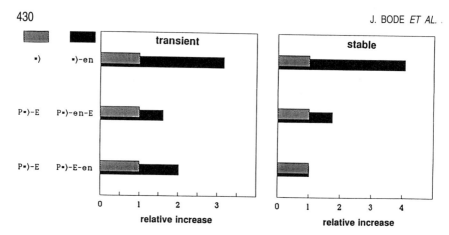

FIG. 18 S/MARs are active in an enhancer blocking test. The luciferase gene controlled by a TATA box (=)) was provided with the CMV enhancer (en) and S/MAR elements P, E. While an enhancer effect can be monitored throughout the transient expressions, it disappears for the stable expression in case it is positioned outside the minidomain (construct P = >-E-en).

Li and Stamatoyannopoulos, 1994a,b; and our own work on the other). Since in standard tests all S/MARs produce effects related to their affinity *in vitro* irrespective of their origin (domain boundary, intronic, enhancer associated), the remaining difference simply concerns length. While extended S/MAR sequences predominate at domain boundaries, intronic S/MARs are at most a few hundred base pairs long. Extending such short S/MARs by oligomerization provides them with the characteristics of longer sequences (Mielke *et al.*, 1990). We cannot rule out, however, that in certain cases the cohabitation or even colocalization of S/MARs and enhancers may explain the differences.

As in the case of copy number dependence, there are potential difficulties associated with the method of enhancer blocking. For standard transfection techniques, multicopy integrates can arise which, due to an interaction between the multimerized constructs, may affect the results. While this is principally overcome by methods (like electroporation) that promote single-copy integration events, the second problem is an intricate one: the experimenter is not in the position to decide whether an observed effect requires the simultaneous presence of the enhancer as for the 5′ S/MAR of the β-globin locus (Yu *et al.*, 1994). Therefore the time has come to develop and apply novel techniques to appreciate fully the capabilities of S/MARs and related elements (see Section VII).

D. Long-Term Stability Mediated by S/MARs

Early observations on the principles of gene order have hinted at an inverse relationship between the size of a chromatin domain and its transcriptional activity (Gasser and Laemmli, 1987). This correlation has been discussed in the framework of various models involving either a mobile polymerase dealing with a fixed domain or a matrix-associated and thereby immobile polymerase through which the coding region is reeled during transcription. Independent of these ideas, it is easy to imagine that small loops are particularly suited to bring the cis regulatory elements into close juxtaposition at the scaffold. In addition, segments of chromatin with multiple points of attachment might physically resist compaction into heterochromatin (Allen *et al.*, 1993), providing an explanation for our observation (Fig. 19) that complete minidomains confer a long-term stability to the enhanced expression levels (construct E-p-W). This is in contrast to a control transgene, EW-p, for which both S/MAR elements (E and W) have been fused to form a single domain border. It is emphasized that the differences elaborated here are ideally traced for single-copy integration events as they will vanish for multicopy tandem arrays, which provide two borders for the majority of copies. This is why for gene transfer electroporation is preferred over transfection while a site-specific recombination approach will be the future method of choice (see Section VII).

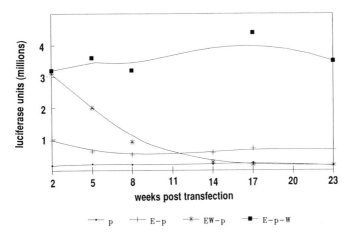

FIG. 19 Minidomains are a means to provide a gene with long-term stability. Expression levels for S/MAR-free (p) and S/MAR constructs (E-p, EW-p, E-p-W) are compared over extended periods of time. Only when the gene is flanked by S/MAR elements E and W do expression levels remain stable (see construct E-p-W).

E. Role of S/MARs in Cultured Cells and in Differentiation Compared

The idea that S/MAR elements participate in nuclear processes is supported by the evolutionary conservation of these sequences from yeasts to humans. If used as part of transgenes for the modification of cultured animal cell lines, S/MARs have consistently been found to increase transcriptional rates due to a nonenhancer mechanism (Blasquez *et al.*, 1989; Stief *et al.*, 1989; Mielke *et al.*, 1990; Phi-Van *et al.*, 1990; Section IV) and sometimes to confer the property of a copy-number-related expression (Stief *et al.*, 1989; Phi-Van *et al.*, 1990; Yu *et al.*, 1994). Additionally, there are three claims that S/MARs are able to limit the action of enhancer elements. While Stief *et al.* (1989) derive this property from episomal copies, Li and Stamatoyannopoulos (1994a) demonstrate such an effect only for a gene at a chromosomal location. The results of our Fig. 18 agree with the latter study as the effect is evident for the stable but not the transient phase of expression.

For plant cells, callus cultures, and complete plants, the situation becomes complicated by the fact that, although an increased and copy-number-related transcription has been described by various authors, these phenomena are only linked in certain cases (Schöffl *et al.*, 1993; Mlynarova *et al.*, 1994; van der Geest *et al.*, 1994). There is another example that for tobacco cells transformed by *Agrobacterium* an S/MAR improves the copy-number-dependent expression without an effect on transcriptional levels (Breyne *et al.*, 1992) and still another one where, upon microprojectile bombardment, levels are dramatically raised while there is no dependence on copy numbers (Allen *et al.*, 1993). These differences may, in part, be due to the gene transfer technique which tends to target transcriptionally competent sites in the first case (see Section VI,A) while random positions are probably hit in the second one (Dietz *et al.*, 1994).

To date, only a few studies have addressed the potential of S/MAR elements to modify or improve the expression of gene constructs in transgenic animals. In most cases, correlations with cells in culture are difficult as S/MARs were used as parts of complete regulatory domains such that possible S/MAR–enhancer and S/MAR–LCR interactions had to be considered. However, fortuitously in this context, there is one very recent report to show that some of the S/MAR activities, which are easily traced in dedifferentiated cells, are overcome during the process of differentiation.

Thompson *et al.* (1994) used a construct from the luciferase gene under the control of the murine HSP 70 promoter with and without the huIFN-β domain borders (Fig. 1), which permits studies of gene expression in a developmental context and in multiple tissues. Preimplantation development of transgenic mice covers periods during which undifferentiated cells divide and it is at these stages up to the blastocyst that the S/MAR⁺ transgenes are transcribed at significantly higher levels and show a positive correlation with copy number. Quite remarkably, both properties largely vanished for newborn and adult mice but were restored in rapidly dividing fibroblast cultures established from tail or ear biopsies. Two important differences between S/MAR⁻ and S/MAR⁺ animals emerged that may be relevant for the potential of S/MARs in transgenic animals:

- Between the newborn and adult stages, S/MAR⁺ lines maintain their respective expression levels while there is a continuous decrease for S/MAR⁺ lines; this is consistent with our observation that S/MARs might prevent chromatin inactivation (Section V,D).
- S/MAR⁺ lines exhibit a tissue-dependent modulation of expression levels while levels are rather uniform for all tissues of a given S/MAR⁻ line. Although, in general, heat shock genes are expressed ubiquitously, these observations hint at a role of S/MAR elements in establishing authentic expression patterns.

The idea that S/MARs enable a developmentally correct expression had already been advanced for two tissue-specific genes, coding for whey acidic protein (WAP) and for lysozyme. The 7-kb WAP gene is an ideal model to this end, since although it contains the specific control elements necessary for developmental and hormonal regulation, its expression is highly position dependent. This gene was cotransferred with S/MARs, in this case the 5′ attachment-element of the chicken lysozyme gene. While all of the S/MAR⁺ lines correctly expressed the transgene in mammary tissue and four out of five lines displayed an accurate hormonal and developmental regulation, only half of the S/MAR⁻ animals showed any expression and this appeared deregulated if compared with the endogenous gene (McKnight *et al.*, 1992). As in the above example, no differences in basal expression were observed between the two types of transgenics, nor were there indications for a copy-number-dependent expression. These results are in line with the above conclusions and also with the recent results on transgenic animals carrying the 20-kb domain of chicken lysozyme with or without the flanking S/MARs (Sippel *et al.*, 1993).

While it has been shown previously that the transfer of the complete lysozyme domain into the germ line of mice leads to a consistent, high-level macrophage-specific expression (Bonifer *et al.*, 1990), more recent experiments have begun to dissect the relative roles of enhancers and bordering elements (Sippel *et al.*, 1993). These studies confirm that S/MARs contribute to suppressing an ectopic gene activity while there was no effect regarding the level or copy-number dependence of expression.

S/MAR elements with a different although probably related function occur in close physical association with established enhancer elements where they are thought to contribute to enhancer function. This "cohabitation" phenomenon was originally observed in *Drosophila* (Gasser and Laemmli, 1986a). For vertebrates, prototype elements have been identified in the Ig light and heavy chain loci. These S/MARs are located adjacent to the intragenic enhancer and support the expression of kappa genes both in plasmacytoma cells and in the spleen of transgenic mice (Blasquez *et al.*, 1989; Xu *et al.*, 1989). The immunoglobulin μ heavy chain locus is regulated by an intronic enhancer that is flanked by an S/MAR on either side. Interestingly, these S/MARs include binding sites for negative regulatory factors that are thought to repress the enhancer in non-B cells by interfering with the S/MAR functions (Scheuermann and Chen, 1989; Dickinson *et al.*, 1992; T. Kohwi-Shigematsu and J. Bode, unpublished). These enhancer-linked S/MARs have now been shown to be essential for μ-gene transcription in transgenic mice which is accompanied by an S/MAR-dependent formation of a DNase I-sensitive domain extending from the hypersensitive enhancer (Forrester *et al.*, 1994). This function may be related to a general property of S/MARs to provide a sink of negative superhelicity that can be transferred in order to unwind or open adjacent genomic districts (see Fig. 8 for a related phenomenon) or to their "domain opening" function in the framework of a model developed by Laemmli and colleagues (Zhao *et al.*, 1993).

The human β-globin gene complex forms one of the most complicated regulatory units known to date and it covers bordering as well as internal S/MARs. S/MARs with a potential bordering function are associated with two constitutive hypersensitive sites, V and VI. The left-hand element meets all criteria that are presently applied to define a prototype domain-bordering element, i.e., it coincides with a decrease of general DNase I sensitivity (Section II,B,4) and acts as an insulator when it is interspersed between hypersensitive site III and the β-globin gene (Li and Stamatoyannopoulos, 1994a). If it is supplemented by enhancer functions, it is also able to confer the property of a copy-number-related expression (Yu *et al.*, 1994).

In addition to these flanking elements, other S/MARs are found within the locus. These intrinsic S/MARs could provide function-related, dynamic boundaries between the transcriptionally active and silent members of the domain. Alternatively or in addition, they could support the function of the enhancers they are associated with (Cunningham *et al.*, 1994). These enhancers had first been identified as developmentally regulated DNase I hypersensitive sites.

- The S/MAR in the large β-intron (IVS2) is associated with an element that is able to confer a β-like expression pattern on the γ-globin gene whereas the second β-S/MAR is closely associated with the tissue-specific enhancer 3' of the β-globin gene (Jarman and Higgs, 1988).

- In addition, a newly discovered S/MAR downstream from the $^A\gamma$-globin gene was first identified as an enhancer and later on as an element that contributes to silencing of the γ-globin gene. These multiple functions may be due to the fact that as in the case of the IgH-enhancer-associated elements, this S/MAR is a target for SATB1, a protein known for its property to disrupt functional S/MAR–scaffold interactions.

The relevance of S/MARs for a position-independent, correctly timed expression has frequently been neglected since, for instance, the property of LCR-β-gene constructs have entirely been ascribed to LCR functions. Recent findings indicate that this picture may be incomplete: Li and Stamatoyannopoulos (1994b) observed that in contrast to β-globin, an LCR-linked γ-globin gene is shielded from position effects only in case the β-gene is present as a tandem partner. Therefore, attention has to be redirected to the above mentioned S/MARs, which were not part of the basic LCR–γ transgenes.

This overview emphasizes that S/MARs have various properties. Some of these may be more important at early developmental stages as they can be overcome in the course of differentiation. S/MARs are found coincident with the borders of certain gene domains and in close association with enhancers, even in transcribed regions. Since these elements can be used interchangeably for creating chromatin minidomains in cultured cells, their functions are likely to be related. S/MARs function as sinks for negative superhelicity providing for a topological separation of the adjacent domains. If the proteins constraining this superhelicity are regulated, this topological status can be transferred to other loci that are primed to open up, for instance, by the presence of an enhancer that has been pre-set by the appropriate factors.

VI. S/MARs as Integration Targets

Two vehicles that are widely used for transferring genes, i.e., the *Agrobacterium tumefaciens*-mediated transfer of T-DNA into the plant genome on one hand and the creation of a provirus upon a retroviral infection of mammalian cells on the other, have one property in common: the integration machinery is in the position of sensing active chromatin structures. This occurs in spite of the fact that the integration steps are totally unrelated from a mechanistic standpoint.

A. Integration of T-DNA

The rearrangements of T-DNA as well as target DNA suggest that the insertion is a multistep process involving different types of recombination, replication, and repair activities that are mostly provided by the host cell. A single-stranded derivative is the most likely integrating form. The attachment to one strand of plant DNA results in a local torsional strain and this may be the reason that the only shared feature between the target sequences is that they are enriched in A and T.

It has been demonstrated for transgenic tobacco plants that genes transferred as part of the T-plasmid are integrated into DNase I-sensitive domains (Weising *et al.,* 1990). Moreover, the high frequency of gene fusions observed for promoterless reporter genes directly shows an integration into loci that are poised for transcription (Mayerhofer *et al.,* 1991).

In an effort to obtain information about the role of integration sites for expression patterns, we have recently analyzed a petunia plant for which the function of an otherwise root-specific promoter was distinctly deregulated (Dietz *et al.,* 1994). At the integration site an AT-rich element was traced that showed properties typical of S/MARs, including across-species borders both *in vitro* and *in vivo* and the cooperation in minidomain constructs composed from a human and the plant element. Considering the unwinding potential of S/MAR sequences, it is intriguing to implicate these elements in the integration process. Such an event would further benefit from the fact that most or all of the required host functions are established matrix enzymes that would be brought into juxtaposition by an S/MAR–matrix or S/MAR–scaffold interaction.

The observed deregulation of tissue specificity indicates that the cis-regulatory elements of the transgene are not shielded but rather are under the control of the respective host domain. This is the likely consequence of an integration adjacent to but not within an S/MAR region (V. Kay, unpublished). If the same S/MAR element is applied to provide the transgene itself with an insulating element at its upstream and downstream positions, all transgenic plants express in the expected root-specific manner (A. Dietz, unpublished).

B. Retroviral Integration

Integration of retroviral DNA occurs by a recombination reaction mediated by the viral integration machinery. Studies on target site specificities have consistently shown that provirus integration can occur at multiple sites without an apparent sequence preference. Like other retroelements, retroviruses show a pronounced tendency to integrate into 5'-noncoding or intron sequences of cellular genes, mostly within a few hundred base pairs of a DNase I-hypersensitive site (Scherdin *et al.,* 1990; Craigie, 1992). That a subgroup of these sites is highly specific has first been reported by Shih *et al.* (1988) who demonstrated that independent integrations could occur at the same nucleotide position.

We have isolated, by inverse PCR techniques, sequences next to 24 integration sites (Mielke, 1993). The amplified DNAs were subjected to a standard scaffold reassociation reaction which led to the surprising conclusion that most of them showed a distinct propensity for binding. Our first assumption that these elements represent S/MARs seemed to gain support from observations that retroviral integration sites contain clusters of strong topoisomerase II sites (Howard and Griffith, 1993).

More detailed analyses, both concerning the sequence features and the affinities, led to the conclusion that the DNAs recovered here are no prototype S/MARs but may represent a subclass thereof. On the average, the fragments were only moderately (50–55%) AT-rich and many of them behaved like bent DNA (C. Mielke and M. Tümmler, unpublished). Since sequences are recognized by the viral integrase protein, the recent analyses of preferred substrates for the enzyme *in vitro* reflects important aspects of retroviral integration. Most of these sites were localized in nucleosomally organized rather than free DNA, i.e., in regions with a widened major groove as they occur

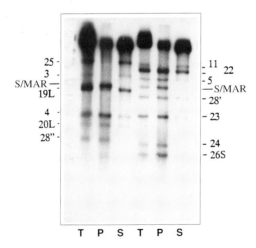

FIG. 20   Host-DNA surrounding the integration site of retroviruses display S/MAR-like properties. MPSV-based constructs have been packaged to form infective virus that was used for an integration into the genome of NIH-3T3 host cells. Inverse PCR techniques have been applied to isolate and sequence host DNA adjacent to the sites of integration. Mixtures of cloned fragments (T) have been used for an *in vitro* scaffold-reassociation experiment analogous to Fig. 9. The control labeled "S/MAR" represents an authentic 800-bp S/MAR fragment.

by wrapping around the histone core. Other substrates were highly bent or kinked DNAs with the same topological features (Mueller *et al.*, 1993).

While bending is a property of many S/MARs, its contribution for binding to the scaffold is not very pronounced. On the other hand, central S/MAR functions may well be supported by adjacent stretches of bent DNA which, in addition, are a feature recognized by DNA topoisomerase I (Caserta *et al.*, 1990). Finally, curved DNA is found for replication origins in yeast, many of which behave like S/MARs.

Among the sequences recovered by us, a $(GAA)_{42}$-tract adjacent to an integration site deserves particular attention since it is recognized by a prototype S/MAR-binding protein, SATB1. The DNA structural analysis revealed a high propensity of unwinding, which is typical for SATB1 substrates. Under certain conditions, the main structure this sequence formed *in vitro* was an intramolecular GGC-like triplex exposing a long stretch of unpaired sequence (T. Kohwi-Shigematsu, unpublished).

VII. Creating and Reusing Artificial Domains

Understanding S/MARs requires their investigation as a single element and in conjunction with another S/MAR at the second domain border. By conventional transfection techniques these alternative configurations are hard to establish due to the known tendency of transgenes to integrate as a tandem array of multiple copies. Moreover, the potential insulating function of S/MARs has almost exclusively been studied by an indirect approach, i.e., by relating the expression copies to expression levels (Section V,A). It is hence assumed that S/MAR constructs behave the same whether they are isolated or parts of a multicopy array. Recent evidence suggests that these assumptions are not necessarily valid, since cells recognize and respond to this situation by DNA methylation/mutagenesis and heterochromatization (Kricker *et al.*, 1992; Allen *et al.*, 1993; Dorer and Henikoff, 1994). Therefore, many of the previous results obtained by such an approach have to be considered with caution.

While the creation of multiple copies during gene transfer is hard to avoid, such a situation is easily recognized by a yeast enzyme, FLP recombinase, if the individual copy is marked by an FLP recognition target (FRT) site (Fig. 21A). FLP recombinase will excise and circularize any sequence between two equally oriented FRT sites, which is of relevance here. On the other hand it will inverse the same segment if orientation of sites is opposed, which is related to the function of the enzyme for the replication of the yeast 2μ circle by which it is encoded (Craig, 1988). Since the reactions are reversible, any circular construct tagged by an FRT site can integrate

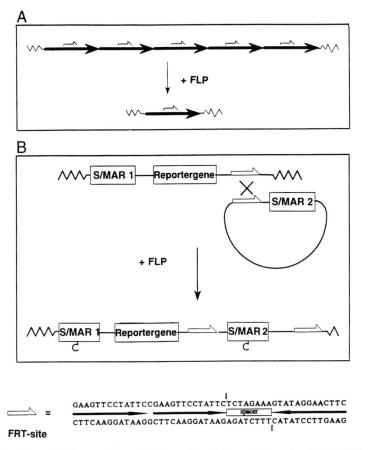

FIG. 21   Uses of FLP-recombinase for the excision (A) and integration (B) of FRT-tagged constructs. A: Reduction of a multicopy array of transgenes to the single-copy level. B: A minidomain is closed *in situ* by adding an FRT-tagged circular S/MAR-vector to a resident single-S/MAR construct.

via a like site of genomic localization and this process can be exploited for a site-directed integration (Fig. 2B).

The established properties of S/MAR elements, like their positive effect on transcriptional levels and their capability to stabilize these high levels over extended periods of time, have made them instruments for the construction of a new generation of expression vectors. Although short and apparently functional S/MAR elements can be constructed by oligomerizing certain sub-S/MAR motifs (core unwinding elements; see Mielke *et al.*, 1990; Bode *et al.*, 1992), the practical use of such synthetic elements is limited by the redundancy of sequence, which gives rise to an intrinsic instability. For practical reasons, therefore, the most favorable and stable situation still is a minidomain with bordering elements of different origin and thereby sequence content.

Following the concept in Fig. 21B, we have transfected a S/MAR–reporter gene construct into recipient cells. Owing to the fact that in case of single-copy integration events (which can be enforced by the approach of Fig. 21A) the reporter is still amenable to influences from the surrounding chromatin, a number of clones with widely differing properties could be selected. For these clones the domain could be closed using a second S/MAR localized on a vector tagged with an FRT site. This approach permitted the evaluation of a new S/MAR element (S/MAR 2) with respect to a reference element (S/MAR 1; T. Schlake and M. Stengert, unpublished). Using integration sites with widely different properties, the still controversial insulating functions are now amenable to experimentation at the single-copy level.

A substantial improvement of this concept has recently been achieved (Schlake and Bode, 1994): flanking a transgene by a FRT site on one end and a mutated FRT′ site on the other allows the entire construct to be replaced by a second cassette carrying the analogous sites. This is the simple consequence of a double-reciprocal crossover mediated in a single enzymatic reaction, i.e., of the fact that recombination involves the pairing of FRT×FRT and FRT′×FRT′ but excludes a combination FRT×FRT′. This approach enables the precise exchange of cassettes with the complete elimination of prokaryotic vector sequences. It is felt that this offers a definite advantage since it eliminates undesired interactions between the unwinding elements typical for S/MARs and those which are known to occur in typical plasmids (Kowalski *et al.*, 1988).

VIII. Understanding S/MAR Functions: How Far Are Results Supported by Theory?

S/MARs are rather unusual *cis*-acting regulatory elements in that they are defined biochemically rather than functionally: S/MAR-DNA defines the base of DNA loops in scaffolds that have been obtained after extracting the bulk of proteins from nuclei. This definition is based on the affinity of a given DNA segment for the protein components of the residual matrix or scaffold but does not address the question if and under what circumstances these contacts are established in the living cell. On balance, the experience from several years of experimentation shows that any fragment isolated by the *in vitro* screening procedure will behave similarly in a number of biological tests.

S/MAR-DNA is detected either at the putative borders of chromatin domains, which are otherwise marked by constitutive DNase I-hypersensitive sites and the decay of general DNase I sensitivity, or in transcribed (almost exclusively intronic) regions and/or in close association with established enhancers. Recent model studies have shown that an S/MAR element of intermediate size (800 bp), positioned within a transcription unit, can be transcribed, although at a largely reduced rate (C. Mielke and J. Bode, unpublished). Since S/MARs from any species or position (flanking, intronic, or enhancer-associated) perform equally in an artificial minidomain (Mielke *et al.,* 1990; Bode *et al.,* 1992) it is concluded that they serve related functions in various contexts.

Like enhancers, S/MARs have to be composed of several modules (here oligo(dA) tracts, certain AT-rich boxes, or unwinding elements) to be active. Although their sequence homology is restricted, the information contained in the modules appears abundant. As a consequence, both enhancers and S/MARs require a minimum length hardly below 200 bp (cf. Schaffner *et al.,* 1988; Dynan, 1989; Mielke *et al.,* 1990). To extend the similarities, both elements act independent of orientation and largely independent of distance. On the other side, enhancers but not S/MARs are active during the phase of transient expression, i.e., at a time where no final chromatin structure has been established. An alternative statement that S/MARs are active at chromosomal but not at episomal locations has to be considered with caution since a report by Hörtnagel *et al.* (1994) demonstrated that S/MAR effects can be reproduced if they are part of a minichromosome based on Epstein–Barr virus sequences. These findings imply that S/MAR functions require an authentic chromatin structure that can only be attained during replication or the potential to introduce torsional strain which is possible for topologically autonomous domains and equally for circular templates.

Enhancers are binding sites for various transcription factors and this property has also been ascribed to S/MARs. In particular, S/MAR sequences have been interpreted as being a mosaic of homeodomain protein-binding elements supporting a role in gene switch and the differential activation of origins of replication in development (Boulikas, 1993b). While such a role appears likely in light of recent findings about the involvement of S/MARs in binding certain positive (Dworetzky *et al.,* 1992) and negative regulatory factors (Dickinson *et al.,* 1992), these questions will be addressed in the parallel articles by Boulikas and Stein. In the following, we will therefore concentrate on a more general role of S/MARs, serving as a constitutive tool that will support the activation of gene domains.

According to a recent suggestion, S/MARs have the potential to act as domain openers. Due to their content of A tracts, they are primary binding sites for histone H1, i.e., they form nucleation centers from which the linker histone can spread into an adjacent domain to inactivate it, possibly by the formation of a 30-nm fiber (Fig. 17). The activation of such a domain will then require a protein with a dystamycin-like action ("D protein") which is capable of displacing H1 from its binding to the minor groove of A-rich DNA. As binding, release is a cooperative process that would then result in opening up the respective domain (Laemmli *et al.*, 1992). A prime candidate for a function as a D proetin is HMG-I/Y (Zhao *et al.*, 1993). In line with such a function is the fact that HMG-I/Y is more abundant in rapidly proliferating rather than differentiated cells (Thompson *et al.*, 1994).

Obviously, the association of H1 reflects only one mode of S/MAR-protein interaction. In striking contrast to other prototype matrix proteins as the lamins, the association of H1 is not dependent on the formation of single strands, which may hence reflect the function of this histone in stabilizing the inactive chromatin state. Our attention has been focused at the propensity of S/MARs to undergo strand separation, which reflects an aspect of a transcriptionally poised or active chromatin domain. We were intrigued by the fact that for a transcriptionally competent domain, huIFN-β, an S/MAR element is a primary target for single-strand-specific probes (Fig. 8). Using similar probes or single-strand-specific nucleases, work by Paul and Ferl (1993), Targa *et al.* (1994), and Iarovaia *et al.* (1995) suggest that this may be a more general property related to the pronounced propensity of S/MAR sequences to separate strands under negative super-helical tension (Fig. 12; and Bode *et al.*, 1992). Upon activation, the under-wound status of DNA is rapidly transferred to the promoter and coding regions where it is thought to support transcriptional initiation and elongation by RNA polymerase II. Such an action requires a progressive rather than a looping-type mechanism and the observation that S/MAR-activities can be blocked by the insertion of GC-rich sequences is certainly in line with such a model (Klehr *et al.*, 1991; Poljak *et al.*, 1994). A related activity may be exerted by S/MAR elements that cohabitate with enhancer elements: the action of the enhancers within the immunoglobulin kappa locus is to a large extent dependent on the presence of an S/MAR (Hörtnagel *et al.*, 1994) and likewise the S/MARs surrounding the intronic enhancer of the immunoglobin μ locus are necessary for establishing an extended DNase I-sensitive chromatin domain and normal rates of transcriptional initiation (Forrester *et al.*, 1994).

Some recent observations have indicated that S/MARs are not necessarily associated with the nuclear matrix *in vivo* as they can be eluted from embedded nuclei (see Eggert and Jack, 1991, for a published report). A possible weakness in the design and interpretation of this experiment is the fact that the transcriptional state of the associated gene was not considered. This argument touches a rather involved discussion of whether one must discriminate between constitutive (permanent) and facultative (activity-related) attachment regions (Jackson *et al.*, 1992; Boulikas, 1993a).

We have demonstrated throughout this chapter that several properties of S/MAR elements originally derived from their presumed function as points of permanent attachment can indeed be verified. The finding of an activity-dependent domain border (Fig. 7) led us to reconsider these findings in light of a more dynamic model of S/MAR-function, which is introduced in Fig. 22. This model describes the state of a domain after it has been released from the inactive state by a D-protein-mediated release of histone H1 and/or by the first round of transcription. Due to scaffold proteins that stabilize the underwound state of DNA, this state will be constrained (overwound stretches, another concomitant of transcription, could be relaxed by the eventual action of topoisomerases). Thereby S/MARs would act as sinks of negative superhelicity and also as topological barriers that prevent superhelical strain from being transmitted to neighboring domains.

As shown for huIFN-β, gene activity is accompanied (or preceded) by the apparent loss of nucleosomes (Fig. 3) and the same process can be mediated by inducing a histone hyperacetylation. While DNA is wrapped in 1.8 left-handed superhelical turns around the histone core, 1 superhelical turn becomes unconstrained after dissolution of the nucleosome (Morse and Simpson, 1988). A similar though smaller effect would be due to the hyperacetylation of the H4 component of a DNA-associated histone core (Norton *et al.*, 1989, 1990), which would result in the generation of under-

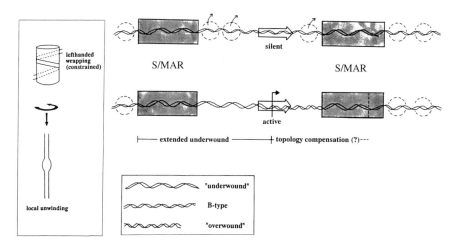

FIG. 22 S/MARs stabilize transcription-dependent topological changes. The removal of nucleosomes upon induction of the huIFN-β gene (Fig. 3) creates underwound or unwound DNA according to the left-hand insert. S/MARs are proposed to stabilize these changes and, in addition, to relieve overwinding of DNA ahead of RNA polymerase II.

wound or locally unwound DNA (Fig. 22). These processes would increase the effect of certain promoter elements (Schlake *et al.*, 1994) and/or reinforce the interaction between he respective regions and the matrix. Finally, the passage of RNA polymerase II, which leaves underwound DNA behind and causes overwinding ahead, might be facilitated by the negative superhelicity stored in the downstream element while it would reinforce the association within the upstream S/MAR.

A final remark concerns the apparently inconsistent interaction between enhancer and S/MAR elements. While in some examples S/MARs delimit the action of enhancers (Li and Stamatoyannopoulos, 1994a,b, Fig. 18), they leave the enhancer action unaffected in another model system (Kellum and Schedl, 1992) and potentiate their action in still other situations (Forrester *et al.*, 1994; Hörtnagel *et al.*, 1994). As a consequence, it is proposed that these interactions are context-dependent. We could envisage a situation where the underwound status of a S/MAR element can be transferred to an enhancer that is preset by the appropriate factors. This requires that unconstrained supercoils be generated, for instance by regulating the proteins involved in S/MAR binding. Such a situation is easy to imagine in a situation where S/MAR and enhancer elements cohabitate and where the S/MARs even reside in the transcribed portion of the gene. If such a regulation cannot be achieved, scaffold attachment would interfere with either the looping or scanning mechanism and restrict an enhancer's action to the stretch of DNA that is interspersed between two S/MAR elements.

During the past decade of active research on S/MAR elements, many ideas about their function have been forwarded. After an initial boom of publications on the localization and basic functions of these elements, new methods had to be developed to evaluate the initial concepts. Some powerful methods for the inspection of S/MAR actions at the single-copy level and at predefined chromosomal sites are now available. Together with the possibilities of creating transgenic organisms, a much more complete picture will predictably emerge in the near future.

References

Adachi, Y., Käs, E., and Laemmli, U. K. (1989). Preferential, cooperative binding of DNA topoisomerase II to scaffold-associated regions. *EMBO J.* **8,** 3997–4006.

Alevy, M. C., Tsai, M. J., and O'Malley, B. W. (1984). DNase I sensitive domain of the gene coding for the glycolytic enzyme glyceraldehyde-3-phosphate dehydrogenase. *Biochemistry,* **23,** 2309–2314.

Allen, G. C., Hall, G. E., Jr., Childs, L. C., Weissinger, A. K., Spiker, S., and Thompson, W. F. (1993). Scaffold attachment regions increase reporter gene expression in stably transformed plant cells. *Plant Cell,* **5,** 603–613.

Anderson, J. N. (1986). Detection, sequence patterns and function of unusual DNA structures. *Nucleic Acids Res.* **21,** 8513–8533.

Antoniou, M., and Grosveld, F. (1990). Beta-globin dominant control region interacts differently with distal and proximal promoter elements. *Gene Dev.* **4,** 1007–1013.

Belgrader, P., Siegal, A. J., and Berezney, R. (1991). A comprehensive study on the isolation and characterization of the HeLa S3 nuclear matrix. *J. Cell Sci.* **98,** 281–291.

Berezney, R., and Coffey, D. S. (1974). Identification of a nuclear protein matrix. *Biochem. Biophys. Res. Commun.* **60,** 1410–1417.

Billiau, A., Edy, V. G., Hermans, van Damme, J., Desmyter, J., Georgiades, J., and de Somer, P. (1977). Human interferon: Mass production in a newly established cell line, MG-63rd *Antimicrob. Agents Chemother.* **12,** 11–15.

Blasquez, V. C., Xu, M., Moses, S. C., and Garrard, W. T. (1989). Immunoglobulin kappa-gene expression after stable integration I: Role of the intronic MAR and enhancer in plasmacytoma cells. *J. Biol. Chem.* **264**(35), 21183–21189.

Bode, J., and Hauser, H. (1992). Interferons: Induction and biological actions. *In* "Biotechnology Focus" (R. K. Finn, P. Präve, M. Schlingman, W. Crueger, K. Esser, R. Thauer, and F. Wagner, eds.), Vol. 3, pp. 57–93. Carl Hanser Verlag, Munich.

Bode, J., and Maass, K. (1988). Chromatin domain surrounding the human interferon-beta gene as defined by scaffold-attached regions. *Biochemistry* **27,** 4706–4711.

Bode, J., Pucher, H. J., and Maass, K. (1986). Chromatin structure and induction-dependent conformational changes of human interferon-β genes in a mouse host cell. *Eur. J. Biochem.* **158,** 393–401.

Bode, J., Kohwi, Y., Dickinson, L., Joh, R. T., Klehr, D., Mielke, C., and Kohwi-Shigematsu, T. (1992). Biological significance of unwinding capability of nuclear matrix-associating DNAs. *Science* **255,** 195–197.

Boffa, L. C., Gruss, R. J., and Allfrey, V. G. (1981). Manifold effects of sodium butyrate on nuclear function. *J. Biol. Chem.* **256,** 9612–9621.

Bohan, C. A., York, D., and Srinivasan, A. (1987). Sodium butyrate activates human immunodeficiency virus long terminal repeat-directed expression. *Biochem. Biophys. Res. Commun.* **148,** 899–905.

Bohan, C. A., Robinson, R. A., Luciw, P. A., and Srinivasan, A. (1989). Mutational analysis of sodium butyrate inducible elements in the human immunodeficiency virus type I long terminal repeat. *Virology* **172,** 573–583.

Bonifer, C., Vidal, M., Grosveld, F., and Sippel, A. E. (1990). Tissue specific and position independent expression of the complete gene domain for chicken lysozyme in transgenic mice. *EMBO J.* **9,** 2843–2848.

Bonifer, C., Hecht, A., Saueressig, H., Winter, D. M., and Sippel, A. E. (1991). Dynamic chromatin: The regulatory domain organization of eukaryotic gene loci. *J. Cell. Biochem.* **47,** 99–108.

Boulikas, T. (1993a). Nature of DNA sequences at the attachment regions of genes to the nuclear matrix. *J. Cell. Biochem.* **52,** 14–22.

Boulikas, T. (1993b). Homeodomain protein binding sites, inverted repeats, and nuclear matrix attachment regions along the human beta-globin gene complex. *J. Cell. Biochem.* **52,** 23–36.

Boulikas, T., and Kong, C. F. (1993). Multitude of inverted repeats characterizes a class of anchorage sites of chromatin loops to the nuclear matrix. *J. Cell. Biochem.* **53,** 1–12.

Boy de la Tour, E., and Laemmli, U. K. (1988). The metaphase scaffold is helically folded: Sister chromatids have predominantly opposite helical handedness. *Cell (Cambridge, Mass.)* **55,** 937–944.

Breyne, P., Van Montagu, M., Depicker, A., and Gheysen, G. (1992). Characterization of a plant scaffold attachment region in a DNA fragment that normalizes transgene expression in tobacco. *Plant Cell* **4,** 463–471.

Brotherton, T., Zenk, D., Kahanic, S., and Reneker, J. (1991). Avian nuclear matrix proteins bind very tightly to cellular DNa of the beta-globin gene enhancer in a tissue-specific fashion. *Biochemistry* **30**, 5845–5850.

Candido, E. P. M., Reeves, R., and Davie, J. R. (1978). Sodium butyrate inhibits histone deacetylation in cultured cells. *Cell (Cambridge, Mass.)* **14**, 105–113.

Cartwright, I. L., and Elgin, S. C. R. (1989). Nonenzymatic cleavage of chromatin. *In* "Methods in Enzymology" (P. Wasserman and R. Kornberg, eds.), Vol. 170, pp. 359–368. Academic Press, San Diego, CA.

Caserta, M., Amadei, A., Camilloni, G., and Di Mauro, E. (1990). Regulation of the function of eukaryotic DNA topoisomerase I: Analysis of the binding step and of the catalytic constants of topoisomerization as a function of DNA topology. *Biochemistry* **29**, 8152–8257.

Charollais, R.-H., Buquet, C., and Mester, J. (1990). Butyrate blocks the accumulation of cdc2 mRNA in late G1 phase but inhibits both the early and late G1 progression in chemically transformed mouse fibroblasts BP-A31st *J. Cell. Physiol.* **145**, 46–52.

Chung, J. H., Whiteley, M., and Felsenfeld, G. (1993). A 5' element of the chicken beta-globin domain serves as an insulator in human erythroid cells and protects against position effect in *Drosophila. Cell (Cambridge, Mass.)* **74**, 505–514.

Cockerill, P. N., and Garrard, W. T. (1986). Chromosomal loop anchorage of the kappan immunoglobulin gene occurs next to the enhancer in a region containing topoisomerase II sites. *Cell (Cambridge, Mass.)* **44**, 273–282.

Collins, J. (1983). Structure and expression of the human interferon genes. *In* "Interferons: From Molecular Biology to Clinical Application" (D. C. Burke and A. G. Morris, eds.), Vol. 35, 35–65. Cambridge Univ. Press, Cambridge, UK.

Collins, J. (1984). Interferon genes: Gene structure and elements involved in gene regulation. *In* "Interferon: Mechanisms of Production and Action" (R. M. Friedman, ed.), Vol. 3, pp. 33–83. Elsevier, Amsterdam.

Cosgrove, D. E., and Cox, G. S. (1990). Effects of sodium butyrate and 5-azacytidine on DNA methylation in human tumor cell lines: Variable response to drug treatment and withdrawal. *Biochim. Biophys. Acta* **1087**, 80–86.

Craig, N. L. (1988). The mechanism of conservative site-specific recombination. *Annu. Rev. Genet.* **22**, 77–105.

Craigie, R. (1992). Hotspots and warm spots: Integration specificity of retroelements. *Trends Genet.* **8**, 187–190.

Cunningham, J. M., Purucker, M. E., Jane, S. M., Safer, B., Vanin, E. F., Ney, P. A., Lowrey, C. H., and Nienhuis, A. W. (1994). The regulatory element 3' to the A gamma-globin gene binds to the nuclear matrix and interacts with special A-T-rich binding protein 1 (SATB1), an SAR/MAR-associating region DNA binding protein. *Blood* **84**, 1298–1308.

de Moura Gallo, C. V., Vassetzky, Y. S., Recillas Targa, F., Georgiev, G. P., Scherrer, K., and Rzain, S. V. (1991). The presence of sequence-specific protein binding sites correlates with replication activity and matrix binding in a 1.7 kb-long DNA fragment of the chicken α-gene domain. *Biochem. Biophys. Res. Commun.* **179**, 512–519.

Dhar, V., Nandi, A., Schildkraut, C. L., and Skoultchi, A. I. (1990). Erythroid-specific nuclease-hypersensitive sites flanking the human beta-globin domain. *Mol. Cell. Biol.* **10**, 4324–4333.

Dickinson, L., Joh, T., Kohwi, Y., and Kohwi-Shigematsu, T. (1992). A tissue-specific MAR/SAR binding protein with unusual binding site recognition. *Cell (Cambridge, Mass.)* **70**, 631–645.

Dietz, A., Kay, V., Schlake, T., Landsmann, J., and Bode, J. (1994). A plant scaffold attached region detected close to a T-DNA integration site is active in mammalian cells. *Nucleic Acids Res.* **22**, 2744–2751.

Dillon, N., and Grosveld, F. (1993). Transcriptional regulation of multigene loci: Multilevel control. *Trends Genet.* **9**, 134–137.

DiMaio, D., Treisman, R., and Maniatis, T. (1982). Bovine papilloma vector that propagates as a plasmid in both mouse and bacterial cells. *Proc. Natl. Acad. Sci. U.S.A.* **79**, 4030–4034.

Dorer, D. R., and Henikoff, S. (1994). Expansions of transgene repeats cause heterochromatin formation and gene silencing in *Drosophila*. *Cell (Cambridge, Mass.)* **77**, 993–1002.

Du, W., Thanos, D., and Maniatis, T. (1993). Mechanisms of transcriptional synergism between distinct virus-inducible enhancer elements. *Cell (Cambridge, Mass.)* **74**, 887–898.

Dworetzky, S. I., Wright, K. L., Fey, E. G., Penman, S., Lian, J. B., Stein, J. L., and Stein, G. S. (1992). Sequence-specific DNA-binding proteins are components of a nuclear matrix-attachment site. *Proc. Natl. Acad. Sci. U.S.A.* **89**, 4178–4182.

Dynan, W. S. (1989). Modularity of promoters and enhancers. *Cell (Cambridge, Mass.)* **58**, 1–4.

Eggert, H., and Jack, R. S. (1991). An ectopic copy of the Drosophila ftz associated SAR neither reorganizes local chromatin structure nor hinders elution of a chromatin fragment from isolated nuclei. *EMBO J.* **10**, 1237–1243.

Fackelmayer, F. O., Dahm, K., Renz, A., Ramsperger, U., and Richter, A. (1994). Nucleic acid binding properties of hnRNP-U/SAF-A: A nuclear matrix protein which binds DNA and RNA in vivo and in vitro. *Eur. J. Biochem.* **221**, 749–757.

Fishel, B. R., Sperry, A. O., and Garrard, W. T. (1993). Yeast calmodulin and a conserved nuclear protein participate in the in vivo binding of a matrix association region. *Proc. Natl. Acad. Sci. U.S.A.* **90**, 5623–5627.

Forrester, W. C., Epner, E., Driscoll, M. C., Enver, T., Brice, M., Papayannopoulou, T., and Groudine, M. (1990). A deletion of the human β-globin locus activation region causes a major alteration in chromatin structure and replication across the entire β-globin locus. *Genes Dev.* **4**, 1637–1649.

Forrester, W. C., Van Genderen, C., Jenuwein, T., and Grosschedl, R. (1994). Dependence of enhancer-mediated transcription of the immunoglobulin Mu gene on nuclear matrix attachment regions. *Science* **265**, 1221–1225.

Fraser, P., Hurst, J., Collis, P., and Grosveld, F. (1990). DNase I hypersensitive sites 1, 2 and 3 of the human β-globin dominant control region direct position-independent expression. *Nucleic Acids Res.* **18**(12), 3503–3508.

Fraser, P., Pruzina, S., Antoniou, M., and Grosveld, F. (1993). Each hypersensitive site of the human beta-globin locus control region confers a different developmental pattern of expression on the globin genes. *Genes Dev.* **7**, 106–113.

Fregeau, C. J., Helgason, C. D., and Bleackley, R. C. (1992). Two cytotoxic cell proteinase genes are differentially sensitive to sodium butyrate. *Nucleic Acids Res.* **20**(12), 3113–3119.

Fritton, H. P., Jantzen, K., Igo-Kemenes, T., Nowock, J., Strech-Jurk, U., Theisen, M., and Sippel, A. E. (1988). Chromatin domains and gene expression—Different chromatin conformations characterize the various functional states of the chicken lysozyme gene. *In* "Architecture of Eukaryotic Genes" (G. Vahl, ed.) Vol. 16, pp. 333–353. VCH, Weinheim.

Froelich-Ammon, S. J., Gale, K. C., and Osheroff, N. (1994). Site-specific cleavage of a DNA hairpin by topoisomerase II. DNA secondary structure as a determinant of enzyme recognition/cleavage. *J. Biol. Chem.* **269**, 7719–7725.

Gasser, S. M., and Laemmli, U. K. (1986a). Cohabitation of scaffold binding regions with upstream-enhancer elements of three developmentally regulated genes of *D. melanogaster*. *Cell (Cambridge, Mass.)* **46**, 521–530.

Gasser, S. M., and Laemmli, U. K. (1986b). The organization of chromatin loops: Characterization of a scaffold attachment site. *EMBO J.* **5**, 511–518.

Gasser, S. M., and Laemmli, U. K. (1987). A glimpse at chromosomal order. *Trends Genet.* **3**, 16–22.

Goodbourn, S., Burstein, H., and Maniatis, T. (1986). The human beta-interferon gene enhancer is under negative control. *Cell (Cambridge, Mass.)* **45**(4), 601–610.

Gromova, I. I., Thomsen, B., and Razin, S. V. (1995). Different topoisomerase II antitumor drugs direct similar specific long-range fragmentation of an amplified c-myc gene locus in living cells and in high-salt extracted nuclei. *Proc. Natl. Acad. Sci. U.S.A.* **92**, 102–106.

Gross, G., Mayr, U., Bruns, W., Grosveld, F., Dahl, H. H., and Collins, J. (1981). The structure of a thirty-six kilobase region of the human chromosome including the fibroblast interferon gene IFN-β. *Nucleic Acids Res.* **9**, 2495–2506.

Grosveld, F., van Assendelft, G. B., Greaves, D. R., and Kollias, G. (1987). Position-independent, high-level expression of the human β-globin gene in transgenic mice. *Cell (Cambridge, Mass.)* **51**, 975–985.

Hakes, D. J., and Berezney, R. (1991). DNA binding properties of the nuclear matrix and individual nuclear matrix proteins. Evidence for salt-resistant DNA binding sites. *J. Biol. Chem.* **266**, 11131–11140.

Hauser, H., Gross, G., Bruns, W., Hochkeppel, H. K., Mayr, U., and Collins, J. (1982). Inducibility of human β-interferon gene in mouse L-cell clones. *Nature (London)* **297**, 650–654.

Hirose, S., and Ohta, T. (1990). DNA supercoiling and eukaryotic transcription—Cause and effect. *Cell Struct. Funct.* **15**, 133–135.

Hirose, S., and Suzuki, Y. (1988). In vitro transcription of eukaryotic genes is affected differently by the degree of DNA supercoiling. *Proc. Natl. Acad. Sci. U.S.A.* **85**, 718–722.

Homberger, H. P. (1989). Bent DNA is a structural feature of scaffold-attached regions in *Drosophila melanogaster* interphase nuclei. *Chromosoma* **98**, 99–104.

Hörtnagel, K., Mautner, J., Strobl, L. J., Wolf, D., Christoph, B., Geltinger, C., and Polack, A. (1994). The role of immunoglobulin kappa elements in c-myc activation. *Oncogene* **10**, 1393–1401.

Hörz, W., and Altenburger, W. (1981). Sequence specific cleavage of DNA by micrococcus nuclease. *Nucleic Acids Res.* **9**, 2643–2658.

Howard, M. T., and Griffith, J. D. (1993). A cluster of strong topoisomerase II cleavage sites is located near an integrated human immunodeficiency virus. *J. Mol. Biol.* **232**, 1060–1068.

Iarovaia, O. V., Lagarkova, M. A., and Razin, S. V. (1995). The specificity of eukaryotic genome long range fragmentation by endogenous topoisomerase II and exogenous Bal 31 nuclease depends on cell proliferation status. *Biochemistry* **34**, 4133–4138.

Izaurralde, E., Mirkovitch, J., and Laemmli, U. K. (1988). Interaction of DNA with nuclear scaffolds in vitro. *J. Mol. Biol.* **200**, 111–126.

Jackson, D. A., Dolle, A., Robertson, G., and Cook, P. R. (1992). The attachments of chromatin loops to the nucleoskeleton. *Cell Biol. Int. Rep.* **16**, 687–696.

Jarman, A. P., and Higgs, D. R. (1988). Nuclear scaffold attachment sites in the human globin gene complexes. *EMBO J.* **7**, 3337–3344.

Johnston, L. A., Tapscott, S. J., and Eisen, H. (1992). Sodium butyrate inhibits myogenesis by interfering with the transcriptional activation function of MyoD and Myogenin. *Mol. Cell. Biol.* **12**(11), 5123–5130.

Kalos, M., and Fournier, R. E. K. (1995). Position-independent transgene expression mediated by boundary elements from the apolipoprotein B chromatin domain. *Mol. Cell. Biol.* **15**, 198–207.

Käs, E., and Chasin, L. A. (1987). Anchorage of the Chinese hamster dihydrofolate reductase gene to the nuclear scaffold occurs in an intragenic region. *J. Mol. Biol.* **198**, 677–692.

Käs, E., and Laemmli, U. K. (1992). In vivo topoisomerase II cleavage of the *Drosophila* histone and satellite III repeats: DNA sequence and structural characteristic. *EMBO J.* **11**, 705–716.

Käs, E., Izaurralde, E., and Laemmli, U. K. (1989). Specific inhibition of DNA binding to nuclear scaffolds and histone H1 by distamycin. The role of oligo(dA)·oligo(dT) tracts. *J. Mol. Biol.* **210**, 587–599.

Käs, E., Poljak, L., Adachi, Y., and Laemmli, U. K. (1993). A model for chromatin opening: Stimulation of topoisomerase II and restriction enzyme cleavage of chromatin by distamycin. *EMBO J.* **12**, 115–126.

Kay, V., and Bode, J. (1994). Binding specificity of a nuclear scaffold: Supercoiled, single-stranded and scaffold-attached-region-DNA. *Biochemistry* **33**, 367–374.

Kay V., and Bode, J. (1995). Detection of scaffold-attached regions (SARs) by in vitro techniques; activities of these elements in vivo. *In* "Methods in Molecular and Cellular Biology: Methods for Studying DNA-Protein Interactions—An Overview" (A. G. Papavassiliou and S. L. King, eds.), Wiley–Liss, New York (in press).

Kellum, R., and Schedl, P. (1991). A position-effect assay for boundaries of higher order chromosomal domains. *Cell (Cambridge, Mass.)* **64**, 941–950.

Kellum, R., and Schedl, P. (1992). A group a scs elements function as domain boundaries in an enhancer-blocking assay. *Mol. Cell. Biol.* **12**, 2424–2431.

Kijima, M., Yoshida, M., Sugita, K., Horinouchi, S., and Beppu, T. (1993). Trapoxin, an antitumor cyclic tetrapeptide, is an irreversible inhibitor of mammalian histone deacetylase. *J. Biol. Chem.* **268**(30), 22429–22435.

Klehr, D., and Bode, J. (1988). Comparative evaluation of bovine papilloma virus (BPV) vectors for the study of gene expression in mammalian cells. *Mol. Genet. (Life Sci. Adv.)* **7**, 47–52.

Klehr, D., Maass, K., and Bode, J. (1991). SAR elements from the human IFN-β domain can be used to enhance the stable expression of genes under the control of various promoters. *Biochemistry* **30**, 1264–1270.

Klehr, D., Schlake, T., Maass, K., and Bode, J. (1992). Scaffold-attached regions (SAR elements) mediate transcriptional effects due to butyrate. *Biochemistry* **31**, 3222–3229.

Kohwi, Y., and Kohwi-Shigematsu, T. (1993). Structural polymorphism of homopurine-homopyrimidine sequences at neutral pH. *J. Mol. Biol.* **231**, 1090–1101.

Kohwi-Shigematsu, T., Gelinas, R., and Weintraub, H. (1983). Detection of an altered DNA conformation at specific sites in chromatin and supercoiled DNA. *Proc. Natl. Acad. Sci. U.S.A.* **80**, 4389–4393.

Kowalski, D., Natale, D. A., and Eddy, M. J. (1988). Stable DNA unwinding, not "breathing" accounts for single-strand-specific nuclease hypersensitivity of specific A+T-rich sequences. *Proc. Natl. Acad. Sci. U.S.A.* **85**, 9464–9468.

Kricker, M. C., Drake, J. W., and Radman, M. (1992). Duplication-targeted DNA methylation and mutagenesis in the evolution of eukaryotic chromosomes. *Proc. Natl. Acad. Sci. U.S.A.* **89**, 1075–1079.

Kruh, J. (1982). Effects of sodium butyrate, a new pharmacological agent, on cell culture. *Mol. Cell. Biol.* **42**, 65–82.

Laemmli, U. K., Käs, E., Poljak, L., and Adachi, Y. (1992). Scaffold-associated regions: cis-acting determinants of chromatin structural loops and functional domains. *Curr. Opin. Genet. Dev.* **2**, 275–285.

Lawson, G. M., Knoll, B. J., March, C. J., Woo, S. L. C., Tsai, M. J., and O'Mally, B. W. (1982). Definition of 5' and 3' structural boundaries of the chromatin domain containing the ovalbumin multigene family. *J. Biol. Chem.* **257**, 1501–1507.

Leder, A., and Leder, P. (1975). Butyric acid, a potent inducer of erythroid differentiation of Friend-virus transformed cells. *Cell (Cambridge, Mass.)* **5**, 319–322.

Lee, M.-S., and Garrard, W. T. (1991a). Positive DNA supercoiling generates a chromatin conformation characteristic of highly active genes. *Proc. Natl. Acad. Sci. U.S.A.* **88**, 9675–9679.

Lee, M.-S., and Garrard, W. T. (1991b). Transcription-induced nucleosome 'splitting': An underlying structure for DNase I sensitive chromatin. *EMBO J.* **10**, 607–615.

Levy-Wilson, B., and Fortier, C. (1989). The limits of the DNAse I-sensitive domain of the human apolipoprotein B gene coincide with the location of chromosomal anchorage loops and define the 5' and 3'-boundaries of the gene. *J. Biol. Chem.* **264**, 21196–21204.

Li, Q., and Stamatoyannopoulos, G. (1994a). Hypersensitive site 5 of the human Beta locus control region functions as a chromatin insulator. *Blood* **84**, 1399–1401.

Li, Q., and Stamatoyannopoulos, J. A. (1994b). Position independence and proper developmental control of gamma-globin gene expression require both a 5' locus control region and a downstream sequence element. *Mol. Cell. Biol.* **14,** 6087–6096.

Lu, Q., Wallrath, L. L., Allan, B. D., Glaser, R. L., Lis, J. T., and Elgin, S. C. R. (1992). Promoter sequence containing (CT)n. (GA)n repeats is critical for the formation of the DNase I hypersensitive sites in the *Drosophila* hsp26 gene. *J. Mol. Biol.* **225,** 985–998.

Lu, Q., Wallrath, L. L., Granok, H., and Elgin, S. C. R. (1993). (CT)n. (GA)n repeats and heat shock elements have distinct roles in chromatin structure and transcriptional activation of the *Drosophila* hsp26 gene. *Mol. Cell. Biol.* **13,** 2802–2814.

Luchnik, A. N., Hisamutdinov, T. A., and Georgiev, G. P. (1988). Inhibition of transcription in eukaryotic cells by X-ray irradiation: Relation to the loss of topological constraint in closed DNA loops. *Nucleic Acids Res.* **16,** 5175–5190.

Ludérus, M. E. E., de Graaf, A., Mattia, E., den Blaauwen, J. L., Grande, M. A., de Jong, L., and van Driel, R. (1992). Binding of matrix attachment regions to lamin B1st. *Cell (Cambridge, Mass.)* **70,** 949–959.

Ludérus, M. E. E., den Blaauwen, J. L., de Smit, O. J. B., Compton, D. A., and van Driel, R. (1994). Binding of matrix attachment regions to lamin polymers involves single-stranded regions and the minor groove. *Mol. Cell. Biol.* **14,** 6297–6305.

Lutfalla, G., Blanc, H., and Bertolotti, R. (1985). Shuttling of integrated vectors from mammalian cells to *E. coli* is mediated by head-to-tail multimeric inserts. *Somatic Cell Mol. Genet.* **11**(3), 223–238.

Marsden, M., and Laemmli, U. K. (1979). Metaphase chromosome structure: Evidence for a radical loop model. *Cell (Cambridge, Mass.)* **17,** 849–858.

Matzke, M. A., and Matzke, A. J. M. (1990). Gene interactions and epigenetic variation in transgenic plants. *Dev. Genet. (Amsterdam)* **11,** 214–223.

Matzke, M. A., Primig, M., Trnovsky, J., and Matzke, A. J. M. (1989). Reversible methylation and inactivation of marker genes in sequentially transformed tobacco plants. *EMBO J.* **8,** 643–649.

Mayerhofer, R., Koncz-Kalman, Z., Nawrath, C., Bakkeren, G., Crameri, A., Angelis, K., Redei, G. P., Schell, J., Hohn, B., and Koncz, C. (1991). T-DNA integration: A mode of illegitimate recombination in plants. *EMBO J.* **10,** 697–704.

McDonagh, K. T., Dover, G. J., Donahue, R. E., Nathan, D. G., Agricola, B., Byrne, E., and Nienhuis, A. W. (1992). Hydroxyurea-induced HbF production in anemic primates: Augmentation by erythropoietin, hematopoietic growth factors, and sodium butyrate. *Exp. Hematol. (Lawrence, Kans.)* **20,** 1156–1164.

McKnight, R. A., Shamay, A., Sankaran, L., Wall, R. J., and Henninghausen, L. (1992). Matrix-attachment regions can impart position-independent regulation of a tissue-specific gene in transgenic mice. *Proc. Natl. Acad. Sci. U.S.A.* **89,** 6943–6947.

Mielke, C. (1993). Einfluss von SAR-Elementen auf die Stabilität und Eigenschaften retroviral infizierter Zellinien. Ph.D. thesis, Universität Braunschweig.

Mielke, C., Kohwi, Y., Kohwi-Shigematsu, T., and Bode, J. (1990). Hierarchical binding of DNA fragments derived from scaffold-attached regions: Correlations of properties in vitro and function in vivo. *Biochemistry* **29,** 7475–7485.

Mirkovitch, J., Mirault, M. E., and Laemmli, U. K. (1984). Organization of the higher-order chromatin loop: Specific DNA attachment sites on nuclear scaffold. *Cell (Cambridge, Mass.)* **39,** 223–232.

Mirkovitch, J., Gasser, S. M., and Laemmli, U. K. (1988). Scaffold attachment of DNA loops in metaphase chromosomes. *J. Mol. Biol.* **200,** 101–110.

Mizutani, M., Ohta, T., Watanabe, H., Handa, H., and Hirose, S. (1991). Negative supercoiling of DNA facilitates an interaction between transcription factor IID and the fibroin gene promoter. *Proc. Natl. Acad. Sci. U.S.A.* **88,** 718–722.

Mlynarova, L., Loonen, A., Heldens, J., Jansen, R. C., Keizer, P., Stiekema, W. J., and Napp, J. P. (1994). Reduced position effect in mature transgenic plants conferred by the chicken lysozyme matrix-associated region. *Plant Cell* **6**, 417–426.

Morse, R. H., and Simpson, R. T. (1988). DNA in the nucleosome. *Cell (Cambridge, Mass.)* **54**, 285–287.

Mueller, H.-P., Pryciak, P. M., and Varmus, H. E. (1993). Retroviral integration machinery as a probe for DNA structure and associated proteins. *Cold Spring Harbor Symp. Quant. Biol.* **58**, 533–541.

Nakagomi, K., Kohwi, Y., Dickinson, L. A., and Kohwi-Shigematsu, T. (1994). A novel DNA-binding motif in the nuclear matrix attachment DNA-binding protein SATB1st *Mol. Cell. Biol.* **14**, 1852–1860.

Nelson, H. C. M., Finch, J. T., Luisi, B. F., and Klug, A. (1987). The structure of an oligo (dA)·oligo(dT) tract and its biological implications. *Nature (London)* **330**, 221–226.

Norton, V. G., Imai, B. S., Yau, P., and Bradbury, E. M. (1989). Histone acetylation reduces nucleosome core particle linking number change. *Cell (Cambridge, Mass.)* **57**, 449–457.

Norton, V. G., Marvin, K. W., Yau, P., and Bradbury, E. M. (1990). Nucleosome linking number change controlled by acetylation of histones H3 and H4th *J. Biol. Chem.* **265**, 19848–19852.

Osheroff, N., Zechiedrich, E. L., and Gale, K. C. (1991). Catalytic function of DNA topoisomerase II. *BioEssays* **13**, 269–275.

Paul, A.-L., and Ferl, R. J. (1993). Osmium tetroxide footprinting of a scaffold attachment region in the maize Adh1 promoter. *Plant Mol. Biol.* **22**, 1145–1151.

Paulson, J. R., and Laemmli, U. K. (1977). The structure of histone-depleted metaphase chromosomes. *Cell (Cambridge, Mass.)* **12**, 817 828.

Phi-Van, L., and Strätling, W. H. (1988). The matrix attachment regions of the chicken lysozyme gene co-map with the boundaries of the chromatin domain. *EMBO J.* **7**, 655–664.

Phi-Van, L., von Kries, J. P., Ostertag, W., and Strätling, W. H. (1990). The chicken lysozyme 5' matrix attachment region increases transcription from a heterologous promoter in heterologous cells and dampens position effects on the expression of transfected genes. *Mol. Cell. Biol.* **10**, 2302–2307.

Pina, B., Truss, M., Ohlenbusch, H., Postma, J., and Beato, M. (1990). DNA rotational positioning in a regulatory nucleosome is determined by base sequence. An algorithm to model the preferred superhelix. *Nucleic Acids Res.* **18**, 6981–6987.

Poljak, L., Seum, C., Mattioni, T., and Laemmli, U. K. (1994). SARs stimulate but do not confer position independent gene expression. *Nucleic Acids Res.* **22**, 4386–4394.

Probst, H., and Herzog, R. (1985). DNA regions associated with the nuclear matrix of Ehrlich ascites cells expose single-stranded sites after deproteinization. *Eur. J. Biochem.* **146**, 167–171.

Puhl, H. L., Gudibande, S. R., and Behe, M. J. (1991). Poly[d(A.T)] and other synthetic polydeoxynucleotides containing oligoadenosine tracts form nucleosomes easily. *J. Mol. Biol.* **222**, 1149–1160.

Razin, S. V., Hancock, R., Iarovaia, O., Westergaard, O., Gromova, I., and Georgiev, G. P. (1993). Structural-functional organization of chromosomal DNA domains. *Cold Spring Harbor Symp. Quant. Biol.* **58**, 25–35.

Razin, S. V., de Moura Gallo, C. V., and Scherrer, K. (1994). Characterization of the chromatin structure in the upstream region of the chicken alpha-globin gene domain. *Mol. Gen. Genet.* **242**, 649–652.

Reitman, M., Lee, E., Westphal, H., and Felsenfeld, G. (1993). An enhancer/locus control region is not sufficient to open chromatin. *Mol. Cell. Biol.* **13**, 3990–3998.

Ríos-Ramírez, M. (1994). Untersuchungen zur Chromatinstruktur von 'Scaffold-attached-regions' (SAR-Elementen). Ph.D. thesis, Universität Braunschweig.

Roberge, M., Dahmus, M. E., and Bradbury, E. M. (1988). Chromosomal loop/nuclear matrix organization of transcriptionally active and inactive RNA polymerase in HeLa nuclei. *J. Mol. Biol.* **201**, 545–556.

Saito, H., Kagawa, T., Tada, S., Tsunematsu, S., Guevara, F. M., Watanabe, T., Morizane, T., and Tsuchiya, M. (1992). Effect of dexamethasone, dimethylsulfoxide and sodium butyrate on a human hepatoma cell line PLC/PRF/5th. *Cancer Biochem. Biophys.* **13**, 75–84.

Saitoh, Y., and Laemmli, U. K. (1993). From the chromosomal loops and the scaffold to the classic bands of metaphase chromosomes. *Cold Spring Harbor Symp. Quant. Biol.* **58**, 755–765.

Saitoh, Y., and Laemmli, U. K. (1994). Metaphase chromosome structure: Bands arise from a differential folding path of the highly AT-rich scaffold. *Cell (Cambridge, Mass.)* **76**, 609–622.

Sander, M., and Hsieh, T. S. (1985). *Drosophila* topoisomerase II double-strand DNA cleavage analysis of DNA sequence homology at the cleavage site. *Nucleic Acids Res.* **13**, 1057–1072.

Sandig, V., Lieber, A., Baehring, S., and Strauss, M. (1993). A phage T7 class-III promoter functions as a polymerase II promoter in mammalian cells. *Gene* **131**, 255–259.

Schaffner, G., Schirm, S., Müller-Baden, B., Weber, F., and Schaffner, W. (1988). Redundancy of information in enhancers as a principle of mammalian transcription control. *J. Mol. Biol.* **201**, 81–90.

Scherdin, U., Rhodes, K., and Breindl, M. (1990). Transcriptionally active genome regions are preferred targets for retrovirus integration. *J. Virol.* **64**, 907–912.

Scheuermann, R. H., and Chen, U. (1989). A developmental-specific factor binds to suppressor sites flanking the immunoglobulin heavy-chain enhancer. *Genes Dev.* **3**, 1255–1266.

Schlake, T. (1994). Entwicklung einer auf SAR-Elementen und sequenzspezifischer Rekombination basierenden Strategie zur stabilen Hochexpression in Mammaliazellen. Ph.D. thesis, Universität Braunschweig.

Schlake, T., and Bode, J. (1994). Use of mutated FLP-recognition-target-(FRT) sites for the exchange of expression cassettes at defined chromosomal loci. *Biochemistry* **33**, 12746–12751.

Schlake, T., Klehr-Wirth, D., Yoshida, M., Beppu, T., and Bode, J. (1994). Gene expression within a chromatin domain: The role of core histone hyperacetylation. *Biochemistry* **33**, 4197–4206.

Schöffl, F., Schröder, G., Kliem, M., and Rieping, M. (1993). An SAR sequence containing 395 bp DNA fragment mediates enhanced, gene-dosage-correlated expression of a chimaeric heat shock gene in transgenic tobacco plants. *Transgenic Res.* **2**, 93–100.

Sealy, L., and Chalkley, R. (1978). Effect of sodium butyrate on histone modification. *Cell (Cambridge, Mass.)* **14**, 115–121.

Shih, C.-C., Stoye, J. P., and Coffin, J. M. (1988). Highly preferred targets for retrovirus integration. *Cell (Cambridge, Mass.)* **53**, 531–537.

Sippel, A. E., Schaefer, G., Faust, N., Saueressig, H., Hecht, A., and Bonifer, C. (1993). Chromatin domains constitute regulatory units for the control of eukaryotic genes. *Cold Spring Harbor Symp. Quant. Biol.* **58**, 37–44.

Stadler, J., Larsen, A., Engel, J. D., Dolan, M., Groudine, M., and Weintraub, H. (1980). Tissue-specific DNA cleavages in the chromatin domain introduced by DNAse I. *Cell (Cambridge, Mass.)* **20**, 451–460.

Stief, A., Winter, D. M., Strätling, W. H., and Sippel, A. E. (1989). A nuclear DNA attachment element mediates elevated and position-independent gene activity. *Nature (London)* **341**, 343–345.

Strätling, W. H., Dölle, A., and Sippel, A. E. (1986). Chromatin structure of the chicken lysozyme gene domain as determined by chromatin fractionation and micrococcal nuclease digestion. *Biochemistry* **196**, 495–502.

Targa, F. R., Razin, S. V., de Moura Gallo, C. V., and Scherrer, K. (1994). Excision close to matrix attachment regions of the entire chicken alpha-globin gene domain by nuclease S1 and characterization of the framing structures. *Proc. Natl. Acad. Sci. U.S.A.* **91**, 4422–4426.

Thomas, D. J., Angle, C. R., Swanson, S. A., and Caffrey, T. C. (1991). Effect of sodium butyrate on metallothionein induction and cadmium cytotoxicity in ROS 17/2.8 cells. *Toxicology* **66**, 35–46.

Thompson, E. M., Christians, E., Stinnakre, M.-G., and Renard, J.-P. (1994). Scaffold attachment regions stimulate HSP70.1 expression in mouse preimplantation embryos but not in differentiated tissues. *Mol. Cell. Biol.* **14,** 4694–4703.

Travers, A. A. (1987). DNA bending and nucleosome positioning. *Trends Biochem. Sci.* **12,** 108–111.

Travers, A. A. (1990). Why bend DNA? *Cell (Cambridge, Mass.)* **60,** 177–180.

Tsuji, N., Kobayashi, M., Nagashima, K., Wakisaka, Y., and Koizumi, K. (1976). A new antifungal antibiotic, Trichostatin. *J. Antibiot.* **29,** 1–6.

Tsutsui, K., Tsutsui, K., and Muller, M. T. (1988). The nuclear scaffold exhibits DNA-binding sites selective for supercoiled DNA. *J. Biol. Chem.* **263,** 7235–7241.

Uvardy, A., Maine, E., and Schedl, P. (1985). The 87A7 chromomere: Identification of novel chromatin structures flanking the heat shock locus that may define the boundaries of higher order domains. *J. Mol. Biol.* **185,** 341–358.

van Assendelft, G. B., Hanscombe, O., Grosveld, F., and Greaves, D. R. (1989). The β-globin dominant control region activates homologous and heterologous promoters in a tissue-specific manner. *Cell (Cambridge, Mass.)* **56,** 969–977.

van der Geest, A. H. M., Hall G. E., Jr., Spiker, S., and Hall, T. C. (1994). The beta-phaseolin gene is flanked by matrix attachment regions. *Plant J.* **6,** 413–423.

van der Krol, A. R., Mur, L. A., Beld, M., and Mol, J. N. M. (1990). Flavonoid genes in petunia: Addition of a limited number of gene copies may lead to a suppression of gene expression. *Plant Cell.* **2,** 291–299.

Vazquez, J., Farkas, G., Gaszner, M., Udvardy, A., Muller, M., Hagstrom, K., Gyurkovics, H., Sipos, L., Gausz, J., Galloni, M., Hogga, I., Karch, F., and Schedl, P. (1993). Genetic and molecular analysis of chromatin domains. *Cold Spring Harbor Symp. Quant. Biol.* **58,** 45–54.

Verreault, A., and Thomas, J. O. (1993). Chromatin structure of the beta-globulin chromosomal domain in adult chicken erythrocytes. *Cold Spring Harbor Symp. Quant. Biol.* **58,** 15–24.

Vidali, G., Boffa, L. C., Bradbury, E. M., and Allfrey, V. G. (1978). Butyrate suppression of histone deacetylation leads to accumulation of multiacetylated forms of histones H3 and H4 and increased DNase I sensitivity of the associated DNA sequences. *Proc. Natl. Acad. Sci. U.S.A.* **75**(5), 2239–2243.

von Kries, J. P., Phi-Van, L., Diekmann, S., and Strätling, W. H. (1990). A non-curved chicken lysozyme 5′ matrix attachment site is 3′ followed by a strongly curved DNA sequence. *Nucleic Acids Res.* **18,** 3881–3885.

von Kries, J. P., Buck, F., and Strätling, W. H. (1994). Chicken MAR binding protein p120 is identical to human heterologous nuclear ribonucleoprotein (hnRNP) U. *Nucleic Acids Res.* **22,** 1215–1220.

Weintraub, H., Larsen, A., and Groudine, M. (1981). α-Globin gene switching during the development of chicken embryos: Expression and chromosome structure. *Cell (Cambridge, Mass.)* **24,** 333–344.

Weising, K., Bohn, H., and Kahl, G. (1990). Chromatin structure of transferred genes in transgenic plants. *Dev. Genet.* **11,** 233–247.

Wolffe, A. (1994). Nucleosome positioning and modification: Chromatin structures that potentiate transcription. *Trends Biochem. Sci.* **19,** 240–244.

Worcel, A., Gargiulo, G., Jessee, B., Udvardy, A., Louis, C., and Schedl, P. (1983). Chromatin fine structure of the histone gene complex of *Drosophila melanogaster*. *Nucleic Acids Res.* **11**(2), 421–439.

Wu, H.-Y., Shyy, S., Wang, J. C., and Liu, L. F. (1988). Transcription generates positively and negatively supercoiled domains in the template. *Cell (Cambridge, Mass.)* **53,** 433–440.

Xu, M., Hammer, R. E., Blasquez, V. C., Jones, S. L., and Garrard, W. T. (1989). Immunoglobulin kappa-gene expression after stable integration II. Role of the intronic MAR and enhancer in transgenic mice. *J. Biol. Chem.* **264**(35), 21190–21195.

Yeivin, A., Tang, D., and Taylor, M. W. (1992). Sodium butyrate selectively induces transcription of promoters adjacent to the MoMSV viral enhancer. *Gene* **116,** 159–164.

Yoshida, M., and Beppu, T. (1988). Reversible arrest of proliferation of rat 3Y1 fibroblasts in both the G1 and G2 phases of trichostatin A. *Exp. Cell Res.* **177,** 122–131.

Yoshida, M., Nomura, S., and Beppu, T. (1987). Effects of trichostatin on differentiation of murine erythroleukemia cells. *Cancer Res.* **47,** 3688–3691.

Yoshida, M., Kijima, M., Akita, M., and Beppu, T. (1990a). Potent and specific inhibition of mammalian histone deacetylase both in vivo and in vitro by trichostatin A. *J. Biol. Chem.* **265**(28), 17174–17179.

Yoshida, M., Hoshikawa, Y., Koseki, K., Mori, K., and Beppu, T. (1990b). Structural specificity for biological activity of trichostatin A, a specific inhibitor of mammalian cell cycle with potent differentiation-inducing activity in Friend leukemia cells. *J. Antibiot.* **43**(9), 1101–1106.

Yu, J., Bock, J. H., Slightom, J. L., and Villeponteau, B. (1994). A 5′ Beta-globin matrix-attachment region and the polyoma enhancer together confer position-independent transcription. *Gene* **139,** 139–145.

Zechiedrich, E. L., and Osheroff, N. (1990). Eukaryotic topoisomerases recognize nucleic acid topology by preferentially interacting with DNA crossovers. *EMBO J.* **9,** 4555–4562.

Zehnbauer, B., and Vogelstein, B. (1986). Supercoiled loops and the organization of replication and transcription in eukaryotes. *BioEssays* **2,** 52–54.

Zhao, K., Käs, E., Gonzalez, E., and Laemmli, U. K. (1993). SAR-dependent mobilization of histone H1 by HMG-I/Y in vitro: HMG-I/Y is enriched in H1-depleted chromatin. *EMBO J.* **12,** 3237–3247.

Origins of Replication and the Nuclear Matrix: The DHFR Domain as a Paradigm

P. A. Dijkwel and J. L. Hamlin
Department of Biochemistry, University of Virginia, School of Medicine,
Charlottesville, Virginia 22908

The eukaryotic genome appears to be organized in a loopwise fashion by periodic attachment to the nuclear matrix. The proposal that a chromatin loop corresponds to a functional domain has stirred interest in the properties of the DNA sequences at the bases of these loops, the matrix-attached regions (MARs). Evidence has been presented suggesting that certain MARs act as boundary elements isolating domains from their chromosomal context. MARs have also been found in the vicinity of promoters and enhancers and they could act by displacing these *cis*-regulatory elements into the proper nuclear subcompartment. Attachment to the matrix might also play a role in DNA replication. A large body of evidence indicates that replication occurs on the nuclear matrix. This implies that any DNA sequence will be attached to the matrix at a certain time during the cell cycle. This *transient* mode of attachment contrasts with the proposed *permanent* attachment of origins of DNA replication with the nuclear matrix. While some data exist that support this suggestion, the current lack of understanding of the mammalian replication origin precludes definitive conclusions regarding the role of MARs in the initiation process.

KEY WORDS: Nuclear matrix, MARs, Replicators, Initiation sites, Two-dimensional replicon mapping, DHFR domain.

I. Introduction

Our understanding of the way in which cells function has increased tremendously due to the explosive development of molecular biological techniques. It is, therefore, not surprising that any cellular process is expected, eventually, to be reproduced in a cell-free system *in vitro*. However, the eukaryotic cell is not a membrane-encapsulated drop of cytoplasm in which enzymes and substrates move around freely. Rather, it is becoming increasingly clear that the cell has a highly ordered three-dimensional architecture. Consequently, linking function, be it cytoplasmic or nuclear, to cellular structure has developed as one of the more interesting and important challenges in molecular and cellular biology. In this chapter, attention will be focused on structure–function relationships in the mammalian cell nucleus as they pertain to the initiation of DNA replication.

II. Functional Organization of the Eukaryotic Genome

A. An Organizing Entity in the Nucleus

The mammalian cell nucleus typically must organize 1–2 m of DNA fiber. This is no mean feat since the diameter of the nucleus rarely exceeds 10 μm. Consequently, even in interphase, DNA has to be compacted several thousandfold, far exceeding what can be achieved by wrapping the DNA around nucleosomes. At the same time, the genome has to be sufficiently dynamic to facilitate a series of important biological processes for survival of the cell. Among these are duplication of the genome, disentanglement of the daughter strands, and their subsequent segregation during mitosis. These processes provide the nucleus with its greatest topological challenge. It is hardly surprising, therefore, that the idea of a nuclear substructure that imparts spatial organization to the genome is not new. Indeed, this idea first surfaced more than a century ago when it was postulated that chromosomes in the interphase nucleus occupy defined zones (Rabl, 1885). However, convincing experimental support for this suggestion was not generated until fairly recently (Hochstrasser *et al.*, 1986).

Immunofluorescence microscopy has shown that the nuclear interior exhibits a high degree of functional organization. When either nascent DNA or RNA is labeled with appropriate fluorescent probes, a punctate (rather than a diffuse) staining pattern is observed throughout the nucleus (Nakamura *et al.*, 1986; Jackson *et al.*, 1993; Wansink *et al.*, 1993). Immunolocalization of the individual molecular components of the replication, transcription, or splicing machineries also reveals a focal distribution (Celis and Celis, 1985; Carmo-Fonseca *et al.*, 1991; Cardoso *et al.*, 1993). Interestingly, the *in situ* hybridization approach has shown that RNA transcripts are concentrated along tracks running from the nuclear interior to the periphery (Lawrence *et al.*, 1989), possibly to specific nuclear pores as proposed earlier in the gene-gating theory (Blobel, 1986). It is tempting to speculate that these tracks outline a nuclear component dedicated to mRNA trafficking.

Not only do chromosomes and regions of nuclear activity appear to occupy more or less defined regions within the nucleus, but chromosomes themselves are highly organized. Cytological staining of metaphase chromosomes by the Giemsa technique results in the appearance of several thousand alternating light and dark bands (Yunis, 1981) that appear to correlate, at the level of resolution of this approach, with genetic activity: light bands replicate early in the S-period and harbor active genes, whereas dark bands are transcriptionally silent and replicate late in S-phase (Yunis et al., 1977; Holmquist et al., 1982).

At the molecular level, several lines of evidence further support the idea that chromatin is organized into topologically separate domains. Relative to bulk chromatin, active and potentially active genes are characterized by increased sensitivity to nucleolytic attack (Weintraub and Groudine, 1976). This open chromatin configuration, however, generally extends well beyond the transcription unit itself in both the 3' and 5' directions (Stalder et al., 1980; Ciejek et al., 1983; Strätling et al., 1986). A rather abrupt transition to a more inaccessible configuration occurs at the edges of these domains and suggests that they could be bordered by specialized sequences that isolate them from the surrounding chromosomal context (Bode and Maass, 1988; Jarmann and Higgs, 1988; Phi-Van and Strätling, 1988; Kellum and Schedl, 1991). This domain organization does not appear to be limited to active chromatin since the kinetics of endonucleolytic digestion of bulk chromatin in isolated nuclei is also consistent with chromatin looping occurring in the majority of the genome (Igo-Kemenes et al., 1977).

Further insight into the nature of the chromosomal domains was provided by an elegant series of studies performed by Cook and colleagues. Nuclei were first isolated and histones and other soluble proteins were then removed by extraction with high concentrations of salt. The residual structures (the shape of which still resembled the original nuclei) were then subjected to sedimentation analyses. It was found that essentially all DNA remained attached to these so-called *nucleoids* (Cook and Brazell, 1975; also Wanka et al., 1977; Berezney and Buchholtz, 1981). Titration with intercalating agents showed that the DNA in these preparations is topologically constrained (Cook and Brazell, 1975, 1976). This quasi-circular behavior immediately suggested that genomic DNA is attached to the residual subnuclear structure to form loops. This arrangement was subsequently corroborated in microscopic studies in which the DNA had the appearance of a halo surrounding the residual nuclear core (Vogelstein et al., 1980). This looped organization of DNA was additionally shown to characterize the DNA in metaphase chromosomes (Paulson and Laemmli, 1977). Interestingly, the average loop size was estimated at 63 kb (Nelson et al., 1986), which is quite similar to the size of the average mammalian replicon as estimated by DNA fiber autoradiography (Huberman and Riggs, 1968).

Although the loop model has gained widespread acceptance, exactly what comprises the subnuclear structure to which the loops are attached is still unclear. It had been repeatedly observed that extraction of the nucleus with high ionic strength buffers generates an insoluble residual structure. Berezney and Coffey (1974) standardized the extraction procedure and operationally defined the "nuclear matrix" as the residual protein-aceous structure that remains after nuclei are depleted of the nuclear membrane, histones, soluble nuclear proteins, and nucleic acids. To rule out the possibility that the matrix is an artifact of the harsh isolation conditions, several alternative protocols for isolation of nuclear matrices have since been developed (Capco et al., 1982; Mirkovitch et al., 1984). Though the residual structures generated by these different protocols differ in detail, they have several important features in common (Belgrader et al., 1991). Most notable are residual nucleoli and the nuclear lamina with its embedded pores. The lamina consists of the lamin proteins (Gerace and Blobel, 1980), which have been shown to be members of the family of intermediate filament proteins (Aebi et al., 1986; McKeon et al., 1986). Another component common to all matrix preparations is an internal fibrillogranular meshwork (Kaufmann et al., 1981; Berezney, 1984), the existence of which was not generally accepted until the advent of resinless embedding techniques for electronmicroscopy (Capco et al., 1984). Though localization studies also strongly argue for the presence of an internal network, this proposed network remains to be characterized.

While the details of the architecture of the internal matrix in the interphase nucleus are still somewhat controversial, the presence of a scaffolding structure in the structurally simpler metaphase chromosomes is now widely accepted (Adolph et al., 1977; Paulson and Laemmli, 1977; Lewis and Laemmli, 1982). Both biochemical and immunological analyses strongly implicate topoisomerase II as an important component of this scaffold (Gasser et al., 1986). Since this enzyme is also present in matrices isolated from interphase nuclei (Berrios et al., 1985), it may be that the chromosome scaffold is derived from the interphase matrix. An informative electron microscopic study on *Physarum* polycephalum supports this idea (Bekers et al., 1981). The details of the composition, ultrastructural appearance, and other properties of the nuclear matrix/chromosomal scaffold structures are presented in several excellent contributions in this volume.

B. Interface of Matrix and DNA Loop:
     Matrix-Attached Regions

The proposal that chromosomal DNA is organized into looped domains by periodic attachment to a nuclear matrix, and furthermore, that these domains have functional significance, has generated interest in the nature of the DNA sequences at the bases of the loops. These matrix-attached regions (MARs, used here for all attachment sites, irrespective of the method by which they were identified) have been identified either by their preferential retention on the matrix after removal of the bulk of genomic DNA with endonucleases (Robinson *et al.*, 1982; Mirkovitch *et al.*, 1984) or by their affinity for isolated matrices *in vitro* (Cockerill and Garrard, 1986). All MARs identified so far contain one or more of the following characteristics: (1) a high A-T content (generally 70%), (2) runs of 10 contiguous As or Ts, (3) one or more matches to the topoisomerase II consensus sequence (Cockerill and Garrard, 1986; Amati and Gasser, 1988; Phi-Van and Strätling, 1988; Dobbs *et al.*, 1994), and (4) a short sequence that has the potential to become unpaired under conditions of superhelical stress (Cockerill *et al.*, 1987; Bode *et al.*, 1992). In spite of these similarities, a MAR consensus sequence has yet to emerge. This has led to the current working hypothesis that MARs display certain sequence motifs (Nakagomi *et al.*, 1994), rather than a specific sequence.

Whether or not the MARs so far identified are representative of attachment sites in general is the subject of some debate. The methods for identifying MARs rely on the specific hybridization of matrix-attached DNA to subfragments of previously cloned DNA sequences. With a few exceptions (Dijkwel and Hamlin, 1988; Mirkovitch *et al.*, 1986; Brun *et al.*, 1990), this has limited searches for specific attachment sites to the vicinity of genes. Consequently, the MARs that have so far been identified in the neighborhood of transcription units are overrepresented relative to those that could potentially be involved in other nuclear processes such as replication.

Recent studies have focused on determining whether MARs might have properties expected of boundary elements. Enhancer blocking assays (Kellum and Schedl, 1991) have shown that insertion of some, but not all, MARs between an enhancer and a promoter can effectively prevent expression of a reporter gene (Kellum and Schedl, 1992; Fishel *et al.*, 1993). This difference among MARs could reflect the existence of classes of attachment sites with different affinities for their target, and might explain why a MAR inserted into an ectopic site could subsequently be eluted from nuclei after *in situ* digestion with a restriction enzyme (Eggert and Jack, 1991). However, in other studies, the inclusion of MARs into a transgene construction has generally been found to decrease the position dependence of the expression of the colinear reporter gene after stable integration into the genome; this is reflected in an increase of the linearity between gene copy number and expression level (Stief *et al.*, 1989; Phi-Van *et al.*, 1990; Klehr *et al.*, 1991).

Recently, MARs have been proposed to act by a rather different mechanism. By serving as nucleation sites for histone H1 by virtue of their high AT content, MARs could act as molecular switches in the H1-dependent condensation and decondensation of chromatin (Käs *et al.*, 1993; Zhao *et al.*, 1993). This proposal is attractive because it would explain why some MARs are positioned in the genome where boundary elements would not seem warranted, i.e., near intragenic enhancers (Cockerill and Garrard, 1986; Gasser and Laemmli, 1986), in the body of genes (Käs and Chasin, 1987), and in the vicinity of replication origins (Amati and Gasser, 1988, 1990; Dijkwel and Hamlin, 1988). Alternatively, as suggested by Mirkovitch *et al.*, 1984), this latter group of MARs could function to localize *cis*-regulatory elements in the proper nuclear subcompartment. In so doing the MAR could actually be an integral part of the genetic signal.

Whatever their role, it seems clear that in order to function, a MAR needs to be recognized by a cognate protein. In recent years several groups have reported the identification of proteins that interact with MARs *in vitro* (von Kries *et al.*, 1991; Dickinson *et al.*, 1992; Romig *et al.*, 1992; Nakagomi *et al.*, 1994). However, the proteins described, at face value, appear to be unrelated to one another. Although this might be expected for factors that recognize an array of sequence motifs rather than a specific sequence, it could also be the unintended consequence of the *in vitro* approaches used to detect MAR-binding proteins. Clearly, extant results have to be supplemented with *in vivo* data.

In summary, MARs probably anchor genomic DNA to a proteinaceous subnuclear structure; however, the exact nature of this interaction *in vivo* remains to be determined. MARs are proposed to function (1) as domain boundaries separating functional chromosomal units from one another, (2) as switches in H1-dependent chromatin opening, or (3) as signals that displace *cis*-regulatory sequences into their appropriate nuclear subcompartment, thus reducing the search volume for transacting factors required for replication and transcription.

C. Nuclear Structure Required for DNA Replication

The present understanding of eukaryotic DNA replication owes much to viral model systems. The development of a cell-free replication system utilizing SV40 genomic DNA (Li and Kelly, 1984), in particular, has led to identification of many key components of the replication reaction. With the notable exception of large T-antigen, SV40 uses the DNA synthetic machinery of the host cell to replicate its genome (DePamphilis and Wassarman, 1982). Since SV40 can replicate in a cell-free system, nuclear architecture seems to be irrelevant to the replication process. However, other data lead to a quite different conclusion. First, SV40 replication *in vitro* systems require unphysiologically high levels of T-antigen (Li and Kelly, 1984), indicating that replication is rather inefficient in this system. It is also true that, with the exception of *Xenopus* egg extracts, no *in vitro* replication systems that have yet been devised are capable of initiating replication on mammalian DNA. Remarkably, *Xenopus* extracts only initiate replication on exogenous templates after assembly of a functional nucleus (Blow and Laskay, 1986; Blow and Sleeman, 1990; Newport *et al.*, 1990). Finally, in quiescent nuclei that are reactivated in cytoplasts of proliferating cells, onset of transcription and replication are preceded by reformation of an internal nuclear matrix (Lafond *et al.*, 1983). These observations argue that nuclear structure is essential for eukaryotic DNA replication and it has been suggested that eukaryotic viruses have also retained this dependency (Buckler-White *et al.*, 1980; Jankelevich *et al.*, 1992).

Cellular localization studies further support the involvement of the nuclear matrix in replication. When replicating DNA is labeled with BrdU and subsequently illuminated with fluorescent anti-BrdU antibodies, highly defined foci of newly made DNA are observed (Nakamura *et al.*, 1986; Nakayasu and Berezney, 1989). This pattern persists in isolated matrices, indicating not only that replication forks are clustered but also that the matrix is probably required to organize these foci (Nakayasu and Berezney, 1989). This punctate pattern was also observed in pseudonuclei that form in *Xenopus* extracts upon introduction of exogenous DNA templates (Newport, 1987; Blow and Sleeman, 1990, Cox and Laskey, 1991). However, interfering in the nuclear assembly process, either by adding inhibitors of topoisomerase II (Newport, 1987) or by depleting extracts of lamins (Newport *et al.*, 1990; Meier *et al.*, 1991), prevents initiation of DNA replication possibly because of the inability of the incomplete nuclei to get replicative enzymes to their appropriate sites or compartments (Meier *et al.*, 1991; Jenkins *et al.*, 1993). The location of these sites, termed prereplicative complexes, appears to be predetermined in the very early G_1 period, but are not fully assembled until the cell arrives at the G_1/S boundary (Adachi and Laemmli, 1994).

Other lines of evidence also support the idea that replication occurs on the nuclear skeleton. Radiolabel incorporated during a short pulse is invariably found in close proximity to the matrix in matrix–halo preparations (Berezney and Coffey, 1975; Dijkwel *et al.*, 1979; McCready *et al.*, 1980; Pardoll *et al.*, 1980). In addition, ultrastructural (Valenzuela *et al.*, 1983) and gel electrophoretic analyses (Vaughn *et al.*, 1990a) have shown that replication forks partition with the nuclear matrix. Finally, enzymes of the so-called replication factories (Jackson and Cook, 1986), as well as the factories themselves (Hozák *et al.*, 1993), have been shown to be associated with the nuclear matrix. These observations fit a model proposed 20 years ago that replication forks as well as origins of replication are attached to a karyoskeleton (Dingman, 1974). This would ensure that physical separation of the daughter strands, which is required for subsequent segregation during mitosis, occurs as a consequence of replication. This model, that origins will be close to or might actually coincide with MARs, has stirred great interest and has served as a starting point for studies aimed at assessing the contribution of matrix attachment to origin function. However, as we will point out, results have been somewhat disappointing owing to the elusive character of origins of replication in metazoan chromosomes.

D. Origins of Replication and the Nuclear Matrix

The first hint that origins of replication might be attached to a karyoskeleton (Dingman, 1974) was provided by the observation that radiolabel incorporated at the onset of the S period remained in close proximity to the matrix, whereas label incorporated later was eventually displaced into the surrounding DNA loops (Aelen et al., 1983; Carri et al., 1986; Dijkwel et al., 1986). While these results support the model in general terms, they do not allow conclusions to be drawn at the molecular level because of the low resolution of the radiolabeling methods. Potentially higher resolution was achieved by analyzing nascent strands extruded from small replication "eyes" at the beginning of the S-period (Zannis-Hadjopoulos et al., 1981; Kaufmann et al., 1985). This DNA fraction, which is assumed to be enriched for origins of replication, was found to contain a small number of sequences that can associate with the nuclear matrix (Mah et al., 1993). In addition, the fraction prepared by a similar extrusion technique contained sequences that hybridized to a restriction fragment harboring the MAR upstream of the chicken α-globin locus, but not to neighboring sequences (Razin et al., 1986). However, it has yet to be demonstrated that the extruded nascent strands in fact behave as replication initiation sites in their natural chromosomal contexts. Additional support for the proposition that origins may be permanently attached to the nuclear matrix has come from attempts to characterize origins of replication in the budding yeast *Saccharomyces cerevisiae* and the amplified dihydrofolate reductase (DHFR) domain of methotrexate-resistant CHO cell lines.

In yeast, progress in identifying origins has been rapid as the consequence of the development of a functional assay to identify autonomously replicating sequences (ARS elements; Stinchcomb et al., 1979). A genetic analysis in *S. cerevisiae* showed that although ARS and MAR functions are separable, plasmid loss in the ARS assay is greatly reduced when the two are juxtaposed in the same plasmid (Amati et al., 1990). Interestingly, a large fraction of the MARs identified in *Drosophila* also show ARS activity when assayed in yeast (Amati and Gasser, 1990; Brun et al., 1990). When these studies were extended to mammalian cells, conflicting results were obtained. One study observed that HeLa DNA sequences that behave as ARS elements in yeast actually reside in the DNA loop fraction (Cook and Lang, 1984), while other studies concluded that such sequences partition with the matrix (Aguinaga *et al.,* 1987; Sykes *et al.,* 1988).

Only recently has it become possible to evaluate the significance of these data. Using two-dimensional replicon mapping techniques (Brewer and Fangman, 1987; Nawotka and Huberman, 1988), which were developed originally to identify chromosomal origins in *Saccharomyces,* it has been shown that many yeast ARS elements do, in fact, function as origins in their natural context (Umek *et al.,* 1989; Dubey *et al.,* 1991). Among the ARS elements that have been shown to function as chromosomal origins, ARS1 (Brewer and Fangman, 1993) and the HMR-E origin (Rivier and Rine, 1992) have been shown to reside close to a MAR (Amati and Gasser, 1988). Only systematic mutagenesis studies will determine whether or not the MAR is indispensable for an ARS to function as an origin in context of the chromosome.

It is still unclear whether any of the *Drosophila* and human sequences that behave as ARS elements in yeast represent bona fide replication origins. In fact, mammalian origins of replication are only now beginning to be understood. Although it is generally accepted that nascent strands start at preferred sites in the genome, it is not known how circumscribed these sites are and whether or not they are genetically defined replicators analogous to those that direct replication in prokaryotic and simple eukaryotic organisms. Obviously the current lack of understanding of the nature of mammalian replication origins has a negative impact on attempts to evaluate the role of MARs in the initiation process.

III. Initiation of DNA Replication in Mammalian Cells

A. General Aspects

As a mammalian cell emerges from mitosis, it is faced with the decision either to continue proliferating or to enter into a quiescent state. If the cell senses adequate nutrient concentrations and receives external stimuli from the appropriate growth factors, the G_1 cyclins will eventually accumulate to levels sufficient for stimulation of the p34^{cdc2} family of protein kinases. The action of these kinases on their targets then allows the cell to pass through the restriction (R) point (Pardee, 1974), the mammalian equivalent of START in yeast, and enter the DNA synthetic phase of the cell cycle. This irrevocably commits the cell to duplicate its entire genome and go through cytokinesis. It is not surprising, therefore, that the cell has developed mechanisms to tightly control this important transition, the deregulation of which results in uncontrolled growth or cancer.

Early investigators of mammalian DNA replication were confronted with a perplexing paradox: the mammalian genome is a thousandfold larger than that of *E. coli,* and mammalian replication forks move at one-tenth the speed of bacterial forks; yet duplication of the mammalian genome requires only about 8 hr—an S-phase only 10 times longer than in bacteria. Fiber autoradiography provided a solution to this apparent dilemma: whereas the bacterial genome consists of one replication unit only, multiple growing points were observed in mammalian chromosomal fibers (Huberman and Riggs, 1968). From this, it was calculated that the mammalian genome is replicated from approximately 50,000 origins. Passage of the cell through the S period is a highly regulated process: different chromosomal domains (as identified by cytological studies) are replicated in a defined order apparently by the activation of clusters of adjacent origins within each domain (Amaldi *et al.,* 1973). Although this pattern is invariable within a given cell type, the replication pattern can be different in cells of different developmental lineages (Hatton *et al.,* 1988). Obviously, a complete understanding of the regulatory processes affecting entry into and orderly passage through S will eventually require a full understanding of developmental control as well as the molecular details of origins of replication.

More than 30 years ago, Jacob and Brenner (1963) proposed the replicon model to explain control over DNA replication in bacteria. The model proposes the existence of a defined genetic *cis*-acting element, termed the replicator, which is activated by interaction with a *trans*-acting factor, the initiator protein. This interaction leads to destabilization of the double helix and subsequent priming of nascent strands at or very close to the replicator itself. A host of observations have shown that the model accurately represents the situation in bacteria, in bacterial and eukaryotic viruses, and in simple eukaryotes such as *S. cerevisiae.* It has also been widely assumed that the model will apply to the mammalian genome. Approaches to define mammalian origins of replication can be divided into two categories: (1) identification of sequences from mammalian genomes that can support autonomous replication of colinear DNA when reintroduced into mammalian cells as plasmids (i.e., replicators) and (2) identification of regions of the genome in which nascent strands initiate (i.e., initiation sites).

B. Identification of Relicators

The identification of chromosomal replicators by assaying for ARS activity is intellectually most appealing because it not only identifies the required genetic element but also results in their isolation. The power of this approach has been amply demonstrated in yeast. As mentioned above, almost all chromosomal origins detected so far were originally identified as ARS elements (Stinchcomb *et al.*, 1979). Systematic mutagenesis has defined the sequences required for origin function (van Houten and Newlon, 1990), which in turn has allowed identification of a multiprotein complex that binds to the replicator (ORC; Bell and Stillman, 1992) and probably represents at least a component of the initiator.

However, the results of the ARS assay, when applied to mammalian cells, have been less than straightforward. Initial reports suggested that origins could be identified by shotgunning digests of the entire mouse genome into a selectable vector that required high copy numbers for detection (Holst *et al.*, 1988). However, the subset of cloned fragments identified by this approach was subsequently shown to be integrated into the genome in multicopy arrays rather than being maintained as bona fide episomes (Grummt, 1989). Although the possibility cannot be formally ruled out that these sequences increased to high copy number because of intrinsic replicator activity, the results cast doubt on the applicability of ARS assays in mammalian systems (Gilbert and Cohen, 1989). In an alternative approach, a library was constructed of density-labeled DNA synthesized in permeabilized cells in the first few minutes after release from an aphidicolin block, i.e., at the onset of the S-period. The genomic DNA was then subjected to high temperature to extrude the small nascent strands centered over initiation sites (Kaufmann *et al.*, 1985). Evidence was presented that several of the extruded fragments can direct replication of pBR322 when transfected into transformed human cells (Frappier and Zannis-Hadjopoulos, 1987). However, the replication levels are low and strict intrinsic controls are lacking. In addition, extensive scrutiny of the DHFR locus of Chinese hamster cells by three different laboratories has failed to produce any evidence for the presence of a mammalian ARS-like element (Burhans *et al.*, 1990; Caddle and Calos, 1992; J. D. Milbrandt, P. K. Foreman, and J. L. Hamlin, unpublished observations) in a region shown to contain a replication origin by several other assays (Heintz and Hamlin, 1982; Burhans *et al.*, 1990; Vaughn *et al.*, 1990b). In marked contrast, when a cloned 15-kb fragment from the same region was transfected into a CHO cell background, it appeared to serve as a replication initiation site regardless of where it had integrated (Handeli *et al.*, 1989). This mixed bag of results is even more puzzling when compared to the results of recent studies in which random restriction fragments were cloned into an EBV-derived vector that provides a nuclear retention function (i.e., a quasi-centromere). Systematic analysis of many individual clones showed that replication ability was linearly related to the length of the inserted fragment (Heinzel *et al.*, 1991). This study actually suggests that potential replicators may occur at very frequent intervals along the chromosome instead of only once per 50–100 kb as suggested by fiberautoradiography (Krysan *et al.*,

1993). In summary, results of assays to identify mammalian ARS elements have been confusing at best, and sequence analysis of potential candidates, unlike the situation in *S. cerevisiae,* does not yield any obvious consensus sequence or sequence arrangement to which an ORC-like multiprotein initiation complex might bind.

C. Identification of Initiation Sites of Nascent Strands

In view of the ambiguities presented by the genetic approach to identifying replication origins, the alternative approach of determining nascent strand start sites has been very much in vogue. In such studies it is assumed that, as in simpler organisms, start sites and replicators will roughly coincide. Determining the start sites of nascent strands in mammalian cells is complicated by (1) the large size and complexity of the mammalian genome and (2) the fact that at any given time in an exponentially growing population of cells, only a minute fraction of any sequence will be in the act of replication.

Initially, two routes were followed to partially obviate these problems: (1) use of synchronization regimens that increase the fraction of cells in the S-period and/or (2) development of model systems in which the region of interest is present in a higher copy number. Though these approaches have helped to open a window on the nature of mammalian origins, neither is without inherent pitfalls. As we will see, the advent of methods based on amplification of signals using the polymerase chain reaction (PCR) has made it possible to avoid the potential pitfalls of induced synchrony and amplification and to study single copy loci in unperturbed cells directly. Most of the work described below will concern the DHFR locus in Chinese hamster cells, since it is the system with which we are most familiar. More importantly, however, this locus is studied by many laboratories around the world and has essentially been subjected to almost all of the methods developed to identify nascent strand initiation sites. As such, it constitutes a paradigm from which to learn the special view that each method offers.

Unambiguous identification of nascent strand start sites would seem to require detection of either the initiation event itself or its imprints. This is beyond the sensitivity of most of the available methods of detection, since at any given moment, less than one in a thousand restriction fragments harboring an origin would contain replication intermediates in an asynchronously growing cell population. If cells could be collected immediately before entry into the S-period, this fraction could be increased (at least for early firing origins; due to decay of synchrony as cells traverse the S-period, firing of later banks is expected to become increasingly more asynchronous), thereby facilitating their analysis. However, suitable inhibitors or mutations to effect synchrony to late G_1 are lacking. This has necessitated the use of compounds that inhibit replication itself, either by lowering pools of deoxyribonucleotides (e.g., hydroxyurea) or by interfering with an enzyme involved in replication fork progression (e.g., aphidicolin). Necessarily, these agents probably allow the event of interest, i.e., initiation, to occur in a large percentage of early firing origins. Commonly, protocols employ either serum or isoleucine starvation to collect cells in a non-S compartment (e.g., the G_1 period). The cells are then transferred to medium containing the appropriate inhibitor, which slows the rate of chain elongation and confines the forks to the vicinity of the start sites. Unfortunately, most of the traditional inhibitors (e.g., ara-C, aphidicolin, hydroxyurea) are moderately leaky allowing forks to move over considerable distances in their presence (Levenson and Hamlin, 1993). However, a novel compound (mimosine) has recently been described that is much more efficacious and might actually prevent forks from being established when present prior to entry of cells into S phase (Dijkwel and Hamlin, 1992; Mosca et al., 1992). Mimosine notwithstanding, results from experiments with synchronized cells have to be viewed with some caution since long-term exposure to inhibitory compounds can induce aberrant initiation patterns under some circumstances (Taylor, 1977; P. A. Dijkwel, unpublished observations) and can provoke nonscheduled DNA synthesis and chromosomal aberrations (van Zeeland et al., 1983).

Identification of a given mammalian initiation site is further compounded by the presence of the "background" of thousands of other initiation events. In order to reduce this unfavorable signal-to-noise ratio, a cell line was developed that contains multiple copies of a single replicon type (Milbrandt et al., 1981). By exposure to stepwise increases of methotrexate, a drug-resistant CHO cell line was eventually selected that has amplified one allele of the DHFR domain 1000-fold.

In CHOC 400 cells, the predominant unit of amplification (amplicon) is 240 kb in size and the copies are arranged in tandem as huge inverted repeats at three chromosomal locations (Fig. 1; Looney and Hamlin, 1987; Ma et al., 1988). In genomic digests of CHOC 400 DNA, amplicon-specific fragments can be seen on agarose gels as prominent bands against a uniform background of single-copy DNA, indicating that amplification has reduced genomic complexity for this domain to that of a bacterial replicon.

To better understand the different approaches taken in identifying initiation sites, a closer look at such a site immediately after activation would be helpful (Fig. 2).

Origin-containing fragments differ from nonorigin fragments in several ways. The most obvious is that when a fragment harbors an initiation site, it is replicated before any other fragment in its vicinity. Immediately after firing, origin fragments have a characteristic physical shape, i.e., they contain a bubble (theta structure). Origin fragments also hybridize to the shortest

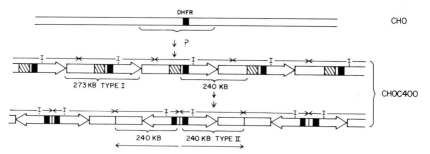

FIG. 1   The amplified DHFR domain in CHO cells. The type I amplicon, accounting for 5% of the total number of amplified units, represents an early type whose sequence is identical to the parental DHFR locus. A 33-kb internal deletion then occurred giving rise to a new 240-kb unit of amplification. These type II amplicons are arranged in head-to-head/tail-to-tail tandem arrays and account for more than 80% of the total number of amplicons. The two closely spaced sites of preferential early labeling are indicated by a single I.

nascent leading strands. Furthermore, inspection of Fig. 1 shows that an initiation bubble is asymmetric: both Okazaki fragments and leading strands switch template strands at the initiation *site*. As we will discuss, each of these characteristics has served as the basis for development of one or more techniques aimed at localizing initiation sites.

The first method employed to characterize replication start sites in the DHFR domain was the straightforward labeling method that was successfully used to pinpoint ori-C in the *E. coli* genome (Marsh and Worcel, 1977). CHOC 400 cells were synchronized by isoleucine starvation followed by a 12-hr treatment with aphidicolin to line cells up near the start of the S-period. The cells were then transferred to drug-free medium containing [^{14}C]thymidine and were harvested at different times after transfer. DNA was isolated, digested with *Eco*RI, separated on an agarose gel, and transferred to a membrane. Autoradiography revealed that bands 11 and 6.2 kb in length were preferentially labeled at the onset of S. Since these bands

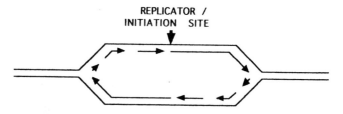

FIG. 2   Schematic representation of a circumscribed origin of bidirectional replication immediately after activation. Long arrows represent leading daughter strands; short arrows represent Okazaki fragments.

were subsequently shown to map to a single cosmid, the DHFR amplicon appears to be replicated from a single initiation site lying downstream of the DHFR gene (Heintz and Hamlin, 1982). To more precisely localize this site, in-gel renaturation (Roninson, 1983) was applied to a digest prepared exactly as described above in order to eliminate the background of single-copy sequences. The autoradiogram could then be subjected to densitometry to estimate the specific radioactivity of individual fragments. When the results were plotted as a function of map position, two peaks of early labeling were detected suggesting the presence of two preferred initiation sites. These sites (termed ori-β and ori-γ) are situated 22 kb apart in the region between the DHFR and 2BE2121 genes (Leu and Hamlin, 1989).

Detection of early-labeled fragments in amplified DNA has also been successfully used to determine the approximate location of the initiation site in the adenosine deaminase amplicon in deoxycoformycin-resistant mouse cells (Carroll et al., 1993).

By exploiting the asymmetry of the replication fork it has been attempted to further narrow down the initiation site in the DHFR domain. When the protein synthesis inhibitor, emetine, is added to proliferating cells, synthesis of lagging strands is preferentially inhibited. Using a density label, leading strand DNA can then be isolated and used to probe separated + and − strands of M13 subclones from various positions within the DHFR locus. By this approach, template strand switches were detected in the vicinities of ori-β and ori-γ with a resolution of several kilobases (Handeli et al., 1989). In a complementary lagging strand assay, which determines the template strand bias of Okazaki fragments labeled in vitro in permeabilized cells, a strand switch was suggested to occur within a 450-bp fragment centered over the ori-β locus. Ori-β was thus defined as an origin of bidirectional replication (OBR; Burhans et al., 1990). In addition, a PCR-based method that has been successfully used to identify initiation loci in the human c-myc (Vassilev et al., 1990a) and in the CHO RPS14 domains (Tasheva and Roufa, 1994) suggests that the shortest nascent leading strands are also centered over ori-β (Vassilev et al., 1990b; unfortunately, in neither the lagging strand assay nor in the nascent strand length analysis was the ori-γ locus analyzed). Thus by several criteria, ori-β/OBR appears to be a genuine origin of replication.

However, two independent two-dimensional (2-D) gel electrophoretic techniques, when applied to the DHFR domain, suggest that the situation is not quite so simple. In the first of these methods, the unique electrophoretic behavior of replication intermediates on neutral/neutral (N/N) 2-D gels is used to localize start sites (Brewer and Fangman, 1987). Genomic DNA from cycling cells is digested with an appropriate restriction enzyme while it is still attached to the nuclear matrix. This prevents destabilization of replication structures and allows for enrichment of replication intermediates by virtue of their attachment to the nuclear matrix. The DNA digest is then separated according to mass in the first dimension on a low-percentage agarose gel. The lane is then excised, turned through 90°, and placed on top of a higher percentage agarose gel. This second dimension gel is run under conditions that maximize the contribution of physical shape to the migration behavior (low temperature, high field strength, presence of intercalating dye). The DNA digest is transferred to a nylon membrane, which is then hybridized sequentially with probes for the genomic fragments of interest.

The diagram in Fig. 3 depicts the patterns expected for a restriction fragment that (1) harbors a centered initiation site (Fig. 3A), (2) is replicated passively by a fork (Fig. 3B), or (3) is replicated from an off-center start site (Fig. 3C). The N/N method was originally developed to analyze origins of replication in budding yeast, an organism that is refractile to most intrinsic labeling approaches. The method was validated using the yeast 2-μm circular replicon as a model, and was subsequently used to show that many ARS elements do indeed function as initiation sites in their normal chromosomal context (Brewer and Fangman, 1987).

Surprisingly, when the amplified DHFR domain in early S-phase CHOC 400 cells was analyzed by this technique, any fragment lying in the 55-kb region between the DHFR and 2BE2121 genes displayed a composite pattern consisting of both a complete bubble arc and a complete fork arc (Fig. 4) (Vaughn *et al.*, 1990b; Dijkwel and Hamlin, 1992; Dijkwel *et al.*, 1994).

In contrast, fragments in the DHFR gene displayed only the single fork arc and no initiations (i.e., bubbles) could be detected. However, small

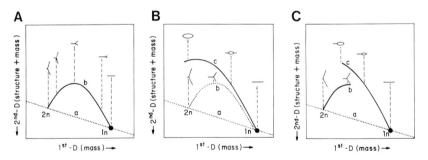

FIG. 3   Principle of the neutral/neutral two-dimensional gel electrophoretic replicon mapping technique. Idealized autoradiographic patterns obtained when a restriction digest of DNA is probed for fragments that contain different replication intermediates. (A) A complete single fork arc (curve b) for a fragment that is replicated passively from an origin elsewhere. Curve a represents the diagonal of nonreplicating fragments. (B) A bubble arc (curve c) for a fragment with a centered origin. (C) A bubble-to-fork arc transition for a fragment that harbors an off-centered origin of replication.

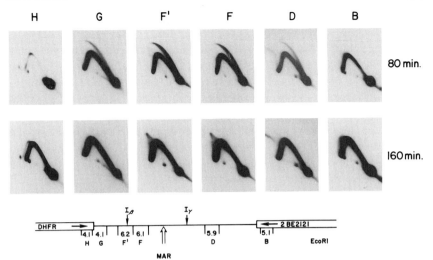

FIG. 4 Replication initiates in the intergenic region. CHOC 400 cells were collected at the G_1/S transition using mimosine and samples were harvested 80 and 160 min after release from the mimosine block. Matrix–halo structures were isolated, nonreplicating DNA was removed by digestion with *Eco*RI, and the matrix-attached DNA fraction was further enriched for replication intermediates by fractionation over BND-cellulose. This double-enriched fraction was then separated on a N/N 2-D gel and transferred to a membrane. This membrane was then hybridized sequentially with probes specific for fragments from the intergenic region (fragments G, F′, F, and D) and from the flanking DHFR (H) and 2BE2121 (B) genes (see map below autoradiograms). In very long exposures, a faint bubble arc was detected in fragment B indicating that small numbers of initiations occur in the body of the 2BE2121 gene (data not shown). Reproduced from Dijkwel and Hamlin (1992) by permission.

numbers of initiations were detected in the coamplified 2BE2121 gene (Dijkwel *et al.*, 1994), possibly because this gene is expressed at a much lower level than the DHFR gene. Since the same results were obtained regardless of the restriction enzyme used to generate the digests, initiation appears to take place at any of a large number of sites scattered throughout the intergenic region. Thus, a picture emerges in which a given restriction fragment sometimes sustains an internal initiation accounting for the bubble arc, but is also sometimes replicated by a fork emanating from an initiation site in a flanking fragment (explaining the presence of the prominent complete single fork arc).

The observation that forks persist in the DHFR domain long after initiation has ceased further suggests that not all amplicons sustain an initiation event and that forks can cross interamplicon junctions and must therefore have arisen from active origins in distant amplicons. This proposal predicts that in exponentially growing cells, the bubble-to-fork-arc ratio will be much lower than in early S-phase cells, which was indeed found to be the case (Vaughn et al., 1990b; Mosca et al., 1992; Dijkwel et al., 1994). In addition, analysis of replication intermediates obtained from asynchronous CHOC 400 cells confirmed that initiation can occur at many sites scattered throughout a broad zone, arguing that this delocalized mode of initiation is not induced by the synchronization procedure (Dijkwel et al., 1994). Finally, to exclude the possibility that delocalized initiation in the DHFR domain could result from amplification per se, the N/N method was applied to an analysis of the DHFR domain in parental CHO cells, which contain only two copies of the DHFR locus (this required modifications that markedly increase the sensitivity of the N/N method; P. A. Dijkwel and J. L. Hamlin, 1995). As in CHOC 400 cells, both bubbles and single replication forks were observed in every fragment from the intergenic region and initiation ceased by the third hour after entry into the S-period (P. A. Dijkwel and J. L. Hamlin, 1995). The persistence of forks for an additional 4 hr suggests that, as in CHOC 400 cells, not every copy of the DHFR domain sustains initiation events in every cell cycle.

N/N 2-D gel electrophoretic studies therefore paint a rather complex picture of initiation in the DHFR domain. Although the data are not incompatible with results from earlier, low-resolution labeling studies, they are difficult to reconcile with results obtained with the lagging strand assay (Burhans et al., 1990) and with the nascent strand size analysis (Vassilev et al., 1990). The existence of broad initiation zones is also suggested by the results of 2-D gel electrophoretic analysis of the human rDNA repeats (Little et al., 1993), multi- and single-copy loci in Drosophila (Shinomiya and Ina, 1993) and in Sciara coprophila (Liang et al., 1993), and in the single copy rhodopsin locus in CHO cells (P. A. Dijkwel, unpublished observations).

Moreover, analysis of the DHFR domain using a second, complementary 2-D gel electrophoretic method also suggests the presence of a broad initiation zone in the intergenic region. By this method, termed the neutral/alkaline (N/A) technique (Nawotka and Huberman, 1988), it is possible to determine the direction in which forks travel through a region of interest by analyzing the size distribution of nascent strands at both ends of any restriction fragment within that region. As expected, early in the S-period, replication forks are observed to travel in both directions in the intergenic region in both CHOC 400 and CHO cells but only outward through the DHFR gene (Vaughn et al., 1990b; Dijkwel et al., 1994; P. A. Dijkwel and J. L. Hamlin, 1995).

This ensemble of results obtained from a host of independent methods for localizing initiation sites suggests that we still do not fully understand the nature of mammalian origins. This contrasts with simpler organisms where genetic approaches, labeling techniques, and methods exploiting physical properties of replicative intermediates have all converged on a coherent picture of fixed initiation sites in chromosomes. Mammalian cells appear to have developed a more complex initiation mechanism of which any single method sees one aspect only.

For instance, methods based on intrinsic labeling of nascent DNA give results that are readily quantified, while data from 2-D gel analyses are not. Labeling approaches, therefore, allow determination of which sites are used most frequently. On the other hand, 2-D methods are extremely sensitive to individual initiation events, while approaches that rely on intrinsically labeled DNA cannot distinguish between true background and contributions of small numbers of initiation events with opposite polarities. Thus, labeling approaches would underestimate the contribution of less favored initiation sites in an initiation zone.

Another important point is that pulse-labeling techniques favor detection of replication intermediates that turn over rapidly, while methods that depend on the physical properties of replicating DNA are skewed toward long-lived structures. Thus, it has been proposed that the long-lived intermediates detected at dispersive sites on 2-D gels actually represent abortive initiations that never mature into full-length daughter strands (DePamphilis, 1993). However, it is equally possible according to this model that the short-lived nascent strands detected around the OBR could comprise the abortive species. Experiments to address this issue are underway, but for now the DHFR "origin," which is the most thoroughly analyzed to date, remains an elusive entity.

IV. Chromosomal Structure and Initiation of DNA Replication in the Chinese Hamster DHFR Domain: An Attempt at Synthesis

From the previous sections it is obvious that a unifying picture of mammalian origins of replication has yet to emerge. However, all methods that have been used to scrutinize the DHFR domain in Chinese hamster cells agree on one thing: the bulk of nascent strands initiate between the DHFR and 2BE2121 genes, and the ori-β/ori-γ region is probably preferred. Thus, the DHFR "origin" provides a useful model in which to ask whether initiation of DNA replication is in some way tied to higher order chromatin structure. In fact, in view of the uncertainty surrounding the nature of origins in mammalian chromosomes, a closer look at the organization of the DHFR domain might suggest different approaches to characterize origins.

In initial studies, the higher order structure of the DHFR domain in CHOC 400 cells was examined using matrix–halo structures isolated by the LIS method (Mirkovitch *et al.*, 1984). The DNA loops were then removed by digestion with combinations of restriction enzymes and the matrix-attached and detached DNA fractions were isolated and purified. Equal amounts (by weight) of both fractions were then separated on agarose gels, transferred to membranes, and probed with a series of recombinant cosmids carrying inserts that span the entire 240-kb DHFR amplicon. By this method, MARs were localized near an interamplicon junction, upstream from the DHFR gene and, most interestingly, approximately midway between ori-β and ori-γ (Dijkwel and Hamlin, 1988).

The *in vitro* binding assay, in which end-labeled restriction fragments are allowed to bind to DNA-depleted matrices (Cockerill and Garrard, 1986), confirmed the presence of a permanent attachment site in the initiation zone. Sequence analysis revealed that this MAR is 70% AT rich. Its most prominent sequence motif is a string of nine almost perfectly repeated AAAT elements (L. Messner and P. A. Dijkwel, unpublished observations). Interestingly, in CHOC 400 only about 15% of amplicons appear to be attached to the nuclear matrix at this MAR (Dijkwel and Hamlin, 1988)— a number corresponding closely to estimates of the fraction of initiation zones active in any single S-period (Vaughn *et al.*, 1990b; Dijkwel and Hamlin, 1992; however, in the parental CHO cell line, virtually all copies of this particular MAR appear to be associated with the matrix; Dijkwel and Hamlin, 1988). Combined with fluorescent *in situ* hybridization data showing that sectors harboring DHFR amplicons bulge out of the nuclear periphery (Trask and Hamlin, 1989), it seems likely that only a minority of amplicons are packaged correctly in CHOC 400 cells.

Northern analysis has revealed a second interesting organizational aspect of the DHFR domain. An additional transcription unit (2BE2121) was identified whose 3' end is 55 kb downstream of the 3' end of the DHFR gene (Foreman and Hamlin, 1989). Therefore, the replication initiation zone, as defined by 2-D gel methods, corresponds to an intergenic region lying between two convergently transcribed genes. This second gene is approximately 35 kb long, is coamplified with DHFR in CHOC 400 cells, and is transcribed preferentially in the late G_1 and early S periods (T.-H. Leu, unpublished observations).

Interestingly, a second early-firing initiation locus situated 200 kb upstream from the DHFR gene in a methotrexate-resistant Chinese hamster lung cell line (A3), is also flanked by two convergently transcribed genes and contains a MAR (Leu and Hamlin, 1992; P. A. Dijkwel, unpublished observations). This leads us to propose a model for early firing origins that could pertain to chromosomal domains harboring housekeeping genes (Fig. 5).

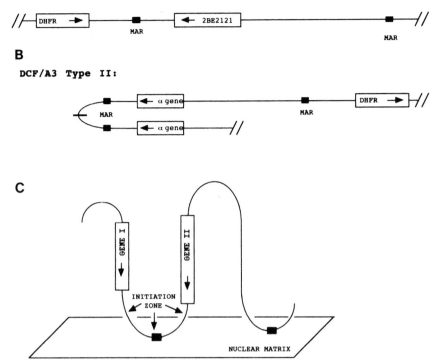

A

FIG. 5   DNA replication initiates in early S-phase in matrix-attached intergenic regions. Early in the S-period, initiation bubbles are detected by N/N 2-D gel electrophoresis in (A) the 55-kb region between the DHFR and 2BE2121 genes in both CHOC 400 and A3 cells and (B) the ~60-kb intergenic region containing the head-to-head junction of the type II amplicons in A3 cells. In addition, both intergenic regions harbor a MAR. (C) Model depicting the organization of early firing initiation zones in chromosomal domains harboring housekeeping genes.

The assumption underlying the model is that nascent strands potentially can initiate at many (possibly random) sites throughout the mammalian genome. However, the actual initiation rate at any one of these sites depends on the local milieu. For instance, according to the twin-supercoiled domain idea (Liu and Wang, 1987), most of the DHFR domain will be negatively supercoiled as the consequence of convergent transcription in the DHFR and 2BE2121 genes. In contrast, the intergenic region will accumulate positive supercoils, which results in nucleosome "splitting" and renders the chromatin relatively open in such a region (Lee and Garrard, 1991). This phenomenon would increase the accessibility of the intergenic region to incoming multienzyme replication complexes, allowing replication forks to be established. In addition, the chance that a positively supercoiled region would become a substrate for the initiation reaction might also depend on where it is located in the nucleus. Only the domains in the proper nuclear subcompartment (i.e., attached to the nuclear matrix) would then be recognized. This would explain why only approximately 15% of the domains actually support initiation (Vaughn *et al.*, 1990b; Dijkwel and Hamlin, 1992; Dijkwel *et al.*, 1994).

The absence of replication initiation in the body of the DHFR gene can also be attributed to chromatin structure. It is well established that active genes have a different chromatin structure than the remainder of the genome (Weintraub and Groudine, 1976). This altered structure might disallow replication forks from being established *de novo* in the body of the genes. Clearly, such modified structure could not prevent forks from entering from flanking regions. Interestingly, in the CHOC 400 DHFR domains, the 2BE2121 gene does sustain a low level of initiations (Dijkwel *et al.*, 1994). Although this observation would appear to negate the model, a possible explanation might be the low level of expression that characterizes this gene. Presumably, the chromatin structure in 2BE2121 is less perturbed than in the DHFR gene, which is transcribed at much higher levels (T. H. Leu, unpublished observations). Alternatively, some copies of the 2BE2121 gene might be inactive.

Recent preliminary data from our laboratory seem to support the model generalized in Fig. 5C. In a N/N 2-D gel analysis of replicating DNA isolated from UA21 (a CHO cell line lacking one copy of the DHFR domain) the initiation pattern in the remaining single DHFR allele was observed to be indistinguishable from that of the parental line carrying two copies of the DHFR locus. However, in a cell line derived from UA21 in which the DHFR promoter has been deleted, replication of the intergenic region did not begin until several hours after entry into the S period. This result indicates that a lesion whose end point in the DHFR gene is more than 20 kb upstream from the intergenic region, can affect the initiation reaction in this region. This is the result that would be predicted by the model in Fig. 5, and warrants further analyses along these lines. The development of replacement strategies based on homologous recombination will now allow systematic screening of any chromosomal locus for potentially important *cis*-acting sequences such as replicators, promoters, and MARs. The results of such studies on the DHFR and other domains will undoubtedly open a new era in the study of mammalian origins.

Acknowledgments

We thank Larry Messner for helpful criticisms on the manuscript. This work was supported by a grant to J.L.H. from the NIH (GM26108).

References

Adachi, Y., and Laemmli, U. K. (1994). Study of the cell cycle-dependent assembly of the DNA pre-replication centers in Xenopus egg extracts. *EMBO J.* **13**, 4153–4164.

Adolph, K. W., Cheng, S. M., Paulson, J. R., and Laemmli, U. K. (1977). Isolation of a protein scaffold from mitotic HeLA cell chromosomes. *Proc. Natl. Acad. Sci. U.S.A.* **74**, 4937–4941.

Aebi, U., Cohn, J., Buhle, L., and Gerace, L. (1986). The nuclear lamina is a meshwork of intermediate filaments. *Nature (London)* **323**, 560–564.

Aelen, J. M. A., Opstelten, R. J. G., and Wanka, F. (1983). Organization of DNA replication in *Physarum:* Attachment of origins of replication and replication forks to the nuclear matrix. *Nucleic Acids Res.* **11**, 1181–1195.

Amaldi, F., Buongiorno, N. M., Carnevali, F., Leoni, L., Mariotti, D., and Pomponi, M. (1973). Replicon origins in Chinese hamster cell DNA. II. Reproducibility. *Exp. Cell Res.* **80**, 79–87.

Amati, B., and Gasser, S. M. (1988). Chromosomal ARS and CEN elements bind specifically to the yeast nuclear scaffold. *Cell (Cambridge, Mass.)* **54**, 967–978.

Amati, B., and Gasser, S. M. (1990). *Drosophila* scaffold attached regions bind nuclear scaffolds and can function as ARS elements in both budding and fission yeasts. *Mol. Cell. Biol.* **10**, 5442–5454.

Amati, B., Pick, L., Laroche, T., and Gasser, S. M. (1990). Nuclear scaffold attachment stimulates, but is not essential for ARS activity in *Saccharomyces cerevisiae:* Analysis of the *Drosophila* ftz SAR. *EMBO J.* **9**, 4007–4016.

Aquinaga, M. P., Kiper, C. E., and Valenzuela, M. S. (1987). Enriched autonomously replicating sequences in a nuclear matrix-DNA complex from synchronized HeLa cells. *Biochem. Biophys. Res. Commun.* **144**, 1018–1024.

Bekers, A. G. M., Gijzen, H. J., Taalman, R. D. F. M., and Wanka, F. (1981). Ultrastructure of the nuclear matrix from *Physarum polycephalum* during the mitotic cycle. *J. Ultrastruct. Res.* **75**, 352–362.

Belgrader, P., Siegel, A. J., and Berezney, R. (1991). A comprehensive study on the isolation and characterization of the HeLa S3 nuclear matrix. *J. Cell Sci.* **98**, 281–291.

Bell, S. P., and Stillman, B. (1992). ATP-dependent recognition of eukaryotic origins of DNA replication by a multiprotein complex. *Nature (London)* **357**, 128–134.

Berezney, R. (1984). Organization and functions of the nuclear matrix. *In* "Chromosomal Nonhistone Proteins" (L. S. Hnilica, ed.), Vol. 4, pp. 119–130. CRC Press, Boca Raton, FL.

Berezney, R., and Buchholtz, L. A. (1981). Isolation and characterization of rat liver nuclear matrices containing high molecular weight deoxyribonucleic acid. *Biochemistry* **20**, 4995–5002.

Berezney, R., and Coffey, D. S. (1974). Identification of a nuclear protein matrix. *Biochem. Biophys. Res. Commun.* **60**, 1410–1417.

Berezney, R., and Coffey, D. S. (1975). Nuclear protein matrix: Association with newly synthesized DNA. *Science* **189**, 291–293.

Berrios, M., Osheroff, N., and Fisher, P. (1985). In situ localization of DNA topoisomerase II, a major polypeptide component of the *Drosophila* nuclear matrix fraction. *Proc. Natl. Acad. Sci. U.S.A.* **82**, 4142–4146.

Blobel, G. (1986). Gene gating: A hypothesis. *Proc. Natl. Acad. Sci. U.S.A.* **82**, 8527–8529.

Blow, J. J., and Sleeman, A. M. (1990). Replication of purified DNA in *Xenopus* egg extract is dependent on nuclear assembly. *J. Cell Sci.* **98**, 383–391.

Blow, J. J., and Laskey, R. A. (1986). Initiation of DNA replication in nuclei and purified DNA by a cell-free extract of *Xenopus* eggs. *Cell (Cambridge, Mass.)* **47**, 577–587.

Bode, J., and Maass, K. (1988). Chromatin domain surrounding the human interferon-b gene as defined by scaffold-attached regions. *Biochemistry* **27**, 4706–4711.

Bode, J., Kohwi, Y., Dickinson, L., Joh, T., Klehr, D., Mielke, C., and Kohwi-Shigematsu, T. (1992). Biological significance of unwinding capability of nuclear matrix-associating DNAs. *Science* **255**, 195–197.

Brewer, B. J., and Fangman, W. L. (1987). The localization of replication origins in ARS plasmids in *S. cerevisiae*. *Cell (Cambridge, Mass.)* **51**, 463–471.

Brewer, B. J., and Fangman, W. L. (1993). Initiation at closely spaced replication origins in a yeast chromosome. *Science* **262**, 1728–1731.

Brun, C., Dang, Q., and Miassod, R. (1990). Studies on a 800 kb stretch of the *Drosophila* X chromosome: comapping of a sub-class of scaffold-attached regions with sequences able to replicate autonomously in yeast. *Mol. Cell. Biol.* **10**, 5455–5463.

Buckler-White, A. J., Humphrey, G. W., and Pigiet, V. (1980). Association of polyoma T antigen and DNA with the nuclear matrix from lytically infected 3T6 cells. *Cell (Cambridge, Mass.)* **22**, 37–46.

Burhans, W. C., Vassilev, L. T., Caddle, M. S., Heintz, N. H., and DePamphilis, M. L. (1990). Identification of an origin of bidirectional replication in mammalian chromosomes. *Cell (Cambridge, Mass.)* **62**, 955–965.

Caddle, M. S., and Calos, M. P. (1992). Analysis of the autonomous replication behavior in human cells of the dihydrofolate reductase putative chromosomal origin of replication. *Nucleic Acids Res.* **20**, 5971–5978.

Capco, D. G., Wan, K. M., and Penman, S. (1982). The nuclear matrix: Three-dimensional architecture and protein composition. *Cell (Cambridge, Mass.)* **29**, 847–858.

Capco, D. G., Krochmalnic, G., and Penman, S. (1984). A new method of preparing embedment-free sections for transmission electron microscopy: Applications to the cytoskeletal framework and other three-dimensional networks. *J. Cell Biol.* **98**, 1878–1885.

Cardoso, M. C., Leonhardt, H., and Nadal-Ginard, B. (1993). Reversal of terminal differentiation and control of DNA replication: Cyclin A and Cdk2 specifically localize at subnuclear sites of DNA replication. *Cell (Cambridge, Mass.)* **74**, 979–992.

Carmo-Fonseca, M., Tollervey, D., Pepperkok, R., Barabino, S. M. L., Merdes, A., Brunner, C., Zamore, P. D., Green, M. R., Hurt, E., and Lamond, A. I. (1991). Mammalian nuclei contain foci which are highly enriched in components of the pre-mRNA splicing machinery. *EMBO J.* **10**, 195–206.

Carri, M. T., Micheli, G., Graziano, E., Pace, T., and Buongiorno-Nardelli, M. (1986). The relationship between chromosomal origins of replication and the nuclear matrix during the cell cycle. *Exp. Cell Res.* **164**, 426–436.

Carroll, S. M., DeRose, M. L., Kolman, J. L., Nonet, G. H., Kelly, R., and Wahl, G. M. (1993). Localization of a bidirectional DNA replication origin in the native locus and in episomally amplified murine adenosine deaminase loci. *Mol. Cell. Biol.* **13**, 2971–1981.

Celis, J. E., and Celis, A. (1985). Cell cycle-dependent variations in the distribution of the nuclear protein cyclin proliferating cell nuclear antigen in cultured cells: Subdivision of S phase. *Proc. Natl. Acad. Sci. U.S.A.* **82**, 3262–3266.

Ciejek, E. M., Tsai, M. J., and O'Malley, B. W. (1983). Actively transcribed genes are associated with the nuclear matrix. *Nature (London)* **306**, 607–609.

Cockerill, P. N., and Garrard, W. T. (1986). Chromosomal loop anchorage of the kappa immunoglobulin gene occurs next to the enhancer in a region containing topoisomerase II sites. *Cell (Cambridge, Mass.)* **44**, 273–282.

Cockerill, P. N., Yuen, M.-H., and Garrard, W. T. (1987). The enhancer of the immunoglobulin heavy chain locus is flanked by presumptive chromosomal loop anchorage elements. *J. Biol. Chem.* **262**, 5394–5397.

Cook, P. R., and Brazell, I. A. (1975). Supercoils in human DNA. *J. Cell Sci.* **19**, 261–279.

Cook, P. R., and Brazell, I. A. (1976). Conformational constraints in nuclear DNA. *J. Cell Sci.* **22**, 287–302.

Cook, P. R., and Lang, J. (1984). The spatial organization of sequences involved in the initiation and termination of eukaryotic DNA replication. *Nucleic Acids Res.* **12**, 1069–1075.

Cox, L. S., and Laskey, R. A. (1991). DNA replication occurs at discrete sites in pseudonuclei assembled from purified DNA in vitro. *Cell (Cambridge, Mass.)* **66**, 271–275.

DePamphilis, M. L. (1993). Eukaryotic DNA replication: Anatomy of an origin. *Annu. Rev. Biochem.* **62**, 29–63.

DePamphilis, M. L., and Wassarman, P. M. (1982). *In* "Organization and Replication of Viral DNA" (A. Kaplan, ed.), pp. 37–114. CRC, Boca Raton, FL.

Dickinson, L. A., Joh, T., Kohwi, Y., and Kohwi-Shigematsu, T. (1992). A tissue specific MAR/SAR DNA-binding protein with unusual binding site recognition. *Cell (Cambridge, Mass.)* **70**, 631–645.

Dijkwel, P. A., and Hamlin, J. L. (1988). Matrix attachment regions are positioned near replication initiation sites, genes, and an interamplicon junction in the amplified dihydrofolate reductase domain of Chinese hamster ovary cells. *Mol. Cell. Biol.* **12**, 5398–5409.

Dijkwel, P. A., and Hamlin, J. L. (1992). Initiation of DNA replication in the dihydrofolate reductase locus is confined to the early S period in CHO cells synchronized with the plant amino acid, mimosine. *Mol. Cell. Biol.* **12**, 3715–3722.

Dijkwel, P. A., and Hamlin, J. L. (1995). The Chinese hamster dihydro-folate reductase origin consists of multiple potential nascent-strand start sites. *Mol. Cell. Biol.* **15**, 3023–3031.

Dijkwel, P. A., Mullenders, L. H. F., and Wanka, F. (1979). Analysis of the attachment of replicating DNA to a nuclear matrix in mammalian interphase nuclei. *Nucleic Acids Res.* **6**, 219–230.

Dijkwel, P. A., Wenink, P. W., and Poddighe, J. (1986). Permanent attachment of replication origins to the nuclear matrix in BHK-cells. *Nucleic Acids Res.* **14**, 3241–3249.

Dijkwel, P. A., Vaughn, J. P., and Hamlin, J. L. (1994). Replication initiation sites are distributed widely in the amplified CHO dihydrofolate reductase domain. *Nucleic Acids Res.* **22**, 4989–4996.

Dingman, C. W. (1974). Bidirectional chromosome replication: Some topological considerations. *J. Theor. Biol.* **43**, 187–195.

Dobbs, D. L., Shaiu, W.-L., and Benbow, R. M. (1994). Modular sequence elements associated with origin regions in eukaryotic chromosomal DNA. *Nucleic Acids Res* **22**, 2479–2489.

Dubey, D. D., Davis, L. R., Greenfelder, S. A., Ong, L. Y., Zhu, J., Broach, J. R., Newlon, C. S., and Huberman, J. A. (1991). Evidence suggesting that the ARS elements associated with silencers of the yeast mating-type locus HML do not function as chromosomal replication origins. *Mol. Cell. Biol.* **11**, 5346–5355.

Eggert, H., and Jack, R. S. (1991). An ectopic copy of the *Drosophila* fitz associated SAR neither reorganizes local chromatin structure nor hinders elution of a chromatin fragment from isolated nuclei. *EMBO J.* **10**, 1237–1243.

Fishel, B. R., Sperry, A. O., and Garrard, W. T. (1993). Yeast calmodulin and a conserved nuclear protein participate in the in vivo binding of a matrix associated region. *Proc. Natl. Acad. Sci. U.S.A.* **90**, 5623–5627.

Foreman, P. K., and Hamlin, J. L. (1989). Identification and characterization of a gene that is coamplified with dihydrofolate reductase in a methotrexate-resistant CHO cell line. *Mol. Cell. Biol.* **9**, 1137–1147.

Frappier, L., and Zannis-Hadjopoulos, M. (1987). Autonomous replication of plasmids bearing monkey DNA origin-enriched sequences. *Proc. Natl. Acad. Sci. U.S.A.* **84**, 6668–6672.

Gasser, S. M., and Laemmli, U. K. (1986). Cohabitation of scaffold binding regions with upstream/enhancer elements of three developmentally regulated genes of *D. melanogaster*. *Cell (Cambridge, Mass.)* **46**, 521–530.

Gasser, S. M., Laroche, T., Falquet, J., Boy de la Tour, E., and Laemmli, U. K. (1986). Metaphase chromosome structure. Involvement of topoisomerase II. *J. Mol. Biol.* **188**, 613–629.

Gerace, L., and Blobel, G. (1980). The nuclear envelope lamina is reversibly depolymerized during mitosis. *Cell (Cambridge, Mass.)* **19**, 277–287.

Gilbert, D., and Cohen, S. N. (1989). Autonomous replication in mouse cells: A correction. *Cell (Cambridge, Mass.)* **56**, 143.

Grummt, F. (1989). Autonomous replication in mouse cells: A correction. *Cell (Cambridge, Mass.)* **56**, 143–144.

Handeli, S., Klar, A., Meuth, M., and Cedar, H. (1989). Mapping replication units in animal cells. *Cell (Cambridge, Mass.)* **57**, 909–918.

Hatton, K. S., Dhar, V., Brown, E. H., Iqbal, M. A., Stuart, S., Didamo, V. T., and Schildkraut, C. L. (1988). Replication program of active and inactive multigene families in mammalian cells. *Mol. Cell. Biol.* **8**, 2149–2158.

Heintz, N. H., and Hamlin, J. L. (1982). An amplified chromosomal sequence that includes the gene for dihydrofolate reductase initiates replication within specific restriction fragments. *Proc. Natl. Acad. Sci. U.S.A.* **79**, 4083–4087.

Heinzel, S. S., Krysan, P. J., Tran, C. T., and Calos, M. P. (1991). Autonomous DNA replication in human cells is affected by the size and source of the DNA. *Mol. Cell. Biol.* **11**, 2263–2271.

Hochstrasser, M. D., Mathog, D., Gruenbaum, Y., Saumweber, H., and Sedat, J. W. (1986). Spatial organization of chromosomes in the salivary gland nuclei of *Drosophila melanogaster*. *J. Cell Biol.* **102**, 112–123.

Holmquist, G., Gray, M., Porter, T., and Jordan, J. (1982). Characterization of Giemsa dark- and light-band DNA. *Cell (Cambridge, Mass.)* **31**, 121–129.

Holst, A., Mueller, F., Zastrow, G., Zentgraf, H., Schwender, S., Dinkl, E., and Grummt, F. (1988). Murine genomic DNA sequences replicating autonomously in mouse L cells. *Cell (Cambridge, Mass.)* **52**, 355–365.

Hozák, P., and Cook, P. R. (1994). Replication factories. *Trends Cell Biol.* **4**, 48–52.

Hozak, P., Massan, A. B., Jackson, D. A., and Cook, P. R. (1993). Visualization of replication factories attached to nucleoskeleton. *Cell* **73**, 361–373.

Huberman, J. A., and Riggs, A. D. (1968). On the mechanism of DNA replication in mammalian chromosomes. *J. Mol. Biol.* **32**, 327–341.

Igo-Kemenes, T., Greil, W., and Zachau, H. G. (1977). Preparation of soluble chromatin and specific chromatin fractions with restriction nucleases. *Nucleic Acids Res.* **4**, 3387–3400.

Jackson, D. A., and Cook, P. R. (1986). A cell cycle-dependent DNA polymerase activity that replicates intact DNA in chromatin. *J. Mol. Biol.* **192**, 65–76.

Jackson, D. A., Hassan, A. B., Errington, R. J., and Cook, P. R. (1993). Visualization of focal sites of transcription within human nuclei. *EMBO J.* **12**, 1059–1065.

Jacob, F., and Brenner, S. (1963). Sur la régulation de la synthèse du AND chez les bactériens: L'hypothèse du replicon. *C. R. Hebd. Seances Acad. Sci.* **256**, 298–300.

Jankelevich, S., Kolman, J. L., Bodnar, J. W., and Miller, G. (1992). A nuclear matrix attachment region organizes the Epstein-Barr viral plasmid in Raji cells into a single DNA domain. *EMBO J.* **11**, 1165–1176.

Jarman, A. P., and Higgs, D. R. (1988). Nuclear scaffold attachment sites in the human globin gene complexes. *EMBO J.* **7**, 3337–3344.

Jenkins, H., Hoelman, T., Lyon, C., Lane, B., Stick, R., and Hutchinson, C. (1993). Nuclei that lack a lamina accumulate karyophylic proteins and assemble a nuclear matrix. *J. Cell Sci.* **106**, 275–285.

Käs, E., and Chasin, L. (1987). Anchorage of the Chinese hamster dihydrofolate reductase gene to the nuclear scaffold occurs in an intragenic region. *J. Mol. Biol.* **198**, 667–692.

Käs, E., Poljak, L., Adachi, Y., and Laemmli, U. K. (1993). A model for chromatin opening: Stimulation of topoisomerase II and restriction enzyme cleavage of chromatin by distamycin. *EMBO J.* **12**, 115–126.

Kaufmann, G., Zannis-Hadjopoulos, M., and Martin, R. G. (1985). Cloning of monkey DNA synthesized early in the cell cycle. *Mol. Cell. Biol.* **5**, 721–727.

Kaufmann, S. H., Coffey, D. S., and Shaper, J. H. (1981). Considerations in the isolation of rat liver nuclear matrix, nuclear envelope, and pore-complex lamina. *Exp. Cell Res.* **132,** 105–123.

Kellum, R., and Shedl, P. (1991). A position-effect assay for boundaries of higher order chromosomal domains. *Cell (Cambridge, Mass.)* **64,** 941–950.

Kellum, R., and Schedl, P. (1992). A group of scs elements function as domain boundaries in an enhancer-blocking assay. *Mol. Cell. Biol.* **12,** 2424–2431.

Klehr, D., Maass, K., and Bode, J. (1991). Scaffold-attached regions from the human interferon-b domain can be used to enhance the stable expression of genes under control of various promoters. *Biochemistry* **30,** 1264–1270.

Krysan, P. J., Smith, J. G., and Calos, M. P. (1993). Autonomous replication in human cells of multimers of specific human and bacterial DNA sequences. *Mol. Cell. Biol.* **13,** 2688–2696.

Lafond, R. E., Woodcock, H., Woodcock, C. L. F., Kundahl, E. R., and Lucas, J. J. (1983). Generation of an internal matrix in mature avian erythrocyte nuclei during reactivation in cytoplasts. *J. Cell Biol.* **96,** 1815–1819.

Lawrence, J. B., Singer, R. H., and Marselle, L. M. (1989). Highly localized tracts of specific transcripts within interphase nuclei visualized by *in situ* hybridization. *Cell* **57,** 493–502.

Lee, M.-S., and Garrard, W. T. (1991). Transcription-induced nucleosome "splitting": an underlying structure for DNaseI sensitive chromatin. *EMBO J.* **10,** 607–615.

Leu, T.-H., and Hamlin, J. L. (1989). High-resolution mapping of replication fork movement through the amplified DHFR domain in CHO cells by in-gel renaturation analysis. *Mol. Cell. Biol.* **9,** 523–531.

Leu, T.-H., and Hamlin, J. L. (1992). Activation of a mammalian origin of replication by chromosomal rearrangement. *Mol. Cell. Biol.* **12,** 2804–2812.

Levenson, V., and Hamlin, J. L. (1993). A general protocol for evaluating the specific effects of DNA replication inhibitors. *Nucleic Acids Res.* **21,** 3997–4004.

Lewis, C. D., and Laemmli, U. K. (1982). Higher order chromosome structure: Evidence for metalloprotein interactions. *Cell (Cambridge, Mass.)* **29,** 171–181.

Li, J. J., and Kelly, T. J. (1984). Simian virus 40 DNa replication in vitro. *Proc. Natl. Acad. Sci. U.S.A.* **81,** 6973–6977.

Liang, C., Spitzer, J. D., Smith, H. S., and Gerbi, S. A. (1993). Replication initiates at a confined region during DNA amplification in Sciara DNA puff II/9A. *Genes Dev.* **7,** 1072–1084.

Little, R. D., Platt, T. H. K., and Schildkraut, C. L. (1993). Initiation and termination of DNA replication in human rRNA genes. *Mol. Cell. Biol.* **13,** 6600–6613.

Liu, L. F., and Wang, J. C. (1987). Supercoiling of the DNA template during transcription. *Proc. Natl. Acad. Sci. U.S.A.* **84,** 7024–7027.

Looney, J. E., and Hamlin, J. L. (1987). Isolation of the amplified dihydrofolate reductase domain from methotrexate-resistant Chinese hamster ovary cells. *Mol. Cell. Biol.* **7,** 569–577.

Ma, C. A., Looney, J. E., Leu, T.-H., and Hamlin, J. L. (1988). Organization and genesis of dihydrofolate reductase amplicons in the genome of a methotrexate-resistant Chinese hamster ovary cell line. *Mol. Cell. Biol.* **8,** 2316–2327.

Mah, D. C. W., Dijkwel, P. A., Todd, A., Klein, V., Price, G. B., and Zannis-Hadjopoulos, M. (1993). ors-12, a mammalian autonomously replicating DNA sequence, associates with the nuclear matrix in a cell cycle-dependent manner. *J. Cell Sci.* **105,** 807–818.

Marsh, R. C., and Worcel, A. (1977). A DNA fragment containing the origin of replication of the *Escherichia coli* chromosome. *Proc. Natl. Acad. Sci. U.S.A.* **74,** 2720–2724.

McCready, S. J., Godwin, J., Mason, D. W., Brazell, I. A., and Cook, P. R. (1980). DNA is replicated at the nuclear cage. *J. Cell Sci.* **46,** 365–386.

McKeon, F. D., Kirschner, M. W., and Caput, D. (1986). Homologies in both the primary and secondary structure between nuclear envelope and intermediate filament proteins. *Nature (London)* **319,** 463–468.

Meier, J., Campbell, K. H. S., Ford, C. C., Stick, R., and Hutchison, C. (1991). The role of lamin LIII in nuclear assembly and DNA replication in cell-free extracts of *Xenopus* eggs. *J. Cell Sci.* **98,** 271–279.

Milbrandt, J. D., Heintz, N. H., White, W. C., Rothman, S. M., and Hamlin, J. L. (1981). Methotrexate-resistant CHO cells have amplified a 135 kb region that includes the dihydrofolate reductase gene. *Proc. Natl. Acad. Sci. U.S.A.* **78,** 6043–6047.

Mirkovitch, J., Mirault, M. E., and Laemmli, U. K. (1984). Organization of the higher-order chromatin loop: Specific DNA attachment sites on the nuclear scaffold. *Cell (Cambridge, Mass.)* **39,** 223–232.

Mirkovitch, J., Spierer, P., and Laemmli, U. K. (1986). Genes and loops in 320,000 base-pairs of the *Drosophila melanogaster* chromosome. *J. Mol. Biol.* **190,** 255–258.

Mosca, P. J., Dijkwel, P. A., and Hamlin, J. L. (1992). The plant amino acid mimosine may inhibit initiation of replication at origins of replication in Chinese hamster cells. *Mol. Cell. Biol.* **12,** 4375–4383.

Nakagomi, K., Kohwi, Y., Dickinson, L. A., and Kohwi-Shigematsu, T. (1994). A novel DNA-binding motif in the nuclear matrix attachment DNA-binding protein SATB1. *Mol. Cell. Biol.* **14,** 1852–1860.

Nakamura, H., Morita, T., and Sato, C. (1986). Structural organization of replicon domains during the DNA synthetic phase in the mammalian nucleus. *Exp. Cell Res.* **165,** 291–297.

Nakayasu, H., and Berezney, R. (1989). Mapping replicational sites in the eukaryotic cell nucleus. *J. Cell Biol.* **103,** 1–11.

Nawotka, K. A., and Huberman, J. A. (1988). Two-dimensional gel electrophoretic method for mapping DNA replicons. *Mol. Cell. Biol.* **8,** 1408–1413.

Nelson, W. G., Pienta, K. J., Barrack, E. R., and Coffey, D. S. (1986). The role of the nuclear matrix in the organization and function of DNA. *Annu. Rev. Biophys. Chem.* **15,** 457–475.

Newport, J. (1987). Nuclear reconstitution in vitro: Stages of assembly around protein-free DNA. *Cell (Cambridge, Mass.)* **48,** 205–217.

Newport, J., Wilson, K. L., and Dunphy, W. G. (1990). A lamin-independent pathway for nuclear envelope assembly. *J. Cell Biol.* **111,** 2247–2259.

Pardee, A. B. (1974). A restriction point for control of normal animal cell proliferation. *Proc. Natl. Acad. Sci. U.S.A.* **71,** 1286–1290.

Pardoll, D. M., Vogelstein, B., and Coffey, D. S. (1980). A fixed site of DNA replication in eukaryotic cells. *Cell (Cambridge, Mass.)* **19,** 527–536.

Paulson, J. R., and Laemmli, U. K. (1977). The structure of histone-depleted metaphase chromosomes. *Cell (Cambridge, Mass.)* **12,** 817–828.

Phi-Van, L., and Strätling, W. H. (1988). The matrix attachment regions of the chicken lysozyme gene co-map with the boundaries of the chromatin domain. *EMBO J.* **7,** 655–664.

Phi-Van, L., von Kries, J. P., Ostertag, W., and Stratling, W. H. (1990). The chicken lysozyme 5' MAR increases transcription from a heterologous promoter in heterologous cells and dampens position effects on the expression of transfected genes. *Mol. Cell Biol.* **10,** 2302–2307.

Rabl, C. (1885). Ueber Zellteilung. *Morphol. Jahrb.* **10,** 214–330.

Razin, S. V., Kekelidze, M. G., Lukanidin, E. M., Scherrer, K., and Georgiev, G. P. (1986). Replication origins are attached to the nuclear skeleton. *Nucleic Acids Res.* **14,** 8189–8207.

Rivier, D. H., and Rine, J. (1992). An origin of DNA replication and a transcription silencer require a common element. *Science* **256,** 659–663.

Robinson, S. I., Nelkin, B., and Vogelstein, B. (1982). The ovalbumin gene is associated with the nuclear matrix of chicken oviduct cells. *Cell (Cambridge, Mass.)* **28,** 99–106.

Romig, H., Fackelmayer, F. O., Renz, A., Ramsperger, U., and Richter, A. (1992). Characterization of SAF-A, a novel nuclear DNA binding protein from HeLa cells with high affinity for nuclear matrix/scaffold attachment DNA elements. *EMBO J.* **11,** 3431–3440.

Roninson, I. (1983). Detection and mapping of homologous, repeated and amplified DNA sequences by DNA renaturation in agarose gels. *Nucleic Acids Res.* **11,** 5413–5431.

Shinomiya, T., and Ina S. (1993). DNA replication of histone gene repeats in *Drosophila melanogaster* tissue culture cells: Multiple initiation sites and replication pause sites. *Mol. Cell. Biol.* **13,** 4098–4106.

Stalder, J., Groudine, M., Dodgson, J. B., Engel, J. D., and Weintraub, H. (1980). Hb switching in chickens. *Cell (Cambridge, Mass.)* **19,** 973–980.

Stief, A., Winter, D. M., Stratling, W. H., and Sippel, A. E. (1989). A nuclear DNA attachment element mediates elevated and position-independent gene activity. *Nature (London)* **341,** 343–345.

Stinchcomb, D. T., Struhl, K., and Davis, R. W. (1979). Isolation and characterization of a yeast chromosomal replicator. *Nature (London)* **282,** 39–43.

Strätling, W. H., Doelle, A., and Sippel, A. E. (1986). Chromatin structure of the chicken lysozyme gene domain as determined by chromatin fractionation and micrococcal nuclease digestion. *Biochemistry* **25,** 495–502.

Sykes, R. C., Lin, D., Hwang, S. J., Framson, P. E., and Chinault, A. C. (1988). Yeast ARS function and nuclear matrix association coincide in a short sequence from the human HPRT locus. *Mol. Gen. Genet.* **212,** 301–309.

Tasheva, E. S., and Roufa, D. J. (1994). A mammalian origin of bidirectional replication within the Chinese hamster RPS14 locus. *Mol. Cell. Biol.* **14,** 5628–5635.

Taylor, J. H. (1977). Increase in DNA replication sites in cells held at the beginning of S phase. *Chromosoma* **62,** 291–300.

Trask, B., and Hamlin, J. L. (1989). Early dihydrofolate reductase gene amplification events in CHO cells usually occur on the same chromosome arm as the original locus. *Genes Dev.* **3,** 1913–1925.

Umek, R. M., Linskens, M. H. K., Kowalski, D., and Huberman, J. A. (1989). New beginnings in the studies of eukaryotic DNA replication origins. *Biochim. Biophys. Acta* **1007,** 1–14.

Valenzuela, M. S., Mueller, G. C., and Dasgupta, S. (1983). Nuclear matrix-DNA complex resulting from EcoRI digests of HeLa nucleoids are enriched for DNA replicating forks. *Nucleic Acids Res.* **11,** 2155–2164.

van Houten, J. V., and Newlon, C. S. (1990). Mutational analysis of the consensus sequence of a replication origin from yeast chromosome III. *Mol. Cell. Biol.* **10,** 3917–3925.

van Zeeland, A. A., Bussman, C. J. M., Degrassi, F., Filon, A. R., van Kesteren-van Leeuwen, A. C., Palitti, F., and Natarajan, A. T. (1983). Effects of aphidicolin on repair replication and induced chromosomal aberrations in mammalian cells. *Mutat. Res.* **92,** 321–332.

Vassilev, L. T., and Johnson, E. M. (1990a). An initiation zone of chromosomal DNA replication located upstream of the c-myc gene in proliferating HeLa cells. *Mol. Cell. Biol.* **10,** 4899–4904.

Vassilev, L. T., Burhans, W. C., and DePamphilis, M. L. (1990b). Mapping an origin of DNA replication at a single-copy locus in exponentially proliferating mammalian cells. *Mol. Cell. Biol.* **10,** 4685–4689.

Vaughn, J. P., Dijkwel, P. A., Mullenders, L. H. F., and Hamlin, J. L. (1990a). Replication forks are attached to the nuclear matrix. *Nucleic Acids Res.* **18,** 1965–1969.

Vaughn, J. P., Dijkwel, P. A., and Hamlin, J. L. (1990b). Replication initiates in a broad zone in the amplified CHO dihydrofolate reductase domain. *Cell (Cambridge, Mass.)* **61,** 1075–1087.

Vogelstein, B., Pardoll, D. M., and Coffey, D. S. (1980). Supercoiled loops and eucaryotic DNA replication. *Cell (Cambridge, Mass.)* **22,** 79–85.

von Kries, J. P., Buhrmester, H., and Strätling, W. H. (1991). A matrix/scaffold attachment region binding protein: Identification, purification and mode of binding. *Cell (Cambridge, Mass.)* **64,** 123–135.

Wanka, F., Mullenders, L. H. F., Bekers, A. G. M., Pennings, L. J., Aelen, J. M. A., and Eygensteyn, J. (1977). Association of nuclear DNA with a rapidly sedimenting structure. *Biochem. Biophys. Res. Commun.* **74,** 739–747.

Wansink, D. G., Schul, W., van der Kraan, I., van Steensel, B., van Driel, R., and de Jong, L. (1993). Fluorescent labelling of nascent RNA reveals transcription by RNA polymerase II in domains scattered throughout the nucleus. *J. Cell Biol.* **122,** 283–293.

Weintraub, H., and Groudine, M. (1976). Chromosomal subunits in active genes have an altered conformation. *Science* **193,** 848–856.

Yunis, J. J. (1981). Mid-prophase human chromosomes. The attainment of 2,000 bands. *Hum. Genet.* **56,** 293–298.

Yunis, J. J., Kuo, M. T., and Saunders, G. F. (1977). Localization of sequences specifying messenger RNA to light-staining G-bands of human chromosomes. *Chromosoma* **61,** 335–344.

Zannis-Hadjopoulos, M., Persico, M., and Martin, R. G. (1981). The remarkable instability of replication loops provides a general method for isolation of origins of DNA replication. *Cell (Cambridge, Mass.)* **27,** 155–163.

Zhao, K., Käs, E., Gonzalez, E., and Laemmli, U. K. (1993). SAR-dependent mobilization of histone H1 by HMG-1/Y in vitro: HMG-1/Y is enriched in H1-depleted chromatin. *EMBO J.* **12,** 3237–3247.

The Nuclear Matrix and Virus Function

W. Deppert* and R. Schirmbeck†
*Heinrich-Pette-Institut für Experimentelle Virologie und Immunologie an der Universität Hamburg, D-20251 Hamburg, Germany, and †Institut für Mikrobiologie, Universität Ulm, Abt. Bakteriologie, D-89089 Ulm, Germany

Replication of the small DNA tumor virus, simian virus 40 (SV40), is largely dependent on host cell functions, because SV40, in addition to virion proteins, codes only for a few regulatory proteins, the most important one being the SV40 large tumor antigen (T-antigen). This renders SV40 an excellent tool for studying complex cellular and viral processes. In this review we summarize and discuss data providing evidence for virtually all major viral processes during the life cycle of SV40 from viral DNA replication to virion formation, being performed at or within structural systems of the nucleus, in particular the chromatin and the nuclear matrix. These data further support the concept that viral replication in the nucleus is structurally organized and demonstrate that viruses are excellent tools for analyzing the underlying cellular processes. The analysis of viral replication at nuclear structures might also provide a means for specifically interfering with viral processes without interfering with the corresponding cellular functions.

KEY WORDS: Simian virus 40, T-antigen, DNA replication, Virion formation, *In situ* cell fractionation, Chromatin, Nuclear matrix, Subcompartmentalization.

I. Introduction

Viruses depend on host cell functions for coming alive. It is obvious that all cellular functions thought to require a structural and functional organization by the nuclear matrix (NM) will also be used by the virus for at least certain viral functions, like transcription and viral replication. So what is of special importance in studying viral functions at the NM? Basically, there is a cellular and a virological aspect of the answer to this question. Viruses can be considered as optimized "Trojan horses" for the introduction of foreign genetic information into cells. Viruses can invade their host cells without having to encounter any resistance, as they use cell surface receptors and internalization pathways developed during evolution for physiological functions of the cells. They use the cellular transcription and translation machinery for the expression of viral genes, and, to a differing extent, the cellular replication machinery for the replication of their genetic information. By doing so, viruses alter cellular pathways for viral replication, with quite varying consequences for the infected cells ranging from no obvious morphological and physiological effects to induction of cellular transformation, or to cell death and lysis for the release of progeny virus. Clearly, viruses encompass an extremely diverse entitiy, so any generalization on viral interactions with their hosts bears the risk of being misleading. Nevertheless, there is one feature common to all viruses—its interactions with its host cell have been optimized during millions of years of evolution for the specific needs of viral replication, and thus for the survival of the particular virus under study. This feature on one hand is responsible for the problems in dealing with viral infections at the cellular level, i.e., by trying to interfere with virus–host cell interactions, especially in fighting off human and animal viruses as major causes of diseases. As a consequence of not being able to prevent effectively the invasion of viruses into, and their replication in, their respective host cells without affecting vital cellular functions, the arsenal of fighting viral infections is still quite limited and with a few exceptions depends on the ability to activate the host organism's immune system to eliminate the virus, or virus-infected cells. On the other hand, as viruses to a large extent depend on host cell functions during their life cycle, they provide an optimal tool for probing complex cellular functions. This, of course, is not a new idea; many basic biological principles have been discovered by studying virus–host interactions, first in prokaryotic systems (e.g., phages λ and T4), and then in eukaryotic systems (e.g., adeno- and polyoma-viruses). The relevant discoveries made by studying these systems now can be found in all modern textbooks on biochemistry and cellular and molecular biology. So the cell biological aspect for analyzing viral functions at the nuclear matrix is that the virus provides specific probes (viral RNA, DNA, or viral proteins), whose functions are already known. These probes can be easily detected and followed up during the viral life cycle. By doing so, we not only learn more about the fate and function of this viral probe, but also about its cellular partners. As viral processes are tightly regulated with respect to the phases of the viral life cycle, temporal and functional aspects of the association of the viral pobe with the nuclear matrix can be correlated. This provides a unique opportunity for analyzing, under closely defined conditions, the function of the NM in the viral processes under study.

The virological aspect of analyzing viral functions at the NM as part of studying virus–host cell interaction is equally intriguing. As stated above, despite all progress in molecular medicine, viral infections still are barely curable at the cellular level. So if the immune system of the host organism is not able to cope with the viral infection, there is very little treatment to offer. Classic biochemical approaches to stop replication, e.g., by inhibiting viral polymerases (if the virus has one of its own), are at their limits. So the only chance for the development of drugs specifically interfering with crucial steps in the viral life cycle is a better understanding not only of the biochemistry of viral transcription and replication, as well as of assembly and release of progeny virus, but even more of the cellular partners involved in these processes. In this regard the analysis of the interaction of viral proteins and of viral genetic information with the nuclear matrix holds great promise. If it is true that the nuclear matrix, or components of it, are involved in fine-tuning the important replicative functions of a cell, then the nuclear matrix should also be the place where cellular functions are modulated to the needs of viral infection. By understanding this modulation, one might be able to devise drugs that specifically interfere with this modulation. Such drugs could possibly prevent an effective replication of the virus, an effective expression of its genetic information, or an effective assembly and release of virus progeny without interfering with the normal cellular function. This can be postulated because such drugs should specifically interfere with the virus-induced modulation of a cellular function, but not with the vital cellular function itself. Therefore, such drugs should provide a tool for specifically interfering with a critical stage in the viral life cycle without doing harm to the cell.

However, despite these obvious advantages of studying viral infections to learn to understand complex cellular functions, the use of viruses as a tool for cell biologists to study nuclear matrix functions is not as widespread as might be anticipated. This seems to be a particular problem in this area of cell biology. Analysis of nuclear matrix functions requires an interdisciplinary approach, combining the expertise of cell biology, biochemistry, and molecular biology. Thus analysis of nuclear matrix functions is not yet the domain of hard-core molecular biologists, as it does not yet allow an easy correlation of data obtained in *in vitro* systems to *in vivo* structure. To provide an example, analysis of the *in vitro* replication of the small minichromosome of the DNA tumor virus simian virus 40 (SV40) during the last decade has unraveled a large part of the biochemistry of eukaryotic DNA replication (Stillman, 1989, 1994; Hurwitz *et al.,* 1990). However, as outlined in a number of chapters in this volume, there is more to eukaryotic DNA replication than just the presence of the appropriate enzymes and substrates. So the analysis of viral DNA replication in conjunction with the participating and organizing cellular structures—the chromatin and the nuclear matrix—should shed light on the functional involvement of these structures in this process. However, such studies are met with quite some skepticism from cell biologists and molecular biologists alike. For the cell biologists, the structural and functional analysis of the nuclear matrix seems to be underrepresented. On the other hand, molecular biologists often tend to negate the need for a structural requirement, e.g., the nuclear matrix structure, for an efficient and coordinated course of nuclear processes such as DNA replication. Anecdotally, the reviews we received on a paper describing the structural topography of SV40 DNA replication (Schirmbeck and Deppert, 1991) emphasize this point. When this paper was submitted to a prime journal of molecular biology, it was rejected with the explanation: "This is one of the nicest studies on nuclear matrix and viral DNA replication I have read. However, SV40 DNA replication works well in soluble *in vitro* systems, so there is no obvious need for a structural organization of this process, and the authors were not able to convincingly demonstrate that an association of this process with the nuclear matrix is required *in vivo*." We then sent out paper to a prime journal in cell biology and received the comment: "This is a nice study on the replication of SV40 DNA at the nuclear matrix. However, it is well known that the nuclear matrix is the site of cellular and viral DNA replication, and this study does not provide any new information on this structure."

This skepticism, however, is not completely unfounded. There are numerous reports on the involvement of the nuclear matrix in viral processes. However, in the majority of these studies, the "nuclear matrix" was simply defined as the insoluble remainder of a rather ill-defined cell fractionation procedure, and the physiological significance of the association of a viral protein or of viral DNA or RNA with this structure remained unclear. Furthermore, in many of these studies the major advantage of working with a viral system, namely the possibility to follow-up the dynamics of the viral life cycle from infection to the release of viral progeny, had not been exploited. This criticism, of course, should not be generalized. There are quite a number of excellent and well-executed studies demonstrating the involvement of the nuclear matrix in viral DNA replication, viral transcription, and virion assembly that have been summarized by others in previous reviews (Hancock, 1982; Berezney, 1984; Nelson *et al.*, 1986; Van der Velden and Wanka, 1987).

In this review, we concentrate on a variety of aspects of nuclear matrix function in the life cycle of SV40 rather than repeating previous reviews that summarized various reports on virus–nuclear matrix preparations. By doing so, we are aware that this is a limitation, as we will not give a complete overview of the whole field. However, by limiting ourselves to this topic, we attempt to describe in an exemplary fashion the crucial role the nuclear matrix plays in several important steps of the life cycle of SV40. SV40 was chosen not only because it is one of the best analyzed tools for studying virus–host interactions, which consequently also implies that the major events of its life cycle are known at the molecular level (Tooze, 1980; DePamphilis and Bradley, 1986), but also, equally important for choosing SV40 as a model system for analyzing the role of the nuclear matrix in viral replication was the fact that SV40 in all of its replicative functions strictly depends on its host. The small genome of SV40 codes for only a very limited number of nonstructural proteins (Fig. 1; DePamphilis and Bradley, 1986; Zerrahn *et al.,* 1993), and with the exception of the helicase activity exerted by the "large" tumor antigen (T-antigen) of SV40 (see below), all of the functions of these proteins are regulatory and not enzymatic. This in turn implies that SV40 must use the host cell machinery throughout its life cycle, and that the regulatory proteins encoded by this virus serve to modulate cellular functions for the needs of viral replication. Thus the association of a viral process with the nuclear matrix most likely is not circumstantial, but reflects the functional involvement of this structure in this viral process. Another advantage for studying SV40 is that this virus has a wide spectrum of possible interactions with host cells: SV40 performs productive infections in cells of its natural host, the rhesus monkey, and in cells of the African green monkey, *Cercopithecus aethiops.* These infections

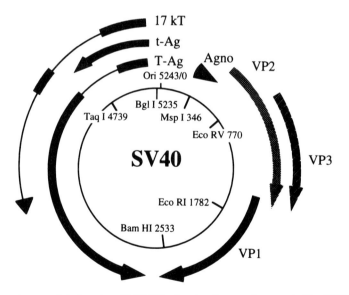

FIG. 1   Genetic map of simian virus 40 (SV40), showing the proteins encoded by SV40. Map coordinates are given according to DePamphilis and Bradley (1986). The SV40 17-kT protein is derived from an alternate splice product of the SV40 large T antigen mRNA and has been described in Zerrahn *et al.* (1993). T-Ag, large tumor antigen; t-Ag: small tumor antigen, VP1-VP3, viral structural proteins; Agno, agnoprotein, a late regulatory protein.

result in the production of SV40 progeny virus and its release from the cells and end with cell death and cell lysis (Tooze, 1980; DePamphilis and Bradley, 1986). SV40 also is able to abortively infect a wide variety of cells of various origins. These infections lead to the transient expression of the major regulatory protein of SV40, the SV40 T-antigen, but no virus progeny is produced. In a few cases, the SV40 DNA gets integrated into the host genome, leading to the permanent expression of T-antigen and to cellular transformation, which has led to the classification of SV40 as a tumor virus. This, however, is a misnomer: SV40 is able to induce tumors rapidly only in newborn hamsters (Eddy *et al.*, 1962), and with a very long latency, also in adult hamsters (Diamandopoulos, 1978) and in mice (Abramczyk *et al.*, 1984).

The majority of the data presented in this review were obtained in this laboratory, and will be summarized and discussed in context. Supporting or controversial data obtained in other laboratories are included.

II. Methodological Aspects

The successful study of virus–host interactions at the cellular level, in addition to the tools of virology, requires a good knowledge cell biology. This seems trivial. However, as indicated above, many studies on viral–cell interactions suffer from too superficial use of the methods of cell biology, especially the methods of cell fractionation. Cell fractionation by biochemical means is, and always will be, a controversial issue. This in principle reflects the fact that it is impossible to separate by biochemical means biological entities that in a living cell form the structural basis for the complex interplay of cellular processes which in their sum provide the basis for life. This dilemma in a rather unproductive way has accompanied in particular the study of the nuclear matrix as the residual proteinaceous framework of the nucleus. Although its existence as such had been rarely disputed, even the "insiders" have spent a lot of time and energy discussing the "correct" method for preparing this structure. However as, by definition, there cannot be a correct preparation of a biological structure by biochemical means, such discussions were then, and still are, in vain. This seems to be especially true for a structure like the nuclear matrix, which is tightly interconnected with other nuclear structures, like the cellular chromatin and the nuclear envelope. However, this caveat is not at all restricted to the preparation of the nuclear matrix, as similar problems arise, e.g., in the isolation and preparation of plasma membranes. Nevertheless, plasma membrane preparations are much less disputed than the preparation of the nuclear matrix. So what is the specific problem in defining the nuclear matrix as a preparative stucture? One reason definitely is that the nuclear matrix is not an easily discernible structure within whole cells or even within intact nuclei, in contrast to, e.g., the plasma membrane. Another reason might be that the fights for the "correct" nuclear matrix structure in a certain way reflect that, from the beginning, nuclear matrices were prepared to demonstrate that they exert certain functions (i.e., the preparation of this nuclear substructure almost always was superimposed onto what the authors thought a nuclear matrix was supposed to do). In contrast to the well-defined function of membranes as biological barriers of permeability, and of "inside" and "outside" of a cell or a cellular compartment, the functions attributed to the nuclear matrix are much less defined. So, at least to critics of the nuclear matrix concept, the nuclear matrix provided an excellent excuse to explain, by the phrase of "structural organization," all we did not (and still do not) understand about the complexity of nuclear functions like DNA replication and transcription. Although this is a valid criticism, negation of the nuclear matrix is as dogmatic as its noncritical appraisal is as the final solution to all the nonresolved problems in the understanding of complex nuclear processes. Finally, there is the problem of whether the nuclear matrix represents an *in vivo* structure like the cytoskeleton and as such is organizing nuclear processes, or whether nuclear matrices, prepared by the various fractionation protocols, reflect the results of the self-organization of nuclear components during the various physiological processes occurring in a cell nucleus, and the fact that the tight and often hydrophobic protein–protein interactions within functional higher-order protein complexes render these complexes insoluble during nuclear extraction. At first sight, there seems to be a fundamental difference between these two views of a nuclear matrix: in the structural concept, a preformed structure serves to organize a complex nuclear process, as opposed to the self-organization of such a process resulting in higher-order protein structures. However, this difference may be mostly semantic considering the dynamics of cellular structures. Also, the cytoskeleton, the plasma membrane, and the internal membranes (i.e., subcellular structures whose existence no one will doubt) are subjected to an enormously dynamic turnover that is the result of functional processes. Even for these well-defined structures the question arises of whether the particular composition of this structure at a given moment and within a certain functional context is structurally predetermined, or whether it reflects self-association of structural subunits and of functional components to form a functional *in vivo* structure.

Given all these caveats, it becomes obvious that the most one can do in the preparation of nuclear matrices is to try to avoid preparational artifacts as much as possible and to devise appropriate controls that verify that functional structures were obtained during preparation of nuclear matrices,

or at least structures that reflect *in vivo* complexes "frozen" in a functional
state. To achieve this goal is difficult enough for "healthy" cells, but it
becomes increasingly critical in the analysis of nuclear matrix function in
viral processes, since at least in cells infected with viruses that replicate
and assemble in the cell nucleus, and which eventually lead to cell death,
one must deal with the problem that the nuclear architecture changes during
the course of viral infection. These changes not only relate to alterations
of nuclear functions due to their modulation by viral processes, but even
more reflect that nuclei of such virus-infected cells become more fragile
during the course of infection. Especially the latter consideration implies
that mechanical forces (encountered, for example, during centrifugation
steps) ought to be avoided.

Another important means to control preparational artifacts is to combine
cell fractionation and analysis of the various nuclear extracts with immunocytochemical or immunofluorescent microscopic analysis of the prepared
structures. This control is necessary because, as indicated above, biochemical cell fractionation procedures hardly yield homogeneous biological structures, as molecules are not extracted by biological criteria but according
to their solubility properties. Consequently, the *in vivo* location of, e.g., a
viral protein, cannot be defined with certainty by analyzing the extracts
alone, and a complementary means is necessary to determine its location
and possible place of function. However, such a complementary analysis
of the structures prepared is only meaningful and revealing if during cell
fractionation cellular structures retain their *in vivo* location. In this regard,
the intermediate filament system of cultured cells poses a real problem:
historically and preparationally, the nuclear matrix is defined as the residual
"insoluble" proteinaceous framework, or skeleton, of the nucleus (Berezney, 1984). A similar preparational definition holds for the cytoplasmic
skeleton of cultured cells, consisting mainly of intermediate filaments and
some actin. This implies that nuclear matrix preparations of cultured cells
always will contain the cytoplasmic skeleton. Because the proteins of the
cytoplasmic skeleton make up for at least 50% of the residual insoluble
cellular structures prepared during nuclear matrix preparation, it is obvious
that it is mandatory to differentiate between an association of a viral protein
with this structure and the nuclear skeleton. This problem is aggrevated
by the fact that isolation of nuclei from cultured cells by conventional means
(e.g., after preparation of single cell suspensions) induces the intermediate
filament system to collapse onto the surface of the nuclei (Staufenbiel and
Deppert, 1982). From then on, it would be extremely difficult to differentiate between nuclear association and intermediate filament association by
any complementary structural analysis.

To circumvent these problems, we have devised an *in situ* cell fractionation protocol for the preparation of *in situ* nuclear matrices. This protocol

is outlined in Fig. 2. Its original version has been described in detail (Staufen-biel and Deppert, 1983, 1984) and its applications and modifications in a number of ensuing publications (Schirmbeck and Deppert, 1987, 1989; Richter and Deppert, 1990).

Basically, the initial step consists of a gentle lysis and extraction of the cells with a nonionic detergent (NP40 or Triton X-100) in a low ionic strength buffer at close to neutral pH. This step is the most critical one, as it decides whether or not the extracted cells (termed "*in situ* nuclei") will remain attached to the substratum. With some experience, it is possible to even retain rounded virus-infected or drug-treated cells in their *in vivo*

In situ Cell Fractionation

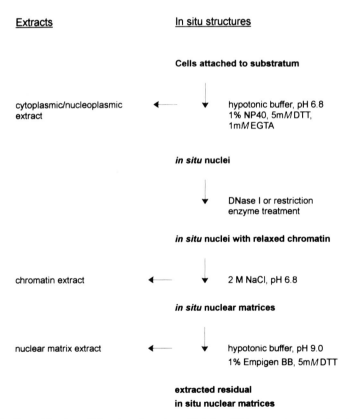

FIG. 2   Outline of *in situ* cell fractionation steps. Details of this procedure are described in Staufenbiel and Deppert (1984) and in the text.

location, with the cytoskeleton still displaying its *in vivo* morphology. The other critical step is the relaxation of the chromatin loops before extracting the cellular chromatin. This step is absolutely required for preserving the integrity of the residual nuclear structures. If this step is omitted, and the chromatin is directly extracted by high-salt treatment, the residual nuclear structures are empty and simply consist of the nuclear lamina. For routine fractionation studies, a short DNase I treatment is sufficient to relax the chromatin loops. In specialized applications (e.g., for the analysis of viral DNA, see below), DNase I treatment can be substituted by treatment of the *in situ* nuclei with appropriate restriction endonucleases. We found it an important prerequisite to test whether the commercial DNase preparations used were free of contaminating protease activity (which actually was very rarely the case). Use of protease-contaminated DNase I preparations not only resulted in at least the partial degradation of the proteins under study, but also in alterations of the structures prepared.

Regardless of whether the cellular DNA is close to quantitatively released by extensive DNase I treatment, or whether only a small percentage of the DNA is released by restriction endonuclease treatment, only very few cellular proteins are released by this step (Staufenbiel and Deppert, 1983, 1984; Schirmbeck *et al.*, 1993). Given the large number of cellular proteins directly interacting with DNA, this finding supports the assumption that a significant percentage of nuclear proteins, i.e., all proteins not extractable by the first extraction step, are organized in higher-order protein complexes, forming the protein constituents of the chromatin and the nuclear matrix.

High-salt treatment then releases what commonly is termed the chromatin fraction. Traditionally, 2 M NaCl is used, but we observed virtually no difference between 1 and 2 M NaCl extractions. The chromatin extraction step can be modified in such a way that consecutive extractions using increasing salt concentrations are performed. Such subfractionations can provide information on the interaction of the molecule in question with a particular subfraction of the cellular chromatin and thus provide a hint on its functional interactions.

The residual nuclear matrices obtained after chromatin extraction can be further freed from residual nucleic acids by treatment with DNase and RNase. However, such treatment no longer releases any additional proteins.

An important prerequisite for the functional analysis of nuclear matrix-associated proteins is that they can be isolated in a native form. As the nuclear matrix preparationally is defined as an "insoluble" structure, this at first glance seems impossible, as solubilization of an otherwise insoluble structure usually requires the use of denaturing agents, like SDS or urea. However, we have found that the zwitterionic detergent Empigen BB provides an excellent means of resolving this problem. Nuclear matrices, like the residual cytoplasmic cytoskeleton, can be dissolved in buffers containing

Empigen BB. Unlike SDS, Empigen BB can be easily removed from proteins by dialysis, or simply "inactivated" by the addition of NP40. Empigen BB-treated proteins thus are not inactivated and retain (or easily regain) their biologically active conformation. This not only allows the immunological analysis of nuclear matrix extracts by immunoprecipitation, but also the purification of nuclear matrix-associated proteins in a native state and thus their further biological and biochemical characterization.

As outlined in Fig. 2, our *in situ* nuclear subfractionation protocol yields nuclear substructures, attached to their substratum, that are amenable to further structural and immunocytochemical or immunofluorescent microscopic analysis, and nuclear extracts, which can be further analyzed for their constituents (e.g., for the presence of a particular protein by immunoprecipitation). By comparing the results obtained with immunocytochemical analyses of the *in situ* extracted cellular structures and of immunoprecipitations of the respective cellular extracts for the protein of interest, a close to unequivocal assignment regarding its subcellular location can be made. As a first example for the successful application of this methodology, the subnuclear distribution of a viral oncogene product, the simian virus 40 (SV40) tumor antigen, has been determined, resulting in the identification of distinct nuclear subclasses of this protein (Staufenbiel and Deppert, 1983).

In addition to the possibility of being able to exclude an artifactual assignment of the molecule of interest to a certain subcellular location as far as possible, the *in situ* cell fractionation protocol outlined above has some additional advantages over conventional cell fractionation procedures. As centrifugation steps are avoided, there is no mechanical stress on the structures prepared. This avoids compression of nuclear structures and results in their easy extraction. Furthermore, because centrifugation steps are time-consuming, lack of these steps results in this cell fractionation protocol being rather rapid. This not only is convenient but further helps in preparing functional structures or functional molecules associated with these structures.

III. Structural Topography of SV40 DNA Replication

A. SV40 as Model System for Eukaryotic DNA Replication

The papova virus SV40 contains a double-stranded, covalently closed circular DNA molecule, which is associated with cellular histones to form a chromatin-like structure, the SV40 minichromosome. An overview of the events occuring during SV40 DNA replication is shown in Fig. 3. Replication of the viral minichromosome starts at specific DNA sequences, the origin

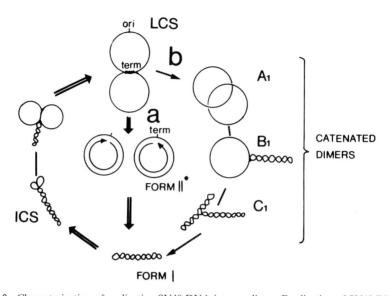

FIG. 3   Characterization of replicating SV40 DNA intermediates. Replication of SV40 DNA starts on specific ORI sequences on supercoiled form I DNA molecules. Bidirectional elongation from the ORI generates intermediate Cairns structures (ICS), containing two relaxed loops of nascent DNA, whereas unreplicated parental DNA preserves its superhelical structure. Elongation of nascent DNA chains on ICS proceeds until replication is about 90% completed. At this point replication forks pause, generating the latest Cairns structure (LCS). The separation of the two daughter molecules from the LCS then proceeds (a) via form II* molecules, containing a short gap in the termination region. The gap is filled up to mature supercoiled minichromosomes. A minor termination pathway (b) can proceed via catenated dimer structures.

of replication (ORI). Bidirectional chain elongation then generates intermediate Cairns structures (ICS), containing two relaxed loops of nascent DNA. Elongation of nascent DNA chains on ICS proceeds until replication of the viral DNA is about 90% completed. At this point replication forks pause, generating the latest Cairns structure (LCS). The separation of the two daughter molecules from the LCS then may occur via two alternative termination pathways: (1) If the remaining parental DNA strands in the LCS were completely unwound during replication, two circular SV40 monomers (SV40 form II* DNA) are separated, each containing a short gap in the termination region of the nascent DNA strand. After the gap is filled up, mature SV40 minichromosomes are detectable, containing superhelical SV40 form I DNA. (2) If not all helical turns of the parental DNA are removed before the replication forks meet, catenated dimers are formed with the two sibling chromosomes intertwined (DePamphilis and Bradley,

1986). As outlined in Fig. 4, the individual steps during replication of the small SV40 minichromosomes bear some similarity to the complex processes necessary for the replication of eukaryotic chromosomes. SV40 DNA replication thus can be considered as a simplified model system for the analysis of these processes (DePamphilis and Bradley, 1986). The usefulness of SV40 minichromosome replication as a model system for studying the more complex molecular processes during eukaryotic DNA replication is underscored by the fact that SV40 DNA replication is almost completely dependent on the cellular DNA replication machinery.

SV40 DNA replication is regulated mainly by a single SV40-encoded protein, the SV40 large tumor antigen (T-antigen), which performs a variety of functions during DNA replication. During the early phase of viral infection, i.e., before the onset of viral DNA replication, T-antigen is responsible for regulating several cellular processes that ultimately activate the cellular DNA replication machinery and lead to the induction of cellular DNA synthesis (Tooze, 1980). During the late phase of infection, which is characterized by the onset of viral DNA replication, T-antigen is required to initiate each round of viral DNA replication (Tegtmeyer, 1972) and is involved in the elongation process of viral DNA synthesis, probably by exerting DNA-helicase activity (Stahl and Knippers, 1987). In addition, T-antigen modulates both early and late viral transcription (Tooze, 1980; Rigby and Lane, 1983). Biochemical activities demonstrated for T-antigen

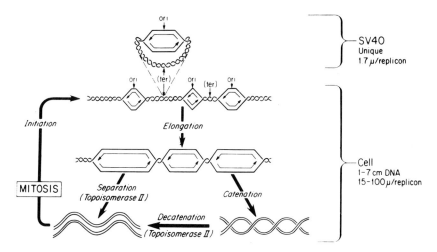

FIG. 4 Comparison of SV40 and of cellular DNA replication. This model has been adapted from DePamphilis and Bradley (1986), and demonstrates the similarities between SV40 and cellular DNA replication.

are its ability to interact with regulatory sequences on the SV40 ORI in a sequence-specific manner, to act as a transactivator or repressor of viral and cellular gene expression, and to exert a helicase activity, which is the only enzymatic activity unequivocally demonstrated for T-antigen. This helicase activity depends on the hydrolysis of ATP; thus the T-antigen intrinsic ATPase activity closely corresponds to this helicase activity and provides a convenient means for measuring it (Stahl and Knippers, 1987).

Considering the limited biochemical activities of T-antigen, one has to postulate that the multitude of functions displayed by this protein during viral replication are the result of its regulatory interactions with (1) regulatory sequences on the viral genome (Rigby and Lane, 1983; DePamphilis and Bradley, 1986) and (2) cellular components involved in transcription and DNA replication, thereby modulating the cellular transcription and DNA replication machinery for the needs of viral replication and transcription. Thus, molecular events during SV40 DNA replication can be probed by analyzing the functions of this regulatory protein in this process.

B. Regulation of Viral Replication by SV40 T-Antigen

1. Characterization of T-Antigen Intrinsic Biochemical Activities

Specific biochemical activities of T-antigen necessary for the regulation of viral replication processes are: (1) Its binding to nucleotide sequences within the SV40 control region (ORI) (Fig. 5A). Specific binding of T-antigen to binding site I within the ORI is necessary (although not sufficient) for controlling autoregulation of early transcription (Rio *et al.*, 1980; Myers *et al.*, 1981), whereas binding to ORI site II is essential for initiating viral DNA replication (Shortle *et al.*, 1979; Wilson *et al.*, 1982; Margolskee and Nathans, 1984). (2) T-antigen exhibits an ATPase activity which is closely associated with and indispensable for its DNA-unwinding function (helicase activity) (Stahl and Knippers, 1987). Considering the different functions exerted by T-antigen during the course of infection, these activities must be closely regulated during the different phases of viral replication. Thus analysis of these activities of T-antigen during the course of infection will shed light on cellular and viral processes regulating SV40 replication.

A prerequisite for performing such analyses is that the biochemical activities of T-antigen can be compared under standardized experimental conditions, as T-antigen is present in different nuclear compartments (see below), requiring drastically different extraction conditions for each individual T-antigen subclass. Furthermore, the amounts of T-antigen change dramatically during the course of infection (Tooze, 1980; Schirmbeck and Deppert,

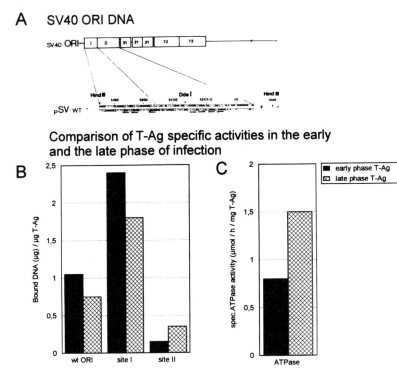

FIG. 5   Comparison of biochemical activities of early- or late-phase large T-antigen. Panel A: Structural organization of the SV40 ORI control region; Panel B: Analysis of DNA-binding and ATPase activities of early phase and late phase T-Ag to the SV40 ORI. About 0.8 μg of T-antigen per assay, obtained by whole-cell extraction at 12 (early phase T-Ag) and 34 hr pi (late phase T-Ag), was analyzed for ORI DNA-binding and ATPase activities. The relative DNA binding activities (μg bound DNA per 1 μg of T-antigen) for the complete ORI, for isolated site I, and for isolated site II (panel B) and the specific ATPase activities (μmol hydrolyzed ATP per hour per mg of T-antigen) (panel C) are shown.

1987). These restrictions render it difficult to standardize reaction conditions for a comparative analysis of T-antigen activities in different cellular ex- tracts. To circumvent this problem, we developed a "target-bound" assay for the analysis of biochemical activities of T-antigen, in which T-antigen was immunoprecipitated from cellular extracts with the T-antigen specific monoclonal antibody (PAb108 [directed against an N-terminal epitope of T-antigen (Gurney et al., 1986)] and protein A–Sepharose. As PAb108 does not interfere with any of the biochemical activities of T-antigen, target- bound, immunopurified T-antigen can be subjected in standardized amounts to various biochemical analyses, providing a convenient means to determine the biochemical properties of T-antigen derived from different sources under standardized conditions (Hinzpeter *et al.,* 1986; Hinzpeter and Deppert, 1987; Schirmbeck and Deppert, 1988, 1989).

a. Characterization of ORI DNA-Binding Activities of T-Antigen   Analysis of the binding affinities and activities of total T-antigen, extracted from TC7 cells lytically infected with SV40 toward isolated binding sites I and II as well as toward combined binding sites I and II on the SV40 wild-type ORI (Fig. 5A) demonstrated that T-antigen bound to these different regulatory DNA sequences with a similar affinity (K_d values between 10 and $50 \times 10^{-10} M$). Scatchard treatment of our binding data also provided the amounts of ORI DNA fragments bound by the amount of input T-antigen under saturating conditions of SV40 DNA, thereby allowing the determination of the fraction of T-antigen molecules binding to a given SV40 ORI DNA fragment under these conditions. In repeated experiments, we found that about 45% of the input T-antigen molecules (calculated per monomers) bound to isolated ORI site I, about 2 to 3% bound to isolated ORI site II, and about 10% bound to combined sites I and II on the SV40 wild-type ORI, indicating that the binding of T-antigen to individual regulatory SV40 ORI sequences and to the complete SV40 ORI is tightly regulated. A detailed analysis of T-antigen binding to these various SV40 ORI fragments has been published (Schirmbeck and Deppert, 1988).

b. Characterization of the ATPase Activity of T-Antigen   To analyze the T-antigen intrinsic ATPase activity, we used a modification of the ATPase assay described by Manos and Gluzman (1985). Defined amounts of target-bound, immunocomplexed T-antigen were subjected to this ATPase assay and the release of free ^{32}P was measured. Thus, ATP hydrolysis could be directly related to the amount of reacted T-antigen. The specificity of this reaction for the T-antigen intrinsic ATPase activity was demonstrated by inhibiting the release of ^{32}P from [γ-^{32}P]ATP with monoclonal antibody PAb204, which specifically blocks the T-antigen intrinsic ATPase-activity (Clark *et al.,* 1981). Again, a detailed analysis has been published (Schirmbeck and Deppert, 1989).

2. Comparison of the ORI DNA-Binding and ATPase Activities of T-Antigen in the Early and Late Phases of Infection

Under our experimental conditions, viral late phase started between 12 and 16 hr postinfection (pi) (Schirmbeck and Deppert, 1987). To characterize biochemical activities of T-antigen during early and during late phases of infection, we analyzed immunocomplexed T-antigen prepared from whole-cell extracts at 12 hr pi ("early phase T-antigen") and 34 hr pi ("late phase T-antigen"), using standardized amounts of T-antigen. Figure 5B shows the DNA-binding activities of early phase and late phase T-antigen toward the various ORI DNA fragments, and Fig. 5C presents their specific ATPase activities. Characteristic differences were found for T-antigen in the early and in the late phases of infection, as T-antigen molecules extracted at the late phase exhibited decreased binding activities toward the complete SV40 ORI (-25%) and toward isolated ORI site I (-15%), compared to T-antigen molecules extracted at the early phase. In contrast, the binding activity of T-antigen toward ORI site II, i.e., the site of initiation of SV40 DNA replication, increased during the late phase ($+130\%$). Similarly, the ATPase activity of late phase T-antigen was considerably higher ($+70\%$) than that of early phase T-antigen. This shift in biochemical activities of T-antigen from early phase to late phase indicates a shift toward an increased replication function of T-antigen during late phase of infection, as initiation of replication requires T-antigen to bind to ORI site II, and elongation requires the T-antigen helicase activity measured by the ATPase activity of T-antigen.

C. Analysis of Biochemical Activities of Nuclear Subclasses of T-Antigen

The results described above demonstrate that T-antigen extracted at early or at late times pi differs significantly in its biochemical activities *in vitro,* possibly reflecting its involvement in different viral processes *in vivo.* Assuming that different biochemical activities of T-antigen are exerted by distinct subpopulations of T-antigen, these results imply that the composition of T-antigen molecules in infected cells changes during the course of infection with regard to such subpopulations.

It previously has been shown that T-antigen exists in various biochemically defined forms that differ in oligomerization and in complex formation with the tumor suppressor p53, as well as in various post-translational modifications (DePamphilis and Bradley, 1986; Stahl and Knippers, 1987; Fanning and Knippers, 1992). In addition, T-antigen interacts with the retinoblastoma gene product, pRB (DeCaprio *et al.,* 1988), the transcription factor AP-1, DNA–polymerase α (Smale and Tjian, 1986; Gannon and Lane, 1987; Mitchell *et al.,* 1987; Mole *et al.,* 1987), and probably a number of other cellular proteins. While the modifications of T-antigen and its interactions with various cellular proteins as such will serve to modulate its functions (Rigby and Lane, 1983; Mole *et al.,* 1987), at a higher level of regulation, modulation of T-antigen functions could be provided (or be reflected) by its interaction with functionally defined structural systems of the nucleus. Complex nuclear functions, such as DNA transcription and replication, are supposed to be performed in conjunction with structural systems of the nucleus, i.e., the chromatin and the nuclear matrix (for details, see specific chapters of this volume). Therefore, at this level of regulation, T-antigen might interact with target molecules at these structures, and, as a consequence, should be found in association with different nuclear substructures. This assumption already had been proven true for T-antigen in transformed cells, which was found as a "soluble" fraction in the nucleoplasm, and in association with the chromatin and the nuclear matrix using the *in situ* cell fractionation procedure described above (Staufenbiel and Deppert, 1983, 1984).

In line with the assumption that such interactions reflect the regulation of T-antigen functions of SV40 infected cells, we found by *in situ* cell fractionation that the subnuclear distribution of T-antigen subclasses (T-antigen in the nucleoplasm, at the cellular chromatin, and associated with the nuclear matrix) followed a defined pattern during the course of viral infection (Schirmbeck and Deppert, 1987) (Fig. 6). T-antigen was barely detectable in any subnuclear fraction until 12 hr pi. At that time (i.e., still during the early phase of infection, Fig. 6A), T-antigen was present mainly in the nucleoplasm and in association with the cellular chromatin. Only a very small amount was found to be associated with the nuclear matrix. During progression of viral infection, T-antigen accumulated mainly at the cellular chromatin and at the nuclear matrix. The beginning of the late phase of infection (around 14 hr pi) was marked by a significant increase of nuclear matrix-associated T-antigen (Fig. 6B). It is important to state that at each time pi the amounts of T-antigen released by *in situ* cell fractionation added up to about the amount of total T-antigen recovered from whole cell lysates, indicating a quantitative recovery of T-antigen from the different nuclear extracts. It thus was possible to analyze the changes in relative mass distribution of T-antigen in each subnuclear fraction during the course of viral infection (Fig. 6C). To do this, the total amount of T-antigen at each time pi was set at 100% and the percentage of T-antigen in each subnuclear fraction was determined. The relative mass distribution of nucleoplasmic and of nuclear matrix T-antigen changed drastically concomitant with the switch from the early (12 hr pi) to the late (16 hr pi) phase of infection. At 12 hr pi, approximately 30% of the total T-antigen had accumulated in the nucleoplasm, 60 to 65% was associated with the chromatin, and only about 5% was bound to the nuclear matrix. At 16 hr pi, the relative amount of nucleoplasmic T-antigen had dropped to about 15%, whereas the relative amount of nuclear matrix-associated T-antigen had increased to about 35%. From 20 hr pi, the relative mass distribution of the three subclasses of T-antigen stayed approximately con-

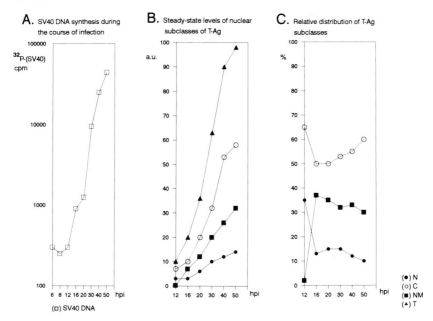

A. SV40 DNA synthesis during the course of infection

B. Steady-state levels of nuclear subclasses of T-Ag

C. Relative distribution of T-Ag subclasses

(□) SV40 DNA

(●) N
(○) C
(■) NM
(▲) T

FIG. 6 Subnuclear distribution of SV40 T-antigen during the course of infection. Panel A: Viral DNA content during the course of infection. TC7 cells were lytically infected with SV40 (MOI of 10). After the time periods indicated, the cells were analyzed for their viral DNA content by Hirt extraction (Hirt, 1967). Steady-state levels of viral DNA were determined by the dot blot procedure and subsequent Southern hybridization using a ^{32}P-labeled SV40 DNA probe. ^{32}P content (cpm) was determined by liquid scintillation counting. Panel B: Steady-state levels of nuclear subclasses of T-antigen. In a parallel experiment, cells were lysed to determine the amount of total large T (T), or subfractionated into a nucleoplasmic fraction (N), a chromatin fraction (C), and a nuclear matrix fraction (NM). T-antigen was immunoprecipitated from the respective extracts and quantitated as described previously (Schirmbeck and Deppert, 1987). Amounts of T-antigen are given in arbitrary units (au). Panel C: Relative distribution of nuclear subclasses of T-antigen. Since the same number of cells was analyzed in whole cell lysates and nuclear extracts, the relative mass distribution of T-antigen subclasses during the course of infection could be determined. The amount of T-antigen (per time point) obtained from whole-cell lysates was designated 100%, and the relative amounts of T-antigen present in the nuclear subclasses were determined.

stant during further course of infection, suggesting a correlation between the association of T-antigen with the nuclear matrix and viral replication.

To further corroborate this correlation, we compared the subnuclear distribution of T-antigen in SV40 wild-type and tsA58 mutant (Tegtmeyer and Ozer, 1971)-infected cells. T-antigen encoded by the tsA58 mutant virus is temperature-sensitive for the initiation of viral DNA replication. At the permissive growth temperature (32°C), viral replication in tsA58 mutant virus-infected cells is similar to that of wild-type virus at this growth temperature, while at the restrictive growth temperature (39°C) tsA58 mutant virus is unable to replicate (Tegtmeyer and Ozer, 1971; Tegtmeyer *et al.*, 1975; Wilson *et al.*, 1982). At the permissive temperature (32°C), viral DNA synthesis both in wild-type SV40 and in tsA58 mutant-infected cells started between 24 and 36 hr pi (Fig. 7A), demonstrating that viral infection proceeded more slowly at 32°C than at 37°C. The subnuclear distribution

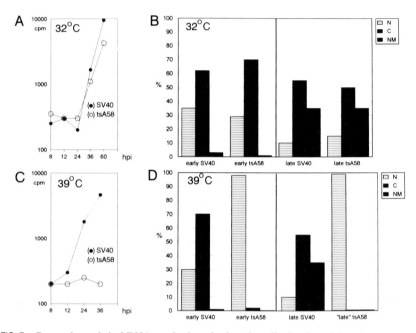

FIG. 7   Comparison of viral DNA synthesis and subnuclear distribution of T-antigen in SV40 and mutant- tsA58-infected cells. TC7 cells were infected with wild-type SV40 or mutant-tsA58 (MOI of 10) at 32°C and either kept at the permissive temperature (32°C) (panels A and B) or shifted to the restrictive temperature for the growth of tsA58 (39°C) at about 2 hr pi (panels C and D). At the indicated time points after infection, the cultures were analyzed for their viral DNA content, as described in the legend to Fig. 6A (panels A and C). To analyze the subnuclear distribution of early-phase or late-phase T-antigen, 32°C cultures (panel B) were subfractionated into a nucleoplasmic fraction (N), a chromatin fraction (C), and a nuclear matrix fraction (NM) at 24 hr pi (early SV40, early tsA58) and at 36 hr pi (late SV40, early tsA58). T-antigen was immunoprecipitated from the respective extracts and quantitated as described previously (Schirmbeck and Deppert, 1987). The relative mass distribution of nulear subclasses of the respective T-antigens was determined as described in the legend to Fig. 6B, C. Furthermore, the relative mass distribution of T-antigen subclasses was determined at 39°C [at 12 hr pi (early SV40, early tsA58) and at 36 hr pi. (late SV40, late tsA58)] (panel D). Note that under the restrictive temperature no late tsA58-T-antigen functions were detectable (panel C), as evidenced by the lack of viral DNA synthesis.

of T-antigen during the course of infection in SV40 wild-type and SV40 tsA58 mutant-infected cells at the permissive temperature resembled that in SV40 wild-type virus-infected cells analyzed at 37°C (Fig. 7B). As in wild-type-infected cells at 37°C, nuclear matrix association of wild-type or tsA mutant T-antigen was barely detectable during the early phase of infection (up to 24 hr pi), but became prominent after the onset of viral DNA replication. During further course of infection, T-antigen accumulated at the chromatin and the nuclear matrix (Fig. 7B). In contrast, if tsA58 mutant virus-infected cells were kept at the restrictive temperature, no viral DNA was synthesized (Fig. 7C). Furthermore, no cytopathic effect was detected in tsA58-infected cells kept at the restrictive temperature for as long as 6 d. The biological differences between SV40 wild-type and SV40 tsA58 mutant virus in cells kept at 39°C were reflected by the subnuclear distribution of the respective T-antigens in these cells (Fig. 7D): Wild-type T-antigen exhibited the subnuclear distribution typical for cells in the early and late phase of infection seen already at 32 and 37°C. This indicates that the overall subnuclear distribution of T-antigen in wild-type-infected cells at early and at late times of infection follows a pattern independent of growth temperature, but strictly correlates with the switch from the early to the late phase of viral replication. In contrast, analysis of the subnuclear distribution of mutant T-antigen in tsA58 mutant-infected cells kept at 39°C showed that this mutant T-antigen barely associated with the cellular chromatin and the nuclear matrix. Thus, most nuclear T-antigen in these cells accumulated in the nucleoplasm. Experiments similar to those described for SV40 tsA58 were performed with another SV40 tsA mutant, tsA28 (Tegtmeyer, 1972), and yielded similar results (data not shown). In summary, these data showed that the subnuclear distribution of T-antigen follows a characteristic pattern during the course of infection (Figs. 6, 7), with a significant change in the relative mass distribution of T-antigen occurring after the switch into the late phase, i.e., concomitant with the onset of viral DNA replication (Schirmbeck and Deppert, 1987).

The strict correlation between the subnuclear distribution of T-antigen and the phases of viral infection suggested that different nuclear subclasses of T-antigen perform different functions during viral replication. If this were the case, the interaction of T-antigen with a certain subnuclear structure would determine its functional profile, or vice versa, the functional profile of T-antigen would determine its association with a certain nuclear structure. Thus, the modulation of functions of T-antigen to the specific needs of viral replication at a given stage of infection would be reflected in a change in the subnuclear distribution of T-antigen. One prediction of this hypothesis is that each nuclear subclass of T-antigen should exhibit a distinct profile of biochemical activities throughout the infectious cycle. Thus the changes in the overall biochemical activities of T-antigen in the early and the late phase of infection would be accommodated by the changes in its subcellular distribution. This assumption could be verified by analysis of the biochemical activities of the individual T-antigen subclasses derived from different subnuclear compartments after *in situ* fractionation (Fig. 8). Under standardized conditions, the analysis of the relative DNA-binding activities revealed that about 9% of the nucleoplasmic T-antigen molecules bound to the complete SV40 ORI, 20% bound to isolated binding site I, and only 0.8% were able to bind to isolated binding site II. Chromatin-associated T-antigen exhibited a significantly higher DNA-binding activity: about 15% bound to the complete ORI, 60% bound to isolated site I, and about 4 to 5% bound to isolated site II. On the other hand, nuclear matrix-associated T-antigen almost completely lacked ORI DNA binding activity (Fig. 8A, left-hand panel). Similarly, the ATPase activities of the different nuclear subclasses of T-antigen also showed a distinct profile: the specific ATPase activities of T-antigen subclasses (expressed as μmoles per hour per milligram) were 0.7 for nucleoplasmic T-antigen, 1.1 for chromatin-associated T-antigen, and about 2.7 for nuclear matrix-associated T-antigen (Fig. 8A, right-hand panel). The fact that nuclear matrix-associated T-antigen exhibited the highest ATPase activity despite being negative for SV40 ORI binding first of all renders it unlikely that the lack of DNA binding activity of this T-antigen subclass is due to preparational artifacts. Furthermore, this finding strongly supports our hypothesis that individual nuclear subclasses of T-antigen exhibit quite distinct profiles of activities.

This conclusion was further corroborated by comparing the experimentally determined biochemical activities of total (unfractionated) T-antigen from whole-cell lysates with the biochemical activities of total T-antigen, calculated from the activities of individual nuclear subclasses of T-antigen. At 34 hr pi, total T-antigen consisted of 15% nucleoplasmic T-antigen, 55% chromatin-associated T-antigen, and 30% nuclear matrix-associated T-antigen (Fig. 8B). Taking into account the relative biochemical activities exerted by each individual subclass of T-antigen (Fig. 8A), the biochemical activities of total T-antigen in these cells were determined. These calculated activities added up to approximately the activities determined experimentally for total (whole cell-extracted) T-antigen at the same time pi (Schirmbeck and Deppert, 1989).

By taking into consideration the actual subnuclear distribution of T-antigen and the relative biochemical activities exerted by the individual subclasses of T-antigen, we were able to determine the absolute distribution of the biochemical activities of T-antigen within the nuclei of infected cells (Fig. 8B). According to this calculation, the bulk of ORI DNA binding activity is performed by chromatin-associated T-antigen molecules, whereas the DNA unwinding activity (ATPase) distributed between chromatin- and nuclear matrix-associated large T molecules. Nucleoplasmic T-antigen was

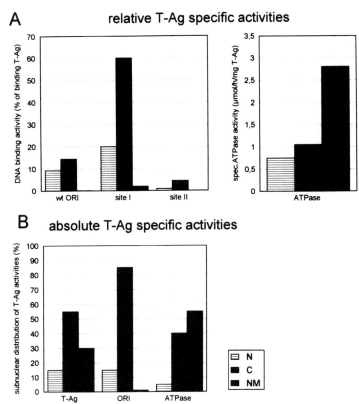

A relative T-Ag specific activities

B absolute T-Ag specific activities

FIG. 8 Quantitation of relative and absolute ORI DNA-binding- and ATPase-activities of nuclear T-antigen subclasses. Panel A: Lytically infected cells were subfractionated at 36 hr pi. Nuclear extracts were immunoprecipitated for T-antigen, and DNA binding of the nucleoplasmic T-antigen (N), of chromatin associated T-antigen (C), and of nuclear matrix associated T-antigen (NM) toward the wild-type ORI and toward isolated binding sites I and II was analyzed. Absolute amounts of T-antigen-bound ORI DNA fragments and of T-antigen present in the respective nuclear subclasses were quantitatively determined as described previously (Schirmbeck and Deppert, 1989) and DNA binding was calculated for 1 μg of T-antigen. Scatchard analysis allowed us to determine the percentage of T-antigen molecules binding to DNA (calculated per T-antigen monomer) in each individual subclass (% binding T-Ag in the left-hand panel A). Furthermore, the specific ATPase activities of the T-antigen subclasses were determined as described previously (Schirmbeck and Deppert, 1989). ATP hydrolysis was calculated for 1 mg of T-antigen (right-hand panel A). Panel B: Considering the actual subnuclear distribution of T-antigen at 36 hr p.i (T-Ag in panel B), and the specific biochemical activities exerted by these individual subclasses, as shown in panel A, the absolute distribution of T-antigen specific biochemical activities could be calculated.

the overall least active one in these assays, providing further support for our hypothesis that replicative functions of T-antigen are structurally organized and require its interaction with structural systems of the nucleus. The different profile of activities of T-antigen in the chromatin and of T-antigen in the nuclear matrix fraction also demonstrates a clear functional distinction between these T-antigen subclasses. This is important for two reasons: (1) functionally, this indicates that the chromatin and the nuclear matrix are functionally distinct entities, exerting different functions in viral replication; and (2) structurally, the possibility to prepare functionally distinct nuclear subclasses of T-antigen supports the concept of cell fractionation as a suitable tool for differentiating between functionally different subcompartments of the cell nucleus.

An important question remains to be resolved: do the biochemical activities of nuclear subclasses of T-antigen differ because this subclass of T-antigen is associated with different nuclear structures, or do subclasses of T-antigen associate with different structural systems of the nucleus because they are biochemically different? In other words: what is the biochemical reason, and what is the driving force for the association of T-antigen with different nuclear substructures and the modulation of this association during the course of infection? So far, this problem has remained refractory to our analyses. Because T-antigen is a phosphoprotein, phosphorylated at multiple sites (Fanning and Knippers, 1992), and as phosphorylation of T-antigen in *in vitro* assays was found to modulate T-antigen functions (Prives, 1990), we compared the phosphorylation patterns of T-antigen extracted at early and at late phase of infection and from different subnuclear compartments by two-dimensional phosphopeptide mapping. The disappointing result was that at least qualitatively, all nuclear T-antigen subclasses were identically phosphorylated (U. Knippschild and W. Deppert, unpublished). In view of the prominent effects of phosphorylation on T-antigen activity and on SV40 DNA replication *in vitro* (Prives, 1990), these results were rather unexpected. The different biochemical activities of nuclear subclasses of T-antigen also cannot be explained by the interaction of T-antigen with cellular partner proteins, like pRB or p53, as only a small subfraction of T-antigen is associated with any of these proteins at any stage of the viral infection.

D. Characterization of SV40 DNA Intermediates

In following up our hypothesis that SV40 replication *in vivo* occurs in association with structural systems of the nucleus, we next analyzed the subnuclear distribution of SV40 DNA and of its replicative intermediates. As shown in Fig. 3, replication of SV40 DNA generates a variety of replicative intermediates. These various replicative forms could be identified by neutral agarose gel electrophoresis as described previously (Sudin and Varshavsky, 1980, 1981; Weaver *et al.*, 1985; Snapka, 1986). The most prominent bands in the ethidium bromide-stained gel (Fig. 9A, lane a), representing steady-state levels of SV40 DNA intermediates, were form I DNA (superhelical), form II + II* DNA (relaxed circles), and catenated dimers [two interlocked SV40 form I molecules (Sudin and Varshavsky, 1981)]. Fluorographic analysis of replicating forms of SV40 DNA (Fig. 9A, lane b) after a 5-min pulse-labeling of the infected cells with [^{3}H]deoxythymidine [^{3}H]dThd) also resulted in radiolabeling of form I DNA, form II + II* DNA, and the LCS. The majority of the radiolabel, however, was present in the ICS. ICS migrate as a smear between form I DNA and the LCS, since they represent a random population of nascent SV40 DNA at various stages of continued replication (Sudin and Varshavsky, 1980; Weaver *et al.*, 1985). Radiolabeled catenated dimers were not detectable after pulse-labeling.

The conversion of transient SV40 DNA intermediates during replication into final products was analyzed in pulse–chase experiments. Figure 9B (lane a) shows the pulse-labeled SV40 DNA intermediates. During the chase period (Fig. 9B, lanes b to e), radiolabeled ICS significantly decreased within 15 min, accompanied by a concomitant increase of radiolabel in form I molecules. LCS molecules also behaved as expected for transient intermediates in the conversion of ICS into form I DNA because they completely disappeared within a 30-min chase period. The form II + II* DNA pool was analyzed for form II* DNA content as described by Weaver *et al.* (1985) and contained about 95% of form II* DNA at the beginning of the chase period (lane b). After the 30-min chase period (lane c), form II* DNA had decreased to 5 to 10%, thus also identifying these molecules as transient intermediates. Following the course of catenated dimers, only 0.5 to 1% of catenated dimers were detectable after a 15-min chase. This form accumulated during a further 15-min chase up to about 2% and remained at this level during a 2-hr chase (lanes b to e). These findings demonstrate that SV40 DNA replication, with regard to termination, in its majority proceeded from LCS via form II* molecules. Thus, these data are in line with the termination pathway proposed by Weaver *et al.* (1985), who showed that under physiological conditions the major termination pathway proceeds via gapped form II* circles into mature form I DNA, and that catenated dimers, representing only a minor SV40 DNA population, are not significantly involved in the replication of SV40 DNA.

To correlate the analysis of the subnuclear location of SV40 DNA with a functional analysis of T-antigen subclasses, we wanted to apply an *in situ* fractionation protocol similar to that used in our analyses of nuclear subclasses of T-antigen. *In situ* fractionation conditions therefore should

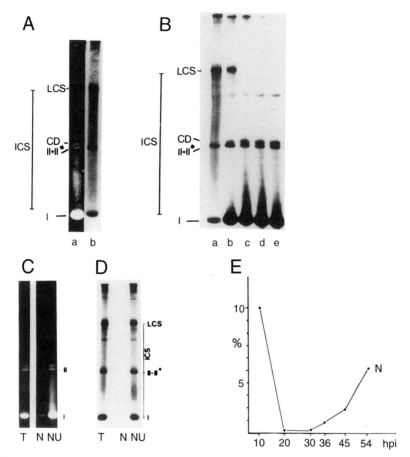

FIG. 9 Analysis of structurally bound SV40 DNA replication intermediates. Panel A: TC7 cells were infected with SV40. At 36 hr pi, cells were labeled with 200 μCi of [³H]dThd for 5 min. Viral DNA was extracted by the method of Hirt (1967), subjected to neutral agarose gel electrophoresis, and visualized by ethidium bromide staining (lane a) or by fluorography (lane b). Panel B: Lytically infected cells were pulse labeled with [³H]dThd for 5 min (lane a), followed by a chase in the presence of unlabeled dThd for 15 min (lane b), 30 min (lane c), 60 min (lane d), and 120 min (lane e). SV40 DNA was extracted and processed as described above. Panels C and D: Lytically infected cells (36 hr pi) were labeled with 200 μCi of [³H]dThd for 10 min. Parallel cultures were either Hirt-extracted for total SV40 DNA (lanes T) or *in situ* fractionated into an NP-40-soluble fraction (lanes N) and a structurally bound fraction by lysing the residual nuclear structures with SDS (Hirt, 1964) (lanes NU) (for details, see Schirmbeck and Deppert, 1991). SV40 DNA was processed for analysis by ethidium bromide staining (panel C) or fluorography (panel D) as described above. Panel E: The NP-40 soluble SV40 DNA (N) and total Hirt-extracted SV40 DNA were analyzed during the course of infection and quantitatively evaluated. The data shown represent the actual proportion of NP-40-soluble SV40 DNA at the respective stages of infection. The positions of SV40 form I DNA (I), SV40 form II + II* DNA (II + II*), catenated dimers (CD), terminating LCS, and elongating ICS are indicated.

preserve the structural integrity of the nuclear substructures prepared and avoid degradation of the SV40 DNA.

1. Characterization of Structurally Bound and Soluble SV40 DNA Intermediates

In a first approach, lytically infected cells were subfractionated into a NP-40 soluble cytoplasmic/nucleoplasmic fraction (N), and a nuclear fraction, termed *in situ* nuclei (NU), reflecting chromatin-associated nuclear matrix structures (Staufenbiel and Deppert, 1983, 1984). The ethidium bromide-stained gel of total viral DNA from unfractionated cells (Fig. 9C, lane T) shows that superhelical form I DNA molecules, derived from *in vivo* unpacked SV40 minichromosomes as well as from minichromosomes encapsidated in mature SV40 virions, were most abundant. Fractionation of total SV40 DNA into soluble SV40 DNA, and into structurally bound SV40 DNA retained in *in situ* nuclei, revealed that, except for about 1% of SV40 form I DNA molecules found in the soluble cytoplasmic/nucleoplasmic fraction (Fig. 9C, lane N), SV40 DNA remained quantitatively associated with nuclear structures (Fig. 9C, lane NU). Most important for supporting our hypothesis of a structural organization of SV40 DNA was that replicating and terminating SV40 DNA intermediates, labeled during a 10-min labeling period, were quantitatively retained within *in situ* nuclei (Fig. 9D). The newly replicated form I DNA molecules found in association with nuclear structures were derived from unpacked minichromosomes (see below).

2. Characterization of Structurally Bound SV40 DNA

To corroborate our finding that all major processes of SV40 replication proceed in association with nuclear structures, soluble and structurally bound SV40 DNA was quantitated during the course of infection. Figure 9E provides the percentage of soluble cytoplasmic/nucleoplasmic SV40 DNA during the course of infection, with the amount of total SV40 DNA in infected cells set as 100% for each time point. During the very early phase of infection (10 hr pi), about 10% of total SV40 DNA (at that time still resulting from infecting virions) was present in the cytoplasm/nucleoplasm. Between 20 and 30 hr pi, i.e., at times of high-level viral DNA replication, only about 0.2% of total SV40 DNA was detectable in this cellular compartment. During the further course of infection, the proportion of soluble SV40 DNA continuously increased to about 7% at 54 hr pi. At that time, the cells started to show cytopathic lysis, thus preventing further analyses. Control experiments, described in detail in Schirmbeck and Deppert (1991), demonstrated that different modifications of the extraction procedure had only minor effects on the qualitative and quantitative association of SV40 DNA with nuclear structures. Thus, the reported release of SV40 nucleoprotein complexes after dounce-homogenization of infected cells (Su and DePamphilis, 1978), was not seen under our *in situ* fractionation conditions, and most likely reflects an artifact, resulting from damage of nuclear structures during preparation. This conclusion was further supported by our finding that under identical buffer conditions, extraction of the nuclei of infected cells released SV40 nucleoprotein complexes when mechanical forces like dounce-homogenization or centrifugation steps were applied (Schirmbeck and Deppert, 1991). We therefore conclude that in all phases of SV40 replication, nuclear structures, i.e., constituents of the cellular chromatin and the nuclear matrix, play an important role in anchoring SV40 chromatin.

3. Subfractionation of *in Situ* Nuclei

The experiment shown in Fig. 9E demonstrated an almost quantitative association of SV40 DNA and its replicative intermediates with nuclear structures, thus emphasizing the need for analyzing the specific interactions of SV40 DNA with the cellular chromatin and the nuclear matrix by subfractionation of *in situ* nuclei. To preserve the integrity of the SV40 DNA and its replicative intermediates, DNase I treatment of isolated nuclei, used in our standard *in situ* fractionation procedure to relax cellular chromatin loops (Staufenbiel and Deppert, 1983, 1984; Schirmbeck and Deppert, 1987, 1989), was replaced by treating *in situ* nuclei with restriction endonucleases that do not cut SV40 DNA (Schirmbeck and Deppert, 1991). A representative *in situ* fractionation experiment is shown in Fig. 10. The chromatin fraction contained a significant proportion (10 to 15%) of the total SV40 form I DNA (panel A; lane C), whereas the majority of this form was associated with the nuclear matrix (panel A; lane NM). Also after a pulse label of 10 min, the chromatin fraction contained a small proportion of radiolabeled SV40 DNA, in its majority consisting of SV40 form I DNA (panel B; lanes C and C*). However, the majority of newly replicated SV40 form I DNA as well as all of SV40-terminating LCS were associated with the nuclear matrix (panel B; compare lane T with lane NM). We found that form II* DNA also was quantitatively associated with this structure (Schirmbeck and Deppert, 1991). The presence of the major termination intermediates at the nuclear matrix strongly suggests that termination of SV40 DNA replication proceeds at this structure. Furthermore, the radiolabeled smear between form I and the LCS in lane NM, representing ICS at various stages of elongation, shows that ICS were also quantitatively associated with the nuclear matrix, thereby also indicating that elongation processes of SV40 DNA replication are quantitatively associated with the

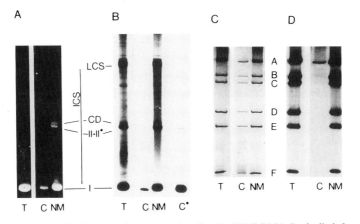

FIG. 10 Subnuclear distribution of mature and replicating SV40 DNA. Lytically infected cells (36 hr pi) were labeled with 200 μCi of [³H]dThd for 10 min, and *in situ* nuclei were prepared by our standard NP-40 lysis of the cells. *In situ* nuclei then were either lysed with SDS for extraction of total DNA (lanes T) or subfractionated into chromatin (lanes C) and nuclear matrix (lanes NM) fractions (Schirmbeck and Deppert, 1991). Nuclear extracts were precipitated for SV40 DNA, and aliquots were subjected to agarose gel electrophoresis, followed by ethidium bromide staining (panel A) or fluorography (panel B). Lane C* represents a long exposure of lane C. The positions of SV40 form I DNA (I), SV40 form II + II* DNA (II + II*); catenated dimers (panels C, D), terminating LCS, and elongating ICS are indicated. Samples of SV40 DNA from the various nuclear fractions were digested with the restriction endonuclease *Hin*dIII and analyzed by SDS–PAGE, followed by silver staining (panel C) for fluorography (panel D).

nuclear matrix. To ensure that the radiolabeled DNA identified as ICS by agarose gel electrophoresis in the nuclear matrix fraction indeed represented authentic SV40 DNA rather than smears of labeled cellular DNA fragments, aliquots of SV40 DNA from the respective nuclear fractions were digested with *Hin*dIII and analyzed by SDS–PAGE. The silver-stained gels (Fig. 10C) revealed the characteristic band pattern of *Hin*dIII-restricted SV40 DNA. Fluorographic analysis of the gels (Fig. 10D) demonstrated a uniform labeling of the SV40 *Hin*dIII fragments in the nuclear matrix fraction (lane NM), unequivocally identifying the radiolabeled SV40 DNA molecules as ICS. In contrast, the chromatin-associated SV40 DNA molecules (lane C) contained the ³H label only in the termination region (SV40 *Hin*dIII-A fragment), thus demonstrating that only newly synthesized, mature form I DNA had been released into the chromatin fraction. In summary, these findings indicate that at least the chain elongation and termination steps of SV40 DNA replication quantitatively proceed at the nuclear matrix. This result is in line with the finding that topoisomerase I was found to be associated both with the nuclear matrix and with SV40 T-antigen in SV40-infected cells, and was present in replicating forks of newly replicating SV40 chromatin (Rainwater and Mann, 1991). It also fits the observation that topoisomerase II, specifically required to decatenate late viral DNA intermediates (Sudin and Varshavsky, 1981; Yang *et al.,* 1987), is tightly bound to the nuclear matrix (Hancock, 1982) and even has been implicated in anchoring SV40 DNA to the nuclear matrix (Pommier *et al.,* 1991). In addition to replicating and terminating SV40 DNA, the majority of newly synthesized, mature SV40 form I DNA also was found to be associated with the nuclear matrix. However, a small proportion of these molecules leave the nuclear matrix and associate with the cellular chromatin.

Because residual nuclear matrices prepared from *in situ* nuclei using restriction enzymes still contain the bulk of the cellular DNA, we wanted to exclude the possibility that replicating SV40 DNA might be associated with the remaining cellular DNA rather than with the residual protein skeleton of the nuclei. Therefore we removed the cellular DNA by repeated restriction endonuclease digests and high-salt extractions. Figure 11 shows that nuclear matrices generated by the first restriction enzyme/high-salt treatment contained about 90% of cellular DNA (NM-1). Since high-salt treatment of nuclear structures removes the histones from the chromatin, cellular DNA then became more accessible to restriction enzymes. Thus, the following restriction enzyme/high-salt treatments resulted in the effective removal of cellular DNA down to about 10% (see lanes NM-2 and NM-3). However, repeated extraction of cellular DNA did not significantly influence the quantitative association of replicating SV40 DNA with the nuclear matrix [compare cellular DNA-rich nuclear matrices (NM-1) and cellular DNA-depleted nuclear matrices (NM-3)]. Considering the stringency of the extraction conditions in this experiment, this finding further supports our conclusion that replicating SV40 DNA is specifically and tightly bound to the residual protein skeleton. Additional control experiments, using a wide spectrum of extraction conditions during *in situ* preparation of nuclei and of nuclear matrices, showed that the specific interactions of mature and replicating SV40 DNA with nuclear structures were only minimally affected by various preparation conditions, including varying salt and Mg^{2+} concentrations, pH, and temperature (Schirmbeck and Deppert, 1991). We interpret these findings to indicate that leaving nuclear structures attached to the substratum, i.e., avoiding mechanical forces, preserves their structural integrity and subsequently renders them highly resistant against extraction-induced artifacts.

Further support for the specific interaction of SV40 DNA with nuclear structures was obtained by analysis of structurally bound SV40 DNA by *in situ* hybridization. SV40 DNA in *in situ* nuclei and at *in situ* matrices is not randomly distributed at isolated nuclear structures but is organized in

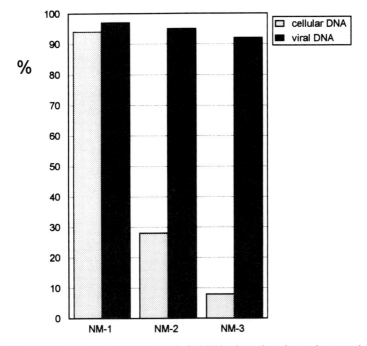

FIG. 11 Quantitative analysis of cellular and viral DNA bound to the nuclear matrix. TC7 cells were labeled with [³H]dThd for 6 hr and either lysed directly in a buffer containing 3% SDS (control value) or subfractionated as described in the text and previously (Schirmbeck and Deppert, 1991), yielding DNA-rich nuclear matrices (NM-1). Restriction digest with SV40 noncutting enzymes and subsequent 1.5 M NaCl extraction was repeated once, or twice, on isolated nuclear matrices prepared in parallel, leading to NM-2 and NM-3 nuclear matrices, respectively. Nuclear matrices then were lysed with SDS and analyzed for remaining cellular DNA by liquid scintillation analysis. In a parallel experiment, lytically infected TC7 cells were extracted as described above. Viral DNA was extracted from the nuclear matrix fractions, purified, and digested with the restriction endonuclease HindIII. HindIII-digested SV40 DNA was analyzed by SDS–PAGE, followed by fluorography and densitometric evaluation of the ³H label. The percentage of cellular DNA and SV40 DNA remaining at the nuclear matrices is shown.

clusters of globular structure (Fig. 12). Similar globular structures previously were reported to be centers of actively replicating viral and cellular DNA (Harper *et al.,* 1985; Voelkerding and Klessig, 1986; DeBruyn-Kops and Knipe, 1988; Nakayasu and Berezney, 1989). SV40 DNA shows a similar pattern of organization, which strongly argues for a functional and highly ordered interaction of SV40 DNA with nuclear matrix components rather than an unspecific entrapment of SV40 DNA within the nuclei during preparation.

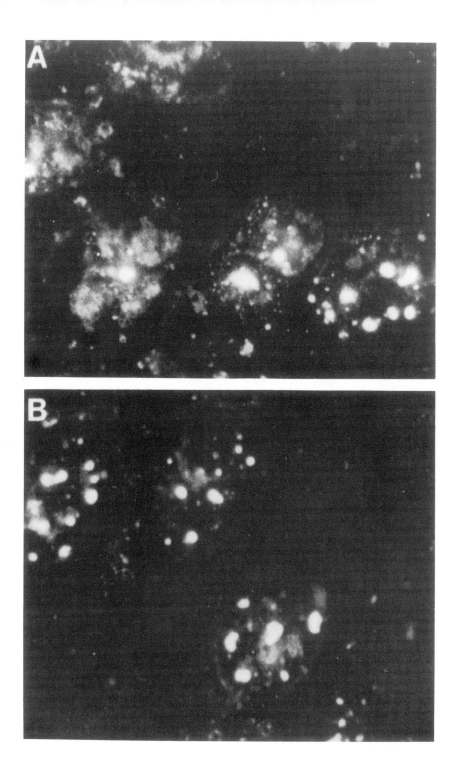

E. Functional and Coordinated Association of SV40 DNA and T-Antigen at Isolated Nuclear Structures

1. SV40 DNA Replication in Isolated *in Situ* Nuclei

The finding that functionally distinct T-antigen populations as well as biologically dissectable SV40 DNA intermediates are specifically associated with the chromatin and the nuclear matrix strongly suggested that T-antigen and SV40 DNA molecules colocalize at nuclear substructures in functional complexes. To test this hypothesis, we first analyzed whether isolated *in situ* nuclei can support SV40 DNA replication in the absence of exogenously added T-antigen. *In situ* nuclei were incubated under replication conditions, without the addition of exogenous T-antigen and exogenous SV40 DNA (Schirmbeck and Deppert, 1991). (Fig. 13A shows that *in situ* nuclei indeed were able to support SV40 DNA replication under these experimental conditions, as is evident from the time-dependent increase in radiolabeled, mature SV40 form I and II DNA, and from the presence of all replicative SV40 DNA intermediates detectable *in vivo*. However, under our *in situ* replication conditions, efficient conversion of mature form II DNA into superhelical form I DNA was decreased, indicating that cellular factors involved in this process were extracted and/or destroyed during preparation of *in situ* nuclei and could not be restored by the addition of cellular extracts (Tapper *et al.*, 1979, 1982). To further exclude the possibility that incorporation into SV40 DNA intermediates represented DNA repair rather than specific SV40 DNA replication, we analyzed newly synthesized SV40 form II DNA, the major end product of *in situ* replication, for the distribution of the radiolabel on the SV40 genome at increasing times of replication. As expected for bidirectional DNA replication, radiolabel was found predominantly (>90%) in the SV40 termination region after brief replication times and migrated toward the origin region during increasing replication times (Nathans and Danna, 1972; Tapper *et al.*, 1979, 1982).

An important aspect of the *in situ* replication system described above was that T-antigen-mediated SV40 DNA replication started immediately after replication conditions were applied. This is in contrast to SV40 replication in soluble *in vitro* assays, where replication starts only after a certain lag period, probably required for the assembly of functional replication

---

FIG. 12   *In situ* hybridization of SV40 DNA at isolated nuclear structures. Lytically infected cells were grown on glass coverslips. At 36 hr pi, *in situ* nuclei (panel A) or *in situ* nuclear matrices (panel B) were prepared, using our standard fractionation procedure (Schirmbeck and Deppert, 1991). Nuclear structures then were processed for *in situ* hybridization by using a biotinylated SV40 probe. SV40 DNA was visualized with a Texas red–streptavidin conjugate.

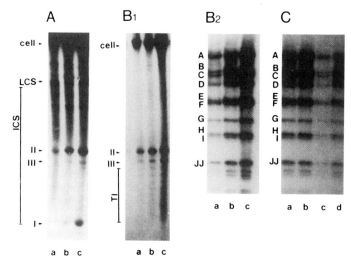

FIG. 13   Detection of elongation-competent SV40 DNA/T-antigen complexes at isolated nuclear structures. *In situ* nuclei (panel A) or *in situ* nuclear matrices (panel B₁) were prepared
36 hr pi. DNA replication conditions were applied as described previously (Schirmbeck and
Deppert, 1991). After 10 (lanes a), 20 (lanes b), and 60 (lanes c) min at 37°C, nuclear structures
still attached to the substratum were lysed with SDS. SV40 DNA was precipitated from the
lysates, subjected to agarose gel electrophoresis, and visualized by autoradiography. The
positions of supercoiled SV40 form I DNA (I), nicked SV40 form II DNA (II), of linear
SV40 form III DNA (III), terminating LCS, elongating ICS, and of cellular DNA (cell) are
indicated. In addition, the position of SV40 topoisomers (TI) is indicated. Panel B₂: Aliquots
of SV40 DNA shown in panel B₁ were digested with *Bst*NI and subjected to SDS–PAGE,
followed by autoradiography. Panel C: Replication conditions were applied to *in situ* nuclear
matrices for 60 min (lane a). Replication was also performed on parallel dishes in the presence
of unspecific mouse control immunoglobulins (lane b), monoclonal anti-T-antigen antibody
PAb204 (lane c), or polyclonal anti-T-antigen antibodies from mouse tumor serum (lane d).
Nuclear matrices were then lysed, and SV40 DNA was precipitated from the extracts, followed
by *Bst*NI digestion of SV40 DNA, SDS–PAGE, and autoradiography. Identical amounts of
SV40 DNA were loaded on the gel, as demonstrated by silver staining of the gel (data
not shown).

complexes (Wobbe *et al.*, 1986). Since SV40 DNA and T-antigen molecules
are immobilized in isolated *in situ* nuclei, these findings strongly argue for
a coordinated and functional association of at least elongation-competent
T-antigen/SV40 DNA complexes (Sewaga *et al.*, 1980; Stahl and Knippers,
1983; Tack and DePamphilis, 1983) at nuclear structures.

2. SV40 DNA Replication at Isolated *in Situ* Nuclear Matrices

The data presented so far strongly indicate that the nuclear matrix plays
an important role in functionally anchoring distinct T-antigen and SV40

DNA molecules in replication complexes, since (1) elongating SV40 DNA intermediates (ICS) were quantitatively associated with the nuclear matrix; (2) the nuclear matrix-associated T-antigen population exhibited the highest ATPase activity, an activity which is closely related to the SV40 DNA elongation function; and (3) more circumstantial, the association of T-antigen with the nuclear matrix in infected cells closely paralleled the onset of viral DNA replication. These correlations strongly suggested that T-antigen-mediated SV40 DNA elongation proceeds at the nuclear matrix. To directly test this hypothesis, we next analyzed isolated *in situ* nuclear matrices for their ability to perform SV40 DNA elongation under our *in situ* replication conditions. Fig. 13B$_1$ shows a time-dependent increase of SV40 form II DNA, the major end product of *in situ* replication, as seen above for *in situ* nuclei. Furthermore, linearized SV40 form III DNA and SV40 form I topoisomers were detectable, whereas elongating and terminating SV40 DNA intermediates were not unequivocally detectable because of the smear of radiolabeled cellular DNA. In addition, a ladder of SV40 DNA topoisomers was generated during replication (Fig. 13B$_1$,TI). The generation of these topoisomers probably reflects the activation of topoisomerases when replication conditions were applied. Since the viral (and the cellular) DNA in the isolated nuclear matrices used for *in situ* replication was histone-depleted, this DNA provided an ideal substrate for these enzymes. The activation of topoisomerases at the nuclear matrix under replication conditions further supports the findings reported above that these enzymes are tightly bound to the nuclear matrix structure and are involved in replication functions.

Restriction of nuclear matrix associated SV40 DNA shown in Fig. 13B$_1$ with BstNI (Fig. 13B$_2$) revealed that fragments representing the SV40 ORI region (fragments G and I) were underrepresented, indicating that isolated nuclear matrices were not able to reinitiate new rounds of SV40 DNA replication to any significant extent. On the other hand, radiolabel preferentially accumulated in the SV40 termination region (fragments A and F), reflecting the fact that termination of SV40 DNA replication is the ratelimiting step during one round of SV40 DNA synthesis (Sudin and Varshavsky, 1980; Weaver *et al.*, 1985). SV40 DNA elongation at isolated nuclear matrices was dependent on the endogenous, i.e., nuclear matrix-associated T-antigen population, since *in situ* SV40 DNA replication (Fig. 13C) was drastically inhibited by antibody PAb204 to about 90 to 95% (lane c) and to about 50 to 60% by polyclonal antibodies against T-antigen from an SV40 tumor serum (lane d), whereas unspecific mouse control immunoglobulins (lane b) did not interfere with SV40 DNA replication. Since PAb204 inhibits the DNA elongation function of T-antigen via inhibition of the T-antigen intrinsic ATPase/helicase activity (Clark *et al.*, 1981; Schirmbeck and Deppert, 1989), these results directly show the T-antigen mediated *in situ* replication of SV40 DNA at isolated nuclear matrices.

IV. Stages of SV40 Virion Assembly

A. Discrimination between Unpacked and Encapsidated SV40 Mini-Chromosomal DNA within the SV40 Form I DNA Pools at Nuclear Substructures

The finding that more than 90% of SV40 DNA remained associated with nuclear structures during all stages of viral infection suggested that, in addition to SV40 DNA replication, all major processes of viral replication, including virion formation, proceed in association with nuclear structures. Therefore we next characterized the events leading from replicating SV40 minichromosomes to encapsidated virions at nuclear substructures. A prerequisite for these studies was to be able to discriminate during *in situ* cell fractionation between unpacked (i.e., not encapsidated) SV40 minichromosomes and SV40 minichromosomes contained in mature virions. Since SV40 DNA in virion minichromosomes is resistant to DNase I treatment, but SV40 DNA in unpacked minichromosomes is not (Ben-Ze'ev *et al.*, 1982), such a discrimination could be achieved by using either non-SV40-cutting restriction endonucleases or DNase I for relaxation of cellular chromatin loops in *in situ* nuclei during cell fractionation. Restriction endonuclease treatment will preserve all SV40 form I DNA (form I DNA in unpacked minichromosomes as well as form I DNA in virion-contained minichromosomes) within the respective nuclear compartments, whereas DNase I treatment will destroy all SV40 DNA in unpacked minichromosomes; and only SV40 form I DNA from encapsidated virion minichromosomes will be recovered (Schirmbeck *et al.*, 1993). The comparison of SV40 form I DNAs derived from the chromatin and the nuclear matrix fractions with either fractionation protocol thus should allow us to quantitate the proportions of unpacked and of encapsidated SV40 minichromosomes, represented by the respective SV40 form I DNA pools.

The validity of this conclusion was tested with SV40 tsB mutant virus-infected cells. SV40 tsB11 (Tegtmeyer *et al.*, 1974) encodes a temperature-sensitive VP1 protein. SV40 tsB11-infected cells thus fail to produce previrion and virion structures at the restrictive growth temperature (39°C). However, viral DNA replication and subsequent accumulation of viral minichromosomes proceed normally under these restrictive growth conditions (Tegtmeyer *et al.*, 1974; Garber *et al.*, 1978, 1980). When SV40 tsB11-infected cells are kept at the permissive growth temperature (32°C), minichromosomes are encapsidated into progeny virions, and a significant amount of SV40 form I DNA, representing tsB11-virion DNA, could be recovered from DNAse I-treated nuclei (Fig. 14). In contrast,

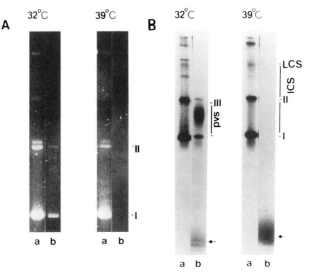

FIG. 14   Characterization of SV40 tsB11 DNA intermediates. TC7 cells were infected with
SV40 tsB11 virus and cultured at 32 or 39°C. At 36 hr pi, cells were labeled with [³H]dThd
for 6 hr. Then *in situ* nuclei were prepared and either Hirt-extracted (lanes a) or treated with
DNase before extraction (lanes b). Purified SV40 DNA obtained from the respective lysates
was analyzed by ethidium bromide staining of the gel (panel A) or by fluorography (panel
B). SV40 tsB11 DNA intermediates are marked as defined in the legend to Fig. 9. In addition,
the positions of linear SV40 tsB11 form III DNA (III) and of SV40 DNA intermediates
released from previrion structures (pvs) are indicated. SV40 tsB11 DNA completely degraded
by DNase is marked by an arrow.

DNase I treatment of *in situ* nuclei from tsB11-infected cells kept at
39°C resulted in a complete degradation of SV40 tsB11 DNA molecules
(Fig. 14).

As extraction of nuclei of infected cells might result in the disruption of
mature virions (Seidman *et al.*, 1979; Boyce *et al.*, 1982), thereby artificially
generating DNase-sensitive SV40 DNA from DNase-insensitive virion
DNA, we performed control experiments taking into consideration those
parameters reported to be critical for maintaining the integrity of SV40
virions during extraction, and found that the fraction of DNase I-resistant
SV40 form I DNA remained constant under all experimental conditions
applied, and that its actual amount depended only on the state of lytic
infection (see below). We thus conclude that the application of the differential extraction procedures in our cell fractionation experiments allowed
us to quantitatively differentiate between unpacked and virion-contained
minichromosomes.

B. SV40 Virion Accumulation during Infection

We next quantitatively analyzed SV40 virion accumulation during the course of infection using our standard fractionation conditions (Schirmbeck and Deppert, 1991). Figure 15 shows that at 8 hr pi (i.e., still in the very early phase of infection), nearly all structurally bound SV40 DNA was DNase sensitive, indicating that infectious virions were almost quantitatively unpacked at this stage of infection. SV40 virion DNA was barely

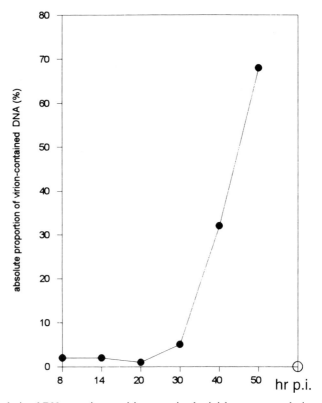

FIG. 15   Analysis of DNase-resistant, virion contained minichromosomes during the course of infection. TC7 cells were lytically infected with SV40. At the indicated time points after infection *in situ* nuclei were either Hirt-extracted (yielding total SV40 minichromosomes) or incubated with DNase before their extraction (yielding only DNase-resistant, virion-contained minichromosomes). SV40 DNA was precipitated from the respective extracts and quantitated as described previously (Schirmbeck *et al.,* 1993). The actual proportion (%) of DNase-resistant, virion-contained minichromosomes at total viral DNA is shown. The start of viral DNA replication (late phase of infection) occurred at around 14 hr pi.

detectable up to 20 hr pi. At 30 hr pi, the proportion of encapsidated SV40 virion minichromosomes had increased and accounted for about 5% of total nuclear SV40 form I DNA. From then on, the proportion of virion minichromosomes drastically increased, and virion minichromosomes accounted for about 70% of the total nuclear SV40 DNA at 50 hr pi. Thus, effective SV40 virion assembly occurs considerably later than high-level viral DNA synthesis, which starts at about 14 hr pi. Considering that about 70% of the total nuclear SV40 DNA became encapsidated between 30 and 50 hr pi, we conclude that virion assembly is very effective at this late stage of infection (Garber *et al.*, 1980; Wang and Roman, 1981).

C. Subnuclear Distribution of Total SV40 Minichromosomes and SV40 Virion-Contained Minichromosomes during Infection

1. Subnuclear Distribution of Total SV40 Minichromosomes

To investigate the involvement of nuclear structures in SV40 virion assembly, we subfractionated *in situ* nuclei of lytically infected cells during the course of infection. The analysis of the subnuclear distribution of total SV40 DNA, isolated after restriction endonuclease treatment of *in situ* nuclei and leading to the recovery of all SV40 DNA (Fig. 16A), shows that at 8 hr pi (panel a), SV40 DNA was mainly recovered from the cellular chromatin fraction, indicating that SV40 minichromosomes released from incoming virions predominantly associated with this structure. The remaining SV40 minichromosomes were about equally distributed between the cytoplasm/nucleoplasm and the nuclear matrix. The subnuclear distribution of SV40 minichromosomes changed significantly after the onset of viral DNA synthesis (20 hr pi), with the bulk of newly replicated minichromosomes now accumulating at the nuclear matrix. Quantitative analysis showed that about 3 to 5% of total SV40 DNA molecules were found in association with the chromatin. Newly replicated minichromosomes continued to accumulate at the nuclear matrix until 30 hr pi, whereas the amount of chromatin-associated minichromosomes increased only slightly. Ongoing viral DNA replication led to a continuous increase in the amount of nuclear matrix-associated SV40 minichromosomes, but a significant proportion of SV40 minichromosomes also associated with the chromatin. SV40 DNA molecules, probably contained within mature virions, could also be extracted from the soluble cytoplasmic/nucleoplasmic fraction.

2. Subnuclear Distribution of Virion-Contained SV40 DNA

In a parallel experiment, *in situ* cell fractionation was performed with DNase I to analyze the subnuclear distribution of DNase-resistant SV40

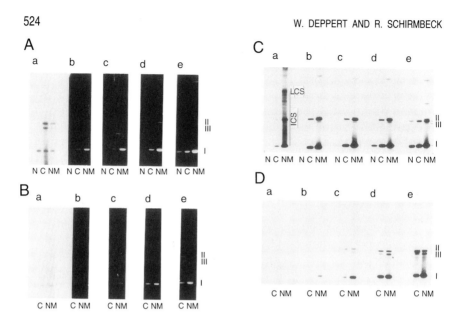

FIG. 16   Subcellular events during SV40 virion assembly A and B: TC7 cells were infected
with SV40 and *in situ* nuclei were prepared at 8 hr pi (panels a), 20 hr pi (panels b), 30 hr
pi (panels c), 40 hr pi (panels d), and 50 hr pi (panels e). To analyze the subnuclear distribution
of total SV40 minichromosomes, *in situ* nuclei were fractionated into chromatin (lanes C)
and nuclear matrix (lanes NM) fractions by using SV40 noncutting enzymes (panel A). In
parallel, fractionation was carried out using DNase (this fractionation condition only detects
virion-contained SV40 minichromosomes) (panel B). Nuclear extracts, including the initial
NP-40-soluble extract (lanes N), were precipitated for SV40 DNA, followed by agarose gel
electrophoresis and ethidium bromide staining (panels b–e) or by Southern hybridization
(panels a). C and D: Lytically infected cells were pulse-labeled with 200 μCi of [³H]dThd at
30 hr pi for 10 min (panels a) and then chased for 1 hr (panels b), 4 hr (panels c), 8 hr (panels
d), or 18 hr (panels e). Pulse and pulse–chase-labeled cells were subfractionated using either
SV40-noncutting enzymes (panel C) or DNase (panel D) as described above. The fluorographic
analysis of the gel is shown. The positions of supercoiled SV40 form I DNA (I), nicked SV40
form II DNA (II), of linear SV40 form III DNA (III), of terminating LCS, and of elongating
ICS are indicated.

virion DNA (Fig. 16B). At 8 hr pi, virion-contained SV40 minichromosomes
were barely detectable in the chromatin and nuclear matrix fractions, reflecting the fact that virtually all incoming infectious virions had been
unpacked. At 20 hr pi, only few virion minichromosomes were detectable
at the nuclear matrix, indicating that at this stage of infection, encapsidation
of newly synthesized SV40 minichromosomes was still inefficient. By 30 hr
pi, the amount of virion-contained minichromosomes had increased slightly
at the nuclear matrix and encapsidated minichromosomes also became
detectable at the chromatin. After 30 hr pi, as virion assembly became

prominent, the amount of virion-contained minichromosomes increased considerably at the nuclear matrix and to a lesser extent at the chromatin.

3. Virion Formation from Nuclear Matrix-Associated Replicating SV40 DNA during the Course of Infection

To directly characterize the course of events leading from nuclear matrix-associated, replicating SV40 minichromosomes to mature virions, pulse–chase experiments were performed. Figure 16C represents the total SV40 DNA in nuclear subfractions of the infected cells, obtained after *in situ* fractionation using restriction endonucleases not cutting in SV40 DNA, whereas Fig. 16D represents SV40 DNA recovered from SV40 virions after *in situ* fractionation using DNase I. Figure 16C shows that after a 10-min pulse-label of SV40-infected cells at 30 hr pi with [³H]dTh, all replicating ICS, terminating LCS intermediates, and about 99% of newly replicated minichromosomes were associated with the nuclear matrix. The chromatin fraction contained only a few newly replicated minichromosomes, with the radiolabel specifically contained within the termination region of the SV40 DNA molecule (Schirmbeck and Deppert, 1991). No radiolabeled SV40 DNA was detectable in the soluble cytoplasmic/nucleoplasmic fraction of these cells. The transient SV40 DNA replicative intermediates (ICS and LCS) completely disappeared during a chase period of 1 hr; their disappearance was accompanied by a concomitant increase of radiolabel in mature SV40 form I DNA molecules. These newly replicated minichromosomes in their majority remained at the nuclear matrix. However, about 10 to 15% of fully replicated molecules became associated with the chromatin. The distribution of radiolabeled SV40 minichromosomes at the chromatin and the nuclear matrix then remained approximately constant during an 18-hr chase period. With longer chase periods, a significant proportion of the SV40 DNA molecules was also recovered from the cytoplasmic/nucleoplasmic fraction.

To determine the actual proportion of newly synthesized mature virions under these conditions, DNase I-resistant, virion-contained minichromosomes were analyzed (Fig. 16D). DNase I treatment resulted in a complete degradation of pulse-labeled viral DNA. However, after the 1-hr chase period, the first newly synthesized, DNase-resistant, SV40 virion-contained minichromosomes were detectable at the nuclear matrix, strongly suggesting that SV40 virion maturation proceeds at this nuclear structure (Fig. 16D). During the 4-hr chase, virion-contained minichromosomes accumulated at the nuclear matrix, but some virions now were also found at the chromatin. Accumulation of SV40 virions at the nuclear matrix and the chromatin continued between the 8- and 18-hr chase periods. Virion-contained minichromosomes therefore accounted for about 80 to 90% of the total radiolabeled SV40 form I DNA recovered from the chromatin and nuclear matrix fractions after the 18-hr chase period. The 8- and 18-hr chase periods corresponded to 38 and 48 hr pi, respectively, and thus to the stage of infection characterized by high-level virion accumulation. Therefore, the increase in SV40 DNA in the soluble cytoplasmic/nucleo-plasmic fraction at these chase times (Fig. 16C) suggests that newly encapsi-dated virions continuously leave nuclear structures. Comparison of total radioactively labeled SV40 minichromosomes (Fig. 16C) with DNase-resistant (i.e., SV40 virion-contained) minichromosomes (Fig. 16D) at the cellular chromatin revealed drastic differences in the respective amounts of SV40 DNA recovered during the 1- and 4-hr chase periods (panels b and c), indicating that during these times after replication, most SV40 DNA at the chromatin is present in unpacked minichromosomes. However, after the 8-hr chase period (panel d), labeled chromatin-associated SV40 mini-chromosomes were almost completely DNase resistant, i.e., present in ma-ture virions.

These experiments reveal that SV40 replication and virion assembly follow a distinct pattern during the course of infection, with a prominent virion maturation phase between 30 and 50 hr pi. Newly replicated SV40 minichromosomes and newly assembled SV40 virions were associated pre-dominantly with the nuclear matrix, indicating a central role for this nuclear structure in these processes. The nuclear matrix-associated elongating and terminating SV40 DNA intermediates fill up a pool of newly replicated SV40 minichromosomes at this nuclear structure. SV40 minichromosomes then could leave this nuclear matrix-associated pool directly and irreversibly after being encapsidated into virions. Alternatively, a small proportion of SV40 minichromosomes can leave this structure to transiently associate with the chromatin, where they might be involved in regulation of SV40 replication, and serve as templates for SV40 early and late transcription. Our data directly show that these newly replicated SV40 minichromosomes leave the nuclear matrix to transiently associate with the cellular chromatin, where they were detectable up to the end of the 4-hr chase period. After that period of time, the bulk of chromatin-associated SV40 minichromosomes became encapsidated into virions. A likely interpretation of this finding is that these chromatin-associated minichromosomes reassociated with the nuclear matrix between the 4- and 8-hr chase periods to become encapsi-dated at this structure, and were exchanged by freshly replicated minichro-mosomes arriving at the chromatin from the nuclear matrix. Alternatively, however, it is also possible that minichromosomes associated with the chro-matin become encapsidated there.

The postulated dynamic association of SV40 minichromosomes with the chromatin is in accordance with previously published data that showed that newly replicated SV40 DNA molecules serve as a template for viral transcription and replication for a similar period of time as we found DNase-sensitive minichromosomes being associated with the cellular chromatin (Green and Brooks, 1978). This further excludes another interpretation, namely, that the presence of virions at the chromatin at later times after infection simply reflects alterations in nuclear structure during course of infection. Our view that the cellular chromatin might be a target for actively transcribed viral minichromosomes is also supported by the observation that, during the early phase of SV40 infection, when SV40 early transcription is prominent (Tooze, 1980), minichromosomes released from infecting virions were preferentially associated with the cellular chromatin (Fig. 16A, panel a). All in all, these data suggest that SV40-specific transcription occurs on chromatin-associated SV40 DNA molecules. Although we favor the interpretation that chromatin-associated SV40 minichromosomes are actively involved in SV40 transcription, there is compelling evidence that SV40 viral transcripts are tightly bound to the nuclear matrix (Ben-Ze'ev et al., 1982). This, however, might not be a contradiction, as it is possible that transcription occurs on chromatin-bound viral minichromsomes, whereas the further processing and transport of these viral transcripts might occur at the nuclear matrix. In analogy, the possibility must be considered that the initiation step of viral DNA synthesis, i.e., the reinitiation of viral replication, occurs at chromatin-bound SV40 minichromosomes, and that initiated minichromsomes at a very early stage of replication are then rapidly translocated to the nuclear matrix. This assumption is supported by our finding that isolated *in situ* nuclear matrices are devoid of reinitiation activity in SV40 viral DNA replication (see above), a finding that correlates with the fact that nuclear matrix associated T-antigen is devoid of SV40 ORI-binding activity. In contrast, the chromatin-associated subclass of T-antigen molecules exhibited the highest level of SV40 ORI DNA-binding activity (Schirmbeck and Deppert, 1989); this activity is a prerequisite for the regulation of viral DNA transcription, and for initiation of viral replication (Rigby and Lane, 1983; DePamphilis and Bradley, 1986; Stahl and Knippers, 1987). These findings further support the concept that the various steps of viral transcription are tightly regulated at the structural level.

SV40 virion formation and release also seem to be dynamic processes, involving a complex spatial organization. The pulse–chase experiments described above revealed that newly assembled SV40 virions were first detectable at the nuclear matrix, and continuously increased in number at that nuclear substructure. Some of these nuclear matrix-associated SV40 virions were released and associated with the chromatin, or were found in the nucleoplasmic/cytoplasmic fraction. The reason for the association of virions with the cellular chromatin is not yet understood. In line with the concept of an active participation of different nuclear structures in virion formation, we favor the possibility that the association of newly formed SV40 virions with the cellular chromatin is transient and might reflect a late step in virion maturation prior to virus release. The presence of SV40 virion-contained minichromosomes in the nucleoplasm and cytoplasm correlated with the presence of mature virions in the growth medium and may be correlated with the continuous release of SV40 virions during the course of infection (von der Weth and Deppert, 1992, 1993).

These findings suggest a highly ordered release of SV40 virions via the chromatin and the nucleoplasm and cytoplasm into the growth medium (Clayson et al., 1989; van der Weth and Deppert, 1992, 1993). The continuous release of SV40 virions led to their accumulation in the growth medium until cell lysis. However, up to that point, virus released into the growth medium accounted only for about 20% of all progeny virions produced (von der Weth and Deppert, 1992). It is important to note that cell lysis resulting from the cytopathic effects of the viral infection did not significantly enhance the amount of SV40 virions in the supernatant, i.e., the majority of the virions produced during infection were not released when the cells died. One thus can assume that virus release in productively infected cells is an active process and not merely the result of cell damage, and thus that cell lysis is a consequence of the viral infection and not a prerequisite for viral release. This view is supported by our finding that rhesus LLC-MK$_2$ cells, in which SV40 establishes a persistent infection, produce significant amounts of SV40 progeny virus, which is released into the growth medium without any obvious cytopathic effect and without cell lysis (von der Weth and Deppert, 1993).

V. Concluding Remarks

Using SV40 as a model system, we have followed up the life cycle of this virus in productively infected TC7 monkey cells and analyzed which viral processes during viral replication are structurally organized. The data summarized above in our opinion provide overwhelming evidence that all prominent processes of SV40 replication proceed in association with nuclear structures. Thus, they support the concept of nuclear subcompartmentalization as a means for an efficient regulation of complex viral processes, ranging from DNA replication to virion assembly and maturation.

An important observation allowing us to make this statement was that our studies provided ample evidence for a dynamic interplay between different nuclear structures in various processes of viral replication. Thus the structural associations observed not simply reflected insolubility of viral proteins and nucleic acids during extraction. "Insolubility" induced by cell fractionation is an argument quite often encountered when the concept of structural organization of complex nuclear processes is faced with the observation that such processes, like SV40 DNA replication, also can be achieved in solution *in vitro,* after reconstitution of the purified constituents (Stillman, 1989, 1994; Hurwitz *et al.,* 1990). However, the fact that structural organization is not absolutely required for this process to work in solution *in vitro* does not necessarily argue against its structural organization *in vivo.* One argument often neglected in such considerations is efficiency. SV40 DNA replication *in vitro* appears to be rather inefficient when compared to SV40 DNA replication in *in situ* nuclei and even to SV40 DNA elongation on isolated *in situ* nuclear matrices. At the moment, this statement is solely supported by the fact that the relatively low amounts of endogenous SV40 T-antigen in *in situ* nuclei and at *in situ* nuclear matrices suffice to perform these functions, as compared to the relatively high amounts of T-antigen that must be added to comparable *in vitro* assays. Thus to substantiate the claim of an enhanced efficiency, comparable *in vitro* and *in situ* assays must be developed that allow a direct analysis of replication efficiency. Another equally important argument is regulation of this process. Our data clearly point to a "division of labor" between different structural components of the nucleus in SV40 DNA replication, with the chromatin being involved in initiation processes and elongation and termination of SV40 DNA replication occurring at the nuclear matrix. This conclusion is supported first by the different biochemical activities exerted by chromatin and nuclear matrix-associated T-antigen. Chromatin-associated T-antigen exhibits the DNA-binding properties required for initiation, i.e., the binding-to-binding site II on the SV40 ORI, and, in addition, ATPase activity reflecting the helicase activity of T-antigen that is also required for the initiation step of SV40 DNA replication (Stahl and Knippers, 1987). Thus chromatin-associated T-antigen performs the function required for site-specific initiation of this process. Nuclear matrix-associated T-antigen, on the other hand, seems to be optimized for the elongation function in SV40 DNA replication, as is indicative from its enhanced ATPase activity and its lack of sequence-specific DNA-binding activity. Thus the T-antigen-specific helicase activity of nuclear matrix-associated T-antigen is freed from the sequence-specific interaction of this molecule with DNA, a fact that might help in speeding up the elongation process. This subnuclear partitioning of "initiation" and "elongation/termination" of SV40 DNA replication might provide a means for regulating this process. In productively infected cells, the rate of SV40 DNA replication is rapidly increasing after the onset of SV40 DNA replication at about 12 to 14 hr pi in TC7 cells. The rate of SV40 DNA replication then reaches a maximum at around 30 hr pi, and then rapidly declines. This decline is not at all due to any deleterious effects of the viral infection on the cells, as cytopathic effects become apparent only at about 54 hr pi (Schirmbeck and Deppert, 1987; von der Weth and Deppert, 1992, 1993). Furthermore, the pattern of SV40 DNA replication is similar, although within a slightly different time scale, in persistently infected LLC-MK$_2$ cells, which do not show any cytopathic effects as a result of the virus infection (Norkin, 1982). Thus the rate of SV40 DNA replication is tightly regulated during the course of viral infection. It is intriguing to speculate that this regulation occurs at the level of initiation, and at the same time provides the signal for encapsidation of the SV40 minichromosomes at the nuclear matrix. The rate-limiting step for initiation of SV40 DNA replication could be provided by a shift in T-antigen activities at the chromatin or by a reduction of templates due to the reassociation of chromatin-bound SV40 minichromosomes with the nuclear matrix. This scenario clearly is still speculative, but it provides a frame for further analyses of the function of different nuclear subcompartments in SV40 DNA replication.

Like SV40 DNA replication, virion formation also is a dynamic process, involving both the nuclear matrix and the chromatin subcompartments of the nucleus. Our data demonstrate that the first DNase-resistant virion structures can be found at the nuclear matrix. However, a certain proportion of SV40 virions also are found in the chromatin fraction. We cannot exclude with certainty that the association of a certain fraction of SV40 virions with the chromatin simply reflects a partial solubilization of these virion structures during chromatin extraction. However, we favor the assumption that the association of SV40 virions with the chromatin represents a necessary step in virion maturation prior to virion release. The argument in favor of this assumption is only indirect but relates to the observation that only a small percentage of the virions formed during a productive infection in TC7 cells actually are released from the cells before cell death. This points to the possibility that SV40 virions at the nuclear matrix are not yet mature, and, therefore, are not released. Clearly, this is a testable hypothesis. It is well known that the virion proteins of SV40, especially the VP1 protein, are subject to multiple post-translational modifications whose biological functions are not yet known (O'Farrell and Goodman, 1976; Ponder et al., 1977; Streuli and Griffin, 1987). A comparative analysis of virion proteins of mature SV40 progeny virus and of virion proteins derived from the chromatin and the nuclear matrix fractions of infected cells may shed light onto this still unresolved question.

The dynamic interplay of the cellular chromatin and the nuclear matrix in various replicative functions of SV40 as demonstrated by the data reviewed above also provides a strong argument for the assumption that the preparatively defined nuclear matrix is not simply an insoluble fraction of the cellular chromatin. Even if the nuclear matrices prepared reflected an arrangement of protein complexes that became insoluble during extraction, rather than a preexisting nuclear substructure, these nuclear matrices are independent functional entities. This is best documented by the finding that the *in situ* nuclear matrices prepared from SV40-infected cells still were able to continue T-antigen-dependent SV40 DNA replication that had been initiated at the chromatin. Thus the concept of chromatin and nuclear matrix reflecting distinct functional entities of nuclear structure appears to be greatly supported by the experiments described in this review.

Because SV40 DNA replication generally is considered to be a model system for eukaryotic DNA replication, the question arises of whether the data described here bear some relevance for cellular DNA replication. Clearly, SV40 DNA replication uses all the machinery used by the cell for cellular DNA replication. Thus our finding that major steps of viral DNA replication occur at the nuclear matrix strongly supports the concept that cellular DNA replication also to a large extent is performed at this nuclear structure. A more difficult question, however, is whether the processes of initiation of viral DNA replication and its regulation, as discussed above, indeed bear relevance to the corresponding processes of cellular DNA replication. Replication of viral DNA is optimized for the needs of viral replication, i.e., to achieve a balance between the synthesis of multiple copies of viral DNA, of proteins that build up the virions, and of virion formation, and as such will require different regulatory elements than cellular DNA replication, which is optimized for a correct passing on of the cell's genetic information to its daughter cells. Thus cellular DNA synthesis occurs only once per cell cycle and is optimized to high fidelity. In contrast, fidelity of DNA replication is drastically reduced in viral DNA replication, generating a significant percentage of mutant viruses during each round of viral infection (Holland, 1992; Morse, 1994). In addition, cellular DNA replication is also controlled at the level of individual chromatin domains (or loops), a fact that seems to be closely related to the regulation of cellular differentiation (Herbomel, 1990; Earnshaw, 1991; Villarreal, 1991). Considering these differences, it appears to be questionable whether comparable regulatory mechanisms as postulated above for SV40 DNA replication also operate in cellular DNA replication. Although this at first glance may seem disappointing from the cell biologist's view, such differences may be of great benefit for the task of selectively interfering with viral replication, as it might provide a means for the development of drugs specifically interfering with viral replication at the level of its regulation by inhibiting viral, but not cellular DNA replication.

The data on SV40 replication in conjunction with nuclear substructures open up a variety of possibilities to further analyze the interactions of this virus with its different host cells. An intriguing example is the persistent infection of SV40 in the established rhesus kidney cell line LLC-MK$_2$ as a model system for the establishment of viral persistence at the cellular level.

So far, virtually nothing is known about what initiates a cytopathic effect after a viral infection. An SV40-specific cytopathic effect is clearly seen in kidney cells of the African green monkey, like in TC7 cells, and also in primary rhesus kidney cells, but is absent in LLC-MK$_2$ cells (von der Weth and Deppert, 1992). Preliminary studies from our laboratory indicate that synthesis of viral proteins and viral DNA are extremely similar in primary rhesus kidney cells and in LLC-MK$_2$ cells, suggesting that the cytopathic effect induced by SV40 does not simply result from toxic effects of viral constituents accumulating in an infected cell (A. von der Weth and W. Deppert, unpublished). It therefore is tempting to speculate that the different fate of an SV40 infection in these two types of cells is determined by different virus–host interactions. Although such a difference has not yet been found, its identification might open the possibility to interfere with viral infections at the cellular level, a possibility also providing great promise for a therapeutical interference in viral infections.

Another important application of this approach to study virus–host interactions is in the analysis of transforming infections of SV40 in nonpermissive cells. SV40 is able to infect a variety of cells of different species in which it cannot replicate. These cells allow the expression of the "early" genes of SV40, in particular the expression of the SV40 T-antigen, which is the major transforming protein of this virus (Livingston and Bradley, 1987). Most of those "abortively" infected cells express this protein only transiently, as the viral DNA becomes degraded over time. In a few cells, however, SV40 DNA becomes stably integrated into the host genome, leading to a permanent expression of T-antigen and induction of stable transformation. As stated at the beginning of this chapter, T-antigen also is subcompartmentalized in SV40-transformed cells (Staufenbiel and Deppert, 1983). By analyzing a variety of cells transformed by mutants in SV40 T-antigen, as well as a cellular revertant of an SV40 wild-type T-antigen-transformed cell line, we were able to demonstrate that the cellular chromatin is an important target for SV40 T-antigen to exert its tranforming functions (Richter and Deppert, 1990; Deppert *et al.*, 1991; Knippschild *et al.*, 1991). This study also revealed that the association of T-antigen with the chromatin, and to a lesser extent with the nuclear matrix, depends on T-antigen being in a biologically active conformation (Richter and Deppert, 1990). As this conformation seems to be controlled by specific phosphorylation events, cellular kinase and phosphatase activities play a major role in controlling the transforming potential of T-antigen (Deppert *et al.*, 1991; Knippschild *et al.*, 1991; Zerrahn and Deppert, 1993). These activities, in turn, can be modulated by T-antigen itself, which seems to be able to induce transformation-specific kinases (Scheidtmann and Harber, 1990), and by another "early" gene product of SV40, the small t-antigen, which is able to associate with phosphatase 2A (PP2A), thereby modulating its activity

(Mumby and Walter, 1991). Thus the studies on nuclear subcompartmentalization of SV40 T-antigen in SV40-transformed cells have opened up another exciting avenue for studying the intricate mechanisms leading to cellular transformation by the expression of viral oncogenes.

In summary, we hope that we were able to demonstrate that the analysis of viral interactions with nuclear structures, as exemplified here for the virus–host interactions of SV40, is an exciting field at the interface of virology and cell biology, with an enormous potential for both disciplines. If this review will get more cell biologists interested in using viruses as tools for studying cellular functions, or vice versa, more virologists interested in applying the knowledge of cell biology to understanding virus–host interactions, then it has served its purpose.

Acknowledgments

Studies performed in the authors' laboratory and summarized in this article were supported by the Deutsche Forschungs gemeiuschaft.

References

Abramczyk, J., Pan, S., Maul, G., and Knowles, B. B. (1984). Tumor induction by simian virus 40 mice is controlled by long-term persistence of the viral genome and the immune response of the host. *J. Virol.* **49,** 540–548.

Ben-Ze'ev, A., Abulafia, R., and Aloni, Y. (1982). SV40 virions and viral RNA metabolism are associated with cellular substructures. *EMBO J.* **1,** 1225–1231.

Berezney, R. (1984). Organization and functions of the nuclear matrix. *In* "Chromosomal Nonhistone Proteins" (L. S. Hnilica, ed.), Vol. 4, pp. 119–180. CRC Press, Boca Raton, FL.

Boyce, F. M., Sudin, O., Barsoum, J., and Varshavsky, A. (1982). New way to isolate simian virus 40 nucleoprotein complexes from infected cells: Use of a thiol-specific reagent. *J. Virol.* **42,** 292–296.

Clark, R., Lane, D. P., and Tjian, R. (1981). Using monoclonal antibodies as probes of simian virus 40 T antigen ATPase activity. *J. Biol. Chem.* **256,** 11854–11858.

Clayson, E. T., Jones-Brando, L. V., and Compans, R. W. (1989). Release of simian virus 40 from epithelial cells is polarized and occurs without cell lysis. *J. Virol* **63,** 2278–2288.

DeBruyn-Kops, A., and Knipe, D. M. (1988). Formation of DNA replication structures in herpes virus-infected cells requires a viral DNA binding protein. *Cell (Cambridge, Mass.)* **55,** 857–868.

DeCaprio, J. A., Ludlow, J. W., Figge, J., Shew, J.-Y., Huang, C.-M., Lee, W.-H., Marsilio, E., Paucha, E., and Livingston, D. M. (1988). SV40 large tumor antigen from a specific complex with the product of retinoblastoma susceptibility gene. *Cell (Cambridge, Mass.)* **54,** 275–283.

DePamphilis, M. L., and Bradley, M. K. (1986). Replication of SV40 and polyoma virus chromosomes. *In* "The Papovaviridae" (N. P. Salzman, ed.), Vol. I, p. 99–246. Plenum, New York.

Deppert, W., Kurth, M., Graessmann, M., Graessmann, A., and Knippschild, U. (1991). Altered phosphorylation at specific sites confers a mutant phenotype to SV40 wild-type large T antigen in a flat revertant of SV40 transformed cells. *Oncogene* **6,** 1931–1938.

Diamandopoulos, G. T. (1978). Incidence, latency, and morphologic types of neoplasms induced by simian virus 40 inoculated intravenously into hamsters of three inbred strains and one outbred stock. *J. Natl. Cancer Inst. (U.S.)* **60,** 445–449.

Earnshaw, W. C. (1991). Large scale chromosome structure and organization. *Curr. Opin. Struct. Biol.* **1,** 237–244.

Eddy, B. E., Borman, G. S., Grubbs, G. E., and Young, R. D. (1962). Identification of the oncogenic substance in rhesus monkey kidney cell cultures as simian virus 40. *Virology* **17,** 65.

Fanning, E., and Knippers, R. (1992). Structure and function of simian virus 40 large tumor antigen. *Annu. Rev. Biochem.* **61,** 55–85.

Gannon, J., and Lane, D. P. (1987). p53 and DNA polymerase α compete for binding to SV40 T antigen. *Nature (London)* **329,** 456–458.

Garber, E. A., Seidman, M., and Levine, A. J. (1978). The detection of multiple forms of SV40 nucleoprotein complexes. *Virology* **90,** 305–316.

Garber, E. A., Seidman, M., and Levine, A. J. (1980). Intracellular SV40 nucleoprotein complexes: Synthesis to encapsidation. *Virology* **107,** 389–401.

Green, M. H., and Brooks, T. L. (1978). Recently replicated simian virus 40 DNA is a preferential template for transcription and replication. *J. Virol.* **26,** 325–334.

Gurney, E. G., Tamowski, S., and Deppert, W. (1986). Antigenic binding sites of monoclonal antibodies specific for simian virus 40 large T antigen. *J. Virol.* **57,** 1168–1172.

Hancock, R. (1982). Topological organisation of interphase DNA: The nuclear matrix and other skeletal structures. *Biol. Cell* **46,** 105–122.

Harper, F., Florentin, Y., and Puvion, E. (1985). Large T antigen-rich viral DNA replication loci in SV40-infected monkey kidney cells. *Exp. Cell Res.* **161,** 434–444.

Herbomel, P. (1990). From gene to chromosome: Organization levels defined by the interplay of transcription and replication in vertebrates. *New Biol.* **2,** 937–945.

Hinzpeter, M., and Deppert, W. (1987). Analysis of biological and biochemical parameters for chromatin and nuclear matrix association of SV40 large T antigen in transformed cells. *Oncogene* **1,** 119–129.

Hinzpeter, M., Fanning, E., and Deppert, W. (1986). A new target bound DNA binding assay for SV40 large T antigen. *Virology* **148,** 159–167.

Hirt, B. (1967). Selective extraction of polyoma DNA from infected mouse cell cultures. *J. Mol. Biol.* **136,** 365–369.

Holland, J. J., ed. (1992). "Current Topics in Microbiology and Immunology," Vol. 176. Springer-Verlag, Berlin and New York.

Hurwitz, J., Dean, F. B., Kwong, A. D., and Lee, S.-H. (1990). The in vitro replication of DNA containing the SV40 origin. *J. Biol. Chem.* **265,** 18043–18046.

Knippschild, U., Kiefer, J., Patschinsky, T., and Deppert, W. (1991). Phenotype-specific phosphorylation of simian virus 40 tsA mutant large T antigens in tsA N-type and A-type transformants. *J. Virol.* **65,** 4414–4423.

Livingston, D. M., and Bradley, M. K. (1987). The simian virus 40 large T antigen. A lot packed into a little. *Mol. Biol. Med.,* **4,** 63–80.

Manos, M., and Gluzman, Y. (1985). Genetic and biochemical analysis of transformation-competent, replication-defective simian virus 40 large T mutants. *J. Virol.* **53,** 120–127.

Margolskee, R. F., and Nathans, D. (1984). Simian virus 40 mutant T antigens with relaxed specifity for the nucleotide sequence at the viral origin of replication. *J. Virol.* **49,** 386–392.

Mitchell, P. J., Wang, C., and Tjian, R. (1987). Positive and negative regulation of transcription in vitro: Enhancer-binding protein AP-2 is inhibited by SV40 T antigen. *Cell (Cambridge, Mass.)* **50,** 847–861.

Mole, S. E., Gannon, J., Ford, M. J., and Lane, D. P. (1987). Structure and function of SV40 T antigen. *Philos. Trans. R. Soc. London, Ser. B* **317**, 455–469.

Morse, S. S., ed. (1994). "The Evolutionary Biology of Viruses." Raven Press, New York.

Mumby, M. C., and Walter, G. (1991). Protein phosphatases and DNA tumor viruses: Transformation through the back door? *Cell Regul.* **2**, 589–598.

Myers, R. M., Rio, D. C., Robbins, A. K., and Tjian, R. (1981). SV40 gene expression is modulated by cooperative binding of T antigen to DNA. *Cell (Cambridge, Mass.)* **25**, 373–384.

Nakayasu, H., and Berezney, R. (1989). Mapping replicational sites in the eukaryotic cell nucleus. *J. Cell Biol.* **108**, 1–11.

Nathans, D., and Danna, K. J. (1972). Specific origin in SV40 DNA replication *Nature (London)*, New Biol. **236**, 200–202.

Nelson, W. G., Pienta, K. J., Barrack, E. R., and Coffey, D. S. (1986). The role of the nuclear matrix in the organization and function of DNA. *Annu. Rev. Biophys. Chem.* **15**, 457–475.

Norkin, L. C. (1982). Papoviral persistent infections. *Microbiol. Rev.* **46**, 384–425.

O'Farrell, P. Z., and Goodman, H. M. (1976). Resolution of simian virus 40 proteins in whole cell extracts by two-dimensional electrophoresis: Heterogeneity of the major capsid protein. *Cell (Cambridge, Mass.)* **9**, 289–298.

Pommier, Y., Capranico, G., Orr, A., and Kohn, K. W. (1991). Distribution of topoisomerase II cleavage sites in simian virus 40 DNA and the effects of drugs. *J. Mol. Biol.* **222**, 909–924.

Ponder, B. A., Robbins, A. K., and Crawford, L. V. (1977). Phosphorylation of polyoma and SV40 virus proteins. *J. Gen. Virol.* **37**, 75–83.

Prives, C. (1990). The replication functions of SV40 T antigen are regulated by phosphorylation. *Cell (Cambridge, Mass.)* **61**, 735–738.

Rainwater, R., and Mann, K. (1991). Association of topoisomerases I and II with the chromatin in SV40-infected monkey cells. *Virology* **181**, 408–411.

Richter, W., and Deppert, W. (1990). The cellular chromatin is an important target for SV40 large T antigen in maintaining the transformed phenotype. *Virology* **174**, 543–556.

Rigby, P. W. J., and Lane, D. P. (1983). Structure and function of simian virus 40 large T antigen. *Adv. Viral Oncol.* **3**, 31–57.

Rio, D. C., Robbins, A. K., Myers, R. M., and Tjian, R. (1980). Regulation of simian virus 40 early transcription in vitro by a purified tumor antigen. *Proc. Natl. Acad. Sci. U.S.A.* **77**, 5706–5710.

Scheidtmann, K. H., and Haber, A. (1990). Simian virus 40 large T antigen induces or activates a protein kinase which phosphorylates the transformation-associated protein p53. *J. Virol.* **64**, 672–679.

Schirmbeck, R., and Deppert, W. (1987). Specific interaction of simian virus 40 large T antigen with cellular chromatin and nuclear matrix during the course of infection. *J. Virol.* **61**, 3561–3569.

Schirmbeck, R., and Deppert, W. (1988). Analysis of mechanisms controlling the interactions of SV40 large T antigen with the SV40 ORI region. *Virology* **165**, 527–538.

Schirmbeck, R., and Deppert, W. (1989). Nuclear subcompartmentalization of simian virus 40 large T antigen: Evidence for an vivo regulation of biochemical activities. *J. Virol.* **63**, 2308–2316.

Schirmbeck, R., and Deppert, W. (1991). Structural topography of simian virus 40 DNA replication. *J. Virol.* **65**, 2578–2588.

Schirmbeck, R., von der Weth, A., and Deppert, W. (1993). Structural requirements for simian virus 40 replication and virion maturation. *J. Virol.* **67**, 894–901.

Seidman, M., Garber, E. A., and Levine, A. J. (1979). Parameters affecting the stability of SV40 virions during the extraction of nucleoprotein complexes *Virology* **95**, 256–259.

Sewaga, M., Sugano, S., and Yamaguchi, N. (1980). Association of simian virus 40 T antigen with replicating nucleoprotein complexes of simian virus 40. *J. Virol.* **35**, 320–330.

Shortle, D. R., Margolskee, R. F., and Nathans, D. (1979). Mutational analysis of the simian virus 40 replicon: Pseudorevertants of mutants with defective replication origin. *Proc. Natl. Acad. Sci. U.S.A.* **76,** 6128–6131.

Smale, S. T., and Tjian, R. (1986). T antigen-DNA polymerase α complex implicated in SV40 DNA replication. *Mol. Cell. Biol.* **6,** 4077–4087.

Snapka, R. M. (1986). Topoisomerase inhibitors can selectively interfere with different stages of simian virus 40 DNA replication. *Mol. Cell. Biol.* **6,** 4221–4227.

Stahl, H., and Knippers, R. (1983). Simian virus 40 large tumor antigen on replicating viral chromatin: Tight binding and localization on the viral genome. *J. Virol.* **47,** 65–76.

Stahl, H., and Knippers, R. (1987). The simian virus 40 large tumor antigen. *Biochim. Biophys. Acta* **910,** 1–10.

Staufenbiel, M., and Deppert, W. (1982). Intermediate filament systems are collapsed onto the nuclear surface after isolation of nuclei from tissue culture cells. *Exp. Cell Res.* **138,** 207–214.

Staufenbiel, M., and Deppert, W. (1983). Different structural systems of the nucleus are targets for SV40 large T antigen. *Cell (Cambridge, Mass.)* **33,** 173–181.

Staufenbiel, M., and Deppert, W. (1984). Preparation of nuclear matrices from cultured cells: Subfractionation of nuclei *in situ. J. Cell Biol.* **98,** 1886–1894.

Stillman, B. (1989). Initiation of eucaryotic DNA replication in vitro. *Annu. Rev. Cell Biol.* **5,** 197–245.

Stillman, B. (1994). Smart machines at the DNA replication fork. *Cell (Cambridge, Mass.)* **78,** 725–728.

Streuli, C. H., and Griffin, B. E. (1987). Myristic acid is coupled to a structural protein of polyoma virus and SV40. *Nature (London)* **326,** 619–622.

Su, R. T., and DePamphilis, M. L. (1978). Simian virus 40 DNA replication in isolated replicating viral chromosomes. *J. Virol.* **28,** 53–65.

Sudin, O., and Varshavsky, A. (1980). Terminal stages of SV40 DNA replication proceed via multiply intertwined catenated dimers. *Cell (Cambridge, Mass.)* **21,** 103–114.

Sudin, O., and Varshavsky, A. (1981). Arrest of segregation leads to accumulation of replication. *Cell (Cambridge, Mass.)* **25,** 659–669.

Tack, L. C., and DePamphilis, M. L. (1983). Analysis of simian virus 40 chromosome-T-antigen complexes: T-antigen is preferentially associated with early replicating DNA intermediates. *J. Virol.* **48,** 281–295.

Tapper, D. P., Anderson, S., and DePamphilis, M. L. (1979). Maturation of replicating simian virus 40 DNA molecules in isolated nuclei by continued bidirectional replication to the normal termination region. *Biochim. Biophs. Acta* **565,** 84–97.

Tapper, D. P., Anderson, S., and DePamphilis, M. L. (1982). Distribution of replicating SV40 DNA in intact cells and its maturation in isolated nuclei. *J. Virol.* **41,** 877–892.

Tegtmeyer, P. (1972). Simian virus 40 desoxyribonucleic acid synthesis: The viral replicon. *J. Virol.* **10,** 591–598.

Tegtmeyer, P., and Ozer, H. L. (1971). Temperature-sensitive mutants of simian virus 40: Infection of permissive cells. *J. Virol.* **8,** 516–524.

Tegtmeyer, P., Robb, J. A., Widmer, C., and Ozer, H. L. (1974). Altered protein metabolism in infection by late tsB11 mutant of simian virus 40. *J. Virol.* **14,** 997–1007.

Tegtmeyer, P., Schwartz, M., Collins, J. K., and Rundell, K. (1975). Regulation of tumor antigen syntheisis by simian virus 40 gene A. *J. Virol.* **16,** 168–178.

Tooze, J., ed. (1980). Molecular biology of tumor viruses. *In* "DNA Tumor viruses," Part 2. Cold Spring Harbor Lab., Cold Spring Harbor, NY.

Van der Velden, H. M. W., and Wanka, F. (1987). The nuclear matrix- Its role in the spatial organization and replication of eukaryotic DNA. *Mol. Biol. Rep.* **12,** 69–77.

Villarreal, L. P. (1991). Relationship of eukaryotic DNA replication to committed gene expression: General theory for gene control. *Microbiol. Rev.* **55,** 512–542.

Voelkerding, K., and Klessig, D. F. (1986). Identification of two nuclear subclasses of adenovirus type 5-encoded DNA- binding protein. *J. Virol.* **60,** 353–362.

von der Weth, A., and Deppert, W. (1992). Lytic infection of primary rhesus kidney cells by simian 40. *Virology* **189,** 334–339.

von der Weth, A., and Deppert, W. (1993). Wild-type p53 is not a negative regulator of simian virus 40 DNA replication in infected monkey cells. *J. Virol.* **67,** 886–893.

Wang, H.-T., and Roman, A. (1981). Cessation of reentry of simian virus 40 DNA into replication and its simultaneous appearance in nucleoprotein complexes of the maturation pathway. *J. Virol.* **39,** 255–262.

Weaver, D. T., Fields-Berry, S. C., and DePamphilis, M. L. (1985). The termination region for SV40 DNA replication directs the mode of separation of the two sibling molecules. *Cell (Cambridge, Mass.)* **41,** 565–575.

Wilson, V. G., Tevethia, M. J., Lewton, A., and Tegtmeyer, P. (1982). DNA binding properties of simian virus 40 temperature sensitive A proteins. *J. Virol.* **44,** 458–466.

Wobbe, C. R., Dean, F. B., Murakami, Y., Weissbach, L., and Hurwitz, J. (1986). Simian virus 40 DNA replication in vitro: Study of events preceeding elongation of chains. *Proc. Natl. Acad. Sci. U.S.A.* **83,** 4612–4616.

Yang, L., Wold, M. S., Li, J. J., Kelly, T. J., and Liu, L. F. (1987). Roles of DNA topoisomerases in simian virus 40 DNA replication in vitro. *Proc. Natl. Acad. Sci. U.S.A.* **84,** 950–954.

Zerrahn, J., and Deppert, W. (1993). Analysis of simian virus 40 small t antigen induces progression of rat F111 cells minimally transformed by large T antigen. *J. Virol.* **67,** 1555–1563.

Zerrahn, J., Knippschild, U., Winkler, T., and Deppert, W. (1993). Independent expression of the transforming amino-terminal domain of SV40 large T antigen from an alternatively spliced third SV40 early mRNA. *EMBO J.* **12,** 4739–4746.

The Nuclear Matrix as a Site of Anticancer Drug Action

D. J. Fernandes and C. V. Catapano

Department of Experimental Oncology, Hollings Cancer Center, Medical University of South Carolina, Charleston, South Carolina 29425

Many nuclear functions, including the organization of the chromatin within the nucleus, depend upon the presence of a nuclear matrix. Nuclear matrix proteins are involved in the formation of chromatin loops, control of DNA supercoiling, and regulation and coordination of transcriptional and replicational activities within individual loops. Various structural and functional components of the nuclear matrix represent potential targets for anticancer agents. Alkylating agents and ionizing radiation interact preferentially with nuclear matrix proteins and matrix-associated DNA. Other chemotherapeutic agents, such as fludarabine phosphate and topoisomerase II-active drugs, interact specifically with matrix-associated enzymes, such as DNA primase and the DNA topoisomerase IIα isozyme. The interactions of these agents at the level of the nuclear matrix may compromise multiple nuclear functions and be relevant to their antitumor activities.

KEY WORDS: Nuclear matrix, Anticancer agents, Alkylating agents, Ionizing radiation, fludarabine phosphate, Mizoribine, Mycophenolic acid, Topoisomerase II-active drugs.

I. Introduction

The presence of a nuclear protein matrix in eukaryotic cells was first reported by Berezney and Coffey (1974). These investigators subjected isolated nuclei from regenerating rat liver to endogenous nuclease digestion and repeated extractions with salt and nonionic detergents. These treatments removed the bulk of the chromatin and the majority of histones leaving a residual nuclear structure, the nuclear matrix. The isolated nuclear matrix contained approximately 10% of the total nuclear proteins, 30% of the total nuclear RNA, 2 to 9% of the total nuclear phospholipid, and only

1 to 3% of the total nuclear DNA (Berezney and Coffey, 1974; Hakes and Berezney, 1991). Alternative procedures for chromatin extraction and isolation of the underlying nuclear framework from nuclei or whole cells were later developed (Fey *et al.*, 1984; Mirkovitch *et al.*, 1984; Gasser and Laemmli, 1986b; Jackson *et al.*, 1988, 1990; He *et al.*, 1990). The nuclear structures isolated by different methods were morphologically and biochemically quite similar and consisted of a subset of nuclear proteins, DNA and RNA.

Studies done within the past two decades have shown that the organization of chromatin and many nuclear functions depends upon the presence of this proteinaceous framework. The nuclear matrix provides the sites for many biochemical reactions that take place inside the nucleus, including replication, transcription, and RNA processing (Berezney, 1984). DNA replicative enzymes and their accessory proteins, various enzymes involved in RNA transcription and splicing, and transcriptional regulatory factors have been reported to be compartmentalized inside the nucleus through their association with the nuclear matrix (Berezney, 1991; Carter *et al.*, 1991; Cook, 1991; Getzenberg *et al.*, 1991; Stein *et al.*, 1991; Xing and Lawrence, 1991). Furthermore, the nuclear matrix has a major role in directing the functional organization of chromatin in eukaryotic cells. The genome of an eukaryotic cell is organized into 50,000 or more replication units and each unit replicates in precise order relative to the others and to the time into S-phase (Jackson, 1990; Berezney, 1991). Actively transcribed genes are thought to be similarly organized in transcriptional units, which are made accessible to the transcription complexes in a specific and coordinate fashion in different cell types (Carter *et al.*, 1991; Getzenberg *et al.*, 1991; Stein *et al.*, 1991; Xing *et al.*, 1993). Replication and transcriptional units may correspond physically to the chromatin loop domains that are formed by binding of DNA to the nuclear matrix. Nuclear matrix proteins that mediate the attachment of the chromatin loops to the nuclear matrix may serve to control and coordinate replication and transcription of genes within the individual chromatin units (Getzenberg *et al.*, 1991; Stein *et al.*, 1991; Dworetsky *et al.*, 1992; Bidewell *et al.*, 1993).

Various structural and functional components of the nuclear matrix represent potential targets for anticancer agents. Recent studies have shown that anticancer agents, such as certain antimetabolites and topoisomerase II-active drugs, inhibit preferentially nuclear matrix-associated enzymes. Ionizing radiation and alkylating agents are also known to interact with matrix-associated DNA and nuclear matrix proteins. In addition to inhibiting specific functions, the interaction of these anticancer agents with nuclear matrix targets may affect the structural integrity of the nuclear matrix and interfere with the organization of chromatin into matrix-attached DNA loops (Fig. 1). Because of the critical role of the nuclear matrix in chromatin organization and nuclear function, loss of structural integrity of the nuclear

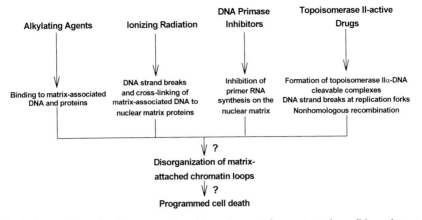

FIG. 1   Interactions of anticancer agents with nuclear matrix targets and possible pathways of activation of programmed cell death.

matrix may represent an important and irreversible step in the process of cell death.

In light of these considerations, we will review certain aspects of chromatin organization and specific matrix-associated activities that are relevant to our understanding of anticancer drug action. The interactions of certain anticancer agents with specific nuclear matrix targets and the relevance of the effects on chromatin organization to the cytotoxic activity of these agents will be discussed.

II. Nuclear Matrix and Chromatin Organization

The basic unit of eukaryotic chromatin is the nucleosome, which consists of a complex of DNA and small basic proteins, the histones (Nelson *et al.,* 1986a). The chromatin fiber is organized into repeating loops with an average size of 50–100 kb (Nelson *et al.,* 1986a). The nuclear matrix is essential to the organization of chromatin loops, which are formed and stabilized by attachment of the bases of the loops with the nuclear matrix (Cook and Brazell, 1976; Vogelstein *et al.,* 1980). The association of chromatin loops with the nuclear matrix maintains the DNA in the loops in a highly supercoiled state (Cook and Brazell, 1976; Vogelstein *et al.,* 1980). The maintenance of a proper level of supercoiling in the DNA loops is important for a number of nuclear processes, including DNA replication, transcription, and recombination (Vogelstein *et al.,* 1980; DiNardo *et al.,* 1984a,b; Wu *et al.*, 1988; Wang *et al.*, 1990). In addition to allowing the formation of chromatin loops and determining the topological state of the DNA in the loops, matrix attachment region (MAR)–matrix interactions are important in regulating the transcriptional and replicational activity within individual loops.

This review will describe first some of the recent studies that have provided insights into the basis of the organization of chromatin into matrix-attached DNA loops. The results of these studies indicate the importance of this precise organization for nuclear function. It is reasonable to assume that any alteration in the organization of chromatin into matrix-attached DNA loops may result in profound effects on cell growth and cell survival. Agents that interfere with the formation and function of the matrix attachment sites may eventually compromise the organization of chromatin loops. As will be described in the following sections, ionizing radiation induces breaks in matrix-associated DNA and cross-linking of DNA to nuclear matrix proteins. Likewise, alkylating agents bind preferentially to matrix-associated DNA and nuclear matrix proteins. Inhibition of specific matrix-associated functions, such as primer RNA synthesis or topoisomerase II activity, may also result in secondary effects on chromatin structure. All these interactions have the potential to interfere with the organization of matrix-attached chromatin loops and may be relevant to the cytotoxic effects of these anticancer agents.

A. Matrix Attachment Sites

The association of chromatin loops to the matrix is not random. Defined DNA sequences (MARs) mediate the attachment of chromatin loops to the nuclear matrix. MARs have been identified in a number of eukaryotic loci and consist of relatively large DNA fragments of 0.3 to 1 kb in length (Mirkovitch *et al.*, 1984; Gasser and Laemmli, 1986a; Bode and Maass, 1988; Dijkwel and Hamlin, 1988; Phi-Van and Strätling, 1988; Levy-Wilson and Fortier, 1990). Characteristic of MARs is the presence of clusters at AT-rich regions. These AT-rich sequences confer unique properties to MAR-containing fragments, including a narrow minor groove (Nelson *et al.*, 1987; Adachi *et al.*, 1989; Käs *et al.*, 1989) and a strong potential for base unpairing under superhelical strain (Kohwi-Shigematsu and Kohwi, 1990; Bode *et al.*, 1992). These structural features, rather than specific DNA sequences, are apparently important for MAR functions.

An AT-rich sequence (AATATATTT), which has been found in the MARs flanking the human β-interferon and immunoglobulin heavy chain gene, has strong base-unpairing property (Kohwi-Shigematsu and Kohwi, 1990; Bode *et al.*, 1992). This core unwinding element may participate in controlling the topological state of chromatin loops and gene expression. Negative supercoiling in the DNA loop might be relieved by DNA unwinding at the level of the core element (Bode et al., 1992). Synthetic oligonucleotides that contain the core unwinding element are capable of unwinding under superhelical strain, exhibit a strong affinity for isolated nuclear matrices, and increase promoter activity when inserted in expression vectors (Bode et al., 1992). Mutations that eliminate the base-unpairing ability and reduce the association of the oligonucleotides with the nuclear scaffold also affect their transcription enhancing activity (Bode et al., 1992).

Studies with the antibiotic distamycin A have shown the importance of the narrow minor groove of AT-rich sequences for the binding of MARs to nuclear matrix proteins. Distamycin A binds specifically to the minor groove of DNA containing (dA)·(dT) sequences and prevents binding of MARs to isolated nuclear matrices (Käs et al., 1989). This agent also inhibited binding of topoisomerase II (Adachi et al., 1989) and lamins (Ludérus et al., 1994) to MARs in vitro. Interestingly, Käs et al. (1993) found that incubation of intact cells or isolated nuclei with distamycin A enhanced the cleavage of MARs mediated by topoisomerase II and restriction enzymes. These results suggested that distamycin may induce a more open conformation of matrix-associated chromatin regions and increase the accessibility to topoisomerase II and restriction endonucleases (Käs et al., 1993). Distamycin A and its derivatives have cytotoxic activity and have been studied as potential anticancer agents (Broggini et al., 1991). The relationship between the effects on nuclear matrix organization and the anticancer activity of these agents needs to be further investigated.

B. MAR-Binding Proteins

Nuclear matrix protein composition is extremely complex as revealed by two-dimensional gel electrophoresis (Fey and Penman, 1988; Stuurman et al., 1990; Nakayasu and Berezney, 1991). Proteins that are potentially involved in anchoring the DNA loops to the nuclear matrix have been isolated based upon their ability to bind MARs (von Kries et al., 1991; Romig et al., 1992; Tsutsui et al., 1993). These proteins are present in large amounts in nuclear matrix preparations from various tissues and species. Two sets of evidence suggest that these proteins are important for the generation of functional chromatin loops in vivo. The DNA binding properties of two of these proteins, ARBP (von Kries et al., 1991) and SAF-A (Romig et al., 1992), mimic exactly those of unfractionated matrix preparations. In addition, both SAF-A and ARBP are able to mediate formation of DNA loops structures in vitro (von Kries et al., 1991; Romig et al., 1992). All the MAR-binding proteins isolated thus far interact with the DNA in a sequence-independent manner, recognizing instead particular structural features of the AT-rich MAR elements (von Kries *et al.*, 1991; Romig *et al.*, 1992; Ludérus *et al.*, 1994). For example, a protein, isolated by Dickinson *et al.* (1992) and named SATB1 (for special AT-rich sequence binding protein 1), binds exclusively to a particular subset of AT-rich sequences with an asymmetric distribution of C and G residues among the two strands of DNA.

Recently, it has been reported that A-type (Ludérus *et al.*, 1994) and B-type lamins (Ludérus *et al.*, 1992) have MAR-binding activity. Attachment of chromatin loops to these components of the peripheral nuclear lamina as opposed to binding to proteins of the internal nuclear matrix may serve to physically segregate transcriptionally active and inactive chromatin domains in these two distinct nuclear subcompartments (Ludérus *et al.*, 1992).

C. Functional Properties of Matrix Attachment Sites

Recent genetic studies have given new insights into the physiological significance of matrix attachment sites in chromatin loop organization and control of nuclear functions (Bonifer *et al.*, 1991). Most of the MARs thus far identified are located at the boundaries of functional transcription units, suggesting that MAR and MAR-binding proteins may participate in transcriptionally regulatory mechanisms (Mirkovitch *et al.*, 1984; Gasser and Laemmli, 1986a; Bode and Maass, 1988; Dijkwel and Hamlin, 1988; Phi-Van and Strätling, 1988; Levy-Wilson and Fortier, 1990). This hypothesis has been tested by constructing artificial chromatin domains in which a reporter gene flanked by MAR elements was inserted into a transfectable expression vector. The level of expression of stably transfected genes is generally variable among different clones carrying identical constructs. The variability depends upon the particular site of integration of the transfected gene in chromosomal DNA and is known as position effect. The presence of MARs in the artificial chromatin domain constructs increased the level of expression of reporter gene compared to constructs without MARs (Stief *et al.*, 1989; Phi-Van *et al.*, 1990; Klehr *et al.*, 1991). Moreover, MAR elements in the constructs decreased the variability in the level of gene expression among the different clones (Stief *et al.*, 1989; Phi-Van *et al.*, 1990; Klehr *et al.*, 1991). MARs exert these effects only after stable integration in the genome and not in transiently transfected cells (Stief *et al.*, 1989; Klehr *et al.*, 1991). Deletion experiments, in which MARs were removed from genes that normally contained these sequences, have confirmed the regulatory role of MARs in gene expression (Blasquez *et al.*, 1989; Xu *et al.*, 1989). MARs also conferred position-independent expression in transgenic mice (Bonifer *et al.,* 1990). Gene constructs containing the flanking MARs functioned as independent genetic loci when introduced into a heterologous genome (Bonifer *et al.,* 1990). These results suggested that MARs acted as locus-controlling regions (LCR) and reduced the dependence of the expression of the transfected genes on the site of chromosomal integration (Bonifer *et al.,* 1991). MARs shielded the transfected genes from the negative or positive influences of long range *cis*-acting transcriptional regulatory elements, like enhancers and silencers.

The studies described above indicate that matrix attachment sites are important structural and functional elements of nuclear organization. Independent, topologically constrained chromatin units are generated by the attachment of DNA loops to the nuclear matrix. Binding of DNA to matrix attachment sites prevents the transmission of superhelical strain from one chromatin domain to another. Each domain is, thus, isolated and independent from the regulatory mechanisms that control transcription and replication in neighboring domains. Cell type- and differentiation stage-dependent gene expression and replication timing depend upon the organization of chromatin in matrix-attached DNA loops. Nuclear matrix proteins that bind to gene regulatory elements appear as major factors in regulating gene expression and replication at the level of entire chromatin units. Damage at the level of matrix attachment sites would interfere with chromatin organization and be a common mechanism of cytotoxicity for certain anticancer agents that interact preferentially with nuclear matrix targets (Fig. 1).

III. Alkylating Agents

A. Chloroethylnitrosoureas

The chloroethylnitrosoureas have clinical activity against lymphomas, malignant melanomas, and brain tumors. Under physiological conditions, chloroethylnitrosoureas spontaneously decompose in two highly reactive species, alkyldiazohydroxide and isocyanate (Montgomery *et al.,* 1967). These products react with nuclear macromolecules and induce a variety of lesions, including monofunctional alkylation of nucleophilic sites of nucleic acids and proteins, DNA inter- and intrastrand crosslinks, DNA–protein crosslinks, and carbamylation of proteins (Kohn, 1977). Induction of DNA interstrand crosslinks is thought to be the major form of damage responsible for the cytotoxic effects of chloroethylnitrosoureas (Kohn, 1977). The amount of DNA interstrand crosslinks is proportional to the cell killing effects of chloroethylnitrosoureas both *in vitro* and *in vivo* (D'Incalci *et al.,* 1988). Moreover, increased levels of the DNA repair enzyme, O[6]-

alkyltransferase, which is able to repair the initial monoadduct on the O^6 of guanine and prevent formation of DNA interstrand crosslinks, have been found to correlate with reduced sensitivity of tumor cells to chloroethylnitrosoureas (D'Incalci et al., 1988).

These correlations suggest that the efficacy of the treatment with chloroethylnitrosoureas is mainly determined by the type and number of drug-induced lesions and the ability of the cell to prevent or repair the damage. However, other factors, such as the localization of drug-induced lesions in certain regions of the DNA or chromatin, may be relevant to the induction of cell death. Certain DNA sequences or regions in the chromatin may be more susceptible to alkylation or are more critical targets for cell survival. In this regard, the nuclear matrix, which is enriched in transcriptionally active chromatin and replication forks, may represent an important target for alkylating agents. Tew (1982) and Tew et al. (1983a) have shown that following incubation of HeLa cells with chloroethylnitrosoureas, such as chlorozotocin and 1-(2-chloroethyl-3-cyclohexyl)-1-nitrosourea (CCNU), approximately 30% of the nuclear-associated drug is covalently bound to the nuclear matrix. This represents a highly preferential interaction of these agents with the nuclear matrix, since HeLa cell matrix preparations contain only 5% of total nuclear protein, 5% of RNA, and 1–2% of the DNA (Tew et al., 1983a). The reasons for the greater susceptibility to alkylation of the nuclear matrix as compared to bulk chromatin may lie in the enrichment of replication and transcription sites on the matrix (Ciejek et al., 1983; Berezney, 1984). Earlier studies indicated that actively transcribed regions of the chromatin were more susceptible to damage by alkylating agents (Tew et al., 1978). Sensitivity to nucleases confirmed that alkylation was more pronounced in regions of active chromatin, which had a more open conformation than inactive chromatin. Furthermore, corticosteroid hormones increased the sensitivity of the cells to chloroethylnitrosoureas by inducing changes in chromatin organization (Tew, 1982). Corticosteroids stimulated transcriptional activity in HeLa S3 cells with consequent induction of a more open chromatin conformation. The transition to an open conformation of the active chromatin was confirmed by the increased sensitivity to micrococcal nuclease. A parallel increase in alkylation and carbamylation was observed in cells incubated with chloroethylnitrosoureas in the presence of corticosteroids, and this resulted in decreased drug binding to nuclear matrix sites and greater cytotoxicity (Tew, 1982). A possible mechanism by which alkylation at the level of the nuclear matrix may contribute to the cytotoxic activity of chloroethylnitrosoureas was suggested by the observation that alkylation of DNA with either chlorozotocin or CCNU reduced the ability of the DNA to bind to HeLa cell matrix preparations in vitro (Tew et al., 1983a). Thus, drug-induced alkylation at the level of matrix attachment sites may disrupt the association of chromatin loop domains with the matrix and have profound consequences on the integrity of various nuclear functions (Fig. 1).

B. Estramustine Phosphate

Estramustine phosphate is an anticancer drug in which the alkylating agent nitrogen mustard is conjugated to the hormone estradiol. Studies have shown that the cytotoxic activity of estramustine does not depend upon decomposition of the conjugate to estradiol and nitrogen mustard (Tew *et al.*, 1983b; Tew and Stearns, 1987). As a result, estramustine acts as neither a steroid nor an alkylating agent in producing its cytotoxic effects (Tew *et al.*, 1983b). The drug is active in estradiol receptor-negative cells and does not induce DNA damage (Tew *et al.*, 1983b). Estramustine binds to nuclear matrix proteins (Tew *et al.*, 1983b; Hartley-Asp and Kruse, 1986). Approximately 40% of the total nuclear estramustine was associated with nuclear matrix proteins, which represented only 5% of the total nuclear proteins (Tew *et al.*, 1983b). The reasons for the preferential binding of estramustine to the nuclear matrix are not known. Drug binding to the matrix was independent of the presence of estradiol receptor sites, since it occurred in HeLa cells that lack of estradiol receptors (Tew *et al.*, 1983b). The importance of the interaction with the nuclear matrix to the therapeutic activity of this agent remains to be demonstrated. Recently, Pienta and Lehr (1993) have shown that binding of estramustine to nuclear matrix sites enhanced the ability of the topoisomerase II-targeted drug, VP-16, to inhibit DNA synthesis. This effect was shown to result in increased cytotoxic activity of VP-16 against prostate cancer cells both *in vitro* and *in vivo* (Pienta and Lehr, 1993). The combination of estramustine and VP-16 is now under investigation in a phase I/II clinical trial for patients with hormone-refractory prostate cancer (Pienta and Lehr, 1993).

IV. Ionizing Radiation

Exposure of cells to ionizing radiation results in the formation of single- and double-strand breaks, various forms of base damage, and DNA–protein crosslinks. Several lines of evidence suggest that ionizing radiation may preferentially damage regions in proximity of matrix-bound replication and transcription sites. Actively transcribed DNA sequences sustain 5 to 6 times more strand breaks after irradiation than does bulk DNA, possibly due to the increased accessibility of active chromatin to locally generated radical species (Chiu and Oleinick, 1982; Chiu *et al.*, 1982; Oleinick *et al.*, 1984).

Unfolding of DNA by removal of magnesium ions or histone H1 from the chromatin, for example, has been shown to result to result in increased susceptibility to radiation damage (Heussen et al., 1987; Nackerdien et al., 1989). DNA replication forks also contain regions of single-stranded DNA that are not protected by histones, which may render them highly sensitive to ionizing radiation (DePamphilis and Wassarman, 1980; Berezney, 1984). Chiu et al. (1986) have reported that, in addition to DNA strand breaks, ionizing radiation induces DNA–protein crosslinks preferentially at the level of the nuclear matrix. Crosslinking between a subset of nuclear matrix proteins and the associated DNA occurs at a frequency 10–16 times higher than that between protein and DNA in the nonmatrix fraction of the nucleus. The reasons for the preferential crosslinking of DNA and protein in the nuclear matrix are not clear. In contrast to nuclear matrix-bound DNA, more of the nonmatrix chromatin is highly condensed and, thus, a poor substrate for the formation of radiation-induced DNA–protein crosslinks. When chromatin from isolated nuclei is expanded in buffers of decreasing ionic strength, the number of DNA–protein crosslinks is greatly increased, suggesting that protected regions of chromatin become exposed to radiation-induced DNA damage in the expanded state (Chiu et al., 1992). In addition, radiation may induce crosslinking of DNA and proteins in the nuclear matrix by stabilizing preexisting interactions between matrix proteins and DNA (Oleinick et al., 1987). It has recently been proposed that copper ions, which may play an important role in the stabilization and maintenance of nuclear matrix organization, mediate radiation-induced damage of nuclear matrix-associated DNA (Chiu et al., 1993). DNA sequences that are bound either directly or indirectly to copper ions, such as the MARs, would be the preferential target of hydroxy radicals generated by copper ions via a Fenton-type reaction (Chiu et al., 1993).

Radiation-induced damage at the level of the MARs is cytotoxic by numerous mechanisms. DNA strand breaks and DNA–protein crosslinks induced by ionizing radiation in matrix-associated DNA may represent critical lesions, since they may block progression of transcription and replication complexes, or induce chromatin disorganization as a result of the detachment of chromatin loops from the nuclear matrix (Fig. 1). Matrix attachment sites isolate chromatin loop domains and prevent transmission of superhelical strain to and from adjacent loops. Alterations of the affinity of MARs to nuclear matrix attachment sites may affect the sensitivity of tumor cells to radiation-induced DNA damage (Vaughan et al., 1991; Milner et al., 1993). Two human squamous cell carcinoma cell lines, which differ in sensitivity to ionizing radiation, have been studied (Milner et al., 1993). The radiosensitive SQ-9G cells exhibit in vivo a weaker association of DNA with the nuclear matrix than the radioresistant SQ-20B cells. The looser association between nuclear matrix proteins and DNA at the matrix attachment sites in SQ-9G cells may allow propagation of torsional energy from a damaged loop to the otherwise intact adjacent loops, thus amplifying the effect of radiation-induced DNA strand breaks. In the case of SQ-20B cells, which exhibit a tighter association between DNA and the nuclear matrix, propagation of superhelical strain would be blocked at the level of the MARs. Interestingly, pretreatment of the cells with the topoisomerase II inhibitor VP-16, which induces formation of a covalent complex between DNA and topoisomerase II, results in an apparent radioprotective effect in both cell lines (Milner *et al.*, 1993). Formation of covalent complexes between matrix-bound topoisomerase II and DNA may block propagation of damage-induced superhelical strain through the MARs.

V. Antimetabolites

A. Association of DNA Replication Sites with the Nuclear Matrix

The nuclear matrix has an important role in the assembly of the replication complexes and in their organization and coordination into multiple replicon clusters. The proposed model of matrix-associated DNA replication sites has provided the basis for the study of the effects of certain anticancer agents (e.g., antimetabolites, topoisomerase II inhibitors) on DNA replication. Berezney and Coffey (1975) were the first to report the preferential association of nascent DNA with the nuclear matrix. These investigators showed that newly replicated DNA was highly enriched on the nuclear matrix of regenerating rat liver following a short pulse with radiolabeled DNA precursors. With increasing pulse time, there was a progressive decrease of the relative specific activity of the nascent DNA attached to the matrix and a corresponding increase of nascent DNA in the nonmatrix fraction of the nucleus. These results have been subsequently confirmed in several different cell types using similar high-salt matrix extraction procedures (Dijkwel *et al.*, 1979; McCready *et al.*, 1980; Pardoll *et al.*, 1980; Vogelstein *et al.*, 1980; Jackson and Cook, 1986; Fernandes *et al.*, 1989) and, recently, a low salt lithium 3,5-diiodosalicylate (LIS) extraction procedure (Vaughn *et al.*, 1990). The results of these pulse-labeling studies suggested that DNA replication sites were fixed on the nuclear matrix, and newly replicated DNA moved progressively away from these matrix-attached replication sites according to the rate of replication fork movement. By autoradiographic analysis of matrix–halo structures prepared from pulse-labeled cells, Vogelstein *et al.* (1980) were able to visualize directly the migration of newly replicated DNA from the central core of the matrix into the halo structure formed by decondensed DNA loops.

The association of replication sites to the nuclear matrix has also been demonstrated using two-dimensional gel electrophoresis followed by blot hybridization to specific radiolabeled probes (Brewer and Fangman, 1988). In the studies carried out by Hamlin and co-workers (Vaughn *et al.*, 1990; Dijkwel *et al.*, 1991), replication forks moving through the amplified dihydrofolate reductase (DHFR) locus were detected only in the matrix-attached DNA fraction, and virtually absent in the soluble nonmatrix DNA of methotrexate-resistant CHOC 400 cells.

Consistent with the model of fixed replication sites associated to the nuclear matrix is the finding that many of the activities involved in DNA replication, including DNA polymerase α (Smith and Berezney, 1980, 1982, 1983; Foster and Collins, 1985), DNA primase (Wood and Collins, 1986; Tubo and Berezney, 1987b; Hirose *et al.*, 1988; Paff and Fernandes, 1990), RNase H (Tubo and Berezney, 1987a), DNA topoisomerase II (Berrios *et al.*, 1985; Earnshaw *et al.*, 1985), and proliferating cell nuclear antigen (PCNA) (Bravo and Macdonald-Bravo, 1987) are present on the nuclear matrix. Tubo and Berezney (1987a) showed that complexes containing all the components of the replicative apparatus, including DNA polymerase α and primase, were assembled on the nuclear matrix prior to the entry of the cells into the S-phase of the cell cycle.

Direct visualization of DNA replication sites in intact or permeabilized cells has shown that individual replicons are organized into groups or clusters of 100 or more replicons (Nakamura *et al.*, 1986; Nakayasu and Berezney, 1989; Hassan and Cook, 1993). Approximately 150 to 300 replication foci were observed using either bromodeoxyuridine or biotinylated-dUTP (Nakamura *et al.*, 1986; Nakayasu and Berezney, 1989; Hassan and Cook, 1993; Hozák *et al.*, 1993; Hassan *et al.*, 1994). Each focus corresponded to a replicon cluster, in which numerous replication complexes were coordinately synthesizing DNA. Interestingly, when the incorporation of bromodeoxyuridine or biotin-dUTP was followed by *in situ* extraction of the nuclear matrix, the size and distribution pattern of the replication foci observed on the nuclear matrix were virtually identical to those in whole cells (Nakayasu and Berezney, 1989; Hozák *et al.*, 1993). These results suggested that replicon clusters were precisely organized in the nucleus and that the nuclear matrix played a role in determining both the structural and functional organization of the replicon clusters.

B. RNA-Primed DNA Synthesis on the Nuclear Matrix

To further evaluate the role of the nuclear matrix in DNA replication, studies were done in our laboratory aimed at determining the intranuclear distribution of RNA-primed DNA in human leukemia cells (Paff and Fernandes, 1990). On the lagging strand of the replication fork, fragments of 100–200 nucleotides in length are synthesized by the DNA polymerase α–primase complex, and then linked together to form a continuous strand of newly replicated DNA (DePamphilis and Wassarman, 1980; Ogawa and Okazaki, 1980; Wang, 1991). DNA primase starts the synthesis of each fragment by making an oligoribonucleotide primer of 8–10 nucleotides (Kaguni and Lehman, 1988). The primer RNA provides the free 3'-OH terminus required by DNA polymerase α to start polymerization of deoxyribonucleotides. It was of interest to determine whether RNA-primed DNA fragments were found in the nuclear matrix, since these fragments would be specific markers of the presence of DNA replication forks.

Previous studies showed that isolated nuclear matrices were capable of supporting the synthesis of putative Okazaki fragments and oligoribonucleotides of 8–50 nucleotides in length (Smith and Berezney, 1982; Tubo and Berezney, 1987b). To determine if the nuclear matrix was the site of synthesis of these replicative intermediates *in vivo*, Paff and Fernandes (1990) incubated whole-cell lysates of human CEM leukemia cells in the presence of radiolabeled precursors, which were incorporated into DNA, RNA, and primer RNA. Following extraction of the nuclear matrix, radiolabeled RNA-primed DNA was separated from bulk RNA by cesium chloride density gradient centrifugation and analyzed by electrophoresis on denaturating polyacrylamide gels. RNA-primed DNA migrated as fragments of 9 to approximately 100 nucleotides in length. After digestion with DNase I, the isolated primer RNA migrated as a product of 8–10 nucleotides in length in agreement with previously reported data (Tseng and Goulian, 1977; Tseng *et al.*, 1979; Kitani *et al.*, 1984; Yamaguchi *et al.*, 1985). Phosphate transfer analysis confirmed that the primer RNA was covalently linked to the 5' end of the nascent DNA (Paff and Fernandes, 1990).

In contrast, neither RNA-primed DNA nor primer RNA was detected in the nonmatrix samples by gel electrophoresis and phosphate transfer analysis (Paff and Fernandes, 1990). Quantitative estimates of the relative amounts of primer RNA in the matrix and nonmatrix fractions indicated that greater than 94% of the primer RNA partitioned with the nuclear matrix fraction of the nucleus. When normalized to the amount of total RNA associated with the nuclear matrix, it was estimated that RNA-primed nascent DNA was enriched at least 80-fold in the nuclear matrix compared to total RNA. Primer RNA is apparently synthesized by DNA primase and extended by DNA polymerase α in the matrix fraction of the nucleus. The primer is then degraded prior to the migration of the nascent DNA into the loops away from the matrix-bound DNA replication sites. Consistent with this hypothesis is the finding that DNA polymerase α (Smith and Berezney, 1980, 1982, 1983; Foster and Collins, 1985) and DNA primase

(Tubo and Berezney, 1987b; Paff and Fernandes, 1990) are both associated with the nuclear matrix. In addition, RNase H, which is thought to catalyze the removal of the primer RNA prior to the ligation of the Okazaki fragments, is associated with the nuclear matrix (Tubo and Berezney, 1987b). Therefore, both synthesis and removal of primer RNA from the Okazaki fragments are likely nuclear matrix-associated functions.

In addition to providing insights into the mechanisms of lagging strand DNA synthesis, the finding of a high degree of enrichment of primer RNA on the nuclear matrix has practical implications in the study of the mechanisms of action of various anticancer agents. Detailed studies of the effects of inhibitors of DNA synthesis on primer RNA formation had been prviously difficult to carry out in intact cells or nuclei because of the very low concentrations of the primer within cells and the relatively large amounts of other RNAs. A new method for quantitative isolation of primer RNA from nuclear matrices has facilitated investigations into the mechanisms of inhibition of DNA and primer RNA synthesis by various anticancer agents (Paff and Fernandes, 1990).

C. Inhibitors of DNA Primase Activity

The synthesis and extension of primer RNA are key reactions in DNA replication (DePamphilis and Wassarman, 1980; Ogawa and Okazaki, 1980; Wang, 1991). Because of the importance of primer RNA formation in DNA synthesis, there has been considerable interest in DNA primase as a potential target for anticancer drugs. The nucleoside triphosphate metabolites of various anticancer agents, including 1-β-D-arabinofuranosylcytosine 5'-triphosphate (araCTP), 9-β-D-arabinofuranosyladenosine 5'-triphosphate (araATP), and 9-β-D-arabinofuranosyl-2-fluoradenosine 5'-triphosphate (FaraATP), were shown to inhibit DNA primase *in vitro* (Tseng and Ahlem, 1983; Yoshida *et al.*, 1985; Parker and Cheng, 1987; Parker *et al.*, 1988; Catapano *et al.*, 1991, 1993; Kuchta and Willhelm, 1991; Kutchta *et al.*, 1992). To confirm that DNA primase was ain intracellular target for an anticancer agent, it was necessary to demonstrate that cytotoxic concentrations of the anticancer agent blocked primer RNA formation and DNA synthesis in intact cells. The demonstration that the nuclear matrix was the subnuclear site of RNA-primed DNA synthesis and that these replicative intermediates could be quantitatively purified from the nuclear matrix (Paff and Fernandes, 1990) made it possible to study the effects of anticancer agents on RNA-primed DNA synthesis in a whole cell system.

1. Fludarabine Phosphate

FaraATP is the active metabolite of the anticancer agent, fludarabine phosphate (9-β-D-arabinofuranosyl-2-fluroadenine 5'-monophosphate, Fludara), which is effective as a single agent against chronic lymphocytic leukemia (Keating *et al.*, 1989; Keating, 1990) and non-Hodgkin's lymphoma (Hochster and Cassileth, 1990). Like many of the available antileukemia agents, fludarabine phosphate appears to exert its cytotoxic effects by inhibiting DNA replication. The nucleoside triphosphate, FaraATP, interferes *in vitro* with the activity of various enzymes involved in DNA replication, including DNA polymerase α (Tseng *et al.*, 1982; Parker *et al.*, 1988, 1991; Huang *et al.*, 1990), DNA primase (Parker and Cheng, 1987; Catapano *et al.*, 1991, 1993), DNA ligase (Yang *et al.*, 1992), and ribonucleotide reductase (Tseng *et al.*, 1982).

When tested in a whole cell lysate system, FaraATP reduced the incorporation of radiolabeled precursors into primer RNA and DNA synthesized on the nuclear matrix (Catapano *et al.*, 1991). The extent of the inhibition of primer RNA synthesis was directly related to the degree of DNA synthesis inhibition over a wide range of FaraATP concentrations. Although it was possible that the effect on primer RNA synthesis was secondary to the inhibition of DNA synthesis, a number of observations indicated that the primary target for FaraATP was the formation of primer RNA. First, FaraATP at concentrations that inhibited DNA primase activity by more than 90%, did not affect DNA polymerase activity when tested in extracts in the presence of saturating concentrations of the natural substrates (Catapano *et al.*, 1991). Comparison of the K_M/K_i ratios for either DNA polymerase α or DNA primase inhibition by FaraATP (3 and 25, respectively) also indicated that FaraATP was a more potent inhibitor of DNA primase than DNA polymerase α (Catapano *et al.*, 1991). Furthermore, other DNA polymerase inhibitors, such as araCTP and aphidicolin, did not block primer RNA formation at concentrations that were effective in inhibiting nascent DNA synthesis in the whole cell lysate system (Catapano *et al.*, 1991).

The mechanism of inhibition of DNA primase activity by FaraATP was investigated in a cell-free system (Catapano *et al.*, 1993). These studies indicated that FaraATP acted as a primer RNA chain terminator. The primase component of the purified DNA polymerase α–primase complex catalyzed the incorporation of ATP into oligomers 2–10 nucleotides in length on a poly(dT) template. The addition of increasing concentrations of FaraATP to the reaction mixtures resulted in a progressive reduction of the amount of full length primers (>7 nucleotides) synthesized by DNA primase. A parallel accumulation of shorter primer RNA intermediates (<6 nucleotides) was observed in the presence of increasing concentrations of FaraATP. These effects were accompanied by the incorporation of FaraATP into oligomers 2–6 nucleotides in length over the entire range of FaraATP concentrations. Sequencing of the truncated primers revealed that FaraAMP was incorporated at the 3'-primer terminus of the truncated primers and no FaraAMP residues were present at internucleotide positions. This provided a direct evidence of the primer RNA chain termination mechanism. DNA primase could not catalyze further elongation of primer RNA after incorporation of a FaraAMP residue at the 3'-primer terminus of a growing primer RNA chain. It was also important that the efficiency of incorporation of FaraATP by DNA primase (expressed as relative k_{cat}/K_M) was 30-fold higher than that of ATP (Catapano *et al.*, 1993).

These studies with the purified DNA polymerase α–primase complex confirmed that DNA polymerase α was less sensitive to FaraATP than DNA primase (Catapano *et al.*, 1993). In primer extension assays, DNA polymerase α was less efficient in extending from a 3'-FaraAMP primer terminus than from a normal 3'-AMP terminus. The observed difference (13-fold), however, was relatively modest when compared to the effect of a 3'-FaraAMP terminus on DNA primase activity. Furthermore, incorporation of FaraATP by DNA polymerase α was unlikely, since this enzyme polymerized FaraATP about 8-fold less efficiently than dATP. These results indicated that the different sensitivity of DNA primase and DNA polymerase α to FaraATP originated from differences in the ability of the two enzymes to discriminate between the analogue and the respective nucleotide substrate, ATP and dATP.

2. IMP Dehydrogenase Inhibitors

This class of metabolic inhibitors of purine biosynthesis has been evaluated recently as potential inhibitors of primer RNA synthesis *in vivo*. Inosine monophosphate (IMP) dehydrogenase (IMPDH) catalyzes the conversion of IMP to GMP in the *de novo* biosynthetic pathway. Incubation of cells in the presence of IMPDH inhibitors, such as mizoribine (Bredenin; 4-carbamoyl-1-β-D-ribofuranosylimidazolium-5-olate) and mycophenolic acid, results in depletion of guanine nucleotide pools (Lowe *et al.*, 1977; Turka *et al.*, 1991). IMPDH inhibitors have been tested as anticancer and immunosuppressive agents (Sakaguchi *et al.*, 1975; Lowe *et al.*, 1977; Lucas *et al.*, 1983; Knight *et al.*, 1987; Kharabanda *et al.*, 1988; Tricott *et al.*, 1989; Turka *et al.*, 1991; Allison *et al.*, 1993). A major metabolic effect of this class of agents is the inhibition of DNA synthesis, which is reversible with repletion of the guanine nucleotide pools (Lowe *et al.*, 1977; Turka *et al.*, 1991). Although both the GTP (guanosine triphosphate) and dGTP pools are reduced after incubation of cells with IMPDH inhibitors, several lines of evidence support a primary role for GTP depletion in the inhibition of DNA synthesis (Cohen *et al.*, 1981; Duan and Sadee, 1982; Cohen and Sadee, 1983). We tested the hypothesis that GTP depletion prevented the formation of primer RNA and, consequently, blocked the synthesis of RNA-primed DNA fragments (Catapano *et al.*, 1995). DNA primase, which is an RNA polymerase, utilizes preferentially guanine and adenine ribonucleosides to initiate primer RNA synthase both *in vitro* (Conaway and Lehman, 1982; Tseng and Ahlem, 1984; Wang *et al.*, 1984; Grosse and Krauss, 1983; Yamaguchi *et al.*, 1985) and *in vivo* (Tseng *et al.*, 1979; Hay *et al.*, 1984; Kitani *et al.*, 1984). Moreover, Sheaff and Kuchta (1993) have shown that DNA primase prefers to incorporate GTP as the second nucleotide of a growing primer RNA chain. Apparently, incorporation of GTP at this position increases the stability of the dinucleotide–template complex because of the additional hydrogen bond present in a G–C base pair compared to an A–T base pair (Sheaff and Kuchta, 1993). This greater stability, in turn, increases the probability that the dinucleotide is elongated to a full-length primer RNA rather than prematurely dissociating from the DNA template.

The dependence of primer RNA synthesis on GTP concentration was tested both *in vitro* and *in vivo* (Catapano *et al.*, 1995). In a cell-free system with purified DNA polymerase α–primase complex and single-stranded M13 DNA as template, primer RNA synthesis decreased progressively as the concentration of GTP in the reaction mixture was reduced. Similarly, in whole cell lysates primer RNA formation was reduced more than 80% in the absence of exogenous GTP. Next, we tested directly the hypothesis that IMPDH inhibitors inhibited DNA synthesis by blocking primer RNA formation by incubating CEM cells with either mizoribine or mycophenolic acid and labeling primer RNA *in vivo* with [^{3}H]adenosine. RNA-primed DNA was then isolated from the nuclear matrices of control and drug-treated cells and purified by cesium chloride density gradient centrifugation. Incubation of the cells for only 2 hr with IC_{50} concentrations of either mizoribine or mycophenolic acid reduced the amount of primer RNA synthesized to 25 and 30% of the untreated controls, respectively. To determine if the inhibition of IMPH and the consequent GTP depletion contributed to the effects of mizoribine and mycophenolic acid on primer RNA synthesis in CEM cells, drug-treated cells were co-incubated with guanosine and 8-aminoguanosine to replete the intracellular GTP. By inhibiting purine nucleoside phosphorylase, 8-aminoguanosine slows the rate of conversion of guanosine to guanine, which results in a more sustained increase in the intracellular levels of GTP compared to guanine alone (Turka *et al.*, 1991). The inhibition of primer RNA synthesis induced by either mizoribine or mycophenolic was prevented by coincubation of the cells with guanosine and 8-aminoguanosine.

These data support the conclusion that inhibition of primer RNA synthesis is an early effect of the GTP depletion induced by IMPDH inhibitors and may be responsible for the antiproliferative effects of these agents. It is interesting that the effects of IMPDH inhibitors on primer RNA and DNA synthesis preceded the induction of internucleosomal DNA fragmentation, which is a feature characteristic of apoptotic cell death (Catapano *et al.*, 1995). A faint nucleosomal ladder was visible only after 4 hr and increased markedly in intensity after 6 hr of incubation with either mizoribine or mycophenolic acid. Both formation of the nucleosomal ladder and loss of cell viability induced by mizoribine and mycophenolic acid were prevented by repletion of intracellular GTP with guanosine and 8-amino-guanosine.

VI. Topoisomerase II-Active Drugs

A. Biological Activities of DNA Topoisomerase II

Chromosomal DNA is organized into loop structures that are constrained in a supercoiled state as a result of their attachment to the nuclear matrix (Cook and Brazell, 1976; Vogelstein et al., 1980). DNA topoisomerases control the level of DNA supercoiling in eukaryotic cells (Wang, 1985). Two major types of eukaryotic topoisomerases have been identified (Wang, 1985). These enzymes modify the topological state of DNA by introducing transient breaks in either one (topoisomerase I) or two (topoisomerase II) strands of a duplex DNA, passing a single- or double-stranded DNA through the break, and finally resealing the breaks (Wang, 1985). Topoisomerase II forms initially an ionic complex with DNA. During catalysis, the enzyme is transiently bound to DNA via a covalent linkage between a tyrosyl residue in the enzyme molecule and the 5'-phosphoryl end of the broken DNA strands (Liu et al., 1983; Wang, 1985; Zechiendrich et al., 1989). After strand passing, the enzyme rejoins the ends of the broken DNA strands and dissociates from the DNA (Wang, 1985; Osheroff, 1986). Topoisomerase I acts in a similar way with the exceptions that the enzyme cleaves a single strand of DNA and is transiently linked to the 3'-end of the broken DNA (Wang, 1985). Both topoisomerase I and II can relax supercoiled DNA. Additionally, topoisomerase II catalyzes unknotting of knotted DNA and decatenation of intertwined DNA molecules (Wang, 1985).

Because of the ability of topoisomerase II to control DNA supercoiling, the enzyme has been implicated in almost every aspect of DNA metabolism, including DNA replication (Nelson et al., 1986b; Brill et al., 1987), RNA transcription (Brill et al., 1987), recombination (Bae et al., 1988), and organization of chromatin loop domains (Earnshaw and Heck, 1985; Gasser et al., 1986). Biochemical and genetic studies favor a direct role of topoisomerase II in various stages of DNA replication. Relaxation of DNA supercoiling by topoisomerase II may allow the initial unwinding at the origins of replication. In addition, the swiveling activity of topoisomerase II may be required for the progression of the replication complex along the DNA template, whereas the decatenating activity of this enzyme is essential for separation of the newly replicated DNA molecules at the end of the replication process (DiNardo et al., 1984a,b; Holm et al., 1985). Topoisomerase II may have a similar role during transcription. The enzyme relaxes DNA supercoiling generated by the progression of the transcription complex and unwinding of the DNA template (Pruss and Drlica, 1986; Liu and Wang, 1987; Wu et al., 1988). Topoisomerase II appears also to be involved in the regulation of homologous recombination and has been implicated in illegitimate recombination events in cells treated with topoisomerase II inhibitors (Sperry et al., 1989; Wang et al., 1990; Han et al., 1993).

B. Nuclear Matrix DNA Topoisomerase II

Topoisomerase II is an abundant nuclear protein (Gasser et al., 1986) and is present in large amounts in the metaphase chromosome scaffolds and the nuclear matrices of interphase cells (Berrios et al., 1985; Earnshaw and Heck, 1985; Earnshaw et al., 1985; Gasser et al., 1986). The high intracellular levels of topoisomerase II and its presence in the chromosome scaffold and nuclear matrix suggest a dual enzymatic and structural role of topoisomerase II in the condensation of interphase chromatin into metaphase chromosomes (Newport and Spann, 1987; Uemura et al., 1987; Wood and Earnshaw, 1990; Adachi et al., 1991) and in the organization of the nuclear matrix in interphase nuclei (Berrios et al., 1985). A number of observations support a direct involvement of topoisomerase II in the attachment of chromatin loops to the nuclear matrix or scaffold. The enzyme is localized at the bases of the loop domains of the metaphase chromosomes (Earnshaw and Heck, 1985; Gasser et al., 1986). MAR or SAR elements, which define the bases of the DNA loops, are characterized by the presence of AT-rich stretches of several hundred base pairs. Within these AT-rich regions, numerous copies of a sequence element related to the consensus binding site for topoisomerase II have been identified (Cockerill and Garrard, 1986; Gasser and Laemmli, 1986b). Furthermore, topoisomerase II is able to bind MAR or SAR-containing DNA fragments in vitro. The binding is cooperative and leads to aggregation of the DNA fragments in looplike structures (Adachi et al., 1989). The interaction of topoisomerase II with MAR or SAR elements has been proposed as an important factor for chromatin loop formation and chromosome condensation in vivo (Adachi et al., 1989, 1991). Drug inhibition studies in cell-free nuclear/chromosome assembly systems also indicate a role for topoisomerase II in chromatin condensation (Newport, 1987; Newport and Spann, 1987).

Topoisomerase II localized at the basis of chromatin loops may represent the form of the enzyme that actively participates in nuclear matrix-associated functions, such as DNA replication, transcription, and recombination. Two isozymes of mammalian topoisomerase II have been identified (Drake *et al.*, 1989). The two isozymes, named α and β, have apparent molecular weights of 170 and 180 kDa, respectively. The α and β isozymes are the products of two distinct genes (Chung *et al.*, 1989), which reside on different chromosomes (Tan *et al.*, 1992). Topoisomerase IIα and β differ in some of their properties, such as thermal stability and sensitivity to certain topoisomerase inhibitors (Drake *et al.*, 1989). The levels of topoisomerase IIα and β appear to be regulated differently depending upon the proliferative state of the cells. The level of the α isozyme is higher in the exponential phase of cell growth than in the stationary phase (Drake *et al.*, 1989; Woessner *et al.*, 1991; Negri *et al.*, 1992). The level of the β isozyme is either constant (Kimura *et al.*, 1994) or increased in cells in stationary phase (Drake *et al.*, 1989; Woessner *et al.*, 1991; Negri *et al.*, 1992). The two isoforms also exhibit different tissue-specific expression patterns (Capranico *et al.*, 1992). Analysis of the cleavage sites induced by purified α and β isozymes in pBR322 DNA indicated a subset of sites that were either preferred or unique for each of the enzymes. The α isozyme preferred AT-rich areas, while the β isozyme preferred GC-rich sequences. An AT-rich oligonucleotide was also able to inhibit the catalytic activity of topoisomerase IIα, but did not affect topoisomerase IIβ. The preferred interaction of topoisomerase IIα with AT-rich region suggested that this form of the enzyme may represent the form associated with the chromosome scaffold and the nuclear matrix, and be involved in nuclear matrix-associated functions (Drake *et al.*, 1989).

In the attempt to provide insights into the specific functions of the two isozymes, various laboratories have recently studied the subnuclear distribution of the α and β isozymes and investigated whether one of the two isozymes exhibited a preferential association with the nuclear matrix. Identification of the specific roles of topoisomerase IIα and β in DNA organization and metabolism may have important implications in cancer chemotherapy. Selective inhibitors of either topoisomerase IIα or β, which would block only the specific functions related to one of the two isozymes, might be useful anticancer agents. Inhibitors of an isozyme involved in DNA replication may be more effective in inhibiting tumor cell growth.

Earlier studies localized topoisomerase II to the nuclear scaffold, but did not distinguish between the two forms of the enzyme (Berrios *et al.*, 1985; Earnshaw and Heck, 1985). The lability of the topoisomerase IIβ has made studies of its subcellular distribution difficult (Woessner *et al.*, 1990; Danks *et al.*, 1994). Danks *et al.* (1994) observed that the 180-kDa band was detected when whole CEM cells were rapidly lysed by a hot SDS method, and the lysates immediately electrophoresed and immunoblotted. In contrast, the 180-kDa band was not detected in preparations of isolated nuclei from the same cells. The isolation procedure apparently induced degradation of the 180-kDa isozyme. Of particular interest was the observation that pretreatment of leukemia cells with the topoisomerase II inhibitor, teniposide, 4'-demethylepipodophyllotoxin 9-(4,6,O-2-thenylidene-β-D-glucopyranoside, (VM-26), which stabilized topoisomerase IIβ–DNA complexes, protected the β isozyme from degradation (Danks *et al.*, 1994). This apparent protective effect was unique to VM-26-stabilized topoisomerase IIβ–DNA complexes, since formation of VM-26-stabilized topoisomerase IIα–DNA complexes did not affect the amount of the α isozyme detected in isolated nuclei of CEM cells.

The stabilizing effect of VM-26 on topoisomerase IIβ was exploited to investigate the distribution of the topoisomerase isozymes in the isolated nuclear matrix and the high-salt-soluble (nonmatrix) fractions of the nucleus. Immunoblotting of matrix and nonmatrix fractions with topoisomerase II-specific antibodies indicated a distinct distribution of the two isozymes. Topoisomerase IIα was present in both the matrix and nonmatrix fractions of nuclei from untreated and VM-26-treated cells. The β isoform was detected in appreciable amounts only in the nonmatrix fraction of nuclei from cells incubated with VM-26. These results indicated that the nuclear matrix form of topoisomerase II is the α isozyme, while both the α and β forms are present in the nonmatrix fraction of the nucleus. Further support for a different nuclear distribution of the two isozymes comes from ultrastructural studies with monoclonal antibodies that were reported to interact selectively with either topoisomerase IIα or β (Zini *et al.*, 1994). These studies showed that topoisomerase IIα was localized primarily in the nucleoplasm, while the β isozyme was present mainly in the nucleolus (Zini *et al.*, 1994). The differences in the subnuclear distribution of topoisomerase IIα and β support the concept that each isozyme has distinct primary functions in the cell. The nuclear matrix-associated topoisomerase IIα may represent the form of the enzyme more directly involved in chromatin loop formation, DNA replication, and transcription. The association of topoisomerase IIβ with nucleoli suggests that this isozyme may be more specifically involved in ribosomal DNA metabolism.

C. Mechanisms of Antitumor Activity of Topoisomerase II-Active Drugs

Many anticancer drugs, including intercalating agents [adriamycin, 4'-(9-acridinylamino)-3-methane-sulfon-*m*-aniside (m-AMSA), ellipticines] and the epipodophyllotoxins [etoposide, 4'-demethyl-epipodophyllotoxin 9-(4,6,O-ethylidene-β-D-glucopyranoside) (VP-16); VM-26] are known to inhibit topoisomerase II activity. The majority of the clinically effective topoisomerase II-active agents allow enzyme binding and DNA cleavage to proceed, but block the resealing of the broken DNA strands (Liu *et al.*, 1983; Osheroff, 1989; Zechiendrich *et al.*, 1989; Robinson and Osheroff, 1990). The result is the stabilization of the transient covalent complex of the enzyme with the cleaved DNA strands. The generation of topoisomerase II–DNA cleavable complexes is thought to be only the initial step in a series of the events that lead to cell death. In fact, most of the topoisomerase II–DNA complexes are reversed and the topoisomerase II-mediated DNA strand breaks resealed after removal of the drug (Tewy *et al.*, 1984; Rowe *et al.*, 1985). Drug-stabilized topoisomerase II–DNA complexes must interact with other cellular processes, such as DNA replication and RNA transcription, and these interactions are ultimately responsible for the cytotoxic effects of topoisomerase II-active drugs.

Topoisomerase II is a major nonhistone protein of the nuclear matrix (Berrios *et al.*, 1985; Gasser *et al.*, 1986). Current data suggest that the matrix-bound topoisomerase II is involved in many nuclear functions, including DNA replication, RNA transcription, recombination, and attachment of chromatin loop domains to the nuclear skeleton. Thus, it seems reasonable to suggest that topoisomerase II-active agents may interfere specifically with some of the nuclear matrix-associated functions by stabilizing covalent topoisomerase II–DNA complexes (Fig. 1). Danks *et al.* (1994) have reported that the α isozyme of topoisomerase II was present in both the matrix and nonmatrix fractions of nuclei from human leukemia CEM cells, whereas the β isozyme was present only in the nonmatrix fraction. The topoisomerase II-active drug VM-26 stimulated the formation of covalent complexes between bulk DNA and both topoisomerase IIα and β. However, only topoisomerase IIα formed complexes with matrix-associated DNA in VM-26-treated cells, suggesting that the α isozyme was the nuclear matrix target for this drug (Danks *et al.*, 1994). This conclusion is further supported by the finding that the selective depletion of topoisomerase IIα in the nuclear matrix of drug-resistant CEM cells (CEM/VM-1) reduces the sensitivity of these cells to VM-26 and other topoisomerase II-active anticancer agents (Fernandes *et al.*, 1990).

1. Interaction of Topoisomerase II-Active Drugs with DNA Replication Forks

Some lines of evidence suggest that the interaction of DNA replication forks with topoisomerase II–DNA covalent complexes may represent a critical event for the conversion of the cleavable complexes into cytotoxic lesions (D'Arpa *et al.*, 1990). Topoisomerase II-active agents, such as the epipodophyllotoxins VP-16 and VM-26, are generally more effective against cells in the S-phase of the cell cycle as opposed to cells in either G_1- or

G_2–M-phases, probably because of the presence of both topoisomerase II–DNA complexes and active replication forks in S-phase cells (Sullivan *et al.*, 1987; D'Arpa and Liu, 1989; D'Arpa *et al.*, 1990, Erba *et al.*, 1992). Pretreatment of tumor cells with the DNA synthesis inhibitor, aphidicolin, protects S-phase cells from topoisomerase II-active agents (Holm *et al.*, 1989; D'Arpa *et al.*, 1990). This protection is achieved without a reduction in the formation of cleavable complexes (Holm *et al.*, 1989; D'Arpa *et al.*, 1990) or protein-associated DNA strand breaks (Holm *et al.*, 1989).

Possible mechanisms by which topoisomerase II-active drugs could block DNA replication have been previously reviewed (Fernandes and Catapano, 1991). Drug-stabilized covalent complexes between DNA and topoisomerase II may act as a physical barrier to the progression of the replication complexes along the DNA template. By alterating the DNA topology at the replication fork, drug–topoisomerase–DNA ternary complexes may also interfere with the binding of the DNA polymerases, primase, and accessory proteins to the template. Furthermore, DNA strand separation at the replication fork may induce disruption of the topoisomerase II–DNA complexes and thereby promotes the transformation of the cleavable complexes into frank DNA strand breaks at the sites of stalled replication forks. DNA breaks may lead to fork breakage and inability to complete DNA replication. Consistent with these hypotheses, studies of SV40 DNA replication both in intact cells and cell-free systems have shown that topoisomerase II inhibitors interfere with the elongation of nascent DNA and induce breaks in replicating DNA (Snapka, 1986; Richter *et al.*, 1987; Richter and Strausfeld, 1988; Snapka *et al.*, 1988). In mammalian cells, topoisomerase II-active agents induce formation of enzyme–DNA complexes or strand breaks preferentially in replicating DNA (Nelson *et al.*, 1986b; Woynarowski *et al.*, 1988) and inhibit elongation of nascent DNA (Kaufmann *et al.*, 1991; Catapano *et al.*, 1994).

Studies from our laboratory suggest that the interaction between replication forks and drug-stabilized topoisomerase II–DNA complexes occur at the level of the nuclear matrix. Compared to the nonmatrix fraction of the nucleus, the nuclear matrix is highly enriched in newly replicated DNA and replication forks. Exposure to relatively low concentrations ($<1\ \mu M$) of either VM-26 or m-AMSA abolished the enrichment of nascent DNA on the nuclear matrix of CEM cells (Fernandes *et al.*, 1989). In contrast, araC and hydroxyurea, at concentrations that induce marked inhibition of total cellular DNA synthesis, do not affect the enrichment of newly replicated DNA on the matrix (Fernandes *et al.*, 1989). These data are consistent with the hypothesis that drug-stabilized topoisomerase II–DNA complexes are converted to DNA strand breaks at replication forks. These breaks may cause detachment of the replication forks and, thus, reduce the enrichment of newly replicated DNA on the nuclear matrix. The progression of

DNA replication forks on the nuclear matrix has also been monitored by following the synthesis of RNA-primed Okazaki fragments, which are present intracellularly only at the nuclear matrix-associated replication sites. In CEM cells, Okazaki fragment synthesis on the nuclear matrix was almost completely inhibited by a 2-hr exposure to 1 μM VM-26 (D. J. Fernandes and C. V. Catapano, unpublished results). The effects of VM-26 on Okazaki fragment synthesis were not due to direct inhibition of either DNA primase or DNA polymerase α, since concentrations of VM-26 as high as 50 μM did not inhibit these enzymes *in vitro*.

2. Other Effects of Topoisomerase II-Active Drugs on Nuclear Matrix-Associated Functions

Other functions of the matrix-bound topoisomerase II may be affected by topoisomerase II-active drugs. Brief incubation of thymocytes with m-AMSA, VP-16, or VM-26 resulted in cleavage of chromosomal DNA into fragments of approximately 50 kb, which may correspond to entire chromatin loop domains (Filipski *et al.*, 1990; Walker *et al.*, 1991). Formation of topoisomerase II-mediated DNA strand breaks in matrix attachment sites was proposed to interfere with the attachment of chromatin loop domains to the nuclear matrix (Filipski *et al.*, 1990; Walker *et al.*, 1991). Interestingly, the appearance of these high-molecular-weight DNA fragments was followed at later times by the appearance of nucleosomal DNA ladders characteristic of apoptotic cells (Walker *et al.*, 1991). Thus, cleavage at the level of the MARs could be an early event in the activation of programmed cell death by topoisomerase II-active drugs (Fig. 1).

Formation of topoisomerase II-mediated DNA strand breaks in matrix-attachment regions may also stimulate nonhomologous recombination in cells treated with topoisomerase II-active drugs (Sperry *et al.*, 1989). These agents are known to stimulate genomic rearrangements resulting in deletions, insertions, and chromosomal translocations in drug-treated cells (Maraschin *et al.*, 1990; Han *et al.*, 1993; Shibuya *et al.*, 1994). Topoisomerase II-mediated nonhomologous recombination may be relevant to both the cytotoxic (Pommier *et al.*, 1985; Kaufmann, 1991) and mutagenic effects (Ratain *et al.*, 1987) of topoisomerase II-active agents. Drug-induced chromosomal translocations, for example, are thought to play a role in the etiology of secondary acute myeloid leukemias in cancer patients treated with topoisomerase II-active drugs (Ratain *et al.*, 1987).

Recent studies have reported a relationship between the sites of deletions or recombination and the sites of topoisomerase II-mediated DNA cleavage in VM-26-treated cells (Bodley *et al.*, 1993; Han *et al.*, 1993). Liu and co-workers (Bodley *et al.*, 1993) reported that the sites of integration of SV40 DNA into cellular DNA corresponded to sites of topoisomerase II-mediated cleavage mapped in SV40 DNA. It is noteworthy that nonhomologous recombination involving SV40 sequences and cellular DNA was affected by the DNA synthesis inhibitor, aphidicolin, suggesting that the process required DNA replication (Bodley *et al.*, 1993). Thus, topoisomerase II-mediated DNA strand breaks at DNA replication forks may be good substrates in the nonhomologous recombination events induced by topoisomerase II-active drugs.

3. Resistance to Topoisomerase II-Active Drugs

Resistance to topoisomerase II-active drugs can be mediated by a variety of mechanisms. A form of resistance to topoisomerase II-active agents is associated with increased expression of the multidrug resistance (MDR)1 gene. The product of the MDR1 gene is a membrane glycoprotein, P-glycoprotein, that acts as an efflux pump and prevent drugs from accumulating within the cells (Endicott and Ling, 1989; Gottesman and Pastan, 1993). Cells that express the MDR phenotype are resistant to multiple topoisomerase II-agents and cross-resistant to a variety of natural products, such as the *Vinca* alkaloids, which have different cellular targets (Endicott and Ling, 1989). Another form of multidrug resistance has been associated with overexpression of a 190-kDa protein, called multidrug resistance-associated protein (MRP; Cole *et al.*, 1992). The pattern of cross-resistance of MRP-expressing cells is similar to that of MDR1-expressing cells. In addition to multidrug-resistant cell lines expressing either MDR1 or MRP phenotypes, other cell lines have been isolated that are resistant exclusively to topoisomerase II-active drugs (Fernandes *et al.*, 1993). These cell lines do not show changes in drug accumulation, MDR1 or MRP gene expression, and sensitivity to *Vinca* alkaloids (Danks *et al.*, 1987; Fernandes *et al.*, 1993). A common feature of these cell lines is a decrease in the amounts of drug-stabilized topoisomerase II–DNA complexes and drug-induced DNA cleavage (Pommier *et al.*, 1986; Danks *et al.*, 1987). Resistance is generally attributed to qualitative or quantitative changes in topoisomerase II activity. This resistant phenotype has been named at-MDR (for altered topoisomerase multidrug resistance).

The role of matrix-bound topoisomerase II in mediating the effects of certain topoisomerase II-active drugs was confirmed by the finding that a form of at-MDR was associated with a specific depletion of the α isozyme of topoisomerase II in the nuclear matrix of resistant cells (Fernandes *et al.*, 1990). Previous studies indicated that VM-26 and m-AMSA reacted preferentially with nuclear matrix topoisomerase II in the parental CEM cells and reduced the enrichment of newly replicated DNA on the nuclear matrix of these cells (Fernandes *et al.*, 1989). In contrast to the results obtained with the drug-sensitive CEM cells, neither VM-26 nor m-AMSA affected the association of newly replicated DNA with the nuclear matrices of the at-MDR CEM/VM-1 cells (Fernandes *et al.*, 1989, 1990). Of particular importance was the observation that nuclear matrices from CEM/VM-1 cells contained about threefold less immunoreactive topoisomerase IIα, whereas the amounts of the enzyme in the nonmatrix were similar in both sensitive and resistant cells (Fernandes *et al.*, 1990). Drug-sensitive (CEM) and -resistant (CEM/VM-1) cells contained also similar amounts of topoisomerase IIβ, which was found exclusively in the nonmatrix nuclear fraction (J. Qiu, C. V. Catapano, and D. J. Fernandes, unpublished results). The unknotting and decatenating activities of topoisomerase II were between six- and sevenfold lower in nuclear matrix preparations from CEM/VM-1 cells compared to the parental cell line. Furthermore, topoisomerase IIα was extracted at lower salt concentrations from nuclear matrices of at-MDR cells compared to drug-sensitive CEM cells (Fernandes *et al.*, 1990). The enzyme from the resistant cells also had altered catalytic activity (ATP requirement) and DNA-binding properties (VM-26-mediated DNA cleavage) (Danks *et al.*, 1988, 1989). Apparently, the selective depletion of matrix-associated topoisomerase IIα was responsible for decreased interaction of VM-26 and m-AMSA with matrix-bound replication sites in CEM/VM-1 cells (Fernandes *et al.*, 1990). Depletion of topoisomerase IIα in the nuclear matrix of resistant cells may be related to mutations (Bugg *et al.*, 1991; Hinds *et al.*, 1991; Danks *et al.*, 1993) or post-translational modifications of the enzyme, such as phosphorylation (Takano *et al.*, 1991), which may impair its association with the matrix. Sequence analysis of the topoisomerase IIα in CEM/VM-1-resistant cells revealed single base changes at positions 449 and 804 in the regions enconding the putative ATP- and DNA-binding domains (Bugg *et al.*, 1991; Danks *et al.*, 1993). It is not yet known if either of these point mutations is primarily responsible for the impaired incorporation of topoisomerase IIα into the nuclear matrices of CEM/VM-1 cells and the high degree of resistance to topoisomerase II-active agents.

VII. Concluding Remarks

Chemotherapeutic agents interfere with the proliferation and survival of tumor cells by a variety of mechanisms. A correlation is observed generally between the amount of damage induced by these agents (e.g., DNA strand breaks for radiation, DNA–interstrand crosslinks for chloroethylnitrosoureas, topoisomerase II–DNA complexes for topoisomerase II-active drugs) and their cell killing effect. These forms of damage are thought to be cytotoxic because of the interference with the template functions of DNA during replication and transcription and/or activation of an apoptotic cell response. However, it is possible that only a fraction of the drug-induced lesions is relevant for the induction of cell death. An important factor is the localization of drug-induced lesions within the nucleus. A number of studies suggest that drug–target interactions at the level of the nuclear matrix are an important, if not the main, event in the induction of cell death by certain anticancer drugs. This can be exemplified by the topoisomerase II-active agents. Most of the drug-stabilized topoisomerase II–DNA complexes in bulk chromatin are reversed when the drug is removed from the cells. It is possible that only the fraction of cleavable complexes formed in the proximity of DNA replication forks may generate a form of irreversible damage. In addition, the topoisomerase II inhibitor, VM-26, forms complexes between DNA and both the α and β isozyme of topoisomerase II. However, studies of drug-sensitive and -resistant CEM cells suggest that only topoisomerase IIα–DNA complexes are relevant to the inhibition of DNA synthesis and, perhaps, to the cytotoxic effects of the drug. This has been related to the differences in the intranuclear distribution of the two isozymes and, consequently, to the location of drug-induced topoisomerase IIα– and β–DNA complexes. Topoisomerase IIα–DNA complexes are formed in the nuclear matrix and are more likely to interact with matrix-bound replication sites.

Other anticancer drugs and ionizing radiation show a similar degree of selectivity toward nuclear matrix components. Preferential binding of alkylating agents and induction of various forms of damage by ionizing radiation in the nuclear matrix has been demonstrated. Antimetabolites, such as fludarabine phosphate and IMPDH inhibitors, may exert their effects by blocking primer RNA synthesis, which occurs exclusively on the nuclear matrix. Selectivity toward nuclear matrix-associated activities is not the only reason why drug–target interactions in the matrix may be relevant for the antitumor effect. The nuclear matrix is a dynamic framework on which DNA is organized into discrete functional units of replication and transcription by virtue of complex interactions with nuclear matrix proteins. The precise organization of chromatin into matrix-attached DNA loops is of extreme importance for virtually all the nuclear functions. It is possible that, in addition to inhibiting specific functions, the interactions of ionizing radiation, alkylating agents, and topoisomerase II-active drugs at the level of the nuclear matrix may affect the integrity of nuclear matrix–DNA associations and the overall functional organization of the DNA. Anticancer drugs would initiate a cascade of events that will compromise multiple nuclear functions and, ultimately, cell survival.

Acknowledgments

The work carried out in the authors' laboratory was supported by Research Grants CA-44597 by the National Cancer Institute and ACS-73049 from the American Cancer Society to D.J.F.

References

Adachi, Y., Käs, E., and Laemmli, U. K. (1989). Preferential cooperative binding of DNA topoisomerase II to scaffold-association regions. *EMBO J.* **8,** 3997–4006.

Adachi, Y., Luke, M., and Laemmli, U. K. (1991). Chromosome assembly in vitro: Topoisomerase II is required for condensation. *Cell (Cambridge, Mass.)* **64,** 137–148.

Allison, A. C., Kowalski, J., Muller, C. D., and Eugui, E. M. (1993). Mechanisms of action of mycophenolic acid. *Ann. N.Y. Acad. Sci.* **696,** 63–87.

Bae, Y.-S., Kawaski, I., Ikeda, H., and Liu, L. F. (1988). Illegitimate recombination mediated by calf thymus DNA topoisomerase II in vitro. *Proc. Natl. Acad. Sci. U.S.A.* **85,** 2076–2080.

Berezney, R. (1984). Organization and functions of the nuclear matrix. *In* "Chromosomal Nonhistone Proteins" (L. S. Hnilica, ed.), pp. 119–180. CRC Press, Boca Raton, FL.

Berezney, R. (1991). The nuclear matrix: A heuristic model for investigating genomic organization and function in the cell nucleus. *J. Cell Biochem.* **47,** 109–123.

Berezney, R., and Coffey, D. S. (1974). Identification of a nuclear protein matrix. *Biochem. Biophys. Res. Commun.* **60,** 1410–1417.

Berezney, R., and Coffey, D. S. (1975). Nuclear protein matrix: Association with newly synthesized DNA. *Science* **189,** 291–293.

Berrios, M., Osheroff, N., and Fisher, P. A. (1985). *In situ* localization of DNA topoisomerase II, a major polypeptide component of the *Drosophila* nuclear matrix fraction. *Proc. Natl. Acad. Sci. U.S.A.* **82,** 4142–4146.

Bidwell, J. P., van Wijnen, A. J., Fey, E. G., Dworetzky, S., Penman, S., Stein, J. L., Lian, J. B., and Stein, G. S. (1993). Osteocalcin gene promoter-binding factors are tissue-specific nuclear matrix components. *Proc. Natl. Acad. Sci. U.S.A.* **90,** 3162–3166.

Blasquez, V. C., Xu, M., Moses, S. C., and Garrard, W. T. (1989). Immunoglobulin kappa gene expression after stable integration. I: Role of intronic matrix attachment region and enhancer in plasmacytoma cells. *J. Biol. Chem.* **264,** 21183–21189.

Bode, J., and Maass, K. (1988). Chromatin domain surrounding the human interferon-β gene as defined by scaffold-attached regions. *Biochemistry* **27,** 4706–4711.

Bode, J., Kohwi, Y., Dickinson, L., Joh, T., Klehr, D., Mielke, C., and Kohwi-Shigematsu, T. (1992). Biological significance of unwinding capability of nuclear matrix-associating DNAs. *Science* **255,** 195–197.

Bodley, A. L., Huang, H.-C., Yu, C., and Liu, L. F. (1993). Integration of simian virus 40 into cellular DNA occurs at or near topoisomerase II cleavage hot spots induced by VM-26 (Teniposide). *Mol. Cell. Biol.* **13,** 6910–6200.

Bonifer, C., Vidal, M., Grosveld, F., and Sippel, A. (1990). Tissue specific and position independent expression of the complete gene domain for chicken lysozyme in transgenic mice. *EMBO J.* **9,** 2843–2848.

Bonifer, C., Hecht, A., Saueressig, H., Winter, D., and Sippel, A. (1991). Dynamic chromatin: The regulatory domain organization of eukaryotic gene loci. *J. Cell Biol.* **47,** 99–108.

Bravo, R., and Macdonald-Bravo, H. (1987). Existence of two populations of cyclin/proliferating cell nuclear antigen during the cell cycle: Association with DNA replication sites. *J. Cell Biol.* **105,** 1549–1554.

Brewer, B. J., and Fangman, W. L. (1988). A replication fork barrier at the 3' end of yeast ribosomal RNA genes. *Cell (Cambridge, Mass.)* **55,** 637–643.

Brill, S. J., DiNardo, S., Voelkel-Meiman, K., and Sternglanz, R. (1987). Need for DNA topoisomerase activity as a swivel for DNA replication and for transcription of ribosomal RNA. *Nature (London)* **326,** 414–416.

Broggini, M., Erba, E., Ponti, M., Ballinari, D., Geroni, C., Spreafico, F., and D'Incalci, M. (1991). Selective DNA interaction of the novel distamycin derivative FCE 24517. *Cancer Res.* **51,** 199–204.

Bugg, B. Y., Danks, M. K., Beck, W. T., and Suttle, D. P. (1991). Expression of a mutant DNA topoisomerase II in CCRF-CEM human leukemic cells selected for resistance to teniposide. *Proc. Natl. Acad. Sci. U.S.A.* **88,** 7654–7658.

Capranico, G., Tinelli, S., Austin, C. A., Fisher, M. L., and Zunino, F. (1992). Different patterns of gene expression of topoisomerase II isoforms in differentiated tissues during murine development. *Biochim. Biophys. Acta* **1132,** 43–48.

Carter, K. C., Taneja, K., and Lawrence, J. B. (1991). Discrete nuclear domains of poly(A)RNA and their relationship to the functional organization of the nucleus. *J. Cell Biol.* **115,** 1191–1202.

Catapano, C. V., Chandler, K. B., and Fernandes, D. J. (1991). Inhibition of primer RNA formation in CCRF-CEM leukemia cells by fludarabine triphosphate. *Cancer Res.* **51,** 1829–1835.

Catapano, C. V., Perrino, F. W., and Fernandes, D. J. (1993). Primer RNA chain termination induced by 9-β-D-arabinofuranosyl-2-fluoroadenine-5′-triphosphate: A mechanism of DNA synthesis inhibition. *J. Biol. Chem.* **268,** 7179–7185.

Catapano, C. V., Pisani, F. D., and Fernandes, D. J. (1994). Topoisomerase II–DNA complexes stabilized by VM-26 block DNA chain elongation in human leukemia CEM cells. *Proc. Am. Assoc. Cancer Res.* **35,** 458.

Catapano, V. C., Dayton, J. S., Mitchell, B. S., and Fernandes, D. J. (1995). GTP-depletion induced by IMP dehydrogenase inhibitors blocks RNA-primed DNA synthesis. *Mol. Pharmacol.* **47,** 948–955.

Chiu, S.-M., and Oleinick, N. (1982). The sensitivity of active and inactive chromatin to ionizing radiation-induced DNA strand breakage. *Int. J. Radiat. Biol.* **41,** 71–77.

Chiu, S.-M., Oleinick, N. L., Friedman, L. R., and Stambrook, P. J. (1982). Hypersensitivity of DNA in transcriptionally active chromatin to ionizing radiation. *Biochim. Biophys. Acta* **699,** 15–21.

Chiu, S.-M., Friedman, L. R., Sokany, N. M., Xue, L.-Y., and Olienick, N. L. (1986). Nuclear matrix proteins are crosslinked to transcriptionally active gene sequences by ionizing radiation. *Radiat. Res.* **107,** 24–38.

Chiu, S.-M., Zue, L. Y., Friedman, L. R., and Oleinick, N. L. (1992). Chromatin compaction and the efficiency of formation of DNA-protein crosslinks in gamma-irradiated mammalian cells. *Radiat. Res.* **129,** 184–191.

Chiu, S.-M., Xue, L.-Y., Friedman, L., and Oleinick, N. (1993). Copper ion-mediated sensitization of nuclear matrix attachment sites to ionizing radiation. *Biochemistry* **32,** 6214–6219.

Chung, T. D. Y., Drake, F. H., Tan, K. B., Per, S. R., Crooke, S. T., and Mirabelli, C. K. (1989). Characterization and immunological identification of cDNA clones encoding two human DNA topoisomerase II isozymes. *Proc. Natl. Acad. Sci. U.S.A.* **86,** 9431–9435.

Ciejek, E. M., Ming-Jer, T., and O'Malley, B. W. (1983). Actively transcribed genes are associated with the nuclear matrix. *Nature (London)* **308,** 607–609.

Cockerill, P. N., and Garrard, W. T. (1986). Chromosomal loop anchorage of the κ immunoglobulin gene occurs next to the enhancer in a region containing topoisomerase II sites. *Cell (Cambridge, Mass.)* **44,** 273–282.

Cohen, M. B., and Sadee, W. (1983). Contributions of the depletions of guanine and adenine nucleotides to the toxicity of purine starvation in the mouse T lymphoma cell line. *Cancer Res.* **43,** 1587–1591.

Cohen, M. B., Maybaum, J., and Sadee, W. (1981). Guanine nucleotide depletion and toxicity in mouse T lymphoma (S-49) cells. *J. Biol. Chem.* **256,** 8713–8717.

Cole, S. P. C., Bhardwaj, G., Gerlach, J. H., Mackie, J. E., Grant, C. E., Almquist, K. C., Stewart, A. J., Kurtz, E. U., Duncan, A. M. V., and Deeley, R. G. (1992). Overexpression of a transporter gene in a multidrug-resistant human lung cancer cell line. *Science* **258,** 1650–1654.

Conaway, R. C., and Lehman, I. R. (1982). A DNA primase activity associated with DNA polymerase α from *Drosophila melanogaster* embryos. *Proc. Natl. Acad. Sci. U.S.A.* **79,** 2523–2527.

Cook, P. R. (1991). The nucleoskeleton and the topology of replication. *Cell (Cambridge, Mass.)* **66,** 627–635.

Cook, P. R., and Brazell, I. A. (1976). Conformational constraints in nuclear DNA. *J. Cell Sci.* **22,** 287–302.

Danks, M. K., Yalowich, J. C., and Beck, W. T. (1987). Atypical multiple drug resistance in a human leukemic cell line selected for resistance to teniposide (VM-26). *Cancer Res.* **47,** 1297–1301.

Danks, M. K., Schmidt, C. A., Cirtain, C. A., Suttle, D. P., and Beck, W. T. (1988). Altered catalytic activity of and DNA cleavage by DNA topoisomerase II from human leukemic cells selected for resistance to VM-26. *Biochemistry* **27,** 8861–8869.

Danks, M. K., Schmidt, C. A., Deneka, D. A., and Beck, W. T. (1989). Increased ATP requirement for activity of and complex formation by DNA topoisomerase II from human leukemic CCRF-CEM cells selected for resistance to teniposide. *Cancer Commun.* **1,** 101–109.

Danks, M. K., Warmoth, M. R., Friche, E., Granzen, B., Bugg, B. Y., Harker, W. G., Zwelling, L. A., Futscher, B. W., Suttle, D. P., and Beck, W. T. (1993). Single-stand conformational polymorphism analysis of the M_r 170,000 isozyme of DNA topoisomerase II in human tumor cells. *Cancer Res.* **53,** 1373–1379.

Danks, M. K., Qiu, J., Catapano, C. V., Schmidt, C. A., Beck, W. T., and Fernandes, D. J. (1994). Subcellular distribution of the α and β topoisomerase II–DNA complexes stabilized by VM-26. *Biochem. Pharmacol.* **48,** 1785–1795.

D'Arpa, P., and Liu, L. F. (1989). Topoisomerase-targeting antitumor drugs. *Biochim. Biophys. Acta* **989,** 163–177.

D'Arpa, P., Beardmore, C., and Liu, L. F. (1990). Involvement of nucleic acid synthesis in cell killing mechanisms of topoisomerase poisons. *Cancer Res.* **50,** 6919–6924.

DePamphilis, M. L., and Wassarman, P. M. (1980). Replication of eukaryotic chromosomes: A close-up of the replication fork. *Annu. Rev. Biochem.* **49,** 627–666.

Dickinson, L. A., Joh, T., Kohwi, Y., and Kohwi-Shigematsu, T. (1992). A tissue-specific MAR/SAR DNA-binding protein with unusual binding site recognition. *Cell (Cambridge, Mass.)* **70,** 631–645.

Dijkwel, P. A., and Hamlin, J. L. (1988). Matrix attachment regions are positioned near replication initiation sites, genes, and an interamplicon junction in the amplified dihydrofolate reductase domain of Chinese hamster ovary cells. *Mol. Cell. Biol.* **8,** 5398–5409.

Dijkwel, P. A., Mullenders, L., and Wanka, F. (1979). Analysis of the attachment of replicating DNA to a nuclear matrix in mammalian interphase nuclei. *Nucleic Acids Res.* **6,** 219–230.

Dijkwel, P. A., Vaughn, J. P., and Hamlin, J. L. (1991). Mapping of replication sites in mammalian genomes by two-dimensional gel analysis: Stabilization and enrichment of replication intermediates by isolation on the nuclear matrix. *Mol. Cell. Biol.* **11,** 3850–3859.

D'Incalci, M., Citti, L., Taverna, P., and Catapano, C. (1988). Importance of the DNA repair enzyme O^6-alkylguanine alkyltransferase (AT) in cancer chemotherapy. *Cancer Treat. Rev.* **15,** 279–292.

DiNardo, S., Voelkel, K., and Sternglanz, R. (1984a). DNA topoisomerase II is required at the time of mitosis in yeast. *Cell (Cambridge, Mass.)* **41,** 553–563.

DiNardo, S., Voelkel, K., and Sternglanz, R. (1984b). DNA topoisomerase II mutant of *S. cerevisiae:* Topoisomerase II is required for segregation of daughter molecules at the termination of DNA replication. *Proc. Natl. Acad. Sci. U.S.A.* **81,** 2616–2620.

Drake, F. H., Hormann, G. A., Bartus, H. F., Mattern, M. R., Crooke, S. T., and Mirabelli, C. K. (1989). Biochemical and pharmacological properties of p170 and p180 forms of topoisomerase II. *Biochemistry* **28,** 8154–8160.

Duan, D., and Sadee, W. (1982). Distinct effects of adenine and guanine starvation of DNA synthesis associated with different pool sizes of nucleotide precursors. *Cancer Res.* **47,** 4047–4051.

Dworetzky, S. I., Wright, K. L., Fey, E. G., Penman, S., Lian, J. B., Stein, J. L., and Stein, G. S. (1992). Sequence-specific DNA-binding proteins are components of a nuclear matrix-attachment site. *Proc. Natl. Acad. Sci. U.S.A.* **89,** 4178–4182.

Earnshaw, W. C., and Heck, M. M. (1985). Localization of topoisomerase II in mitotic chromosomes. *J. Cell Biol.* **100,** 1716–1725.

Earnshaw, W. C., Halligan, W. B., Cooke, C. A., Heck, M. M. S., and Fiu, L. F. (1985). Topoisomerase II is a structural component of mitotic chromosome scaffolds. *J. Cell Biol.* **100,** 1706–1715.

Endicott, J. A., and Ling, V. (1989). The biochemistry of p-glycoprotein-mediate multidrug resistance. *Annu. Rev. Biochem.* **58,** 137–171.

Erba, E., Sen, S., Lorico, A., and D'Incalci, M. (1992). Potentiation of etoposide cytotoxicity against a human ovarian cancer cell line by pretreatment with non-toxic concentrations of methotrexate or aphidicolin. *Eur. J. Cancer* **28,** 66–71.

Fernandes, D. J., and Catapano, C. V. (1991). Nuclear matrix targets for anticancer agents. *Cancer Cells* **262,** 5857–5865.

Fernandes, D. J., Smith-Nanni, C., Paff, M. T., and Neff, T.-A. M. (1989). Effects of antileukemic agents on nuclear matrix-bound DNA replication in CCRF-CEM leukemia cell. *Cancer Res.* **48,** 1850–1855.

Fernandes, D. J., Danks, M. K., and Beck, W. T. (1990). Decreased nuclear matrix DNA topoisomerase II in human leukemia cells resistant to VM-26 and m-AMSA. *Biochemistry* **29,** 4235–4241.

Fernandes, D. J., Catapano, C. V., and Townsend, A. J. (1993). Topoisomerase-related mechanisms of drug resistance. *In* "Drug Resistance in Oncology" (B. A. Teicher, ed.), pp. 479–498. Dekker, New York.

Fey, E. G., and Penman, S. (1988). Nuclear matrix proteins reflect cell type of origin in cultured human cells. *Proc. Natl. Acad. Sci. U.S.A.* **85,** 121–125.

Fey, E. G., Wan, K. M., and Penman, S. (1984). Epithelial cytoskeletal framework and nuclear matrix-intermediate filament scaffold: Three-dimensional organization and protein composition. *J. Cell Biol.* **98,** 1973–1984.

Filipski, J., Leblanc, J., Youdale, T., Sikorska, M., and Walker, P. R. (1990). Periodicity of DNA folding in higher order chromatin structures. *EMBO J.* **9,** 1319–1327.

Foster, K. A., and Collins, J. M. (1985). The interrelation between DNA synthesis rates and DNA polymerases bound to the nuclear matrix in synchronized HeLa cells. *J. Biol. Chem.* **7,** 4229–4235.

Gasser, S. M., and Laemmli, U. K. (1986a). Cohabitation of scaffold binding regions with upstream/enhancer elements of three developmentally regulated genes of *D. melanogaster. Cell (Cambridge, Mass.)* **46,** 521–530.

Gasser, S. M., and Laemmli, U. K. (1986b). The organization of chromatin loops: Characterization of a scaffold attachment site. *EMBO J.* **5,** 511–518.

Gasser, S. M., Laroche, R., Falquet, J., Boy de la Tour, E., and Laemmli, U. K. (1986). Metaphase chromosome structure: Involvement of topoisomerase II. *J. Mol. Biol.* **188,** 613–629.

Getzenberg, R. H., Pienta, K. J., Ward, W. S., and Coffey, D. S. (1991). Nuclear structure and three-dimensional organization of DNA. *J. Cell. Biochem.* **47,** 289–299.

Gottesman, M. M., and Pastan, I. (1993). Biochemistry of multidrug resistance mediated by the multidrug transporter. *Annu. Rev. Biochem.* **62,** 385–427.

Grosse, F., and Krauss, G. (1985). Primase activity of DNA polymerase α from calf thymus. *J. Biol. Chem.* **260,** 1881–1888.

Hakes, D. J., and Berezney, R. (1991). DNA binding properties of the nuclear matrix and individual nuclear matrix proteins: Evidence for salt-resistant DNA binding sites. *J. Biol. Chem.* **266**, 11131–11140.

Han, Y.-H., Finley-Austin, M. J., Pommier, Y., and Povirk, L. F. (1993). Small deletion and insertion mutations induced by the topoisomerase II inhibitor teniposide in CHO cells and comparison with sites of drug-stimulated DNA cleavage in vitro. *J. Mol. Biol.* **229**, 52–66.

Hartley-Asp, B., and Kruse, E. (1986). Nuclear protein matrix as a target for estramustine-induced cell death. *Prostate* **9**, 387.

Hassan, A. B., and Cook, P. R. (1993). Visualization of replication sites in unfixed human cells. *J. Cell Sci.* **105**, 541–550.

Hassan, A. B., Errington, R. J., White, N. S., Jackson, D. A., and Cook, P. R. (1994). Replication and transcription sites are colocalized in human cells. *J. Cell Sci.* **107**, 425–434.

Hay, R. T., Hendrickson, E. A., and DePamphilis, M. L. (1984). Sequence specificity for the initiation of RNA-primed simian virus 40 DNA synthesis *in vivo*. *J. Mol. Biol.* **175**, 131–157.

He, D., Nickerson, J. A., and Penman, S. (1990). Core filaments of the nuclear matrix. *J. Cell Biol.* **110**, 569–580.

Heussen, C., Nackerdien, A., Smit, B., and Bohm, L. (1987). Irradiation damage in chromatin isolated from V-79 chinese hamster lung fibroblast. *Radiat. Res.* **110**, 84–94.

Hinds, M., Deisseroth, K., Mayes, J., Altschuler, E., Jansen, R., Ledley, F. D., and Zwelling, L. A. (1991). Identification of a point mutation in the topoisomerase II gene from a human leukemia cell line containing an amsacrine-resistant form of topoisomerase II. *Cancer Res.* **51**, 4729–4731.

Hirose, F., Yamamoto, S., Yamaguchi, M., and Matsukage, A. (1988). Identification and subcellular localization of the polypeptide for chick DNA primase with a specific monoclonal antibody. *J. Biol. Chem.* **263**, 2925–2933.

Hochster, H., and Cassileth, P. (1990). Fludarabine phosphate therapy of non-Hodgkin's lymphoma. *Semin. Oncol.* **17**, 63–65.

Holm, C., Goto, T., Wang, J. C., and Botstein, D. (1985). DNA topoisomerase II is required at the time of mitosis in yeast. *Cell (Cambridge, Mass.)* **41**, 553–563.

Holm, C., Covey, J. M., Kerrigan, D., and Pommier, Y. (1989). Differential requirement of DNA replication for the cytotoxicity of DNA topoisomerase I and II inhibitors in chinese hamster DC3F cells. *Cancer Res.* **49**, 6365–6368.

Hozák, P., Hassan, A. B., Jackson, D. A., and Cook, P. R. (1993). Visualization of replication factors attached to a nucleoskeleton. *Cell (Cambridge, Mass.)* **73**, 361–373.

Huang, P., Chubb, S., and Plunkett, W. (1990). Termination of DNA synthesis by 9-β-D-arabinofuranosyl 2-fluoroadenine: A mechanism for cytotoxicity. *J. Biol. Chem.* **265**, 16617–16625.

Jackson, D. A. (1990). The organization of replication centres in higher eukaryotes. *BioEssays* **12**, 87–89.

Jackson, D. A., and Cook, P. R. (1986). Replication occurs at the nucleoskeleton. *EMBO J.* **5**, 1403–1410.

Jackson, D. A., Yuan, J., and Cook, P. R. (1988). A gentle method for preparing cyto-and nucleoskeletons and associated chromatin. *J. Cell Sci.* **90**, 365–378.

Jackson, D. A., Dickson, P., and Cook, P. R. (1990). Attachment of DNA to the nucleoskeleton of HeLa cells examined using physiological conditions. *Nucleic Acids Res.* **18**, 4385–4393.

Kaguni, L. S., and Lehman, I. R. (1988). Eukaryotic DNA polymerase-primase: structure, mechanism and function. *Biochim. Biophys. Acta* **950**, 87–101.

Käs, E., Izaurralde, E., and Laemmli, U. K. (1989). Specific inhibition of DNA binding to nuclear scaffolds and histone H1 by distamycin. The role of oligo(dA) · oligo(dT) tracts. *J. Mol. Biol.* **210**, 587–599.

Käs, E., Poljak, L., Adachi, Y., and Laemmli, U. K. (1993). A model for chromatin opening: Stimulation of topoisomerase II and restriction enzyme cleavage by chromatin by distamycin. *EMBO J.* **12**, 115–126.

Kaufmann, S. H. (1991). Antagonism between camptothecin and topoisomerase II-directed chemotherapeutic agents in a human leukemia cell line. *Cancer Res.* **51**, 1129–1136.

Kaufmann, W. K., Boyer, J. C., Estabrooks, L. L., and Wilson, S. J. (1991). Inhibition of replicon initiation in human cells following stabilization of topoisomerase-DNA cleavable complexes. *Mol. Cell. Biol.* **11**, 3711–3718.

Keating, J. J. (1990). Fludarabine phosphate in the treatment of chronic lymphocytic leukemia. *Semin. Oncol.* **17**, 49–62.

Keating, M. M., Kantarjian, J., Talpaz, M., Redman, J., Koller, C., Barlogie, B., Velasquez, W., Plunkett, W., Freireich, E. J., and McCredie, K. B. (1989). Fludarabine: A new agent with major activity against chronic lymphocytic leukemia. *Blood* **74**, 19–25.

Kharabanda, S. M., Sherman, M. L., Spriggs, D. R., and Kufe, D. W. (1988). Effects of tiazofurin on protooncogene expression during HL60 cell differentiation. *Cancer Res.* **48**, 5965–5968.

Kimura, K., Saijo, M., Ui, M., and Enomoto, T. (1994). Growth state- and cell cycle-dependent fluctuation in the expression of two forms of DNA topoisomerase II and possible specific modification of the higher molecular weight form in the M phase. *J. Biol. Chem.* **269**, 1173–1176.

Kitani, T., Yoda, K.-Y., and Okazaki, T. (1984). Discontinuous DNA replication of *Drosophila melanogaster* is primed by an octaribonucleotide primer. *Mol. Cell. Biol.* **4**, 1591–1596.

Klehr, D., Maass, K., and Bode, J. (1991). Scaffold-attachment regions from the human interferon β domain can be used to enhance the stable expression of genes under the control of various promoters. *Biochemistry* **30**, 1264–1270.

Knight, R. D., Mangum, J., Lucas, D. L., Cooney, D. A., Kahn, E. C., and Wright, D. G. (1987). Inosine monophosphate dehydrogenase and myeloid cell maturation. *Blood* **69**, 634–639.

Kohn, K. W. (1977). Interstand cross-linking of DNA by 1,3-bis(2-chlorethyl)-1-nitrosourea and other 1-(2-haloethy)-1-nitrosoureas. *Cancer Res.* **37**, 1450–1454.

Kohwi-Shigematsu, T., and Kohwi, Y. (1990). Torsional stress stabilizes extended base unpairing in suppressor sites flanking immunoglobulin heavy chain enhancer. *Biochemistry* **29**, 9551–9560.

Kuchta, R. D., and Willhelm, L. (1991). Inhibition of DNA primase by 9-β-D-arabinofuranosyladenosine triphosphate. *Biochemistry* **30**, 797–803.

Kuchta, R. D., Ilsley, D., Kravig, K. D., Schubert, S., and Harris, B. (1992). Inhibition of DNA primase and polymerase α by arabinofuranosylnucleoside triphosphates and related compounds. *Biochemistry* **31**, 4720–4728.

Levy-Wilson, B., and Fortier, C. (1990). The limits of the DNase I-sensitive domain of the human apolipoprotein B gene coincide with the locations of the chromosomal anchorage loops and define the 5′ and 3′ boundaries of the gene. *J. Biol. Chem.* **264**, 21196–21204.

Liu, L. F., and Wang, J. C. (1987). Supercoiling of the DNA template during transcription. *Proc. Natl. Acad. Sci. U.S.A.* **84**, 7024–7027.

Liu, L. F., Yang, T. C., Yang, L., Tewey, K. M., and Chen, G. L. (1983). Cleavage of DNA by mammalian DNA topoisomerase II. *J. Biol. Chem.* **256**, 4805–4809.

Lowe, J. K., Brox, L., and Henderson, J. F. (1977). Consequences of inhibition of guanine nucleotide synthesis by mycophenolic acid and virazole. *Cancer Res.* **37**, 736–743.

Lucas, D. L., Robins, R. K., Knight, R. D., and Wright, D. G. (1983). Induced maturation of the human promyelocytic leukemia cell line HL60 by 2-β-D-ribofuranozylselenazole-4-carboxamide. *Biochem. Biophys. Res. Commun.* **1115**, 971–980.

Ludérus, M. E. E., de Graaf, A., Mattia, E., den Blaauwen, J. L., Grande, M. A., de Jong, L., and van Driel, R. (1992). Binding of matrix attachment regions to lamin B. *Cell (Cambridge, Mass.)* **70**, 949–959.

Ludérus, M. E. E., den Blaauwen, J. L., de Smit, O. J. B., Compton, D. A., and van Driel, R. (1994). Binding of matrix attachment regions of lamin polymers involves single-stranded regions and the minor groove. *Mol. Cell. Biol.* **14**, 6297–6305.

Maraschin, J., Dutrillaux, B., and Aurias, A. (1990). Chromosome aberrations induced by etoposide (VP-16) are not random. *Int. J. Cancer* **46,** 808–812.

McCready, S. J., Godwin, J., Mason, D. W., Brazell, I. A., and Cook, P. R. (1980). DNA is replicated at the nuclear cage. *J. Cell Sci.* **46,** 365–386.

Milner, A., Gordon, D., Turner, B., and Vaughan, A. (1993). A correlation between DNA-nuclear matrix binding and relative radiosensitivity in two human squamous cell carcinoma cell line. *Int. J. Radiat. Biol.* **63,** 13–20.

Mirkovitch, J., Mirault, M.-E., and Laemmli, U. K. (1984). Organization of the higher order chromatin loop: Specific DNA attachment sites on nuclear scaffold. *Cell (Cambridge, Mass.)* **39,** 223–232.

Montgomery, J. A., James, R., McCaleb, G. S., and Johnston, T. P. (1967). The modes of decomposition of 1,3-bis-2-chlorethyl)-1-nitrosourea and related compounds. *J. Med. Chem.* **10,** 668–674.

Nackerdien, Z., Miche, J., and Bohm, J. (1989). Chromatin decondensed by acetylation shows an elevated radiation response. *Radiat. Res.* **117,** 234–244.

Nakamura, H., Morita, T., and Sato, C. (1986). Structural organization of replicon domains during the DNA synthetic phase in the mammalian nucleus. *Exp. Cell Res.* **165,** 291–297.

Nakayasu, H., and Berezney, R. (1989). Mapping replicational sites in the eucaryotic cell nucleus. *J. Cell Biol.* **108,** 1–11.

Nakayasu, H., and Berezney, R. (1991). Nuclear matrins: Identification of the major nuclear matrix proteins. *Proc. Natl. Acad. Sci. U.S.A.* **88,** 10312–10316.

Negri, C., Chiesa, R., Cerino, A., Bestagno, M., Sala, C., Zini, N., Miraldi, N. M., and Astaldi-Ricotti, G. C. (1992). Monoclonal antibodies to human DNA topoisomerase I and the two isoforms of DNA topoisomerase II: 170- and 180-kDa isozymes. *Exp. Cell Res.* **200,** 452–459.

Nelson, H. C. M., Finch, J. T., Luisi, B. F., and Klug, A. (1987). The structure of an oligo(dA). oligo(dT) tract and its biological implications. *Nature (London)* **330,** 211–226.

Nelson, W. G., Pienta, K. J., Barrack, E. R., and Coffey, D. S. (1986a). The role of the nuclear matrix in the organization and function of DNA. *Annu. Rev. Biophys. Biophys. Chem.* **15,** 457–475.

Nelson, W. G., Liu, L. F., and Coffey, D. S. (1986b). Newly replicated DNA is associated with DNA topoisomerase II in cultured rat prostatic adenocarcinoma cells. *Nature (London)* **322,** 187–189.

Newport, J. (1987). Nuclear reconstitution in vitro: Stages of assembly around protein-free DNA. *Cell (Cambridge, Mass.)* **48,** 205–217.

Newport, J., and Spann, T. (1987). Disassembly of the nucleus in mitotic extracts: Membrane vesicularization, lamin disassembly, and chromosome condensation are independent processes. *Cell (Cambridge, Mass.)* **48,** 219–230.

Ogawa, T., and Okazaki, T. (1980). Discontinuous DNA replication. *Annu. Rev. Biochem.* **49,** 421–457.

Oleinick, N. L., Chiu, S.-M., and Friedman, L. R. (1984). Gamma radiation as a probe of chromatin structure: Damage to and repair of active chromatin in the metaphase chromosome. *Radiat. Res.* **98,** 629–641.

Oleinick, N. L., Chiu, S.-M., Ramakrishnan, N., and Xue, L. Y. (1987). The formation, identification, and significance of DNA-protein cross-links in mammalian cells. *Br. J. Cancer* **8,** 135–140.

Osheroff, N. (1986). Eukaryotic topoisomerase II: Characterization of enzyme turnover. *J. Biol. Chem.* **261,** 9944–9950.

Osheroff, N. (1989). Effect of antineoplastic agents on the DNA cleavage/religation reaction of eukaryotic topoisomerase II: Inhibition of DNA religation by etoposide. *Biochemistry* **28,** 6157–6160.

Paff, M. T., and Fernandes, D. J. (1990). Synthesis and distribution of primer RNA in nuclei of CCRF-CEM leukemia cells. *Biochemistry* **29,** 3442–3450.

Pardoll, D. M., Vogelstein, B., and Coffey, D. S. (1980). A fixed site of DNA replication in eucaryotic cells. *Cell (Cambridge, Mass.)* **19**, 527–536.

Parker, W. B., and Cheng, Y.-C. (1987). Inhibition of DNA primase by nucleoside triphosphates and their arabinofuranosyl analogs. *Mol. Pharmacol.* **31**, 146–151.

Parker, W. B., Bapat, A. R., Shen, J. X., Townsend, A. J., and Cheng, Y.-C. (1988). Interaction of 2-halogenated dATP analogs (F, Cl, and Br) with human DNA polymerases, DNA primase, and ribonucleotide reductase. *Mol. Pharmacol.* **34**, 485–491.

Parker, W. B., Shaddix, S. C., Change, C. H., White, E. L., Rose, L. M., Brockman, R. W., Shortnacy, A. T., Montgomery, J. A., III, and Bennett, L. L., Jr. (1991). Effects of 2-Chloro-9-(D-deoxy-2-fluoro-β-D-arabinofuranosyl)adenine on K562 cellular metabolism and the inhibition of human ribonucleotide reductase and DNA polymerases by its 5'-triphosphate. *Cancer Res.* **51**, 2386–2394.

Phi-Van, L., and Strätling, W. H. (1988). The matrix attachment regions of the chicken lysozyme gene co-map with the boundaries of the chromatin domain. *EMBO J.* **7**, 655–664.

Phi-Van, L., von Kries, J. P., Ostertag, W., and Strätling, W. H. (1990). The chicken lysozyme 5' matrix attachment region increases transcription from a heterologous promoter in heterologous cells and dampens position effects on the expression of transfected genes. *Mol. Cell. Biol.* **10**, 2303–2307.

Pienta, K., and Lehr, J. (1993). Inhibition of prostate cancer growth by estramustine and etoposide: Evidence for interaction at the nuclear matrix. *J. Urol.* **149**, 1622–1625.

Pommier, Y., Zwelling, L. A., Kao-Shan, C.-S., Whang-Peng, J., and Bradley, M. O. (1985). Correlations between intercalator-induced DNA strand breaks and sister chromatid exchanges, mutations, and cytotoxicity in Chinese hamster cells. *Cancer Res.* **45**, 3143–3149.

Pommier, Y., Kerrigan, D., Schwartx, R. E., Swack, J. A., and McCurdy, A. (1986). Altered DNA topoisomerase II activity in Chinese hamster cells resistant to topoisomerase II inhibitors. *Cancer Res.* **46**, 3075–3081.

Pruss, G. J., and Drlica, K. (1986). Topoisomerase I mutants: The gene on pBR322 that encodes resistance to tetracycline affects plasmid DNA supercoiling. *Proc. Natl. Acad. Sci. U.S.A.* **83**, 8952–8956.

Ratain, M. J., Kaminer, L. S., Bitrain, J. D., Larson, R. A., LeBeau, M. M., Skosey, C., Purl, S., Hoffman, P. C., Wade, J. Vardiman, J. W., Daly, K., Rowley, J. D., and Golomb, H. M. (1987). Acute nonlymphocytic leukemia following etoposide and cisplatin combination chemotherapy for advanced non-small carcinoma of the lung. *Blood* **70**, 1412–1417.

Richter, A., and Strausfeld, U. (1988). Effects of VM-26, a specific inhibitor of type II DNA topoisomerase, on SV40 chromatin replication in vitro. *Nucleic Acids Res.* **16**, 10119–10129.

Richter, A., Strausfeld, U., and Knippers, R. (1987). Effects of VM-26 (teniposide), a specific inhibitor of type II DNA topoisomerase, on SV40 DNA replication in vivo. *Nucleic Acids Res.* **15**, 3455–3468.

Robinson, M. J., and Osheroff, N. (1990). Stabilization of the topoisomerase II–DNA cleavage complex by antineoplastic drugs: Inhibition of enzyme-mediated DNA religation by 4'-(9-acridinylamino)methanesulfon-*m*-anisidide. *Biochemistry* **29**, 2511–2515.

Romig, H., Fackelmayer, F. O., Renz, A., Ramsperger, U., and Richter, A. (1992). Characterization of SAF-A, a novel nuclear DNA binding protein from HeLa cells with high affinity for nuclear matrix/scaffold attachment DNA elements. *EMBO J.* **11**, 3431–3440.

Rowe, T., Kupfer, G., and Ross, W. (1985). Inhibition of epipodophyllotoxin cytotoxicity by interference with topoisomerase-mediated DNA cleavage. *Biochem. Pharmacol.* **34**, 2483–2487.

Sakaguchi, K., Tsujino, M., Yoshizawa, M., Mizuno, K., and Hayano, K. (1975). Action of Bredenin on mammalian cells. *Cancer Res.* **35**, 1643–1648.

Sheaff, R. J., and Kuchta, R. D. (1993). Mechanism of calf thymus DNA primase: Slow initiation, rapid polymerization, and intelligent termination. *Biochemistry* **32**, 3027–3037.

Shibuya, M. L., Ueno, A. M., Vannais, D. B., Craven, P. A., and Waldren, C. A. (1994). Megabase pair deletions in mutant mammalian cells following exposure to amsacrine, an inhibitor of DNA topoisomerase II. *Cancer Res.* **54,** 1092–1097.

Smith, H. C., and Berezney, R. (1980). DNA polymerase alpha is tightly bound to the nuclear matrix of actively replicating liver. *Biochem. Biophys. Res. Commun.* **97,** 1541–1547.

Smith, H. C., and Berezney, R. (1982). Nuclear matrix-bound deoxyribonucleic acid synthesis: An in vitro system. *Biochemistry* **21,** 6751–6761.

Smith, H. C., and Berezney, R. (1983). Dynamic domains of DNA polymerase alpha in regenerating rat liver. *Biochemistry* **22,** 3042–3046.

Snapka, R. M. (1986). Topoisomerase inhibitors can selectively interfere with different states of simian virus 40 DNA replication. *Mol. Cell. Biol.* **6,** 4221–4227.

Snapka, R. M., Powelson, M. A., and Strayer, J. M. (1988). Swiveling and decatenation of replicating simian virus genomes in vivo. *Mol. Cell. Biol.* **8,** 515–521.

Sperry, A. O., Blasquez, V. C., and Garrard, W. T. (1989). Dysfunction of chromosomal loop attachment sites: Illegitimate recombination linked to matrix association regions and topoisomerase II. *Proc. Natl. Acad. Sci. U.S.A.* **86,** 5497–5501.

Stein, G. S., Lian, J. B., Dworetzky, S. I., Owen, T. A., Bortell, R., Bidwell, J. P., and van Wijnen, A. J. (1991). Regulation of transcription–factor activity during growth and differentiation: Involvement of the nuclear matrix in concentration and localization of promoter binding proteins. *J. Cell. Biochem.* **47,** 300–305.

Stief, A., Winter, D. M., Strätling, W. H., and Sippel, A. E. (1989). A nuclear DNA attachment element mediates elevated and position-independent gene activity. *Nature (London)* **341,** 343–345.

Stuurman, N., Meijne, A. M., van der Pol, A. J., de Jong, L., van Driel, R., and van Renswoude, J. (1990). The nuclear matrix from cells of different origin. *J. Biol. Chem.* **265,** 5460–5465.

Sullivan, D. M., Chow, K.-C., Glisson, B. S., and Ross, W. E. (1987). Role of proliferation in determining sensitivity to topoisomerase II-active chemotherapy agents. *Natl. Cancer Inst.* **4,** 73–78.

Takano, H., Kohno, K., Ono, M., Uchida, Y., and Kuwano, M. (1991). Increased phosphorylation of DNA topoisomerase II in etoposide-resistant mutants of human cancer KB cells. *Cancer Res.* **51,** 3951–5957.

Tan, K. B., Dorman, T. E., Falls, K. M., Chung, T. D. Y., Mirabelli, C. K., Crooke, S. T., and Mao, J.-I. (1992). Topoisomerase IIα and topoisomerase IIβ genes: Characterization and mapping to human chromosome 17 and 3, respectively. *Cancer Res.* **52,** 231–234.

Tew, K. D. (1982). The interaction of nuclear reactant drugs with the nuclear membrane and nuclear matrix. In "The Nuclear Envelope and the Nuclear Matrix" (G. G. Maul, ed.), pp. 279–292. Alan R. Liss, New York.

Tew, K. D., and Stearns, M. E. (1987). Hormone-independent, non alkylating mechanism of cytotoxicity for estramustine. *Urol. Res.* **15,** 155–160.

Tew, K. D., Sudhakar, S., Schein, P., and Smulson, M. (1978). Binding of chlorozotocin and 1-(2-chlorethyl)-3-cyclohexyl-1-nitrosourea to chromatin and nucleosomal fractions HeLa cells. *Cancer Res.* **38,** 3371–3378.

Tew, K. D., Wang, A.L., and Schein, P. S. (1983a). Alkylating agent interactions with the nuclear matrix. *Biochem. Pharmacol.* **32,** 3509–3516.

Tew, K. D., Erickson, L., White, G., Wang, A. L., Schein, P. S., and Hartley-Asp, B. (1983b). Cytotoxicity of estramustine, a steroid-nitrogen mustard derivative, through non-DNA targets. *Mol. Pharmacol.* **24,** 324–328.

Tewey, K. M., Chen, G. L., Nelson, E. M., and Liu, L. F. (1984). Intercalative antitumor drugs interfere with the breakage-reunion of mammalian DNA topoisomerase II. *J. Biol. Chem.* **259,** 9182–9187.

Tricott, G. J., Jayaram, H. N., Lapis, E., Natsumeda, Y., Nichols, C. R., Kneebone, P., Heerema, N., Weber, G., and Hoffman, R. (1989). Biochemically directed therapy of leukemia with tiazofurin, selective blocker of inosine 5'-phosphate dehydrogenase activity. *Cancer Res.* **49**, 3696–3701.

Tseng, B. Y., and Ahlem, C. N. (1983). A DNA primase from mouse cells: Purification and partial characterization. *J. Biol. Chem.* **258**, 9845–9849.

Tseng, B. Y., and Ahlem, C. N. (1984). Mouse primase initiation sites in the origin region of simian virus 40. *J. Biol. Chem.* **81**, 2342–2346.

Tseng, B. Y., and Goulian, M. (1977). Initiator RNA of discontinuous DNA synthesis in human lymphocytes. *Cell (Cambridge, Mass.)* **12**, 483–489.

Tseng, B. Y., Erickson, J. M., and Goulian, M. (1979). Initiator RNA of nascent DNA from animal cells. *J. Mol. Biol.* **129**, 531–545.

Tseng, W. C., Derse, D., Cheng, Y. C., Brockman, R. W., and Bennett, L. L., Jr. (1982). In vitro biological activity of 9-β-D-arabinofuranosyl-2-fluoroadenine and the biochemical actions of its triphosphate on DNA polymerases and ribonucleotide reductase from HeLa cells. *Mol. Pharmacol.* **21**, 474–477.

Tsutsui, K., Tsutsui, K., Okada, S., Watarai, S., Seki, S., Yasuda, R., and Shohmori, T. (1993). Identification and characterization of a nuclear scaffold protein that binds the matrix attachment region DNA. *J. Biol. Chem.* **268**, 12886–12894.

Tubo, R. A., and Berezney, R. (1987a). Pre-replicative association of multiple replicative enzyme activities with the nuclear matrix during rat liver regeneration. *J. Biol. Chem.* **262**, 1148–1154.

Tubo, R. A., and Berezney, R. (1987b). Nuclear matrix-bound DNA primase. *J. Biol. Chem.* **262**, 6637–6642.

Turka, L. A., Dayton, J., Sinclair, G., Thompson, C. B., and Mitchell, B. S. (1991). Guanine ribonucleotide depletion inhibits T cell activation: Mechanism of action of the immunosuppressive drug mizoribine. *J. Clin. Invest.* **87**, 940–948.

Uemura, T., Ohkura, H., Adachi, Y., Morino, K., Shiozaki, K., and Yanagida, M. (1987). DNA topoisomerase II is required for condensation and separation of mitotic chromosomes in *S. pombe. Cell (Cambridge, Mass.)* **50**, 917–925.

Vaughan, A., Milner, T., Gordon, D., and Schwartz, J. (1991). Interaction between ionizing radiation and supercoiled DNA within human tumour cells. *Cancer Res.* **51**, 3857–3861.

Vaughn, J. P., Dijkwel, P. A., Mullenders, L. H. F., and Hamlin, J. L. (1990). Replication forks are associated with the nuclear matrix. *Nucleic Acids Res.* **18**, 1965–1969.

Vogelstein, B., Pardoll, D. M., and Coffey, D. S. (1980). Supercoiled loops and eucaryotic DNA replication. *Cell (Cambridge, Mass.)* **22**, 79–85.

von Kries, P., Bugrnester, H., and Sträting, W. H. (1991). A matrix/scaffold attachment region binding protein: Identification, purification, and mode of binding. *Cell (Cambridge, Mass.)* **64**, 123–135.

Walker, P. R., Smith, C., Youdale, T., Leblanc, J., Whitfield, J. F., and Sikorska, M. (1991). Topoisomerase II-reactive chemotherapeutic drugs induce apoptosis in thymocytes. *Cancer Res.* **51**, 1078–1085.

Wang, J. C. (1985). DNA topoisomerase. *Annu. Rev. Biochem.* **54**, 665–697.

Wang, J. C., Caron, P. R., and Kim, R. A. (1990). The role of DNA topoisomerase in recombination and genome stability: A double-edged sword? *Cell (Cambridge, Mass.)* **62**, 403–406.

Wang, T. S.-F. (1991). Eukaryotic DNA polymerases. *Annu. Rev. Biochem.* **60**, 513–552.

Wang, T. S.-F., Hu, S.-Z., and Korn, D. (1984). DNA primase from KB cells. Characterization of a primase activity tightly associated with immunoaffinity-purified DNA polymeraseα. *J. Biol. Chem.* **259**, 1854–1865.

Woessner, R. D., Chung, T. D. Y., Hoffmann, G. A., Mattern, M. R., Mirabelli, C. K., Drake, F. H., and Johnson, R. K. (1990). Differences between normal and *ras*-transformed NIH-3T3 cells in expression of the 170 kD and 180 kD forms of topoisomerase II. *Cancer Res.* **50**, 2901–2908.

Woessner, R. D., Mattern, M. R., Mirabelli, C. K., Johnson, R. K., and Drake, F. H. (1991). Proliferation- and cell cycle-dependent differences in expression of the 170 kilodalton and 180 kilodalton forms of topoisomerase II in NIH-3T3 cell. *Cell Growth Differ.* **2,** 209–214.

Wood, E. R., and Earnshaw, W. C. (1990). Mitotic chromatin condensation in vitro using somatic cell extracts and nuclei with variable levels of endogenous topoisomerase II. *J. Cell Biol.* **111,** 2839–2850.

Wood, S. H., and Collins, J. M. (1986). Preferential binding of DNA primase to the nuclear matrix in HeLa cells. *J. Biol. Chem.* **261,** 7119–7122.

Woynarowski, J. M., Sigmund, R. D., and Beerman, T. A. (1988). Topoisomerase-II-mediated lesions in nascent DNA: Comparison of the effects of epipodophyllotoxin derivatives, VM-26 and VP-16, and 9-anilinoacridine derivatives, m-AMSA and o-AMSA. *Biochim. Biophys. Acta* **950,** 21–29.

Wu, H.-Y., Shyy, S., Wang, J. C., and Liu, L. F. (1988). Transcription generates positively and negative supercoiled domains in the template. *Cell (Cambridge, Mass.)* **53,** 433–440.

Xing, Y., and Lawrence, J. B. (1991). Preservation of specific RNA distribution within chromatin-depleted nuclear substructure demonstrated by in situ hybridization coupled with biochemical fractionation. *J. Cell Biol.* **112,** 1055–1063.

Xing, Y., Johnson, C. V., Dobner, P. R., and Lawrence, J. B. (1993). Higher level organization of individual gene transcription and RNA splicing. *Science* **259,** 1326–1330.

Xu, M., Hammer, R. E., Blasquez, V. C., Jones, S. L., and Garrard, W. T. (1989). Immunoglobulin kappa gene expression after stable integration. II: Role of intronic matrix attachment region and enhancer in transgenic mice. *J. Biol. Chem.* **264,** 21190–21195.

Yamaguchi, M., Hendrickson, E. A., and DePamphilis, M. (1985). DNA primase–DNA polymerase α from simian cells: Sequence specificity of initiation sites on simian virus 40 DNA. *Mol. Cell. Biol.* **5,** 1170–1183.

Yang, S. W., Huang, P., Plunkett, W., Becker, F. F., and Chan, J. Y. H. (1992). Dual mode of inhibition of purified DNA ligase I from human cells by 9-β-D-arabinofuranosyl-2-fluoroadenine triphosphate. *J. Biol. Chem.* **267,** 2345–2349.

Yoshida, S., Suzuki, R., Masaki, S., and Koiwai, O. (1985). Aranbinosylnucleoside 5'-triphosphate inhibits DNA primase of calf thymus. *J. Biochem. (Tokyo)* **98,** 427–433.

Zechiendrich, E. L., Christiansen, K., Andersen, A. H., Westergaard, O., and Osheroff, N. (1989). Double-strand DNA cleavage/religation reaction of eukaryotic topoisomerase II: Evidence for a nicked DNA intermediate. *Biochemistry* **28,** 6229–6236.

Zini, N., Santi, S., Ognibene, A., Bavelloni, A., Neri, L. N., Valmori, A., Mariani, E., Negri, C., Astaldi-Ricotti, B., and Maraldi, N. M. (1994). Discrete localization of different topoisomerases in HeLa and K562 cell nuclei and subnuclear fractions. *Exp. Cell Res.* **210,** 336–348.

INDEX

ISBN 0-12-364565-4

90018